GEOTECHNICAL ENGINEER'S PORTABLE HANDBOOK

GEOTECHNICAL ENGINEER'S PORTABLE HANDBOOK

Robert W. Day

McGraw-Hill

New York San Francisco Washington, D.C. Auckland Bogotá
Caracas Lisbon London Madrid Mexico City Milan
Montreal New Delhi San Juan Singapore
Sydney Tokyo Toronto

McGraw-Hill

A Division of The ***McGraw·Hill*** *Companies*

9 10 11 12 13 14 15 QDB QDB 14 13 12 11

ISBN-13: 978-0-07-135111-9
ISBN-10: 0-07-135111-6

The sponsoring editor of this book was Larry S. Hager, the editing supervisor was David E. Fogarty, and the production supervisor was Sherri Souffrance. It was set in the HB1 design in Times by Paul Scozzari, Deirdre Sheean, Joanne Morbit, and Kim Sheran of McGraw-Hill's Professional Book Group composition unit.

Printed and bound by R. R. Donnelley & Sons Company.

This book was printed on recycled, acid-free paper containing a minimum of 50% recycled, de-inked fiber.

Dedicated with love to
my daughter Rachel

CONTENTS

Preface xiii
Acknowledgments xv

Part 1 Geotechnical Engineering

Chapter 1. Introduction 1.3

1.1. Definitions / *1.3*
1.2. Qualifying Experience / *1.3*
1.3. Project Requirements / *1.4*
1.4. Proposals and Contracts / *1.4*

Chapter 2. Field Exploration 2.1

2.1. Document Review / *2.1*
2.2. Subsurface Exploration / *2.1*
2.3. Soil Sampling / *2.2*
2.4. Field Testing / *2.2*
2.5. Exploratory Logs / *2.2*
2.6. Geophysical Techniques / *2.2*
2.7. Subsoil Profile / *2.3*

Chapter 3. Laboratory Testing 3.1

3.1. Index Tests / *3.1*
3.2. Oedometer Apparatus / *3.1*
3.3. Shear Strength / *3.1*
3.4. Permeability / *3.2*

3.5. Laboratory Compaction Tests / *3.2*
3.6. Laboratory Tests for Pavement Design / *3.2*

Chapter 4. Soil and Rock Classification 4.1

4.1. Soil Classification Systems / *4.1*
  4.1.1. Unified Soil Classification System (USCS) / *4.1*
  4.1.2. Inorganic Soil Classification System Based on Plasticity (ISBP) / *4.2*
  4.1.3. AASHTO Soil Classification System / *4.2*
  4.1.4. USDA Textural Classification / *4.2*
  4.1.5. Organic Soil Classification / *4.2*
4.2. Unusual Soil / *4.3*
4.3. Rock Classification / *4.3*

Chapter 5. Phase Relationships 5.1

5.1. Soil Element / *5.1*
5.2. Phase Relationships Directly from Laboratory Testing / *5.2*
5.3. Indirect Phase Relationships / *5.2*
5.4. Example of the Calculation of Phase Relationships / *5.3*

Chapter 6. Effective Stress and Stress Distribution 6.1

6.1. Introduction / *6.1*
6.2. Effective Stress, Total Stress, and Pore Water Pressure / *6.2*
6.3. Stress Distribution / *6.2*
  6.3.1. One-Dimensional Loading / *6.2*
  6.3.2. 2:1 Approximation / *6.2*
  6.3.3. Equations Based on the Theory of Elasticity / *6.3*
  6.3.4. Charts Based on the Theory of Elasticity / *6.3*
6.4. Example of Calculations Using Various Stress Distributions / *6.3*
6.5. Mohr Circle / *6.3*

Chapter 7. Shear Strength 7.1

7.1. Introduction / *7.1*
7.2. Cohesionless Soil / *7.2*
7.3. Cohesive Soil / *7.2*
  7.3.1. Undrained Shear Strength / *7.2*
  7.3.2. Drained Shear Strength / *7.3*
  7.3.3. Drained Residual Shear Strength / *7.3*
  7.3.4. Factors that Effect the Shear Strength of Cohesive Soil / *7.3*
7.4. Total Stress and Effective Stress Analyses / *7.3*

Chapter 8. Permeability and Seepage 8.1

8.1. Introduction / *8.1*
8.2. Permeability / *8.2*
8.3. Superficial Velocity and Seepage Velocity / *8.2*
8.4. Seepage Forces / *8.2*
8.5. Two-Dimensional Flow Nets / *8.2*
8.5.1. Using Flow Nets to Calculate Pore Water Pressures and Exit Gradients / *8.3*

Chapter 9. Settlement Analyses 9.1

9.1. Introduction / *9.1*
9.2. Allowable Settlement / *9.1*
9.3. Collapsible Soil / *9.2*
9.4. Settlement of Cohesive and Organic Soils / *9.2*
9.4.1. Immediate Settlement / *9.2*
9.4.2. Consolidation / *9.2*
9.4.3. Secondary Compression / *9.2*
9.5. Settlement of Cohesionless Soil / *9.3*
9.6. Other Common Causes of Settlement / *9.3*

Chapter 10. Bearing Capacity Analyses 10.1

10.1. Introduction / *10.1*
10.2. Bearing Capacity for Shallow Foundations / *10.1*
10.3. Bearing Capacity for Deep Foundations / *10.2*

Chapter 11. Pavement and Pipeline Design 11.1

11.1. Pavement Design / *11.1*
11.1.1. California Method of Flexible Pavement Design / *11.2*
11.2. Pipeline Design / *11.2*
11.2.1. Rigid Pipelines and Flexible Pipeline Design / *11.2*

Chapter 12. Expansive Soil 12.1

12.1. Introduction / *12.1*
12.2. Laboratory Testing / *12.1*
12.3. Swelling of Desiccated Clay / *12.1*
12.4. Types of Expansive Soil Movement / *12.2*
12.5. Calculating Foundation Heave / *12.2*
12.6. Soil Treatment and Foundation Design / *12.2*
12.7. Pavements and Flatwork on Expansive Soil / *12.2*
12.8. Expansive Rock / *12.3*

Chapter 13. Slope Stability 13.1

13.1. Typical Types of Slope Movement / *13.1*
13.2. Allowable Lateral Movement / *13.1*
13.3. Rockfall / *13.1*
13.4. Surficial Slope Stability Analysis / *13.2*
13.5. Gross Slope Stability Analysis / *13.2*
13.6. Landslide Stability Analysis / *13.2*
13.7. Debris Flow / *13.2*
13.8. Slope Softening and Creep / *13.3*
13.9. Analysis for Earth Dams / *13.3*

Chapter 14. Earthquake Analyses 14.1

14.1. Introduction / *14.1*
14.2. Liquefaction / *14.1*
14.3. Earthquake-Induced Settlement and Slope Movement / *14.1*

Chapter 15. Erosion Analyses 15.1

15.1. Introduction / *15.1*
15.2. Universal Soil Loss Equation / *15.1*
15.3. Design and Construction / *15.1*

Chapter 16. Retaining Walls 16.1

16.1. Introduction / *16.1*
16.2. Retaining Wall Analyses / *16.1*
16.3. Design and Construction of Retaining Walls / *16.1*
16.4. Restrained Retaining Walls / *16.2*
16.5. Mechanically Stabilized Earth Retaining Walls / *16.2*
16.6. Sheet Pile Walls / *16.2*
16.7. Temporary Retaining Walls / *16.3*
16.8. Pier Walls / *16.3*

Chapter 17. Deterioration 17.1

17.1. Introduction / *17.1*
17.2. Sulfate Attack of Concrete / *17.1*
17.3. Pavement Deterioration / *17.2*
17.4. Frost / *17.2*

Chapter 18. Foundations **18.1**

18.1. Introduction / *18.1*
18.2. Shallow Foundations / *18.1*
18.3. Deep Foundations / *18.1*

Part 2 Construction

Chapter 19. Grading and Other Site Improvement Methods **19.3**

19.1. Introduction / *19.3*
19.2. Grading Specifications / *19.3*
19.3. Compaction Fundamentals / *19.4*
19.4. Site Improvement Methods / *19.4*
19.4.1. Soil Replacement / *19.4*
19.4.2. Water Removal / *19.4*
19.4.3. Site Strengthening / *19.5*
19.4.4. Grouting / *19.5*
19.4.5. Thermal / *19.5*
19.4.6. Summary / *19.5*
19.5. Observational Method / *19.6*

Chapter 20. Groundwater and Percolation Tests **20.1**

20.1. Introduction / *20.1*
20.2. Groundwater / *20.1*
20.2.1. Groundwater Control / *20.2*
20.2.2. Pavements / *20.2*
20.2.3. Slopes / *20.3*
20.3. Moisture Migration through Floor Slabs and Basement Walls / *20.4*
20.4. Percolation Tests for Sewage Disposal System / *20.4*
20.5. Surface Drainage and Pipe Breaks / *20.5*

Chapter 21. Excavation, Underpinning, and Field Load Tests **21.1**

21.1. Introduction / *21.1*
21.2. Excavations / *21.1*
21.2.1. Footing Excavations / *21.2*
21.2.2. Excavation of Piers / *21.2*
21.2.3. Open Excavations / *21.2*
21.2.4. Braced Excavations / *21.2*
21.2.5. Tunnels / *21.2*

21.3. Underpinning / *21.3*
21.4. Field Load Tests / *21.3*

Chapter 22. Geosynthetics 22.1

22.1. Introduction / *22.1*
22.2. Geogrids / *22.1*
22.3. Geotextiles / *22.1*
22.4. Geomembranes / *22.2*
22.5. Geonets and Geocomposites / *22.2*
22.6. Geosynthetic Clay Liners / *22.2*

Chapter 23. Instrumentation 23.1

23.1. Introduction / *23.1*
23.2. Commonly Used Monitoring Devices / *23.1*

Appendix A. Glossary A.1

Glossary 1. Engineering Geology and Subsurface Exploration Terminology / *A.4*
Glossary 2. Laboratory Testing Terminology / *A.10*
Glossary 3. Terminology for Engineering Analysis and Computations / *A.15*
Glossary 4. Construction and Grading Terminology / *A.22*
Glossary 5. Terminology for Sewage Disposal and Percolation Tests / *A.30*

Appendix B. Example of Grading Specifications B.1

Appendix C. Percolation Test Procedures C.1

Appendix D. Conversion Factors D.1

Appendix E. References E.1

Index follows Appendix E

PREFACE

This book is a reference geotechnical handbook for the engineer, student, or contractor. The unique feature of the *Geotechnical Engineer's Portable Handbook* is that it provides quick access to important geotechnical and foundation information. The goal of the book is to reduce the task of searching through numerous geotechnical and foundation engineering publications in order to find the desired information.

The book should be primarily used for reference. The book contains easy-to-use tables and charts that enable the geotechnical engineer, student, and contractor to quickly locate the desired reference material.

The book is divided into two separate parts. Part 1 (Chaps. 1 to 18) provides information concerning geotechnical engineering, including field exploration guidelines, laboratory testing procedures, soil and rock classification, basic phase relationships, stress distribution tables and charts, and pavement and pipeline design charts. Also included is information on all types of geotechnical analyses including settlement, bearing capacity, expansive soil, slope stability, earthquake, erosion, and deterioration analyses. The first part also covers retaining walls and building foundations.

Part 2 (Chaps. 19 to 23) provides information for construction-related aspects of geotechnical and foundation engineering. Topics covered include grading, instrumentation, excavation, underpinning, groundwater control, and geosynthetics.

The appendixes consist of a glossary, typical grading specifications, an example of percolation test procedures, conversion factors, and references.

The book presents the practical aspects of geotechnical and foundation engineering in an easy-to-use format. It provides frequently used equations, charts, professional guidelines, common laboratory procedures, and a glossary. It is a portable, compact, and indispensable ready-reference source for the field and office.

Robert W. Day

ACKNOWLEDGMENTS

I am grateful for the contributions of the many people who helped make this book. Special thanks is due Larry Hager, senior editor at McGraw-Hill Book Company. It was his idea to create a geotechnical portable handbook that could be used as a source of reference material.

Specific parts of the book are derived from *Geotechnical and Foundation Engineering: Design and Construction*. I would like to thank Professor Charles C. Ladd, at the Massachusetts Institute of Technology, who reviewed that material and offered many helpful suggestions during its initial preparation for publication.

Numerous practicing engineers reviewed portions of the text and provided valuable assistance during its initial development. In particular, I am indebted to Robert Brown, Edred Marsh, Rick Walsh, and Scott Thoeny. Thanks also to Dennis Poland, Ralph Jeffery, and Todd Page for their help with the geologic aspects of the book, and Rick Dorrah and Eric Noether for drafting the figures for the book.

I would also like to thank Professor Timothy Stark, at the University of Illinois, who performed the ring shear tests, provided a discussion of the test procedures, and prepared the ring shear test plots. Thanks also to Kean Tan who performed the triaxial compression tests and prepared the shear strength data plots. I am also indebted to Gregory Axten, president of American Geotechnical, who provided valuable support during the review and preparation of the book, and to Carl Bonura, for the many conversations we have had over the years about his unique experience in engineering geology.

Tables and figures taken from other sources are acknowledged where they occur in the text. Finally, I wish to thank David Fogarty, Sherri Souffrance, George Watson, and others on the McGraw-Hill editorial staff, who made this book possible and refined my rough draft into this finished product.

P · A · R · T · 1

GEOTECHNICAL ENGINEERING

CHAPTER 1
INTRODUCTION

Chapter 1 contains basic geotechnical information, such as definitions, qualifying experience, project requirements, and information on proposals and contracts.

1.1 DEFINITIONS

Table 1.1 presents definitions of geotechnical engineering and foundation engineering. Foundations are commonly divided into two categories: shallow and deep foundations. Table 1.2 presents a list of common types of foundations. Additional terms and definitions are presented in the Glossary (App. A).

1.2 QUALIFYING EXPERIENCE

Table 1.3 indicates those items that are considered to be qualifying experience for geotechnical engineers. Table 1.3 also lists the five basic aspects of geotechnical engineering and the typical types of analyses performed by geotechnical engineers. Table 1.4 presents a summary of the fields of expertise for the engineering geologist and geotechnical engineer, with the last column indicating the areas of overlapping expertise. The individual areas of responsibility of the engineering geologist and the geotechnical engineer are summarized in Table 1.5.

1.3 PROJECT REQUIREMENTS

Table 1.6 indicates typical types of projects that involve geotechnical engineers and Table 1.7 lists the general requirements for geotechnical and foundation projects.

1.4 PROPOSALS AND CONTRACTS

Often the first steps in a project are to plan the work, prepare a cost estimate, and provide a proposal that is acceptable to the client. Table 1.8 presents an example of a schedule of fees, Table 1.9 provides an example of a cost estimating sheet, and Table 1.10 summarizes those items that should be included in the contract between the geotechnical engineer and the client.

TABLE 1.1 Definitions

Term (1)	Definition (2)
Geotechnical engineering	In a broad sense, the definition of a geotechnical engineer is an individual who performs an engineering evaluation of earth materials. This typically includes soil, rock, and groundwater and their interaction with earth retention systems, structural foundations, and other civil engineering works. Geotechnical engineering is a subdiscipline of civil engineering and requires a knowledge of basic engineering principles, such as statics, dynamics, fluid mechanics, and the behavior of engineering materials. An understanding of construction techniques and the performance of civil engineering works influenced by earth materials is also required. Geotechnical engineering is often divided into two categories: soil mechanics and rock mechanics.
	Soil mechanics: The majority of geotechnical engineering deals with soil mechanics and, in practice, the term "soils engineer" is synonymous with "geotechnical engineer." Soil has many different meanings, depending on the field of study. For example, in agronomy (application of science to farming) soil is defined as a surface deposit that contains mineral matter that originated from the original weathering of rock and also contains organic matter that has accumulated through the decomposition of plants and animals. To an agronomist, soil is that material that has been sufficiently altered and supplied with nutrients that it can support the growth of plant roots. But to a geotechnical engineer, soil has a much broader meaning and can include not only agronomic material, but also broken-up fragments of rock, volcanic ash, alluvium, aeolian sand, glacial material, and any other residual or transported product of rock weathering. Difficulties naturally arise because there is not a distinct dividing line between rock and soil. For example, to a geologist a given material may be classified as a formational rock because it belongs to a definite geologic environment, but to a geotechnical engineer it may be sufficiently weathered or friable that it should be classified as a soil.

TABLE 1.1 Definitions (*Continued*)

Term (1)	Definition (2)
Geotechnical engineering (*Continued*)	*Rock mechanics:* To the geotechnical engineer, rock is a relatively solid mass that has permanent and strong bonds between the minerals. Rocks can be classified as being either sedimentary, igneous, or metamorphic. There are significant differences in the behavior of soil versus rock, and there is not much overlap between soil mechanics and rock mechanics.
Foundation engineering	A foundation is defined as that part of the structure that supports the weight of the structure and transmits the load to underlying soil or rock. Some engineers consider foundation engineering to be a part of geotechnical engineering (e.g., Cernica 1995a), while others consider it to be a separate field of study (e.g., Holtz and Kovacs 1981). In general, foundation engineering applies the knowledge of geology, soil mechanics, rock mechanics, and structural engineering to the design and construction of foundations for buildings and other structures. The most basic aspect of foundation engineering deals with the selection of the type of foundation, such as using a shallow or deep foundation system. Another important aspect of foundation engineering involves the development of design parameters, such as the bearing capacity of the foundation. Foundation engineering could also include the actual foundation design, such as determining the type and spacing of steel reinforcement in concrete footings.
Engineering geologist	An engineering geologist is defined as an individual who applies geologic data, principles, and interpretation so that geologic factors affecting planning, design, construction, and maintenance of civil engineering works are properly recognized and utilized (*Geologist and Geophysicist Act* 1986).

Note: See App. A for additional terms and definitions.

TABLE 1.2 Common Types of Foundations

Category (1)	Common types (2)	Comments (3)
Shallow foundations	Spread footings (also called pad footings)	Spread footings are often square in plan view, are of uniform reinforced concrete thickness, and are used to support a single column load located directly in the center of the footing.
	Strip footings (also called wall footings)	Strip footings are often used for load-bearing walls. They are usually long, reinforced concrete members of uniform width and shallow depth.
	Combined footings	Reinforced-concrete combined footings are often rectangular or trapezoidal in plan view, and carry more than one column load.
	Conventional slab-on-grade	A continuous reinforced-concrete foundation consisting of bearing wall footings and a slab-on-grade. Concrete reinforcement often consists of steel rebar in the footings and wire mesh in the concrete slab.
	Posttensioned slab-on-grade	A continuous posttensioned concrete foundation. The posttensioning effect is created by tensioning steel tendons or cables embedded within the concrete. Common posttensioned foundations are the ribbed foundation, California slab, and PTI foundation.
	Raised wood floor	Perimeter footings that support wood beams and a floor system. Interior support is provided by pad or strip footings. There is a crawl space below the wood floor.
	Mat foundation	A large and thick reinforced-concrete foundation, often of uniform thickness, that is continuous and supports the entire structure. A mat foundation is considered to be a shallow foundation if it is constructed at or near ground surface.

TABLE 1.2 Common Types of Foundations (*Continued*)

Category (1)	Common types (2)	Comments (3)
Deep foundations	Driven piles	Driven piles are slender members, made of wood, steel, or precast concrete, that are driven into place by pile-driving equipment.
	Other types of piles	There are many other types of piles, such as bored piles, cast-in-place piles, and composite piles.
	Piers	Similar to cast-in-place piles, piers are often of large diameter and contain reinforced concrete. Pier and grade beam support is often used for foundation support on expansive soil.
	Caissons	Large piers are sometimes referred to as *caissons*. A caisson can also be a watertight underground structure within which construction work is carried on.
	Mat or raft foundation	If a mat or raft foundation is constructed below ground surface or if the mat or raft foundation is supported by piles or piers, then it should be considered to be a deep foundation system.
	Floating foundation	A special foundation type where the weight of the structure is balanced by the removal of soil and construction of an underground basement.
	Basement-type foundation	A common foundation for houses and other buildings in frost-prone areas. The foundation consists of perimeter footings and basement walls that support a wood floor system. The basement floor is usually a concrete slab.

Note: Classification of foundations as shallow or deep in this table is based on the depth of the soil or rock support of the foundation.

TABLE 1.3 Qualifying Experience for Geotechnical Engineers

Qualifying experience (1)	Typical items (2)
Development of programs of geotechnical investigation	Communication with other design consultants to determine their geotechnical input needs. Performance of literature searches and site history analyses related to surface and subsurface conditions. Formulation or engineering evaluation of field exploration and laboratory testing programs to accomplish the scope of the investigation. Preparation or engineering evaluation of proposals.
Geotechnical field and laboratory studies	Direction and/or modification of field exploration programs, as required, upon evaluation of the conditions being encountered. Classification and evaluation of subsurface conditions. Understanding the purposes for and being qualified to perform routine field and laboratory tests for soil strength, bearing capacity, expansion properties, consolidation, soil collapse potential, erosion potential, compaction characteristics, material acceptability for use in fill, pavement support qualities, freeze-thaw properties, grain size, permeability/percolation properties, groundwater conditions, and soil dynamic properties.
Analysis of geotechnical data and engineering computations	Analysis of field and laboratory data. Performance of computations using test results and available data regarding bearing capacity; foundation type, depth, and dimensions; allowable soil bearing pressures; potential settlement; slope stability; retaining systems; soil treatment; dewatering and drainage; floor support; pavement design; site preparation; fill construction; liquefaction potential; ground response to seismic forces; groundwater problems and seepage; and underpinning.
Performance or engineering evaluation of construction	Performance or supervision of geotechnical testing and observation of site grading. Analysis, design, and evaluation of instrumentation programs to evaluate or monitor various phenomena in the field, such as settlement, slope creep, pore water pressures, and groundwater variations.
Preparation or engineering evaluation of geotechnical reports	Preparation of plans, logs, and test results. Documentation testing and observation. Preparation of written reports which present findings, conclusions, and recommendations of the investigation. Preparation of specifications and guidelines.

Note: Adapted from the *California Plain Language Pamphlet of the Professional Engineers Act and Board Rules*, 1995.

TABLE 1.4 Fields of Expertise

Topic (1)	Engineering geologist (2)	Geotechnical engineer (3)	Overlapping areas of expertise (4)
Project planning	Development of geologic parameters Geologic feasibility	Design Material analysis Economics	Planning investigations Urban planning Environmental factors
Mapping	Geologic mapping Aerial photography Air photo interpretation Landforms Subsurface configurations	Topographic survey Surveying	Soil mapping Site selections
Exploration	Geologic aspects (fault studies, etc.)	Engineering aspects	Conducting field exploration Planning, observation, etc. Selecting samples for testing Describing and explaining site conditions
Engineering geophysics	Soil and rock hardness Mechanical properties Depth determinations	Engineering applications	Minimal overlapping of expertise
Classification and physical properties	Rock description Soil description (Modified Wentworth system)	Soil testing Earth materials Soil classification (USCS)	Soil description
Earthquakes	Location of faults Evaluation of active and inactive faults. Historic record of earthquakes	Response of soil and rock materials to seismic activity Seismic design of structures	Seismicity Seismic considerations Earthquake probability

TABLE 1.4 Fields of Expertise (*Continued*)

Topic (1)	Engineering geologist (2)	Geotechnical engineer (3)	Overlapping areas of expertise (4)
Rock mechanics	Rock mechanics Description of rock Rock structure, performance, and configuration	Rock testing Stability analysis Stress distribution	*In situ* studies Regional or local studies
Slope stability	Interpretative Geologic analyses and geometrics Spatial relationship	Engineering aspects of slope stability analysis and testing	Stability analyses Grading in mountainous terrain
Surface waters	Geologic aspects during design	Design of drainage systems Coastal and river engineering Hydrology	Volume of runoff Stream description Silting and erosion potential Source of material and flow Sedimentary processes
Groundwater	Occurrence Structural controls Direction of movement	Mathematical treatment of well systems Development concepts	Hydrology
Drainage	Underflow studies Storage computation Soil characteristics	Regulation of supply Economic factors Lab permeability	Well design, specific yield Field permeability Transmissibility

Note: Adapted from *Fields of Expertise,* undated.

TABLE 1.5 Areas of Responsibility

Responsibilities of the engineering geologist (1)	Responsibilities of the geotechnical engineer (2)
Description of the geologic environment pertaining to the engineering project.	Directing and coordinating the team efforts where engineering is a predominant factor.
Description of earth materials, such as their distribution and general physical and chemical characteristics.	Controlling the project in terms of time and money requirements and degree of safety desired.
Deduction of the history of pertinent events affecting the earth materials.	Deciding on optimum procedures.
Forecast of future events and conditions that may develop.	Making final judgments on economy and safety matters.
Recommendation of materials for representative sampling and testing.	Engineering testing and analysis.
Recommendation of ways of handling and treating various earth materials and processes.	Developing designs consistent with data and recommendations of team members.
Recommending or providing criteria for excavation (particularly angle of cut slopes) in materials where engineering testing is inappropriate or where geologic elements control stability.	Reviewing and evaluating data, conclusions, and recommendations of the team members.
Inspection during construction to confirm conditions.	Inspection during construction to assure compliance.

Note: Adapted from *Fields of Expertise*, undated.

TABLE 1.6 Types of Projects that Involve Geotechnical Engineers

Type of project (1)	Discussion (2)
Single-family dwellings and condominiums	Because of urban sprawl, the most numerous types of structures being built are single-family dwellings or condominiums and their associated roads and utilities. Especially for large housing tracts, the geotechnical engineer will usually have significant involvement with the project. Such projects are often divided into two basic categories: flatland and hillside. A more rigorous geotechnical and geologic investigation is often required for a hillside than a flatland site. A common feature of single-family dwellings and low-rise condominiums is the use of lightweight construction, such as wood framing or even aluminum framing. Usually footing widths and depths of single-family dwellings are governed by minimum building code requirements, rather than the loads applied to the foundation.
Commercial and industrial sites	The most common types of commercial projects are office buildings, including skyscrapers, that are either built specifically for the use of the client or rented out to various tenants. Common types of office buildings are steel-framed, reinforced-concrete, combined reinforced-concrete and steel buildings, and tilt-up concrete exterior panel buildings. Industrial sites can contain a variety of projects such as factories and refineries. Commercial and industrial projects frequently have a variety of loading and performance criteria that require special geotechnical investigation and foundation design.
Other projects in the private sector	There are many other types of private sector projects besides dwellings, commercial and industrial sites. Examples include the construction of small private dams, power plants, and energy transmission facilities, and transportation projects, such as privately owned roads.
Public works projects	This category of projects is very broad and includes all types of projects built with public money. Examples include levees and dams, harbors, airports, stadiums, and publicly owned buildings. This category also includes public transportation facilities, such as roads, highways, train beds, highway overpasses, bridges, and tunnels. Military projects are also included in this category, such as armories, waterway projects, military housing projects, and other military base facilities.
Essential facilities	Essential facilities are defined as those structures or buildings that must be safe and usable for emergency purposes after an earthquake or other natural disaster in order to preserve the health and safety of the general public. Typical examples of essential facilities are as follows (*Uniform Building Code* 1997): hospitals and other medical facilities having surgery or emergency treatment areas, fire and police stations, municipal government disaster operations, and communication centers deemed to be vital in emergencies.

TABLE 1.7 General Requirements for Foundation Engineering Projects

General requirement (1)	Discussion (2)
Topography	Knowledge of the general topography of the site as it affects foundation design and construction, e.g., surface configuration; adjacent property; the presence of watercourses, ponds, hedges, trees, rock outcrops, etc.; and the available access for construction vehicles and materials.
Utilities	The location of buried utilities such as electric power and telephone cables, water mains, and sewers.
Geology	The general geology of the area with particular reference to the main geologic formations underlying the site and the possibility of subsidence from mineral extraction or other causes.
Site history	The previous history and use of the site including information on any defects or failures of existing or former buildings attributable to foundation conditions.
Special features	Any special features such as the possibility of earthquakes and climate factors such as flooding, seasonal swelling and shrinkage, permafrost, or soil erosion.
Available materials	The availability and quality of local construction materials such as concrete aggregates, building and road stone, and water for construction purposes.
Marine structures	For maritime or river structures, information on tidal ranges and river levels, velocity of tidal and river currents, and other hydrographic and meteorological data.
Subsurface	A detailed record of the soil and rock strata and groundwater conditions within the zones affected by foundation bearing pressures and construction operations, or of any deeper strata affecting the site conditions in any way.
Laboratory testing	Results of laboratory tests on soil and rock samples appropriate to the particular foundation design or construction problems.
Chemical analysis	Results of chemical analyses on soil or groundwater to determine possible deleterious effects of foundation structures.

Source: Tomlinson 1986.

TABLE 1.8 Example of a Schedule of Fees

Professional and staff hourly rates	
Principal geotechnical engineer or principal engineering geologist	$/h
Chief geotechnical engineer or chief engineering geologist	$/h
Senior geotechnical engineer or senior engineering geologist	$/h
Project engineer or project geologist	$/h
Staff engineer or staff geologist	$/h
Associate engineer or associate geologist	$/h
Compaction testing technician	$/h
Drafting or CAD services	$/h
Office services	$/h
Subsurface exploration, compaction testing, and monitoring	
Drill rig rental costs (24-in.-diameter bucket auger boring)	$/h
Drill rig rental costs (solid- or hollow-stem auger)	$/h
Drill rig rental costs (rotary coring)	$/h
Drill rig rental costs (air track)	$/h
Bulldozer for construction of drill rig access roads	$/h
Test pit excavation costs	$/h
Trench excavation costs (backhoe)	$/h
Mobile laboratory equipment for field compaction	$/h
Inclinometer pipe and materials	$/ft
Piezometer pipe and materials	$/ft
Laboratory testing	
Moisture content (ASTM D 2216)	$/test
Wet density	$/test
Atterberg limits (ASTM D 4318) liquid/plastic	$/test
Particle size analysis (ASTM D 422)	$/test
Specific gravity—soils (ASTM D 854)	$/test
Specific gravity—oversize particles (ASTM C 127)	$/test
Sand equivalent (ASTM D 2419)	$/test
Collapse test (ASTM D 5333)	$/test
Swell test (ASTM D 4546)	$/test
Expansion index (UBC Std. 18-2)	$/test
Modified Proctor compaction test (ASTM D 1557)	$/test
R-value (ASTM D 2844)	$/test
Unconfined compression (ASTM D 2166)	$/test
Direct shear test (ASTM D 3080)	$/test
Triaxial compression test (ASTM D 4767)	$/test
Consolidation test (ASTM D 2435)	$/test
Hydraulic conductivity (permeability, ASTM D 2434 or D 5084)	$/test
Special handling, storage, and/or disposal	Hourly rates
Outside laboratory	Cost + 20%

TABLE 1.9 Example of a Cost Estimating Sheet

Category (1)	Description (2)	Hours (3)	Rate (4)	Cost (5)
Proposal	Planning and preparation of proposal			
Field exploration	In-house and agency research Review client's or other engineers' documents Subsurface exploration (drilling, test pits, trenches) Engineering geologist work (mapping, aerial photos, etc.) Preparation of logs and field paperwork Preparation of soil profile			
Laboratory testing	Soil classification tests (particle size, Atterberg limits) Moisture content and wet density determinations Settlement potential (consolidation, collapse tests) Expansion potential (expansion index, swell tests) Shear strength (direct shear, triaxial, etc.) Erosion and deterioration potential Compaction tests (Modified or Standard Proctor) Miscellaneous (specific gravity, sand equivalent, *R*-value)			
Analysis of data and engineering computations	Laboratory data reduction and analysis Engineering calculations (settlement, bearing capacity, expansive soil, slope stability, seismic analysis, etc.) Development of design parameters for foundations, retaining walls, effect of groundwater, etc. Computer analyses (slope stability, etc.) Design of geotechnical elements (foundations, etc.) Engineering geology analyses and recommendations			
Compaction testing and other construction services	Compaction testing (technician) Observations during grading by engineer and geologist Sampling and testing during grading operations Other construction services			

TABLE 1.9 Example of a Cost Estimating Sheet (*Continued*)

Category (1)	Description (2)	Hours (3)	Rate (4)	Cost (5)
Report preparation	Report writing, editing, and review Preparation of laboratory and field data Drafting and graphics for report Word processing and report production Blueprinting and production of plans			
Subcontract expenses	Subcontract expenses for drill rig rental Subcontract expenses for test pits and trenches Expenses for monitoring equipment Other subcontract expenses			

Total estimate = $______ Price quoted = $______

TABLE 1.10 Typical Items Included in the Contract between the Geotechnical Engineer and the Client

Typical items (1)	Discussion (2)
Contract title and introductory wording	The contract should contain a title and introductory wording indicating that the document is the contract between the geotechnical engineer and the client.
Project and client information	The section of the contract where the project name and address and client's name and address are inserted.
Type of services	The section of the contract where a brief summary of the scope of services is inserted.
Cost of services	The section of the contract where the cost estimate or not-to-exceed dollar amount is inserted.
Signature page	A final section of the contract which states that both the geotechnical engineer and client have read the contract and agree to all the terms and conditions. Spaces should be provided for both the geotechnical engineer and the client to sign and date the contract.
Extras	The contract should indicate that work requested by responsible parties outside the scope will be billed as "extras" on a time-and-expense basis under purview of this proposal, unless another proposal is specifically requested.
Safety	A statement indicating that the geotechnical engineer will not be responsible for the general safety on the job or the work of other contractors and third parties.
Termination and modification of agreement	The contract could indicate the procedure to terminate or modify the agreement, such as requiring that any termination or modification of the agreement must be in writing and signed by all parties.
Retainers and payments	Prompt payment for engineering services is always desirable. It may be appropriate to state that payments are due within 30 days upon receipt of the invoice for engineering services. An interest charge for payments beyond the due date could also be listed.
Limitation of liability	The contract could include a limitation of liability clause. Geotechnical engineering is often described as a risky profession, and these clauses are inserted in order to reduce the potential liability of the geotechnical engineer.

Note: It is always best to have an attorney prepare or review the contract. Other items which could be included in the contract include statements concerning the ownership of documents, disclaimer of warranties, the contract jurisdiction, and the time limit for the signing of the contract.

CHAPTER 2
FIELD EXPLORATION

The goal of the field exploration is to obtain a detailed understanding of the engineering and geologic properties of the soil and rock strata and groundwater conditions that could impact the proposed development.

2.1 DOCUMENT REVIEW

Often the first step in the field exploration is to perform a document review. Table 2.1 presents a list of typical documents that may need to be reviewed for the field exploration, and Table 2.2 provides a discussion of these types of documents. Examples are presented in Figs. 2.1 to 2.4.

2.2 SUBSURFACE EXPLORATION

There are different types of subsurface exploration, such as borings, test pits, and trenches. Table 2.3 summarizes the boring, core drilling, sampling, and other exploratory techniques that can be used by the geotechnical engineer. The type and number of borings selected for the project depends on many factors including experience and judgment. Table 2.4 lists various factors to be considered in the selection of borings, Table 2.5 presents details on the boring layout, and Table 2.6 presents guidelines on the depth of subsurface exploration. Table 2.7 presents the uses, capabilities, and limitations of test pits and trenches.

2.3 SOIL SAMPLING

Table 2.8 summarizes the three types of soil samples that can be recovered from borings, test pits, and trenches, and Fig. 2.5 presents a definition of terms for sampling tubes. Table 2.9 presents a list of various factors that can cause sample disturbance and Table 2.10 provides a discussion of radiographs that are used to detect sample disturbance. Figures 2.6 to 2.8 present three examples of radiographs.

2.4 FIELD TESTING

There are numerous types of field tests that can be performed during the subsurface exploration. The most common field test is the standard penetration test (SPT) (see Table 2.11). Table 2.12 presents a correlation between the measured SPT N-value (blows per foot) and the density condition of a clean sand deposit. Other types of field tests are discussed in Table 2.13 and Figs. 2.9 to 2.12.

2.5 EXPLORATORY LOGS

A log is defined as a written record prepared during the subsurface excavation of borings, test pits, or trenches that documents the observed conditions. Table 2.14 lists items that should be included on the excavation log. An important part of the preparation of logs is to determine the geologic or man-made process that created the soil deposit. Table 2.15 presents a list of common soil deposits encountered during subsurface exploration and Table 2.16 discusses soil and rock types that require special consideration. An example of a boring log is shown in Fig. 2.13.

2.6 GEOPHYSICAL TECHNIQUES

Geophysical techniques can be employed by the engineering geologist to obtain data on the subsurface conditions (see Table 2.17). Probably the most commonly used geophysical technique is the seismic refraction method, which can be used to determine the seismic velocity of the soil or rock strata. The seismic velocity can be used to determine whether the soil and rock strata are rippable with conventional excavation equipment (Figs. 2.14 to 2.17).

2.7 SUBSOIL PROFILE

The final section of this chapter presents examples of subsoil profiles. The results of the subsurface exploration are often summarized on a subsoil profile. Usually the engineering geologist is the person most qualified to develop the subsoil profile on the basis of experience and judgment in extrapolating conditions between the borings, test pits, and trenches.

Figures 2.18 to 2.21 show four examples of subsoil profiles. As shown in Figures 2.18 to 2.21, the results of field and laboratory tests are often included on the subsoil profiles. The development of a subsoil profile is often a required element for geotechnical and foundation engineering analyses. For example, subsoil profiles are used to determine the foundation type (shallow versus deep foundation), to calculate the amount of settlement or heave of the structure, to evaluate the effect of groundwater on the project and develop recommendations for dewatering of underground structures, to perform slope stability analyses for projects having sloping topography, and to prepare site development recommendations.

FIGURE 2.1 Geologic map (see Table 2.2). (*From Kennedy 1975.*)

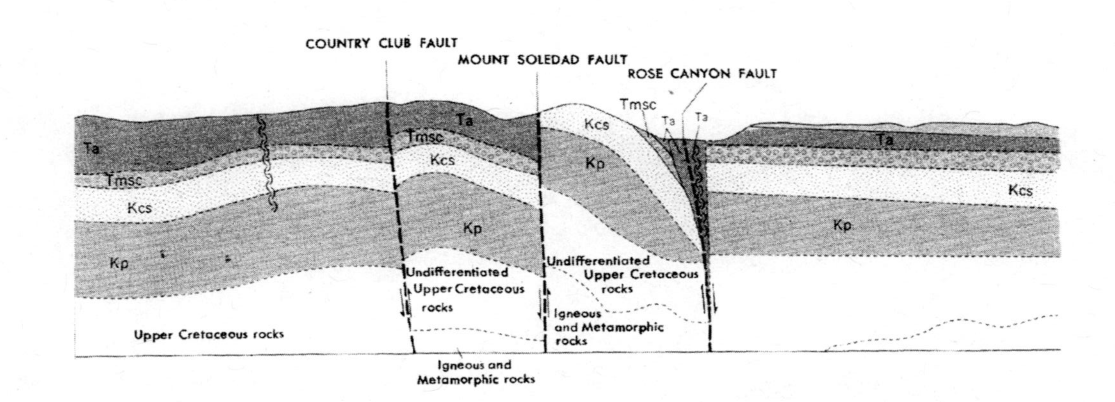

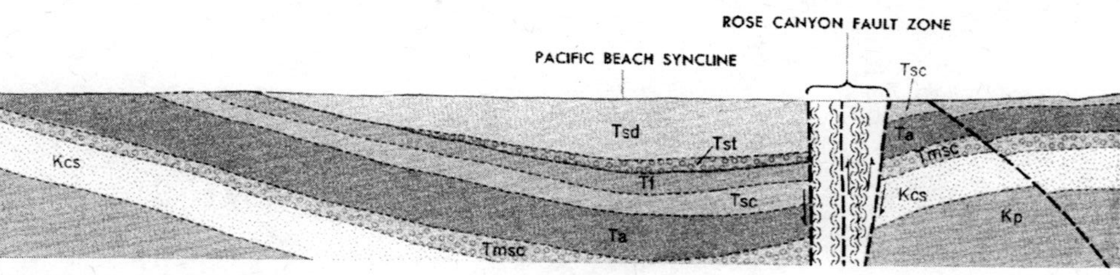

FIGURE 2.2 Geologic cross-sections (see Table 2.2). (*From Kennedy 1975.*)

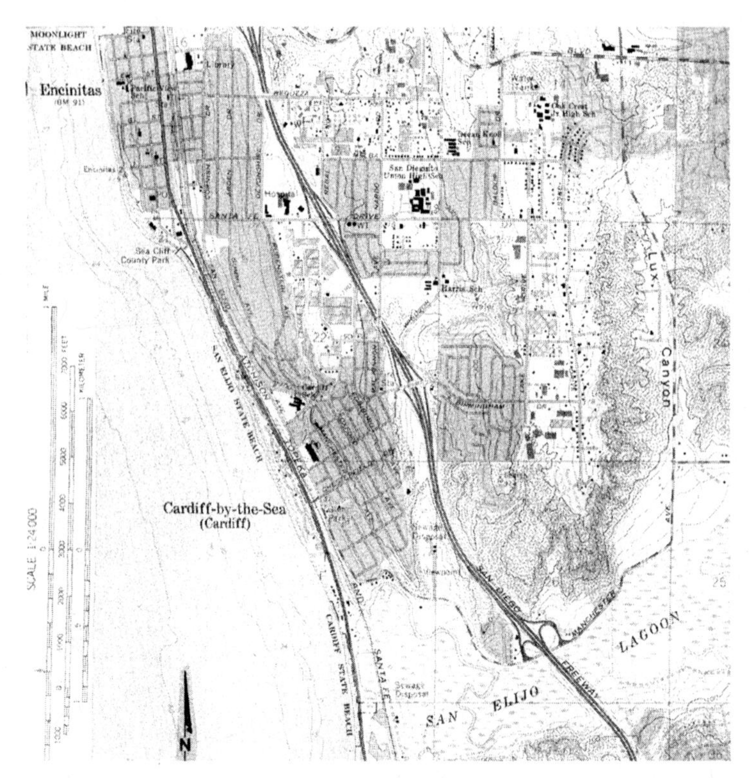

FIGURE 2.3 Topographic map (see Table 2.2). (*From USGS 1975.*)

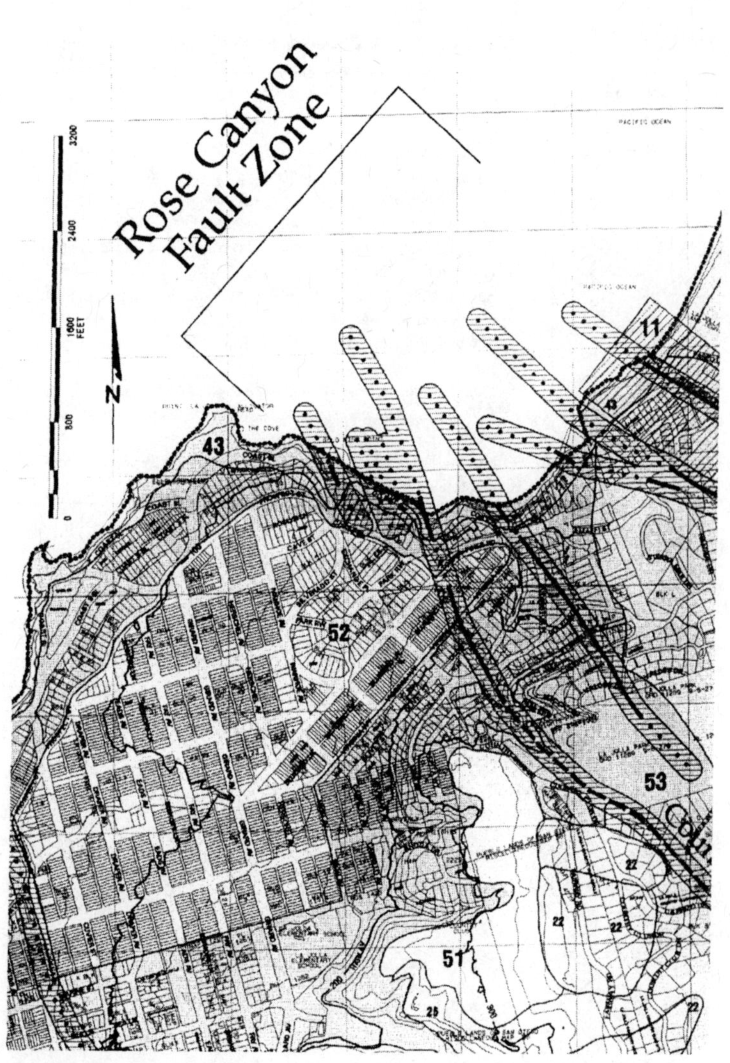

FIGURE 2.4 Portion of *Seismic Safety Study* 1995 (see Table 2.2). (*Developed by the City of San Diego.*)

TABLE 2.1 Typical Documents That May Need to Be Reviewed for the Project

Project phase (1)	Type of documents (2)
Design phase of the project	1. Available design information, such as preliminary data on the type of project to be built at the site and typical foundation design loads. 2. If applicable, data on the history of the site, such as information on prior fill placement or construction at the site. 3. Data (if available) on the design and construction of adjacent property. 4. Local building code. 5. Special study data, developed by the local building department or other governing agency. 6. Standard drawings issued by the local building department or other governing agency. 7. Standard specifications that may be applicable to the project, such as *Standard Specifications for Public Works Construction* or *Standard Specifications for Highway Bridges.* 8. Other reference material, such as seismic activity records, geologic and topographic maps, aerial photographs.
Construction phase of the project	1. Reports and plans developed during the design phase 2. Construction specifications. 3. Field change orders. 4. Information bulletins used during construction. 5. Project correspondence between different parties. 6. Building department reports or permits.

TABLE 2.2 Discussion of the Types of Documents That May Need to Be Reviewed for the Project

Type of document (1)	Discussion (2)
Preliminary design information	The documents dealing with preliminary design and proposed construction of the project should be reviewed. For example, the structural engineer or architect may have design information, such as the building size, height, loads, and details on proposed construction materials and methods. Preliminary plans may even have been developed that show the proposed construction.
History of the site	If the site had prior development, it is also important to obtain information on the history of the site. The site could contain old deposits of fill, abandoned septic systems and leach fields, buried storage tanks, seepage pits, cisterns, mining shafts, tunnels, and other man-made surface and subsurface works that could impact the new proposed development. There may also be information concerning on-site utilities and underground pipelines, which may need to be capped or rerouted around the project.
Aerial photographs	Aerial photographs are taken from an aircraft flying at a prescribed altitude along preestablished lines. Interpretation of aerial photographs takes considerable judgment and it is usually the engineering geologist who interprets the aerial photographs. By viewing a pair of aerial photographs, with the aid of a stereoscope, a three-dimensional view of the land surface is provided. This view may reveal important geologic information at the site, such as the presence of landslides, fault scarps, types of landforms (e.g., dunes, alluvial fans, glacial deposits such as moraines and eskers), erosional features, general type and approximate thickness of vegetation, and drainage patterns. By comparing older versus newer aerial photographs, any man-made or natural changes that have occurred at the site can be observed.
Geologic maps	Geologic maps can be especially useful because they often indicate potential geologic hazards (e.g., faults, landslides) as well as the type of near-surface soil or rock at the site. A major source for geologic maps in the United States is the United States Geological Survey (USGS). The USGS prepares many different geologic maps, books, and charts, and a list of the USGS publications is provided in their *Index of Publications of the Geological Survey* (USGS 1997). The USGS also provides an *Index to Geologic Mapping in the United States*, which shows a map of each state and indicates the areas where a geologic map has been published. Figure 2.1 presents a portion of a geologic map and Fig. 2.2 shows cross sections through the area shown in Figure 2.1 (from Kennedy 1975). Geologic maps and cross sections indicate the location of faults and the width of fault zones, and often state whether the faults are active or inactive. For example, in Fig. 2.2, the Rose Canyon Fault Zone is shown, which is an active fault having a wide shear zone that has displaced rock layers. Geologic symbols are used to identify different soil and rock layers.

TABLE 2.2 Discussion of the Types of Documents That May Need to Be Reviewed for the Project (*Continued*)

Type of document (1)	Discussion (2)
Topographic maps	Both old and recent topographic maps can provide valuable site information. The topographic map is usually to scale and shows the locations of buildings, roads, freeways, train tracks, and other civil engineering works as well as natural features such as canyons, rivers, lagoons, sea cliffs, and beaches. The topographic map can even show the locations of sewage disposal ponds, and water tanks, and, by using different colors and shading, it indicates older versus newer development. A main purpose of the topographic map is to indicate ground surface elevations or elevations of the sea floor. This information can be used to determine the major topographic features at the site and for the planning of subsurface exploration, such as determining available access to the site for drilling rigs. Figure 2.3 presents a portion of a topography map from the Encinitas Quadrangle, Encinitas, California (from USGS 1975).
Special study maps	For some areas, special study maps may have been developed that indicate local hazards. For example, Fig. 2.4 presents a portion of the *Seismic Safety Study* (1995) that shows the location of the Rose Canyon Fault Zone. Special study maps may also indicate other hazards, such as potentially liquefiable soil, landslides, and abandoned mines.
Building code and other specifications	A copy of the most recently adopted local building code should be reviewed. Usually only a few sections of the building code will be directly applicable to geotechnical projects. For example, the main applicable geotechnical sections in the *Uniform Building Code* (1997) are Chap. 18, "Foundations and Retaining Walls," and Chap. 33, "Excavation and Grading." Depending on the type of project, there may be other specifications that are applicable for the project and will need to be reviewed. Documents that may be needed for public works projects include the *Standard Specifications for Public Works Construction* (1997) or the *Standard Specifications for Highway Bridges* (AASHTO 1996).
Documents at the local building department	Other useful technical documents include geotechnical reports for adjacent properties, which can provide an idea of possible subsurface conditions. A copy of geotechnical reports on adjacent properties can often be obtained at the archives of public agencies, such as the local building department. Other valuable reference materials are standard drawings or standard specifications, which can also be obtained from the local building department.

TABLE 2.3 Boring, Core Drilling, Sampling, and Other Exploratory Techniques

Method (1)	Procedure (2)	Type of sample (3)	Applications (4)	Limitations (5)
Auger boring, ASTM D 1452	Dry hole drilled with hand or power auger; samples preferably recovered from auger flutes	Auger cuttings, disturbed, ground up, partially dried from drill heat in hard materials	In soil and soft rock; to identify geologic units and water content above water table	Soil and rock stratification destroyed; sample mixed with water below the water table
Test boring, ASTM D 1586	Hole drilled with auger or rotary drill; at intervals samples taken 36-mm ID and 50-mm OD driven 0.45 m in three 150-mm increments by 64-kg hammer falling 0.76 m; hydrostatic balance of fluid maintained below water level	Intact but partially disturbed (number of hammer blows for second plus third increment of driving is standard penetration resistance or *N*)	To identify soil or soft rock; to determine water content; in classification tests and crude shear test of sample (*N*-value a crude index to density of cohesionless soil and undrained shear strength of cohesive soil)	Gaps between samples, 30 to 120 cm; sample too distorted for accurate shear and consolidation tests; sample limited by gravel; *N*-value subject to variations, depending on free fall of hammer
Test boring of large samples	50- to 75-mm ID and 63- to 89-mm OD samplers driven by hammers up to 160 kg	Intact but partially disturbed (number of hammer blows for second plus third increment of driving is penetration resistance)	In gravelly soils	Sample limited by larger gravel
Test boring through hollow-stem auger	Hole advanced by hollow-stem auger; soil sampled below auger as in test boring above	Intact but partially disturbed (number of hammer blows for second plus third increment of driving is *N*-value)	In gravelly soils (not well adapted to harder soils or soft rock)	Sample limited by larger gravel; maintaining hydrostatic balance in hole below water table is difficult
Rotary coring of soil or soft rock	Outer tube with teeth rotated; soil protected and held stationary in inner tube; cuttings flushed upward by drill fluid (examples: Denison, Pitcher, and Acker samplers)	Relatively undisturbed sample, 50 to 200 mm wide and 0.3 to 1.5 m long in liner tube	In firm to stiff cohesive soils and soft but coherent rock	Sample may twist in soft clays; sampling loose sand below water table is difficult; success in gravel seldom occurs
Rotary coring of swelling clay, soft rock	Similar to rotary coring of rock; swelling core retained by third inner plastic liner	Soil cylinder 28.5 to 53.2 mm wide and 600 to 1500 mm long, encased in plastic tube	In soils and soft rocks that swell or disintegrate rapidly in air (protected by plastic tube)	Sample smaller; equipment more complex

TABLE 2.3 Boring, Core Drilling, Sampling, and Other Exploratory Techniques (*Continued*)

Method (1)	Procedure (2)	Type of sample (3)	Applications (4)	Limitations (5)
Rotary coring of rock, ASTM D 2113	Outer tube with diamond bit on lower end rotated to cut annular hole in rock; core protected by stationary inner tube; cuttings flushed upward by drill fluid	Rock cylinder 22 to 100 mm wide and as long as 6 m, depending on rock soundness	To obtain continuous core in sound rock (percent of core recovered depends on fractures, rock variability, equipment, and driller skill)	Core lost in fractured or variable rock; blockage prevents drilling in badly fractured rock; dip of bedding and joint evident but not strike
Rotary coring of rock, oriented core	Similar to rotary coring of rock above; continuous grooves scribed on rock core with compass direction	Rock cylinder, typically 54 mm wide and 1.5 m long with compass orientation	To determine strike of joints and bedding	Method may not be effective in fractured rock
Rotary coring of rock, wire line	Outer tube with diamond bit on lower end rotated to cut annular hole in rock; core protected by stationary inner tube; cuttings flushed upward by drill fluid; core and stationary inner tube retrieved from outer core barrel by lifting device or "overshot" suspended on thin cable (wire line) through special large-diameter drill rods and outer core barrel	Rock cylinder 36.5 to 85 mm wide and 1.5 to 4.6 m long	To recover core better in fractured rock, which has less tendency for caving during core removal; to obtain much faster cycle of core recovery and resumption of drilling in deep holes	Same as ASTM D 2113 but to lesser degree
Rotary coring of rock, integral sampling method	22-mm hole drilled for length of proposed core; steel rod grouted into hole; core drilled around grouted rod with 100- to 150-mm rock coring drill (same as for ASTM D 2113)	Continuous core reinforced by grouted steel rod	To obtain continuous core in badly fractured, soft, or weathered rock in which recovery is low by ASTM D 2113	Grout may not adhere in some badly weathered rock; fractures sometimes cause drift of diamond bit and cutting rod
Thin-wall tube, ASTM D 1587	75- to 1250-mm thin-wall tube forced into soil with static force (or driven in soft rock); retention of sample helped by drilling mud	Relatively undisturbed sample, length 10 to 20 diameters	In soft to firm clays, short (5-diameter) samples of stiff cohesive soil, soft rock, and, with aid of drilling mud, in firm to dense sands	Cutting edge wrinkled by gravel; samples lost in loose sand or very soft clay below water table; more disturbance occurs if driven with hammer

Thin-wall tube, fixed piston	75- to 1250-mm thin-wall tube, which has internal piston controlled by rod and keeps loose cuttings from tube, remains stationary while outer thin-wall tube forced ahead into soil; sample in tube is held in tube by aid of piston	Relatively undisturbed sample, length 10 to 20 diameters	To minimize disturbance of very soft clays (drilling mud aids in holding samples in loose sand below water table)	Method is slow and cumbersome
Swedish foil	Samples surrounded by thin strips of stainless steel, stored above cutter, to prevent contact of soil with tube as it is forced into soil	Continuous samples 50 mm wide and as long as 12 m	In soft, sensitive clays	Samples sometimes damaged by coarse sand and fine gravel
Dynamic sounding	Enlarged disposable point on end of rod driven by weight falling fixed distance in increments of 100 to 300 mm	None	To identify significant differences in soil strength or density	Misleading in gravel or loose saturated fine cohesionless soils
Static penetration	Enlarged cone, 36 mm diameter and 60° angle, forced into soil; force measured at regular intervals	None	To identify significant differences in soil strength or density; to identify soil by resistance of friction sleeve	Stopped by gravel or hard seams
Borehole camera	Inside of core hole viewed by circular photograph or scan	Visual representation	To examine stratification, fractures, and cavities in hole walls	Best above water table or when hole can be stabilized by clear water
Pits and trenches	Pit or trench excavated to expose soils and rocks	Chunks cut from walls of trench; size not limited	To determine structure of complex formations; to obtain samples of thin critical seams such as failure surface	Moving excavation equipment to site, stabilizing excavation walls, and controlling groundwater may be difficult

TABLE 2.3 Boring, Core Drilling, Sampling, and Other Exploratory Techniques (*Continued*)

Method (1)	Procedure (2)	Type of sample (3)	Applications (4)	Limitations (5)
Rotary or cable tool well drill	Toothed cutter rotated or chisel bit pounded and churned	Ground	To penetrate boulders, coarse gravel; to identify hardness from drilling rates	Identifying soils or rocks difficult
Percussion drilling (jack-hammer or air track)	Impact drill used; cuttings removed by compressed air	Rock dust	To locate rock, soft seams, or cavities in sound rock	Drill becomes plugged by wet soil

Note: Reprinted with permission from *Landslides: Analysis and Control, Special Report 176*. Copyright 1978 by the National Academy of Sciences. Courtesy of the National Academy Press, Washington, D.C.

Source: Sowers and Royster (1978), based on the work by ASTM (1970, 1971), Lambe (1951), Sanglerat (1972), and Sowers and Sowers (1970). Also see ASCE (1972, 1976, and 1978).

TABLE 2.4 Guidelines for Number of Borings

Item (1)	Discussion (2)
Relative costs of the investigation	The cost of additional borings must be weighed against the value of additional subsurface information.
Type of project	A more detailed and extensive subsurface investigation is required for an essential facility as compared to a single-family dwelling.
Topography	A hillside project usually requires more subsurface investigation than a flatland project because of the slope stability requirements.
Nature of soil deposits	Fewer borings may be needed when the soil deposits are uniform as compared to erratic deposits.
Geologic hazards	The more known or potential geologic hazards at the site, the greater the need for subsurface exploration.
Access	In many cases, the site may be inaccessible and access roads will have to be constructed. Creating access roads throughout the site can be expensive and disruptive and may influence decisions on the number and spacing of borings.
Governmental or local building department requirements	For some projects, there may be specifications on the required number and spacing of borings. For example, the *Standard Specifications for Highway Bridges* (AASHTO 1996) states: "A minimum of one soil boring shall be made for each substructure unit [note: a substructure unit is defined as every pier, abutment, retaining wall, foundation, or similar item]. For substructure units over 30 m (100 ft) in width, a minimum of two borings shall be required."
Preliminary subsurface plan	Often a preliminary subsurface plan is developed to perform a limited number of exploratory borings. The purpose is just to obtain a rough idea of the soil, rock, and groundwater conditions at the site. Then once the preliminary subsurface data is analyzed, additional borings as part of a detailed exploration are performed. The detailed subsurface exploration can be used to better define the soil profile, explore geologic hazards, and obtain further data on the critical subsurface conditions that will likely have the most impact on the design and construction of the project.

TABLE 2.5 Guidelines for Boring Layout

Areas of investigation (1)	Boring layout (2)
New site of wide extent	Space preliminary borings 60 to 150 m (200 to 500 ft) apart so that area between any four borings includes approximately 10% of total area. In detailed exploration, add borings to establish geological sections at the most useful orientations.
Development of site on soft compressible soil	Space borings 30 to 60 m (100 to 200 ft) at possible building locations. Add intermediate borings when building site is determined.
Large structure with separate closely spaced footings	Space borings approximately 15 m (50 ft) in both directions, including borings at possible exterior foundation walls, at machinery or elevator pits, and at the most useful orientations to establish geologic sections.
Low-load warehouse building of large area	Minimum of four borings at corners plus intermediate borings at interior foundations sufficient to define subsoil profile.
Isolated rigid foundation	For foundation 230 to 930 m^2 (2500 to 10,000 ft^2) in area, minimum of three borings around perimeter. Add interior borings, depending on initial results.
Isolated rigid foundation	For foundation less than 230 m^2 (2500 ft^2) in area, minimum of two borings at opposite corners. Add more for erratic conditions.
Major waterfront structures, such as dry docks	If definite site is established, space borings generally not farther than 15 m (50 ft), adding intermediate borings at critical locations, such as deep pump well, gate seat, tunnel, or culverts.
Long bulkhead or wharf wall	Preliminary borings on line of wall at 60-m (200-ft) spacing. Add intermediate borings to decrease spacing to 15 m (50 ft). Place certain intermediate borings inboard and outboard of wall line to determine materials in scour zone at toe and in active wedge behind wall.
Cut stability, deep cuts, and high embankments	Provide three to five borings on line in the critical direction to provide geological section for analysis. Number of geologic sections depends on extent of stability problem. For an active slide, place at least one boring upslope of sliding area.
Dams and water-retention structures	Space preliminary borings approximately 60 m (200 ft) over foundation area. Decrease spacing on centerline to 30 m (100 ft) by intermediate borings. Include borings at location of cutoff, critical spots in abutment, spillway, and outlet works.

Source: NAVFAC DM-7.1 (1982).

TABLE 2.6 Guidelines for Boring Depths

Topic (1)	Discussion (2)
Large structure with separate closely spaced footings	Extend to depth where increase in vertical stress from combined footings is less than 10% of effective overburden stress. Generally all borings should extend to no less than 9 m (30 ft) below lowest part of foundation unless rock is encountered at shallower depth.
Isolated rigid foundations	Extend to depth where vertical stress decreases to 10% of bearing pressure. Generally all borings should extend no less than 9 m (30 ft) below lowest part of foundation unless rock is encountered at shallower depth.
Long bulkhead or wharf wall	Extend to depth below dredge line between 0.75 and 1.5 times unbalanced height of wall. Where stratification indicates possible deep stability problem, selected borings should reach top of hard stratum.
Slope stability	Extend to an elevation below active or potential failure surface and into hard stratum, or to a depth for which failure is unlikely because of geometry of cross section.
Deep cuts	Extend to depth between 0.75 and 1 times base width of narrow cuts. Where cut is above groundwater in stable materials, depth of 1.2 to 2.4 m (4 to 8 ft) below base may suffice. Where base is below groundwater, determine extent of pervious strata below base.
High embankments	Extend to depth between 0.5 to 1.25 times horizontal length of side slope in relatively homogeneous foundation. Where soft strata are encountered, borings should reach hard materials.
Dams and water-retention structures	Extend to depth of 0.5 base width of earth dams or 1 to 1.5 times height of small concrete dams in relatively homogeneous foundations. Borings may terminate after penetration of 3 to 6 m (10 to 20 ft) in hard and impervious stratum if continuity of this stratum is known from reconnaissance.
Minimum depth of borings	Borings should always be extended through unsuitable foundation bearing material, such as uncompacted fill, peat, soft clays and organic soil, and loose sands, and into dense soil or hard rock of adequate bearing capacity.
Required specifications	For some projects, there may be specifications on the required depth of borings. For example, the *Standard Specifications for Highway Bridges* (AASHTO 1996) states: "When substructure units will be supported on deep foundations, the depth of subsurface exploration shall extend a minimum of 6 m (20 ft) below the anticipated pile or shaft tip elevation. Where pile or shaft groups will be used, the subsurface exploration shall extend at least two times the maximum pile group dimension below the anticipated tip elevation, unless the foundation will be end bearing on or in rock. For piles bearing on rock, a minimum of 3 m (10 ft) of rock core shall be obtained at each exploration location to insure the exploration has not been terminated on a boulder."

TABLE 2.6 Guidelines for Boring Depths (*Continued*)

Topic (1)	Discussion (2)
Backfill	At the completion of each boring, it should immediately be backfilled with on-site soil and compacted by using the drill rig equipment. In certain cases, the holes may need to be filled with a cement slurry or grout. For example, if the borehole is to be converted to an inclinometer (slope monitoring device), then it should be filled with a weak cement slurry. Likewise, if the hole is to be converted to a piezometer (pore water pressure monitoring device), then special backfill materials, such as a bentonite seal, will be required. It may also be necessary to seal the hole with grout or bentonite if there is the possibility of water movement from one stratum to another. For example, holes may need to be filled with cement or grout if they are excavated at the proposed locations of dams, levees, or reservoirs.

Sources: NAVFAC DM-7.1 (1982) and Day 1999.

TABLE 2.7 Use, Capabilities, and Limitations of Test Pits and Trenches

Exploration method (1)	General use (2)	Capabilities (3)	Limitations (4)
Hand-excavated test pits	Bulk sampling, *in situ* testing, visual inspection.	Provides data in inaccessible areas, less mechanical disturbance of surrounding ground.	Expensive, time-consuming, limited to depths above groundwater level.
Backhoe-excavated test pits and trenches	Bulk sampling, *in situ* testing, visual inspection, excavation rates, depth of bedrock and groundwater.	Fast, economical, generally less than 4.6 m (15 ft) deep, can be up to 9 m (30 ft) deep.	Equipment access, generally limited to depths above groundwater level, limited undisturbed sampling.
Dozer cuts	Bedrock characteristics, depth of bedrock and groundwater level, rippability, increase depth capability of backhoe, level area for other exploration equipment.	Relatively low cost, exposures for geologic mapping.	Exploration limited to depth above the groundwater table.
Trenches for fault investigations	Evaluation of presence and activity of faulting and sometimes landslide features.	Definitive location of faulting, subsurface observation up to 9 m (30 ft) deep.	Costly, time-consuming, requires shoring, only useful where datable materials are present, depth limited to zone above the groundwater level.

Source: NAVFAC DM-7.1 (1982).

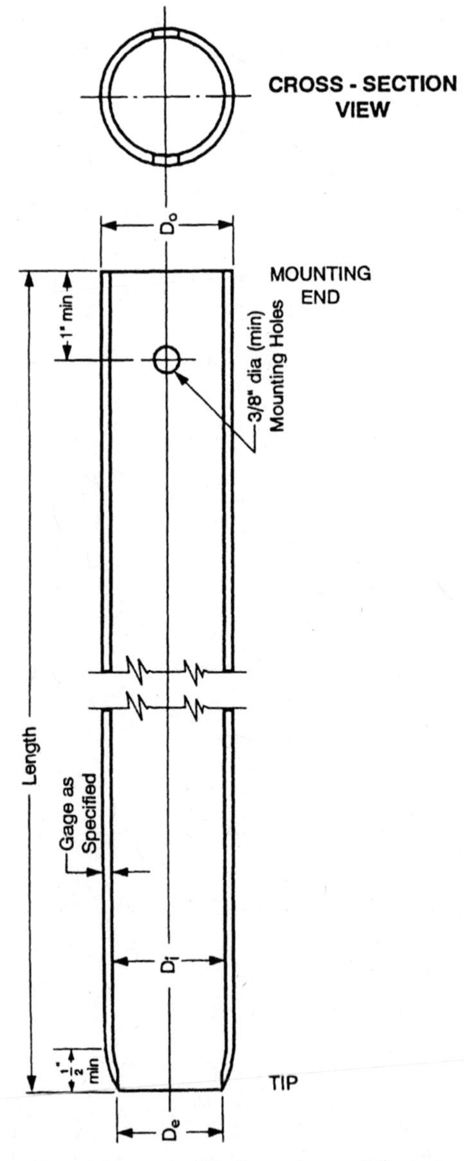

FIGURE 2.5 Definition of terms for sampling tube (see Table 2.8).

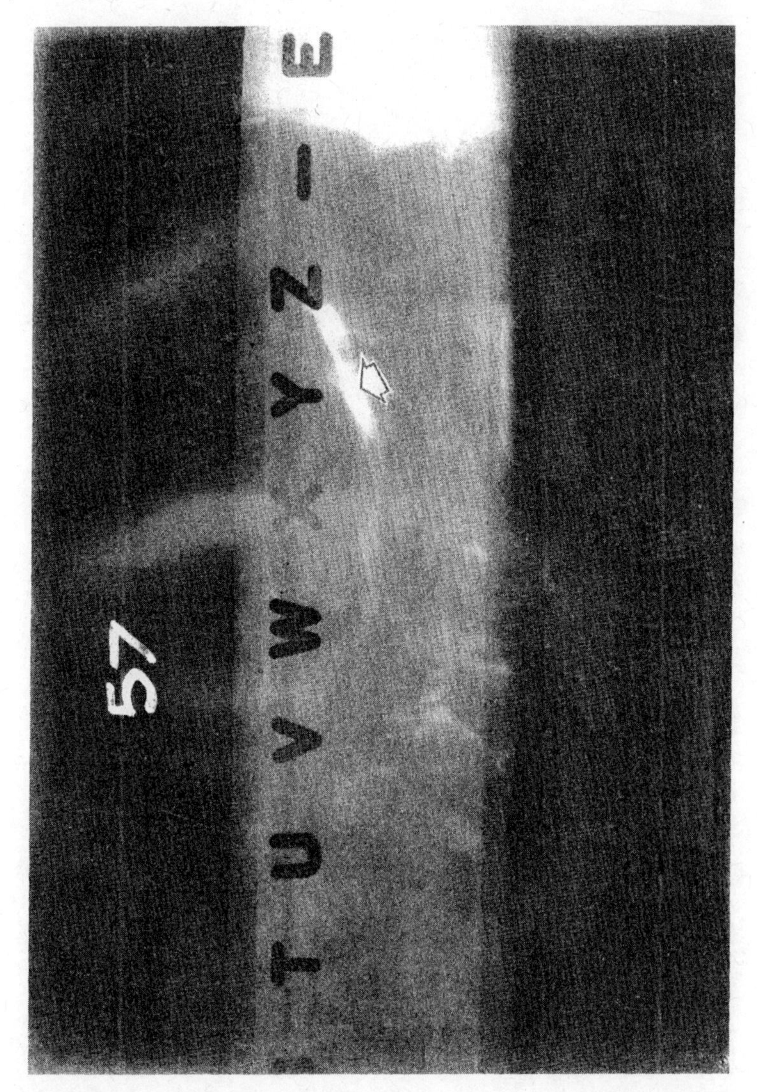

FIGURE 2.6 Radiograph of Orinoco clay within a Shelby tube (see Table 2.10). (*From Day 1980 and Ladd et al. 1980.*)

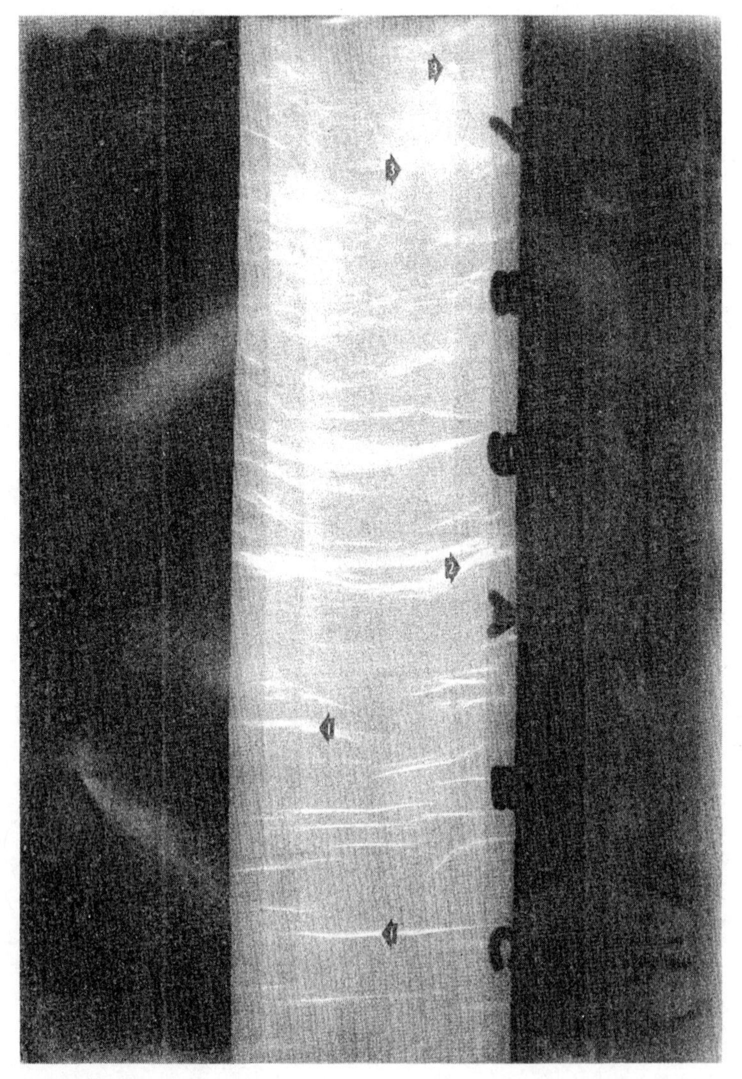

FIGURE 2.7 Radiograph of Orinoco clay within a Shelby tube (see Table 2.10). (*From Day 1980 and Ladd et al. 1980.*)

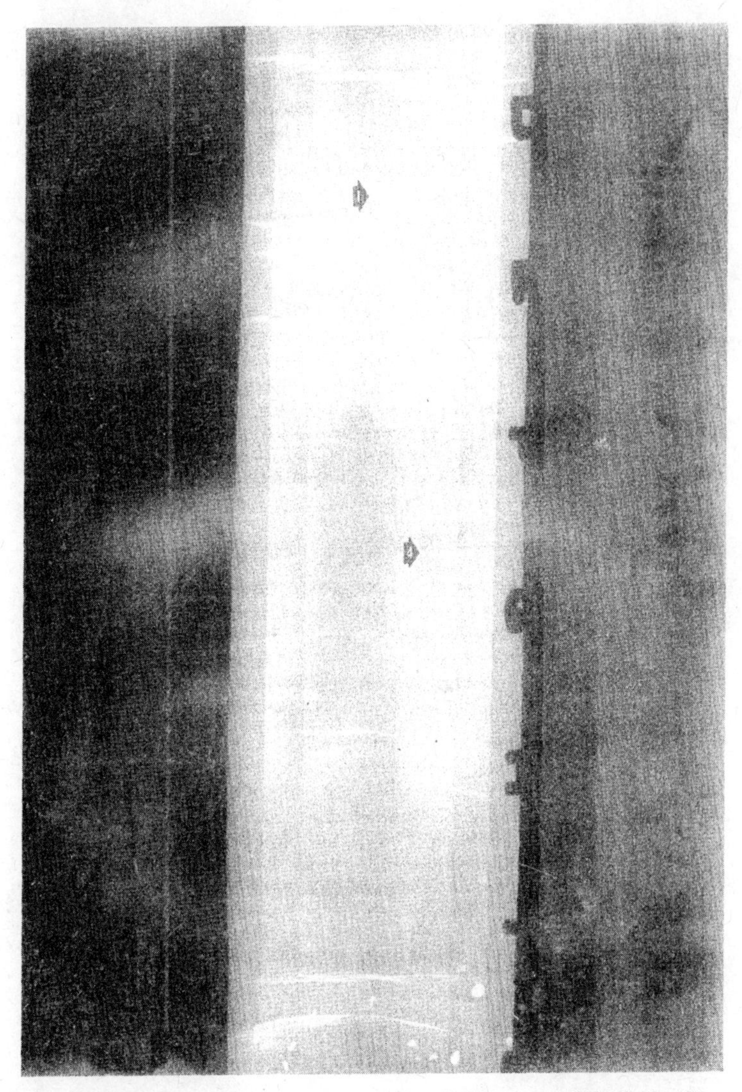

FIGURE 2.8 Radiograph of Orinoco clay within a Shelby tube (see Table 2.10). (*From Day 1980 and Ladd et al. 1980.*)

TABLE 2.8 Three Types of Soil Samples Recovered during Subsurface Exploration

Type of soil sample (1)	Discussion (2)
Altered soil	During the boring operations, soil can be altered by mixing or contamination. For example, if the boring is not cleaned out prior to sampling, a soil sample taken from the bottom of the borehole may actually consist of cuttings from the side of the borehole. These borehole cuttings, which have fallen to the bottom of the borehole, will not represent *in situ* conditions at the depth sampled. In other cases, the soil sample may become contaminated with drilling fluid, which is used for wash-type borings. These types of soil samples that have been mixed or contaminated by the drilling process should not be used for laboratory tests because they will lead to incorrect conclusions regarding subsurface conditions. Soil that has a change in water content due to the drilling fluid or from heat generated during the drilling operations should also be classified as altered soil. Soil that has been densified by overpushing or overdriving the soil sampler should also be considered as altered because the process of overpushing or overdriving could squeeze water from the soil.
Disturbed samples	Disturbed soil is defined as soil that has been remolded during the sampling process. For example, soil obtained from driven samplers, such as the standard penetration test (SPT), split spoon sampler, or chunks of intact soil brought to the surface in an auger bucket (i.e., bulk samples) are considered disturbed soil. Disturbed soil can be used for numerous types of laboratory tests, such as water content, specific gravity, soil classification tests, and Proctor compaction tests.
Undisturbed sample	It should be recognized that no soil sample can be taken from the ground and be in a perfectly undisturbed state. But this terminology has been applied to those soil samples taken by certain sampling methods. Undisturbed samples are often defined as those samples obtained by slowly pushing thin-walled tubes, having sharp cutting ends and tip relief, into the soil. Two parameters, the inside clearance ratio and area ratio, are often used to evaluate the disturbance potential of different samplers, and they are defined as follows: $$\text{Inside clearance ratio (\%)} = 100\,\frac{D_i - D_e}{D_e}$$ $$\text{Area ratio (\%)} = 100\,\frac{D_o^2 - D_i^2}{D_i^2}$$ where D_e = diameter at the sampler cutting tip, D_i = inside diameter of the sampling tube, and D_o = outside diameter of the sampling tube as shown in Fig. 2.5. In general, a sampling tube for undisturbed soil specimens should have an inside clearance ratio of about 1% and an

TABLE 2.8 Three Types of Soil Samples Recovered during Subsurface Exploration (*Continued*)

Type of soil sample (1)	Discussion (2)
Undisturbed sample (*Continued*)	area ratio of about 10% or less. Having an inside clearance ratio of about 1% provides for tip relief of the soil and reduces the friction between the soil and inside of the sampling tube during the sampling process. A thin film of oil can be applied at the cutting edge to also reduce the friction between the soil and metal tube during sampling operations. The purpose of having a low area ratio and a sharp cutting end is to slice into the soil with as little disruption and displacement of the soil as possible. Shelby tubes are manufactured to meet these specifications and are considered to be undisturbed soil samplers. As a comparison, the California sampler has an area ratio of 44% and is considered to be a thick-walled sampler.

TABLE 2.9 Factors That Can Cause Sample Disturbance

Factors (1)	Discussion (2)
Sample disturbance	An undisturbed soil specimen will have little rearrangement of the soil particles and perhaps no disturbance except that caused by stress relief where there is a change from the *in situ* k_o (at rest) condition to an isotropic "perfect sample" stress condition (Ladd and Lambe 1963). A disturbed soil specimen will have a decrease in effective stress, a reduction in the interparticle bonds, and a rearrangement of the soil particles. In measuring the shear strength or deformation characteristics of the soil, the results of laboratory tests run on undisturbed specimens obviously better represent *in situ* properties than laboratory tests run on disturbed specimens.
Gravel-size particles	Pieces of hard gravel or shell fragments in the soil can cause voids to develop along the sides of the sampling tube during the sampling process.
Stress relief	The soil in the ground is confined by the overburden pressure. The soil specimen can become disturbed during the reduction in confining pressure during the sampling process.
Sampling process	There can be sample disturbance and disruption of the soil structure due to hammering or pushing the sampling tube into the soil stratum.
Expansion of gas	The pore water may contain dissolved gas. When the confining pressure is removed during the sampling process, this gas can come out of solution and cause sample disturbance.
Transportation of soil samples	Sample disturbance can occur during the transportation of the soil samples to the laboratory. Banging or jarring the soil within the sampling tube can disrupt the soil structure.
Removing the sample from the sampling tube	If there is significant binding between the soil and interior of the sampling tube, the soil sample can be densified as it is pushed out of the sampling tube. If soil samples have been left in the sampler tube for a long time, there may be sample disturbance due to drying or moisture migration within the soil sample.
Preparation of soil specimen for laboratory testing	The soil specimen can be disturbed during the trimming, handling, and setup process of preparing the soil specimen for laboratory testing.

Note: Soil samples recovered from the subsurface exploration should be kept within the sampling tube or sampling rings. The soil sampling tube should be tightly sealed with end caps or the sampling rings thoroughly sealed in containers to prevent a loss of moisture during transportation to the laboratory. The soil samples should be marked with the file or project number, date of sampling, name of engineer or geologist who performed the sampling, and boring number and depth (e.g., B-1 @ 20–21 ft).

TABLE 2.10 Use of Radiographs to Detect Sample Disturbance

Topic (1)	Discussion (2)
Definition	A radiograph is a photographic record produced by the passage of x-rays through an object and onto photographic film. One method of assessing the quality of soil samples is to obtain an x-ray radiograph of the soil contained in the sampling tube (ASTM D 4452-95, 1998). Denser objects absorb the x-rays and can appear as dark areas on the radiograph. Worm holes, coral fragments, cracks, gravel inclusions, and sand or silt seams can easily be identified by using radiography (Allen et al. 1978).
Examples	Figures 2.6 to 2.8 (from Day 1980, Ladd et al. 1980) present three radiographs taken of Orinoco clay contained within Shelby tubes. These three radiographs illustrate the common types of soil disturbance.
Types of observed disturbance	Typical types of observed disturbance are as follows: 1. *Voids:* The arrow in Fig. 2.6 points to a soil void. Such voids are often caused by the sampling and transporting process. The open voids can be caused by many different factors, such as gravel or shells which impact with the cutting end of the sampling tube and/or scrape along the inside of the sampling tube and create voids. The voids and highly disturbed clay shown in Fig. 2.6 are possibly caused by cuttings inadvertently left at the bottom of the borehole. Some of the disturbance could also be caused by tube friction during sampling as the clay near the tube wall becomes remolded as it travels up the tube. 2. *Soil cracks*: Figures 2.7 and 2.8 show numerous cracks in the clay. For example, the arrows labeled 1 point to some of the soil cracks in Figs. 2.7 and 2.8. The soil cracks probably developed during the sampling process. A contributing factor in the development of the soil cracks may have been gas coming out of solution which fractured the clay. 3. *Turning of edges*: Turning or bending of edges of various thin layers shows as curved-down edges on the sides of the specimen. This effect is caused by friction between the soil and sampler. There could be other distortion-type effects at the soil/sampling tube interface. For example, the arrow labeled 2 in Fig. 2.7 shows cracking and distortion of the soil at the tube interface. 4. *Gas-related voids*: The circular voids (labeled 3) in Fig. 2.7 were caused by gas coming out of solution during the sampling process when the confining pressures were essentially reduced to zero. In contrast to soil disturbance, the arrow labeled 4 in Fig. 2.8 indicates an undisturbed section of the soil sample. Note in Fig. 2.8 that the individual fine layering of the soil sample can even be observed.

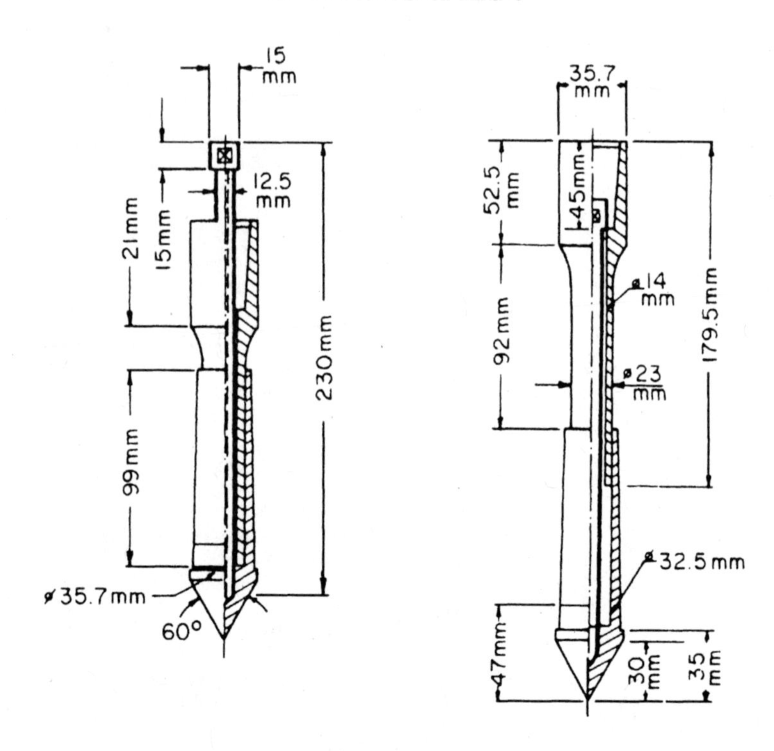

FIGURE 2.9 Example of mechanical cone penetrometer tip (Dutch mantle cone) (see Table 2.13). (*Reprinted with permission from the American Society for Testing and Materials, 1998.*)

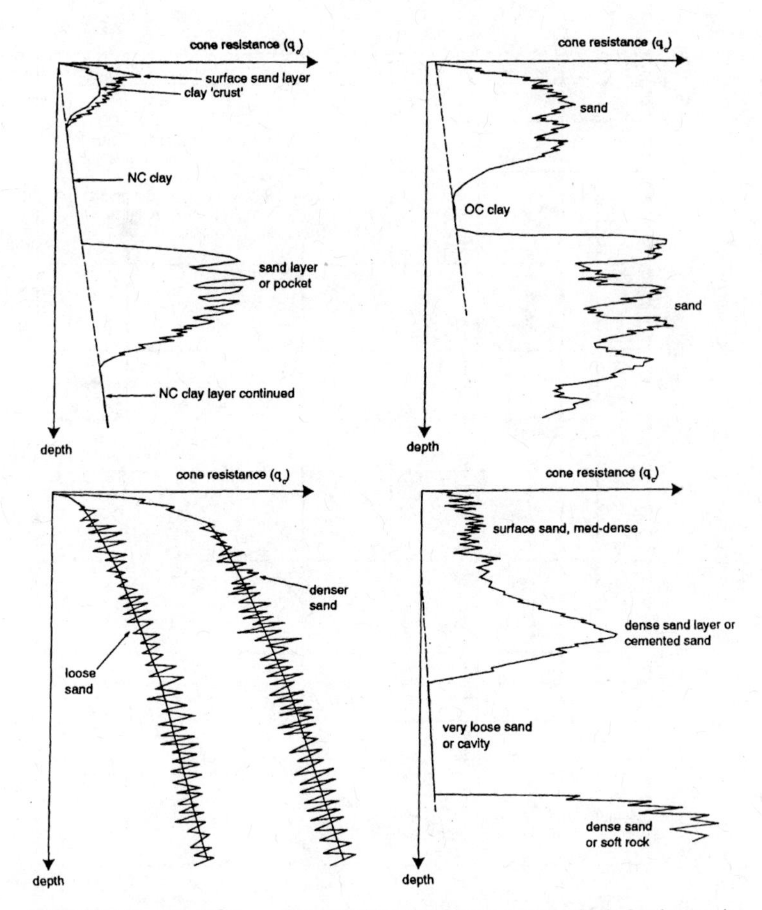

FIGURE 2.10 Simplified examples of CPT cone resistance q_c versus depth showing possible interpretations of soil types and conditions (see Table 2.13). (*From Schmertmann 1977.*)

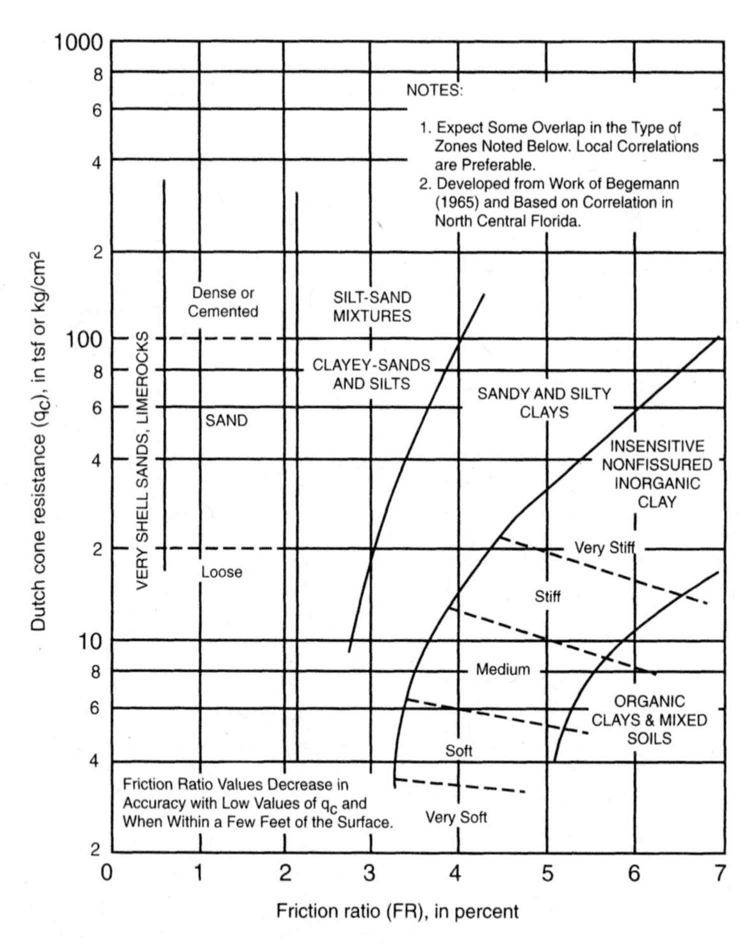

FIGURE 2.11 Guide for estimating soil type from Dutch mantle cone [enter chart with cone resistance q_c and friction ratio (FR = sleeve friction divided by cone resistance = 100 f_s/q_c); see Table 2.13]. (*From Schmertmann 1977.*)

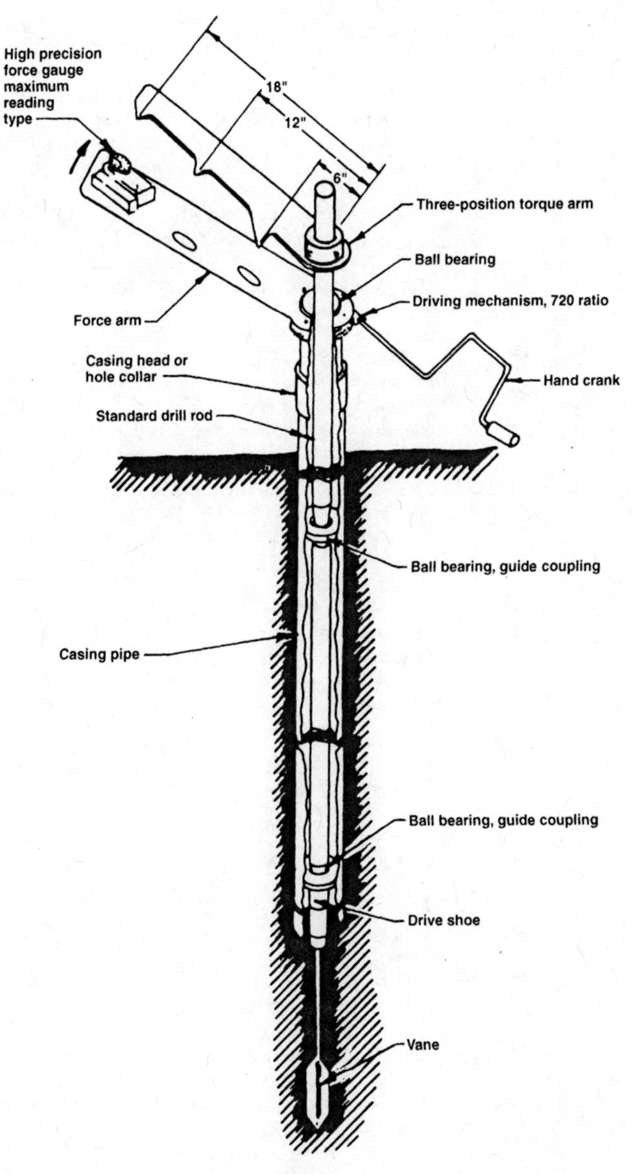

FIGURE 2.12 Diagram illustrating the field vane test (see Table 2.13). (*From NAVFAC DM-7.1, 1982.*)

TABLE 2.11 Standard Penetration Test

Topic (1)	Discussion (2)
Definition	The standard penetration test (SPT) consists of driving a thick-walled sampler into the sand deposit. According to ASTM D 1586-92 (1998), the SPT sampler must have an inside barrel diameter (D_i) = 3.81 cm (1.5 in.) and an outside diameter (D_o) = 5.08 cm (2 in.). The SPT sampler is driven into the sand by using a 63.5-kg (140-lb) hammer falling a distance of 0.76 m (30 in.). The SPT sampler is driven a total of 45 cm (18 in.), with the number of blows recorded for each 15-cm (6-in.) interval. The "measured SPT *N*-value" (blows per foot) is defined as the penetration resistance of the sand, which equals the sum of the number of blows required to drive the SPT sampler over the depth interval of 15 cm to 45 cm (6 in. to 18 in.). The reason the number of blows required to drive the SPT sampler for the first 15 cm (6 in.) is not included in the *N*-value is because the drilling process often disturbs the soil at the bottom of the borehole and the readings at 15 cm to 45 cm (6 in. to 18 in.) are believed to be more representative of the *in situ* penetration resistance of the sand.
Uses	The standard penetration test (SPT) can be used for all types of soil, but, in general, the SPT should be used only for sand deposits (Coduto 1994). The SPT can be especially of value for clean sand deposits where the sand falls or flows out from the sampler when retrieved from the ground. Without a soil sample, other types of tests, such as the standard penetration test, must be used to assess the engineering properties of the sand. Often in drilling a borehole, if subsurface conditions indicate a sand stratum and sampling tubes come up empty, the sampling gear can be quickly changed to perform standard penetration tests.
Correction for testing factors	Testing factors can affect the measured SPT *N*-value, and the following equation can be used to compensate for these testing factors (Skempton 1986): $$N_{60} = 1.67\, E_m\, C_b\, C_r\, N$$ where N_{60} = SPT *N*-value corrected for field testing procedures E_m = Hammer efficiency (for U.S. equipment, E_m = 0.6 for a safety hammer and E_m = 0.45 for a donut hammer) C_b = Borehole diameter correction (C_b = 1.0 for boreholes of 65 to 115 mm diameter, 1.05 for 150 mm diameter, and 1.15 for 200 mm diameter hole) C_r = Rod length correction (C_r = 0.75 for up to 4 m of drill rods, 0.85 for 4 to 6 m of drill rods, 0.95 for 6 to 10 m of drill rods, and 1.00 for drill rods in excess of 10 m) N = measured SPT *N*-value

TABLE 2.11 Standard Penetration Test (*Continued*)

Topic (1)	Discussion (2)
General factors that influence *N*-value	The measured SPT *N*-value can be influenced by the type of soil, such as the amount of fine and gravel-size particles in the soil. Saturated sands that contain appreciable fine soil particles, such as silty or clayey sands, could give abnormally high *N*-values if they have a tendency to dilate or abnormally low *N*-values if they have a tendency to contract during the undrained shear conditions associated with driving the SPT sampler. Gravel-size particles increase the driving resistance (hence increase *N*-value) by becoming stuck in the SPT sampler tip or barrel. Another factor that could influence the measured SPT *N*-value is groundwater. It is important to maintain a level of water in the borehole at or above the *in situ* groundwater level. This is to prevent groundwater from rushing into the bottom of the borehole which could loosen the sand and result in low measured *N*-values.
Specific factors that influence *N*-value	The following factors may affect the measured *N*-value (from NAVFAC DM-7.1, 1982): 1. *Inadequate cleaning of the borehole:* SPT is only partially made in original soil. Sludge may be trapped in the sampler and compressed as the sampler is driven, increasing the blow count. This may also prevent sample recovery. 2. *Not seating the sampler spoon on undisturbed material:* Incorrect *N*-value is obtained. 3. *Driving of the sample spoon above the bottom of the casing: N*-value is increased in sands and reduced in cohesive soil. 4. *Failure to maintain sufficient hydrostatic head in boring:* The water table in the borehole must be at least equal to the piezometric level in the sand, otherwise the sand at the bottom of the borehole may be transformed into a loose state. 5. *Attitude of operators:* Blow counts for the same soil using the same rig can vary, depending on who is operating the rig, and perhaps the mood of operator and time of drilling. 6. *Overdriven sample:* Higher blow counts usually result from overdriven sampler. 7. *Sampler plugged by gravel:* Higher blow counts result when gravel plugs the sampler. The resistance of loose sand could be highly overestimated. 8. *Plugged casing:* High *N*-values may be recorded for loose sand in sampling below the groundwater table. Hydrostatic pressure causes sand to rise and plug the casing. 9. *Overwashing ahead of casing*: Low blow count may result for dense sand, since sand is loosened by overwashing. 10. *Drilling method*: Drilling technique (e.g., cased holes versus mud-stabilized holes) may result in different *N*-values for the same soil. 11. *Not using the standard hammer drop:* Energy delivered per blow is not uniform. European countries have adopted an automatic trip hammer not currently in use in North America.

TABLE 2.11 Standard Penetration Test (*Continued*)

Topic (1)	Discussion (2)
Specific factors that influence *N*-value (*Continued*)	12. *Free fall of the drive weight is not attained:* Using more than 1.5 turns of rope around the drum and/or using wire cable will restrict the fall of the drive weight. 13. *Not using the correct weight:* Driller frequently supplies drive hammers with weights varying from the standard by as much as 10 lb. 14. *Weight does not strike the drive cap concentrically:* Impact energy is reduced, increasing the *N*-value. 15. *Not using a guide rod:* Incorrect *N*-value is obtained. 16. *Not using a good tip on the sampling spoon:* If the tip is damaged and reduces the opening or increases the end area, the *N*-value can be increased. 17. *Use of drill rods heavier than standard:* With heavier rods, more energy is absorbed by the rods, causing an increase in the blow count. 18. *Not recording blow counts and penetration accurately:* Incorrect *N*-values are obtained. 19. *Incorrect drilling procedures:* The standard penetration test was originally developed from wash-boring techniques. Drilling procedures which seriously disturb the soil will affect the *N*-value, for example, drilling with cable tool equipment. 20. *Using large drill holes:* A borehole correction is required for large-diameter boreholes. This is because larger diameters often result in a decrease in the blow count. 21. *Inadequate supervision:* Frequently a sampler will be impeded by gravel or cobbles, causing a sudden increase in blow count. This is often not recognized by an inexperienced observer. Accurate recording of drilling, sampling, and depth is always required. 22. *Improper logging of soils:* Not describing the sample correctly. 23. *Using too large a pump:* Too high a pump capacity will loosen the soil at the base of the hole, causing a decrease in blow count.
Popularity	Even with the limitations and all of the corrections that must be applied to the measured SPT *N*-value, the standard penetration test is probably the most widely used field test in the United States. This is because it is relatively easy to use, the test is economical compared to other types of field testing, and the SPT equipment can be quickly adapted and included as part of almost any type of drilling rig.

TABLE 2.12 Correlation between SPT N-value and Density of Clean Sand

N-value, blows per foot (1)	Sand density (2)	Relative density (D_r) (3)
0 to 4	Very loose condition	0 to 15%
4 to 10	Loose condition	15 to 35%
10 to 30	Medium condition	35 to 65%
30 to 50	Dense condition	65 to 85%
Over 50	Very dense condition	85 to 100%

Sources: Terzaghi and Peck 1967 and Lambe and Whitman 1969.

Notes:

1. The correlation between N-value and sand density presented in this table is approximate and the boundaries between different density conditions are not as distinct as implied by the table. If it only takes four blows or less to drive the SPT sampler, then the sand should be considered to be very loose and could be subjected to significant settlement due to the weight of a structure or due to earthquake shaking. On the other hand, if it takes over 50 blows to drive the SPT sampler, then the sand is considered to be in a very dense condition and would be able to support high bearing loads and would be resistant to settlement from earthquake shaking.
2. See Table 3.16 for further details on relative density (D_r).

TABLE 2.13 Other Types of Field Tests

Type of field test (1)	Discussion (2)
Cone penetration test (CPT)	The cone penetration test (CPT) is similar to the standard penetration test (SPT) except that instead of driving a thick-walled sampler into the soil, a steel cone is pushed into the soil. There are many different types of cone penetration devices, such as the mechanical cone, mechanical-friction cone, electric cone, and piezocone (see App. A, Glossary 1, for descriptions). The simplest type of cone is shown in Fig. 2.9 (from ASTM D 3441-94, 1998). The cone is first pushed into the soil to the desired depth (initial position) and then a force is applied to the inner rods which moves the cone downward into the extended position. The force required to move the cone into the extended position (Fig. 2.9) divided by the horizontally projected area of the cone is defined as the cone resistance q_c. By continually repeating the two-step process shown in Fig. 2.9, the cone resistance data is obtained at increments of depth. A continuous record of the cone resistance versus depth can be obtained by using the electric cone, where the cone is pushed into the soil at a rate of 10 to 20 mm/s (2 to 4 ft/min). A considerable amount of work has been performed in correlating cone resistance q_c with subsurface conditions. Figure 2.10 (from Schmertmann 1977) presents four examples, where the cone resistance q_c has been plotted versus depth below ground surface. The shape of the cone resistance q_c plots versus depth can be used to identify sands, clays, cavities, or rock. Some cones can be adapted to measure the frictional resistance f_s along a side sleeve or the development and dissipation of excess pore water pressure during penetration of the cone, and this data can also be useful in identifying different types of soil, such as sand or clay (see Fig. 2.11). The cone can even be equipped with a video camera (Raschke and Hryciw 1997) to enable the type of soil to be viewed during the cone penetration test. A major advantage of the cone penetration test is that by using the electric cone, a continuous subsurface record of the cone resistance q_c, such as shown in Fig. 2.10, can be obtained. This is in contrast to the standard penetration test (SPT), which obtains data at intervals in the soil deposit. Disadvantages of the cone penetration test are that soil samples can not be recovered and special equipment is required to produce a steady and slow penetration of the cone. Unlike the SPT, the ability to obtain a steady and slow penetration of the cone is not included as part of conventional drilling rigs. Because of these factors, in the United States, the CPT is used less frequently than the SPT.
Vane shear test (VST)	The SPT and CPT are used to correlate the resistance of driving a sampler (*N*-value) or pushing a cone (q_c) with the engineering properties (such as density condition) of the soil. In contrast, the vane test is a different *in situ* field test because it directly measures a specific soil property, the undrained shear strength (s_u) of clay.

TABLE 2.13 Other Types of Field Tests (*Continued*)

Type of field test (1)	Discussion (2)
Vane shear test (VST) (*Continued*)	The vane test consists of inserting a four-bladed vane, such as shown in Fig. 2.12, into the borehole and then pushing the vane into the clay deposit located at the bottom of the borehole. Once inserted into the clay, the maximum torque T_{max} required to rotate the vane and shear the clay is measured. The undrained shear strength s_u of the clay can then be calculated by using the following equation, which assumes uniform end shear for a rectangular vane. (*Note:* The following equation is valid only for a rectangular vane with shear failure along the entire perimeter and at both ends of the vane): $$s_u = \frac{T_{max}}{\pi\,(0.5\,D^2H + 0.167\,D^3)}$$ where T_{max} = maximum torque required to rotate the rod which shears the clay; H = height of the vane; D = diameter of the vane Standard field specifications, such as ASTM D 2573-94 (1998), have been developed to provide uniformity of the field testing procedure. The vane can provide an undrained shear strength s_u that is too high if the vane is rotated too rapidly. The vane test also gives unreliable results for clay strata that contain sand layers or lenses or varved clay, or if the clay contains gravel or gravel-size shell fragments. There are other types of vane testing equipment, such as the tapered vane (see ASTM D 2573-94, 1998). There are even miniature vane devices, which can be used in the field or laboratory. The miniature vane device is not inserted down a borehole, but rather the test is performed on clay specimens brought to the surface from undisturbed samplers. An example of a miniature vane is the torvane device, which is a handheld vane that is manually inserted into the clay surface and then rotated to induce a shear failure of the clay. On top of the torvane there is a calibrated scale that directly indicates the undrained shear strength s_u of the clay. The torvane device has a quick failure rate which could overestimate the undrained shear strength s_u. Because the miniature vane tests only a very small portion of the clay, the strength could be overestimated for fissured clay, varved clay, or clay containing slickensides. Also, ASTM D 4648-94 (1998) indicates that the miniature vane provides unreliable readings for clays having an undrained shear strength s_u in excess of 100 kPa (1.0 tsf) because the actual failure surface deviates from the assumed cylindrical failure surface, resulting in an overestimation of the undrained shear strength s_u.
Other types of field tests performed in boreholes	There are many other types of field tests that can be performed in boreholes. Examples include the pressuremeter test (PMT), screw plate compressometer (SPC), and Iowa borehole shear test (BST) (Holtz and Kovacs 1981, Mitchell 1978). These types of tests are used much less frequently than the SPT, CPT, and VST.

TABLE 2.13 Other Types of Field Tests (*Continued*)

Type of field test (1)	Discussion (2)
Other types of field tests performed in boreholes (*Continued*)	The pressuremeter test (PMT) is a field test that involves the expansion of a cylindrical probe within an uncased borehole. The screw plate compressometer (SPC) is a field test where a plate is screwed down to the desired depth, and then as pressure is applied, the settlement of the plate is measured. The Iowa borehole shear test (BST) is a field test where the device is lowered into an uncased borehole and then expanded against the sidewalls. The force required to pull the device toward ground surface is measured and much as in a direct shear device, the shear strength properties of the *in situ* soil can then be determined.
California bearing ratio	The California bearing ratio (CBR) and the plate load test are two types of field tests that can be performed at ground surface, or on a level surface excavated in a test pit, trench, or bulldozer cut. The California bearing ratio (CBR) test consists of pushing a 2.0-in.-diameter piston into the soil. The CBR is calculated as follows (see ASTM D 4429-93, 1998): (1) Record the load (pounds) versus depth of penetration (inches) of the piston as it is being pushed into the soil. (2) Calculate the stress (psi) on the piston by dividing the load by the area of the piston. (3) Plot the stress on the piston (psi) versus the depth of penetration (inches). (4) The bearing value is normally the stress (psi) corresponding to a depth of penetration of 0.10 in. (5) This bearing value is converted to a ratio by dividing it by 1000 psi, which represents the penetration resistance of compacted crushed rock. (6) This bearing ratio, which is a fraction, is multiplied by 100 and reported as the California bearing ratio (CBR). Typical values of California bearing ratio range from less than 5 for soft clays up to 80 for dense sandy gravel. When interpreting the CBR data, the geotechnical engineer must consider a possible increase in field moisture content of the soil which could soften fine-grained soils and lead to a lower CBR than originally measured. There is the possibility that the soil might be disturbed and loosened during the construction operations, also resulting in a lower CBR than originally measured.
Plate load test	The plate load test is similar to the California bearing ratio, except that instead of pushing a piston into the soil, a steel plate is pushed into the soil (ASTM D 1196-97, 1998). The usual procedure is to push a square or round steel plate into the soil and record the load versus depth of penetration. The load is converted to a stress by dividing the load by the area of the plate. The stress versus depth of penetration data is then used to calculate the subgrade modulus K_v, which is also known as the modulus of subgrade reaction. The procedure (per NAVFAC DM-7.1, 1982) is to plot the stress on the plate versus the penetration of the plate. Using this plot, the yield point at which the penetration rapidly

TABLE 2.13 Other Types of Field Tests (*Continued*)

Type of field test (1)	Discussion (2)
Plate load test (*Continued*)	increases is estimated. Then the stress q and depth of penetration of the plate (δ) corresponding to one-half the yield point are determined from the plot, and the subgrade modulus K_v is equal to q divided by δ. The plate load test can also be used to directly estimate the settlement potential of a footing. For settlement of sands caused by an applied surface loading, an empirical equation that relates the depth of penetration of the steel plate (S_1) to the settlement of the actual footing (S) is as follows (Terzaghi and Peck 1967): $$S = \frac{4S_1}{(1 + D_1/D)^2}$$ where D_1 = smallest dimension of the steel plate and D = smallest dimension of the actual footing. In order to use this equation, the stress exerted by the steel plate corresponding to S_1 must be the same bearing stress exerted by the actual footing. The calculated settlement from this equation can significantly underestimate the actual settlement at the site in cases where there is settlement due to secondary influences (such as collapsible soil) or in cases where there is a deep looser or softer layer that is not affected by the small plate, but is loaded by the much larger footing.

TEST EXCAVATION LOG No. **LB-1** F.N. **60210.02**

Project Name: Sheet: **1** of **2**

Location: Start: **7 OCT 94**

Estimated Surface Elevation: **1050' ±** Total Depth: **35.0'** Rig Type: **24" Limited Access** End: **7 OCT 94**

Field Description By: **DMP**

Surface Conditions:
Grass landscaping approximately 8 feet from the northeast corner of the house.

Subsurface Conditions: **FORMATION: Classification, color, moisture, tightness, etc.**

Depth-Feet	Sample Type (Intact / Bulk)	Blow Counts Per Half Foot	Dry Unit Weight pcf	Moisture Content %	Relative Compaction	Saturation %	USCS Symbol	Graphic Log	Layer depth	Field Description
0										FILL: From 0.0'-3.5', Silty CLAY, dark gray/brown, moist, slightly stiff. Semi-abundant to abundant roots to 1/2" in diameter, slightly porous.
			107	18		87			3.50	
5										From 3.5'-6.5', becomes dark brown, moist, slightly stiff to stiff. Semi-abundant Siltstone gravel near basal contact with Landslide Debris (Qls). At 6.5', undulating near horizontal contact with Qls.
			73	28		59			6.50	
									9.00	LANDSLIDE DEBRIS (Qls): From 6.5'-9.0', SILTSTONE, light gray/brown, dry to damp, loose to very loose. Very fractured to angular gravel to cobble (3") sized pieces. Occasional infilled animal burrow, near-vertical infilling of topsoil (6" to 8" in width) within very fractured zone indicates ripping of rock during grading of pad. At 9.0', approximately 1" thick fine sand lense, light gray, oriented N25E 50B.
10			76	34		77			12.00	From 9.0'-12.0', increasing moisture, fractured rock generally cobble (to 4") sized.
			76	36		81			13.40	From 12.0'-13.4', brecciated zone with randomly oriented sub-angular cobbles (6" to 18") of laminated Siltstone, light gray/brown, within clayey matrix.
15			81	37		94			16.00	From 13.4'-16.0', Fine Sandy SILTSTONE, gray/chocolate brown, moist to very moist, hard. Fractured.
									18.60	From 16.0'-18.6', Fine Sandy Clayey SILTSTONE, gray/chocolate brown, moist to very moist, hard. Fractured, occasional rootlet to 1/16" in diameter, fractured surfaces sub-polished and stained orange, slight clay infilling of fractures which are near-vertical to sub-horizontal. At 18.6', SHEAR ZONE, poorly developed, approximately 1/2" to 1" thick brecciated clay, stained orange/brown, moist, slightly soft to slightly stiff, oriented N60W 32N.
20										From 18.6'-24.6', Clayey SILTSTONE, reddish/gray/brown, moist, hard to very hard. At 24.5', Fine Sandstone lense approximately 1/2" thick, light brown, dry to moist, soft to slightly hard. Oriented N70W 15N

NOTES: **Kelly Weights 0-12 feet: 850 lbs; 12-20.5 feet: 650 lbs; 20.5+ feet: 450 lbs.**

FIGURE 2.13 Boring log (see Table 2.14). (*From Day and Poland 1996; reproduced with permission of ASCE.*)

TEST EXCAVATION LOG No. **LB-1** F.N. 60210.02

Project Name: Sheet: 2 of 2

Field Description

By: DMP

Surface Conditions:

Grass landscaping approximately 8 feet from the northeast corner of the house.

Subsurface Conditions: **FORMATION: Classification, color, moisture, tightness, etc.**

Depth-Feet	Sample Type (Intact / Bulk)	Blow Counts Per Half Foot	Dry Unit Weight pcf	Moisture Content %	Relative Compaction	Saturation %	USCS Symbol	Graphic Log	Field Description
			78	41		97			
24.60									From 24.6'-25.1', Clayey SILTSTONE, reddish/gray/brown, moist, hard to very hard. Slightly fractured.
25.10									From 25.1'-25.5', SHEAR ZONE, Silty CLAY, gray/brown, moist, slightly soft to slightly stiff, diatomaceous, brecciated. Basal shear zone approximately 1/2" thick oriented N70W 15NE.
25.50			83	33		88			**BEDROCK-UNNAMED SHALE (Tush):** From 25.5'-28.5', Clayey SILTSTONE, reddish/gray/brown, moist, hard to very hard. Slightly fractured. At 28.5' contact with bedrock.
28.50									From 28.5'-32.0', Clayey SILTSTONE, reddish/gray/brown, moist, very hard. Occasional sub-horizontal 1/2" gypsum vein, occasional dark gray/black staining to 4" in diameter.
32.00									From 32.0'-35.0', Fine Sandy SILTSTONE, dark gray/black, moist, very hard.
35.00									Total Depth = 35.0' No Water No Caving

Depth scale: 25, 30, 35, 40, 45

FIGURE 2.13 (*Continued*)

TABLE 2.14 Types of Information That Should Be Recorded on Exploratory Logs

Item (1)	Description (2)
Excavation number	Each boring, test pit, or trench excavated at the site should be assigned an excavation number. For example, in Fig. 2.13, the boring has been labeled LB-1, for large diameter boring number 1.
Project information	Project information should include the project name, file number, client, and site address. The individual preparing the log should also be noted.
Type of equipment	Include on the log the type of excavation (i.e., hand-dug pit, backhoe trench, etc.) and the total depth and size of the excavation. For borings, indicate type of drilling equipment, use of drilling fluid, and kelly bar weights. Also indicate if casing was used.
Site-specific information	The exploratory log should list the surface elevation, date(s) of excavation, and ground surface conditions.
Type of field tests	For borings, list all field tests, such as SPT, CPT, or vane test. Also indicate if the boring was converted to a monitoring device, such as a piezometer.
Type of sampler	Indicate type of sampler and depth of each soil or rock sample recovered from the excavation. For driven samplers, indicate type and weight of hammer and number of blows per foot to drive the sampler. Indicate sample recovery and RQD for rock strata.
Soil and rock descriptions	Classify the soil and rock exposed in the excavation (see Chap. 4). Also indicate moisture and density condition of the soil and rock.
Excavation problems	List excavation problems, such as instability (sloughing, groundwater-induced caving, squeezing of the hole, etc.), hard drilling, or boring termination due to refusal.
Groundwater	Indicate depth to groundwater or seepage zones. At the end of the subsurface exploration, indicate the depth of free-standing water in the excavation.
Geologic features and hazards	Identify geologic features and hazards. Geologic features include type of deposit (see Table 2.15), formation name, fracture condition of rock, etc. Geologic hazards include landslides, active fault shear zones, liquefaction-prone sand, bedding shear surfaces or slickensides, and underground voids or caverns.
Unusual conditions	Any unusual subsurface condition should be noted. Examples include artesian groundwater, boulders or other obstructions, or loss of drilling fluid which could indicate an underground void or cavity.
Laboratory tests and results	The types and locations of laboratory tests are usually included on the log. Test results, such as the unit weight and water content of the soil and rock, are also included on the boring log.

Note: Additional information may be required for subsurface explorations for mining or agricultural purposes, for the investigation of hazardous waste, or other special types of subsurface exploration. Although logs are often prepared by technicians or even the driller, the most appropriate individuals to log the subsurface conditions are geotechnical engineers or engineering geologists who have considerable experience and judgment acquired by many years of field practice. It is especially important that the subsurface conditions likely to have the most impact on the proposed project be adequately described. See Fig. 2.13 for an example of a boring log.

TABLE 2.15 Common Man-Made and Geologic Soil Deposits

Main category (1)	Common types of soil deposits (2)	Possible engineering problems (3)
Structural fill	Dense or hard fill. Often the individual fill lifts can be identified.	Upper surface of structural fill may have become loose or weathered.
Uncompacted fill	Random soil deposit that can contain chunks of different types and sizes of rock fragments.	Susceptible to compression and collapse.
Debris fill	Contains pieces of debris, such as concrete, brick, and wood fragments.	Susceptible to compression and collapse.
Municipal dump	Contains debris and waste products such as household garbage or yard trimmings.	Significant compression and gas from organic decomposition.
Residual soil deposit	Soil deposits formed by in-place weathering of rock.	Engineering properties are highly variable.
Organic deposit	Examples include peat and muck which form in bogs, marshes, and swamps.	Very compressible and unsuitable for foundation support.
Alluvial deposit	Soil transported and deposited by flowing water, such as streams and rivers.	All types of grain sizes, loose sandy deposits susceptible to liquefaction.
Aeolian deposit	Soil transported and deposited by wind. Examples include loess and dune sands.	Can have unstable soil structure that may be susceptible to collapse.
Glacial deposit	Soil transported and deposited by glaciers or their melt water. Examples include till.	Erratic till deposits and soft clay deposited by glacial melt water.
Lacustrine deposit	Soil deposited in lakes or other inland bodies of water.	Unusual soil deposits can form, such as varved silts or varved clays.
Marine deposit	Soil deposited in the ocean, often from rivers that empty into the ocean.	Granular shore deposits but offshore areas can contain soft clay deposits.
Colluvial deposit	Soil transported and deposited by gravity, such as talus, hill-wash, or landslide deposits	Can be geologically unstable deposit.
Pyroclastic deposit	Material ejected from volcanoes. Examples include ash, lapilli, and bombs.	Weathering can result in plastic clay. Ash can be susceptible to erosion.

Note: The first four soil deposits are man-made; all others are due to geologic processes. Usually the engineering geologist is most qualified to determine the type of soil deposit. The different soil deposits can have unique geotechnical and foundation implications, and it is always of value to determine the geologic or man-made process that created the soil deposit.

TABLE 2.16 Problem Conditions Requiring Special Consideration

Problem type (1)	Description (2)	Comments (3)
Soil	Organic soil, highly plastic soil	Low strength and high compressibility
	Sensitive clay	Potentially large strength loss upon large straining
	Micaceous soil	Potentially high compressibility
	Expansive clay, silt, or slag	Potentially large expansion upon wetting
	Liquefiable soil	Complete strength loss and high deformations caused by earthquakes
	Collapsible soil	Potentially large deformations upon wetting
	Pyritic soil	Potentially large expansion upon oxidation
Rock	Laminated rock	Low strength when loaded parallel to bedding
	Expansive shale	Potentially large expansion upon wetting; degrades readily upon exposure to air and water
	Pyritic shale	Expands upon exposure to air and water
	Soluble rock	Rock such as limestone, limerock, and gypsum that is soluble in flowing and standing water
	Cretaceous shale	Indicator of potentially corrosive groundwater
	Weak claystone	Low strength and readily degradable upon exposure to air and water
	Gneiss and schist	Highly distorted with irregular weathering profiles and steep discontinuities
	Subsidence	Typical in areas of underground mining or high groundwater extraction
	Sinkholes	Areas underlain by carbonate rock (karst topography)
Condition	Negative skin friction	Additional compressive load on deep foundations due to settlement of soil
	Expansion loading	Additional uplift load on foundation due to swelling of soil
	Corrosive environment	Acid mine drainage and degradation of soil and rock
	Frost and permafrost	Typical in northern climates
	Capillary water	Rise in water level which leads to strength loss for silts and fine sands

Source: Reproduced with permission from *Standard Specifications for Highway Bridges,* 16th ed., AASHTO 1996.

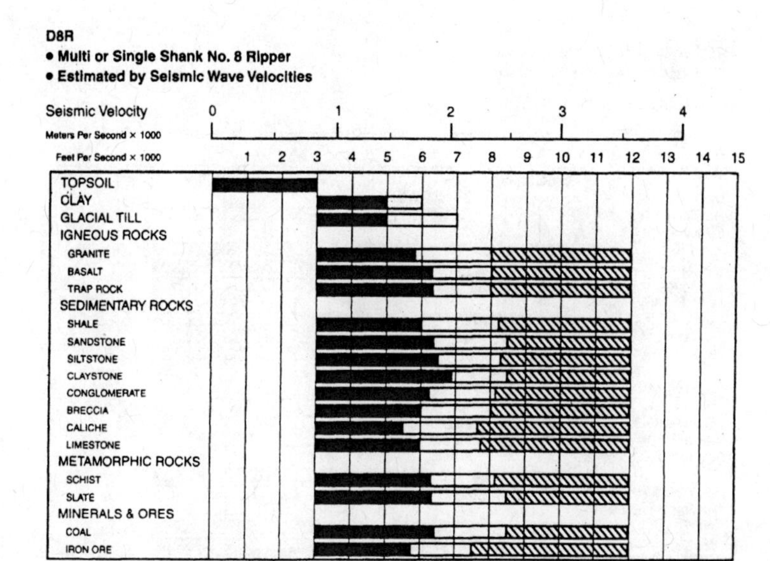

FIGURE 2.14 Rippability of rock versus seismic velocity for a Caterpillar D8R tractor/ripper. (*From Caterpillar Performance Handbook 1997.*)

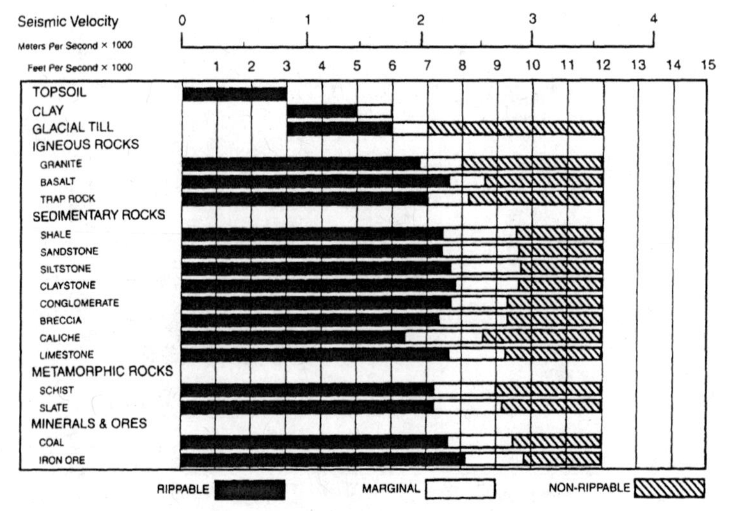

FIGURE 2.15 Rippability of rock versus seismic velocity for a Caterpillar D9R tractor/ripper. (*From Caterpillar Performance Handbook 1997.*)

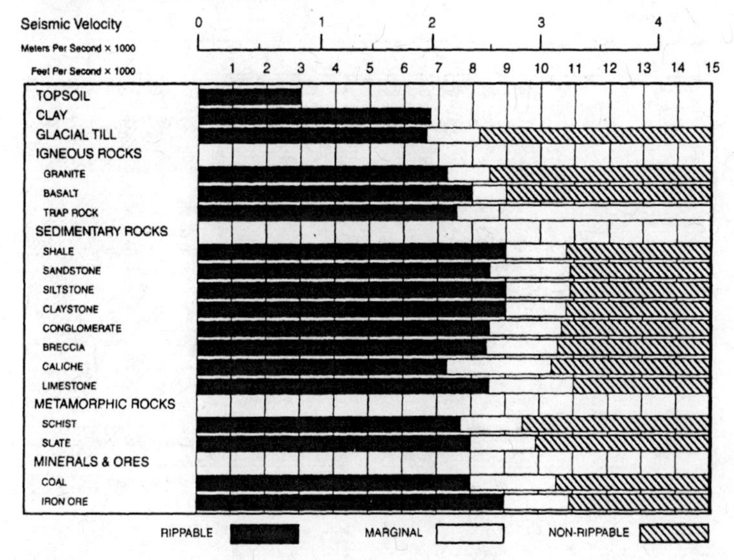

FIGURE 2.16 Rippability of rock versus seismic velocity for a Caterpillar D10R tractor/ripper. (*From Caterpillar Performance Handbook 1997.*)

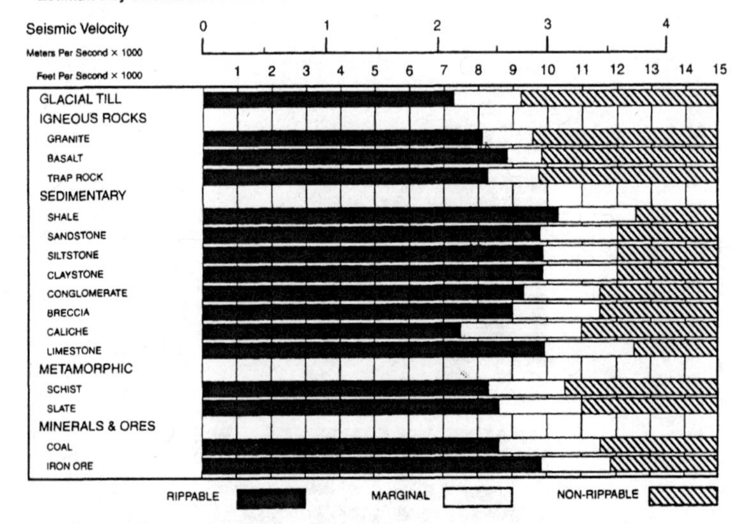

FIGURE 2.17 Rippability of rock versus seismic velocity for a Caterpillar D11R tractor/ripper. (*From Caterpillar Performance Handbook 1997.*)

TABLE 2.17 Geophysical Techniques

Name of method (1)	Procedure or principle utilized (2)	Applicability and limitations (3)
	Seismic methods	
1. Refraction	Based on time required for seismic waves to travel from source of energy to points on ground surface, as measured by geophones spaced at intervals on a line at the surface. Refraction of seismic waves at the interface between different strata gives a pattern of arrival times at the geophones versus distance to the source of seismic waves. Seismic velocity can be obtained from a single geophone and recorder with the impact of a sledge hammer on a steel plate as a source of seismic waves.	Utilized for preliminary site investigation to determine rippability, faulting, and depth to rock or other lower stratum substantially different in wave velocity than the overlying material. Generally limited to depths up to 30 m (100 ft) of a single stratum. Used only where wave velocity in successive layers becomes greater with depth.
2. High-resolution reflection	Geophones record travel time for the arrival of seismic waves reflected from the interface of adjoining strata.	Suitable for determining depths to deep rock strata. Generally applies to depths of a few thousand feet. Without special signal enhancement techniques, reflected impulses are weak and easily obscured by the direct surface and shallow refraction impulses. Method is useful for locating groundwater.
3. Vibration	The travel time of transverse or shear waves generated by a mechanical vibrator consisting of a pair of eccentrically weighted disks is recorded by seismic detectors placed at specific distances from the vibrator.	Velocity of wave travel and natural period of vibration gives some indication of soil type. Travel time plotted as a function of distance indicates depths or thickness of surface strata. Useful in determining dynamic modulus of subgrade reaction and obtaining information on the natural period of vibration for the design of foundations of vibrating structures.
4. Uphole, downhole, and crosshole surveys	*a. Uphole or downhole*: Geophones on surface, energy source in borehole at various locations starting from hole bottom. Procedure can be revised with energy source on surface, detectors moved up or down the hole. *b. Downhole*: Energy source at the surface (e.g., wooden plank struck by hammer), geophone probe in borehole. *c. Crosshole*: Energy source in central hole, detectors in surrounding holes.	Obtain dynamic soil properties at very small strains, rock mass quality, and cavity detection. Unreliable for irregular strata or soft strata with large gravel content. Also unreliable for velocities decreasing with depth. Crosshole measurements best suited for *in situ* modulus determination.

TABLE 2.17 Geophysical Techniques (*Continued*)

Name of method (1)	Procedure or principle utilized (2)	Applicability and limitations (3)
	Electrical methods	
1. Resistivity	Based on the difference in electrical conductivity or resistivity of strata. Resistivity is correlated to material type.	Used to determine horizontal extent and depths up to 30 m (100 ft) of subsurface strata. Principal applications are for investigating foundations of dams and other large structures, particularly in exploring granular river channel deposits or bedrock surfaces. Also used for locating fresh/saltwater boundaries.
2. Drop in potential	Based on the determination of the drop in electrical potential.	Similar to resistivity methods but gives sharper indication of vertical or steeply inclined boundaries and more accurate depth determinations. More susceptible than resistivity method to surface interference and minor irregularities in surface soils.
3. E-logs	Based on differences in resistivity and conductivity measured in borings as the probe is lowered or raised.	Useful in correlating units between borings, and has been used to correlate materials having similar seismic velocities. Generally not suited to civil engineering exploration but valuable in geologic investigations.
	Other methods	
Magnetic measurements	Highly sensitive proton magnetometer is used to measure the Earth's magnetic field at closely spaced stations along a traverse.	Difficult to interpret in quantitative terms but indicates the outline of faults, bedrock, buried utilities, or metallic trash in fills.
Gravity measurements	Based on differences in density of subsurface materials which affects the gravitational field at the various points being investigated.	Useful in tracing boundaries of steeply inclined subsurface irregularities such as faults, intrusions, or domes. Methods not suitable for shallow depth determination but useful in regional studies. Some application in locating limestone caverns.

Source: NAVFAC DM-7.1, 1982.
Note: Also see AGI Data Sheets 59.1 to 60.2 (American Geological Institute 1982) for a summary of the applications of geophysical methods.

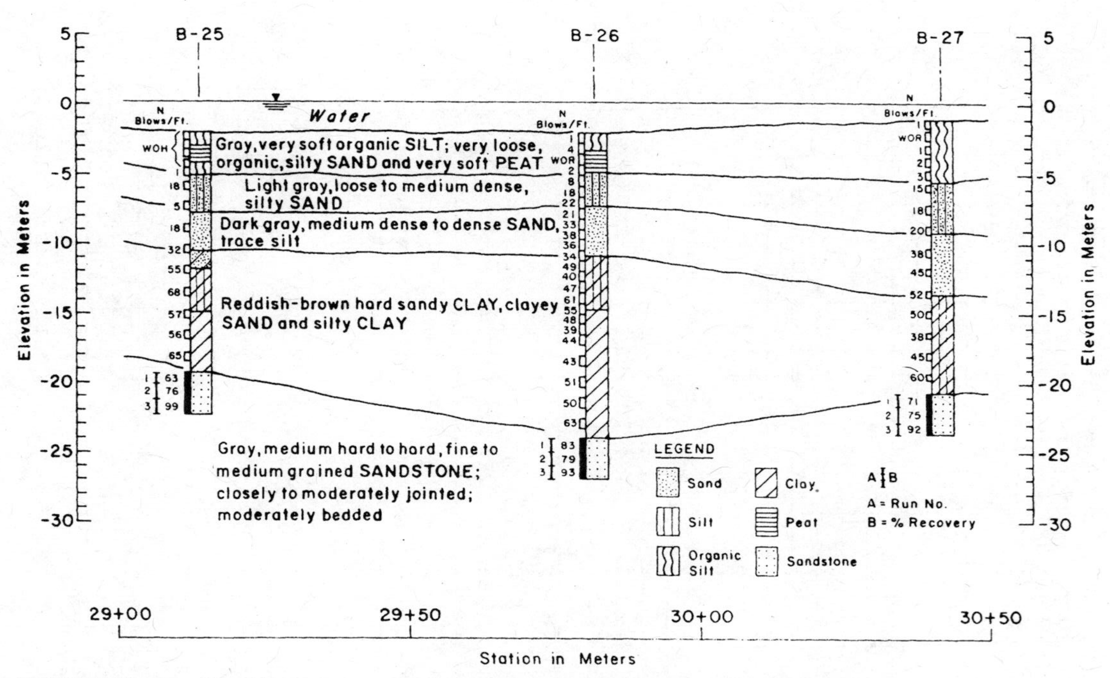

FIGURE 2.18 Subsoil profile. (*From Lowe and Zaccheo 1975; copyright Van Nostrand Reinhold Company.*)

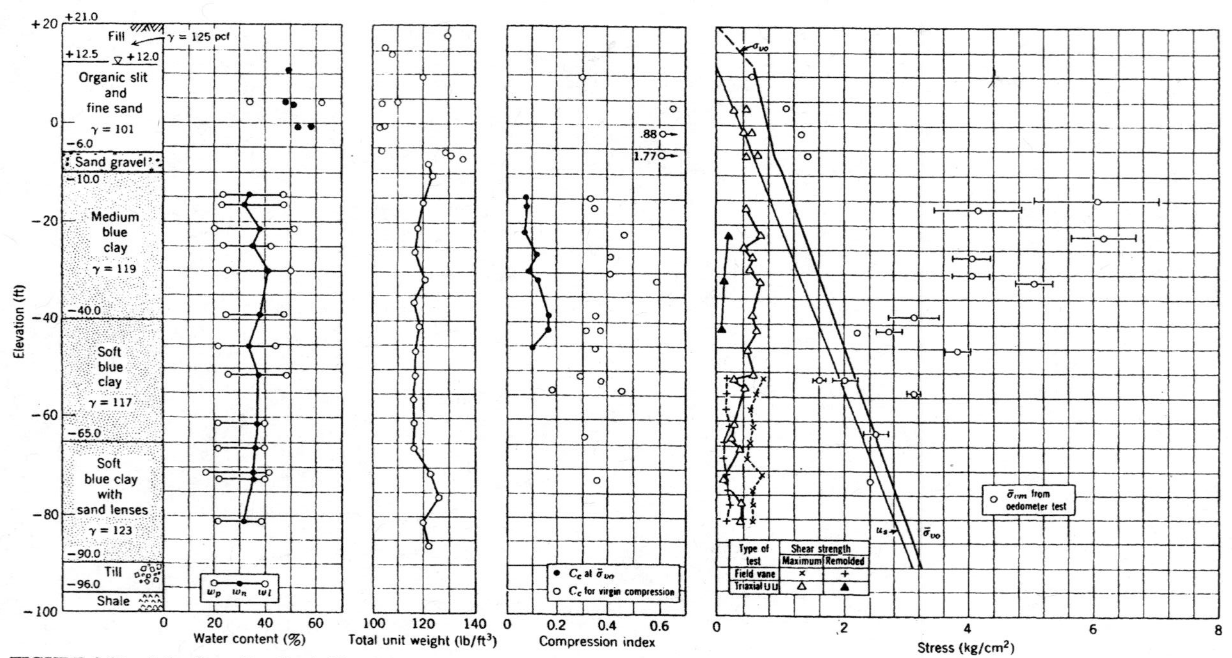

FIGURE 2.19 Subsoil profile, Cambridge, Massachusetts. (*From Lambe and Whitman 1969; reprinted with permission of John Wiley & Sons.*)

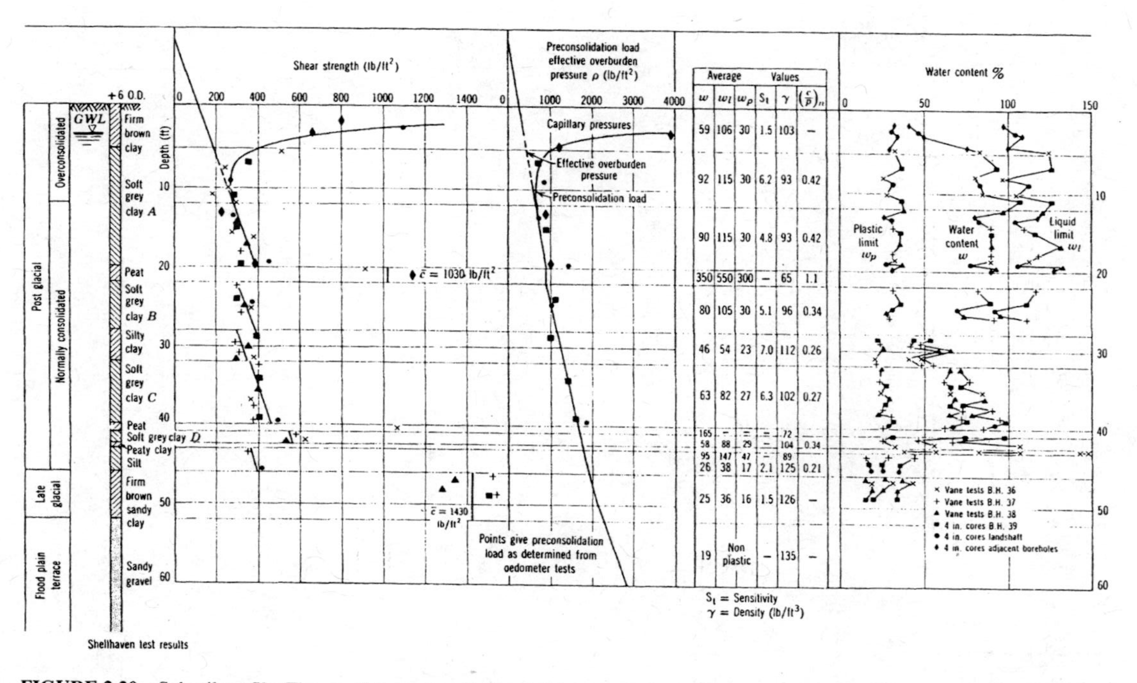

FIGURE 2.20 Subsoil profile, Thames Estuary clay, England. (*From Skempton and Henkel 1953, reprinted from Lambe and Whitman 1969; reprinted with permission of John Wiley & Sons.*)

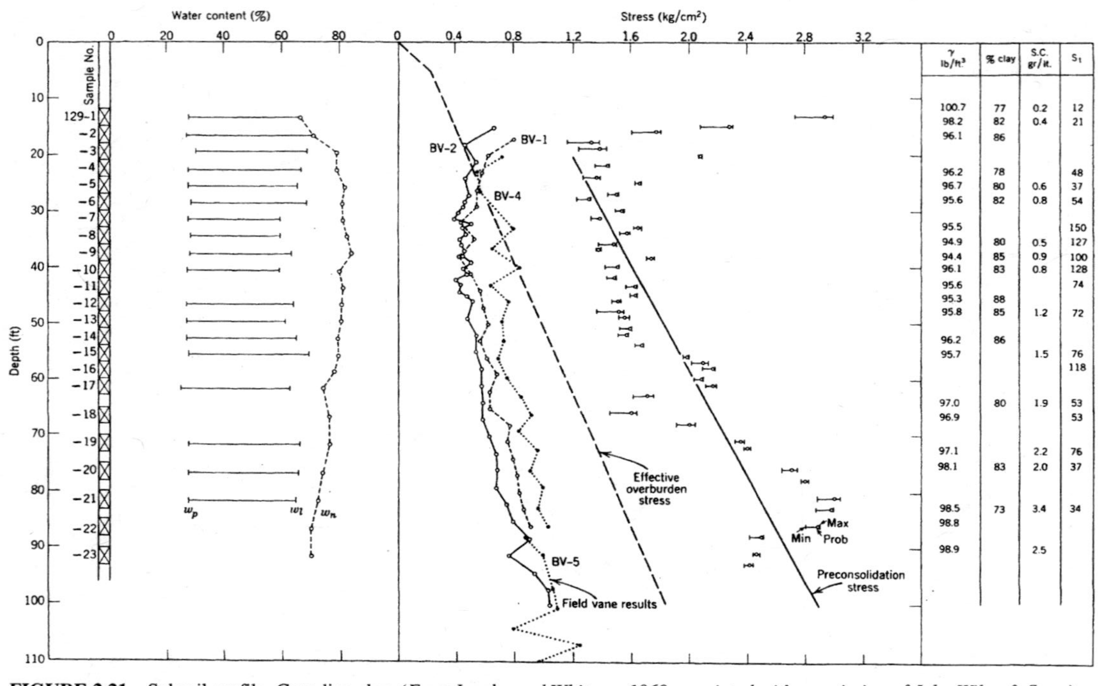

FIGURE 2.21 Subsoil profile, Canadian clay. (*From Lambe and Whitman 1969; reprinted with permission of John Wiley & Sons.*)

CHAPTER 3
LABORATORY TESTING

The soil, rock, or groundwater samples recovered during subsurface exploration can be tested in the laboratory. This chapter discusses the most common types of laboratory tests. The step-by-step procedures for performing the laboratory tests are not presented in this chapter and standardized testing procedures [such as those stated by the American Society for Testing and Materials (ASTM)] should be reviewed prior to performing the laboratory tests.

Table 3.1 presents a general discussion of laboratory tests and Table 3.2 lists common laboratory tests used in geotechnical engineering.

3.1 INDEX TESTS

As indicated in Table 3.2, one group of soil tests is known as "index tests." These types of tests are discussed in Tables 3.3 to 3.7.

3.2 OEDOMETER APPARATUS

The oedometer apparatus can be used to study the settlement or expansive soil behavior of soil. Table 3.8 discusses the oedometer apparatus and Fig. 3.1 presents a diagram of the fixed-ring and floating-ring oedometer apparatus.

3.3 SHEAR STRENGTH

The shear strength of a soil is a basic geotechnical engineering parameter and is required for the analysis of foundations, earthwork, and slope stability problems. Tables 3.9 to 3.13 summarize the different types of laboratory tests used to determine the shear strength of soil and Figs. 3.2 and 3.3 illustrate the triaxial and direct shear laboratory equipment used for shear strength tests. The shear strength of soil is further discussed in Chap. 7.

3.4 PERMEABILITY

Permeability is defined as the ability of water to flow through a saturated soil. A high permeability indicates that water flows rapidly through the void spaces, and vice versa. A measure of the soil's permeability is the coefficient of permeability, also known as the hydraulic conductivity k. The coefficient of permeability can be measured in the laboratory by using the constant-head (Fig. 3.4) or falling-head (Fig. 3.5) permeameter (see Table 3.14). Figure 3.6 summarizes the coefficient of permeability versus soil type and drainage property and Fig. 3.7 presents data on the coefficient of permeability versus void ratio for various soil types. Permeability and seepage are further discussed in Chap. 8.

3.5 LABORATORY COMPACTION TESTS

The most common laboratory compaction tests are the Modified Proctor and Standard Proctor (see Table 3.15). Figure 3.8 shows laboratory compaction test results for a silty sand. These laboratory compaction tests are used to determine the relative compaction of fill (not to be confused with relative density, Table 3.16). The placement and compaction of fill will be further discussed in Chap. 19.

3.6 LABORATORY TESTS FOR PAVEMENT DESIGN

The final section of Chap. 3 (i.e., Table 3.17) presents a summary of laboratory tests commonly used for the design of pavements.

TABLE 3.1 General Discussion of Laboratory Tests

Topic (1)	Discussion (2)
Laboratory testing program	Laboratory testing usually begins once the subsurface exploration is complete. The first step in laboratory testing is to log in all of the materials (soil, rock, or groundwater) recovered from the subsurface exploration. Then the geotechnical engineer and engineering geologist prepare a laboratory testing program, which basically consists of assigning specific laboratory tests for the soil specimens. The actual laboratory testing of the soil specimens is often performed by experienced technicians, who are under the supervision of the geotechnical engineer. Because the soil samples can dry out or there could be changes in the soil structure with time, it is important to perform the laboratory tests as soon as possible. Usually at the time of the laboratory testing, the geotechnical engineer and engineering geologist will have located the critical soil layers or subsurface conditions that will have the most impact on the design and construction of the project. The laboratory testing program should be oriented toward the testing of those critical soil layers or subsurface conditions. For many geotechnical projects, it is also important to determine the amount of ground surface movement due to construction of the project. In these cases, laboratory testing should model future expected conditions so that the amount of movement or stability of the ground can be analyzed. During the planning stage, specific types of laboratory tests may have been selected, but based on the results of the subsurface exploration, additional tests or a modification of the planned testing program may be required.
Laboratory testing specifications	Laboratory tests should be performed in accordance with standard procedures, such as those recommended by the American Society for Testing and Materials (ASTM) or those procedures listed in standard textbooks or specification manuals (e.g., Lambe 1951, Bishop and Henkel 1962, Department of the Army 1970, *Standard Specifications* 1997).
Assessing test results	Tomlinson (1986) states: "It is important to keep in mind that natural soil deposits are variable in composition and state of consolidation; therefore it is necessary to use considerable judgment based on common sense and practical experience in assessing test results and knowing where reliance can be placed on the data and when they should be discarded. It is dangerous to put blind faith in laboratory tests, especially when they are few in number. The test data should be studied in conjunction with the borehole records and the site observations, and any estimations of bearing pressures or other engineering design data obtained from them should be checked as far as possible with known conditions and past experience. "Laboratory testing should be as simple as possible. Tests using elaborate equipment are time-consuming and therefore costly, and are liable to serious error unless carefully and conscientiously carried out by highly experienced technicians. Such methods may be quite unjustified if the samples are few in number, or if the cost is high in relation to the cost of the project. Elaborate and costly tests are justified only if the increased accuracy of the data will give worthwhile savings in design or will eliminate the risk of a costly failure."

TABLE 3.2 Common Soil Laboratory Tests Used in Geotechnical Engineering

Type of condition (1)	Soil properties (2)	Specification (3)
Index tests	Water content (moisture content)	ASTM D 2216-92
	Unit weight	Block samples or sampling tubes.
	Specific gravity	ASTM D 854-92
	Particle size (sieve and hydrometer)	ASTM D 422-90
	Atterberg limits	ASTM D 4318-95
	Sand equivalent (SE)	ASTM D 2419-95*
Permeability (Chap. 8)	Constant head	ASTM D 2434-94
	Falling head	ASTM D 5084-97†
Settlement (Chap. 9)	Consolidation	ASTM D 2435-96
	Collapse	ASTM D 5333-96†
	Organic content	ASTM D 2974-95
	Fill compaction: Standard Proctor	ASTM D 698-91
	Fill compaction: Modified Proctor	ASTM D 1557-91
Expansive soil (Chap. 12)	Swell	ASTM D 4546-96
	Expansion index test	ASTM D 4829-95 or UBC 18-2
Shear strength for slope stability (Chap. 13)	Unconfined compressive strength	ASTM D 2166-91
	Unconsolidated undrained	ASTM D 2850-95
	Consolidated undrained	ASTM D 4767-95
	Direct shear	ASTM D 3080-90
	Ring shear	Stark and Eid 1994
	Miniature vane (i.e., torvane)	ASTM D 4648-94
Erosion (Chap. 15)	Dispersive clay	ASTM D 4647-93
	Erosion potential	Day 1990b
Pavements and deterioration (Chaps. 11, 17)	Pavements: CBR	ASTM D 1883-94
	Pavements: *R*-value	ASTM D 2844-94
	Sulfate	Chemical Analysis

Notes: *This specification is in the ASTM Standards Volume 04.03. †These specifications are in the ASTM Standards Volume 04.09. All other ASTM standards are in Volume 04.08.

TABLE 3.3 Index Test (Water Content)

Topic (1)	Discussion (2)
General discussion	The water content w is also known as the moisture content. The water content test is probably the most common and simplest type of laboratory test. This test can be performed on disturbed or undisturbed soil specimens. The water content test consists of determining the mass of the wet soil specimen and then drying the soil in an oven overnight (12 to 16 hours) at a temperature of 110°C (ASTM D 2216-92, 1998). The water content w of a soil is defined as the mass of water (M_w) in the soil divided by the dry mass of the soil (M_s), expressed as a percentage, or: $$w = 100\ M_w / M_s$$ Water content values are often reported to the nearest 0.1% or 1% of measured value. Values of water content w can vary from essentially 0% up to 1200%. A water content of 0% indicates a dry soil. An example of a dry soil would be near-surface rubble, gravel, or clean sand located in a hot and dry climate, such as Death Valley, California. Soil having the highest water content is organic soil, such as fibrous peat, which has been reported to have a water content as high as 1200% (NAFVAC DM-7.1, 1982). The water content data is often plotted with depth on the subsoil profile.
Common laboratory errors	According to Rollings and Rollings (1996), common laboratory errors are: 1. Inaccurate or poorly calibrated balance. 2. Loss of soil between first and second weighing. 3. Loss of moisture from sample before first weighing. 4. Addition of moisture to sample after drying and before second weighing. 5. Incorrect oven temperature, specimen too small, or incorrect tare weight. 6. Specimen removed from the oven before reaching a constant oven-dry mass. 7. Weighing the dry specimen while still hot.

Minimum wet soil specimen mass for water content determination	Largest particle dimension	Sieve size corresponding to particle dimension	Minimum mass for values reported to the nearest 0.1%	Minimum mass for values reported to the nearest 1%
	2 mm or less	No. 10	20 grams	—
	4.75 mm	No. 4	100 grams	20 grams
	9.5 mm	3/8 inch	500 grams	50 grams
	19.0 mm	3/4 inch	2,500 grams	250 grams

Note: Minimum mass for water content determination based on ASTM D 2216-92 (1998). For soil containing particles larger than 3/4 in., first sieve the sample on the 3/4-in. sieve, record the mass of oversize particles (plus 3/4-in. sieve material), and then determine the water content of the soil matrix (minus 3/4-in. sieve soil).

TABLE 3.4 Index Test (Unit Weight)

Topic (1)	Discussion (2)
Total unit weight γ_t	The total density ρ_t, also known as wet density, must be obtained from undisturbed block samples or soil samples extruded from thin-walled samplers (Table 2.8). The total density is defined as: $$\rho_t = M/V$$ where M = total mass of the soil, which is the sum of the mass of water (M_w) and mass of solids (M_s), and V = total volume of the soil sample. Since most laboratories use balances that record in grams, and the gram is a unit of mass in the International System of Units (SI), the correct terminology for the above equation is density (mass per unit volume). When dealing with engineering calculations, it is usually easier to deal with total unit weight γ_t, i.e., weight per unit volume. In order to convert total density to total unit weight in the International System of Units (SI), the density is multiplied by g (where g = acceleration of gravity = 9.81 m/s^2) to obtain the unit weight, which has units of kN/m^3. For example, in SI units, the density of water (ρ_w) = 1.0 g/cm^3 or 1.0 Mg/m^3, while the unit weight of water (γ_w) = 9.81 kN/m^3. In the United States Customary System, density and unit weight have exactly the same value. Thus the density of water and the unit weight of water are 62.4 pcf (pounds per cubic foot). However, for the density of water (ρ_w), the units should be thought of as pounds-mass (lbm) per cubic foot, while for unit weight (γ_w), the units are pounds-force (lbf) per cubic foot. In the United States Customary System, it is common to assume that 1 lbm is equal to 1 lbf. Common errors in the determination of total unit weight (γ_t) include (Rollings and Rollings 1996): (1) inaccurate measurement of the sample, (2) voids on the sides of the specimen from improper trimming or from sampling of soil having oversize particles, and (3) loss of material while extruding the specimen from the sampling tube.
Dry unit weight γ_d	By knowing the total unit weight γ_t and the water content w of the soil, the dry unit weight γ_d can be calculated as follows: $$\gamma_d = \gamma_t/(1 + w)$$ where the water content w of the soil is expressed as a decimal.
Saturated unit weight γ_{sat}	The saturated unit weight γ_{sat} of the soil represents the condition where all the void spaces are filled with water (i.e., saturated soil, S = 100%). The saturated unit weight can be calculated by using the following equation: $$\gamma_{sat} = \gamma_w(G + e)/(1 + e)$$ where γ_w = unit weight of water, G = specific gravity of solids (Table 3.5), and e = void ratio (Chap. 5).
Buoyant unit weight γ_b	The buoyant unit weight γ_b, also known as the submerged unit weight, can be calculated as follows: $$\gamma_b = \gamma_{sat} - \gamma_w$$ where γ_{sat} = saturated unit weight and γ_w = unit weight of water.

TABLE 3.5 Index Test (Specific Gravity)

Topic (1)	Discussion (2)
General discussion	The specific gravity G is a dimensionless parameter that relates the density of dry soil particles to the density of water, or: $$G = \rho_s/\rho_w$$ where ρ_s = dry density of the soil particles (i.e., $\rho_s = M_s/V_s$) and ρ_w = density of water (1 g/cm^3 or 62.4 pcf). For soil, the specific gravity is obtained by measuring the dry mass of the soil particles and then using a pycnometer to obtain the volume of the soil particles (see ASTM D 854-92, 1998). Typical values and ranges of specific gravity versus different types of soil minerals are presented below. Because quartz is the most abundant type of soil mineral, the specific gravity for inorganic soil is often assumed to be 2.65. For clays, the specific gravity is often assumed to be 2.70 because common clay particles, such as montmorillonite and illite, have slightly higher specific gravity values. Oversize particles are often defined as gravel and cobbles that are retained on the ¾-in. sieve (Day 1989). For these large particles, there may be internal rock fractures, voids, or moisture trapped within the gravel and cobbles. The specific gravity test procedures and calculations can include these features and the test result is referred to as the *bulk specific gravity* (ASTM C 127-93, 1998).
Common laboratory errors	According to Rollings and Rollings (1996) common laboratory errors are: 1. Imprecise weights of pycnometer flask and its contents (e.g., the flask is not clean). 2. Moisture on the outside of the pycnometer flask or on inside of its neck. 3. Meniscus is not level with the mark on the neck of the pycnometer flask. 4. Use of tap water or other water that is not distilled or demineralized. 5. Incomplete removal of entrapped air from soil suspension. This is often the most serious error and will tend to lower the computed value. 6. Gain in moisture of the oven-dried soil before it is weighed. 7. Loss of material from the oven-dried soil specimen.

Type of mineral	Formula	Specific gravity	Comments
Quartz	SiO_2	2.65	Silicate, most common type of soil mineral
K feldspar	$KAlSi_3O_8$	2.54–2.57	Feldspars are also silicates and are the second most common type of mineral
Na feldspar	$NaAlSi_3O_8$	2.62–2.76	
Calcite	$CaCO_3$	2.71	Basic constituent of carbonate rocks
Dolomite	$CaMg(CO_3)_2$	2.85	Basic constituent of carbonate rocks
Muscovite	Varies	2.76–3.0	Silicate sheet-type mineral (mica group)

TABLE 3.5 Index Test (Specific Gravity) (*Continued*)

Type of mineral	Formula	Specific gravity	Comments
Biotite	Complex	2.8–3.2	Silicate sheet-type mineral (mica group)
Hematite	Fe_2O_3	5.2–5.3	Frequent cause of reddish-brown color in soil
Gypsum	$CaSO_4 \cdot 2H_2O$	2.35	Can lead to sulfate attack of concrete
Serpentine	$Mg_3Si_2O_5(OH)_4$	2.5–2.6	Silicate sheet or fibrous-type mineral
Kaolinite	$Al_2Si_2O_5(OH)_4$	2.61–2.66	Silicate clay mineral, low activity
Illite	Complex	2.60–2.86	Silicate clay mineral, intermediate activity
Montmorillonite	Complex	2.74–2.78	Silicate clay mineral, highest activity

Notes: Silicates are very common and account for about 80% of the minerals at the earth's surface. Data accumulated from the following sources: Lambe and Whitman (1969) and Mottana et al. (1978).

TABLE 3.6 Index Tests (Sieve and Hydrometer)

Topic (1)	Discussion (2)
General discussion	The particle sizes of a specimen of soil are determined from sieve analysis and hydrometer analysis. The size analysis is used to determine the particle size distribution for those particles larger than the No. 200 sieve (0.075 mm) and the hydrometer analysis is used to determine the particle size distribution for those soil particles finer than the No. 200 sieve (i.e., silt and clay size particles).
Particle size analysis—sieve analysis	*Sieve analysis:* A sieve is a piece of laboratory equipment that consists of a pan with a screen (square woven wire mesh) at the bottom. U.S. standard sieves are used to separate particles of a soil sample into various sizes. A sieve analysis is performed on dry soil particles that are larger than the No. 200 U.S. standard sieve (i.e., sand-size, gravel-size, and cobble-size particles). The first step in a sieve analysis is to oven-dry the soil and obtain its initial mass. The soil is then washed on the No. 200 sieve to remove all the fines (silt- and clay-size particles). The portion retained on the No. 200 sieve is oven-dried. A series of sieves, arranged in descending size of openings, are stacked and the dry soil is poured into the top. The sieves are shaken (usually with a mechanical shaker) and the amount of dry soil retained on each sieve is recorded. Based on this data, the percent finer is calculated as the dry mass of soil passing through the sieve divided by the initial dry mass of the soil specimen, expressed as a percentage. Commonly used United States Standard sieve numbers and openings are listed below: No. 4 sieve opening = 4.75 mm; No. 60 sieve opening = 0.25 mm No. 10 sieve opening = 2.00 mm; No. 100 sieve opening = 0.15 mm No. 20 sieve opening = 0.85 mm; No. 140 sieve opening = 0.106 mm No. 40 sieve opening = 0.425 mm; No. 200 sieve opening = 0.075 mm According to Rollings and Rollings (1996) common laboratory errors are: (1) failure to break down agglomerations of material into individual grains, (2) loss of material during testing, (3) overloading of sieves, (4) broken or distorted sieves, and (5) inadequate shaking of sieves.
Particle size analysis—hydrometer analysis	*Hydrometer analysis:* The particle size distribution for fines (silt- and clay-size particles finer than the No. 200 sieve) are determined by a sedimentation process. A hydrometer is used to obtain the necessary data during the sedimentation process (ASTM D 422-90, 1998). The test procedure consists of mixing a known mass (usually about 50 g) of soil with a dispersing agent (sodium hexametaphosphate). The dispersing agent prevents the clay size particles from forming flocs during the hydrometer test. The soil specimen, dispersing agent, and distilled water are thoroughly mixed and then transferred to a 1000-mL glass cylinder. Hydrometer readings versus time from the beginning of sedimentation are then recorded. The hydrometer is calibrated so that it records the actual mass of soil particles and dispersing agent that is in suspension. The hydrometer test is based on Stokes' law, which relates the diameter of a single sphere to the time required for the sphere to fall a certain distance in a liquid of known viscosity. The idea for the hydrometer analysis is that a larger, and hence heavier, soil particle will fall faster through distilled water than a smaller, and hence lighter, soil particle. The test procedure is approximate because many fine soil particles are not spheres, but are rather of a platelike shape. According to Rollings and Rollings (1996) common laboratory errors are: (1) inadequate type or quantity of dispersing agent, (2) incomplete dispersion of soil, (3) too much soil in suspension, and (4) disturbance of suspension during insertion of hydrometer.

TABLE 3.7 Index Tests (Atterberg Limits)

Topic (1)	Discussion (2)
General discussion	The Atterberg limits are defined as the water content corresponding to different behavior conditions of silts and clays. Although originally six limits were defined by Albert Atterberg (1911), in geotechnical engineering, the term *Atterberg limits* typically refers only to the liquid limit (LL), plastic limit (PL), and shrinkage limit (SL). In accordance with ASTM, the liquid limit, plastic limit, and shrinkage limit are performed on that portion of the soil that passes the No. 40 sieve (0.425 mm). For many soils, a significant part of the soil specimen (i.e., those soil particles larger than the No. 40 sieve) will be excluded during testing. This can cause problems in classifying the soil (see Chap. 4).
Liquid limit (LL)	The liquid limit (LL) is defined as the water content corresponding to the behavior change between the liquid and plastic state of a silt or clay. The liquid limit is arbitrarily defined as the water content at which a pat of soil, cut by a groove of standard dimensions, will flow together for a distance of 12.7 mm (0.5 in.) under the impact of 25 blows in a standard liquid limit device (ASTM D 4318-95, 1998).
Plastic limit (PL)	The plastic limit (PL) is defined as the water content corresponding to the behavior change between the plastic and semisolid state of a silt or clay. The plastic limit is arbitrarily defined as the water content at which a silt or clay will just begin to crumble when rolled into a thread approximately 3.2 mm ($^1/_8$ in.) in diameter (ASTM D 4318-95, 1998).
Shrinkage limit (SL)	The shrinkage limit (SL) is defined as the water content corresponding to the behavior change between the semisolid and solid states of a silt or clay. The shrinkage limit is also defined as the water content at which any further reduction in water content will not result in a decrease in volume of the soil mass (ASTM D 427-93 or D 4943-95, 1998). The shrinkage limit is rarely obtained in practice because of laboratory testing difficulties and limited use of the data.
Plasticity index (PI)	A measure of a soil's plasticity is the plasticity index (PI), defined as: $$PI = LL - PL$$ where LL = liquid limit and PL = plastic limit. The PI is often expressed as a whole number. The plasticity index is important because it has been correlated with numerous soil engineering properties.
Common laboratory errors	According to Rollings and Rollings (1996), common laboratory errors are: 1. Nonrepresentative sample (the soil sample must be the same for both the liquid and plastic limit tests). 2. Improperly prepared and cured samples. 3. Incorrect water content determination. Additional common laboratory errors for the liquid limit test include: (1) improper, unadjusted, or uncalibrated test device, (2) worn parts on the test device, (3) soil at the contact point between the cup and rubber base of the test device, and (4) loss of moisture during the test. Additional common laboratory errors for the plastic limit test include: (1) incorrect final thread diameter and (2) stopping the rolling process too soon.

TABLE 3.8 Oedometer Apparatus

Topic (1)	Discussion (2)
Purpose	The oedometer (also known as a consolidometer) is the primary laboratory equipment used to study the settlement or expansion behavior of soil. The purpose of the oedometer is to laterally confine the soil specimen and allow for a vertical pressure to be applied to the soil specimen (see Fig. 3.1). The oedometer test should be performed only on undisturbed soil specimens, or in the case of studies of fill behavior, on specimens compacted to anticipated field and moisture conditions.
Types of oedometer apparatus	Figure 3.1 shows a diagram of the two basic types of oedometers. The equipment can vary, but in general, an oedometer consists of the following: 1. A metal ring that is used to laterally confine the soil specimen. 2. Porous plates placed on the top and bottom of the soil specimen. 3. A loading device that is used to apply a vertical stress to the soil specimen. 4. A dial gauge to measure the vertical deformation of the soil specimen as it is loaded or unloaded. 5. A surrounding container to allow the soil specimen to be submerged in distilled water. For the fixed-ring oedometer, the soil specimen will be compressed from the top downward as the load is applied, while for the floating-ring oedometer, the soil specimen is compressed inward from the top and bottom. An advantage of the floating ring oedometer is that there is less friction between the confining ring and the soil. Disadvantages of the floating ring are that it is often more difficult to set up and soil may squeeze or fall out of the junction of the bottom porous plate and ring. Because of these disadvantages, the fixed ring oedometer is the most popular testing setup.
Popularity	The oedometer test is popular because of its simplicity and it can be used to model and predict the behavior of the *in situ* soil. For example, a soil specimen can be placed in the oedometer and then subjected to an increase in pressure equivalent to the weight of the proposed structure. By analyzing the settlement-versus-load data, the geotechnical engineer can calculate the amount of expected settlement due to the weight of the proposed structure. As will be discussed in Chaps. 9 and 12, the oedometer can also be used to predict the consolidation behavior of soft saturated clay, settlement potential of collapsible soil, and the amount of heave of expansive soils.
Common laboratory errors	Common laboratory errors using the oedometer apparatus are: (1) the specimen contains voids or oversize particles, (2) the applied vertical load is inaccurate or not applied concentrically, (3) the dial gauge is inaccurate or not read correctly, (4) the correct loading sequence is not followed, and (5) specimen is submerged in tap water, rather than distilled water. According to Rollings and Rollings (1996), other common laboratory errors are as follows: 1. Disturbance of the specimen during trimming. 2. Specimen not snugly fitting into and filling the confining ring. 3. Permeability of the porous plates is too low (e.g., the plates are clogged with soil). 4. Excessive friction between the specimen and confining ring. 5. Inappropriate or inaccurate measurement of the specimen height.

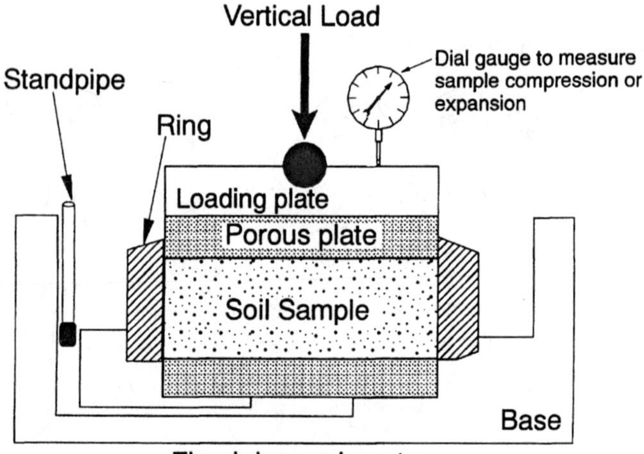

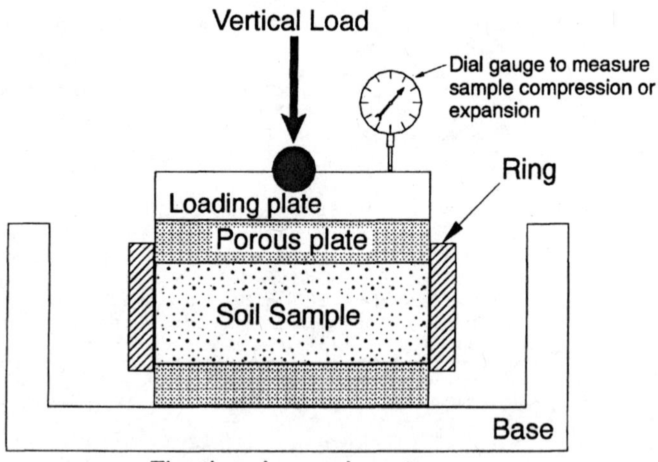

Note: **Although not shown above, the dial gauge should be positioned at the center of the loading plate.**

FIGURE 3.1 Fixed- and floating-ring oedometer apparatus (see Table 3.8).

TABLE 3.9 Laboratory Shear Strength Tests (General Discussion)

Types (1)	Discussion (2)
General discussion	There are many types of shear strength tests that can be performed in the laboratory. The objective is to obtain the shear strength of the soil. Laboratory tests are generally divided into two categories: 1. *Shear strength tests based on total stress:* The purpose of these laboratory tests is to obtain the undrained shear strength s_u of the soil or the failure envelope in terms of total stresses (total cohesion c and total friction angle ϕ). These types of shear strength tests are often referred to as *undrained* shear strength tests. 2. *Shear strength tests based on effective stress:* The purpose of these laboratory tests is to obtain the effective shear strength of the soil based on the failure envelope in terms of effective stress (effective cohesion c' and effective friction angle ϕ'). These types of shear strength tests are often referred to as *drained* shear strength tests. For a further discussion of the concept of total and effective stress, see Chap. 6.
Type of shear strength test	1. *Unconsolidated undrained (UU):* The unconsolidated undrained shear strength test is used to obtain the undrained shear strength s_u of the soil. Typical types of laboratory tests are the laboratory vane test and the unconfined compression test (see Table 3.10). 2. *Consolidated undrained (CU):* The consolidated undrained shear strength test is used to obtain the failure envelope in terms of total stresses, i.e., total cohesion c and total friction angle ϕ. The triaxial apparatus is used to perform this test (see Table 3.11). 3. *Consolidated undrained with pore pressure measurements (CU′):* The consolidated undrained shear strength test with pore pressure measurements is used to obtain the failure envelope in terms of effective stress, i.e., effective cohesion c' and effective friction angle ϕ'. The triaxial apparatus is used to perform this test (see Table 3.11). 4. *Consolidated drained (CD):* The consolidated drained test is also used to obtain the failure envelope in terms of effective stress, i.e., effective cohesion c' and effective friction angle ϕ'. The consolidated drained test can be performed in the direct shear apparatus (see Table 3.12) or in the triaxial apparatus (see Table 3.11). 5. *Drained residual shear strength:* The drained residual shear strength is used to obtain the residual failure envelope in terms of effective stress, i.e., residual friction angle ϕ'_r. The drained residual shear strength can be obtained by shearing back and forth a specimen in the direct shear apparatus or by using the ring shear apparatus (see Table 3.13).

TABLE 3.10 Laboratory Shear Strength Tests (Vane Tests and the Unconfined Compression Test)

Types (1)	Discussion (2)
Laboratory vane tests	As discussed in Chap. 2 (e.g., see Fig. 2.12), the vane shear test (VST) can be performed in the field to obtain the undrained shear strength s_u of clay. The miniature vane or torvane device could also be used in the laboratory to obtain the undrained shear strength (s_u) of clay.
Unconfined compression test	The unconfined compression test is a very simple type of test that consists of applying a vertical compressive pressure to a cylinder of laterally unconfined clay. The vertical compressive pressure that causes shear failure of the clay is termed the unconfined compressive strength q_u. Some clays are so plastic that instead of shearing, they deform and bulge, and in these cases the unconfined compressive strength q_u is defined as the vertical compressive pressure at 15% axial strain. The unconfined compressive test is a shear strength test based on total stress, and it can be used to determine the undrained shear strength (s_u) of the clay, defined as: $$s_u = q_u/2$$ The unconfined compression test should not be performed on dry and crumbly soils, fissured or varved clays, silts, peat, and all types of granular material. The height to diameter of the cylinder of clay should be between 2 and 2.5. The clay specimen should also be tested in standard laboratory equipment that can accurately record the force applied to the top of the clay specimen as well as record the vertical deformation of the specimen as it is loaded. Typical loading rates of the clay specimen are 0.5 to 2% of the initial height L_o per minute (ASTM D 2166-91, 1998). Because the area of the clay specimen changes as it is loaded, corrections should be made to the unconfined compressive strength, as follows: $$\epsilon_f = \Delta L/L_o$$ where ϵ_f = axial strain at failure, ΔL = change in height of the clay specimen at failure, and L_o = original height of the test specimen. From this data, the cross-sectional area of the specimen at failure (A_f) can be calculated, and is: $$A_f = A_o/(1 - \epsilon_f)$$ where A_o = initial cross-sectional area of the cylinder of clay. The unconfined compressive strength q_u of the clay can then be calculated as follows: $$q_u = P_f/A_f$$ where P_f is the recorded vertical force that caused a shear failure of the cylinder of clay. To determine the *in situ* undrained shear strength of the clay deposit, undisturbed soil specimens are required. The unconfined compressive test could also be performed on compacted fill specimens to determine the shear strength of fill. According to Rollings and Rollings (1996), the following are the most common errors with this test: (1) the test is run on inappropriate soil, (2) the specimen is disturbed during trimming, (3) there is a loss of soil moisture through evaporation during specimen preparation, and (4) the rate of strain or rate of loading is too fast.

TABLE 3.11 Laboratory Shear Strength Tests (Triaxial Compression Tests)

Types (1)	Discussion (2)
Unconsolidated undrained triaxial compression test	The unconsolidated undrained triaxial compression test is similar to the unconfined compression test, except that clay specimens are subjected to a confining fluid pressure in a triaxial chamber (ASTM D 2850-95, 1998). Figure 3.2 shows a diagram of the triaxial apparatus. The cylindrical clay specimen is placed in the center of the chamber, sealed with a rubber membrane, and subjected to a confining fluid pressure. A vertical pressure is slowly applied to the top of the specimen by slowly loading the piston. No drainage of the clay specimen is allowed during the test. The clay specimen is sheared in compression without drainage at a constant rate of axial deformation (strain-controlled). If the clay specimens are saturated, then the confining pressure will not change the wet density of the specimen. Therefore, if several identical clay specimens are tested, they will all have approximately the same undrained shear strength (failure envelope will be a horizontal line, i.e., $\phi = 0$ and $c =$ constant). Since this test does not provide any additional information for saturated clay specimens, it is used much less frequently than the unconfined compression test. According to Rollings and Rollings (1996), the following are the most common errors with the triaxial test: (1) leakage of chamber fluid into the specimen; (2) leakage of pore water out of the specimen; (3) the rubber membrane is too thick, i.e., too strong for use on a particular soil, thus making the soil appear stronger; (4) there is friction on the loading piston, making the load at failure appear larger than it really is; and (5) the specimen is disturbed during trimming or while putting on the rubber membrane.
Consolidated undrained triaxial compression test	The consolidated undrained triaxial compression test is similar to the unconsolidated undrained compression test, except that the clay specimen is first saturated and consolidated prior to shearing (ASTM D 4767-95, 1998). The process of saturation is achieved by using a back pressure. This test can also be used to measure the excess pore water pressures that develop within the clay as it is sheared. The shear strength based on total stress (c and ϕ) or based on effective stress (c' and ϕ' by measuring the pore water pressure during shear) can be obtained from this laboratory test. The test is quite complicated and should be performed only by experienced personnel. This test has been used to study the shear behavior of all types of soil, and further details of this test will be presented in Chap. 7. According to Rollings and Rollings (1996), additional common errors with this test include: (1) the application of back pressure is done in increments that are too large, (2) the back pressure is applied too rapidly, (3) the back pressure and chamber pressure are not precisely maintained during consolidation, (4) the specimen is not completely consolidated prior to shearing, (5) the rate of strain is too fast, (6) there are excessive temperature variations during shear testing, and (7) the permeability of the porous plates is too low, which affects the pore water pressure measurements during shearing.

TABLE 3.11 Laboratory Shear Strength Tests (Triaxial Compression Tests) (*Continued*)

Types (1)	Discussion (2)
Consolidated drained triaxial compression test	The consolidated drained triaxial compression test is similar to the consolidated undrained triaxial compression test, except that during shearing, drainage is allowed and the soil specimen is sheared slow enough so that excess (positive or negative) pore water pressure do not develop. For silts and clays, the shearing portion of this test will require a considerable length of time to ensure that the excess pore water pressures are allowed to dissipate. This test is used to obtain the shear strength parameters in terms of effective stress (c' and ϕ').

TABLE 3.12 Laboratory Shear Strength Tests (Direct Shear)

Topic (1)	Discussion (2)
General discussion	The direct shear test, first used by Coulomb in 1776, is the oldest type of shear testing equipment. The direct shear test is also the most common laboratory equipment used to obtain the drained shear strength (shear strength based on effective stress) of a soil. Figure 3.3 shows the direct shear testing device.
Testing procedure	The testing procedure is as follows (ASTM D 3080-90, 1998): 1. The soil specimen (typical diameter = 63.5 mm and height = 25.4 mm) is placed on top of a porous plate in the center of the direct shear apparatus. 2. A porous plate is placed on the top of the specimen and a dial gauge is set up to measure vertical deformation. 3. A loading device is used to apply a vertical force to the soil specimen. This vertical force, also known as the normal force, is converted to a stress by dividing the vertical force by the cross-sectional area of the soil specimen. 4. A surrounding container (not shown in Fig. 3.3) allows the soil specimen to be submerged in distilled water. 5. The soil specimen is allowed to equilibrate or fully consolidate under the applied vertical stress. 6. A dial gauge is set up to measure horizontal deformation, and then an increasing force is applied to the upper half of the direct shear apparatus. The shear force can be applied by using deadweights placed on a hanger (stress-controlled test) or by a motor acting through gears (strain-controlled test). 7. The soil specimen is slowly sheared in half (a single shear plane). The maximum shear force required to shear the sample is recorded and the shear strength is defined as this shear force divided by the cross-sectional area of the soil specimen. 8. The shear strength versus the vertical stress can be plotted and a failure envelope (defined by c' and ϕ') is obtained.
Rate of shearing	It is important to shear the soil specimen slowly enough so that excess pore water pressures do not develop within the soil specimen. For clean sands, the total elapsed time to failure should be about 10 minutes, while for silty sands the total elapsed time to failure should be about 60 minutes. For clean sands and silty sands, the setup and actual testing operation is relatively quick and simple, which is the main reason for the test's popularity. If sands contain gravel size particles, then the gravel can artificially increase the shear strength of the sand because of the usual small height (25.4 mm) of the specimen. Other limitations of the device are that the shear stresses and displacements are nonuniformly distributed, and the failure plane may not be perfectly horizontal within the specimen (Dounias and Potts 1993, Day 1995a). Even with these limitations, the direct shear apparatus generally provides reasonably accurate values of the effective friction angle (ϕ') for cohesionless soil.

TABLE 3.12 Laboratory Shear Strength Tests (Direct Shear) (*Continued*)

Topic (1)	Discussion (2)
Rate of shearing (*Continued*)	For silts and clays, the drained shear strength (c' and ϕ') can also be determined, but the rate of shearing must be much slower. According to ASTM D 3080-90 (1998), the total elapsed time to failure (t_f) should be: $$t_f = 50\ t_{50}$$ where t_{50} = time for the silt or clay to achieve 50% consolidation under the vertical stress. This time (t_{50}) is obtained from a time versus vertical deformation plot (Chap. 9). For silt or clay, the value of t_f can be quite high. For example, for a clay that reaches 50% consolidation in 20 minutes, the total elapsed time to failure must be at least 1000 minutes (16.7 hours). This may require the installation of special equipment (special gears to slow down the strain rate) to allow the direct shear device to shear very slowly.
Common laboratory errors	Probably the two most common problems with the direct shear testing of a silt or clay are: 1. The soil is not saturated prior to shearing 2. The soil is sheared too quickly Both of these conditions can result in the shear strength parameters being overestimated. Other common errors with this test include the following (Rollings and Rollings, 1996): 1. Loss of moisture through evaporation during specimen preparation 2. The top and bottom of the soil specimen are not flat and parallel 3. The gap between the top and bottom of the shear box is too large or too small 4. There is an inaccurate measurement of small shear stresses 5. The porous plates are not permeable enough 6. The test is stopped too soon (i.e., the data was not plotted during the shear testing to establish when the test can be stopped)

TABLE 3.13 Laboratory Shear Strength Tests (Drained Residual Shear Strength)

Topic (1)	Discussion (2)
General discussion	A common purpose of shear strength tests is to obtain the undrained shear strength s_u or the shear strength parameters (c or c' and ϕ or ϕ') that define the failure envelope. These tests normally determine the shear strength of the soil, which is defined as the shear stress at failure (i.e., the peak shear strength).
Use of the drained residual shear strength	For some projects, it may be important to obtain the residual shear strength ϕ'_r of fine-grained soil, which is defined as the remaining (or residual) shear strength after a considerable amount of shear deformation has occurred. For example, a clay specimen could be placed in the direct shear box and then sheared back and forth several times to develop a well-defined shear failure surface. Once the shear surface is developed, the drained residual shear strength would be obtained by performing a final, slow shear of the specimen. The drained residual shear strength can be applicable to many types of soil or rock conditions where a considerable amount of shear deformation has already occurred. For example, the stability analysis of ancient landslides, slopes in overconsolidated fissured clays, and slopes in fissured shales will often be based on the drained residual shear strength of the failure surface (Bjerrum 1967a, Skempton and Hutchinson 1969, Skempton 1985, Hawkins and Privett 1985, Ehlig 1992). Skempton (1964) stated that the residual shear strength (friction angle) is independent of the original shear strength, water content, and liquidity index, and depends only on the size, shape, and mineralogical composition of the constituent particles. Besides the direct shear equipment, the drained residual shear strength can be determined by using a modified Bromhead ring shear apparatus (Stark and Eid 1994). Back calculations of landslide shear strength indicate that the residual shear strength from ring shear tests are reasonably representative of the slip surface (Watry and Ehlig 1995). The ring shear specimen is annular with an inside diameter of 7 cm (2.8 in.) and an outside diameter of 10 cm (4 in.). Drainage is provided by annular bronze porous plates secured to the bottom of the specimen container and the loading platen.
Testing procedure	Remolded specimens are often used for the ring shear testing. The typical testing procedure is a follows: 1. The remolded specimen is obtained by air drying the soil (such as slide plane clay seam material), crushing it with a mortar and pestle, ball-milling the crushed material, and processing it through the U.S. Standard Sieve No. 200. 2. Distilled water is added to the processed soil until a water content approximately equal to the liquid limit is obtained. The specimen is then allowed to rehydrate for 7 days in a moist room. A spatula is used to place the remolded soil paste into the annular specimen container. To measure the drained residual shear strength, the ring shear specimen is often consolidated at a high vertical pressure (such as 700 kPa) and then the specimen is unloaded to a much lower vertical pressure (such as 50 kPa). 3. The specimen is then presheared by slowly rotating the ring shear base for one complete revolution by using the handwheel.

TABLE 3.13 Laboratory Shear Strength Tests (Drained Residual Shear Strength) (*Continued*)

Topic (1)	Discussion (2)
Testing procedure (*Continued*)	4. After preshearing, the specimen is sheared at a slow drained displacement rate (such as 0.02 mm/min). This slow shear displacement rate has been successfully used to test soils that are very plastic. 5. After a drained residual strength condition is obtained at the low vertical pressure, shearing is stopped and the normal stress is increased to a higher pressure (such as 100 kPa). After consolidation at this higher pressure, the specimen is sheared again until a drained residual condition is obtained. 6. This procedure is also repeated for other effective normal stresses. The slow shear displacement rate (0.02 mm/min) should be used for all stages of the multistage test. 7. The maximum shear stress is plotted versus normal stress in order to obtain the drained residual shear strength envelope (see Chap. 7 for examples).

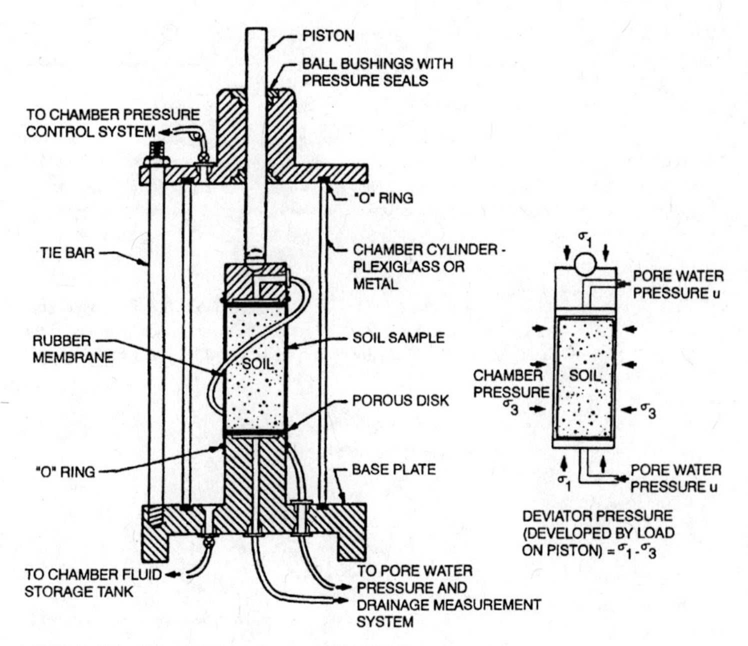

FIGURE 3.2 Triaxial apparatus (see Table 3.11).

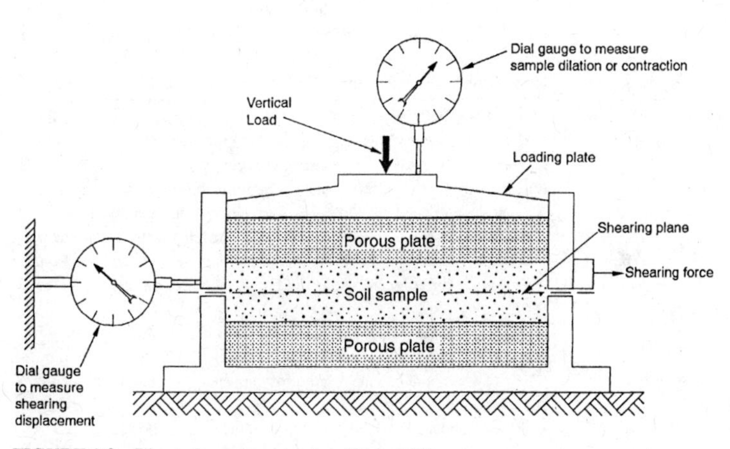

FIGURE 3.3 Direct shear apparatus (see Table 3.12).

TABLE 3.14 Laboratory Tests to Determine the Permeability of Soil

Topic (1)	Discussion (2)
Constant-head permeameter	Figure 3.4 shows a constant-head permeameter apparatus. A saturated soil specimen is placed in the permeameter and then a head of water (Δh) is maintained. The hydraulic conductivity is based on Darcy's law, which states that the velocity v of flow in soil is proportional to the hydraulic gradient i, or: $$v = ki$$ where k = hydraulic conductivity (also known as coefficient of permeability), with units of cm/s, and i = hydraulic gradient, which is defined as the change in total head Δh divided by the length of the soil specimen L, or: $$i = \Delta h / L$$ From Darcy's law, the coefficient of permeability (k) from the constant head permeameter can be calculated as follows: $$k = \frac{QL}{\Delta h A t}$$ where Q = total discharged volume (mL) in a given time t A = area of the soil specimen L = length of the soil specimen Δh = total head loss for the constant-head permeameter as defined in Fig. 3.4 The constant-head permeameter is often used for sandy soil that has a high permeability. It is important that the porous plate which supports the soil specimen (Fig. 3.4) has a very high permeability. As an alternative, a reinforced permeable screen can be used in place of the porous plate. Another important consideration is that the soil specimen diameter should be at least 10 times larger than the size of the largest soil particle.
Falling-head permeameter	Figure 3.5 (from Department of the Army 1970) shows a laboratory falling-head permeameter. This equipment is often used to determine the hydraulic conductivity of a saturated silt or clay specimen. Filter paper is often placed over the porous plate to prevent the migration of soil fines through the porous plate. Also, a frequent cause of inaccurate results is the inability to obtain a seal between the soil specimen and the side of the permeameter. Because of these factors (migration of fines and inadequate sealing), a greater degree of skill is required to perform the falling-head permeability test. The objective of the falling-head permeability test is to allow the water level in a small-diameter tube to fall from an initial position h_o to a final position h_f. The amount of time it takes for the water level to fall from h_o to h_f is recorded. According to Darcy's law, the equation to determine the hydraulic conductivity k for a falling-head test is as follows: $$k = \frac{2.3\, aL}{At} \log_{10}(h_o/h_f)$$

TABLE 3.14 Laboratory Tests to Determine the Permeability of Soil (*Continued*)

Topic (1)	Discussion (2)
Falling-head permeameter (*Continued*)	where a = area of the standpipe L = length of the soil specimen A = area of the soil specimen t = time it takes for the water level in the standpipe to fall from h_o to h_f
Permeability data	The bold lines in Fig. 3.6 indicate major divisions in the hydraulic conductivity. A hydraulic conductivity of about 1 cm/s is the approximate boundary between laminar and turbulent flow. A hydraulic conductivity of about 1×10^{-4} cm/s is the approximate dividing line between good drainage and poorly drained soils. In general, those soils that contain more fines, such as clays, will have the lowest hydraulic conductivity. This is because clay particles provide very small drainage paths, with a resultant large resistance to fluid flow, even though a clay will often have significantly more void space than a sand. According to Terzaghi and Peck (1967), the classification of soil according to hydraulic conductivity k is as follows: • High degree of permeability: k is over 0.1 cm/s • Medium degree of permeability: k is between 0.1 and 0.001 cm/s • Low permeability: k is between 0.001 and 1×10^{-5} cm/s • Very low permeability: k is between 1×10^{-5} and 1×10^{-7} cm/s • Practically impermeable: k is less than 1×10^{-7} cm/s In terms of engineering practice, the hydraulic conductivity is often reported to one or, at most, two significant figures. This is because in many cases the laboratory permeability will not represent *in situ* conditions. For example, Tomlinson (1986) states: "There is a difference between the horizontal and vertical permeability of natural soil deposits due to the effects of stratification with alternating beds of finer or coarser grained soils. Thus the results of laboratory tests on a few samples from a vertical borehole are of rather doubtful value in assessing the representative permeability of the soil for calculating the quantity of water to be pumped from a foundation excavation."
Permeability versus void ratio	The type of soil (i.e., open graded gravel versus fat clay) is the most important factor governing the hydraulic conductivity. Another important factor is the void ratio (defined in Chap. 5). Figure 3.7 shows the relationship between void ratio e and the coefficient of permeability (hydraulic conductivity) for several different types of soil.

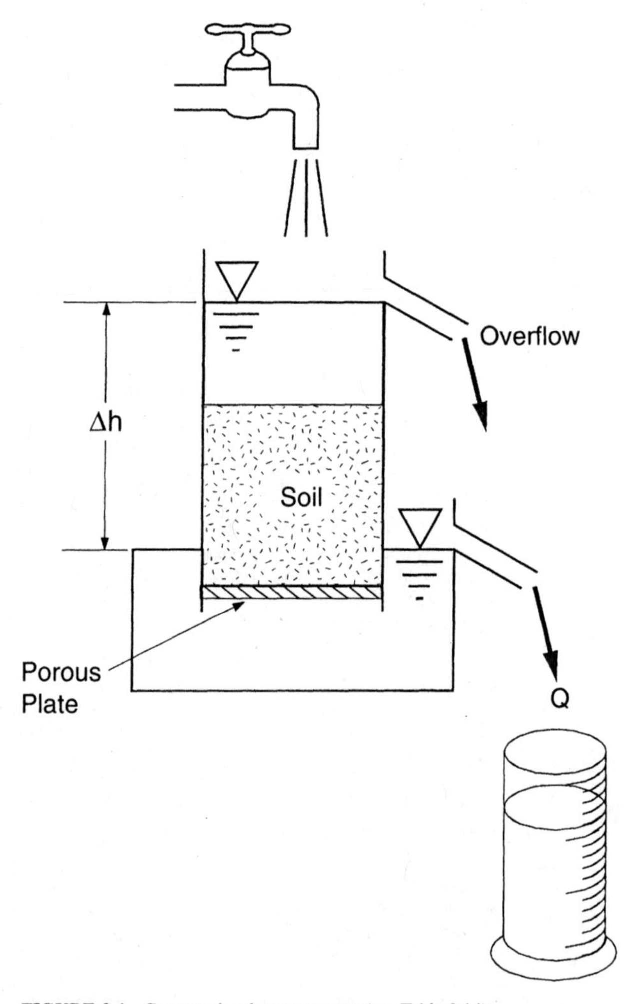

FIGURE 3.4 Constant-head permeameter (see Table 3.14).

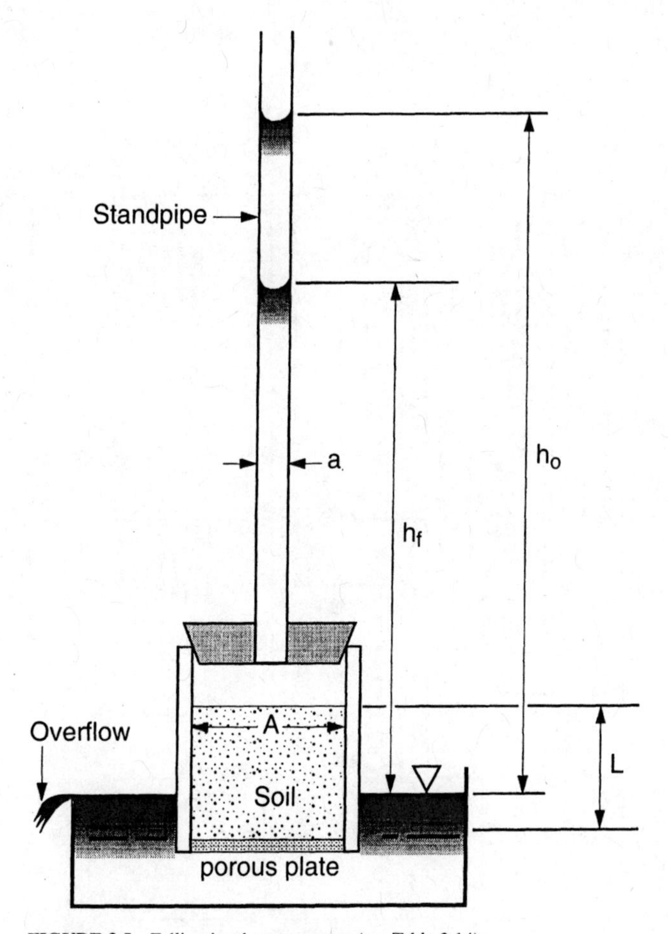

FIGURE 3.5 Falling-head permeameter (see Table 3.14).

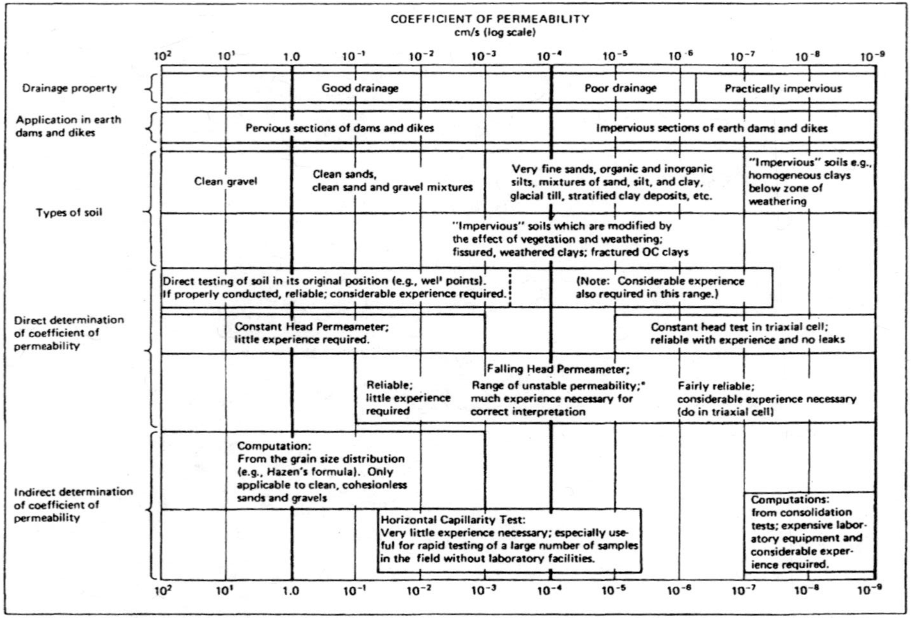

COEFFICIENT OF PERMEABILITY
cm/s (log scale)

10^2 10^1 1.0 10^{-1} 10^{-2} 10^{-3} 10^{-4} 10^{-5} 10^{-6} 10^{-7} 10^{-8} 10^{-9}

Drainage property
Good drainage
Poor drainage
Practically impervious

Application in earth dams and dikes
Pervious sections of dams and dikes
Impervious sections of earth dams and dikes

Types of soil
Clean gravel
Clean sands, clean sand and gravel mixtures
Very fine sands, organic and inorganic silts, mixtures of sand, silt, and clay, glacial till, stratified clay deposits, etc.
"Impervious" soils e.g., homogeneous clays below zone of weathering
"Impervious" soils which are modified by the effect of vegetation and weathering; fissured, weathered clays; fractured OC clays

Direct determination of coefficient of permeability
Direct testing of soil in its original position (e.g., well points). If properly conducted, reliable; considerable experience required.
(Note: Considerable experience also required in this range.)
Constant Head Permeameter; little experience required.
Constant head test in triaxial cell; reliable with experience and no leaks
Falling Head Permeameter;
Reliable; little experience required
Range of unstable permeability;* much experience necessary for correct interpretation
Fairly reliable; considerable experience necessary (do in triaxial cell)

Indirect determination of coefficient of permeability
Computation: From the grain size distribution (e.g., Hazen's formula). Only applicable to clean, cohesionless sands and gravels
Horizontal Capillarity Test: Very little experience necessary; especially useful for rapid testing of a large number of samples in the field without laboratory facilities.
Computations: from consolidation tests; expensive laboratory equipment and considerable experience required.

10^2 10^1 1.0 10^{-1} 10^{-2} 10^{-3} 10^{-4} 10^{-5} 10^{-6} 10^{-7} 10^{-8} 10^{-9}

*Due to migration of fines, channels, and air in voids.

FIGURE 3.6 Coefficient of permeability versus drainage property, soil type, and method of determination (see Table 3.14). (*Developed by Casagrande, with minor additions by Holtz and Kovacs 1981; reproduced from Holtz and Kovacs 1981.*)

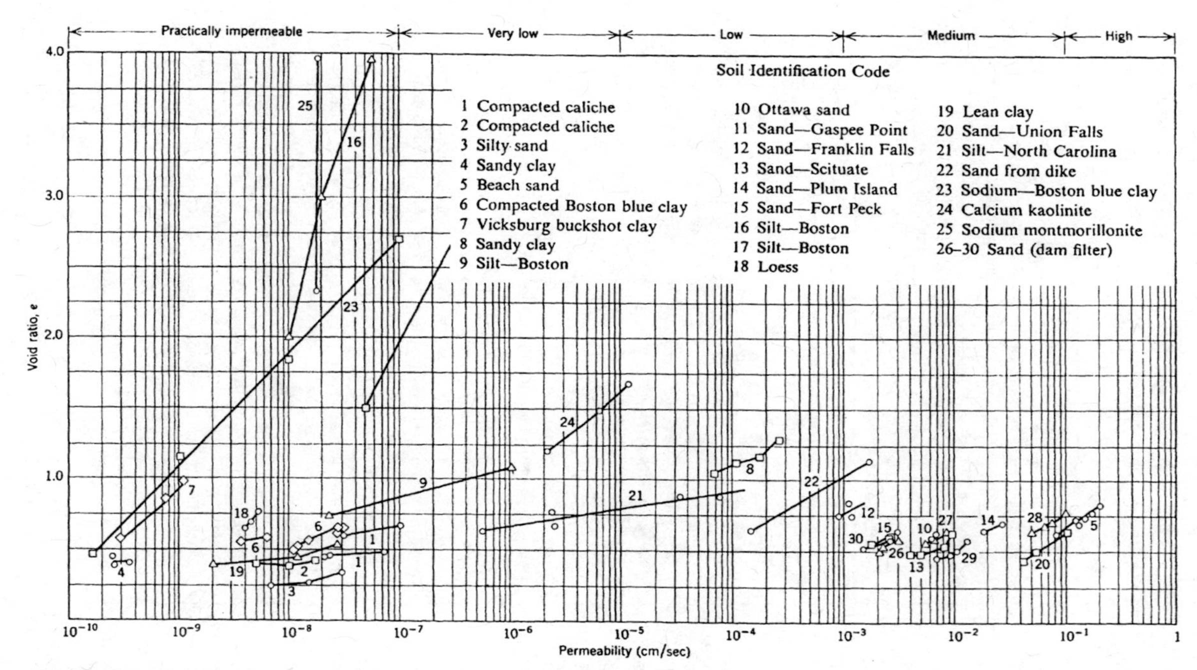

FIGURE 3.7 Coefficient of permeability k versus void ratio e (see Table 3.14) (*From Lambe and Whitman 1969; reproduced with permission from John Wiley & Sons.*)

TABLE 3.15 Modified Proctor and Standard Proctor Laboratory Compaction Tests

Topic (1)	Discussion (2)
General discussion	The laboratory compaction test consists of compacting a soil at a known water content into a mold of specific dimensions using a certain compaction energy. The procedure is repeated for various water contents to establish the compaction curve. The most common testing procedures (compaction energy, number of soil layers in the mold, etc.) are the Modified Proctor (ASTM D 1557-91, 1998) and the Standard Proctor (ASTM D 698-91, 1998). The term *Proctor* is in honor of R. R. Proctor, who in 1933 showed that the dry density of a soil for a given compactive effort depends on the amount of water the soil contains during compaction.
Modified Proctor	In California, there is nearly exclusive use of the Modified Proctor compaction specifications. For the Modified Proctor (ASTM D 1557-91, 1998, procedure A), the soil is compacted into a 10.2-cm- (4-in.-) diameter mold that has a volume of 944 cm^3 ($\frac{1}{30}$ ft^3), where five layers of soil are compacted into the mold with each layer receiving 25 blows from a 44.5-N (10-lbf) hammer that has a 0.46-m (18-in.) drop. The Modified Proctor has a compaction energy of 2700 kN-m/m^3 (56,000 ft-lbf/ft^3). The test procedure is to prepare soil at a certain water content, compact the soil into the mold, and then by recording the mass of soil within the mold, the wet density of the compacted soil is obtained. By knowing the water content of the compacted soil, the dry density can be calculated (i.e., divide the wet density by $1 + w$, where w is in a decimal form). This compaction procedure is repeated for the soil at different water contents and then the data is plotted on a graph in order to obtain the compaction curve.
Standard Proctor	For the Standard Proctor (ASTM D 698-91, 1998, procedure A), the soil is compacted into a 10.2-cm- (4-in.-) diameter mold that has a volume of 944 cm^3 ($\frac{1}{30}$ ft^3), where three layers of soil are compacted into the mold, with each layer receiving 25 blows from a 24.4-N (5.5-lbf) hammer that has a 0.31-m (12-in.) drop. The Standard Proctor has a compaction energy of 600 kN-m/m^3 (12,400 ft-lbf/ft^3).
Compaction curve	Figure 3.8 shows the compaction curve for a nonplastic silty sand. The compaction curve shows the relationship between the dry density and water content for a given compaction effort. The compaction data presented in Fig. 3.8 was obtained by using the Modified Proctor specifications. Note in Fig. 3.8 that the vertical axis is dry density and the horizontal axis is water content. The density of the dry soil particles is of interest, which is why the dry density, and not wet density, is the vertical axis. Four data points, representing four individual soil specimens compacted into the Proctor mold, are shown in Fig. 3.8. The four data points are connected in order to create the compaction curve.

TABLE 3.15 Modified Proctor and Standard Proctor Laboratory Compaction Tests (*Continued*)

Topic (1)	Discussion (2)
Zero air voids curve	The three lines to the right of the compaction curve (see Figure 3.8) are each known as a *zero air voids curve.* These curves represent the relationship between water content and dry density for a condition of saturation (S = 100%) for a specified specific gravity. Note how the right side of the compaction curve is approximately parallel to the zero air voids curve. This is often the case for many soil types and can be used as a check on the laboratory test results.
Maximum dry density and optimum moisture content	The peak point of the compaction curve is the laboratory maximum dry density. In Fig. 3.8, the laboratory maximum dry density is 122.5 pcf (1.96 Mg/m^3). The water content corresponding to the laboratory maximum dry density is known as the optimum moisture content. In Fig. 3.8, the optimum moisture content w_{opt} for this soil is 11%. This laboratory data is important because it tells the grading contractor that the moisture content of the soil should be about 11% for the most efficient compaction of the soil. Grading specifications (see App. B) often require compaction "near optimum," which means that fill should be compacted at a water content that is from 1% below to 3% above optimum (e.g., for the data in Fig. 3.8, w = 10 to 14%).

TABLE 3.16 Relative Density D_r of Granular, Nonplastic Soil

Topic (1)	Discussion (2)
Definition of relative density	The relative density is a measure of the density state of a nonplastic soil. The relative density can be used only for soil that is nonplastic, such as sands and gravels. The relative density (D_r in %) is defined as: $$D_r\,(\%) = 100\,\frac{e_{max} - e}{e_{max} - e_{min}}$$ where e_{max} = void ratio corresponding to the loosest possible state of the soil, usually obtained by pouring the soil into a mold of known volume (ASTM D 4254-96, 1998); e_{min} = void ratio corresponding to the densest possible state of the soil, usually obtained by vibrating the soil particles into a dense state (ASTM D 4253-96, 1998); and e = the natural void ratio of the soil. For the definition of void ratio e, see Chap. 5.
Relative density (D_r) state	The density state of the natural soil can be described as the following: • D_r = 0% to 15% Very loose condition • D_r = 15% to 35% Loose condition • D_r = 35% to 65% Medium condition • D_r = 65% to 85% Dense condition • D_r = 85% to 100% Very dense condition
Use of relative density D_r	The relative density D_r should not be confused with the relative compaction (RC) which will be discussed in Chap. 19. The relative density D_r of an *in situ* granular soil can be estimated from the results of standard penetration tests (SPT, see Table 2.12). If the relative density of a granular soil is known, the effective shear strength ϕ' can be estimated from Fig. 7.5. The relative density D_r of sands and gravels is important because it is a primary factor in the amount of settlement due to applied foundation loads or the liquefaction potential of submerged soil.

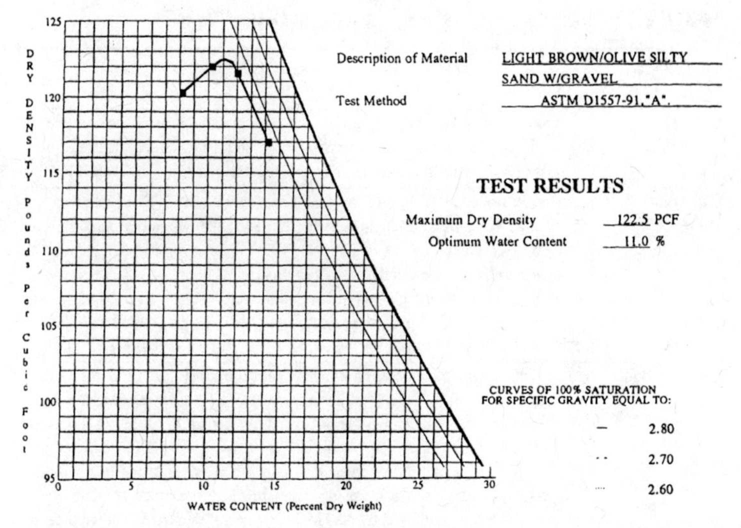

FIGURE 3.8 Compaction curve for a silty sand, determined by the Modified Proctor test specification (see Table 3.15).

TABLE 3.17 Laboratory Tests Used for Pavement Design

Type of Test (1)	Discussion (2)
General discussion	Two common types of laboratory tests for the design of pavements are the laboratory California bearing ratio (CBR) and the *R*-value. The use of these laboratory tests in the design of pavements is discussed in Chap. 11.
California bearing ratio (CBR)	As mentioned in Table 2.13, the California bearing ratio (CBR) can be determined in the field. It can also be performed in the laboratory on compacted soil specimens (ASTM D 1883-94, 1998). The laboratory CBR could be used to assess the condition of base material. It can also be used for the design of roads and airfields where compacted fill is to be used as the subgrade. The testing procedure and calculations for the laboratory CBR are essentially the same as for the field CBR discussed in Table 2.13.
R-value	In California, the most common type of laboratory test used for flexible pavement design is the *R*-value (ASTM D 2844-94, 1998). The *R*-value is the resistance of the subgrade or base and refers to the ability of the soil to resist lateral deformation when acted upon by a vertical load. In essence, the *R*-value is a relative measure of the shear strength of the soil. The *R*-value is determined by using standardized test procedures and equipment, such as specified by ASTM D 2844-94, "Standard Test Method for Resistance *R*-Value and Expansion Pressure of Compacted Soils" (1998). The *R*-value is determined on specimens compacted to anticipated field conditions. Prior to testing, the specimens are soaked in water to achieve saturation. By soaking the soil prior to testing, the *R*-value is supposed to represent the worst possible state that the soil might attain in the field. A crushed-rock base can have an *R*-value of 75 to 85, while a clay subgrade will usually have an *R*-value of less than 10.

CHAPTER 4
SOIL AND ROCK CLASSIFICATION

The purpose of a soil classification system is to provide the geotechnical engineer with a way to predict the behavior of the soil for engineering projects. There are many soil classification systems that are in use and each has different particle size dimensions and terminology. The geotechnical engineer should be aware that terms or soil descriptions used by geologists or soil scientists may not match the engineer's description because they are using a different classification system.

4.1 SOIL CLASSIFICATION SYSTEMS

Important aspects of soil classification systems are clay mineralogy and soil plasticity (Table 4.1). The plasticity chart is shown in Fig. 4.1 and the plasticity characteristics of common clay minerals as they plot on the plasticity chart are indicated in Fig. 4.2. In addition to the classification of a soil, Table 4.2 indicates other items that should be included in the field or laboratory description of a soil.

The following classification systems are presented in this chapter:

4.1.1 Unified Soil Classification System (USCS)

For geotechnical engineers in the United States, the most widely used soil classification system is the Unified Soil Classification System (see Tables 4.3 and 4.4).

4.1.2 Inorganic Soil Classification System Based on Plasticity (ISBP)

This classification system is similar to the Unified Soil Classification System except that the two major divisions of soil are in terms of plastic and non-plastic soil (see Table 4.5).

4.1.3 AASHTO Soil Classification System

This classification system was developed by the American Association of State Highway and Transportation Officials (see Table 4.6). Inorganic soils are divided into 7 groups (A-1 through A-7), with the eighth group (A-8) reserved for highly organic soils. Soil types A-1, A-2, and A-7 have subgroups as indicated in Table 4.6. Those soils having plastic fines can be further categorized by using the group index (defined in Table 4.6).

Figure 4.3 shows the plasticity chart and a comparison of the classifications of fine-grained soil (by the USCS) to silt-clay materials (by AASHTO).

4.1.4 USDA Textural Classification

Figure 4.4 presents the simple textural classification system according to the U.S. Department of Agriculture (1975), where "percent sand" is defined as the percentage of soil particles between 2.0 mm and 0.050 mm in size, "percent silt" is defined as the percentage of soil particles between 0.050 mm and 0.002 mm, and "percent clay" is defined as the percentage of soil particles finer than 0.002 mm. Note that the definitions of sand- and silt-size particles differ from those used in the other classification systems. If a soil contains a significant soil fraction larger than 2.0 mm, the textural classification is preceded with the term "gravelly" for soil fragments up to 76 mm in diameter, "cobbly" for rock fragments between 76 mm and 250 mm, and "stony" or "bouldery" for rock fragments larger than 250 mm.

Figure 4.5 shows an example of the grain size distribution and summarizes the Atterberg limits data (performed on soil passing the No. 40 sieve) for a soil. Table 4.7 summarizes the different classifications of this soil using the USCS, ISBP, AASHTO, and USDA classification systems.

4.1.5 Organic Soil Classification

Table 4.8 presents a classification system for organic soils.

4.2 UNUSUAL SOIL

An unusual soil can be defined as a soil that has rare or unconventional engineering behavior. Such soils frequently cause damage and distress or result in construction cost overruns. This is often due to their unusual properties that were not identified during the subsurface exploration or properly evaluated during the design phase. Some unusual soils are discussed in Table 4.9.

4.3 ROCK CLASSIFICATION

The purpose of this last section of Chapter 4 is to provide a brief introduction to rock classification. There are three basic types of rocks: igneous, sedimentary, and metamorphic. Because of special education and training, usually the best person to classify rock is the engineering geologist. Table 4.10 presents a simplified rock classification and common rock types. In addition to determining the type of rock, it is often important to identify the rock hardness (Table 4.11) and describe the physical character of the rock (Table 4.12).

Table 4.13 presents a summary of the typical engineering properties of rocks.

TABLE 4.1 Clay Mineralogy and Related Topics

Topic (1)	Discussion (2)
Clay mineral structure	The amount and type of clay minerals present in a soil have a significant effect on soil engineering properties such as plasticity, swelling, shrinkage, shear strength, consolidation, and permeability. This is due in large part to their very small flat or platelike shape which enables them to attract water to their surfaces, also known as the *double layer* effect. The double layer is a grossly simplified interpretation of the positively charged water layer and the negatively charged surface of the clay particle itself. Two reasons for the attraction of water to the clay particle (double layer) are: 1. The dipolar structure of the water molecule which causes it to be electrostatically attracted to the surface of the clay particle, and 2. The clay particles attract cations which contribute to the attraction of water by the hydration process. Ion exchange can occur in the double layer, where under certain conditions, sodium, potassium, and calcium cations can be replaced by other cations present in the water. This property is known as *cation exchange capacity.* In addition to the double layer, there is an "absorbed water layer" that consists of water molecules that are tightly held to the clay particle face, for example, by the process of hydrogen bonding. The presence of the very small clay particles surrounded by water helps explain their impact on the engineering properties of soil. For example, clays that have been deposited in lakes or marine environments often have a very high water content and are very compressible with a low shear strength because of this attracted and bonded water. Another example is desiccated clays, which have been dried, but have a strong desire for water and will swell significantly upon wetting (Chap. 12).
Plasticity	The term *plasticity* is applied to silts and clays and indicates an ability to be rolled and molded without breaking apart. A measure of the plasticity of a silt or clay is the plasticity index (PI), defined as the liquid limit (LL) minus the plastic limit (PL); see Table 3.7.
Liquidity index (LI)	A useful parameter is the liquidity index, defined as: $$\mathrm{LI} = \frac{w - \mathrm{PL}}{\mathrm{PI}}$$ The liquidity index can be used to identify sensitive clays. For example, quick clays often have an *in situ* water content w that is greater than the liquid limit, and thus the liquidity index is greater than 1.0. At the other extreme are clays that have liquidity index values that are zero or even negative. These liquidity index values indicate a soil that is desiccated and could have significant expansion potential (Chap. 12). In accordance with ASTM, the Atterberg limits are performed on soil that is finer than the No. 40 sieve, but the water content can be performed on soil containing larger

TABLE 4.1 Clay Mineralogy and Related Topics (*Continued*)

Topic (1)	Discussion (2)
Liquidity index (LI) (*Continued*)	soil particles (Table 3.3), and thus the liquidity index should be calculated only for soil that has all its particles finer than the No. 40 sieve.
Plasticity chart	Based on the Atterberg limits, the plasticity chart was developed by Casagrande (1932, 1948) and is used in the Unified Soil Classification System to classify soils. As shown in Fig. 4.1, the plasticity chart is a plot of the liquid limit (LL) versus the plasticity index (PI). Figure 4.2 shows the locations where common clay minerals plot on the plasticity chart (from Mitchell 1976 and Holtz and Kovacs 1981). Casagrande (1932) defined two basic dividing lines on the plasticity chart, as follows: 1. LL = 50 line: This line is used to divide silts and clays into high-plasticity (LL > 50) and low- to medium-plasticity (LL < 50) categories. 2. A-line: The A-line is defined as: $$PI = 0.73\ (LL - 20)$$ where PI = plasticity index and LL = liquid limit. The A-line is used to separate clays, which plot above the A-line, from silts, which plot below the A-line. An additional line has been added to the Casagrande plasticity chart, known as the U-line (see Fig. 4.1). The U-line (or upper-limit line) is defined as: $$PI = 0.90\ (LL - 8)$$ The U-line is valuable because it represents the uppermost boundary of test data found thus far for natural soils. The U-line is a good check on erroneous data, and any test results that plot above the U-line should be rechecked. There have been other minor changes proposed for Casagrande's original plasticity chart (e.g., note the hatched zone in Fig. 4.1).
Activity	In 1953, Skempton defined the activity A of a clay as: $$A = \frac{PI}{\text{clay fraction}}$$ where the clay fraction = that part of the soil specimen finer than 0.002 mm, based on dry weight. Clays that are *inactive* are defined as those clays that have an activity less than 0.75, *normal activity* is defined as those clays having an activity between 0.75 and 1.25, and an *active* clay is defined as those clays having an activity greater than 1.25. Quartz has an activity of zero, while at the other extreme is sodium montmorillonite, which can have an activity from 4 to 7. Because the PI is determined from Atterberg limits that are performed on soil that passes the No. 40 sieve (0.425 mm), a correction is required for soils that contain a large fraction of particles coarser than the No.

TABLE 4.1 Clay Mineralogy and Related Topics (*Continued*)

Topic (1)	Discussion (2)
Activity (*Continued*)	40 sieve. For example, suppose a clayey gravel contains 70% gravel particles (particles coarser than No. 40 sieve), 20% silt-size particles, and 10% clay-size particles (silt and clay size particles are finer than No. 40 sieve). If the PI = 40 for the soil particles finer than the No. 40 sieve, then the activity for the clayey gravel would be 1.2 (i.e., 40/33.3).
Common types of clay minerals	Clay minerals present in a soil can be identified by their x-ray diffraction patterns. This process is rather complicated, expensive, and involves special equipment which is not readily available to the geotechnical engineer. A more common approach is to use the location of clay particles as they plot on the plasticity chart (Fig. 4.2) to estimate the type of clay mineral in the soil. This approach is often inaccurate, because soil can contain more than one type of clay mineral. The three most common clay minerals are listed below, with their respective activity (*A*) values (from Skempton 1953 and Mitchell 1976): 1. *Kaolinite* (*A* = 0.3 to 0.5). The kaolin minerals are a group of clay minerals consisting of hydrous aluminum silicates. A common kaolin mineral is kaolinite, having the general formula $Al_2\ Si_2(OH)_4$. Kaolinite is usually formed by alteration of feldspars and other aluminum-bearing minerals. Kaolinite is usually a large clay mineral of low activity and often plots below the *A*-line (Fig. 4.2). Holtz and Kovacs (1981) state that kaolinite is a relatively inactive clay mineral and even though it is technically a clay, it behaves more like a silt material. Kaolinite has many industrial uses, including the production of china, medicines, and cosmetics. 2. *Montmorillonite* (Na-montmorillonite, *A* = 4 to 7 and Ca-montmorillonite, *A* = 1.5). A group of clay minerals that are characterized by weakly bonded layers. Each layer consists of two silica sheets with an aluminum (gibbsite) sheet in the middle. Water and exchangeable cations (Na, Ca, etc.) can enter and separate the layers, creating a very small crystal that has a strong attraction for water. Montmorillonite has the highest activity and it can have the highest water content, greatest compressibility, and lowest shear strength of all the clay minerals. As shown in Fig. 4.2, montmorillonite plots just below the U-line. Montmorillonite often forms as the result of the weathering of ferromagnesian minerals, calcic feldspars, and volcanic materials (Coduto 1994). For example, sodium montmorillonite is often formed from the weathering of volcanic ash. Other environments that are likely to form montmorillonite are alkaline conditions with a supply of magnesium ions and a lack of leaching (Coduto 1994). Such conditions are often present in semiarid and arid regions. 3. *Illite* (*A* = 0.5 to 1.3). This clay mineral has a structure similar to montmorillonite, but the layers are more strongly bonded together. In cation exchange capacity, in ability to absorb and retain water,

TABLE 4.1 Clay Mineralogy and Related Topics (*Continued*)

Topic (1)	Discussion (2)
Common types of clay minerals (*Continued*)	and in physical characteristics such as plasticity index, illite is intermediate in activity between clays of the kaolin and montmorillonite groups. As shown in Fig. 4.2, illite often plots just above the A-line. There are many other types of clay minerals. Even within a clay mineral category, there can be different crystal components because of isomorphous substitution. This is the process where ions of approximately the same size are substituted in the crystalline framework. Also shown on Fig. 4.2 are two other less common clay minerals, chlorite and halloysite, both of which have less activity than the three clay minerals previously described. Although not very common, halloysite is an interesting clay mineral because instead of the usual flat particle shape, it is tubular in shape, which can effect engineering properties in unusual ways. It has been observed that classification and compaction tests made on air-dried halloysite samples give markedly different results than tests on samples at their natural water content (Holtz and Kovacs 1981).

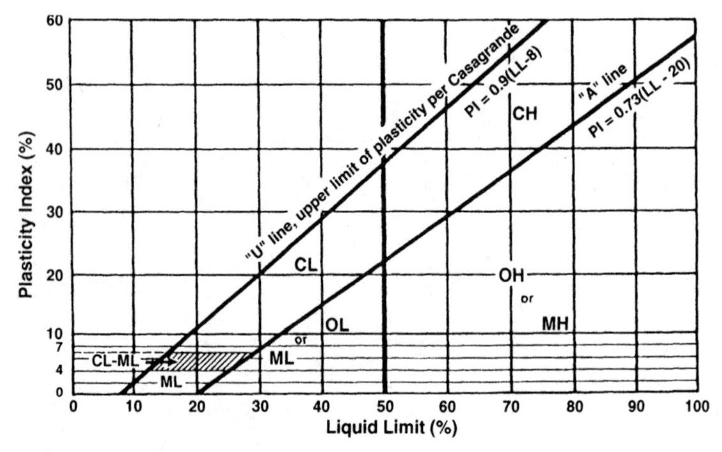

FIGURE 4.1 Plasticity chart.

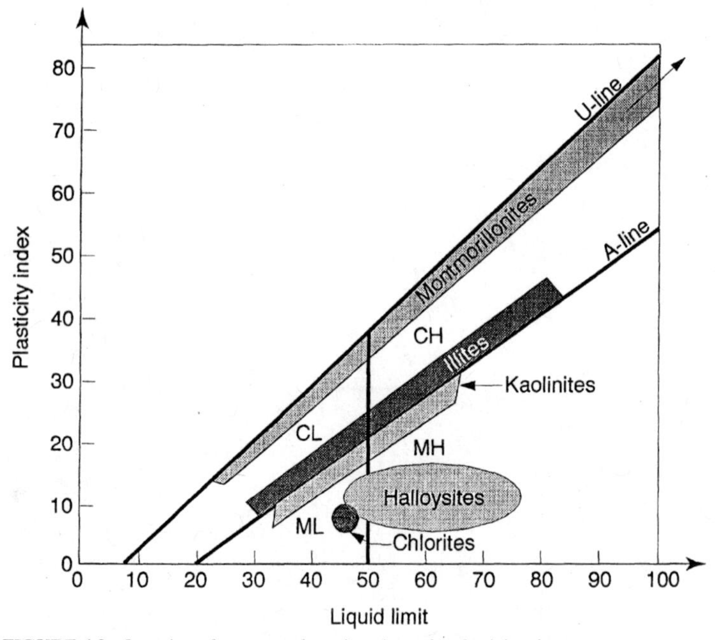

FIGURE 4.2 Location of common clay minerals on the plasticity chart.

TABLE 4.2 Various Items Used to Describe Soil

Topic (1)	Discussion (2)
Soil color	Usually the standard primary color (red, orange, yellow, etc.) of the soil is listed. Although not frequently used in geotechnical engineering, color charts have been developed. For example, the *Munsell Soil Color Charts* (1975) display 199 different standard color chips systematically arranged according to their Munsell notations, on cards carried in a loose leaf notebook. The arrangement is by the three variables that combine to describe all colors and are known in the Munsell system as hue, value, and chroma. Color can be very important in identifying different types of soil. For example, the Friars formation, which is a stiff-fissured clay and is a frequent cause of geotechnical problems such as landslides and expansive soil, can often be identified by its dark green color. Another example is the Sweetwater formation, which is also a stiff-fissured clay, and has a bright pink color due to the presence of montmorillonite.
Soil structure	In some cases the structure of the soil may be evident. Definitions vary, but in general, the soil structure refers to both the geometric arrangement of the soil particles and the interparticle forces which may act between them (Holtz and Kovacs 1981). There are many different types of soil structure, such as cluster, dispersed, flocculated, honeycomb, single-grained, and skeleton (see App. A for definitions). In some cases, the soil structure may be visible under a magnifying glass, or in other cases the soil structure may be reasonably inferred from laboratory testing results.
Soil texture	The texture of a soil refers to the degree of fineness of the soil. For example, terms such as smooth, gritty, or sharp can be used to describe the texture of the soil when it is rubbed between the fingers.
Soil porosity	The soil description should also include the *in situ* condition of the soil. For example, numerous small voids may be observed in the soil, and this is referred to as *pinhole porosity.*
Clay consistency	For clays, the consistency (i.e., degree of firmness) should be listed. The consistency of a clay varies from "very soft" to "hard," depending on the undrained shear strength of the clay. The undrained shear strength s_u can be determined from the vane tests or the unconfined compression test (Table 3.10). Values of the undrained shear strength s_u corresponding to various degrees of consistency are as follows: • *Very soft:* $s_u < 12$ kPa ($s_u < 250$ psf). The clay is easily penetrated several centimeters by the thumb. The clay oozes out between the fingers when squeezed in the hand. • *Soft:* 12 kPa $\leq s_u <$ 25 kPa (250 psf $\leq s_u <$ 500 psf). The clay is easily penetrated 2 to 3 cm (1 in.) by the thumb. The clay can be molded by slight finger pressure. • *Medium:* 25 kPa $\leq s_u <$ 50 kPa (500 psf $\leq s_u <$ 1000 psf). The clay can be penetrated about 1 cm (0.4 in.) by the thumb with moderate effort. The clay can be molded by strong finger pressure.

TABLE 4.2 Various Items Used to Describe Soil (*Continued*)

Topic (1)	Discussion (2)
Clay consistency (*Continued*)	• *Stiff:* 50 kPa $\le s_u <$ 100 kPa (1000 psf $\le s_u <$ 2000 psf). The clay can be indented about 0.5 cm (0.2 in.) by the thumb with great effort. • *Very stiff:* 100 kPa $\le s_u <$ 200 kPa (2000 psf $\le s_u <$ 4000 psf). The clay cannot be indented by the thumb, but can be readily indented with the thumbnail. • *Hard:* $s_u \ge$ 200 kPa ($s_u \ge$ 4000 psf). With great difficulty, the clay can only be indented with the thumbnail.
Sand density condition	For sands, the density state of the soil varies from "very loose" to "very dense" (see Table 3.16).
Soil moisture condition	The moisture condition of the soil should also be listed. Moisture conditions vary from a "dry" soil to a "saturated" soil. The degree of saturation indicates the degree to which the soil voids are filled with water (see Chap. 5). A totally dry soil will have a degree of saturation S of 0%, while a saturated soil, such as a soil below the groundwater table, will have a degree of saturation S of 100%. Typical ranges of degree of saturation versus soil condition are as follows (Terzaghi and Peck 1967): • Dry: S = 0% • Humid: S = 1 to 25% • Damp: S = 26 to 50% • Moist: S = 51 to 75% • Wet: S = 76 to 99% • Saturated: S = 100% Especially in the arid climate of the southwestern United States, soil may be in a dry and powdery state. Often these soils are misclassified as silts, when in fact they are highly plastic clays. It is always important to add water to dry or powdery soils in order to assess their plasticity characteristics.
Additional descriptive items	Most classification systems are applicable only for soil and rock particles passing the 75 mm (3 in.) sieve. Cobbles and boulders are larger than 75 mm (3 in.), and if applicable, the words "with cobbles" or "with boulders" should be added to the soil classification. Other descriptive terminology includes the presence of rock fragments, such as "crushed shale, claystone, sandstone, siltstone, or mudstone fragments," and unusual constituents such as "shells, slag, glass fragments, and construction debris."

TABLE 4.3 Unified Soil Classification System (Background Information)

Topic (1)	Discussion (2)
General information	The Unified Soil Classification System is abbreviated USCS (not to be confused with the United States Customary System units which has the same abbreviation). The Unified Soil Classification System was initially developed by Casagrande (1948) and then later modified by Casagrande in 1952.
Two main groups	The Unified Soil Classification System (USCS) separates soils into two main groups: coarse-grained soils and fine-grained soils. The basis of the USCS is that the engineering behavior of coarse-grained soils is related to their grain size distributions and the engineering behavior of fine-grained soil is related to their plasticity characteristics. Table 4.4 (adapted from ASTM D 2487-93, 1998) presents a summary of the Unified Soil Classification System. As indicated in Table 4.4, the two main groups of soil are defined as follows: • *Coarse-grained soils:* Defined as having more than 50% (by dry mass) of soil particles retained on the No. 200 sieve. • *Fine-grained soils:* Defined as having 50% or more (by dry mass) of soil particles passing the No. 200 sieve. As indicated in Table 4.4, the coarse-grained soils are divided into gravels and sands. Both gravels and sands are further subdivided into four secondary groups as indicated in Table 4.4. The four secondary classifications are based on whether the soil is well graded, is poorly graded, contains silt-size particles, or contains clay-size particles. As indicated in Table 4.4, the fine-grained soils are divided into soils of low or high plasticity. The three secondary classifications are based on liquid limit (LL) and plasticity characteristics (PI).
Distribution of particle sizes	The distribution of particle sizes larger than 0.075 mm (No. 200 sieve) is determined by sieving, while the distribution of particle sizes smaller than 0.075 mm is determined by a sedimentation process (hydrometer). For the USCS, the rock fragments or soil particles versus size are defined as follows (from largest to smallest particle sizes): • *Boulders.* Rocks that have an average diameter greater than 300 mm (12 in.). • *Cobbles.* Rocks that are smaller than 300 mm (12 in.) and are retained on the 75-mm (3-in.) U.S. standard sieve. • *Gravel-size particles.* Rock fragments or soil particles that will pass a 75-mm (3-in.) sieve and be retained on a No. 4 (4.75-mm) U.S. standard sieve. Gravel-size particles are subdivided into coarse gravel sizes and fine gravel sizes. • *Sand-size particles.* Soil particles that will pass a No. 4 (4.75-mm) sieve and be retained on a No. 200 (0.075-mm) U.S. standard sieve. Sand-size particles are subdivided into coarse sand size, medium sand size, and fine sand size. • *Silt-size particles.* Fine soil particles that pass the No. 200 (0.075-mm) U.S. standard sieve and are larger than 0.002 mm. • *Clay-size particles.* Fine soil particles that are smaller than 0.002 mm.

TABLE 4.3 Unified Soil Classification System (Background Information) (*Continued*)

Topic (1)	Discussion (2)
Particle size versus soil description	It is very important to distinguish between the size of a soil particle and the classification of the soil. For example, a soil could have a certain fraction of particles that are of "clay size." The same soil could also be classified as a "clay." But the classification of a "clay" does not necessarily mean that the majority of the soil particles are of clay size (smaller than 0.002 mm). In fact, it is not unusual for a soil to be classified as a "clay" and have a larger mass of silt-size particles than clay-size particles. When reference is given to particle size, the terminology *clay-size particles* or *silt-size particles* should be used. When reference is given to a particular soil, then the terms such as *silt* or *clay* should be used.
Coarse-grained-soil coefficients	On the basis of particle sizes, the coefficient of uniformity C_u and coefficient of curvature C_c can be calculated as follows: $$C_u = \frac{D_{60}}{D_{10}}$$ $$C_c = \frac{(D_{30})^2}{(D_{10})(D_{60})}$$ where D_{10} = particle size corresponding to 10% finer soil-size particles, D_{30} = particle size corresponding to 30% finer soil-size particles, and D_{60} = particle size corresponding to 60% finer soil-size particles. The coefficient of uniformity C_u and the coefficient of curvature C_c are used in the Unified Soil Classification System to determine whether a coarse-grained soil is well graded (many different particle sizes) or poorly graded (many particles of about the same size).
Group symbols	Note in column 3 of Table 4.4 that symbols (known as *group symbols*) are used to identify different soil types. The group symbols consist of two capital letters. The first letter indicates the following: G Gravel S Sand M Silt C Clay O Organic The second letter indicates the following: W Well graded, which indicates that a coarse-grained soil has particles of all sizes. P Poorly graded, which indicates a coarse-grained soil has particles of the same size, or the soil is skip-graded or gap-graded. M Indicates a coarse-grained soil that has silt-size particles. C Indicates a coarse-grained soil that has clay-size particles. L Indicates a fine-grained soil of low plasticity. H Indicates a fine-grained soil of high plasticity. An exception is peat, where the group symbol is PT. Also note in Table 4.4 that certain soils require the use of dual symbols.

TABLE 4.4 Unified Soil Classification System (USCS)

Major divisions (1)	Subdivisions (2)	USCS symbol (3)	Typical names (4)	Laboratory classification criteria (5)	
Coarse-grained soils (more than 50% retained on No. 200 sieve)	Gravels (more than 50% of coarse fraction retained on No. 4 sieve)	GW	Well-graded gravels or gravel-sand mixtures, little or no fines	Less than 5% fines*	$C_u \geq 4$ and $1 \leq C_c \leq 3$
		GP	Poorly graded gravels or gravelly sands, little or no fines	Less than 5% fines*	Does not meet C_u and/or C_c criteria listed above
		GM	Silty gravels, gravel-sand-silt mixtures	More than 12% fines*	Minus No. 40 soil plots below the A-line
		GC	Clayey gravels, gravel-sand-clay mixtures	More than 12% fines*	Minus No. 40 soil plots on or above the A-line (Fig. 4.1)
	Sands (50% or more of coarse fraction passes No. 4 sieve)	SW	Well-graded sands or gravelly sands, little or no fines	Less than 5% fines*	$C_u \geq 6$ and $1 \leq C_c \leq 3$
		SP	Poorly graded sands or gravelly sands, little or no fines	Less than 5% fines*	Does not meet C_u and/or C_c criteria listed above
		SM	Silty sands, sand-silt mixtures	More than 12% fines*	Minus No. 40 soil plots below the A-line (Fig. 4.1)
		SC	Clayey sands, sand-clay mixtures	More than 12% fines*	Minus No. 40 soil plots on or above the A-line (Fig. 4.1)

TABLE 4.4 Unified Soil Classification System (USCS) (*Continued*)

Major divisions (1)	Subdivisions (2)	USCS symbol (3)	Typical names (4)	Laboratory classification criteria (5)	
Fine-grained soils (50% or more passes the No. 200 sieve)	Silts and clays (liquid limit less than 50)	ML	Inorganic silts, rock flour, silts of low plasticity	Inorganic soil	PI < 4 or plots below A-line (Fig. 4.1)
		CL	Inorganic clays of low plasticity, gravelly clays, sandy clays, etc.	Inorganic soil	PI > 7 and plots on or above A-line†
		OL	Organic silts and organic clays of low plasticity	Organic soil	LL (oven dried)/LL (not dried) < 0.75
	Silts and clays (liquid limit 50 or more)	MH	Inorganic silts, micaceous silts, silts of high plasticity	Inorganic soil	Plots below A-line (Fig. 4.1)
		CH	Inorganic highly plastic clays, fat clays, silty clays, etc.	Inorganic soil	Plots on or above A-line (Fig. 4.1)
		OH	Organic silts and organic clays of high plasticity	Organic soil	LL (oven dried)/LL (not dried) < 0.75
Peat	Highly organic	PT	Peat and other highly organic soils	Primarily organic matter, dark in color, and organic odor	

**Fines* are those soil particles that pass the No. 200 sieve. For gravels and sands with between 5 and 12% fines, use of dual symbols is required (e.g., GW-GM, GW-GC, GP-GM, or GP-GC).

†If $4 \le PI \le 7$ and PI plots above A-line, then dual symbols (e.g., CL-ML) are required.

TABLE 4.5 Inorganic Soil Classification Based on Plasticity (ISBP)

Major divisions (1)	Subdivisions (2)	ISBP symbol (3)	Typical names (4)	Laboratory classification criteria (5)
Nonplastic soils	Gravels (Greater fraction of total sample is retained on No. 4 sieve)	GW	Well-graded gravels, sandy gravels, silty-sandy gravels	$C_u \geq 4$ and $1 \leq C_c \leq 3$
		GP	Poorly graded gravels, gravel-sand-silt mixtures	Does not meet C_u and/or C_c criteria listed above. In addition, the % passing No. 200 sieve < 15%
		GM	Poorly graded, nonplastic silty gravels, gravel-silt mixtures	Does not meet C_u and/or C_c criteria listed above. In addition, the % passing No. 200 sieve ≥ 15%
	Sands (Greater fraction of total sample is between No. 4 and No. 200 sieves)	SW	Well-graded sands and gravelly sands	$C_u \geq 6$ and $1 \leq C_c \leq 3$
		SP	Poorly graded sands or sand-gravel-silt mixtures	Does not meet C_u and/or C_c criteria listed above. In addition, the % passing No. 200 sieve < 15%
		SM	Poorly graded, nonplastic silty sands, sand-silt mixtures	Does not meet C_u and/or C_c criteria listed above. In addition, the % passing No. 200 sieve ≥ 15%
	NP silt	MN	Nonplastic silts, rock flour. Gravelly silts and sandy nonplastic silts	Greater fraction of the total sample passes the No. 200 sieve. Silts are nonplastic
Plastic soils	Plastic silts (Minus No. 40 fraction plots below A-line, Fig. 4.1)	GM*	Plastic silty gravels, gravel-silt mixtures	50% or more particles retained on the No. 200 sieve with the greater fraction of gravel size
		SM*	Plastic silty sands, sand-silt mixtures	50% or more particles retained on the No. 200 sieve with the greater fraction of sand size
		ML MI MH	Plastic silts, sandy silts, and clayey silts	For silt of low plasticity (ML) PI ≤ 10 For silt of intermediate plasticity (MI) 10 < PI ≤ 30 For silt of high plasticity (MH) PI > 30
	Clays (Minus No. 40 fraction plots on or above A-line, Fig. 4.1)	GC*	Clayey gravels, gravel-clay mixtures	50% or more particles retained on the No. 200 sieve with the greater fraction of gravel size
		SC*	Clayey sands, sand-clay mixtures	50% or more particles retained on the No. 200 sieve with the greater fraction of sand size
		CL CI CH	Clay, sandy clays, and silty clays	For clay of low plasticity (CL) PI ≤ 10 For clay of intermediate plasticity (CI) 10 < PI ≤ 30 For clay of high plasticity (CH) PI > 30

*Must state high, intermediate, or low plasticity where the PI (for this table) = PI times the fraction of soil passing No. 40 sieve.

Note: Dual symbols are not used in the ISBP classification system. See Day (1994a, 1999) for further details.

TABLE 4.6 AASHTO Soil Classification System

Major divisions (1)	Group (2)	AASHTO symbol (3)	Typical names (4)	Sieve analysis (percent passing) (5)	Atterberg limits (6)
Granular materials (35% or less passing No. 200 sieve)	Group A-1	A-1-a	Stone or gravel fragments	Percent passing: No. 10 ≤ 50%, No. 40 ≤ 30%, No. 200 ≤ 15%	PI ≤ 6
		A-1-b	Gravel and sand mixtures	No. 40 ≤ 50%, No. 200 ≤ 25%	PI ≤ 6
	Group A-3	A-3	Fine sand that is nonplastic	No. 40 > 50%, No. 200 ≤ 10%	PI = 0 (nonplastic)
	Group A-2	A-2-4	Silty or clayey gravel and sand	Percent passing No. 200 sieve ≤ 35%	LL ≤ 40, PI ≤ 10
		A-2-5	Silty or clayey gravel and sand	Percent passing No. 200 sieve ≤ 35%	LL > 40, PI ≤ 10
		A-2-6	Silty or clayey gravel and sand	Percent passing No. 200 sieve ≤ 35%	LL ≤ 40, PI > 10
		A-2-7	Silty or clayey gravel and sand	Percent passing No. 200 sieve ≤ 35%	LL > 40, PI > 10
Silt-clay materials (more than 35% passing No. 200 sieve)	Group A-4	A-4	Silty soils	Percent passing No. 200 sieve > 35%	LL ≤ 40, PI ≤ 10
	Group A-5	A-5	Silty soils	Percent passing No. 200 sieve > 35%	LL > 40, PI ≤ 10
	Group A-6	A-6	Clayey soils	Percent passing No. 200 sieve > 35%	LL ≤ 40, PI > 10
	Group A-7	A-7-5	Clayey soils	Percent passing No. 200 sieve > 35%	LL > 40, PI ≤ LL − 30, PI > 10
		A-7-6	Clayey soils	Percent passing No. 200 sieve > 35%	LL > 40, PI > LL − 30, PI > 10
Highly organic	Group A-8	A-8	Peat and other highly organic soils	Primarily organic matter, dark in color, and organic odor	

Notes:

1. Classification procedure: First decide which of the three main categories (granular materials, silt-clay materials, or highly organic) the soil belongs. Then proceed from the top to the bottom of the chart and the first group that meets the particle size and Atterberg limits criteria is the correct classification.

TABLE 4.6 AASHTO Soil Classification System (*Continued*)

Notes:

2. Group index = $(F - 35)[0.2 + 0.005(\mathrm{LL} - 40)] + 0.01(F - 15)(\mathrm{PI} - 10)$, where F = percent passing No. 200 sieve, LL = liquid limit, and PI = plasticity index. Report group index to nearest whole number. For negative group index, report as zero. When working with A-2-6 and A-2-7 subgroups, use only the PI portion of the group index equation.
3. Atterberg limits are performed on soil passing the No. 40 sieve. LL = liquid limit, PL = plastic limit, and PI = plasticity index (see Table 3.7).
4. AASHTO definitions of particle sizes are as follows: (*a*) boulders: above 75 mm, (*b*) gravel: 75 mm to No. 10 sieve, (*c*) coarse sand: No. 10 to No. 40 sieve, (*d*) fine sand: No. 40 to No. 200 sieve, and (*e*) silt-clay–size particles: material passing No. 200 sieve.
5. Example: An example of an AASHTO classification for a clay is A-7-6 (30), or group A-7, subgroup 6, group index 30.

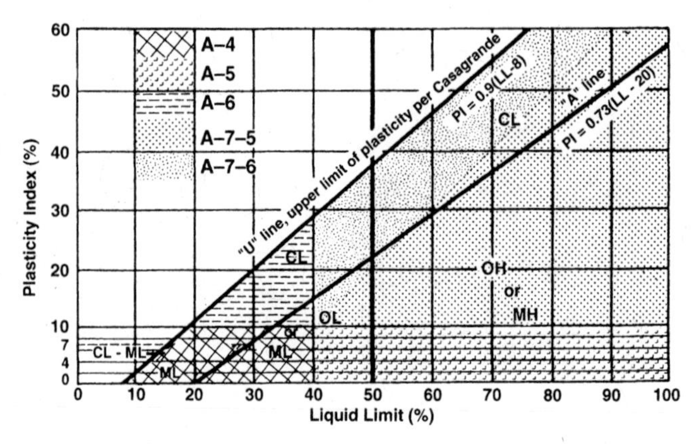

FIGURE 4.3 Plasticity chart showing the location of fine-grained soil per USCS and the location of silt-clay materials (shaded areas) per AASHTO.

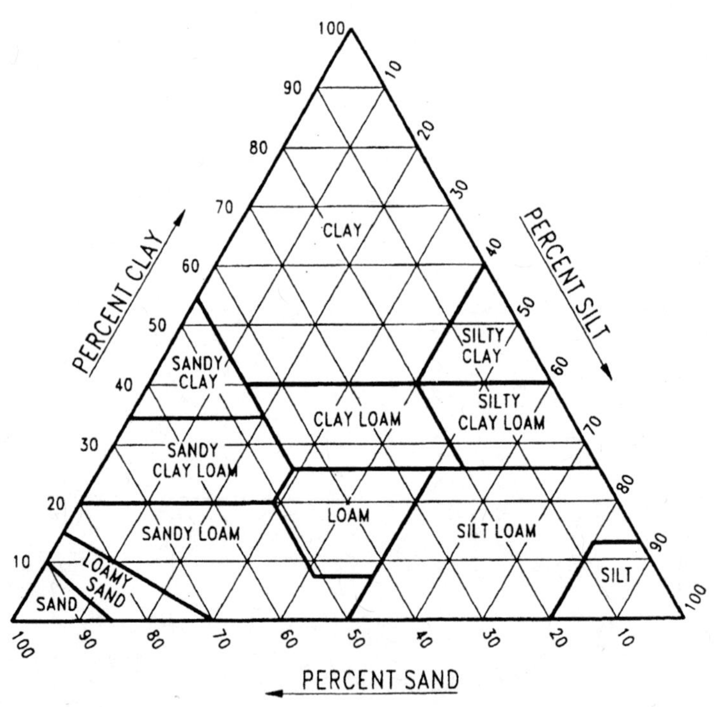

FIGURE 4.4 USDA Textural Classification System. (*Note:* Percent sand = 2.0 mm to 0.050 mm, percent silt = 0.050 mm to 0.002 mm, and percent clay is finer than 0.002 mm.)

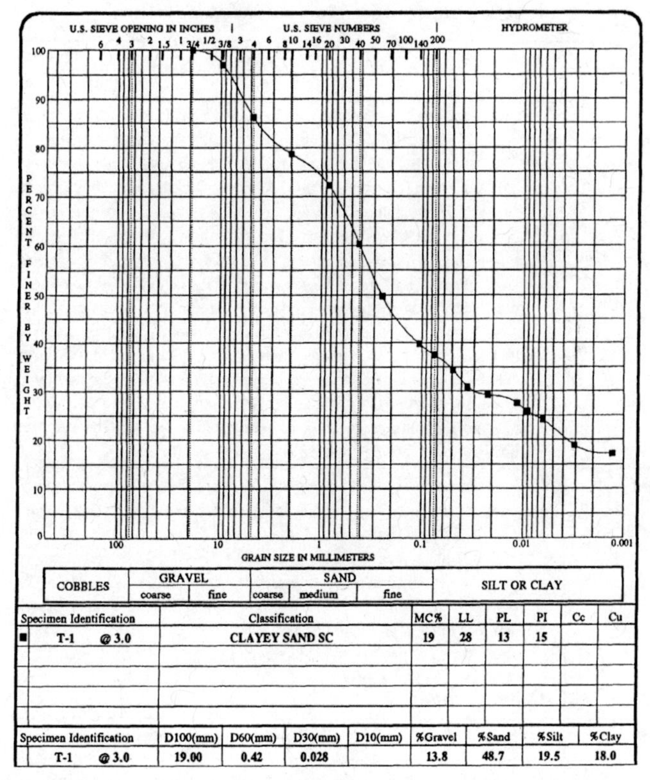

Specimen Identification	Classification	MC%	LL	PL	PI	Cc	Cu
■ T-1 @ 3.0	CLAYEY SAND SC	19	28	13	15		

Specimen Identification	D100(mm)	D60(mm)	D30(mm)	D10(mm)	%Gravel	%Sand	%Silt	%Clay
T-1 @ 3.0	19.00	0.42	0.028		13.8	48.7	19.5	18.0

FIGURE 4.5 Grain size curve and Atterberg limit test data (see Table 4.7).

TABLE 4.7 Example of Soil Classification (for Data Shown in Fig. 4.5)

Topic (1)	Discussion (2)
General discussion	Figure 4.5 shows the results of a grain-size analysis (sieve and hydrometer tests) for a tested soil. The sieve portion of the analysis is shown at the top of the graph. Note the items labeled "U.S. sieve openings in inches" and "U.S. sieve numbers." The U.S. sieve openings in inches refer to sieves that are manufactured with specific-size openings. For example, the 3-in. U.S. standard size sieve has square openings that are 3 in. (75 mm) wide. The label "Hydrometer" indicates that part of the grain size curve determined from the hydrometer analysis. If the sieve and the hydrometer tests are performed correctly, the portion of the grain size curve from the sieve analysis should merge smoothly into the portion of the curve from the hydrometer analysis, as shown in Fig. 4.5. A large and abrupt jump in the grain size curve from the sieve to the hydrometer test indicates errors in the laboratory testing procedure. The Atterberg limits test data is also summarized on Fig. 4.5. The Atterberg limits were performed on soil passing the No. 40 sieve.
Unified Soil Classification System (USCS)	The classification of the soil (e.g., clayey sand, SC) and percent gravel, sand, silt, and clay-size particles indicated at the bottom of Fig. 4.5 are according to the Unified Soil Classification System. In Fig. 4.5, the percent passing the No. 200 sieve is 37.5%, and therefore the soil is classified as a coarse-grained soil (see Table 4.4). The greater fraction of the soil retained on the No. 200 sieve is of sand size, and therefore the soil is a sand. Because there is more than 12% fines (i.e., more than 12% passing No. 200 sieve) and the Atterberg limits (PI = 15, LL = 28) plot above the A-line in Fig. 4.1, the soil is classified as a clayey sand (SC).
Inorganic Soil Classification System Based on Plasticity (ISBP)	Since the soil has a plasticity index, the soil is classified as a plastic soil (see Table 4.5). The Atterberg limits (PI = 15, LL = 28) plot above the A-line in Fig. 4.1, and therefore the soil is within the clay subcategory of plastic soil. Because more than 50% of the particles are retained on the No. 200 sieve and the greater fraction of these particles are sand-size, the soil is classified as a clayey sand. Adjusting the plasticity index for soil particles larger than the No. 40 sieve, the corrected plasticity index = (15) (fraction of soil passing No. 40 sieve) = (15) (0.60) = 9. Since the corrected PI is less than 10, the classification is a clayey sand (SC) of low plasticity (SC of low plasticity).
AASHTO Soil Classification System	In Fig. 4.5, the percent passing the No. 200 sieve is 37.5%, and therefore the soil is classified as a silt-clay material (see Table 4.6). From the Atterberg limits (PI = 15, LL = 28), the LL is less than 40 and the PI is greater than 10, and therefore the soil is classified as Group A-6 (clayey soil). The group index (see Table 4.6) can be calculated by using $F = 37.5$, LL = 28, and PI = 15, and the result is a group index = 1.48. The final AASHTO classification is A-6 (1).

TABLE 4.7 Example of Soil Classification (for Data Shown in Fig. 4.5) (*Continued*)

Topic (1)	Discussion (2)
USDA Textural Classification System	The USDA Textural Classification System (see Fig. 4.4) does not have the same definitions for particle size as the other classification systems. In order to use the USDA system, the sand-, silt-, and clay-size particles from Fig. 4.5 must be adjusted for the fraction larger than 2.0 mm. Therefore, since 79% of the particles are finer than 2.0 mm and 35% of the particles are finer than 0.050 mm, % sand-size particles = (79% − 35%)/0.79 = 55.7% % silt-size particles = (35% − 18%)/0.79 = 21.5% % clay-size particles = 18%/0.79 = 22.8% Entering Fig. 4.4 with % sand = 55.7, % silt = 21.5, and % clay = 22.8, find the resulting classification is "sandy clay loam."
Summary of classification systems	The following is a summary of the soil classification systems for the soil data shown in Fig. 4.5: USCS: Coarse-grained soil classified as a clayey sand (SC) ISBP: Plastic soil classified as a clayey sand of low plasticity (SC of low plasticity) AASHTO: Silt-clay material further classified as A-6 (1), i.e., a clayey soil USDA: Sandy-clay loam
Complete classification	The complete classification of the soil is as follows: According to the ISBP, clayey sand of low plasticity (SC of low plasticity). Based on dry mass, the soil contains 13.8% gravel-size particles, 48.7% sand-size particles, 19.5% silt-size particles, and 18.0% clay-size particles. The gravel-size particles are predominantly hard and angular rock fragments. The sand-size particles are predominantly composed of angular quartz grains. Atterberg limits performed on the soil passing the No. 40 sieve indicate a LL = 28 and a PI = 15. The *in situ* soil has a reddish-brown color, soft consistency, gritty texture, and is wet. The *in situ* soil is a residual soil and is part of an old vernal pool.

TABLE 4.8 Soil Classification for Organic Soil

Major divisions (1)	Organic content (2)	USCS symbol (3)	Typical names (4)	Distinguishing characteristics for visual identification (5)	Typical range of laboratory test results (6)
Organic matter	75% to 100% organics (either visible or inferred)	PT	Fibrous peat (woody, mats, etc.)	Lightweight and spongy. Shrinks considerably on air drying. Much water squeezes from sample.	w = 500% to 1200% γ_t = 9.4 to 11 kN/m^3 (60 to 70 pcf) G = 1.2 to 1.8, $C_c/(1 + e_o) \geq 0.40$
		PT	Fine-grained peat (amorphous)	Lightweight and spongy. Shrinks considerably on air drying. Much water squeezes from sample.	w = 400% to 800%, PI = 200 to 500 γ_t = 9.4 to 11 kN/m^3 (60 to 70 pcf) G = 1.2 to 1.8, $C_c/(1 + e_o) \geq 0.35$
Highly organic soils	30% to 75% organics (either visible or inferred)	PT	Silty peat	Relatively lightweight, spongy. Shrinks on air drying. Usually can readily squeeze water from sample.	w = 250% to 500%, PI = 150 to 350 γ_t = 10 to 14 kN/m^3 (65 to 90 pcf) G = 1.8 to 2.3, $C_c/(1 + e_o)$ = 0.3 to 0.4
		PT	Sandy peat	Sand fraction visible. Shrinks on air drying. Often a gritty texture. Usually can squeeze water from sample.	w = 100% to 400%, PI = 50 to 150 γ_t = 11 to 16 kN/m^3 (70 to 100 pcf) G = 1.8 to 2.4, $C_c/(1 + e_o)$ = 0.2 to 0.3

Organic soils (*Continued*)	5% to 30% organics (either visible or inferred)	OH	Clayey organic silt	Often has strong hydrogen sulfide (H_2S) odor. Medium dry strength and slow dilatency.	w = 65% to 200%, PI = 50 to 150 γ_t = 11 to 16 kN/m^3 (70 to 100 pcf) G = 2.3 to 2.6, $C_c/(1 + e_o)$ = 0.2 to 0.35
		OL	Organic sand or silt	Threads weak and friable near plastic limit, or will not roll at all. Low dry strength, medium to high dilatency.	w = 30% to 125%, PI = NP to 40 γ_t = 14 to 17 kN/m^3 (90 to 110 pcf) G = 2.4 to 2.6, $C_c/(1 + e_o)$ = 0.1 to 0.25
Slightly organic soils	Less than 5% organics	Use Table 4.4	Soil with slight organic fraction	Depends on the characteristics of the inorganic fraction.	Depends on the characteristics of the inorganic fraction.

Source: NAVFAC DM-7.1 (1982), based on unpublished work by Ayers and Plum.

Notes: w = *in situ* water content, PI = plasticity index, NP = nonplastic, γ_t = total unit weight, G = specific gravity, C_c = compression index, e_o = initial void ratio, and $C_c/(1 + e_o)$ = modified compression index.

TABLE 4.9 Unusual Soil

Examples of unusual soil (1)	Discussion (2)
Rock flour (or bull's liver)	This soil consists predominantly of silt-size particles, but has little or no plasticity. Nonplastic rock flour contains particles of quartz, ground to a very fine state by the abrasive action of glaciers. Terzaghi and Peck (1967) state that because of its fine particle size, this soil is often mistaken as clay. In describing this soil, the term "bull's liver" apparently comes from its *in situ* appearance. It has been observed that in a saturated state, it quakes like jelly from shock or vibration and can even flow like a liquid (Sowers and Sowers 1970).
Peat	As indicated in Table 4.8, peat is composed of partially decayed organic matter, where the remains of leaves, stems, twigs, and roots can be identified. The places where peat accumulates are known as peat bogs or peat moors. Its color ranges from light brown to black. Peat is unusual because it has a very high water content, which makes it extremely compressible. This almost always makes it unsuitable for supporting foundations (Terzaghi and Peck 1967).
Nonwelded tuff and volcanic ash	Tuff is a pyroclastic rock, originating as airborne debris from explosive volcanic eruptions. The largest fragments (in excess of 64 mm) are called *blocks* and *bombs,* fragments between 4 mm and 64 mm are called *lapilli,* fragments between 4 mm and 0.25 mm are *ash,* and the finest fragments (less than 0.25 mm) are *volcanic dust* (Compton 1962). An important aspect of tuff is the degree of welding, which can be described as either nonwelded, partially welded to varying degrees, or densely welded. Welding is generally caused by fragments that are hot when deposited, and because of this heat, the sticky glassy fragments may actually fuse together (Best 1982). There are distinct changes in the original shards and pumice fragments, such as the union and elongation of the glassy shards and flattening of the pumice fragments, which is characteristic of completely welded tuff (Ross and Smith 1961). The degree of welding depends on many factors, such as type of fragments, plasticity of the fragments (which depends on the emplacement temperature and chemical composition), thickness of the resulting deposit, and rate of cooling (Smith 1960). Deposition of volcanic ash directly from the air may result in an unconsolidated (geologically speaking) deposit, which would then be called ash; but indurated deposits are called tuff. Nonwelded tuff has an engineering behavior similar to volcanic ash. These materials have been used as mineral filler in highways and other earth-rock construction. Some types of volcanic ash have been used as pozzolanic cement and as admixtures in concrete to retard undesired reactions between cement alkalies and aggregates.

TABLE 4.9 Unusual Soil (*Continued*)

Examples of unusual soil (1)	Discussion (2)
Nonwelded tuff and volcanic ash (*Continued*)	Natural deposits of nonwelded tuff and volcanic ash are unusual because they have very low dry densities (e.g., 1 Mg/m^3) because of the presence of lightweight glass and pumice. The materials are also highly susceptible to erosion, which can cause the development of unusual eroded landforms known as "pinnacles."
Loess	Loess is widespread in the central portion of the United States. It consists of uniform cohesive wind-blown silt, commonly light brown, yellow, or gray in color, with most of the particle sizes between 0.01 and 0.05 mm (Terzaghi and Peck 1967). The cohesion is commonly due to calcareous cement which binds the particles together. An unusual feature of loess is the presence of vertical root holes and fractures that make it much more permeable in the vertical direction than the horizontal direction. Another unusual feature of loess is that it can form near vertical slopes, but when saturated, the cohesion is lost and the slope will fail or the ground surface settle.
Caliche	This type of material is common in arid or semi-arid parts of the southwestern United States and consists of soil that is normally cemented together by calcium carbonate. When water evaporates near or at ground surface, the calcium carbonate is deposited in the void spaces between soil particles. Caliche is generally strong and stable in an undisturbed state, but it can become unstable if the cementing agents are leached away by water from leaky pipes or sewers or from the infiltration of irrigation water.
Debris flow and alluvial fan deposits	A debris flow can transport a wide variety of soil particle sizes, including boulders and cobbles (see Sec. 13.7). Boulders, cobbles, and coarse gravel are typically described as oversize particles, and the finer soil particles are described as the soil matrix. Such debris flow and alluvial fan deposits are often unstable because of the erratic and unsteady arrangement of oversize particles and loose matrix soil.
Varved clay	Varved clays ordinarily form as lake deposits and consist of alternating layers of soil. Each varve represents the deposition during a year, where the lower sandy part is deposited during the summer, and the upper clayey part is then deposited during the winter when the surface of the lake is frozen and the water is tranquil. This causes an unusual variation in shear strength in the soil, where the horizontal shear strength along the clay portion of the varve is much less than the vertical shear strength. This can cause the stability of structures founded on varved clay to be overestimated, resulting in a bearing capacity–type failure.
Bentonite	Bentonite is a deposit consisting mainly of montmorillonite clay particles. It is derived from the alteration of volcanic tuff or ash. Bentonite is

TABLE 4.9 Unusual Soil (*Continued*)

Examples of unusual soil (1)	Discussion (2)
Bentonite (*Continued*)	mined to make products that are used as impermeable barriers, such as geosynthetic clay liners (GCL) which are bentonite/geosynthetic composites. Because bentonite consists almost exclusively of montmorillonite, it will swell, shrink, and cause more expansive soil–related damage than any other type of soil.
Sensitive or quick clays	The sensitivity S_t of a clay is defined as the undisturbed or natural undrained shear strength divided by its remolded shear strength, or: $$S_t = \frac{\text{undrained shear strength (undisturbed state)}}{\text{undrained shear strength (remolded state)}}$$ On the basis of this value, the clay can be rated as follows (Holtz and Kovacs 1981): • Soil with a "low" sensitivity: S_t is less than 4 • Soil with a "medium" sensitivity: S_t is from 4 to 8 • Soil with a "high" sensitivity: S_t is from 8 to 16 • Soil with a "quick" sensitivity: S_t is greater than 16 An unusual feature of highly sensitive or quick clays is that the *in situ* water content is often greater than the liquid limit (liquidity index greater than one). Figure 2.21 shows sensitive Canadian clay that has a water content greater than its liquid limit. Sensitive clays have unstable bonds between particles. As long as these unstable bonds are not broken, the clay can support a heavy load. But once remolded, the bonding is destroyed and the shear strength is substantially reduced. For example, sensitive Leda clay, from Ottawa, Ontario, has a high shear strength in the undisturbed state, but once remolded, the clay is essentially a fluid (no shear strength). There are reports of entire hillsides of quick clays becoming unstable and then simply flowing away (Lambe and Whitman 1969).
Diatomaceous earth	Diatoms are microscopic, single-celled plants of the class Bacillariophyceae, which grow in both marine and fresh water (Bates and Jackson 1980). Diatoms secrete outer shells of silica, called frustules, in a great variety of forms which can accumulate in sediments in enormous amounts (Bates and Jackson 1980). Deposits of diatoms have low dry density and high moisture content because the structure of the diatom is an outer shell of silica that can contain water. Common shapes of diatoms are rodlike, spherical, or circular disks having a typical length or diameter of about 0.03 to 0.11 mm (Spencer 1972). Diatoms typically have rough surface features, such as protrusions or indentations. A natural deposit of diatoms is commonly referred to as diatomaceous earth or diatomite. Diatomaceous earth usually consists of fine, white,

TABLE 4.9 Unusual Soil (*Continued*)

Examples of unusual soil (1)	Discussion (2)
Diatomaceous earth (*Continued*)	siliceous powder, composed mainly of diatoms or their remains (Terzaghi and Peck 1967, Stokes and Varnes 1955). Diatomite is a organogenetic sedimentary rock containing frustules of diatoms and sometimes mixed with shells of radiolarians, spicules of sponges, and foraminifera (Mottana et al. 1978). Industrial uses of diatomaceous earth or diatomite are as filters to remove impurities, as abrasives to polish soft metals, and when mixed with nitroglycerin, as an absorbent in the production of dynamite (Mottana et al. 1978). Diatomaceous earth can be very compressible when used as fill. At high pressures, the diatoms (which are essentially hollow shells of silica) can be crushed together, resulting in a high compressibility (Day 1995e).

TABLE 4.10 Simplified Rock Classification

Common igneous rocks		
Major division (1)	Secondary divisions (2)	Rock types (3)
Extrusive	Volcanic explosion debris (fragmental)	Tuff (lithified ash) and volcanic breccia
	Lava flows and hot siliceous clouds	Obsidian (glass), pumice, and scoria
	Lava flows (fine-grained texture)	Basalt, andesite, and rhyolite
Intrusive	Dark minerals dominant	Gabbro
	Intermediate (25–50% dark minerals)	Diorite
	Light color (quartz and feldspar minerals)	Granite
Common sedimentary rocks		
Major division (1)	Texture (grain size) or chemical composition (2)	Rock types (3)
Clastic rocks	Grain sizes larger than 2 mm (pebbles, gravel, cobbles, and boulders)	Conglomerate (rounded cobbles) or breccia (angular rock fragments)
	Sand-size grains, 0.062 to 2 mm	Sandstone
	Silt-size grains, 0.004 to 0.062 mm	Siltstone
	Clay-size grains, less than 0.004 mm	Claystone and shale
Chemical and organic rocks	Carbonate minerals (e.g., calcite)	Limestone
	Halite minerals	Rock salt
	Sulfate minerals	Gypsum
	Iron-rich minerals	Hematite
	Siliceous minerals	Chert
	Organic products	Coal
Common metamorphic rocks		
Major division (1)	Structure (foliated or massive) (2)	Rock types (3)
Coarse crystalline	Foliated	Gneiss
	Massive	Metaquartzite
Medium crystalline	Foliated	Schist
	Massive	Marble, quartzite, serpentine, soapstone
Fine to microscopic	Foliated	Phyllite, slate
	Massive	Hornfels, anthracite coal

***Note*:** Grain sizes correspond to the Modified Wentworth scale.

TABLE 4.11 Hardness of Rock Versus Unconfined Compressive Strength

Hardness (1)	q_u (2)	Rock description (3)
Very soft	10 to 250 tsf	The rock can be readily indented, grooved, or gouged with fingernail, or carved with a knife. Breaks with light manual pressure. The rock disintegrates upon the single blow of a geologic hammer.
Soft	250 to 500 tsf	The rock can be grooved or gouged easily by a knife or sharp pick with light pressure. Can be scratched with a fingernail. Breaks with light to moderate manual pressure.
Hard	500 to 1000 tsf	The rock can be scratched with a knife or sharp pick with great difficulty (heavy pressure is needed). A heavy hammer blow is required to break the rock.
Very hard	1000 to 2000 tsf	The rock cannot be scratched with a knife or sharp pick. The rock can be broken with several solid blows of a geologic hammer.
Extremely hard	>2000 tsf	The rock cannot be scratched with a knife or sharp pick. The rock can only be chipped with repeated heavy hammer blows.

Notes:

1. One measure of the quality of rock is its hardness, which has been correlated with the unconfined compressive strength of rock specimens. This table lists hardness of rock as a function of the unconfined compressive strength q_u. Because the unconfined compressive strength is performed on small rock specimens, in most cases, it will not represent the actual condition of *in situ* rock. The reason is the presence of joints, fractures, fissures, and planes of weakness in the actual rock mass which govern its engineering properties, such as deformation characteristics, shear strength, and permeability. The unconfined compressive test also does not consider other rock quality factors, such as its resistance to weathering or behavior when submerged in water.

2. q_u = unconfined compressive strength (tsf) of the rock. 1 tsf is approximately equal to 100 kPa.

3. Sources: *Basic Soils Engineering* (Hough 1969) and *Engineering Geology Field Manual* (1987).

TABLE 4.12 Physical Description of Rock

Topic (1)	Discussion (2)
Weathering characteristics	The following are commonly used weathering descriptors: • *Fresh:* No discoloration or oxidation. • *Slightly weathered:* Discoloration or oxidation is limited to surface of, or short distance from, fractures; some feldspar crystals are dull. • *Moderately weathered:* Discoloration or oxidation extends from fractures, usually throughout; Fe and Mg minerals are "rusty" and feldspar crystals are "cloudy." • *Intensely weathered:* Discoloration or oxidation throughout; all feldspars and Fe and Mg minerals are altered to clay to some extent; or chemical alteration produces *in situ* disaggregation. • *Decomposed:* Discolored or oxidized throughout, but resistant minerals such as quartz may be unaltered; all feldspars and Fe and Mg minerals are completely altered to clay.
Joint spacing	The following are commonly used joint spacing descriptors: • *Massive:* Greater than 3 m (10 ft) spacing of joints or fractures. • *Very thickly (bedded, foliated, or banded):* The spacing of joints or fractures is 1 to 3 m (3 to 10 ft). • *Thickly:* The spacing of joints or fractures is 300 mm to 1 m (1 to 3 ft). • *Moderately:* The spacing of joints or fractures is 100 to 300 mm (0.3 to 1 ft). • *Thinly:* The spacing of joints or fractures is 30 to 100 mm (0.1 to 0.3 ft). • *Very thinly:* The spacing of joints or fractures is 10 to 30 mm (0.03 to 0.1 ft). • *Laminated (intensely foliated or banded):* Less than 10 mm (0.03 ft) spacing of joints or fractures.
Rock quality	A measure of the quality of the rock is the RQD (rock quality designation), which is computed by summing the lengths of all pieces of the core (NX size) equal to or longer than 100 mm (4 in.) and dividing by the total length of the core run. The RQD is multiplied by 100 and expressed as a percentage. The mass rock quality can be defined as follows: • RQD = 0 to 25%, rock quality is defined as "very poor" • RQD = 25 to 50%, rock quality is defined as "poor" • RQD = 50 to 75%, rock quality is defined as "fair" • RQD = 75 to 90%, rock quality is defined as "good" • RQD = 90 to 100%, rock quality is defined as "excellent" In calculating the RQD, only the natural fractures should be counted, and any fresh fractures due to the sampling process should be ignored. RQD measurements can provide valuable data on the quality of the *in situ* rock mass, and can be used to locate zones of extensively fractured or weathered rock.

Source: *Engineering Geology Field Manual* (1987)

TABLE 4.13 Summary of Typical Engineering Properties of Rock

Rock type (1)	Mechanical strength (2)	Durability (3)	Chemical stability (4)	Surface (5)	Impurities (6)	Crushed shape (7)
Igneous						
Granite, syenite, and diorite	Good	Good	Good	Good	Possible	Good
Felsite	Good	Good	Questionable	Fair	Possible	Fair
Basalt, diabase, and gabbro	Good	Good	Good	Good	Seldom	Fair
Peridotite	Good	Fair	Questionable	Good	Possible	Good
Sedimentary						
Limestone and dolomite	Good	Fair	Good	Good	Possible	Good
Sandstone	Fair	Fair	Good	Good	Seldom	Good
Chert	Good	Poor	Poor	Fair	Likely	Poor
Conglomerate and breccia	Fair	Fair	Good	Good	Seldom	Fair
Shale	Poor	Poor	—	Good	Possible	Fair to poor
Metamorphic						
Gneiss and schist	Good	Good	Good	Good	Seldom	Good to poor
Quartzite	Good	Good	Good	Good	Seldom	Fair
Marble	Fair	Good	Good	Good	Possible	Good
Serpentine	Fair	Fair	Good	Fair to poor	Possible	Fair
Amphibolite	Good	Good	Good	Good	Seldom	Fair
Slate	Good	Good	Good	Poor	Seldom	Poor

Source: *Concrete Construction Handbook* (Waddell and Dobrowolski 1993). Reproduced with permission of McGraw-Hill, Inc.

CHAPTER 5
PHASE RELATIONSHIPS

Chapters 2 and 3 deal with field exploration and laboratory testing. The remainder of the chapters in Part I of the book deal with engineering analyses that utilize this data. It is important to recognize that, without adequate and meaningful data from subsurface exploration and laboratory testing, the engineering analyses presented in the following chapters will be of doubtful value and may even lead to erroneous conclusions.

Phase relationships are the basic soil relationships used in geotechnical engineering. The phase relationships are also known as *mass-volume relationships.*

5.1 SOIL ELEMENT

Figure 5.1 shows an element of soil that can be divided into three basic parts:

1. *Solids.* The mineral soil particles
2. *Liquids.* Usually water that is contained in the void spaces between the solid mineral particles
3. *Gas.* For example, air that is contained in the void spaces between the solid mineral particles

As indicated on the right side of Fig. 5.1, the three basic parts of soil can be rearranged into their relative proportions based on volume and mass. Note that the symbols as defined in Fig. 5.1 will be used throughout this chapter.

5.2 PHASE RELATIONSHIPS DIRECTLY FROM LABORATORY TESTING

Certain phase relationships can be determined directly from laboratory testing. These phase relationships, discussed in Chap. 3, include the following:

1. *Water content* w. Also known as moisture content (see Table 3.3), it is defined as the mass of water divided by the mass of dry soil, usually expressed as a percentage ($w = 100\ M_w/M_s$).
2. *Total density* ρ_t. As indicated in Table 3.4, the total density is defined as the mass per unit volume ($\rho_t = M/V$). In the United States Customary System, total density ρ_t and total unit weight γ_t have exactly the same value. In the International System of Units (SI), the total density (in Mg/m^3) is multiplied by the acceleration of gravity ($a = 9.81$ m/s^2) in order to obtain the total unit weight (kN/m^3).
3. *Specific gravity* G. As indicated in Table 3.5, the specific gravity G is defined as the density of solids (ρ_s) divided by the density of water (ρ_w), or $G = \rho_s/\rho_w$. The density of solids (ρ_s) is defined as the mass of solids (M_s) divided by the volume of solids (V_s). The density of water (ρ_w) is equal to 1 g/cm^3 (or 1 Mg/m^3) and 62.4 pcf.

5.3 INDIRECT PHASE RELATIONSHIPS

Most phase relationships can not be determined in a laboratory, but instead must be calculated. Table 5.1 presents a discussion of various indirect phase relationships, Table 5.2 presents various equations for mass and volume, and Table 5.3 presents various equations for unit weight.

5.4 EXAMPLE OF THE CALCULATION OF PHASE RELATIONSHIPS

Table 5.4 presents examples of the calculation of phase relationships.

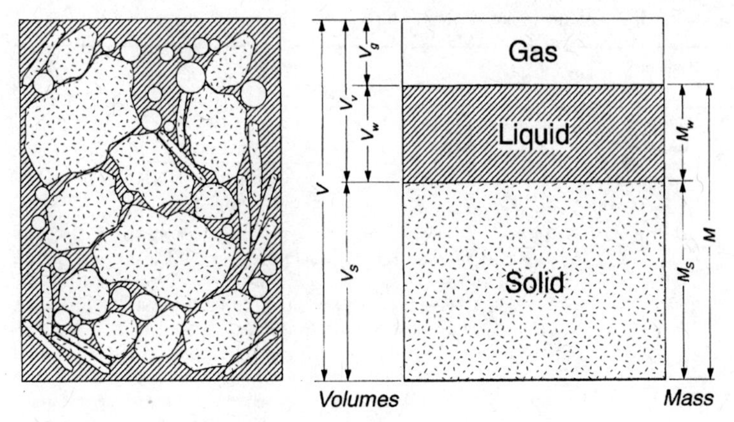

FIGURE 5.1 Soil element and the soil element separated into phases.

TABLE 5.1 Indirect Phase Relationships

Parameter (1)	Relationships (2)
Volume of voids (V_v)	$V_v = V_g + V_w$, where V_g = volume of gas (air) and V_w = volume of water (see Fig. 5.1)
Void ratio e and porosity n	$e = V_v/V_s$ where V_s = volume of solids (see Fig. 5.1) $n = V_v/V$ *Note:* $e = n/(1 - n)$ and $n = e/(1 + e)$ The void ratio and porosity indicate the relative amount of void space in a soil. The lower the void ratio and porosity, the denser the soil (and vice versa). The natural soil having the lowest void ratio is probably till. For example, a typical value of dry density for till is 2.34 Mg/m^3 (146 pcf), which corresponds to a void ratio of 0.14 (NAVFAC DM-7.1, 1982). A typical till consists of a well-graded soil ranging in particle sizes from clay to gravel and boulders. The high density and low void ratio is due to the extremely high stress exerted by glaciers (Winterkorn and Fang 1975). For compacted soil, the soil type with typically the lowest void ratio is a well-graded decomposed granite (DG). A typical value of maximum dry density (Modified Proctor) for a well-graded DG is 2.20 Mg/m^3 (137 pcf), which corresponds to a void ratio of 0.21. In general, the factors needed for a very low void ratio for compacted and naturally deposited soil are (Aberg 1996, Day 1997a): 1. A well-graded grain-size distribution 2. A high ratio of D_{100}/D_0 (ratio of the largest and smallest grain sizes) 3. Clay particles (having low activity) to fill in the smallest void spaces 4. A process, such as compaction or the weight of glaciers, to compress the soil particles into dense arrangements At the other extreme are clays, such as sodium montmorillonite, which at low confining pressures can have a void ratio of more than 25. Highly organic soil, such as peat, can have even higher void ratios.
Degree of saturation S	$S\ (\%) = 100\ V_w / V_v$ If a soil is obtained below the groundwater table or after submergence in the laboratory, the degree of saturation is often assumed to be 100%, and then phase relationships (such as the void ratio) are back-calculated. However, for soil below the groundwater table, a better approach is to use the degree of saturation as a final check on the accuracy of the laboratory test data (i.e., γ_t, w, and G). See Table 4.2 (Soil Moisture Condition) for values of the degree of saturation versus descriptive terminology.
Useful relationship	A frequently used method of solving phase relationships is to first in the phase diagram shown in Fig. 5.1. Once the different mass and volumes are known, then the various phase relationships can be determined. Another approach is to use equations that relate different parameters. An example of a commonly used equation is: $Gw = Se$

TABLE 5.2 Mass and Volume Relationships*

Parameter (1)	Relationships (2)
Mass	Mass of solids (M_s): $M_s = \dfrac{M}{1+w} = \dfrac{M_w G}{eS} = GV\rho_w(1-n)$
	Mass of water (M_w): $M_w = \dfrac{eM_s S}{G} = wM_s = S\rho_w V_v$
	Total mass (M): $M = M_s + M_w = M_s(1+w)$
Volume	Volume of solids (V_s): $V_s = \dfrac{M_s}{G\rho_w} = \dfrac{V}{1+e} = \dfrac{V_v}{e} = V(1-n) = V-(V_g+V_w)$
	Volume of water (V_w): $V_w = \dfrac{M_w}{\rho_w} = \dfrac{SVe}{1+e} = SV_s e = SV_v = V_v - V_g$
	Volume of gas (V_g): $V_g = \dfrac{(1-S)\,Ve}{1+e} = (1-S)\,V_s e = V-(V_s+V_w) = V_v - V_w$
	Volume of voids (V_v): $V_v = \dfrac{V_s n}{1-n} = V - \dfrac{M_s}{G\rho_w} = \dfrac{Ve}{1+e} = V_s e = V - V_s$
	Total volume (V): $V = \dfrac{V_s}{1-n} = \dfrac{V_v(1+e)}{e} = V_s(1+e) = V_s + V_g + V_w$

*See Fig. 5.1 and Table 5.1 for definitions of terms.

TABLE 5.3 Unit Weight Relationships*

Parameter (1)	Relationships (2)
Total unit weight γ_t	$\gamma_t = \dfrac{W_s + W_w}{V} = \dfrac{G\gamma_w (1 + w)}{1 + e}$
Dry unit weight γ_d	$\gamma_d = \dfrac{W_s}{V} = \dfrac{G\gamma_w}{1 + e} = \dfrac{\gamma_t}{1 + w}$
Saturated unit weight γ_{sat}	$\gamma_{sat} = \dfrac{W_s + V_v\gamma_w}{V} = \dfrac{(G + e)\,\gamma_w}{1 + e} = \dfrac{G\gamma_w (1 + w)}{1 + Gw}$ ***Note:*** The total unit weight γ_t is equal to the saturated unit weight γ_{sat} when all the void spaces are filled with water (i.e., $S = 100\%$).
Buoyant unit weight γ_b	$\gamma_b = \gamma_{sat} - \gamma_w$ $\gamma_b = \dfrac{\gamma_w (G - 1)}{1 + e} = \dfrac{\gamma_w (G - 1)}{1 + Gw}$ ***Note:*** The buoyant unit weight is also known as the *submerged unit weight.*

*See Fig. 5.1 and Tables 3.4 and 5.1 for definitions of terms.

Notes:

1. In the equations listed in Tables 5.2 and 5.3, water content w and degree of saturation S must be expressed as a decimal (not as a percentage).

2. ρ_w = density of water (1.0 Mg/m^3, 62.4 pcf) and γ_w = unit weight of water (9.81 kN/m^3, 62.4 pcf).

3. In the United States Customary System, it is common to concurrently use pounds to represent both a unit of mass (M) and a unit of force (W). Thus the density of water (mass per unit volume: ρ_w = 62.4 pcf) has the same numerical value as the unit weight of water (weight per unit volume: γ_w = 62.4 pcf). In the International System of Units (SI), the mass (M) is multiplied by the acceleration of gravity (a = 9.81 m/s^2) in order to obtain the weight (i.e., $W = Ma$). Thus the density of water (ρ_w) equals 1.0 Mg/m^3 or 1.0 g/cm^3 and the unit weight of water (γ_w) equals 9.81 kN/m^3.

TABLE 5.4 Example

Example (1)	Discussion (2)
General discussion	As mentioned in Chap. 4, the soil classification provides a written description of the soil. In essence, the phase relationships provide a mathematical description of the soil which is used in engineering analyses. For the soil data shown in Fig. 4.5, the written and mathematical descriptions of the soil are as follows: *Written description (soil classification):* Per the ISBP, clayey sand of low plasticity (SC of low plasticity). Based on dry mass, the soil contains 13.8% gravel-size particles, 48.7% sand-size particles, 19.5% silt-size particles, and 18.0% clay-size particles. The gravel-size particles are predominantly hard and angular rock fragments. The sand-size particles are predominantly composed of angular quartz grains. Atterberg limits performed on the soil passing the No. 40 sieve indicate a LL = 28 and a PI = 15. The *in situ* soil has a reddish-brown color, soft consistency, and gritty texture, and is wet. The *in situ* soil is a residual soil and is part of an old vernal pool. *Mathematical description (phase relationships):* Water content (w) = 19%, total unit weight (γ_t) = 19 kN/m^3 (120 pcf), dry unit weight (γ_d) = 16 kN/m^3 (100 pcf), specific gravity (G) = 2.65, void ratio (e) = 0.63, porosity (n) = 0.39 (or 39%), and degree of saturation (S) = 80%.
Example problem	A soil specimen has the following measured properties: Total unit weight = 19.0 kN/m^3 (121 pcf) Water content = 18.0% Specific gravity = 2.65 Calculate the parameters shown in Fig. 5.1 assuming that the total volume (V) = 1 m^3 and 1 ft^3. Also calculate void ratio e, porosity n, and degree of saturation S. *Solution:* For SI units, V = 1 m^3. $M_s = M/(1 + w) = 19/[(1 + 0.18)(9.81)] = 1.64$ Mg $V_s = M_s/(G\rho_w) = 1.64/[(2.65)(1.0)] = 0.62$ m^3 $V_w = M_w/\rho_w = [19.0-(19/1.18)]/[(9.81)(1.0)] = 0.30$ m^3 $V_g = V - (V_s + V_w) = 1 - (0.62 + 0.30) = 0.08$ m^3 $M_w = V_w\rho_w = (0.30)(1.0) = 0.3$ Mg *Solution:* For United States Customary System Units, V = 1 ft^3. $M_s = M/(1 + w) = 121/(1 + 0.18) = 103$ lb $V_s = M_s/(G\,\rho_w) = 103/[(2.65)(62.4)] = 0.62$ ft^3 $V_w = M_w/\rho_w = [121 - (121/1.18)]/(62.4) = 0.30$ ft^3 $V_g = V - (V_s + V_w) = 1 - (0.62 + 0.30) = 0.08$ ft^3 $M_w = V_w\,\rho_w = (0.30)(62.4) = 18$ lb Void ratio (e) = $V_v/V_s = (1 - 0.62) / 0.62 = 0.61$ Porosity (n) = $V_v/V = (1 - 0.62) / 1.0 = 0.38$ (or 38%) Degree of saturation (S) = $V_w/V_v = 0.30/(1 - 0.62) = 0.78$ (or 78%)

CHAPTER 6
EFFECTIVE STRESS AND STRESS DISTRIBUTION

6.1 INTRODUCTION

After characterizing the soil in terms of a written description (soil classification, Chap. 4) and mathematical description (phase relationships, Chap. 5), the next step in the analysis is often to determine the stresses acting on the soil. This is important because most geotechnical projects deal with a change in stress of the soil. For example, the construction of a building applies an additional stress onto the soil supporting the foundation, which results in settlement of the building.

Stress is defined as the load divided by the area over which it acts. In geotechnical engineering, a compressive stress is considered positive and tensile stress is negative. Stress and pressure are often used interchangeably in geotechnical engineering. When using the International System of Units (SI), the units for stress are kPa. In the United States Customary System, the units for stress are psf (pounds-force per square foot). Stress expressed in units of kg/cm^2 have been used in the past (e.g., see Figs. 2.19 and 2.21). One kg/cm^2 is approximately equal to one ton per square foot (tsf).

6.2 EFFECTIVE STRESS, TOTAL STRESS, AND PORE WATER PRESSURE

Table 6.1 presents a discussion and the equations for effective stress σ', total stress σ, and pore water pressure u. Related topics (capillarity and coefficient of earth pressure at rest) are discussed in Tables 6.2 and 6.3.

6.3 STRESS DISTRIBUTION

The previous section presented equations that are used to determine the existing stresses within the soil mass. This section describes commonly used methods to determine the increase in stress in the soil deposit due to applied loads. This is naturally important in settlement analysis because the settlement of the structure is due directly to its weight, which causes an increase in stress in the underlying soil. In most cases, it is the increase in vertical stress that is of most importance in settlement analyses. The symbol σ_z is often used to denote an increase in vertical stress in the soil, although $\Delta\sigma_v$ (change in total vertical stress) is also used.

The following stress distribution methods are presented in this chapter:

6.3.1 One-Dimensional Loading

In dealing with stress distribution, a distinction must be made between one-dimensional and two- or three-dimensional loading. A one-dimensional loading applies a stress increase at depth that is 100 percent of the applied surface stress. Table 6.4 presents a discussion of common one-dimensional loading conditions.

6.3.2 2:1 Approximation

Surface loadings can cause both vertical and horizontal strains, and this is referred to as two- or three-dimensional loading. Common examples of two-dimensional loading are from strip footings or long embankments (i.e., plane strain conditions). Examples of three-dimensional loading would be square and rectangular footings (spread footings) and round storage tanks. The following subsections describe methods that can be used to determine the change in vertical stress for two-dimensional (strip footings and long embankments) and three-dimensional (spread footings and round storage tanks) loading conditions. In these cases, the load usually dissipates rapidly with depth. One commonly used method to determine the increase in vertical stress due to a two- or three-dimensional surface loading is the 2:1 approximation (see Table 6.5 and Fig. 6.1).

6.3.3 Equations Based on the Theory of Elasticity

Equations and charts have been developed to determine the change in stress due to applied loads from the theory of elasticity. Table 6.6 and Fig. 6.2 present equations by Boussinesq (1885) that can be used to determine the increase in vertical stress due to a point or line load.

6.3.4 Charts Based on the Theory of Elasticity

On the basis of the theory of elasticity, numerous charts (Figs. 6.3 to 6.9) have been developed to determine the increase in vertical stress due to different types of applied loads. Table 6.7 presents a discussion of these charts. In addition to the equations and charts described above, the theory of elasticity has been applied to many other types of loading conditions (e.g., see Poulos and Davis 1974).

6.4 EXAMPLE OF CALCULATIONS USING VARIOUS STRESS DISTRIBUTIONS

Table 6.8 presents an example of the calculations using various stress distributions presented in this chapter.

6.5 MOHR CIRCLE

The Mohr circle is a graphical representation of the state of stress on different planes at a particular point in the soil mass. By knowing the magnitude and direction of the major principal stress σ_1 and the minor principal stress σ_3, it is possible to determine the stresses in any other direction by using the Mohr circle (see Fig. 6.10).

TABLE 6.1 Effective Stress, Total Stress, and Pore Water Pressure

Topic (1)	Discussion (2)
Effective stress σ'	An important concept in geotechnical engineering is effective stress. The effective stress σ' is defined as follows: $$\sigma' = \sigma - u$$ where σ = total stress and u = pore water pressure. Many engineering analysis use the vertical effective stress, also known as the effective overburden stress, which is designated σ'_v or σ'_{vo}.
Total stress σ	For the condition of a uniform soil and a level ground surface (geostatic condition), the total vertical stress σ_v at a depth z below the ground surface is: $$\sigma_v = \gamma_t z$$ where γ_t = total unit weight of the soil (see Table 5.3). For soil deposits having layers with different total unit weights, the total vertical stress is the sum of the vertical stress for each individual soil layer. For some projects, total pressure cells (Chap. 20) can be installed to measure the total stress in the soil or the total stress of the soil acting on an earth structure, such as a retaining wall.
Pore water pressure u	For the condition of a hydrostatic groundwater table (i.e., no groundwater flow or excess pore water pressures), the static pore water pressure (u or u_s) is: $$u = \gamma_w z_w$$ where γ_w = unit weight of water and z_w = depth below the groundwater table. By knowing the total unit weight of the soil (γ_t) and the pore water pressure u, the vertical effective stress σ'_v can be calculated. An alternative method is to use the buoyant unit weight γ_b to calculate the vertical effective stress. For example, suppose that a groundwater table corresponds with the ground surface. In this case, the vertical effective stress σ'_v is simply the buoyant unit weight γ_b times the depth below the ground surface. More often, the groundwater table is below the ground surface, in which case the vertical total stress of the soil layer above the groundwater table must be added to the buoyant unit weight calculations. The vertical effective stress σ'_v is often plotted versus depth and included with the subsoil profile. For example, in Fig. 2.19, the vertical effective stress σ'_{vo} and the static pore water pressure u_s are plotted versus depth. Likewise in Figs. 2.20 and 2.21, the vertical effective stress (also known as effective overburden stress) have been plotted versus depth. For cases where there is flowing groundwater or excess pore water pressure due to the consolidation of clay, the pore water will not be hydrostatic. Engineering analyses, such as seepage analyses (Chap. 8) or the theory of consolidation (Chap. 9) can be used to predict the pore water pressure. For some projects, piezometers (Chap. 23) can be installed to measure the pore water pressure u in the ground.

TABLE 6.2 Capillarity

Topic (1)	Discussion (2)
Capillarity	Soil above the groundwater table can be subjected to negative pore water pressure. This is known as *capillarity,* also known as *capillary action,* which is the rise of water through soil due to the fluid property known as surface tension. Because of capillarity, the pore water pressures are less-than-atmospheric values produced by the surface tension of pore water acting on the meniscus formed in the void spaces between the soil particles. If the soil is saturated above the groundwater table, then the pore water pressure u is negative and can be calculated as the distance above the groundwater table h times the unit weight of water γ_w, or: $$u = \gamma_w h$$ Capillary is important in the understanding of soil behavior. Because of the negative value of pore water pressure due to capillarity, it essentially holds together the soil particles. For large-size soil particles, such as gravel and coarse sand-size particles, the effect of capillarity is negligible and the soil particles simply fall apart (hence they are cohesionless). Medium to fine sands do have a low capillarity, which is enough to build a sand castle at the beach, but this small capillarity is lost when the sand becomes submerged in water or the sand completely dries. Silt- and clay-size particles are so small that they are strongly influenced by capillarity. These fine soil size particles can be strongly held together by capillarity, which give the soil the ability to be remolded and rolled without falling apart (hence they are cohesive). Capillary is the mechanism that gives a silt or clay its plasticity, the ability to be remolded and rolled without falling apart. When the remolded silt or clay is submerged in water, the capillary tension is slowly eliminated and the soil particles will often disperse. As will be discussed in Chap. 7, the undrained shear strength of silts and clays is strongly influenced by capillarity.
Height of capillary rise	The height of capillary rise h_c is related to the pore size of the soil, as follows (Hansbo 1975): • Open-graded gravel: $h_c = 0$ • Coarse sand: $h_c = 0.03$ to 0.15 m (0.1 to 0.5 ft) • Medium sand: $h_c = 0.12$ to 1.1 m (0.4 to 3.6 ft) • Fine sand: $h_c = 0.3$ to 3.5 m (1.0 to 12 ft) • Silt: $h_c = 1.5$ to 12 m (5 to 40 ft) • Clay: $h_c \geq 10$ m ($\geq$ 33 ft) As the above data indicates, there will be no capillary rise for open-graded gravel because of the large void spaces between the individual gravel size particles. But for clay, which has very small void spaces, the capillary rise can be in excess of 10 m (33 ft).

TABLE 6.3 Coefficient of Earth Pressure at Rest

Topic (1)	Discussion (2)
Coefficient of lateral earth pressure at rest (k_o)	The preceding tables discuss the vertical effective stress σ'_v for soil deposits. For many geotechnical projects, it may be important to determine the *in situ* horizontal stress σ_h. Like the vertical stress, the horizontal total stress σ_h can be measured by using total pressure cells, and by knowing the pore water pressure, the horizontal effective stress σ'_h can be calculated (i.e., $\sigma'_h = \sigma_h - u$). The horizontal effective stress σ'_h can also be calculated as follows: $$\sigma'_h = k_o \sigma'_v$$ where k_o = coefficient of lateral earth pressure at rest. The value of this coefficient depends on many factors, such as the soil type, density condition (loose versus dense), geological depositional environment (i.e., alluvial, glacial, etc.), and the stress history of the site (Massarsch et al. 1975, Massarsch 1979). The value of k_o in natural soils can be as low as 0.4 for soils formed by sedimentation and never preloaded, up to 3.0 or greater for some heavily preloaded soil deposits (Holtz and Kovacs 1981).
k_o for normally consolidated soils	For soil deposits that have not been significantly preloaded (such as normally consolidated soils), a value of k_o = 0.5 is often assumed in practice, or the following equation is used (Jaky 1944, 1948, Brooker and Ireland 1965): $$k_o = 1 - \sin \phi'$$ where ϕ' = effective friction angle of the soil.
k_o for overconsolidated soils	As an approximation, the value of k_o for preloaded soil can be determined from the following equation (adapted from Alpan 1967, Schmertmann 1975, Ladd et al. 1977): $$k_o = 0.5(\text{OCR})^{0.5}$$ where OCR = the overconsolidation ratio, defined as the largest vertical effective stress ever experienced by the soil deposit (σ'_p) divided by the existing vertical effective stress σ'_v. The overconsolidation ratio OCR and the preconsolidation pressure σ'_p are discussed in Chap. 9.

TABLE 6.4 One-Dimensional Loading Conditions

Topic (1)	Discussion (2)
One-dimensional loading (fill of uniform thickness)	A common example of a one-dimensional loading is the placement of a fill layer of uniform thickness and large areal extent at ground surface. Beneath the center of the uniform fill, the *in situ* soil is subjected to an increase in vertical stress that equals the following: $$\sigma_z = \Delta\sigma_v = h\gamma_t$$ where h = thickness of the fill layer and γ_t = total unit weight of the fill. In this case of one-dimensional loading, the soil would be compressed only in the vertical direction (i.e., strain only in the vertical direction).
One-dimensional loading (lowering of the groundwater table)	Another common instance of a one-dimensional loading is the uniform lowering of a groundwater table. If the total unit weight of the soil does not change as the groundwater table is lowered, then the one-dimensional increase in vertical stress for the *in situ* soil located below the groundwater table would equal the following: $$\sigma_z = \Delta\sigma_v = h\gamma_w$$ where h = vertical distance that the groundwater table is uniformly lowered and γ_w = unit weight of water.
Examples	1. *Fill surcharge.* A uniform fill of large areal extent will be placed at the ground surface. The thickness of the fill layer will be 3 m (10 ft). The total unit weight of the fill is equal to 18.7 kN/m^3 (119 pcf). If the groundwater table is below the ground surface, calculate the increase in vertical stress ($\Delta\sigma_v$) beneath the center of the constructed fill mass. *Solution.* Let h = the thickness of the fill layer. Since the loading condition is one-dimensional: $$\Delta\sigma_v = h\gamma_t = (3)(18.7) = 56 \text{ kPa or } 1190 \text{ psf}$$ 2. *Lowering of the groundwater table.* Assume that the groundwater table is permanently and uniformly lowered 3 m (10 ft). Also assume that after the lowering of the groundwater table, the pore water pressures are hydrostatic and the soil located in the zone of groundwater table lowering remains saturated. Determine the increase in vertical stress ($\Delta\sigma_v$) for the soil located below the lowered groundwater table. *Solution.* Let h = the vertical distance that the groundwater table is uniformly lowered. Since the loading is one-dimensional: $$\Delta\sigma_v = h\gamma_w = (3)(9.81) = 29 \text{ kPa or } 610 \text{ psf}$$

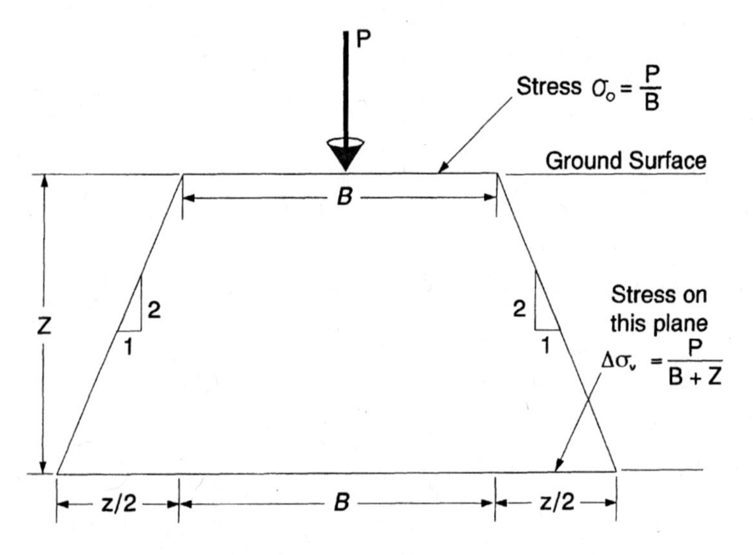

FIGURE 6.1 The approximation for the calculation of the increase in vertical stress at depth due to an applied load *P*; see Table 6.5.

TABLE 6.5 2:1 Approximation

Topic (1)	Discussion (2)
General discussion	A simple method to determine the increase in vertical stress with depth is the 2:1 approximation (also known as the 2:1 method). Figure 6.1 illustrates the basic principle of the 2:1 approximation. This method assumes that the stress dissipates with depth in the form of a trapezoid that has 2:1 (vertical:horizontal) inclined sides as shown in Fig. 6.1. The purpose of this method is to approximate the actual "pressure bulb" stress increase beneath a footing.
Strip footings (two-dimensional loading)	If there is a strip footing of width B that has a vertical load P per unit length of footing, then, as indicated in Fig. 6.1, the stress applied by the footing (σ_o) would be: $$\sigma_o = \frac{P}{B}$$ As indicated in Fig. 6.1, at a depth z below the footing, the vertical stress increase σ_z due to the strip footing load would be: $$\sigma_z = \Delta\sigma_v = \frac{P}{B + z}$$
Square or rectangular footings (three-dimensional loading)	If the footing is a rectangular spread footing having a length = L and a width = B, then the stress applied by the rectangular footing (σ_o) would be: $$\sigma_o = \frac{P}{BL}$$ where P = entire load of the rectangular spread footing. According to the 2:1 approximation, the vertical stress increase (σ_z) at a depth = z below the rectangular spread footing would be: $$\sigma_z = \Delta\sigma_v = \frac{P}{(B + z)\,(L + z)}$$
Simplicity of use	A major advantage of the 2:1 approximation is its simplicity, and for this reason, it is probably used more often than any other type of stress distribution method. The main disadvantage with the 2:1 approximation is that the stress increase under the center of the loaded area equals the stress increase under the corner or side of the loaded area. The actual situation is that the soil underlying the center of the loaded area is subjected to a higher vertical stress increase than the soil underneath a corner or edge of a uniformly loaded area. Thus the 2:1 approximation is often only used to estimate the average settlement of the loaded area. Different methods, such as stress distribution based on the theory of elasticity, can be used to calculate the change in vertical stress between the center and corner of the loaded area.

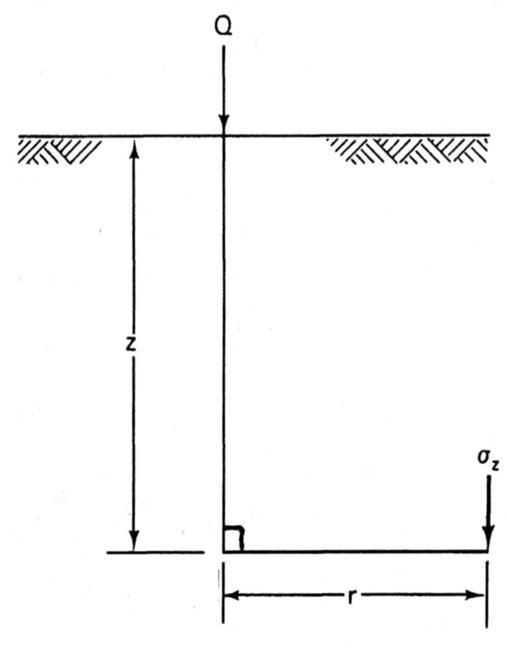

FIGURE 6.2 Definition of terms for Boussinesq equations; see Table 6.6.

TABLE 6.6 Boussinesq Equations

Topic (1)	Discussion (2)
General discussion	In 1885, Boussinesq published equations based on the theory of elasticity. The solutions assume an elastic and homogeneous soil that is continuous and in static equilibrium. The elastic solutions also use a specific type of applied load, such as a point load, uniform load, or linearly increasing load (triangular distribution). For loads where the length of the footing is greater than 5 times the width, such as for strip footings, the stress distribution is considered to be plane strain. This means that the horizontal strain of the elastic soil occurs only in the direction perpendicular to the long axis of the footing. Although equations and charts based on the theory of elasticity are often used to determine the change in soil stress, soil is not an elastic material. For example, if a heavy foundation load is applied to a soil deposit, there will be vertical deformation of the soil in response to this load. If this heavy load is removed, the soil will rebound but not return to its original height because soil is not elastic. However, it has been stated that as long as the factor of safety against shear failure exceeds about 3, then stresses imposed by the foundation load are roughly equal to the values computed from elastic theory (NAVFAC DM-7.1 1982).
Point load	For a surface point load Q applied at the ground surface such as shown in Fig. 6.2, the vertical stress increase at any depth z and distance r from the point load can be calculated by using the following Boussinesq (1885) equation: $$\sigma_z = \Delta\sigma_v = \frac{3Qz^3}{2\pi(r^2 + z^2)^{5/2}}$$
Uniform line load	If there is a uniform line load Q (force per unit length), the vertical stress increase at a depth z and distance r from the line load would be: $$\sigma_z = \Delta\sigma_v = \frac{2Qz^3}{\pi(r^2 + z^2)^2}$$ For the same load Q, the line load equation will usually give a higher value of σ_z than the point load equation.

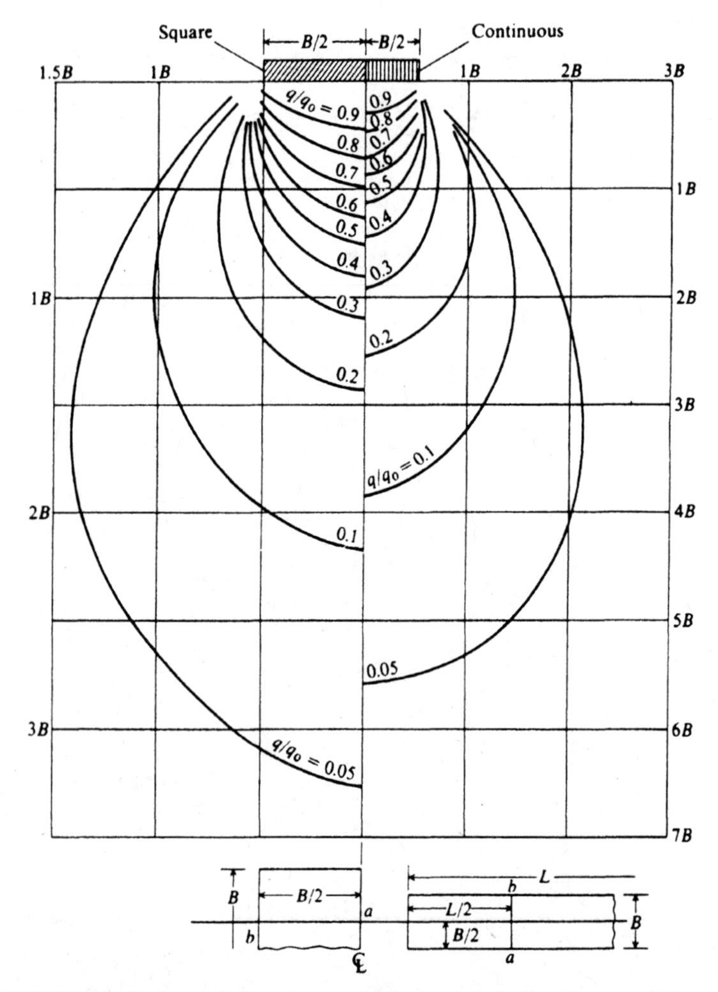

FIGURE 6.3 Pressure bulb beneath square footing and strip footing. The curves indicate the values of $\Delta\sigma_v/q_o$ beneath the footings, where q_o = uniform footing pressure; see Table 6.7. *Note:* applicable only along the line *ab* from center to edge of base. (*From Bowles 1982; reproduced with permission of McGraw-Hill, Inc.*)

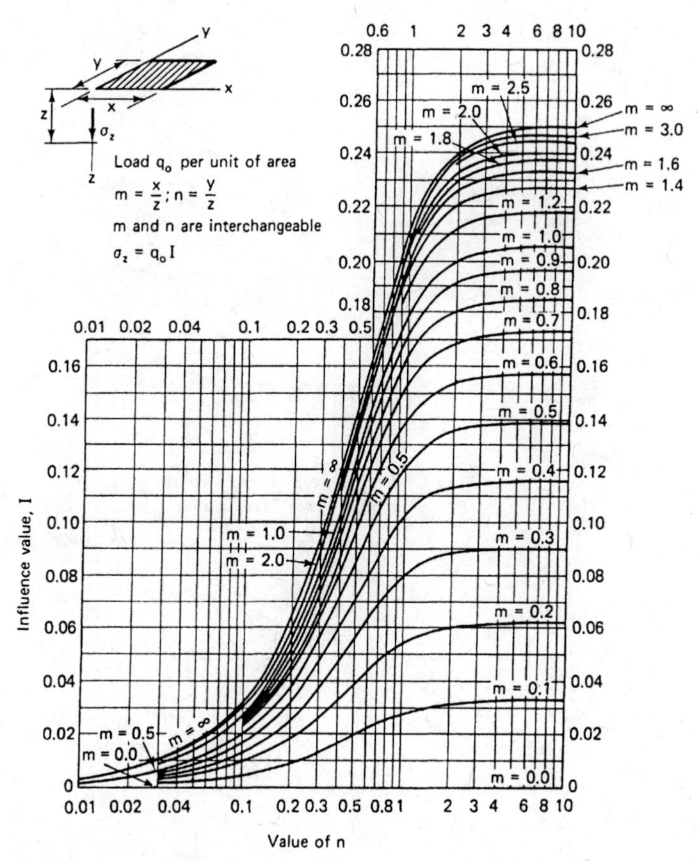

FIGURE 6.4 Chart for calculating the increase in vertical stress beneath the corner of a uniformly loaded rectangular area; see Table 6.7. (*From NAVFAC DM-7.1 1982*; *reproduced from Holtz and Kovacs 1981.*)

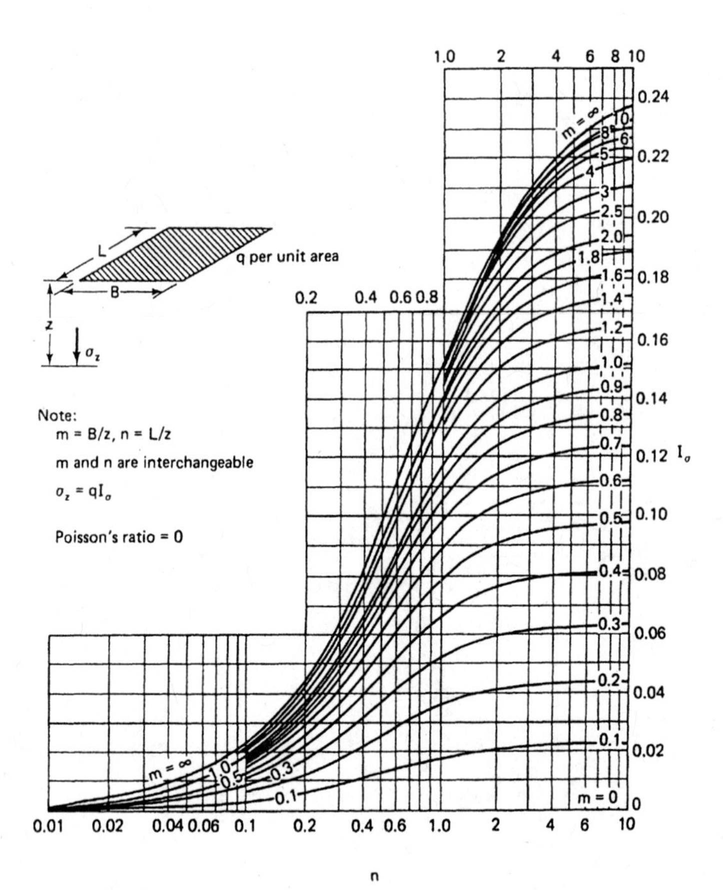

FIGURE 6.5 Chart based on Westergaard theory for calculating the increase in vertical stress beneath the corner of a uniformly loaded rectangular area; see Table 6.7. (*From Duncan and Buchignani 1976*; *reproduced from Holtz and Kovacs 1981.*)

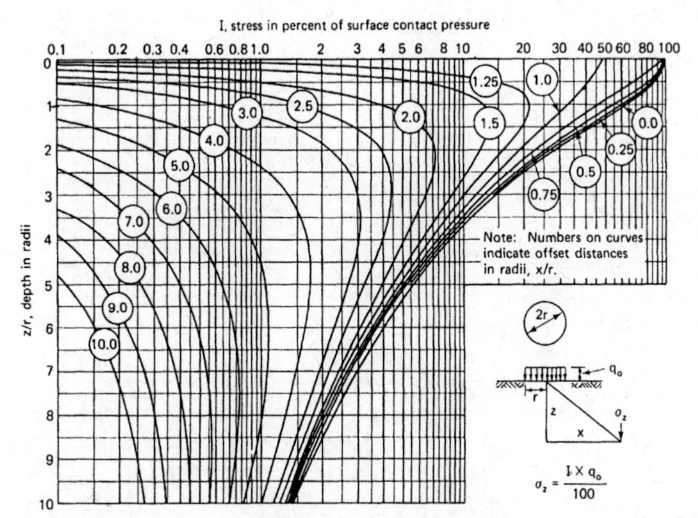

FIGURE 6.6 Chart for calculating the increase in vertical stress beneath a uniformly loaded circular area; see Table 6.7. (*From NAVFAC DM-7.1 1982 and Foster and Ahlvin 1954; reproduced from Holtz and Kovacs 1981.*)

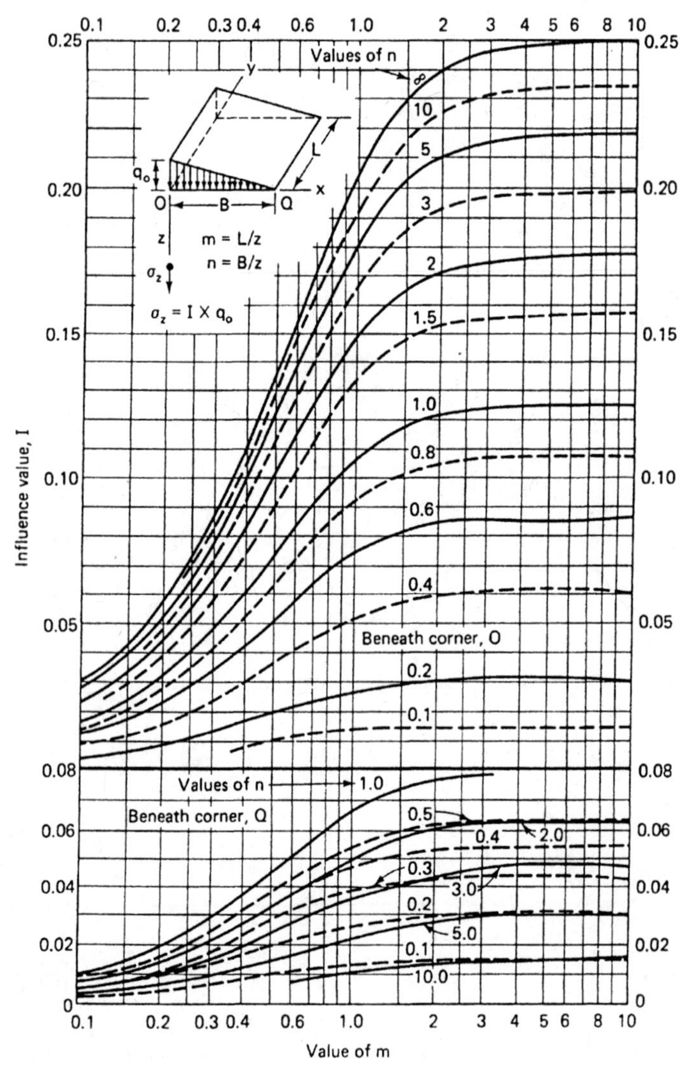

FIGURE 6.7 Chart for calculating the increase in vertical stress beneath the corner of a rectangular area that has a triangular load; see Table 6.7. (*From NAVFAC DM-7.1 1982; reproduced from Holtz and Kovacs 1981.*)

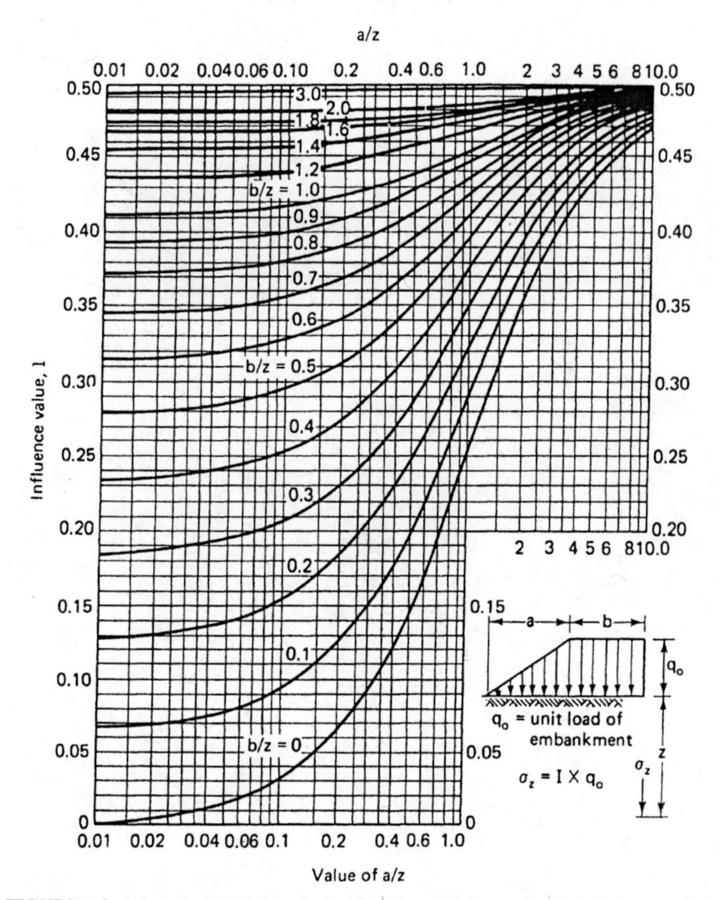

FIGURE 6.8 Chart for calculating the increase in vertical stress beneath the center of a very long embankment; see Table 6.7. (*From NAVFAC DM-7.1 1982 and Osterberg 1957; reproduced from Holtz and Kovacs 1981.*)

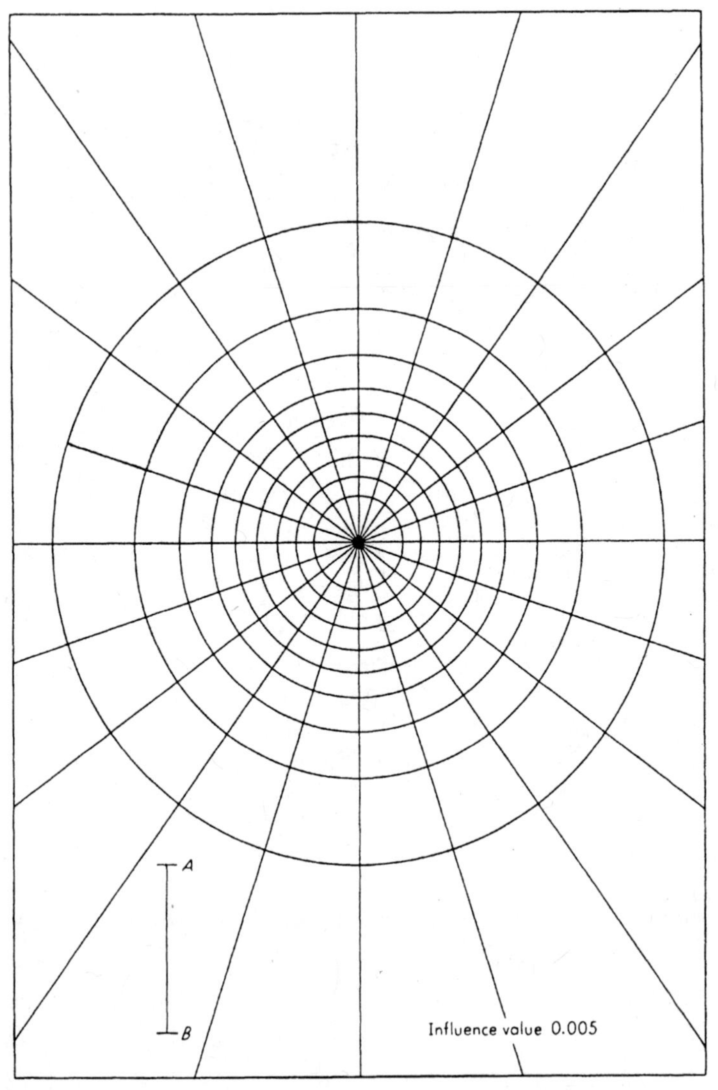

FIGURE 6.9 Newmark chart for calculating the increase in vertical stress beneath a uniformly loaded area of any shape; see Table 6.7. (*From Newmark 1942; reproduced from Bowles 1982.*)

TABLE 6.7 Charts Based on the Theory of Elasticity

Figure number (1)	Discussion (2)
Figure 6.3	Figure 6.3 shows the "pressure bulbs" (also known as *isobars*) beneath a square footing and a strip footing. In Fig. 6.3, $\sigma_z = \Delta\sigma_v$ = value from curve times q_o, where q_o = uniform applied footing pressure.
Figure 6.4	In 1935, Newmark performed an integration of the Boussinesq equation and derived a new equation to determine the vertical stress increase σ_z under the corner of a loaded area. Convenient charts have been developed, based on the Newmark (1935) equation. For example, the chart shown in Fig. 6.4 is easy to use and consists of first calculating m and n. The value m is defined as the width of the loaded area (x) divided by the depth z to where the vertical stress increase σ_z is to be calculated. The value n is defined as the length of the loaded area (y) divided by the depth z. The chart is entered with the value of n and, at the intersection with the desired m curve, the influence value I is then obtained from the vertical axis. As indicted in Fig. 6.4, vertical stress increase σ_z is then calculated as the uniform applied foundation pressure q_o times the influence value I. Figure 6.4 can also be used to determine the vertical stress increase σ_z below the center of a rectangular loaded area. In this case, the rectangular loaded area would be divided into four parts and then Fig. 6.4 would be used to find the stress increase below the corner of one of the parts. By multiplying this stress by 4 (i.e., 4 parts), the vertical stress increase σ_z below the center of the total loaded area is obtained. This type of analysis is possible because of the principle of superposition for elastic materials. To find the vertical stress increase σ_z outside the loaded area, additional rectangular areas can be added and subtracted as needed in order to model the loading condition.
Figure 6.5	Figure 6.5 presents the Westergaard (1938) analysis for a soft elastic material reinforced by numerous strong horizontal sheets. This chart was based on an elastic material that contains numerous thin and perfectly rigid layers that allow for only vertical strain but not horizontal strain. This chart may represent a better model for the increase in vertical stress σ_z for layered soils, such as a soft clay deposit that contains numerous horizontal layers of sand.
Figures 6.6 to 6.8	Figures 6.6 to 6.8 present additional charts for different types of loading conditions, based on the integration of the original Boussinesq equations. Note that Fig. 6.8 can be used to determine the vertical stress increase σ_z beneath the center of a long embankment by splitting the embankment down the middle and then multiplying the final result by 2 (i.e., two parts of the embankment).
Figure 6.9	Figure 6.9 presents a Newmark (1942) chart which can be used to determine the vertical stress increase σ_z beneath a uniformly loaded area of any shape. There are numerous influence charts, each having a different influence value. Note that the chart in Fig. 6.9 has an influence value I of 0.005. The first step

TABLE 6.7 Charts Based on the Theory of Elasticity (*Continued*)

Figure number (1)	Discussion (2)
Figure 6.9 (*Continued*)	is to draw the loaded area onto the chart, using a scale where *AB* equals the depth z. The center of the chart must correspond to the point where the increase in vertical stress (σ_z) is desired. The increase in vertical stress σ_z is then calculated as: $\sigma_z = q_o IN$ where q_o = uniform applied pressure from the irregular area, I = influene value (0.005 for Fig. 6.9) and N = number of blocks within the irregular shaped area plotted on Fig. 6.9. When obtaining the value of N, portions of blocks are also counted. Note that the entire procedure must be repeated if the increase in vertical stress σ_z is needed at a different depth.

TABLE 6.8 Example

Topic (1)	Discussion (2)
Problem statement	A proposed building will be constructed with a mat foundation. Assume that the mat foundation will be constructed at ground surface and the dimensions of the foundation are 6 m (20 ft) wide and 10 m (33 ft) long. Also assume that the proposed foundation will impose a uniform vertical pressure = 56 kPa (1170 psf). Calculate the increase in vertical stress $\Delta\sigma_v$ beneath the center of the mat foundation at a depth of 12 m (39 ft).
2:1 Approximation (Table 6.5)	For the loaded area, $B = 6$ m and $L = 10$ m. At a depth $z = 12$ m, therefore: $P = BL\sigma_o = (6)(10)(56) = 3360$ kN; $\Delta\sigma_v = P/[(B + z)(L + z)] = 3360/[(6 + 12)(10 + 12)] = 8.5$ kPa or 180 psf
Boussinesq equation (Table 6.6)	Assuming the foundation pressure is a concentrated load: $P = Q = 3360$ kN $r = 0$ $z = 12$ m; $\Delta\sigma_v = 3Qz^3/[2\pi(r^2 + z^2)^{5/2}] = 3(3360)(12)^3/[2\pi(12^2)^{5/2}] = 11$ kPa or 250 psf
Newmark chart (Fig. 6.4)	$q_o = 56$ kPa $m = x/z = 3/12 = 0.25$ $n = y/z = 5/12 = 0.42$; From Fig. 6.4 and the m and n values, $I = 0.044$. Then $\Delta\sigma_v = \sigma_z = 4q_oI = (4)(56)(0.044) = 9.9$ kPa or 210 psf
Westergaard chart (Fig. 6.5)	$q_o = 56$ kPa $m = 0.25$ $n = 0.42$; From Fig. 6.5 and the m and n values, $I = 0.027$. Then $\Delta\sigma_v = \sigma_z = 4q_oI = (4)(56)(0.027) = 6.1$ kPa or 130 psf
Newmark chart (Fig. 6.9)	Using Fig. 6.9 with the distance AB = 12 m, draw a rectangle with width= 6/12 = 0.5 *AB* and length = 10/12 = 0.83 *AB*, with the center of the rectangle at the center of Fig. 6.9. The number of blocks within the rectangle = 34, therefore: $\Delta\sigma_v = \sigma_z = q_oIN = (56)(0.005)(34) = 9.5$ kPa or 200 psf
Summary of values	2:1 approximation — 8.5 kPa or 180 psf Concentrated load — 11 kPa or 230 psf Newmark chart (Fig. 6.4) — 9.9 kPa or 210 psf Westergaard chart — 6.1 kPa or 130 psf Newmark chart (Fig. 6.9) — 9.5 kPa or 200 psf

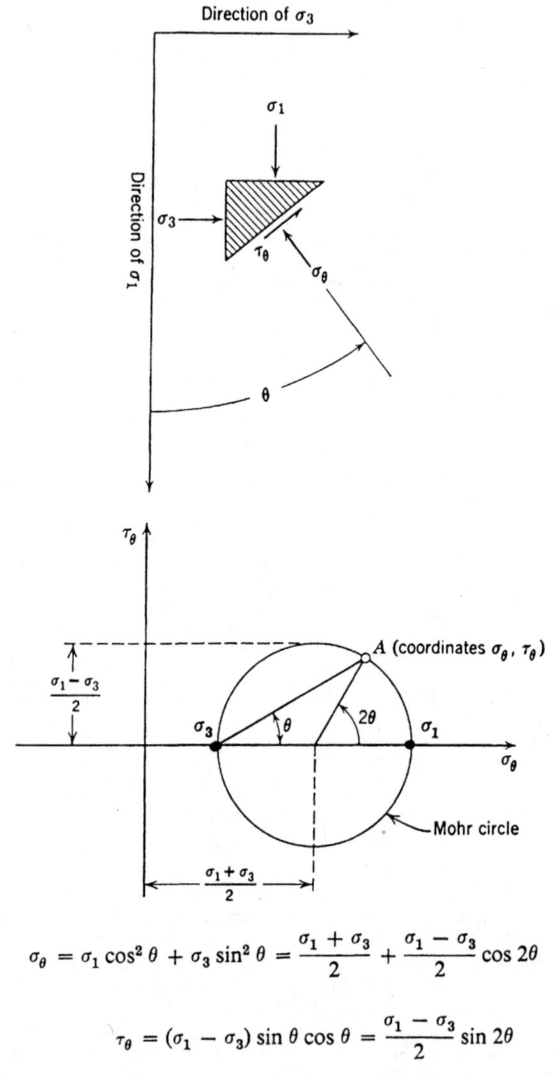

$$\sigma_\theta = \sigma_1 \cos^2 \theta + \sigma_3 \sin^2 \theta = \frac{\sigma_1 + \sigma_3}{2} + \frac{\sigma_1 - \sigma_3}{2} \cos 2\theta$$

$$\tau_\theta = (\sigma_1 - \sigma_3) \sin \theta \cos \theta = \frac{\sigma_1 - \sigma_3}{2} \sin 2\theta$$

FIGURE 6.10 Mohr circle. Upper diagram shows the definition of terms and the lower diagram shows the graphical construction of Mohr circle. The equations for the calculation of normal stress σ_θ and shear stress τ_θ on a given plane are also indicated in this figure. (*Adapted from Lambe and Whitman 1969.*)

CHAPTER 7
SHEAR STRENGTH

7.1 INTRODUCTION

An understanding of shear strength is essential in geotechnical engineering, because most geotechnical failures involve a shear-type failure of the soil. This is due to the nature of soil, which is composed of individual soil particles that slide (i.e., shear past each other) when the soil is loaded.

Section 3.3 presents a discussion of shear strength tests performed in the laboratory. As discussed in Sec. 3.3, there are many types of shear strength tests that can be performed in the laboratory. Laboratory tests are generally divided into two categories:

1. *Shear strength tests based on total stress.* The purpose of these laboratory tests is to obtain the undrained shear strength s_u of the soil or the failure envelope in terms of total stresses (total cohesion c and total friction angle ϕ). These types of shear strength tests are often referred to as "undrained" shear strength tests. Examples include the laboratory vane test, unconfined compression test, unconsolidated undrained triaxial test, and the consolidated undrained triaxial test (see Sec. 3.3). These types of laboratory tests are performed exclusively on plastic (cohesive) soils.
2. *Shear strength tests based on effective stress.* The purpose of these laboratory tests is to obtain the effective shear strength of the soil based on the failure envelope in terms of effective stress (effective cohesion c' and effective friction angle ϕ'). These types of shear strength tests are often referred to as "drained" shear strength tests. Examples include the direct shear test, consolidated drained triaxial test, consolidated undrained triaxial test with pore water pressure measurements, and the drained residual

shear strength (see Sec. 3.3). These types of laboratory tests can be performed on plastic (cohesive) soil and nonplastic (cohesionless) soil.

The mechanisms that control the shear strength of soil are complex, but in simple terms the shear strength of soils can be divided into two broad categories: cohesionless (nonplastic) soils and cohesive (plastic) soils.

7.2 COHESIONLESS SOIL

A general discussion of the shear strength of cohesionless (nonplastic) soil is provided in Table 7.1. An example of laboratory shear strength test data for a cohesionless silty sand is presented in Table 7.2 and Figs. 7.1 and 7.2. Factors that affect the shear strength of cohesionless soil and typical values of the effective friction angle (ϕ') are listed in Table 7.3.

For many projects, the effective friction angle ϕ' of a sand deposit is determined from indirect means, such as the standard penetration test (SPT) and the cone penetration test (CPT, see Sec. 2.4). As indicated in Figs. 7.3 and 7.4, the effective friction angles ϕ' for clean quartz sand can be estimated from the results of standard penetration tests or cone penetration tests for various values of effective overburden pressure σ'_v. Another useful chart is presented in Fig. 7.5, which can be used to estimate the effective friction angle ϕ' on the basis of the soil classification and the relative density (Table 3.16).

7.3 COHESIVE SOIL

The shear strength of cohesive (plastic) soil is much more complicated than the shear strength of cohesionless soils. Also, in general the shear strength of cohesive (plastic) soils tends to be lower than the shear strength of granular soils. As a result, more shear-induced failures occur in cohesive soils, such as clays, than in granular (nonplastic) soils. Cohesive soils have "fines," which are silt- and clay-size particles that give the soil a plasticity or ability to be molded and rolled. The shear strength of cohesive (plastic) soil can be divided into three broad groups: undrained shear strength, drained shear strength, and drained residual shear strength.

7.3.1 Undrained Shear Strength

A general discussion of the undrained shear strength of cohesive soil is provided in Table 7.4 and the undrained shear strength s_u for a highly plastic clay is shown in Fig. 7.6. An alternative approach to determining the undrained shear strength s_u of the plastic soil is to obtain the undrained shear strength in terms of the total shear strength parameters c and ϕ (see Tables 7.5 and 7.6 and Figs. 7.7 and 7.8).

7.3.2 Drained Shear Strength

Drained shear strength tests (such as drained direct shear tests) can be used to obtain the effective shear strength parameters c' and ϕ'. Because of the low permeability of cohesive soil, it is often more convenient to perform consolidated undrained triaxial compression tests with pore water pressure measurements in order to determine the effective shear strength parameters (see Table 7.7 and Fig. 7.9). Factors that affect the shear strength of cohesive soil and typical values of c' and ϕ' are discussed in Table 7.8 and Figs. 7.10 and 7.11.

Calculating the pore water pressure parameters and plotting the stress paths are often useful in analyzing the data from triaxial tests on cohesive soil (see Tables 7.9 to 7.12 and Figs. 7.12 to 7.15).

7.3.3 Drained Residual Shear Strength

The drained residual shear strength is discussed in Table 3.13. The residual shear strength is defined as the remaining (or residual) shear strength of plastic soil after a considerable amount of shear deformation has occurred. As previously mentioned in Table 3.13, the drained residual shear strength ϕ'_r can be very low. For example, Fig. 7.16 shows shear stress versus deformation plots and Fig. 7.17 shows the drained residual shear strength envelope for ring shear tests performed on slide plane material. As shown in Fig. 7.17, the drained residual shear strength envelope is often nonlinear, and the drained residual friction angle ϕ'_r equals 6° at high effective normal stresses.

7.3.4 Factors that Affect the Shear Strength of Cohesive Soil

Factors that affect the shear strength of cohesive soil are discussed in Table 7.13.

7.4 TOTAL STRESS AND EFFECTIVE STRESS ANALYSES

The final section of Chap. 7 presents a discussion of total stress and effective stress analyses (see Table 7.14). Example problems are presented in Table 7.15.

TABLE 7.1 Shear Strength of Cohesionless Soil

Topic (1)	Discussion (2)
General discussion	Cohesionless soils are also known as nonplastic soils. As indicated in Table 4.5, nonplastic soils include gravels, sands, and nonplastic silt such as rock flour. A cohesionless soil develops its shear strength as a result of the frictional and interlocking resistance between the individual soil particles. Cohesionless soils can be held together only by confining pressure, and will fall apart when the confining pressure is released.
Shear strength equation	The shear strength of a soil can be defined as (Mohr-Coulomb failure law): $$\tau_f = c' + \sigma'_n \tan \phi'$$ where τ_f = shear strength of the soil, c' = effective cohesion, σ'_n = effective normal stress acting on the shear surface, and ϕ' = effective friction angle. For cohesionless soil, the effective cohesion c' is zero. An exception is the testing of cohesionless soil at high normal pressures, where the shear strength envelope may actually be curved because of particle crushing (Holtz and Gibbs 1956). In this case, a straight line approximation at high normal stresses may indicate a cohesion intercept, but this value should be regarded as an extrapolated value that is not representative of the shear strength of cohesionless soils at lower values of σ'_n.
Dilation and contraction	During shear of cohesionless soils, those soils in a dense state will tend to dilate (increase in volume), while those soils in a loose state tend to contract (decrease in volume). As indicated in Fig. 3.6, cohesionless soils have a high permeability and for the shear strength testing of saturated cohesionless soils, water usually flows quickly into the soil when it dilates or out of the soil when it contracts. Thus the drained shear strength (i.e., ϕ') is of most importance for cohesionless soils. An important exception is the liquefaction of saturated, loose cohesionless soils, which are discussed in Chap. 14.
Saturated versus dry state	The shear strength of cohesionless soils can be measured in the direct shear apparatus (see Table 3.12). As discussed in Table 6.2, there can be a small capillary tension in cohesionless soils and thus the soil specimen is saturated prior to shearing (ASTM D 3080-90, 1998). Because these test specifications require the direct shear testing of soil in a saturated and drained state, the shear strength of the soil is expressed in terms of the effective friction angle ϕ'. Cohesionless soils can also be tested in a dry state, and the shear strength of the soil is then expressed in terms of the friction angle (ϕ). In a comparison of the effective friction angle ϕ' from drained direct shear tests on saturated cohesionless soil and the friction angle ϕ from direct shear tests on the same soil in a dry state, it has been determined that ϕ' is only 1 to 2° lower than ϕ (Terzaghi and Peck 1967, Holtz and Kovacs 1981). This slight difference is usually ignored and the friction angle ϕ and effective friction angle ϕ' are typically considered to mean the same thing for cohesionless (nonplastic) soils.

TABLE 7.2 Example of Laboratory Shear Strength Tests on Cohesionless Soil

Topic (1)	Discussion (2)
General discussion	Figure 7.1 presents the results of drained direct shear tests performed on undisturbed samples of silty sand. Undisturbed samples were obtained from two borings (B-1 and B-2) excavated in the silty sand deposit. A total of six direct shear tests were performed. For each direct shear test, the effective normal pressure (i.e., the vertical load divided by the specimen area; see Fig. 3.3) was based on the *in situ* vertical overburden pressure.
Shear strength	There is considerable scatter of the data in Fig. 7.1, but this is not unusual for soil which usually has variable engineering properties. The straight line drawn in Fig. 7.1 is the *shear strength envelope,* also known as the *failure envelope,* for the silty sand deposit. Considerable experience is needed in determining this envelope. For example, taking a conservative approach, the envelope was drawn by ignoring the two high peak shear strength points (B-2 at 2.5 ft and B-2 at 10 ft) and then a best-fit line was drawn through the rest of the data points. In addition, the angle of inclination of the failure envelope, which is known as the effective friction angle ϕ', is 30°, which is a typical value for silty sands. Thus the failure envelope drawn in Fig. 7.1 was based on engineering judgment and experience. Note in Fig. 7.1 that the failure envelope (straight line) passes through the origin of the plot. This means that the effective cohesion c' is zero. If the line had intersected the vertical axis, then that value would be defined as the effective cohesion value c'. From the shear strength equation listed in Table 7.1: $\tau_f = c' + \sigma'_n \tan \phi' = 0 + \sigma'_n \tan 30°$ $= 0.577\ \sigma'_n$ In essence, the laboratory direct shear test results indicate that the shear strength (maximum shearing stress the soil can sustain) is 57.7% of the vertical pressure acting on the horizontal shear surface.
Stress-deformation plots	During the shearing of the silty sand specimens in the direct shear apparatus, the shear stress (which equals the shearing force divided by the specimen area; see Fig. 3.3) versus lateral deformation was recorded and is plotted for two of the tests in Fig. 7.2. Note in Fig. 7.2 that the drained direct shear test performed on the silty sand at B-2 at 10 ft has a distinct peak in the curve, while the drained direct shear test on the silty sand at B-1 at 6 ft does not exhibit this distinctive peaking of the curve. This is due to the dry unit weight of the soil specimens, where the specimen at 10 ft is in a denser state (γ_d = 104 pcf, 16.4 kN/m^3) than the silty sand at a depth of 6 ft (γ_d = 93.4 pcf, 14.7 kN/m^3). Both the peak (highest point) and ultimate (final value) from Fig. 7.2 have been plotted in Fig. 7.1 and are designated as the *ultimate shear strength* (squares) and the *peak shear strength* (circles). In those cases where the peak and ultimate values coincide, only one point is shown.

TABLE 7.3 Factors That Affect the Shear Strength of Cohesionless Soil and Typical Values of Effective Friction Angle

Topic (1)	Discussion (2)
Factors that affect the shear strength	For cohesionless soils, $c' = 0$ and the effective friction angle ϕ' depends on: **1.** *Soil type.* As indicated below, sand and gravel mixtures have a higher effective friction angle than nonplastic silts. **2.** *Soil density.* For a given cohesionless soil, the denser the soil, the higher the effective friction angle. This is because of the interlocking of soil particles, where at a denser state the soil particles are interlocked to a higher degree and hence the effective friction angle is greater than in a loose state. It has been observed that in the ultimate shear strength state, the shear strength and density of a loose and dense sand tend to approach each other. **3.** *Grain size distribution.* A well-graded cohesionless soil will usually have a higher friction angle than a uniform soil. With more soil particles to fill in the small spaces between soil particles, there is more interlocking and frictional resistance developed for a well-graded than for a uniform cohesionless soil. **4.** *Mineral type, angularity, and particle size.* Soil particles composed of quartz tend to have a higher friction angle than soil particles composed of weak carbonate. Angular soil particles tend to have rougher surfaces and better interlocking ability. Larger-size particles, such as gravel-size particles, typically have higher friction angles than sand. **5.** *Deposit variability.* Because of variations in soil types, gradations, particle arrangements, and dry density values, the effective friction angle is rarely uniform with depth. It takes considerable judgment and experience in selecting an effective friction angle based on an analysis of data such as shown in Fig. 7.1. **6.** *Indirect methods.* For many projects, the effective friction angle of a sand deposit is determined from indirect means, such as the standard penetration test (SPT) and the cone penetration test (CPT; see Sec. 2.4). As indicated in Figs. 7.3 and 7.4, the effective friction angles ϕ' for clean quartz sand can be estimated from the results of standard penetration tests or cone penetration tests for various values of effective overburden pressure σ'_v. Another useful chart is presented in Fig. 7.5, which can be used to estimate the effective friction angle ϕ' on the basis of the dry unit weight γ_d and soil classification (ISBP, Table 4.5).
Typical effective friction angles ϕ'	Values of effective friction angles for different types of cohesionless (nonplastic) soils are presented below (data from Hough 1969). An exception to the values presented below is cohesionless soils that contain appreciable mica flakes. A micaceous sand will often have a high void ratio and hence little interlocking and a lower friction angle (Horn and Deere 1962).

TABLE 7.3 Factors That Affect the Shear Strength of Cohesionless Soil and Typical Values of Effective Friction Angle (*Continued*)

Typical effective friction angles ϕ' (*Continued*)	Soil types	Effective friction angles ϕ' at peak strength		Effective friction angle ϕ'_u at ultimate strength
		Medium dense	Dense	
	1. Silt (non-plastic)	28–32°	30–34°	26–30°
	2. Uniform fine to medium sand	30–34°	32–36°	26–30°
	3. Well-graded sand	34–40°	38–46°	30–34°
	4. Sand and gravel mixtures	36–42°	40–48°	32–36°

Note: The effective friction angle ϕ'_u at the ultimate shear strength state could be considered to be the same as the friction angle ϕ' for the same soil in a loose state.

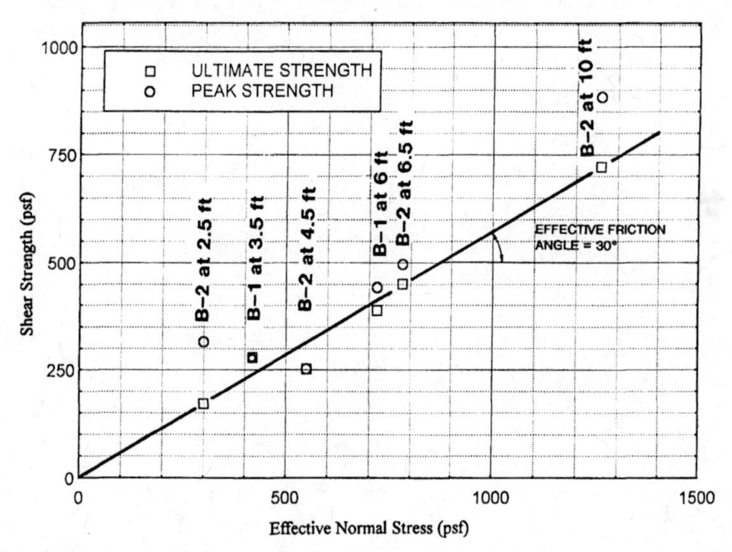

FIGURE 7.1 Shear strength versus effective normal stress for drained direct shear tests on silty sand specimens (see Table 7.2).

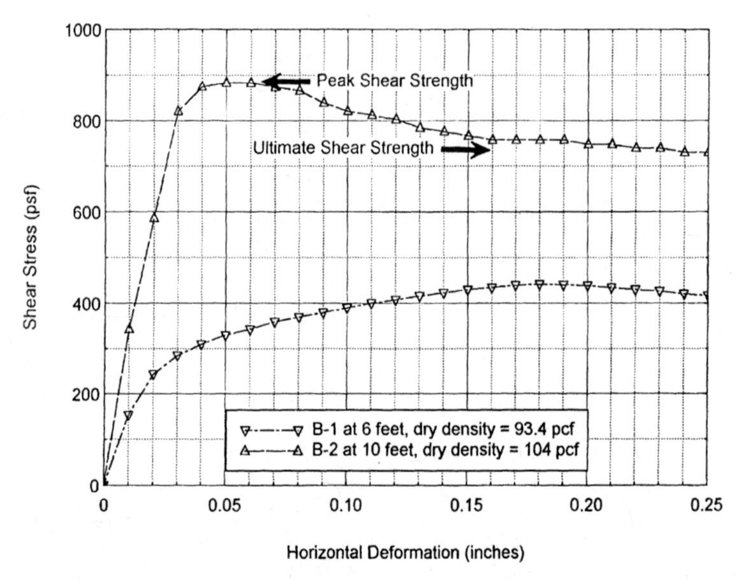

FIGURE 7.2 Shear stress versus horizontal deformation from the drained direct shear tests on silty sand specimens from a depth of 6 ft and 10 ft (see Table 7.2).

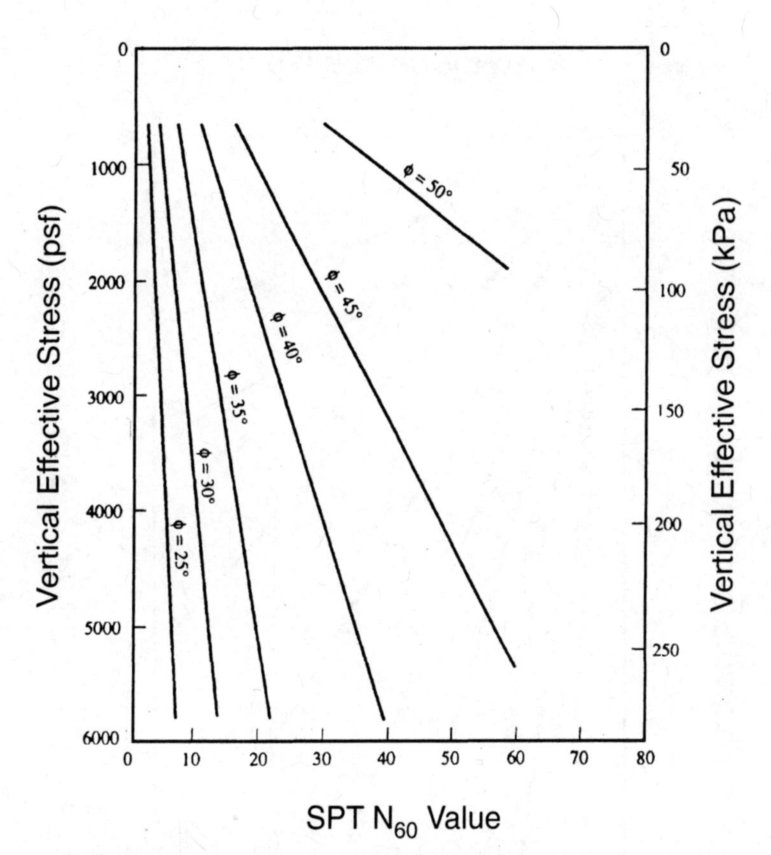

FIGURE 7.3 Empirical correlation between SPT N_{60} value (see Table 2.11), vertical effective stress, and friction angle for clean quartz sand deposits. See Table 7.3. (*Adapted from de Mello 1971, reproduced from Coduto 1994.*)

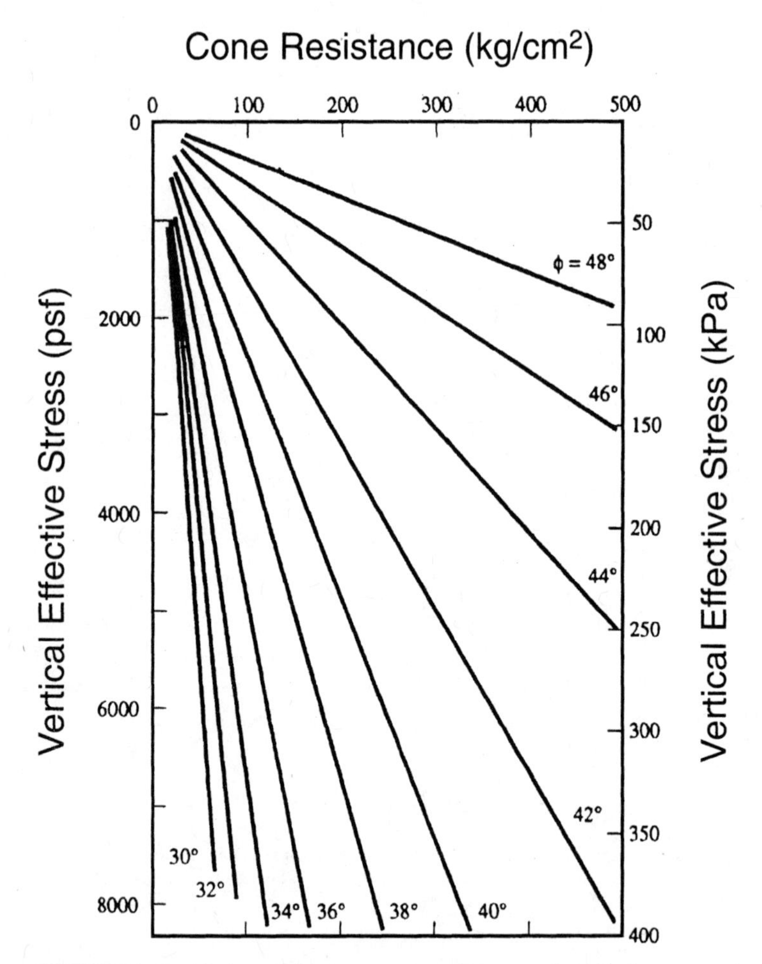

FIGURE 7.4 Empirical correlation between cone resistance, vertical effective stress, and friction angle for clean quartz sand deposits. See Table 7.3. *Note:* 1 kg/cm^2 approximately equals 1 tsf and 100 kPa. (*Adapted from Robertson and Campanella 1983, reproduced from Coduto 1994.*)

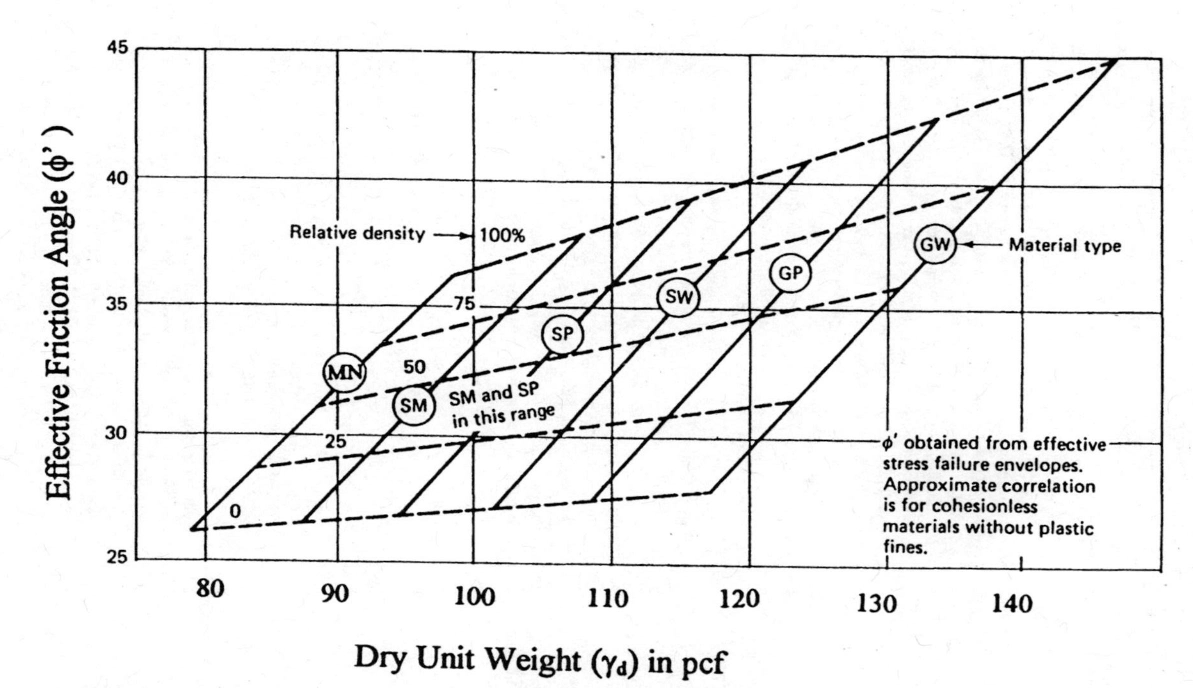

FIGURE 7.5 Approximate correlations to determine the effective friction angle ϕ' for cohesionless (nonplastic) soil. Enter the figure with the dry unit weight (γ_d, see Table 5.3), intersect the soil type (ISBP symbols, Table 4.5) to determine ϕ'. As an alternative, enter the chart with the relative density (see Table 3.16) and soil type to determine ϕ'. See Table 7.3 for further discussion. (*Adapted from NAVFAC DM-7.1 1982.*)

TABLE 7.4 Undrained Shear Strength s_u of Cohesive Soil

Topic (1)	Discussion (2)
General discussion	As the name implies, the undrained shear strength s_u refers to a shear condition where water does not enter or leave the cohesive soil during the shearing process. In essence, the water content of the soil must remain constant during the shearing process. There are many projects where the undrained shear strength is used in the design analysis. In general, these field situations must involve loading or unloading of the plastic soil at a rate that is much faster than the shear-induced pore water pressures can dissipate. During rapid loading of saturated plastic soils, the shear-induced pore water pressures can only dissipate by the flow of water into (negative shear-induced pore water pressures) or out of (positive shear-induced pore water pressures) the soil. But as indicated in Fig. 3.6, cohesive (plastic) soil has a low permeability, and if the load is applied quick enough, there will not be enough time for water to enter or leave the plastic soil.
Undrained shear strength s_u	Section 3.3 describes various laboratory tests that can be performed to determine the undrained shear strength s_u of plastic soil. For example, the undrained shear strength could be determined from the unconfined compression test. The undrained shear strength could also be determined by using the miniature lab vane or the torvane device. The undrained shear strength could even be determined in the field, by using the vane shear test (VST; see Fig. 2.12).
Normalized undrained shear strength s_u/σ_v'	The parameter s_u/σ'_v is known as the normalized undrained shear strength. It is a valuable parameter for the analysis of the undrained shear strength of saturated clay deposits. For example, note in Fig. 2.20 that the values of s_u/σ'_v (labeled *c/p* in Fig. 2.20) have been listed for the various clay deposits. The normalized shear strength s_u/σ'_v is related to the OCR of a clay as follows (Jamiolkowski et al. 1985): $$s_u/\sigma'_v = (0.23 \pm 0.04)\ \mathrm{OCR}^{0.8}$$ where OCR = the overconsolidation ratio, defined as the largest vertical effective stress ever experienced by the soil deposit (σ'_p) divided by the existing vertical effective stress σ'_v. The overconsolidation ratio and the preconsolidation pressure σ'_p are discussed in Sec. 9.4.
Example	Figure 7.6 (from Day 1980 and Ladd et al. 1980) shows the undrained shear strength s_u versus depth for borings E1 and F1 excavated in an offshore deposit of highly plastic Orinoco clay (created by sediments from the Orinoco River, Venezuela). The undrained shear strength s_u was determined in the laboratory by using the torvane device, laboratory vane, and unconfined compression test (UUC). Note in Fig. 7.6 that there is a distinct discontinuity in the undrained shear strength s_u at a depth of 60 ft (boring E1) and 40 ft (boring F1). This discontinuity was a result of different sampling

TABLE 7.4 Undrained Shear Strength s_u of Cohesive Soil (*Continued*)

Topic (1)	Discussion (2)
Example (*Continued*)	procedures. Above a depth of 60 ft at boring E1 and 40 ft at boring F1, samplers were hammered into the clay deposit, causing sample disturbance and a lower shear strength value for the upper zone of clay. For the deeper zone of clay, a WIP sampling procedure was utilized which produced less sample disturbance and hence a higher undrained shear strength. There is a considerable amount of scatter in the undrained shear strength s_u shown in Fig. 7.6. This is not unusual, for soil deposits and nonuniform deposits often have much larger scatter in data than shown in Fig. 7.6. For average values of shear strength from Fig. 7.6, the normalized undrained shear strength is $s_u/\sigma'_v = 0.20$ for boring E1 and $s_u/\sigma'_v = 0.22$ for boring F1.

TABLE 7.5 Undrained Shear Strength in Terms of Total Stress Parameters c and ϕ, Based on the Unconsolidated Undrained Triaxial Compression Test

Topic (1)	Discussion (2)
General discussion	An alternative approach to determining the undrained shear strength s_u of the plastic soil is to obtain the undrained shear strength in terms of the total shear strength parameters c and ϕ. The usual process is to perform triaxial compression tests on specimens of the plastic soil. The triaxial test has been discussed in Sec. 3.3. For the unconsolidated undrained triaxial compression test (ASTM D 2850-95, 1998), the cylindrical specimen of cohesive soil is placed in the center of the triaxial apparatus, sealed in a rubber membrane, and then subjected to a confining pressure without allowing the soil specimen to have a change in water content at any time during the triaxial testing. The soil is sheared in the triaxial apparatus (see Fig. 3.2) by increasing the vertical stress on the soil specimen.
Mohr circle	The results of a triaxial test can be graphically represented by the Mohr circle, which is illustrated in Fig. 6.10. In the triaxial compression test, the vertical stress at failure is the major principal stress σ_1 and is equal to the cell pressure plus the vertical load from the piston divided by the area of the specimen (using an area correction, see Table 3.10). The horizontal stress is the minor principal stress σ_3 and is equal to the cell pressure. With the major (σ_1) and minor (σ_3) principal stresses at failure known, a Mohr circle can be drawn, such as that shown in Fig. 6.10. The major principal stress minus the minor principal stress (i.e., $\sigma_1 - \sigma_3$) is referred to as the *deviator, deviatoric stress,* or *stress difference.* Note that the maximum shear stress τ_{max} is equal to one-half the deviator stress and equals the radius of the Mohr circle. In geotechnical engineering, usually only the upper half of the Mohr circle is used in the analysis.
Test data	Figure 7.7 shows the results of four (Tests No. 1–4) unconsolidated undrained triaxial compression tests (ASTM D 2850-95, 1998) performed on a plastic soil (LL = 64, PI = 33) having a dry unit weight (γ_d) = 11.0 kN/m^3 (70 pcf), a water content (w) = 49.3%, and a degree of saturation (S) = 95%. Plots of the deviator stress versus axial strain for these four tests are also shown in Fig. 7.7. Note that this stress-strain plot is similar to the shear stress-deformation plot shown in Fig. 7.2. The stress-strain plots shown in Fig. 7.7 do not have a peak point and are indicative of normally consolidated (OCR = 1) clays. Note that Tests No. 2 through 4 have essentially the same shear strength (peak point on the Mohr circle). The reason is because the increased cell pressure for Tests No. 3 and 4 did not densify the soil specimen, but rather only increased the pore water pressure of the clay specimens. In essence, the effective stress did not change for Tests No. 2 through 4 and thus the shear strength could not increase. This is an important concept in geotechnical engineering, that the shear strength of saturated soil cannot increase unless the effective stress of the soil first increases. In Fig. 7.7, the failure envelope has been drawn for Tests No. 2 through 4. Using the Mohr-Coulomb failure law, the total stress parameters are $\phi = 0$ and c = 2.1 psi (14.5 kPa). These types of shear strength results have been

TABLE 7.5 Undrained Shear Strength in Terms of Total Stress Parameters c and ϕ, Based on the Unconsolidated Undrained Triaxial Compression Test (*Continued*)

Topic (1)	Discussion (2)
Test data (*Continued*)	termed the "$\phi = 0$ concept." It is important to recognize that this concept does not say that the clay has zero frictional resistance (i.e., $\phi = 0$). Instead, this concept is indicating that there is no shear strength increase in the clay for the condition of a rapidly applied load. The clay must first consolidate (flow of water from the clay) before it can increase its shear strength.
Example	An example of the use of the shear strength data shown in Fig. 7.7 would be a quick loading condition. Examples are the fast construction of a building or embankment fill on top of this clay. Because the construction is quick, there would be no drainage (i.e., consolidation) of the clay during construction. To model this quick loading condition, it would be appropriate to use the undrained shear strength parameters from Fig. 7.7, i.e., $\phi = 0$ and $c = 2.1$ psi (14.5 kPa).

TABLE 7.6 Undrained Shear Strength in Terms of Total Stress Parameters c and ϕ, Based on the Consolidated Undrained Triaxial Compression Test

Topic (1)	Discussion (2)
General discussion	In some cases, it may be appropriate to obtain the total stress shear strength parameters c and ϕ from consolidated undrained triaxial compression tests (ASTM D 4767-95, 1998). The procedure consists of placing a cylindrical specimen of plastic (cohesive) soil in the center of the triaxial apparatus (Fig. 3.2), sealing the specimen in a rubber membrane, applying a confining pressure, and then allowing enough time for the soil specimen to consolidate. At the completion of consolidation, an axial load is applied to the soil specimen without allowing a change in water content (i.e., undrained loading).
Shear strength envelope	For the triaxial compression tests, the minor principal stress (σ_3) = the confining pressure and the major principal stress (σ_1) = the vertical pressure at failure. With σ_3 and σ_1 known, the Mohr circle (Fig. 6.10) at failure can be drawn. By performing a series of triaxial tests on soil specimens, a failure envelope (straight line) can be drawn tangent to the Mohr circles and the cohesion (c = y axis intercept) and friction angle (ϕ = angle of inclination of the straight line) can be determined. The shear strength τ_f from the triaxial compression tests can be expressed as follows (Mohr-Coulomb failure law): $$\tau_f = c + \sigma_n \tan \phi$$ where σ_n = total normal stress acting on the shear surface. Note that this equation is identical to the equation listed in Table 7.1, except that one equation is expressed in terms of effective stress and the other is expressed in terms of total stress.
Test data	Figure 7.8 presents data from three triaxial tests (Nos. 1–3) performed in accordance with ASTM D 4767-95 (1998) test specifications (Day and Marsh 1995). The three specimens used for the triaxial tests were composed of silty clay (LL = 54, PI = 28, classified as a clay of intermediate plasticity, CI). In order to create the three triaxial specimens, the silty clay was compacted into a cylindrical mold to a dry unit weight of 13.4 kN/m^3 (85 pcf, relative compaction = 80%) at a water content of 20%. The test procedure first consisted of saturating and consolidating the three soil specimens in the triaxial apparatus (for Tests No. 1–3, consolidation pressure = 3.5 psi, 13.9 psi, and 27.8 psi respectively). After consolidation, the clay specimens were sheared in the triaxial apparatus (see Fig. 3.2) by increasing the vertical stress and not allowing drainage during the loading process. By knowing the major (σ_1) and minor (σ_3) principal stresses at failure, a Mohr circle can be drawn such as shown in Fig. 7.8. The failure envelope (straight line) was drawn tangent to the Mohr circles, and the cohesion (c = y axis intercept = 1.2 psi) and friction angle (ϕ = angle of inclination of the straight line = 16°) were determined. From the shear strength equation listed above:

TABLE 7.6 Undrained Shear Strength in Terms of Total Stress Parameters c and ϕ, Based on the Consolidated Undrained Triaxial Compression Test (*Continued*)

Topic (1)	Discussion (2)
Test data (*Continued*)	$\tau_f = c + \sigma_n \tan \phi = 1.2 + \sigma_n \tan 16° = 1.2 + 0.287\ \sigma_n$ In essence, the laboratory test results indicate that the shear strength (maximum shear stress on the failure plane) is 1.2 psi plus 28.7% of the total pressure acting perpendicular to the failure plane.
Example	An example would be the construction of a structure (such as an oil tank or grain elevator) where sufficient time elapses so that the saturated plastic soil consolidates under this load. If the oil tank or grain elevator is then quickly filled, the saturated plastic soil would be subjected to an undrained loading. The consolidation followed by the quick loading can be modeled by performing consolidated undrained triaxial compression tests (ASTM D 4767-95, 1998) in order to determine the total stress parameters c and ϕ.

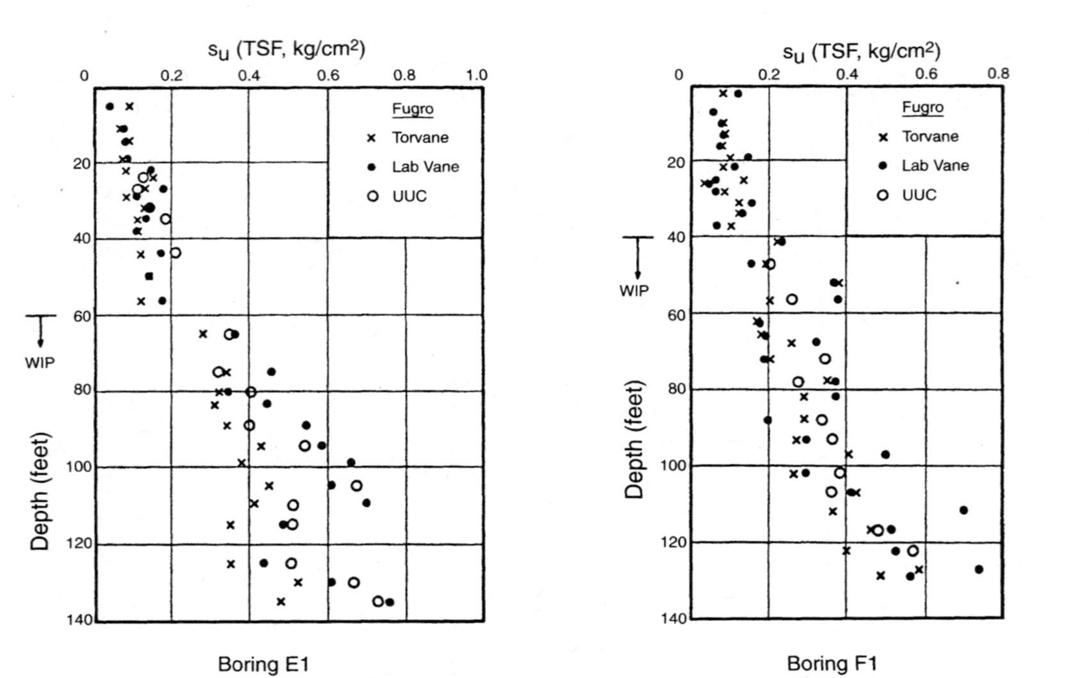

FIGURE 7.6 Undrained shear strength versus depth for Orinoco clay at Borings E1 and F1. (*From Day 1980 and Ladd et al. 1980.*)

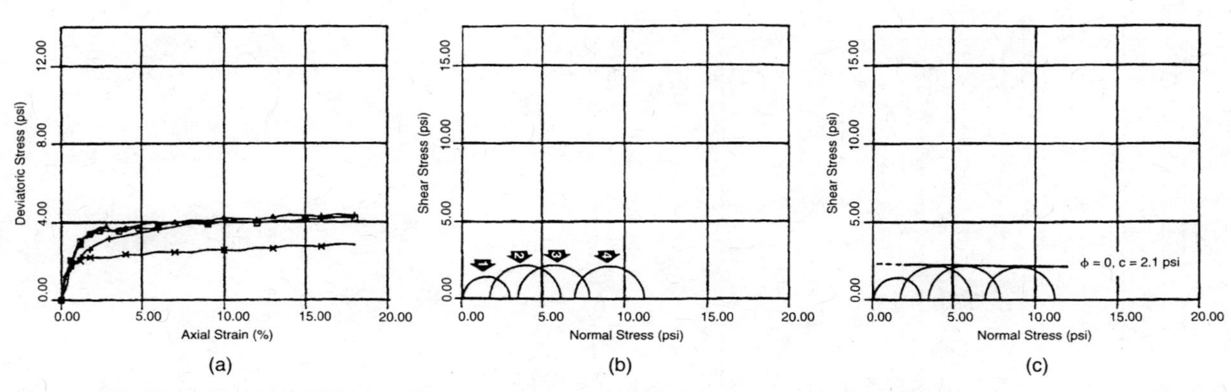

FIGURE 7.7 Unconsolidated undrained triaxial compression test data performed on a clay. (*a*) Stress-strain curves for the four triaxial tests; (*b*) the Mohr circles for the four triaxial tests; (*c*) the failure envelope for the clay. See Table 7.5.

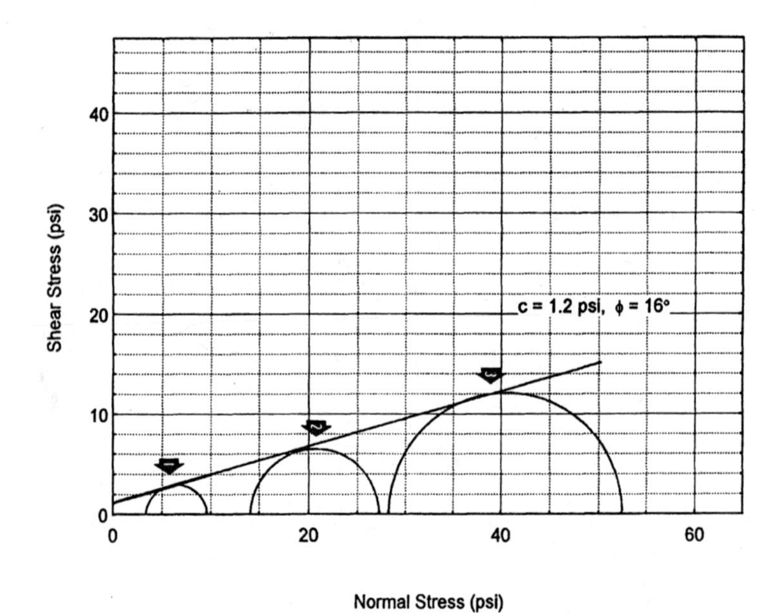

FIGURE 7.8 Consolidated undrained triaxial compression tests performed on a silty clay. See Table 7.6.

TABLE 7.7 Consolidated Undrained Triaxial Compression Test with Pore Water Pressure Measurements

Topic (1)	Discussion (2)
General discussion	The effective friction angle ϕ' and effective cohesion c' of a plastic (cohesive) soil could be obtained by performing drained shear strength tests (such as drained direct shear tests). But, as indicated in Fig. 3.6, plastic soils usually have a very low permeability and it often takes a considerable amount of time to perform drained shear strength tests. An alternative procedure that is routinely used in practice is to perform triaxial compression tests with pore pressure measurements. As discussed in Table 7.6, consolidated undrained triaxial compression tests can be used to obtain the total stress shear strength parameters (ASTM D 4767-95, 1998). If the pore water pressures are measured during the shearing of the saturated cohesive soil, then the effective stress shear strength parameters c' and ϕ' could also be determined (also ASTM D 4767-95, 1998). This type of triaxial test is very practical because it can be used to determine the undrained shear strength in terms of the total stress parameters c and ϕ as well as the shear strength in terms of the effective stress parameters c' and ϕ' when the pore water pressures are measured during the shearing of the saturated cohesive soil specimen.
Mohr circles	In order to plot the Mohr circles at failure in terms of effective stress, the following equations (which are identical to the effective stress equation in Table 6.1) are used: $$\sigma'_1 = \sigma_1 - u \qquad \sigma'_3 = \sigma_3 - u$$ where σ'_1 and σ'_3 are the major and minor principal effective stresses at failure. Failure is often defined as the largest vertical load (maximum σ_1) that the cohesive specimen can sustain. Since σ_3 is constant during the triaxial compression test, the maximum σ_1 will occur at exactly the same time as the "maximum deviator stress," also known as "maximum stress difference" or maximum value of $\sigma_1 - \sigma_3$. In dealing with effective stresses, failure could also be defined as the maximum deviator stress (maximum value of $\sigma'_1 - \sigma'_3$) or as the maximum obliquity (maximum value of σ'_1/σ'_3).
Example	Figure 7.9 presents the Mohr circles at failure in terms of effective stress for the three triaxial tests on compacted silty clay (i.e., same triaxial tests used to produce Fig. 7.8). During the undrained shearing of the three saturated silty clay specimens, the pore water pressures u were measured and the above equations were used to calculate σ'_1 and σ'_3. In order to determine the effective shear strength envelope, a straight line was drawn tangent to the Mohr circles in Fig. 7.9. The effective shear strength envelope can be defined by $c' = 0.5$ psi (3.4 kPa) and $\phi' = 29°$.

TABLE 7.7 Consolidated Undrained Triaxial Compression Test with Pore Water Pressure Measurements (*Continued*)

Topic (1)	Discussion (2)
Example (*Continued*)	By using the triaxial test with pore water pressure measurements (ASTM D 4767-95, 1998), both the shear strength parameters in terms of total stress (Fig. 7.8) and the shear strength parameters in terms of effective stress (Fig. 7.9) were obtained for the compacted silty clay. For this plastic soil, the shear strength parameters are as follows: Total stress analysis (Fig. 7.8): $c = 1.2$ psi (8.3 kPa) and $\phi = 16°$ Effective stress analysis (Fig. 7.9): $c' = 0.5$ psi (3.4 kPa) and $\phi' = 29°$ In Figs. 7.8 and 7.9, note that the size of the Mohr circles for each test number are identical (same radius). In comparing Fig. 7.8 with 7.9, there has been a shifting to the left of the Mohr circles. The amount of shifting of the Mohr circles is equal to the pore water pressure developed during the triaxial shearing of the three specimens. This can be deduced from the above equations, which indicates that both σ'_1 and σ'_3 (i.e., the x-axis coordinates of the Mohr circle) are adjusted by the pore water pressure u.

TABLE 7.8 Factors That Affect the Shear Strength of Cohesive Soil and Typical Values of c' and ϕ'

Topic (1)	Discussion (2)
Factors that affect the shear strength	For cohesive soil, the values of the effective cohesion c' and effective friction angle ϕ' depend on: **1.** *Soil type.* Clays will usually have lower values of effective shear strength parameters c' and ϕ' than silts. The more plastic the soil (higher the PI), the lower the values of c' and ϕ'. For example, Fig. 7.10 presents an empirical correlation between the effective friction angle ϕ' and the plasticity index of the soil. **2.** *Particle bonding.* Naturally cemented clays, where there is an actual bonding between clay particles, can have relatively high values of c' and ϕ'. An example of such soils is "quick clays," which are relatively strong in an undisturbed state, but if they are sheared and remolded, the structure of the soil breaks down and they can be transformed into a liquid. **3.** *Stress history.* The overconsolidation ratio (OCR) is defined as the maximum past pressure (also known as preconsolidation pressure) divided by the existing vertical effective stress. An overconsolidation ratio of 1.0 indicates that the soil is normally consolidated while an OCR greater than 1.0 indicates that the cohesive soil is overconsolidated (i.e., densified). For a given soil, the higher the overconsolidation ratio, the greater the values of c' and ϕ'. **4.** *Peak and ultimate shear strength.* Similar to dense cohesionless soils, cohesive soil that is overconsolidated often has a peak shear strength that is greater than the ultimate shear strength. **5.** *Other factors.* Other factors that affect shear strength are sample disturbance, strain rate, and anisotropy. Table 7.13 presents a discussion of these factors.
Typical values of c' and ϕ'	For undisturbed natural clays, the values of ϕ' range from around 20° for normally consolidated highly plastic clays, up to 30° or more for other types of plastic (cohesive) soil (Holtz and Kovacs 1981). The value of ϕ' for compacted clay is typically in the range of 25° to 30° and occasionally as high as 35°. In terms of effective cohesion for plastic soil, Holtz and Kovacs (1981) indicate that the value of c' for normally consolidated noncemented clays is very small and can be assumed to be zero for practical work.
Curved failure envelope	If the plastic soil is overconsolidated (from such a cause as compaction, higher load in the past, or desiccation), then there may be an effective cohesion intercept (c'), as shown in Fig. 7.11. Often the higher the overconsolidation ratio (i.e., the higher the preloaded condition of the plastic soil), the higher the effective cohesion intercept. It should be mentioned that the failure envelope at

TABLE 7.8 Factors That Affect the Shear Strength of Cohesive Soil and Typical Values of c' and ϕ' (*Continued*)

Topic (1)	Discussion (2)
Curved failure envelope (*Continued*)	low effective stresses is often curved and passes through the origin, and the effective cohesion c' is usually an extrapolated value that does not represent the actual effective shear strength at low effective stresses (see Fig. 7.11). This is very important in engineering analyses where the effective confining pressure is very low, as analyses of surficial stability (Chap. 13).

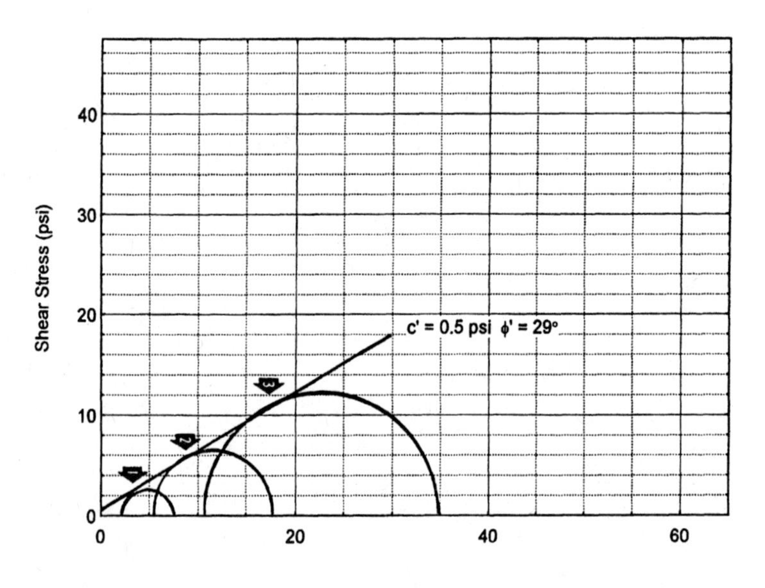

FIGURE 7.9 Mohr circles in terms of effective stress (same triaxial tests used to produce Fig. 7.8). See Table 7.7.

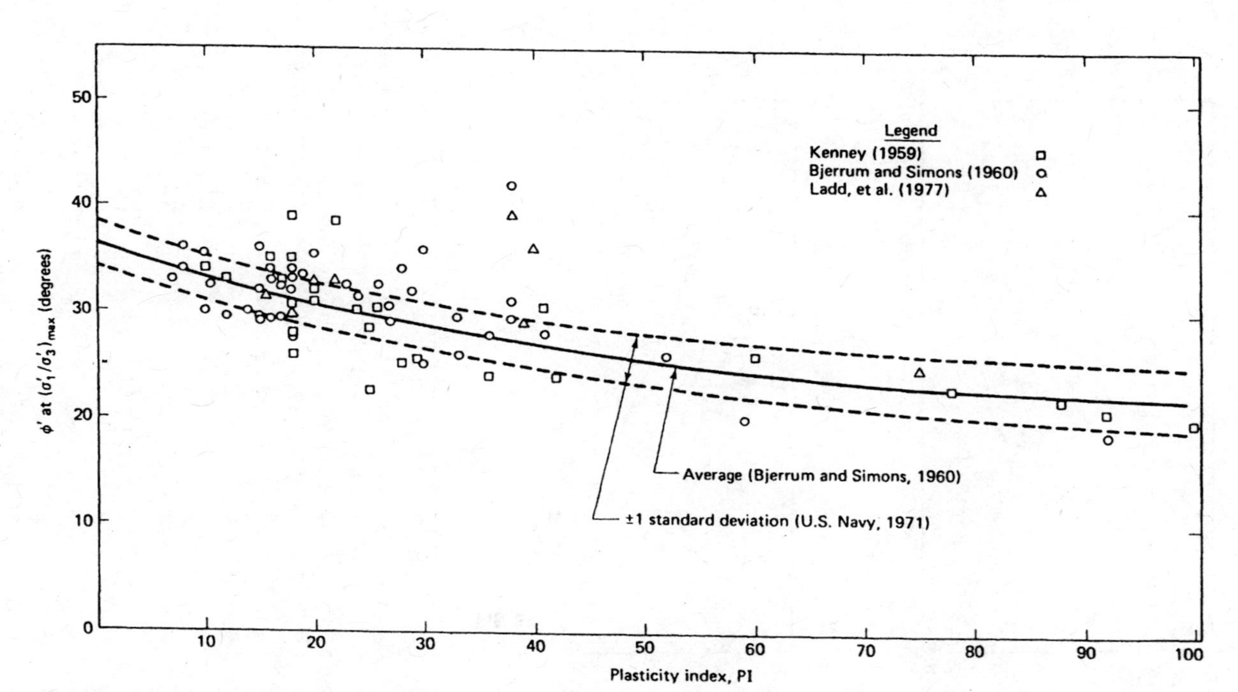

FIGURE 7.10 Empirical correlation between ϕ' and PI from triaxial compression tests on normally consolidated undisturbed clays. (*From NAVFAC DM-7, 1971 and Ladd et al., 1977. Reproduced from Holtz and Kovacs, 1981.*)

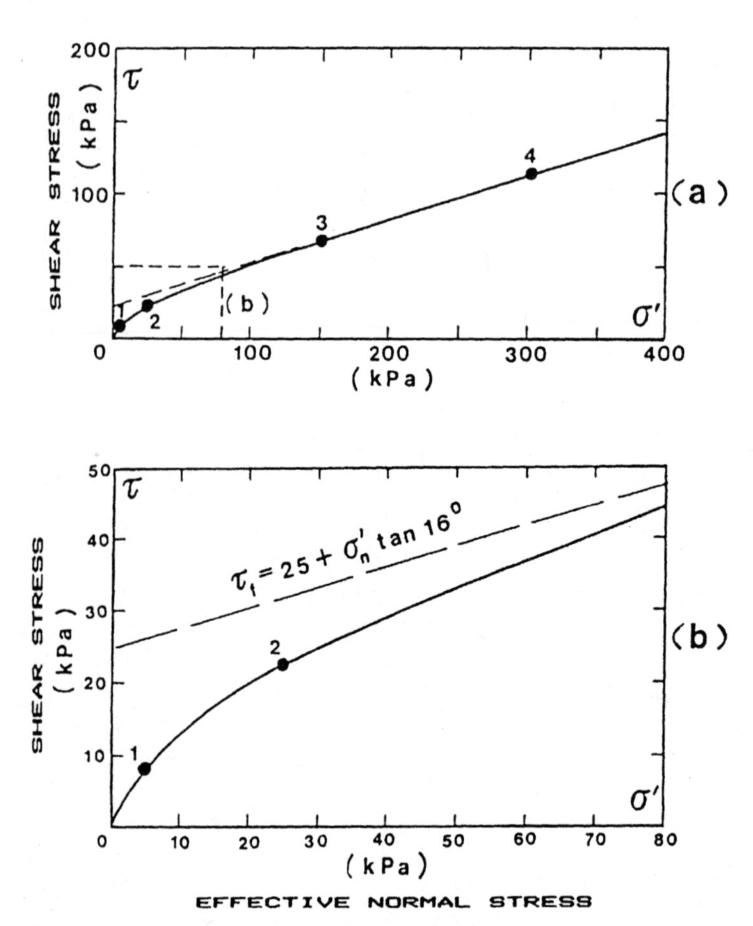

FIGURE 7.11 Failure envelope for compacted London clay. (*a*) Investigated stress range with detail; (*b*) low-stress range. (*From Maksimovic 1989b; reprinted with permission from the American Society of Civil Engineers.*)

TABLE 7.9 Skempton Pore Water Pressure Parameters

Topic (1)	Discussion (2)
General discussion	The Skempton (1954) pore water pressure parameters relate the change in total stresses for an undrained loading. For an undrained triaxial test, the general expression is as follows: $$\Delta u = B[\Delta\sigma_3 + A(\Delta\sigma_1 - \Delta\sigma_3)]$$ where Δu = change in pore water pressure generated during undrained shear, kPa or psf $\Delta\sigma_1$ = change in total major principal stress during undrained shear, kPa or psf $\Delta\sigma_3$ = change in total minor principal stress during undrained shear, kPa or psf A, B = Skempton pore water pressure coefficients (dimensionless)
B-value	In performing a consolidated undrained triaxial compression test on a saturated plastic soil (ASTM D 4767-95, 1998), a high backpressure is used to be sure of saturation of the triaxial specimen. After the cohesive soil has consolidated and prior to triaxial shearing, the B-value should be calculated. The process consists of preventing drainage of the plastic soil and then applying a confining pressure ($\Delta\sigma_c$) to the specimen by increasing the cell pressure. In this case, $\Delta\sigma_c = \Delta\sigma_1 = \Delta\sigma_3$ and the above equation reduces to: $$B = \Delta u/\Delta\sigma_3$$ From the increase in confining pressure ($\Delta\sigma_c = \Delta\sigma_3$), and by measuring the increase in pore water pressure (Δu) during application of the confining pressure, the B-value can be calculated. Table 7.10 lists values of measured B-values for various types of clay.
A-value	While the B-value is calculated before the triaxial shearing part of the test is performed, the A-value is calculated during the actual shearing of the cohesive specimen. For the conventional triaxial compression test in which the confining pressure is held constant throughout shearing, $\Delta\sigma_3 = 0$, and if the degree of saturation is 100% ($B = 1.0$), then the above equation reduces to the following: $$A = \Delta u/\Delta\sigma_1$$ On the basis of this equation and triaxial compression tests, Skempton and Bjerrum (1957) provided typical A-values at failure for various plastic soils, as follows: • Very sensitive soft clay: A-value at failure is greater than 1 • Normally consolidated clay: A-value at failure is 0.5 to 1 • Overconsolidated clay: A-value at failure is 0.25 to 0.5 • Heavily overconsolidated clay: A-value at failure is 0 to 0.25

TABLE 7.9 Skempton Pore Water Pressure Parameters (*Continued*)

Topic (1)	Discussion (2)
A-value (*Continued*)	The reason for high *A*-values for very soft, sensitive clay and normally consolidated clay is the contraction of the soil structure during shear, and because it is an undrained loading, positive pore water pressures will develop. The reason for zero or even negative *A*-values for highly overconsolidated soil is the dilation of the soil structure during shear, and because it is an undrained loading, negative pore water pressures will develop. The *A*-value can indicate a normally consolidated versus a heavily overconsolidated soil. For soil compacted at optimum moisture content, the *A*-value at failure from a consolidated undrained triaxial compression test (ASTM D 4767-95, 1998) can indicate the type of soil behavior. For example, Fig. 7.12 shows a plot of the *A*-value at failure versus the percentage of swell or collapse measured in an oedometer apparatus (Day and Marsh 1995). Table 7.11 was developed from the data in Fig. 7.12 to indicate the type of soil behavior in terms of swell or collapse versus the triaxial *A*-value. As indicated in Fig. 7.12, soil that collapsed in the oedometer apparatus had a high *A*-value from the triaxial test, which is indicative of contractive behavior, and soil that tended to swell conversely had a low to negative *A*-value from the triaxial test, which is indicative of dilative behavior.

TABLE 7.10 *B*-Values for Various Clays

Types of soil (1)	Degree of saturation, % (2)	*B*-value (3)
London clay (OC)	100	0.9981
London clay (NC)	100	0.9998
Vicksburg clay	100	0.9990
Kawasaki clay	100	0.999
Boulder clay	93	0.69
Boulder clay	87	0.33
Boulder clay	76	0.10

Notes: OC = overconsolidated clay, NC = normally consolidated clay. Note that if the clay is saturated (S = 100%), the *B*-value is essentially equal to 1.0. This data shows that a stress applied to a saturated cohesive soil will be carried by an increase in pore water pressure and not the soil particle skeleton (i.e., $\Delta u = \Delta\sigma_c$). The measurements of the *B*-value on saturated clays are further confirmation that the shear strength of a saturated soil cannot increase unless the effective stress of the soil first increases. If the soil is not saturated, the *B*-value is less than 1.0. The lower the degree of saturation, the lower the *B*-value. Thus the lower the degree of saturation, the greater the portion of the load that will be immediately carried by the soil particle skeleton. In the case of a completely dry soil (S = 0%), the *B*-value is zero and all of the load will be immediately carried by the soil particle skeleton.

Sources: Skempton (1954, 1961) and Lambe and Whitman (1969).

TABLE 7.11 Soil Behavior versus *A*-value

	Triaxial *A*-value at failure*			
Soil type (1)	*A*-value less than 0 (2)	*A*-value = 0 to 0.4 (3)	*A*-value = 0.4 to 0.6 (4)	*A*-value greater than 0.6 (5)
Compacted granular soil	No collapse	No appreciable collapse	Transition: possible collapse	Significant collapse
Compacted clay	Very high swell	Significant swell	Transition: some swell or collapse	Significant collapse

*Shear failure defined as maximum obliquity (σ'_1 / σ'_3).

Note: This table is valid for soil compacted at optimum moisture content (optimum moisture content determined per ASTM D 1557-91, 1998). *A*-values at failure were determined from consolidated undrained triaxial compression tests (ASTM D 4767-95, 1998).

TABLE 7.12 Stress Paths

Topic (1)	Discussion (2)
General discussion	A stress path is defined as a series of stress points that are connected together to form a line or curve. Figure 7.13 illustrates the stress path for a triaxial compression test. The state of stress for the triaxial test starts at point *A* where the cohesive specimen is subjected to isotropic compression (i.e., $\sigma_c = \sigma_1 = \sigma_3$). Then, as the specimen is axially loaded, σ_1 increases while σ_3 remains constant. This results in a series of ever larger Mohr circles, such as shown in Fig. 7.13. Mohr circle *E* represents stress conditions at failure.
Stress path	Instead of drawing a series of Mohr circles to represent the change in stress during the triaxial shear test, a stress path can be drawn as shown in Fig. 7.14. The peak point of each Mohr circle (i.e., points *A, B, C, D,* and *E*) are used to draw the stress path on a *p-q* plot. The peak point of each Mohr circle is defined as: In terms of total stresses: $p = \frac{\sigma_1 + \sigma_3}{2}$ and $q = \frac{\sigma_1 - \sigma_3}{2}$ In terms of effective stresses: $p' = \frac{\sigma'_1 + \sigma'_3}{2}$ and $q = \frac{\sigma'_1 - \sigma'_3}{2}$ Note that p represents the center of the Mohr circle and q is the radius of the Mohr circle. For a given stress state defined by σ_1, σ_3, and u, the value of q is the same in terms of total or effective stresses. As previously discussed for Figs. 7.8 and 7.9, the Mohr circles for a given total and effective stress state are only shifted and the radius (i.e., value of q) of the Mohr circle does not change.
Shear strength parameters	The shear strength parameters from the conventional plot (Figs. 7.8 and 7.9) are related to the shear strength parameters from the *p-q* plot by the following relationships: In terms of total stresses: $\sin \phi = \tan \alpha$ and $c \cos \phi = a$ In terms of effective stresses: $\sin \phi' = \tan \alpha'$ and $c' \cos \phi' = a'$ Total stress paths, such as shown in Fig. 7.14, or effective stress paths can be drawn for the consolidated undrained triaxial compression test (ASTM D 4767-95, 1998). In drawing effective stress paths, the effective principal stresses are first determined during the actual shearing process of the specimen. Then the above equations are used to calculate the *p-q* points that are connected together to form the effective stress paths.
Summary	Figure 7.15 presents a summary of the triaxial testing of cohesive soil.

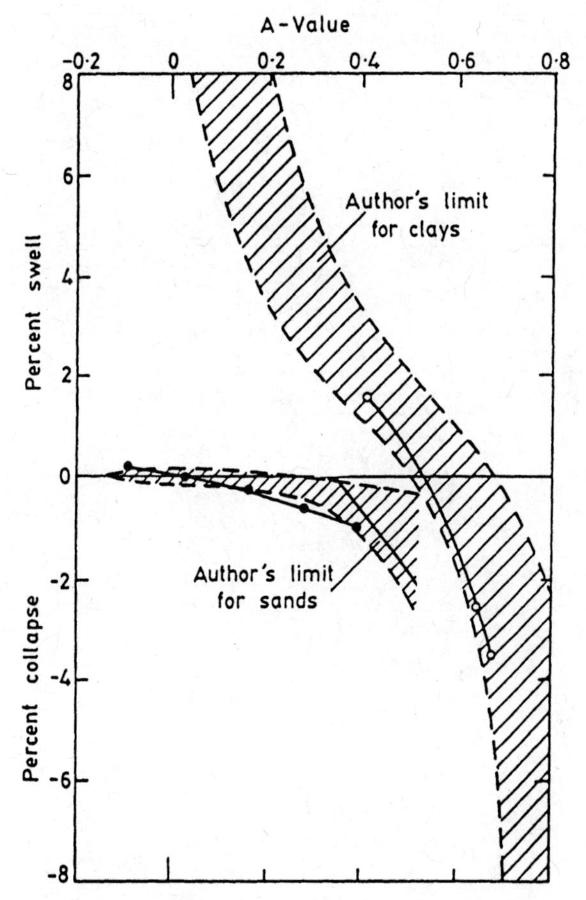

FIGURE 7.12 Percentage swell or collapse versus triaxial A-value at failure. See Tables 7.9 and 7.11.

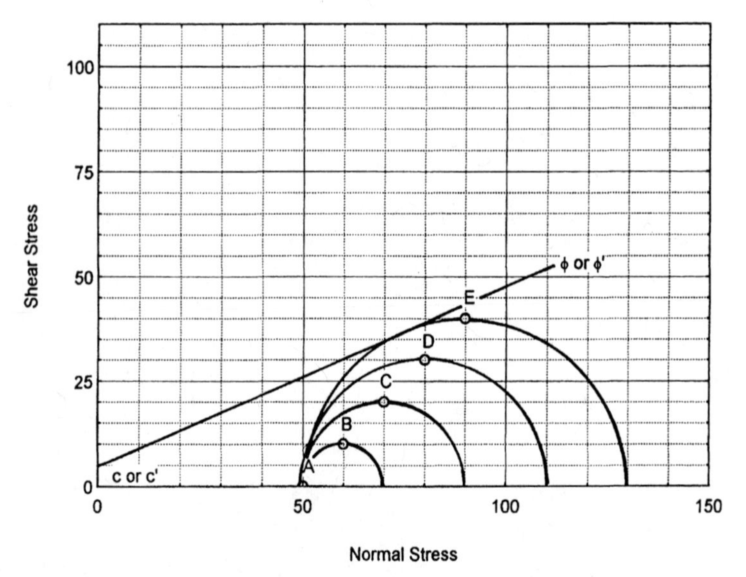

FIGURE 7.13 Series of Mohr circles representing the state of stress during a triaxial compression test (Mohr circle *E* represents the Mohr circle at failure). Figure 7.14 shows the stress path for these Mohr circles. See Table 7.12.

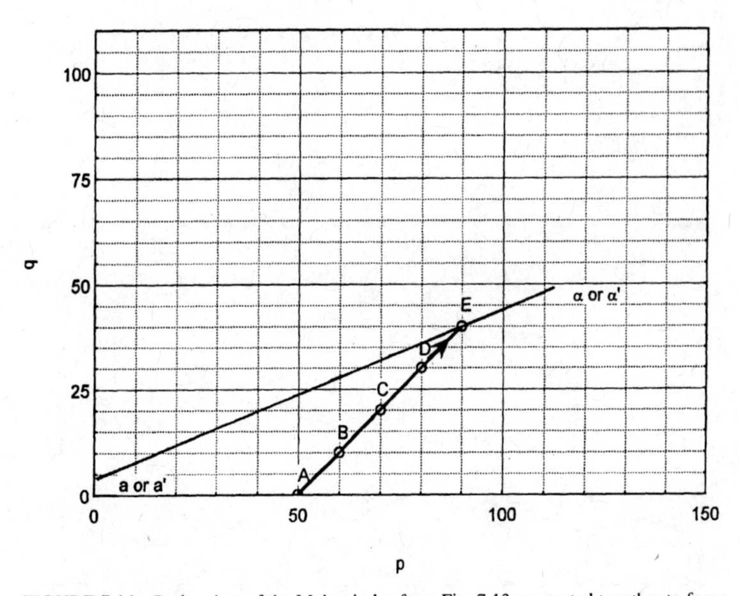

FIGURE 7.14 Peak points of the Mohr circles from Fig. 7.13 connected together to form a stress path. See Table 7.12.

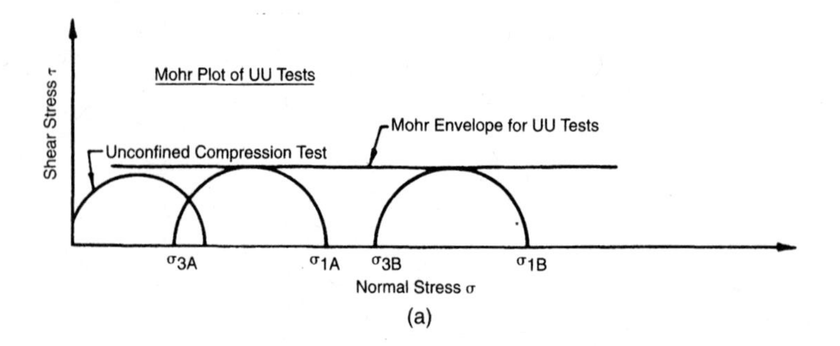

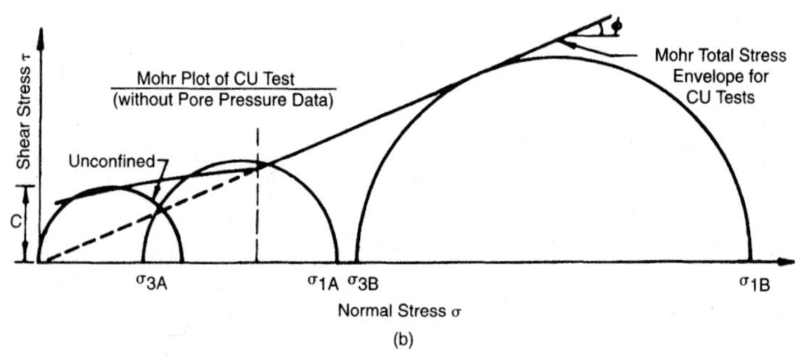

FIGURE 7.15 Summary of triaxial testing for cohesive soil. (*a*) Data from unconsolidated undrained triaxial compression test (see Table 7.5); (*b*) consolidated undrained triaxial compression test (see Table 7.6); (*c*) consolidated undrained triaxial compression test with pore water pressure measurements (see Table 7.7); (*d*) stress paths (see Table 7.12). (*Reproduced from NAVFAC DM-7.2 1982.*)

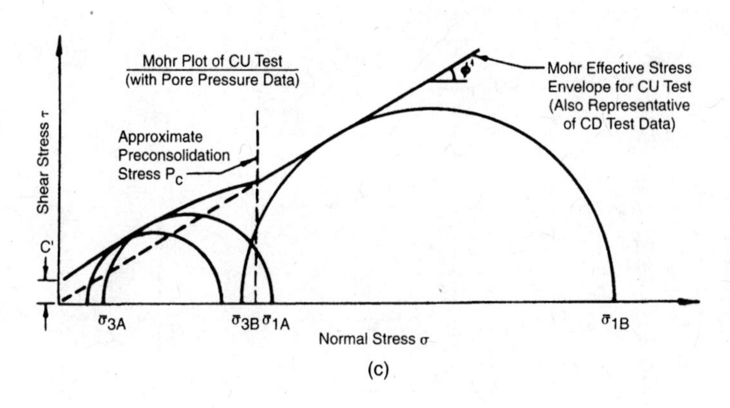

(c)

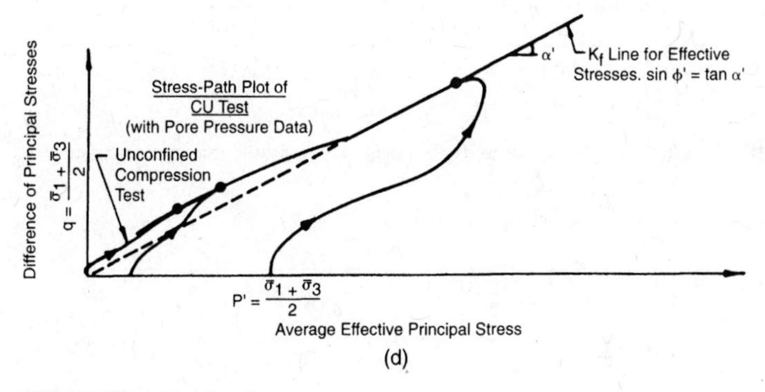

(d)

FIGURE 7.15 (*Continued*)

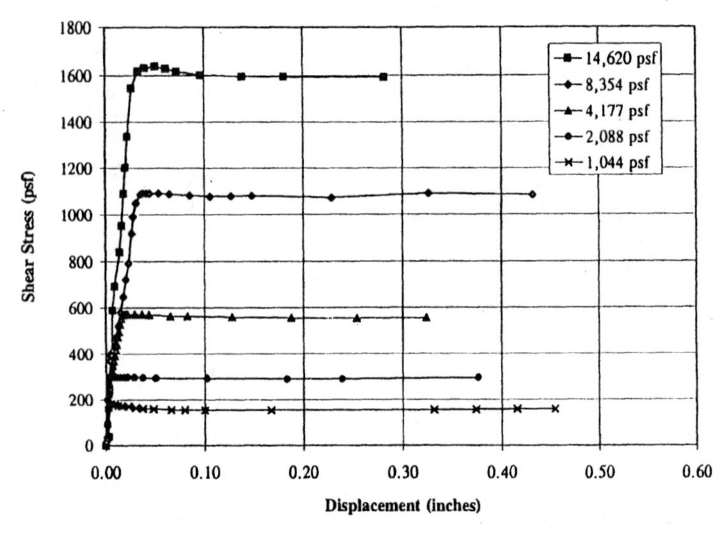

FIGURE 7.16 Shear stress versus displacement from ring shear test on slide plane material. (See Table 3.13.)

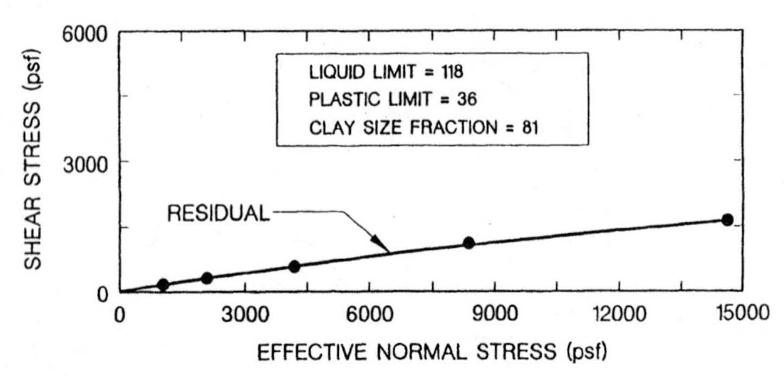

FIGURE 7.17 Drained residual shear strength envelope from ring shear test on slide plane material. (See Table 3.13.)

TABLE 7.13 Factors That Affect the Shear Strength of Cohesive Soil

Topic (1)	Discussion (2)
Sample disturbance	As discussed in Sec. 2.3, a soil sample cannot be taken from the ground and be in a perfectly undisturbed state. Disturbance causes a decrease in effective stress, a reduction in the interparticle bonds, and a rearrangement of the soil particles. Sample disturbance can cause the greatest reduction in shear strength than any other factor. As an example of the effects of sample disturbance, Fig. 7.18 shows the undrained shear strength s_u of "undisturbed" and remolded Orinoco clay and indicates about a 75% reduction in the undrained shear strength (Ladd et al. 1980, Day 1980). For clays having a high sensitivity, such as quick clays, disturbance can severely affect the shear strength. In instances of severe disturbance, quick clays can be disturbed to such an extent that they even become liquid (essentially no shear strength).
Strain rate	The faster a soil specimen is sheared, i.e., the faster the strain rate, the higher the value of the undrained shear strength s_u. For torvane, lab vane, and unconfined compression tests, the strain rate is very fast with failure occurring in only a few minutes or less. In Fig. 7.18, torvane tests were performed on the Orinoco clay with the tests having different times to failure. Note in this figure that for both the undisturbed specimens and the remolded specimens, a slower strain rate results in a lower undrained shear strength. This is because a slower strain rate allows the soil particles more time to slide, deform, and creep around each other, resulting in a lower undrained shear strength.
Anisotropy	Soils, especially clays, have a natural strength variation, where s_u depends on the orientation of the failure plane; thus s_u along a horizontal failure plane will not equal s_u along a vertical failure plane. For example, note the very fine layering of the Orinoco clay that is visible from the radiograph in Fig. 2.8. Lab vane and torvane tests have simultaneous horizontal and vertical failure planes with an undrained shear strength that rarely equals s_u from an unconfined compression test, which has an oblique failure plane.
Field vane tests	The three factors listed above can affect soil in different ways. For example, it has been stated that for the vane shear test, the values of the undrained shear strength from field vane tests are likely to be higher than can be mobilized in practice (Bjerrum 1972, 1973). This has been attributed to a combination of anisotropy of the soil and the fast rate of shearing involved with the vane shear test. Because of these factors, Bjerrum (1972) proposed that, for field vane tests performed on saturated normally consolidated clays, the undrained shear strength s_u be reduced according to the plasticity index of the clay. Figure 7.19 shows Bjerrum's (1972) recommendation, where the *in situ* undrained shear strength s_u is equal to the shear strength determined from the field vane test times the correction factor determined from Fig. 7.19.
Unconfined compression test	Like the vane shear test, the unconfined compression test can provide values of undrained shear strength that are too high because the saturated plastic soil is sheared too quickly. In many cases, the unconfined compression test

TABLE 7.13 Factors That Affect the Shear Strength of Cohesive Soil (*Continued*)

Topic (1)	Discussion (2)
Unconfined compression test (*Continued*)	has provided reasonable values of undrained shear strength s_u because of a compensation of factors: too high an undrained shear strength, due to the fast strain rate, that is compensated by a reduction in undrained shear strength, due to sample disturbance (Ladd 1971).
Experience and judgment	Because of all of these factors (sample disturbance, strain rate, and anisotropy), considerable experience and judgment are needed for the selection of shear strength parameters for cohesive soil.

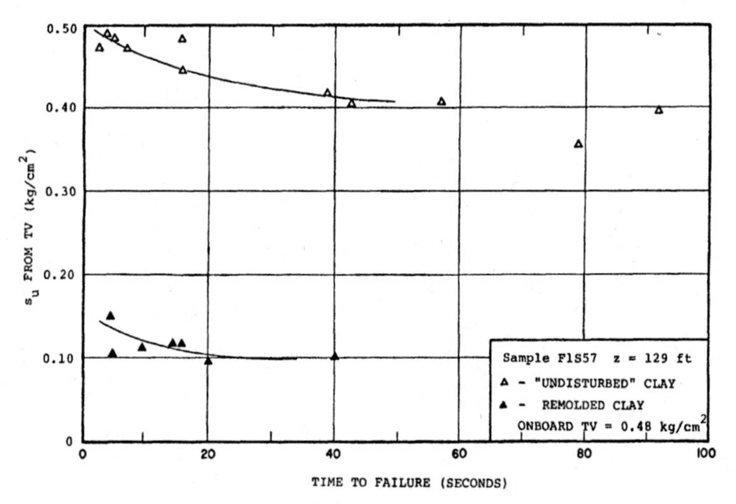

FIGURE 7.18 Effects of sample disturbance and strain rate on the undrained shear strength of Orinoco clay. See Table 7.13. (*From Day 1980.*)

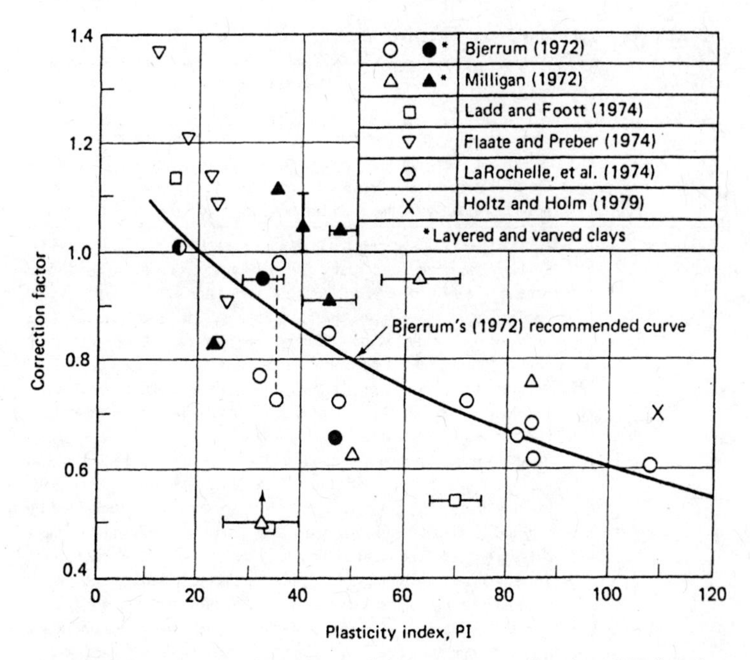

FIGURE 7.19 Correction factor for the field vane test as a function of PI, based on embankment failures. *Note: In situ* s_u = s_u from the field vane test times the correction factor. See Table 7.13. (*After Ladd 1975, Ladd et. al. 1977, reproduced from Holtz and Kovacs 1981.*)

TABLE 7.14 Total Stress and Effective Stress Analyses

Topic (1)	Discussion (2)
General discussion	An understanding of the total stress analysis and the effective stress analysis is essential in geotechnical engineering, and these two methods are used in the evaluation of all types of earth projects, such as slope stability, earth pressure calculations, and foundation design.
Total stress analysis	The total stress analysis uses the undrained shear strength of the soil and is typically used only for cohesive soil. This analysis is applicable to field situations where there is a change in shear stress that occurs quickly enough so that soft cohesive soil does not have time to consolidate, or in the case of heavily overconsolidated cohesive soils, the negative pore water pressures do not have time to dissipate. For this reason, the total stress analysis is often termed a *short-term analysis*. The total stress analysis uses the total unit weight γ_t of the soil, and the location of the groundwater table is not considered in the analysis. The total stress analysis is also applicable when there is a sudden change in loading condition of a cohesive soil. Examples include wind or earthquake loadings that exert compression or tension forces on piles supported by clay strata, or the rapid drawdown of a reservoir which induces shear stresses upon the clay core of the dam. The total stress analysis uses the undrained shear strength s_u of the plastic soil obtained from vane tests or unconfined compression tests (see Table 7.4). An alternative approach is to use the total stress parameters c and ϕ from unconsolidated undrained triaxial compression tests (see Table 7.5) or the consolidated undrained triaxial test (see Table 7.6). An advantage of the total stress analysis is that the undrained shear strength is obtained from tests (such as the vane test or unconfined compression test) that are easy to perform. A major disadvantage of this approach is that the accuracy of the undrained shear strength is always in doubt because it depends on the shear-induced pore water pressures (which are not measured) and factors such as sample disturbance, strain rate effects, and anisotropy.
Effective stress analysis	The effective stress analysis uses the drained shear strength parameters c' and ϕ'. Except for earthquake loading (i.e., liquefaction), essentially all analyses of cohesionless soils are made by using the effective stress analysis. For cohesionless soil, $c' = 0$ and the effective friction angle ϕ' is often obtained from drained direct shear tests, drained triaxial tests, or from empirical correlations, such as the standard penetration test (Fig. 7.3) or the cone penetration test (Fig. 7.4). See Tables 7.1 to 7.3 for further details. For cohesive soils, the effective shear strength could be obtained from drained shear strength tests, although it is more common to perform consolidated undrained triaxial compression tests with pore water pressure measurements performed on saturated cohesive soil in order to determine the effective shear strength parameters c' and ϕ' (see Table 7.7). Effective stress parameters are used for all long-term analyses where conditions are relatively constant. Examples include the long-term stability of slopes,

TABLE 7.14 Total Stress and Effective Stress Analyses (*Continued*)

Topic (1)	Discussion (2)
Effective stress analysis (*Continued*)	embankments, earth supporting structures, and bearing capacity of foundations for cohesionless and cohesive soil. The effective stress analysis can also be used for any situation where the pore water pressures induced by loading can be estimated or measured. The effective stress analysis can use either one of the following approaches: (1) total unit weight of the soil γ_t and the boundary pore water pressures or (2) buoyant unit weight of the soil γ_b and the seepage forces. An advantage of the effective stress analysis is that it more fundamentally models the shear strength of the soil, because shear strength is directly related to effective stress. A major disadvantage of the effective stress analysis is that the pore water pressures must be included in the analysis. The accuracy of the pore water pressure is often in doubt because of the many factors which affect the magnitude of pore water pressure changes, such as the determination of changes in pore water pressure resulting from changes in external loads. For effective stress analysis, assumptions are frequently required concerning the pore water pressures that will be used in the analysis.

TABLE 7.15 Example Problems

Topic (1)	Discussion (2)
Effective friction ϕ' estimated from the SPT test	*Problem:* A clean sand deposit has a level ground surface, a total unit weight γ_t above the groundwater table of 18.9 kN/m^3 (120 pcf), and a submerged unit weight γ_b of 9.84 kN/m^3 (62.6 pcf). The groundwater table is located 1.5 m (5 ft) below ground surface. Standard penetration tests were performed in a 10-cm- (4-in.-) diameter borehole. At a depth of 3 m (10 ft) below ground surface, a standard penetration test (SPT) was performed with a donut hammer with a blow count of 3 blows for the first 15 cm (6 in.), 4 blows for the second 15 cm (6 in.), and 5 blows for the third 15 cm (6 in.) of driving penetration. Assuming hydrostatic pore water pressures, determine the vertical effective stress σ'_v at a depth of 3 m (10 ft), the corrected *N*-value (i.e., N_{60}, Table 2.11), and the friction angle ϕ of the sand. *Solution:* $\sigma'_v = \sigma_v - u = \gamma_t\,(1.5) + \gamma_b\,(1.5) = (18.9)(1.5) + (9.84)(1.5) =$ 43 kPa. $N_{60} = 1.67\,E_m C_b C_r N$ (Table 2.11), where $E_m = 0.45$ (donut hammer), $C_b = 1.0$ (100 mm diameter hole), $C_r = 0.75$ (3 m length of drill rods), $N = 4 + 5 = 9$ (*N*-value). Substituting these values into the equation: $N_{60} = 1.67\,(0.45)(1.0)(0.75)(9) = 5$. In Fig. 7.3, for $N_{60} = 5$ and $\sigma'_v = 43$ kPa, $\phi' = 30°$.
Direct shear test	*Problem:* A drained direct shear test was performed on a nonplastic cohesionless soil. The specimen diameter = 6.35 cm (2.5 in.). At a vertical load of 150 N (34 lb), the peak shear force = 94 N (21 lb). For a second specimen tested at a vertical load of 300 N (68 lb), the peak shear force = 188 N (42 lb). Determine the effective friction angle ϕ' for the cohesionless soil. *Solution:* The soil is nonplastic and cohesionless, therefore $c' = 0$. The specimen diameter is 6.35 cm, therefore area $(A) = 0.00317$ m^2. *First test:* $\sigma_n = N/A = 150/0.00317 = 47{,}000 \text{ Pa} = 47.3 \text{ kPa}$ $\tau = T/A = 94/0.00317 = 29{,}700 \text{ Pa} = 29.7 \text{ kPa}$ Therefore, since $\tau = \sigma_n \tan \phi'$, $\tan \phi' = 29.7/47.3$ or $\phi' = 32°$ *Second test:* $\sigma_n = N/A = 300/0.00317 = 94{,}600 \text{ Pa} = 94.6 \text{ kPa}$ $\tau = T/A = 188/0.00317 = 59{,}300 \text{ Pa} = 59.3 \text{ kPa}$ Therefore, since $\tau = \sigma_n \tan \phi'$, $\tan \phi' = 59.3/94.6$ or $\phi' = 32°$

TABLE 7.15 Example Problems (*Continued*)

Topic (1)	Discussion (2)
Consolidated undrained triaxial compression test	*Problem:* A consolidated undrained triaxial compression test was performed on a saturated cohesive soil specimen. The soil specimen was first consolidated at an effective confining pressure = 100 kPa (i.e., $\sigma'_1 = \sigma'_3$). At the end of consolidation, the soil specimen had an area = 9.68 cm^2 (1.50 in^2) and a height = 11.7 cm (4.60 in.). The specimen was then subjected to an axial load, and at failure, the axial deformation = 1.48 cm (0.583 in.), the axial load = 48.4 N (10.9 lb), and the change in pore water pressure (Δu) = 45.6 kPa (950 psf). Calculate the area of the specimen at failure (A_f), the major and minor principal effective stresses at failure (σ'_1 and σ'_3), q and p' (in terms of effective stresses), the effective friction angle (ϕ') assuming $c' = 0$, and the A-value at failure. *Solution:* $A_f = A_o/(1 - \epsilon_f) = 9.68/[1 - (1.48/11.7)] = 11.1\ cm^2$ $\sigma'_1 = \sigma'_v = \sigma_1 - u = [100 + (0.0484/0.00111)] - 45.6 = 98$ kPa $\sigma'_3 = \sigma'_h = \sigma_3 - u = 100 - 45.6 = 54.4$ kPa $p' = 0.5(\sigma'_1 + \sigma'_3) = (0.5)(98 + 54.4) = 76.2$ kPa $q = 0.5(\sigma'_1 - \sigma'_3) = (0.5)(98 - 54.4) = 21.8$ kPa for $c' = 0$, $a' = 0$ Since $q = p' \tan \alpha'$, $\tan \alpha' = q/p' = 21.8/76.2 = 0.286$ Therefore $\alpha' = 16°$ $\sin \phi' = \tan \alpha' = \tan 16°$ Then the friction angle is $\phi' = 17°$ The A-value at failure is $A = \Delta u/\Delta\sigma_1 = 45.6/(0.0484/0.00111) = 1.05$

CHAPTER 8
PERMEABILITY AND SEEPAGE

8.1 INTRODUCTION

This chapter presents the basic principles of permeability and groundwater seepage through soil. In the previous chapters, the discussion of groundwater has included the following:

- *Subsurface exploration (Sec. 2.2).* One of the main purposes of subsurface exploration is to determine the location of the groundwater table.
- *Laboratory testing (Sec. 3.4).* The rate of flow through soil is dependent on its permeability. The constant-head permeameter (Fig. 3.4) or the falling-head permeameter (Fig. 3.5) can be used to determine the coefficient of permeability (also known as the hydraulic conductivity). Equations to calculate the coefficient of permeability in the laboratory are based on Darcy's law ($v = ki$). Figure 3.6 summarizes the coefficient of permeability versus soil type and drainage property, and Fig. 3.7 presents data on the coefficient of permeability versus void ratio for various soil types.
- *Pore water pressure (Sec. 6.2).* The pore water pressure u for a level groundwater table with no flow and hydrostatic conditions is calculated by using the equation in Table 6.1 for soil below the groundwater table and the equation in Table 6.2 for soil above the groundwater table that is saturated because of capillary rise.

- *Shear strength (Chap. 7).* The shear strength of saturated soil is directly related to the water that fills the soil pores. Normally consolidated plastic soils have an increase in pore water pressure as the soil structure contracts, while heavily overconsolidated plastic soils have a decrease in pore water pressure as the soil structure dilates during shear.

The main items that will be discussed in this chapter are permeability, seepage velocity, seepage force, and two-dimensional flow nets. In the following chapters, these principles will be used for all types of engineering analyses, such as consolidation, slope stability, and the design of earth dams. Chapter 20 will specifically discuss groundwater control.

8.2 PERMEABILITY

Table 8.1 presents a discussion of the factors affecting the permeability of soil. Various standpipe conditions and the equations used to calculate the field permeability k are shown in Fig. 8.1.

8.3 SUPERFICIAL VELOCITY AND SEEPAGE VELOCITY

A discussion and the equations needed to calculate the superficial velocity v and the seepage velocity v_s for the flow of water in soil are presented in Table 8.2.

8.4 SEEPAGE FORCES

Table 8.3 presents a discussion of seepage force and the equation needed to calculate seepage forces within the soil mass.

8.5 TWO-DIMENSIONAL FLOW NETS

A flow net is defined as a graphical representation used to study the flow of water through soil (see Table 8.4). The flow net can be used to determine the quantity of water (Q) flowing through the soil. Examples of two-dimensional flow nets are presented in Figs. 8.2 to 8.4.

8.5.1 Using Flow Nets to Calculate Pore Water Pressures and Exit Gradients

The flow net can also be used to determine the pore water pressures *u* in the ground and the exit gradients, which are important in the analysis of piping (see Table 8.5). Example problems are presented in Table 8.6.

TABLE 8.1 Factors Affecting the Coefficient of Permeability

Topic (1)	Discussion (2)
General discussion	As previously mentioned in Sec. 3.4, cohesive soils such as clays have very low values of the coefficient of permeability *k,* while cohesionless soils such as clean sands and gravels can have a coefficient of permeability *k* that is a billion times larger than cohesive soils (see Fig. 3.6). The reason is because of the size of the drainage paths through the soil. Open graded gravel has very large, interconnected void spaces. At the other extreme are clay-size particles, which produce minuscule void spaces with a large resistance to flow and hence a very low permeability.
Factors affecting the coefficient of permeability *k*	Although the type of soil is the most important parameter that governs the coefficient of permeability *k* of saturated soil, there are many other factors that affect *k,* such as: **1.** *Void ratio.* For a given soil, the higher the void ratio, the higher the coefficient of permeability. For example, Fig. 3.7 presents data from various soils and shows how the coefficient of permeability *k* increases as the void ratio increases. **2.** *Particle size distribution.* A well-graded soil will tend to have a lower coefficient of permeability *k* than a uniform soil. This is because a well-graded soil will have more particles to fill in the void spaces and make the flow paths smaller. **3.** *Soil structure.* Some types of soil structure, such as the honeycomb structure or cardhouse structure, have larger interconnected void spaces with a corresponding higher permeability. **4.** *Layering of soil.* Natural soils are often stratified and have layers or lenses of permeable soil, resulting in a much higher horizontal permeability than vertical permeability. This is important because oftentimes it is only the vertical permeability that is measured in the constant-head and falling-head permeameters (see Figs. 3.4 and 3.5). **5.** *Soil imperfections or discontinuities.* Natural soil has numerous imperfections, such as root holes, animal burrows, joints, fissures, seams, and soil cracks, that significantly increase the permeability of the soil mass. For compacted clay liners, there may be incomplete bonding between fill lifts, which allows water to infiltrate through the fill mass.
Field tests	In many cases, the determination of the coefficient of permeability *k* from laboratory tests on small soil specimens may not be representative of the overall field condition. A common method of determining the field coefficient of permeability is through the measurements of the change in water levels in open standpipes. Figure 8.1 presents various standpipe conditions and the equations used to calculate the field permeability *k.*

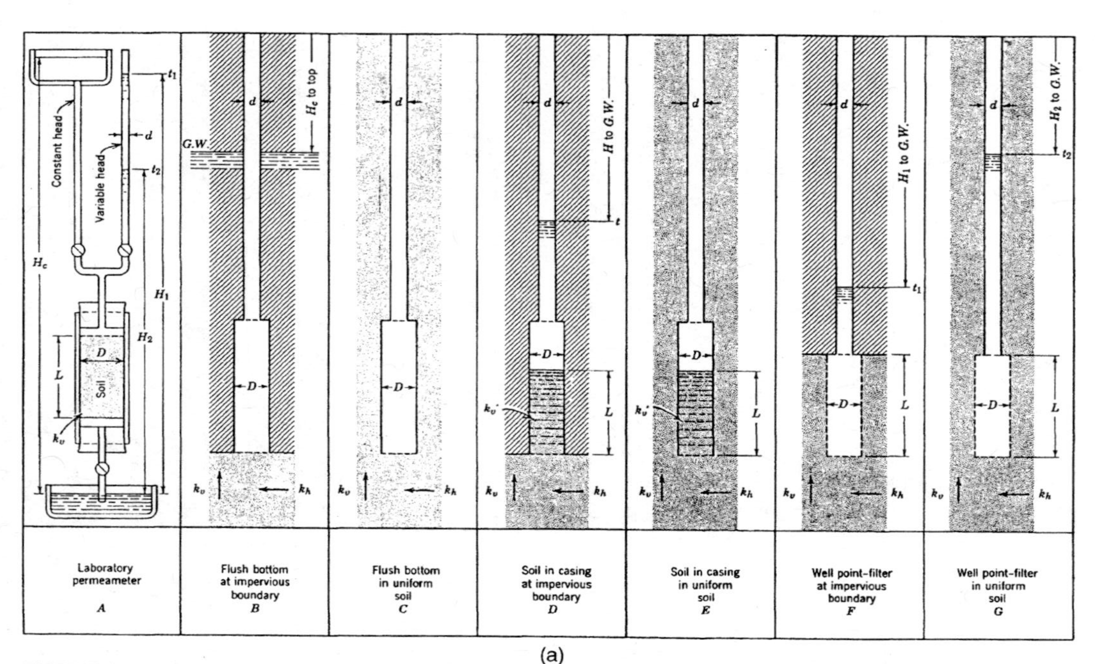

FIGURE 8.1a Methods for determining permeability. (*From Lambe and Whitman 1969; reproduced with permission from John Wiley & Sons.*)

Case	Constant Head	Variable Head
A	$k_v = \dfrac{4 \cdot q \cdot L}{\pi \cdot D^2 \cdot H_c}$	$k_v = \dfrac{d^2 \cdot L}{D^2 \cdot (t_2 - t_1)} \ln \dfrac{H_1}{H_2}$ $k_v = \dfrac{L}{t_2 - t_1} \ln \dfrac{H_1}{H_2}$ for $d = D$
B	$k_m = \dfrac{q}{2 \cdot D \cdot H_c}$	$k_m = \dfrac{\pi \cdot d^2}{8 \cdot D \cdot (t_2 - t_1)} \ln \dfrac{H_1}{H_2}$ $k_m = \dfrac{\pi \cdot D}{8 \cdot (t_2 - t_1)} \ln \dfrac{H_1}{H_2}$ for $d = D$
C	$k_m = \dfrac{q}{2.75 \cdot D \cdot H_c}$	$k_m = \dfrac{\pi \cdot d^2}{11 \cdot D \cdot (t_2 - t_1)} \ln \dfrac{H_1}{H_2}$ $k_m = \dfrac{\pi \cdot D}{11 \cdot (t_2 - t_1)} \ln \dfrac{H_1}{H_2}$ for $d = D$
D	$k_v' = \dfrac{4 \cdot q \left(\dfrac{\pi}{8} \cdot \dfrac{k_v'}{k_v} \cdot \dfrac{D}{m} + L\right)}{\pi \cdot D^2 \cdot H_c}$	$k_v' = \dfrac{d^2 \cdot \left(\dfrac{\pi}{8} \cdot \dfrac{k_v'}{k_v} \cdot \dfrac{D}{m} + L\right)}{D^2 \cdot (t_2 - t_1)} \ln \dfrac{H_1}{H_2}$ $k_v = \dfrac{\dfrac{\pi}{8} \cdot \dfrac{D}{m} + L}{t_2 - t_1} \ln \dfrac{H_1}{H_2}$ for $\begin{cases} k_v' = k_v \\ d = D \end{cases}$
E	$k_v' = \dfrac{4 \cdot q \cdot \left(\dfrac{\pi}{11} \cdot \dfrac{k_v'}{k_v} \cdot \dfrac{D}{m} + L\right)}{\pi \cdot D^2 \cdot H_c}$	$k_v' = \dfrac{d^2 \cdot \left(\dfrac{\pi}{11} \cdot \dfrac{k_v'}{k_v} \cdot \dfrac{D}{m} + L\right)}{D^2 \cdot (t_2 - t_1)} \ln \dfrac{H_1}{H_2}$ $k_v = \dfrac{\dfrac{\pi}{11} \cdot \dfrac{D}{m} + L}{t_2 - t_1} \ln \dfrac{H_1}{H_2}$ for $\begin{cases} k_v' = k_v \\ d = D \end{cases}$
F	$k_h = \dfrac{q \cdot \ln \left[\dfrac{2mL}{D} + \sqrt{1 + \left(\dfrac{2mL}{D}\right)^2}\right]}{2 \cdot \pi \cdot L \cdot H_c}$	$k_h = \dfrac{d^2 \cdot \ln \left[\dfrac{2mL}{D} + \sqrt{1 + \left(\dfrac{2mL}{D}\right)^2}\right]}{8 \cdot L \cdot (t_2 - t_1)} \ln \dfrac{H_1}{H_2}$ $k_h = \dfrac{d^2 \cdot \ln \left(\dfrac{4mL}{D}\right)}{8 \cdot L \cdot (t_2 - t_1)} \ln \dfrac{H_1}{H_2}$ for $\dfrac{2mL}{D} > 4$
G	$k_h = \dfrac{q \cdot \ln \left[\dfrac{mL}{D} + \sqrt{1 + \left(\dfrac{mL}{D}\right)^2}\right]}{2 \cdot \pi \cdot L \cdot H_c}$	$k_h = \dfrac{d^2 \cdot \ln \left[\dfrac{mL}{D} + \sqrt{1 + \left(\dfrac{mL}{D}\right)^2}\right]}{8 \cdot L \cdot (t_2 - t_1)} \ln \dfrac{H_1}{H_2}$ $k_h = \dfrac{d^2 \cdot \ln \left(\dfrac{2mL}{D}\right)}{8 \cdot L \cdot (t_2 - t_1)} \ln \dfrac{H_1}{H_2}$ for $\dfrac{mL}{D} > 4$

ASSUMPTIONS

Soil at intake, infinite depth, and directional isotropy (k_v and k_h constant). No disturbance, segregation, swelling, or consolidation of soil. No sedimentation or leakage. No air or gas in soil, well point, or pipe. Hydraulic losses in pipes, well point, or filter negligible.

FIGURE 8.1b Formulas for determination of permeability. (*From Hvorslev 1951; reproduced from Lambe and Whitman 1969, with permission from John Wiley & Sons.*)

Case	Basic Time Lag	Notation
A	$k_v = \frac{d^2 \cdot L}{D^2 \cdot T}$ $k_v = \frac{L}{T}$ for $d = D$	D = Diam, intake, sample (cm) d = Diameter, standpipe (cm) L = Length, intake, sample (cm)
B	$k_m = \frac{\pi d^2}{8 \cdot D \cdot T}$ $k_m = \frac{\pi \cdot D}{8 \cdot T}$ for $d = D$	H_c = Constant piez. head (cm) H_1 = Piez. head for $t = t_1$ (cm) H_2 = Piez. head for $t = t_2$ (cm)
C	$k_m = \frac{\pi \cdot d^2}{11 \cdot D \cdot T}$ $k_m = \frac{\pi \cdot D}{11 \cdot T}$ for $d = D$	q = Flow of water (cm³/sec) t = Time (sec) T = Basic time lag (sec) k_v' = Vert. perm. casing (cm/sec)
D	$k_v' = \frac{d^2 \cdot \left(\frac{\pi}{8} \cdot \frac{k_v'}{k_v} \cdot \frac{D}{m}\right) + L}{D^2 \cdot T}$ $k_v = \frac{\frac{\pi}{8} \cdot \frac{D}{m} + L}{T}$ for $\begin{cases} k_v' = k_v \\ d = D \end{cases}$	k_v = Vert. perm. ground (cm/sec) k_h = Horz. perm. ground (cm/sec) k_m = Mean coeff. perm. (cm/sec) m = Transformation ratio $k_m = \sqrt{k_h \cdot k_v}$ $m = \sqrt{k_h/k_v}$ $\ln = \log_e = 2.3 \log_{10}$
E	$k_v' = \frac{d^2 \cdot \left(\frac{\pi}{11} \cdot \frac{k_v'}{k_v} \cdot \frac{D}{m} + L\right)}{D^2 \cdot T}$ $k_v' = \frac{\frac{\pi}{11} \cdot \frac{D}{m} + L}{T}$ for $\begin{cases} k_v' = k_v \\ d = D \end{cases}$	
F	$k_h = \frac{d^2 \cdot \ln\left[\frac{2mL}{D} + \sqrt{1 + \left(\frac{2mL}{D}\right)^2}\right]}{8 \cdot L \cdot T}$ $k_h = \frac{d^2 \cdot \ln\left(\frac{4mL}{D}\right)}{8 \cdot L \cdot T}$ for $\frac{2mL}{D} > 4$	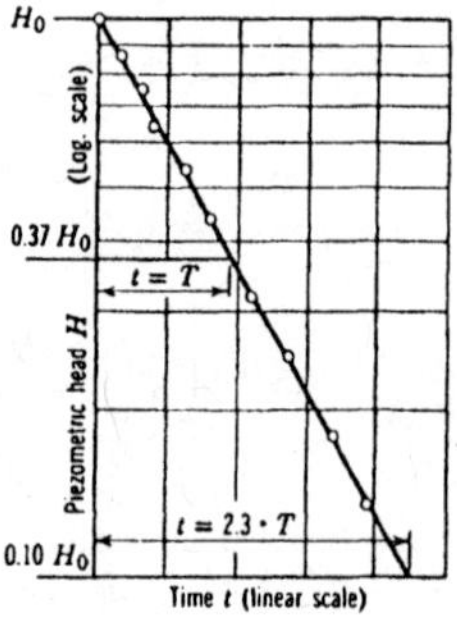 H_0 0.37 H_0 $t = T$ Piezometric head H (Log. scale)
G	$k_h = \frac{d^2 \cdot \ln\left[\frac{mL}{D} + \sqrt{1 + \left(\frac{mL}{D}\right)^2}\right]}{8 \cdot L \cdot T}$ $k_h = \frac{d^2 \cdot \ln\left(\frac{2mL}{D}\right)}{8 \cdot L \cdot T}$ for $\frac{mL}{D} > 4$	$t = 2.3 \cdot T$ 0.10 H_0 Time t (linear scale) Determination basic time lag T

FIGURE 8.1b *(Continued)*

TABLE 8.2 Superficial Velocity and Seepage Velocity

Topic (1)	Discussion (2)
Bernoulli's equation	From Bernoulli's energy equation, the total head h is equal to the sum of the velocity head, pressure head, and elevation head. Head has units of length. For seepage problems in soil, the velocity head is usually small enough to be neglected, and thus for laminar flow in soil, the total head h is equal to the sum of the pressure head h_p and the elevation head h_e. In order for there to be flow of water through soil, there must be a change in total head (Δh). For example, given a level groundwater table with hydrostatic pore water pressures, the total head will be identical at all points below the groundwater table. Since there is no change in total head ($\Delta h = 0$), there can be no flow of water.
Superficial velocity v	Darcy's law ($v = ki$) is used for the analysis of laminar flow in soil, and it relates the velocity v to the change in total head ($i = \Delta h/L$, where L = length between the total head differences). Since the coefficient of permeability k is measured in a permeameter where the quantity of water is measured flowing out of (constant-head permeameter) or into (falling-head permeameter) the soil, the velocity v is actually a superficial velocity and not the actual velocity of flow in the soil voids v_s.
Seepage velocity v_s	The seepage velocity v_s is the velocity of flow of water in the soil. The seepage velocity v_s is related to the superficial velocity v, by the following equation: $$v_s = \frac{v}{n} = \frac{ki}{n} = \frac{k\Delta h}{nL}$$ where v_s = seepage velocity, i.e., velocity of flow of water in the soil (cm/s or in./min) v = superficial velocity, i.e., velocity of flow of water into or out of the soil (cm/s or in./min) k = the coefficient of permeability, also known as the hydraulic conductivity (cm/s or in./min) i = hydraulic gradient ($i = \Delta h/L$), a dimensionless parameter n = porosity of the soil, Table 5.1, expressed as a decimal (dimensionless parameter) Δh = change in total head between two points in the soil mass (m or ft) L = length between the two points in the soil mass (m or ft) Between any two points in a saturated soil mass below a sloping groundwater table or below a level groundwater table that does not have hydrostatic pore water pressure (such as an artesian condition), this equation can be used to determine the velocity of flow v_s in the soil.

TABLE 8.3 Seepage Force

Topic (1)	Discussion (2)
General discussion	When water flows through a soil, it exerts a drag force on the individual soil particles. This force has been termed the *seepage force.* For example, an artesian condition can cause an upward flow of water to the ground surface. If this upward flow of water exerts enough of a seepage force upon the individual soil particles, they can actually lose contact with each other. This fluid condition of a cohesionless soil has been termed *quicksand.* In order to create this fluid condition, the vertical effective stress σ'_v in the soil must equal zero.
Seepage force	There are two basic approaches to evaluating a seepage condition, and each approach gives the same answer. The first approach is to use the total unit weight of the soil γ_t and the boundary pore water pressures. The second approach is to use the buoyant unit weight of the soil γ_b and the seepage force j, which is equal to: $$j = \gamma_w i$$ where γ_w = unit weight of water (9.81 kN/m^3 or 62.4 pcf) and i = hydraulic gradient ($i = \Delta h/L$). The seepage force j has units of force per unit volume of soil (kN/m^3 or pcf).
Most common analysis	Because it is often easier to measure or predict the pore water pressure, the first approach, which uses the total unit weight of the soil and the boundary pore water pressures, is used more often in engineering analyses. The most common method used to predict the pore water pressures in the ground due to flowing groundwater is to use a flow net. An example of an engineering analysis that may need to include the seepage effects of groundwater is the slope stability of an earth dam (Chap. 13).

TABLE 8.4 Two-Dimensional Flow Nets

Topic (1)	Discussion (2)
General discussion	As shown in Figs. 3.4 and 3.5, the constant-head and falling-head permeameters have a one-dimensional flow condition (i.e., water flows vertically through the soil). For two-dimensional flow conditions, where, for example, the groundwater is flowing horizontally and vertically, the situation is more complex. The most common method of estimating the quantity of water (Q) and pore water conditions in the soil for a two-dimensional case is to construct a flow net.
Flow lines and equipotential lines	A flow net is defined as a graphical representation used to study the flow of water through soil. A flow net is composed of two types of lines: (1) flow lines, which are lines that indicate the path of flow of water through the soil, and (2) equipotential lines, which are lines that connect points of equal total head; i.e., all along an equipotential line, the numerical value of the total head h is constant.
Flow net example—foundation pit	Figure 8.2 presents a cross section depicting the seepage of water underneath a sheet pile wall and into a foundation pit. On one side of the sheet pile wall, the level of water is at a height = h_w. On the other side of the sheet pile wall (i.e., the foundation pit), the water has been pumped out, and the groundwater table corresponds to the ground surface. In this case, the change in total head (Δh) between the outside water level and the inside foundation pit water level is equal to h_w (i.e., $\Delta h = h_w$). Because there is a difference in total head, there will be the flow of water from the outside to the inside of the foundation pit. Note in Fig. 8.2 that the sheet pile wall is embedded a depth of 12 m into the soil. The site consists of 25 m of pervious sand overlying a relatively impervious clay stratum. Assuming isotropic and homogeneous soil, the flow net has been drawn as shown in Fig. 8.2. The flow lines are those lines that have arrows which indicate the direction of flow. The equipotential lines are those lines that are generally perpendicular to the flow lines. Note in Fig. 8.2 that the flow lines and equipotential lines intersect at right angles and tend to form squares.
Quantity of seepage (Q)	By using the flow net, the quantity of seepage (Q) entering the foundation pit, per unit length of sheet pile wall, can be estimated by using Darcy's law, or: $$Q = kiAt = k\Delta ht(A/L) = k\Delta ht\,(n_f/n_d)$$ where Q = quantity of water that enters the foundation pit per unit length of sheet pile wall (m^3 for a 1-m length of wall or ft^3 for a 1-ft length of wall) k = coefficient of permeability of the pervious stratum (m/day or ft/day) Δh = change in total head (m or ft) t = time (days) A/L = cross-sectional area of flow divided by the length of flow which is equal to n_f / n_d, i.e., the number of flow channels n_f divided by the number of equipotential drops n_d.

TABLE 8.4 Two-Dimensional Flow Nets (*Continued*)

Topic (1)	Discussion (2)
Flow net example—earth dam	Figure 8.3 shows an earth dam with the seepage of water through the dam and into a longitudinal drainage filter. In Fig. 8.3, the flow lines are those lines that have arrows which indicate the direction of flow. The equipotential lines are those lines that are generally perpendicular to the flow lines. The vertical difference between the water level on the upstream side of the earth dam and the level of water in the drainage filter (i.e., dashed line) is the change in total head (Δh). Note in Fig. 8.3 that a series of horizontal lines have been drawn adjacent to the line representing the groundwater table in the earth dam. The distance between these horizontal lines are all equal and represent equal drops in total head for successive equipotential lines. The quantity of flow (Q) collected by the drainage filter can be determined by using the previous equation. Another use for the flow net shown in Fig. 8.3 would be to determine the stability of earth dam slopes. From the flow net, the pore water pressures u could be determined and then the effective shear strength could be used in conjunction with a slope stability analysis to determine the stability of the earth dam (Chap. 13).
Flow net construction	The usual procedure in the preparation of a flow net is trial-and-error sketching. A flow net is first sketched in pencil with a selected number of flow channels, and then the equipotential lines are inserted. If the flow net does not have flow lines and equipotential lines intersecting at right angles, or if the flow net is not composed of approximate squares, then the lines are adjusted. Computer programs have been developed to aid in the construction of a flow net. For example, the SEEP/W (Geo-Slope 1992) computer program can be used to determine the total head h distributions and the flow vectors (i.e., flow lines) for both simple and highly complex seepage problems. The flow nets presented in Figs. 8.2 and 8.3 are based on the assumption of isotropic and homogeneous soil. Flow nets can also be drawn for the flow of water from one stratum of soil to another and for anisotropic soil where the coefficient of permeability is higher in the horizontal direction than the vertical direction (see the portions labeled "Transfer conditions at interface between strata" and "For anisotropic soil" in Fig. 8.4).
Rules for flow net construction	The rules for a flow net construction are as follows (NAVFAC DM-7.1, 1982): **1.** When the soil is isotropic with respect to permeability, the pattern of flow lines and equipotential lines intersect at right angles. Draw a pattern in which squares are formed between flow lines and equipotential lines. **2.** Usually it is expedient to start with an integer number of equipotential drops, dividing the total head loss by a whole number and drawing flow lines to conform to these equipotential drops. In a general sense, the outer flow path will form a rectangle rather than a square. The shape of these rectangles must be constant (i.e., constant ratio of B/L) **3.** The upper boundary of the flow net that is at atmospheric pressure is a "free water surface." Equipotential lines intersect the free water surface at points spaced at equal vertical intervals (for example, see Fig. 8.3).

TABLE 8.4 Two-Dimensional Flow Nets (*Continued*)

Topic (1)	Discussion (2)
Rules for flow net construction (*Continued*)	**4.** A discharge face through which seepage passes is an equipotential line if the discharge is submerged, or a free water surface if the discharge is not submerged. If it is a free water surface, the flow net directly adjacent to the discharge face will not be composed of squares.
	5. In a stratified soil profile where the ratio of permeability of different layers exceeds a factor of 10, the flow in the more permeable layer controls. In this case, the flow net may be drawn for the more permeable layer, assuming the less permeable layer is impervious.
	6. When materials are anisotropic with respect to permeability, the cross section may be transformed by changing the scale as indicated in Fig. 8.4, and the flow net is then drawn as for an isotropic soil. In computing the quantity of seepage Q, the total head difference Δh is not changed for the transformation.
	7. Where only the quantity of seepage is to be determined, an approximate flow net suffices. If pore water pressures are to be determined, the flow net must be accurately drawn.

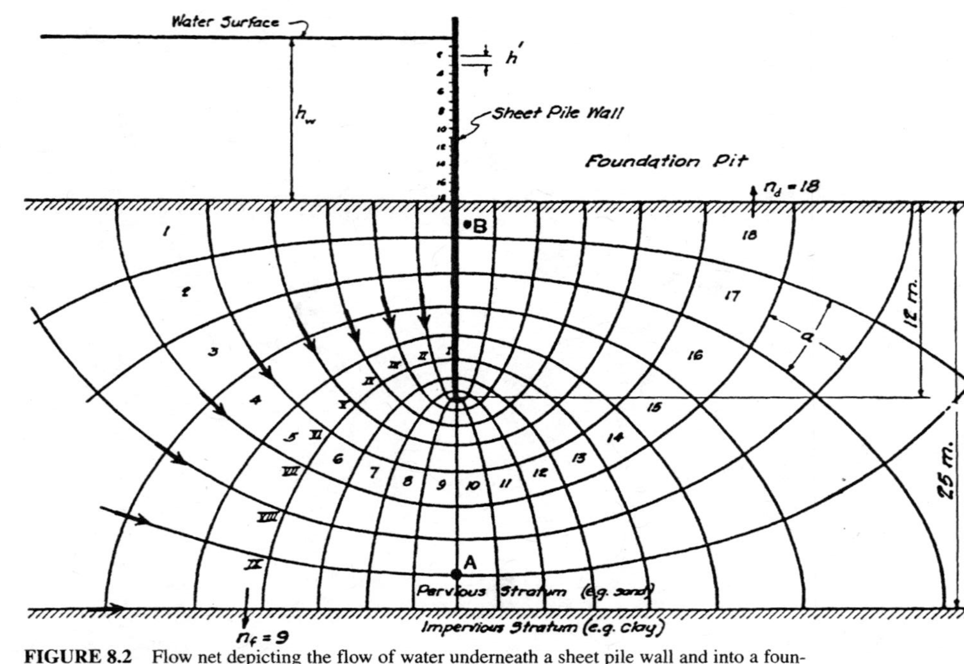

FIGURE 8.2 Flow net depicting the flow of water underneath a sheet pile wall and into a foundation pit. See Table 8.4. (*Adapted from Casagrande 1940.*)

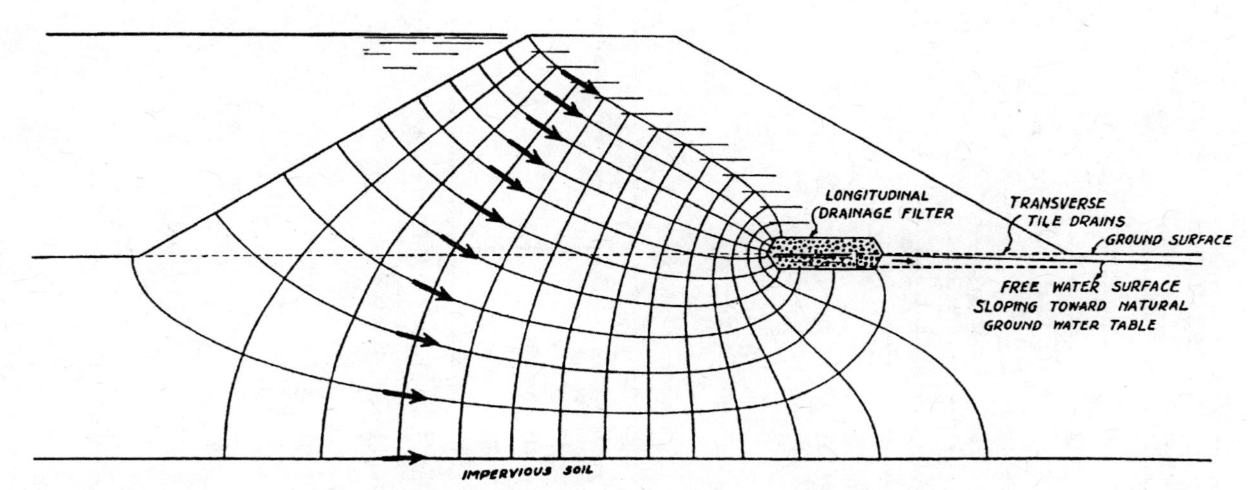

FIGURE 8.3 Flow net depicting the flow of water through an earth dam. See Table 8.4. (*Adapted from Casagrande 1940.*)

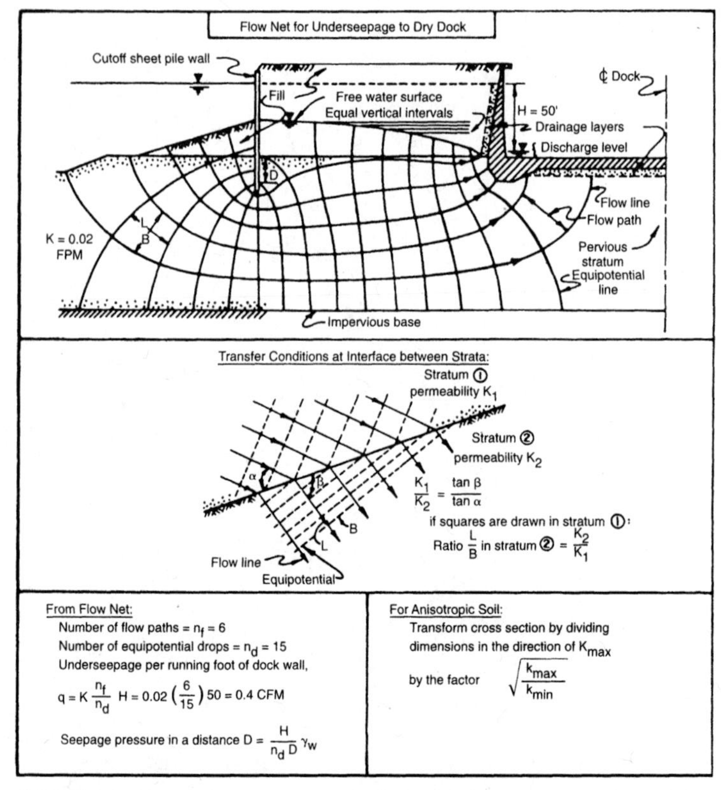

FIGURE 8.4 Flow net for seepage into a dry dock. Also shown are the transfer conditions for the flow of water from one soil stratum to another and the procedure for construction of a flow net for anisotropic soil. See Table 8.4. (*From NAVFAC DM-7.1, 1982.*)

TABLE 8.5 Using the Flow Net to Calculate Pore Water Pressure and Exit Gradients

Topic (1)	Discussion (2)
Pore water pressure u	The flow net can be used to determine the pore water pressures u in the ground. The equipotential lines are used to determine the pore water pressure in the soil. In Fig. 8.2, the value of h' is shown. This value represents the drop in total head between each successive equipotential line. Because there are 18 equipotential drops for the flow net shown in Fig. 8.2, $h' = \Delta h/18 = h_w/18$. Thus each successive equipotential line represents an additional drop in total head $= h'$. By establishing a datum to obtain the elevation head h_e, and knowing the value of the total head h along each equipotential line, one can calculate the pressure head ($h_p = h - h_e$) and hence pore water pressure ($u = h_p \gamma_w$).
Piping	In a sense, the reduction in total head (i.e., loss of energy) as the water flows through the soil is absorbed by the frictional drag resistance of the soil particles. If this frictional drag resistance of the soil particles becomes too great, piping may occur. Piping is the physical transport of soil particles by the groundwater seepage. Piping can result in the opening of underground voids, which could lead to the sudden failure. Piping is further discussed in Sec. 13.9.
Exit hydraulic gradient	The exit hydraulic gradient i_e can be used to assess the piping potential of soil. For the foundation pit (Fig. 8.2), the most likely location for a piping failure is at point B because this is where the flow of groundwater is directly upward and the distance between equipotential lines is least. If the distance between the two equipotential lines at point B is 2 m, then the exit hydraulic gradient at point B (i_e) would equal: $$h_w = 10 \text{ m}$$ Therefore $$h' = 10/18 = 0.556 \text{ m and } i_e = h'/L = 0.556 \text{ m}/2 \text{ m} = 0.28$$
Quicksand condition	It has been determined that for a quicksand condition (also known as a *blowout* or *boiling* condition), the critical hydraulic gradient i_c is equal to the buoyant unit weight γ_b divided by the unit weight of water γ_w, or $i_c = \gamma_b/\gamma_w$. The buoyant unit weight is often about equal to the unit weight of water, and thus the critical hydraulic gradient i_c is about equal to 1. Certainly if the soil at point B in Fig. 8.2 should turn to quicksand, then there would be significant piping (loss of soil particles) at this location.
Factor of safety for quicksand condition	It is often useful to compare the exit hydraulic gradient i_e to the critical hydraulic gradient which will cause quicksand i_c, or: $$F = i_c/i_e$$ where F = the factor of safety for the development of a quicksand condition. For the foundation pit (Fig. 8.2), i_e (point B) = 0.28, and for $i_c = 1.0$, the factor of safety = 3.6. Because of the catastrophic effects of a piping failure, usually a high factor of safety is required for the exit hydraulic gradient (according to Cedergren 1989, factor of safety should be at least 2 to 2.5).

TABLE 8.6 Example Problems

Topic (1)	Problem and solution (2)
Field permeability tests	*Problem:* In a uniform soil deposit having a level ground surface, an open standpipe with a constant interior diameter of 5.0 cm is installed to a depth of 10 m. The standpipe has a flush bottom (Case C, Fig. 8.1*a*). The groundwater table is located 4.0 m below the ground surface. The standpipe is filled with water and the water level is maintained at ground surface by adding 1.0 liter per every 33 seconds. Calculate the mean coefficient of permeability k for the soil deposit. *Solution:* $d = D = 5.0$ cm $q = 1000/33 = 30.3\ \text{cm}^3/\text{s}$ $H_c = 4\ \text{m} = 400$ cm For Case C (Fig. 8.1*b*) and a constant head condition: $k = q/[2.75DH_c] = 30.3/[(2.75)(5.0)(400)] = 0.006$ cm/s.
Quantity of seepage Q	*Problem:* Using the flow net shown in Fig. 8.2, assume the coefficient of permeability (k) = 0.1 m/day (0.33 ft/day), the height of the water level outside the foundation pit (h_w) = 10 m (33 ft), and the length of the sheet pile wall is 100 m (328 ft). Calculate the quantity of water that enters the foundation pit per week (t = 7 days). *Solution:* Use the following values: k = 0.1 m/day (0.33 ft/day), $\Delta h = h_w$ = 10 m (33 ft), t = 7 days, n_f = number of flow channels from the flow net = 9 (see Fig. 8.2), and n_d = number of equipotential drops from the flow net = 18 (see Fig. 8.2). Inserting these values into the equation in Table 8.4, the quantity of water (Q) that enters the foundation pit = 3.5 m^3 per linear meter of wall length (38 ft^3 per linear foot of wall length). For a 100-m-long sheet pile wall, the total amount of water that enters the foundation pit per week = 350 m^3 (12,500 ft^3). This is a considerable amount of water and extensive pumping would be required to keep the foundation pit dry.
Pore water pressure u	*Problem:* Assume the same conditions as the previous example. Calculate the pore water pressure u at point A in Fig. 8.2, if point A is located 2 m (6.6 ft) above the impervious stratum. *Solution:* Let the datum be at the top of the impervious stratum. Then the total head h corresponding to the ground surface on the left side of the sheet pile wall is equal to the elevation head (h_e = 25 m) plus the pressure head ($h_p = h_w$ = 10 m) or a total head h of 35 m (115 ft). In Fig. 8.2, the value of $h' = \Delta h/n_d = h_w/18 = 0.556$ m (1.82 ft). Since there are 9 equipotential drops for the flow of water to point A (Fig. 8.2), the reduction in total head = 9 times 0.556 m, or 5.0 m (16.5 ft). The total head along the equipotential line that contains point A can now be calculated. The total head h at point A would equal 35 m (115 ft) minus the reduction in total head due to the 9 equipotential drops

TABLE 8.6 Example Problems (*Continued*)

Topic (1)	Problem and solution (2)
Pore water pressure u (*Continued*)	(5.0 m), or 30 m (98.5 ft). Since the elevation head at point A = 2 m (6.6 ft), and since the total head h is equal to the sum of the pressure head h_p and the elevation head h_e, the pressure head is equal to 28 m (i.e., $h_p = h - h_e$ = 30 m − 2 m = 28 m). Converting head to pore water pressure ($u = h_p \gamma_w$), the pore water pressure u at point A = 275 kPa (5730 psf). *Check:* Because point A is located at exactly the midpoint in the flow net, a quick check can be performed on the above calculations. At the midpoint, the drop in head of water would equal ½ h_w, or 5 m. According to a piezometer measurement at point A, the water level would rise 5 m above ground surface, or a total of 28 m (23 m plus 5 m).
Exit gradient	*Problem:* Calculate the exit gradient i_e and the factor of safety for piping failure F for the flow net square labeled 18 in Fig. 18.2. Assume h_w = 10 m and γ_t = 19.8 kN/m^3. *Solution:* $L = 4$ m; $i_e = h'/L = (10/18)/4 = 0.14$; $i_c = \gamma_b/\gamma_w = (19.8 - 9.81)/9.81 = 1.02$; $F = i_c/i_e = 1.02/0.14 = 7$

CHAPTER 9
SETTLEMENT ANALYSES

9.1 INTRODUCTION

Settlement can be defined as the permanent downward displacement of the foundation. There are two basic types of settlement, as discussed in Table 9.1. Determining the settlement of the structure is one of the primary obligations of the geotechnical engineer. In general, three parameters are required: the total settlement ρ_{max}, differential settlement Δ, and rate of settlement. One approach for the design of the foundation is to first obtain the allowable settlement (total and differential) of the structure (Sec. 9.2). Then, on the basis of the results of the field exploration and laboratory testing, the total and differential settlement of the structure can be calculated (Secs. 9.3 to 9.6). As discussed in Chap. 18, the foundation type may have to be changed or mitigation measures adopted if the calculated settlement of the structure exceeds the allowable settlement.

9.2 ALLOWABLE SETTLEMENT

The allowable settlement is defined as the acceptable amount of settlement, and it usually includes a factor of safety. In many cases the allowable settlement of the project will be provided by the structural engineer. In other cases, the allowable settlement can be estimated from prior studies of damaged structures (see Tables 9.2 to 9.4 and Figs. 9.1 to 9.4).

9.3 COLLAPSIBLE SOIL

Collapsible soil can be broadly classified as soil that is susceptible to a large and sudden reduction in volume upon wetting. Collapsible soil is discussed in Tables 9.5 to 9.8 and Figs. 9.5 and 9.6.

9.4 SETTLEMENT OF COHESIVE AND ORGANIC SOILS

Cohesive and organic soils can be susceptible to a large amount of settlement from structural loads. It is usually the direct weight of the structure that causes settlement of the cohesive or organic soil (i.e., first basic category of settlement; see Table 9.1). The settlement of saturated clay or organic soil can have three different components: immediate (also known as "initial" settlement), consolidation, and secondary compression.

9.4.1 Immediate Settlement

Immediate settlement s_i is due to undrained shear deformations, or in some cases contained plastic flow, caused by two- or three-dimensional loading (see Table 9.9 and Figs. 9.7 and 9.8).

9.4.2 Consolidation

A load applied to a saturated cohesive soil will cause the development of excess pore water pressures which slowly dissipate as water flows out of the soil, and this process is known as *primary consolidation,* or simply *consolidation* (s_c). As the water slowly flows from the cohesive soil, the load is transferred to the soil particle skeleton, thereby increasing the effective stress of the soil. Consolidation is a time-dependent process that may take many years to complete (see Tables 9.10 to 9.15 and Figs. 9.9 to 9.13).

9.4.3 Secondary Compression

The final component of settlement is a result of secondary compression s_s, which is that part of the settlement that occurs after essentially all of the excess pore water pressures have dissipated (i.e., settlement that occurs at constant effective stress). The usual assumption is that secondary compression does not start until after primary consolidation is complete (see Table 9.16).

The total settlement ρ_{max} is equal to the sum of the immediate settlement s_i, consolidation settlement s_c, and the secondary compression s_s, or:

$$\rho_{max} = s_i + s_c + s_s$$

9.5 SETTLEMENT OF COHESIONLESS SOIL

This section deals with the settlement of cohesionless (nonplastic) soil caused by the structural load. A major difference between saturated cohesive soil and cohesionless soil is that the settlement of cohesionless soil is not time-dependent. Because of the generally high permeability of cohesionless soil, the settlement of a saturated cohesionless soil usually occurs as the load is applied during the construction of the building. Commonly used methods to estimate the settlement of cohesionless soil are presented in Table 9.17 and Figs. 9.14 and 9.15.

9.6 OTHER COMMON CAUSES OF SETTLEMENT

There are many other causes of settlement of structures. Some of the more common causes are discussed in Table 9.18 and Figs. 9.16 and 9.17. Example problems are presented in Table 9.19.

TABLE 9.1 Two Basic Types of Settlement

Topic (1)	Discussion (2)
Settlement due directly to the weight of the structure	The first basic type of settlement is that due directly to the weight of the structure. For example, the weight of a building may cause compression of an underlying sand deposit (Sec. 9.5) or consolidation of an underlying clay layer (Sec. 9.4). Often the settlement analysis is based on the actual dead load of the structure. The dead load is defined as the structural weight due to beams, columns, floors, roofs, and other fixed members. The dead load does not include nonstructural items. Live loads are defined as the weight of nonstructural members, such as furniture, occupants, inventory, and snow. Live loads can also result in settlement of the structure. For example, if the proposed structure is a library, then the actual weight of the books (a live load) should be included in the settlement analyses. Likewise, for a proposed warehouse, it may be appropriate to include the actual weight of anticipated stored items in the settlement analyses. In other projects where the live loads represent a significant part of the loading, such as large electrical transmission towers that will be subjected to wind loads, the live load (wind) may also be included in the settlement analysis. As mentioned, the load used for settlement analyses should include the actual dead weight of the structure, and in some cases, can include live loads. Considerable experience and judgment are required to determine the load that is to be used in the settlement analyses.
Settlement due to secondary influences	The second basic type of settlement of a building is caused by secondary influence, which may develop at a time long after the completion of the structure. This type of settlement is not directly caused by the weight of the structure. For example, the foundation may settle as water infiltrates the ground and causes unstable soils to collapse (i.e., collapsible soil, Sec. 9.3). The foundation may also settle because of yielding of adjacent excavations or the collapse of limestone cavities or underground mines and tunnels (Sec. 9.6). Other causes of settlement that would be included in this category are natural disasters, such as settlement caused by earthquakes (Chap. 14) or undermining of the foundation from floods.
Subsidence	Subsidence is usually defined as a sinking down of a large area of the ground surface. Subsidence could be caused by the extraction of oil or groundwater which leads to a compression of the underlying porous soil or rock structure. Since subsidence is caused by a secondary influence (extraction of oil or groundwater), its effect on the structure would be included in the second basic type of settlement described above.
Expansive soil	A special case is the downward displacement of the foundation due to the drying of underlying wet clays, discussed in Chap. 12. Often this downward displacement of the foundation caused by the desiccation of clays is referred to as *settlement.* But upon the introduction of moisture,

TABLE 9.1 Two Basic Types of Settlement (*Continued*)

Topic (1)	Discussion (2)
Expansive soil (*Continued*)	for example, during the rainy season, the desiccated clay will swell and the downward displacement will be reversed. The foundation could even heave more than it initially settled. In dealing with expansive clays, it is best to consider the downward displacement of the foundation as part of the cyclic heave and shrinkage of expansive soil and not permanent settlement.
Structural design analyses	The structural design of foundations, such as determining the thickness of the concrete foundation and type and spacing of steel reinforcement, is not covered in this book. There are many excellent references, such as *Foundation Analysis and Design* (Bowles 1982) and *Foundation Design, Principles and Practices* (Coduto 1994), that present the methods and procedures for the analysis and design of foundations.

TABLE 9.2 Allowable Settlement

Topic (1)	Discussion (2)
General Discussion	At the start of a project, the structural engineer should be consulted about the anticipated loading conditions, allowable settlement of the structure, and preliminary thoughts on the type of foundation. The ideal situation would be to obtain this information directly from the structural engineer. If this information is unavailable, then the geotechnical engineer will have to estimate the allowable settlement of the structure in order to determine an appropriate type of foundation. There is a considerable amount of data available on the allowable settlement of structures (e.g., Leonards 1961, ASCE 1964, Feld 1965, Peck et al. 1974, Bromhead 1984, Wahls 1994). For example, it has been stated that the allowable differential and total settlement should depend on the flexibility and complexity of the structure including the construction materials and type of connections (*Foundation Engineering Handbook*, Winterkorn and Fang 1975).
Allowable settlement per Coduto (1994)	In terms of the allowable settlement, Coduto (1994) states that it depends on many factors, including the following: 1. The Type of Construction: For example, wood-frame buildings with wood siding would be much more tolerant than unreinforced brick buildings. 2. The Use of the Structure: Even small cracks in a house might be considered unacceptable, whereas much larger cracks in an industrial building might not even be noticed. 3. The Presence of Sensitive Finishes: Tile or other sensitive finishes are much less tolerant of movements. 4. The Rigidity of the Structure: If a footing beneath part of a very rigid structure settles more than the others, the structure will transfer some of the load away from the footing. However, footings beneath flexible structures must settle much more before any significant load transfer occurs. Therefore, a rigid structure will have less differential settlement than a flexible one. Coduto (1994) also states that the allowable settlement for most structures, especially buildings, will be governed by aesthetic and serviceability requirements, not structural requirements. Unsightly cracks, jamming doors and windows, and other similar problems will develop long before the integrity of the structure is in danger,
Allowable settlement per Skempton and MacDonald (1956)	A major reference for the allowable settlement of structures is the paper by Skempton and MacDonald (1956) titled "The Allowable Settlements of Buildings." As shown in Fig. 9.1, Skempton and MacDonald defined the maximum angular distortion (δ/L) and the maximum differential settlement (Δ) for a building with no tilt. The angular distortion (δ/L) is defined as the differential settlement between two points divided by the distance between them less the tilt, where tilt equals rotation of the entire building. As shown in Fig. 9.1, the maximum angular distortion (δ/L) does not necessarily occur at the location of maximum differential settlement (Δ).

TABLE 9.2 Allowable Settlement (*Continued*)

Topic (1)	Discussion (2)
Allowable settlement per Skempton and MacDonald (1956) (*Continued*)	Skempton and MacDonald (1956) studied 98 buildings, where 58 had suffered no damage and 40 had been damaged in varying degrees as a consequence of settlement. From a study of these 98 buildings, Skempton and MacDonald (1956) in part concluded the following: 1. The cracking of the brick panels in frame buildings or load-bearing brick walls is likely to occur if the angular distortion of the foundation exceeds 1/300. Structural damage to columns and beams is likely to occur if the angular distortion of the foundation exceeds 1/150. 2. By plotting the maximum angular distortion (δ/L) versus the maximum differential settlement (Δ) such as shown in Figure 9.2, a correlation was obtained that is defined as $\Delta = 350\ \delta/L$ (note: Δ is in inches). Using this relationship and an angular distortion (δ/L) of 1/300, cracking of brick panels in frame buildings or load bearing brick walls is likely to occur if the maximum differential settlement (Δ) exceeds 32 mm ($1\frac{1}{4}$ inches). 3. The angular distortion criteria of 1/150 and 1/300 were derived from an observational study of buildings of load-bearing-wall construction, and steel and reinforced concrete-frame buildings with conventional brick panel walls but without diagonal bracing. The criteria are intended as no more than a guide for day-to-day work in designing typical foundations for such buildings. In certain cases they may be overruled by visual or other considerations.
Grant et al. (1974)	The paper by Grant et al. (1974) updated the Skempton and MacDonald data pool and also evaluated the rate of settlement with respect to the amount of damage incurred. Grant et al. (1974) in part concluded the following: 1. A building foundation that experiences a maximum value of deflection slope (δ/L) greater than 1/300 will probably suffer some damage. However, damage does not necessarily occur at the point where the local deflection slope exceeds 1/300. 2. For any type of foundation on sand or fill, new data tend to support Skempton and MacDonald's suggested correlation of $\Delta = 350\ \delta/L$ (see Fig. 9.2). 3. Consideration of the rate of settlement is important only for the extreme situations of either very slow or very rapid settlement. Based on the limited data available, the values of maximum δ/L corresponding to building damage appear to be essentially the same for cases involving slow and fast settlements.
Slab-on-grade foundations	Data concerning the behavior of lightly reinforced, conventional slab-on-grade foundations have also been included in Fig. 9.2. This data indicates that cracking of gypsum wall board panels is likely to occur if the angular distortion of the slab-on-grade foundation exceeds 1/300 (Day 1990a). The data plotted in Fig. 9.2 would indicate that the relationship $\Delta = 350\ \delta/L$ can also be used for buildings supported by lightly reinforced slab-on-grade foundations. Using δ/L = 1/300 as the boundary where cracking

TABLE 9.2 Allowable Settlement (*Continued*)

Topic (1)	Discussion (2)
Slab-on-grade foundations (*Continued*)	of panels in wood-frame residences supported by concrete slab-on-grade is likely to occur and substituting this value into the relationship $\Delta = 350\ \delta/L$ (Fig. 9.2), the calculated differential slab displacement is 32 mm (1¼ inches). For buildings on lightly reinforced slabs-on-grade, cracking of gypsum wallboard panels is likely to occur when the maximum slab differential exceeds 32 mm (1¼ inches).
Allowable settlement per Terzaghi (1938)	Concerning allowable settlement Terzaghi (1938) stated: "Differential settle ment must be considered inevitable for every foundation, unless the foundation is supported by solid rock. The effect of the differential settlement on the building depends to a large extent on the type of construction." Terzaghi (1938) summarized his studies on several buildings in Europe where he found that walls 18 m (60 ft) and 23 m (75 ft) long with differential settlements over 2.5 cm (1 in.) were all cracked, but four buildings with walls 12 m (40 ft) to 30 m (100 ft) long were undamaged when the differential settlement was 2 cm (¾ in.) or less. This is probably the basis for the general design guide that building foundations should be designed so that the differential settlement is 2 cm (¾ in.) or less.
Allowable settlement per Sowers (1962)	Another example of allowable settlements for buildings is Table 9.3 (from Sowers 1962). In this table, the allowable foundation displacement has been divided into three categories: total settlement, tilting, and differential movement. Table 9.3 indicates that those structures that are more flexible (such as simple steel frame buildings) or have more rigid foundations (such as mat foundations) can sustain larger values of total settlement and differential movement.
Allowable settlement per Bjerrum (1963)	Figure 9.3 presents data from Bjerrum (1963). Similar to the studies previously mentioned, this figure indicates that cracking in panel walls is to be expected at an angular distortion (δ/L) of 1/300 and that structural damage of buildings is to be expected at an angular distortion (δ/L) of 1/150. This figure also provides other limiting values of angular distortion (δ/L), such as for buildings containing sensitive machinery or overhead cranes.
Settlement versus cracking damage	Table 9.4 summarizes the severity of cracking damage versus approximate crack widths, typical values of maximum differential movement Δ, and maximum angular distortion δ/L of the foundation (Burland et al. 1977, Boone 1996, Day 1998). The relationship between differential settlement Δ and angular distortion δ/L) was based on the equation $\Delta = 350\ \delta/L$ (from Fig. 9.2). When assessing the severity of damage for an existing structure, the damage category (Table 9.4) should be based on multiple factors, including crack widths, differential settlement, and the angular distortion of the foundation. Relying on only one parameter, such as crack width, can be inaccurate in cases where cracking has been hidden or patched, or in cases where other factors (such as concrete shrinkage) contribute to crack widths.

TABLE 9.2 Allowable Settlement (*Continued*)

Topic (1)	Discussion (2)
Component of lateral movement	Foundations subjected to settlement can be damaged by a combination of both vertical and horizontal movements. For example, a common cause of foundation damage is fill settlement. Figure 9.4 shows an illustration of the settlement of fill in a canyon environment. Over the sidewalls of the canyon, there tends to be a pulling or stretching of the ground surface (tensional features), with compression effects near the canyon centerline. This type of damage is due to two-dimensional settlement, where the fill compresses in both the vertical and horizontal directions (Lawton et al. 1991, Day 1991a). Another common situation where both vertical and horizontal foundation displacement occurs is at cut-fill transitions. A cut-fill transition occurs when a building pad has some rock removed (the cut portion), with a level building pad being created by filling in (with soil) the remaining portion. If the cut side of the building pad contains non-expansive rock that is dense and unweathered, then very little settlement would be expected for that part of the building on cut. But the fill portion could settle under its own weight and cause damage. For example, a slab crack will typically open at the location of the cut-fill transition. The building is damaged by both the vertical foundation movement (settlement) and the horizontal movement, which manifests itself as a slab crack and drag effect on the structure. A typical measure to prevent this type of damage is shown in Standard Detail No. 6 (Appendix B), where the cut portion of the building pad in undercut and replaced with fill to eliminate the abrupt change in bearing resistance at a cut-fill transition. In the cases described above, the lateral movement is a secondary result of the primary vertical movement due to settlement of the foundation. Therefore, Table 9.4 can be used as a guide to correlate damage category with Δ and δ/L. In cases where lateral movement is the most predominant or critical mode of foundation displacement, Table 9.4 may underestimate the severity of cracking damage for values of Δ and δ/L (Day 1998, Boone 1998).

TABLE 9.3 Allowable Settlement

Type of movement (1)	Limiting factor (2)	Maximum settlement (3)
Total settlement	Drainage	15 to 30 cm (6 to 12 in.)
	Access	30 to 60 cm (12 to 24 in.)
	Probability of nonuniform settlement:	
	Masonry walled structure	2.5 to 5 cm (1 to 2 in.)
	Framed structures	5 to 10 cm (2 to 4 in.)
	Smokestacks, silos, mats	8 to 30 cm (3 to 12 in.)
Tilting	Stability against overturning	Depends on H and W
	Tilting of smokestacks, towers	$0.004L$
	Rolling of trucks, etc.	$0.01L$
	Stacking of goods	$0.01L$
	Machine operation—cotton loom	$0.003L$
	Machine operation—turbogenerator	$0.0002L$
	Crane rails	$0.003L$
	Drainage of floors	$0.01L$ to $0.02L$
Differential movement	High continuous brick walls	$0.0005L$ to $0.001L$
	One-story brick mill building, wall cracking	$0.001L$ to $0.002L$
	Plaster cracking (gypsum)	$0.001L$
	Reinforced concrete building frame	$0.0025L$ to $0.004L$
	Reinforced concrete building curtain walls	$0.003L$
	Steel frame, continuous	$0.002L$
	Simple steel frame	$0.005L$

Notes: L = distance between adjacent columns that settle different amounts, or between any two points that settle differently. Higher values are for regular settlements and more tolerant structures. Lower values are for irregular settlement and critical structures. H = height and W = width of structure.

Source: From Sowers (1962).

TABLE 9.4 Severity of Cracking Damage

Damage category (1)	Description of typical damage (2)	Approx. crack width (3)	Δ (4)	δ/L (5)
Negligible	Hairline cracks	< 0.1 mm	< 3 cm (<1.2 in.)	< 1/300
Very slight	Very slight damage includes fine cracks that can be easily treated during normal decoration, perhaps an isolated slight fracture in building, and cracks in external brickwork visible on close inspection	1 mm	3–4 cm (1.2–1.5 in.)	1/300 to 1/240
Slight	Slight damage includes cracks that can be easily filled and redecoration would probably be required; several slight fractures may appear showing on the inside of the building; cracks that are visible externally and some repointing may be required; doors and windows may stick	3 mm	4–5 cm (1.5–2.0 in.)	1/240 to 1/175
Moderate	Moderate damage includes cracks that require some opening up and can be patched by a mason; recurrent cracks that can be masked by suitable linings; repointing of external brickwork and possibly a small amount of brickwork replacement may be required; doors and windows stick; service pipes may fracture; weather-tightness is often impaired	5 to 15 mm or a number of cracks > 3 mm	5–8 cm (2.0–3.0 in.)	1/175 to 1/120
Severe	Severe damage includes large cracks requiring extensive repair work involving breaking out and replacing sections of walls (especially over doors and windows); distorted windows and door frames; noticeably sloping floors; leaning or bulging walls; some loss of bearing in beams; and disrupted service pipes	15 to 25 mm but also depends on number of cracks	8–13 cm (3.0–5.0 in.)	1/120 to 1/70
Very severe	Very severe damage often requires a major repair job involving partial or complete rebuilding; beams lose bearing; walls lean and require shoring; windows are broken with distortion; and there is danger of structural instability	Usually > 25 mm but also depends on number of cracks	> 13 cm (> 5 in.)	> 1/70

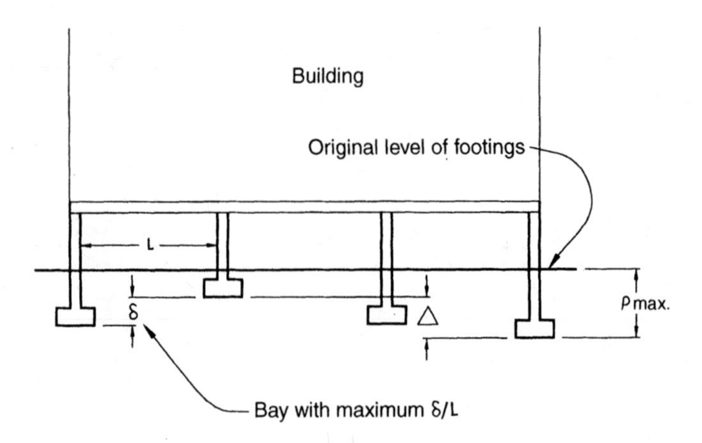

FIGURE 9.1 Diagram illustrating the definitions of maximum angular distortion and maximum differential settlement. See Table 9.2.

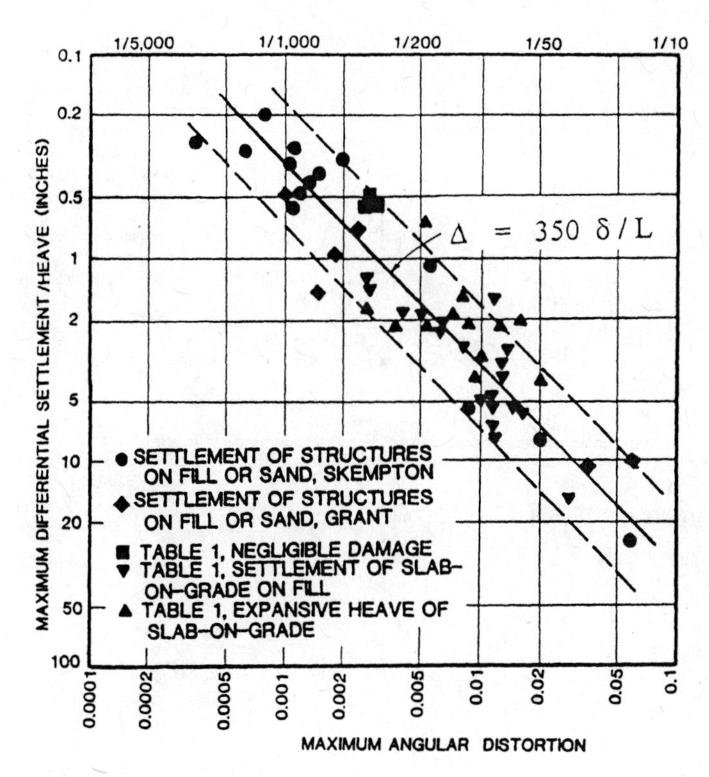

FIGURE 9.2 Maximum differential settlement versus maximum angular distortion. See Table 9.2. (*Data from Skempton and MacDonald 1956 and Day 1990a.*)

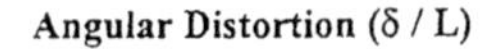

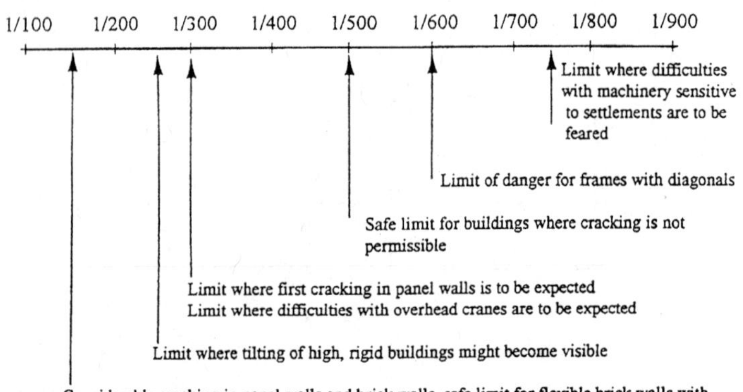

FIGURE 9.3 Damage criteria. See Table 9.2. (*After Bjerrum 1963.*)

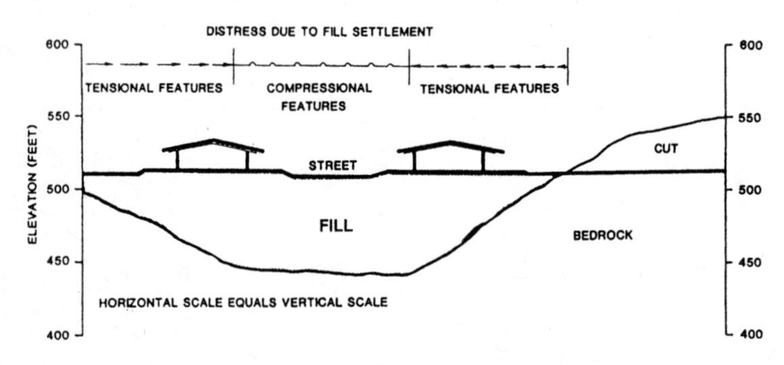

FIGURE 9.4 Fill settlement in a canyon environment. See Table 9.2.

TABLE 9.5 Collapsible Soil

Topic (1)	Discussion (2)
General Discussion	In the southwestern United States, probably the most common cause of settlement is collapsible soil. For example, Johnpeer (1986) states that ground subsidence from collapsing soils is a common occurrence in New Mexico. The category of collapsible soil would include settled debris, uncontrolled fill, deep fill, or natural soil, such as alluvium or colluvium. Collapsible soil can be broadly classified as soil that is susceptible to a large and sudden reduction in volume upon wetting. Collapsible soil usually has a low dry density and low moisture content. Such soil can withstand a large applied vertical stress with a small compression, but then experience much larger settlements after wetting, with no increase in vertical pressure (Jennings and Knight 1957). As such, collapsible soil falls within the second basic category of settlement (see Table 9.1), which is settlement of the structure due to secondary influences. In general, there has been an increase in damage due to collapsible soil, probably because of the lack of available land in many urban areas. This causes development of marginal land, which may contain deposits of dumped fill or deposits of natural collapsible soil. Also, substantial grading can be required to develop level building pads, which results in more areas having deep fill.
Collapsible fill	Deep fill has been defined as fill that has a thickness greater than 6 m (20 ft), (Greenfield and Shen 1992). Uncontrolled fills include fills that were not documented with compaction testing as they were placed; these include dumped fills, fills dumped under water, hydraulically placed fills, and fills that may have been compacted but for which there is no documentation of testing or the amount of effort that was used to perform the compaction (Greenfield and Shen 1992). These conditions may exist in rural areas where inspections are lax or for structures built many years ago when the standards for fill compaction were less rigorous. For collapsible fill, compression will occur as the overburden pressure increases. The increase in overburden pressure could be due to the placement of overlying fill or the construction of a building on top of the fill. The compression due to this increase in overburden pressure involves a decrease in void ratio of the fill due to expulsion of air. The compression usually occurs at constant moisture content. After completion of the fill mass, water may infiltrate the soil because of irrigation, rainfall, or leaky water pipes. The mechanism that usually causes the collapse of the loose soil structure is a decrease in negative pore water pressure (capillary tension) as the fill becomes wet. For a fill specimen submerged in distilled water, the main variables that govern the amount of one-dimensional collapse are the soil type, compacted moisture content, compacted dry density, and the vertical pressure (Dudley 1970, Lawton et al. 1989, 1991, 1992, Tadepalli and Fredlund 1991, Day 1994b). In general, the one-dimensional collapse of fill will increase as the dry density decreases, the moisture content decreases, or the vertical pressure increases.

TABLE 9.5 Collapsible Soil (*Continued*)

Topic (1)	Discussion (2)
Collapsible fill (*Continued*)	For a constant dry density and moisture content, the one-dimensional collapse will decrease as the clay fraction increases once the optimum clay content (usually a low percentage) is exceeded (Rollins et al. 1994).
Collapsible alluvium and colluvium	For natural deposits of collapsible soil in the arid climate of the southwest, a common mechanism involved in rapid volume reduction entails breaking of bonds at coarse particle contacts by weakening of fine-grained materials brought there by surface tension in evaporating water. In other cases, the alluvium or colluvium may have an unstable soil structure which collapses as the wetting front passes through the soil.

TABLE 9.6 Laboratory Testing of Collapsible Soil

Topic (1)	Discussion (2)
General discussion	If the results of field exploration indicate the possible presence of collapsible soil at the site, then undisturbed soil specimens should be obtained and tested in the laboratory. One-dimensional collapse is usually measured in the oedometer (ASTM D 5333-96 "Standard Test Method for Measurement of Collapse Potential of Soils" 1998).
Laboratory testing procedure	The laboratory testing procedure consists of placing the soil specimen in an oedometer (see Table 3.8) and then increasing the vertical pressure until it approximately equals the anticipated overburden pressure after completion of the structure. At this vertical pressure, distilled water is added to the oedometer to measure the amount of collapse of the soil specimen. Percent collapse (%*C*) is defined as the change in height of the specimen due to inundation divided by the initial height of the specimen (ASTM D 5333-96, 1998). For projects where it is anticipated that there will be fill placement, the fill specimens can be prepared by compacting them to the anticipated field as-compacted density and moisture condition. The fill specimens would then be subjected to the anticipated overburden pressure. The procedure for measuring the one-dimensional collapse of the fill specimen would then be the same as outlined above (ASTM D 5333-96, 1998).
Laboratory test results	Figures 9.5 and 9.6 present the results of a one-dimensional collapse test performed on a fill specimen. The fill specimen contains 60% sand-size particles, 30% silt-size particles, and 10% clay-size particles and is classified as a silty sand (SM). To model field conditions, the silty sand was compacted at a dry unit weight γ_d of 14.5 kN/m^3 (92.4 pcf) and water content w of 14.8%. The silty sand specimen, having an initial height of 25.4 mm (1.0 in.), was subjected to a vertical stress of 144 kPa (3000 psf) and then inundated with distilled water (see Fig. 9.5). Figure 9.6 shows the amount of vertical deformation (collapse) as a function of time after inundation. Percent collapse (%*C*) is defined as: $\%C = \frac{100\ \Delta e_c}{1 + e_o}$ or $\%C = \frac{100\ \Delta H_c}{H_o}$ where Δe_c = change in void ratio upon wetting e_o = initial void ratio ΔH_c = change in height upon wetting H_o = initial height For the collapse test on the silty sand, the percent collapse (%*C*) = (2.62 mm/25.4 mm) × 100 = 10.3%.
Collapse potential	The collapse potential (CP) of a soil can be determined by applying a vertical stress of 200 kPa (2 tsf) to the soil specimen, submerging it in distilled water, and then determining the percent collapse, which is designated the collapse potential (CP). This collapse potential can be considered as an index test to compare the susceptibility of collapse for different soils. The collapse potential versus severity of the problem is listed below:

TABLE 9.6 Laboratory Testing of Collapsible Soil (*Continued*)

Topic (1)	Discussion (2)	
Collapse potential (*Continued*)	Collapse potential (CP)	Severity of problem
	0	No problem
	0.1 to 2%	Slight
	2.1 to 6%	Moderate
	6.1 to 10%	Moderately severe
	> 10%	Severe

Note: For collapse potential (CP), the vertical stress upon inundation must equal 200 kPa. Collapse potential values are based on ASTM D 5333-96, 1998.

TABLE 9.7 Settlement Analyses for Collapsible Soil

Topic (1)	Discussion (2)
Settlement analyses	For collapsible soil, the settlement analysis is usually as follows: **1.** The amount of collapse for the different soil layers underlying the site are obtained from the laboratory oedometer tests on undisturbed samples (ASTM D 5333-96, 1998). The vertical stress used in the laboratory testing should equal the overburden pressure plus the weight of the structure (using stress distribution, Sec. 6.3). **2.** To obtain the settlement of each collapsible soil layer, the percent collapse is multiplied by the thickness of the soil layer. **3.** The total settlement is the sum of the collapse value from the different soil layers.
Example	As an example, suppose the data shown in Figs. 9.5 and 9.6 represents a collapse test on an undisturbed soil specimen taken from the middle of a 3-m (10-ft) uniform layer of collapsible soil. Then the estimated settlement for this layer of collapsible soil would be 0.103 (10.3%) times 3 m (10 ft), or a settlement of 0.31 m (1.0 ft).
Maximum and differential settlement	The above analysis could be performed for each boring excavated within the proposed footprint of the building. The settlement could then be calculated for each boring and the largest value would be the maximum settlement (ρ_{max}). The maximum differential settlement Δ is more difficult to determine because it depends to a large extent on how water infiltrates the collapsible soil. For example, for the wetting of the collapsible soil only under a corner or portion of a building, the maximum differential settlement Δ could approach in value the maximum total settlement ρ_{max}. Usually because of the unknown wetting conditions that will prevail in the field, the maximum differential settlement Δ is assumed to be from 50 to 75% of the maximum total settlement ρ_{max}. The amount of time for this settlement to occur depends on the availability of water and the rate of infiltration of the water. Usually the collapse process in the field is slow and may take many years to complete.
Design and construction for collapsible soil	There are many different methods to deal with collapsible soil. If there is a shallow deposit of natural collapsible soil, then the deposit can be removed and recompacted during the grading of the site. In some cases, the soil can be densified (for example, by compaction grouting) to reduce the collapse potential of the soil. Another method to deal with collapsible soil is to flood the building footprint or force water into the collapsible soil stratum by using wells. As the wetting front moves through the ground, the collapsible soil will densify and reach an equilibrium state. Flooding or forcing water into collapsible soil should not be performed if there are adjacent buildings because of the possibility of damaging these structures. Also, after the completion of the flooding process, subsurface exploration and laboratory testing should be performed to evaluate the effectiveness of the process.

TABLE 9.7 Settlement Analyses for Collapsible Soil (*Continued*)

Topic (1)	Discussion (2)
Design and construction for collapsible soil (*Continued*)	There are also foundation options that can be used for sites containing collapsible soil. A deep foundation system, that derives support from strata below the collapsible soil, could be constructed. Also, posttensioned foundations or mat slabs can be designed and installed to resist the larger anticipated settlement from the collapsible soil. As previously mentioned, the triggering mechanism for the collapse of fill or natural soil is the introduction of moisture. Common reasons for the infiltration of moisture include water from irrigation, broken or leaky water lines, and an altering of surface drainage which allows rainwater to pond near the foundation. For sites with collapsible soil, it is good practice to emphasize the importance of positive drainage (no ponding of water at the site) and an immediate repair of any leaking utilities.

TABLE 9.8 Collapse and Settlement Caused by Soluble Soil Particles

Topic (1)	Discussion (2)
General discussion	Especially in arid parts of the world, the soil may contain soluble soil particles, such as halite (salt). Deposits of halite can form in salt playas, sabkhas (coastal salt marshes), and salinas (Bell 1983). Besides halite, the soil may contain other minerals that are soluble, such as magnesium or calcium carbonate (caliche) and gypsum (gypsiferous soil). These soluble soil particles are dense and hard enough to carry the overburden pressure. But after the site is developed, there can be infiltration of water into the ground from irrigation or leaky water pipes. As this water penetrates the soil containing soluble minerals, two types of settlement can occur: (1) the collapse of the soil structure due to weakening of salt cemented bonds at particle contacts and (2) the water can dissolve away the soluble minerals (i.e., a loss of solids) resulting in ground surface settlement.
Percent soluble soil particles	A simple method to determine the presence of soluble minerals is to perform a permeameter test. A specimen of the soil, having a dry mass (M_o) of about 100g, is placed in the permeameter apparatus. Filter paper should be used to prevent the loss of fines during the test. Distilled water is slowly flushed through the soil specimen. Usually about 2 liters of distilled water is flushed through the soil specimen. After flushing, the soil is dried and the percent soluble soil particles (% soluble) is determined as the initial (M_o) minus final (M_f) dry mass divided by the initial dry mass of soil (M_o), expressed as a percentage, i.e., $$\%\ \text{soluble}\ (S_L) = 100\ (M_o - M_f)/M_o$$ As a check on the amount of soluble minerals, the water flushed through the soil specimen can be collected, placed in a sedimentation cylinder (1000 mL), and a hydrometer can be used to determine the amount (grams) of dissolved minerals. As an alternative, the water flushed through the soil specimen can be boiled and the residue collected and weighed.
Settlement analysis	Ground surface settlement due to the dissolution of soluble soil particles can be calculated from the following equation: $$s = S_L H (G_s/G_{sol})$$ where s = settlement of the soil layer due to loss of soluble soil particles S_L = soluble soil particles in the soil (above equation expressed as a fraction) H = thickness of the soluble soil layer G_s = specific gravity of the insoluble soil minerals G_{sol} = specific gravity of the soluble soil minerals The above calculated settlements should be added to the collapse settlement determined from oedometer tests.
Example	Suppose a 2-ft-thick soil layer has an average value of 10% soluble gypsum soil particles (G_{sol} = 2.35) and a specific gravity G_s of the insoluble soil minerals = 2.70. The amount of settlement due to loss of soluble soil particles is as follows: $$s = S_L H\ (G_s/G_{sol}) = (0.10)(2\ \text{ft})(2.70/2.35) = 0.23\ \text{ft or } 2.8\ \text{in.}$$

TABLE 9.8 Collapse and Settlement Caused by Soluble Soil Particles (*Continued*)

Topic (1)	Discussion (2)	
Design and construction for soluble soil	A common recommendation is that soil which contains above 6% soluble soil particles can not be used as structural fill (Converse Consultants Southwest, Inc. 1990). Soil having between 2 and 6% soluble minerals can be blended with nonsoluble soil in accordance with the following ratios (per Converse Consultants Southwest, Inc. 1990):	
	Percent soluble soil particles	Blending proportions (nonsoluble soil:soluble soil)
	2 to 4	1:1
	4 to 6	2:1
	above 6	Can not be used as structural fill

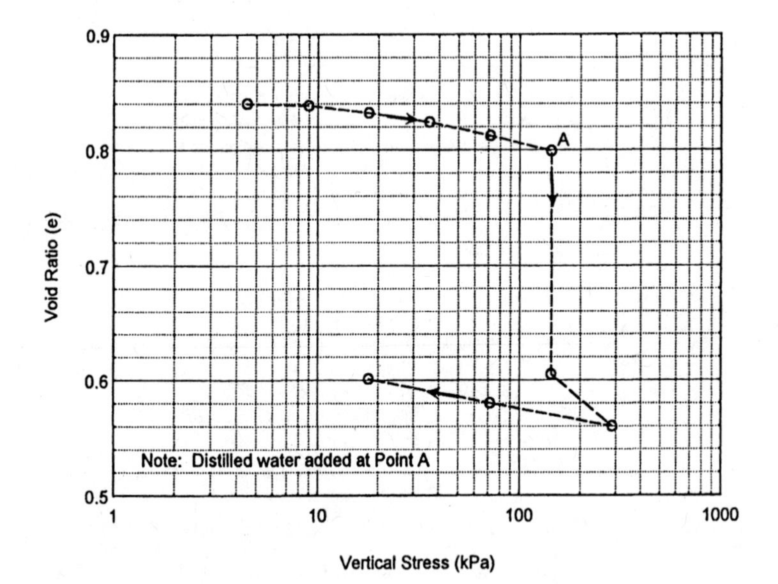

FIGURE 9.5 Vertical pressure versus void ratio for a silty sand. See Table 9.6.

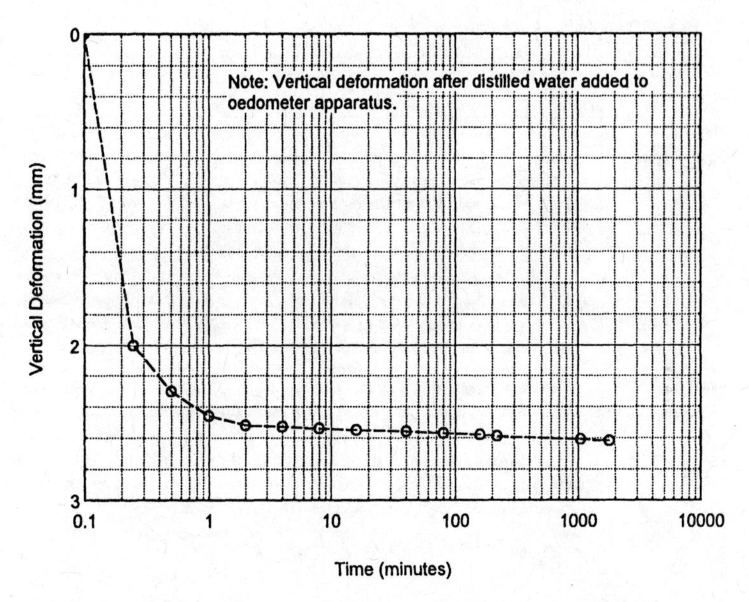

FIGURE 9.6 Vertical deformation versus time for a silty sand. See Table 9.6.

TABLE 9.9 Analyses for Immediate Settlement

Topic (1)	Discussion (2)
General discussion	In most situations, surface loading causes both vertical and horizontal strains, and this is referred to as two- or three-dimensional loading. Immediate settlement is due to undrained shear deformations, or in some cases contained plastic flow, caused by the two- or three-dimensional loading (Ladd et al. 1977, Foott and Ladd 1981, Day 1995b). Common examples of three-dimensional loadings are those from square footings and round storage tanks. A field condition that often leads to a substantial amount of immediate settlement is a thick near- or at-surface clay layer that is normally consolidated (OCR = 1) or is subjected to a quick loading ($\Delta\sigma_v$) that results in a vertical stress exceeding the maximum past pressure σ'_p, as illustrated in Fig. 9.7.
Theory of elasticity	There are many different methods available to determine the amount of immediate settlement for two- or three-dimensional loadings. One approach is to use the undrained modulus E_u (also known as the *stress-strain modulus* or *modulus of elasticity of the soil*) from triaxial compression tests in order to estimate the immediate settlement based on elastic theory. A commonly used equation based on the theory of elasticity is as follows: $$s_i = qBI\frac{1-\nu^2}{E_u}$$ where s_i = immediate settlement of the loaded area (m or ft) q = uniform pressure applied by the foundation (kPa or psf) B = width of the foundation (m or ft) I = dimensionless parameter derived from the theory of elasticity to account for the thickness of the compressible layer, shape of the foundation, and flexibility of the foundation. Theoretical influence values have been obtained for the situation of an underlying very thick compressible layer relative to the size of the foundation or for the situation of a relatively incompressible layer at a depth equal to H (Fig. 9.8). ν = Poisson's ratio, which is often assumed to be 0.5 for saturated plastic soil subjected to undrained loading (dimensionless parameter) E_u = undrained modulus of the clay (kPa or psf). This is often obtained from an undrained triaxial compression test performed on an undisturbed specimen where the stress-strain curve is plotted (i.e., plot on the left side of Fig. 7.7). The initial tangent modulus to the stress-strain curve is often assumed to be equal to E_u. This approach based on the theory of elasticity will often provide approximate results, provided the entire soil deposit underlying the foundation is overconsolidated and the load does not cause the vertical effective stress to exceed the maximum past pressure. However, using the theory of elasticity could significantly underestimate the amount of immediate settlement for situations where there is plastic flow of the soil (e.g., the example shown in Fig. 9.7).

TABLE 9.9 Analyses for Immediate Settlement (*Continued*)

Topic (1)	Discussion (2)
Plate load tests	A second approach is to perform field plate load tests (see Table 2.13) to measure the amount of immediate settlement due to an applied load. The plate load test could significantly underestimate the immediate settlement if the test is performed on a near surface sandy layer or surface crust of clay that is overconsolidated. Low values of immediate settlement would be recorded because the pressure bulb of the plate load test is very small. But when a large structure is built, the pressure bulb is much larger, which could result in significant plastic flow if there is a normally consolidated clay layer underlying the stiff surface layer.
Stress path method	A third approach is to determine the amount of immediate settlement from actual laboratory tests that model the field loading conditions. An undisturbed soil specimen could be set up in the triaxial apparatus, and then the specimen could be subjected to vertical and horizontal stresses that are equivalent to the anticipated loading condition. The undrained vertical deformation (i.e., immediate settlement) due to the applied loading could then be measured. By measuring the amount of vertical deformation from a series of specimens taken from various depths below the proposed structure, the total amount of immediate settlement could be calculated. This approach has been termed the *stress-path method* (Lambe 1967).

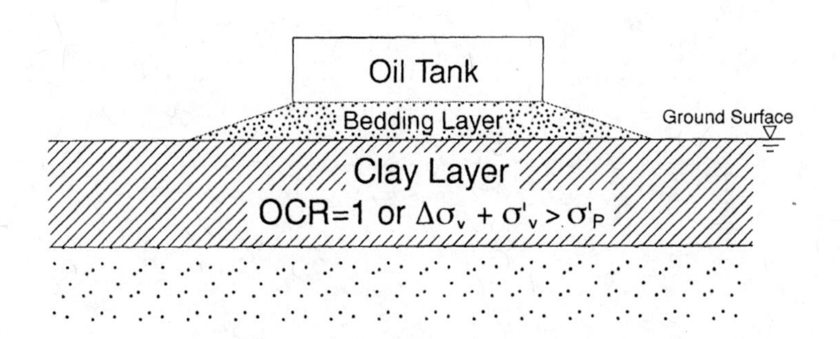

FIGURE 9.7 Example of a condition causing significant immediate settlement. See Table 9.9.

Shape and Rigidity Factor I for Loaded Areas on an Elastic Half-Space of Infinite Depth

Shape and Rigidity	Center	Corner	Edge/Middle of Long Side	Average
Circle (flexible)	1.00		0.64	0.85
Circle (rigid)	0.79		0.79	0.79
Square (flexible)	1.12	0.56	0.76	0.95
Square (rigid)	0.82	0.82	0.82	0.82
Rectangle: (flexible) length/width				
2	1.53	0.76	1.12	1.30
5	2.10	1.05	1.68	1.82
10	2.56	1.28	2.10	2.24
Rectangle: (rigid) length/width				
2	1.12	1.12	1.12	1.12
5	1.6	1.6	1.6	1.6
10	2.0	2.0	2.0	2.0

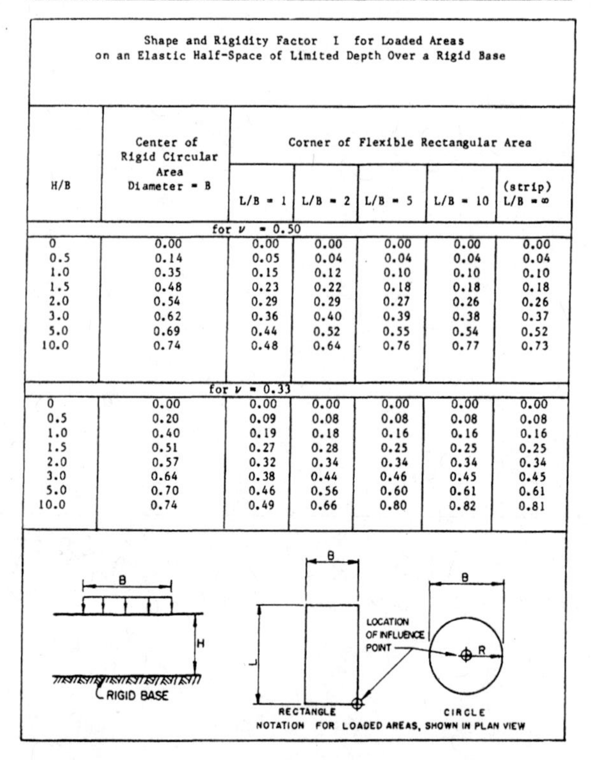

Shape and Rigidity Factor I for Loaded Areas on an Elastic Half-Space of Limited Depth Over a Rigid Base

H/B	Center of Rigid Circular Area Diameter = B	Corner of Flexible Rectangular Area				
		L/B = 1	L/B = 2	L/B = 5	L/B = 10	(strip) L/B = ∞
for ν = 0.50						
0	0.00	0.00	0.00	0.00	0.00	0.00
0.5	0.14	0.05	0.04	0.04	0.04	0.04
1.0	0.35	0.15	0.12	0.10	0.10	0.10
1.5	0.48	0.23	0.22	0.18	0.18	0.18
2.0	0.54	0.29	0.29	0.27	0.26	0.26
3.0	0.62	0.36	0.40	0.39	0.38	0.37
5.0	0.69	0.44	0.52	0.55	0.54	0.52
10.0	0.74	0.48	0.64	0.76	0.77	0.73
for ν = 0.33						
0	0.00	0.00	0.00	0.00	0.00	0.00
0.5	0.20	0.09	0.08	0.08	0.08	0.08
1.0	0.40	0.19	0.18	0.16	0.16	0.16
1.5	0.51	0.27	0.28	0.25	0.25	0.25
2.0	0.57	0.32	0.34	0.34	0.34	0.34
3.0	0.64	0.38	0.44	0.46	0.45	0.45
5.0	0.70	0.46	0.56	0.60	0.61	0.61
10.0	0.74	0.49	0.66	0.80	0.82	0.81

FIGURE 9.8 Shape and rigidity factors *I* for calculating immediate settlement from the theory of elasticity. See Table 9.9. (*From NAVFAC DM-7.1, 1982.*)

TABLE 9.10 Primary Consolidation and Overconsolidation Ratio (OCR)

Topic (1)	Discussion (2)
Definition of primary consolidation	The increase in vertical pressure due to the weight of the structure constructed on top of saturated soft clays and organic soil will initially be carried by the pore water in the soil. This increase in pore water pressure is known as an excess pore water pressure u_e. The excess pore water pressure will decrease with time, as water slowly flows out of the cohesive soil. This flow of water from cohesive soil (which has a low permeability) as the excess pore water pressures slowly dissipate is known as *primary consolidation*, or simply *consolidation*. As the water slowly flows from the cohesive soil, the structure settles as the load is transferred to the soil particle skeleton, thereby increasing the effective stress of the soil. Consolidation is a time-dependent process that may take many years to complete. The typical one-dimensional case of settlement involves strain in only the vertical direction. Common examples of one-dimensional loading include the lowering of the groundwater table or a uniform fill surcharge applied over a very large area. In the case of a one-dimensional loading, both the strain and flow of water from the cohesive soil as it consolidates will be only in the vertical direction. Consolidation can also occur for two- or three-dimensional loadings, in which case there would be both horizontal and vertical flow of water from the cohesive soil as it consolidates.
Definition of overconsolidation ratio (OCR)	Depending on the stress history of saturated cohesive soils, they are considered to be either underconsolidated, normally consolidated, or overconsolidated. The overconsolidation ratio (OCR) is used to describe the stress history of cohesive soil, and it is defined as: $OCR = \sigma'_{vm}/\sigma'_{vo}$ where OCR = overconsolidation ratio (dimensionless parameter); σ'_{vm} or σ'_p = maximum past pressure (σ'_{vm}), also known as the preconsolidation pressure σ'_p, which is equal to the highest previous vertical effective stress that the cohesive soil was subjected to and consolidated under (kPa or psf); and σ'_{vo} or σ'_v = existing vertical effective stress (kPa or psf).
Stress history of saturated cohesive soil	In terms of the stress history of a cohesive soil, there are three possible conditions: *Underconsolidated* (*OCR* < 1). A saturated cohesive soil is considered underconsolidated if the soil is not fully consolidated under the existing overburden pressure and excess pore water pressures u_e exist within the soil. Underconsolidation occurs in areas where a cohesive soil is being deposited very rapidly and not enough time has elapsed for the soil to consolidate under its own weight.

TABLE 9.10 Primary Consolidation and Overconsolidation Ratio (OCR) (*Continued*)

Topic (1)	Discussion (2)
Stress history of saturated cohesive soil (*Continued*)	*Normally consolidated* (*OCR* = *1*). A saturated cohesive soil is considered normally consolidated if it has never been subjected to an effective stress greater than the existing overburden pressure and if the deposit is completely consolidated under the existing overburden pressure. *Overconsolidated or preconsolidated* (*OCR* > *1*). A saturated cohesive soil is considered overconsolidated if it has been subjected in the past to a vertical effective stress greater than the existing vertical effective stress. An example of a situation that creates an overconsolidated soil is where a thick overburden layer of soil has been removed by erosion over time. As indicated in Table 9.11, there are other mechanisms that can cause a cohesive soil to become overconsolidated.
Importance of the overconsolidation ratio	If the cohesive soil is underconsolidated, then considerable settlement due to continued consolidation caused by the soil's own weight as well as the applied structural load would be expected. On the other hand, if the cohesive soil is highly overconsolidated, then a load can often be applied to the cohesive soil without significant settlement.

TABLE 9.11 Mechanisms That Can Create an Overconsolidated Soil

Main mechanism (1)	Item creating overconsolidated soil (2)	Remarks or references (3)
Change in total stress	Removal of overburden	Soil erosion
	Past structures	Human-induced factors
	Weight of past glaciers	Melting of glaciers
Change in pore water pressure	Change in groundwater table elevation	Sea-level changes (Kenney 1964)
	Artesian pressures	Common in glaciated areas
	Deep pumping of groundwater	Common in many cities
	Desiccation due to drying	May have occurred during deposition
	Desiccation due to plant life	
Change in soil structure	Secondary compression (aging)	
	Changes in environment, such as pH, temperature, and salt concentration	Lambe 1958a, 1958b; Bjerrum 1967b; Leonards and Altschaeffl 1964
	Chemical alterations due to weathering, precipitation of cementing agents, and ion exchange	Bjerrum 1967b; Cox, 1968

Source: Ladd (1973).

TABLE 9.12 Laboratory Consolidation Test

Topic (1)	Discussion (2)
Oedometer apparatus	The oedometer apparatus (Fig. 3.1) is used to determine the consolidation properties of saturated cohesive soil (ASTM D 2435-96, 1998). The typical testing procedure consists of placing an undisturbed specimen (usual diameter = 2.5 in., height = 1.0 in.) within the apparatus, applying a vertical seating pressure to the laterally confined specimen, and then submerging the specimen in distilled water. The specimen is then subjected to an incremental increase in vertical pressure, with each pressure remaining on the specimen for a period of 24 hours. Because of the small thickness of the soil specimen, a time period of 24 hours is ordinarily sufficient to allow the cohesive soil to consolidate and come to equilibrium with little or no further deformation and with the excess pore water pressures u_e within the specimen approximately equal to zero. Dial readings of the vertical deformation versus time can be recorded for each vertical pressure applied to the specimen.
Consolidation curve	Figure 9.9 presents results of consolidation tests performed on two specimens of Orinoco clay at a depth of 40 m (128 ft) (Ladd et al. 1980, Day 1980). The upper plot is known as the *consolidation curve* for the two Orinoco clay soil specimens. The horizontal axis is the effective consolidation stress (i.e., the applied vertical pressure) which is designated σ'_{vc}, with the subscripts *vc* referring to vertical consolidation pressure and the prime mark indicating that it is an effective stress. Note that the horizontal axis is a logarithm scale. In Fig. 9.9, the vertical axis is percent vertical strain ϵ_v. The vertical axis could also be in terms of the void ratio *e*. The two consolidation tests shown in Fig. 9.9 start out at zero vertical strain. Then, as the vertical pressure is increased in increments, the soil consolidates and the vertical strain increases. The usual procedure is to apply a vertical pressure that is double the previously applied pressure (load increment ratio = 1.0). However, as shown in Fig. 9.9, in order to better define the consolidation curve, the loading increments can be adjusted to better define the breaking point of the consolidation curve. For the two consolidation tests performed on Orinoco clay (Fig. 9.9), the specimens were loaded up to a vertical pressure of 12 kg/cm^2 (24,600 psf, 1180 kPa) and then unloaded.
Maximum past pressure σ'_{vm} (also known as the preconsolidation pressure, σ'_p)	In Fig. 9.9, the arrows indicate the numerical values of the vertical effective stress σ'_{vo} and the maximum past pressure σ'_{vm}, for both consolidation tests. The vertical effective stress σ'_{vo} was determined by using the effective stress equation listed in Table 6.1. The most commonly used procedure for determining the maximum past pressure σ'_{vm} is to use the Casagrande construction technique (1936). Figure 9.10 illustrates this procedure, which is performed as follows: **1.** Locate the point of minimum radius of the consolidation curve (point *A*).

TABLE 9.12 Laboratory Consolidation Test (*Continued*)

Topic (1)	Discussion (2)
Maximum past pressure σ'_{vm} (also known as the preconsolidation pressure, σ'_p) (*Continued*)	**2.** Draw a line tangent to the consolidation curve at point *A*. **3.** Draw a horizontal line from point *A*. **4.** Bisect the angle made by steps 2 and 3. **5.** Extend the straight line portion of the virgin compression curve up to where it meets the bisect line. The point of intersection (point *B*) of these two lines is the maximum past pressure (also known as preconsolidation pressure, σ'_p). This construction procedure will give the same results if the vertical axis is in terms of vertical strain ϵ_v or void ratio *e*. Since the Casagrande technique is only an approximate method of determining the maximum past pressure, it is often useful to obtain a range in values. Note in Fig. 9.10 that the range in possible values has been determined as varying from point *D* to point *E*.
Sample disturbance	By the Casagrande construction technique, the maximum past pressure σ'_{vm} was determined for the two consolidation tests on Orinoco clay (Fig. 9.9). Note that these two tests have significantly different values of maximum past pressure (2.75 versus 1.35 kg/cm^2). The reason for the difference is sample disturbance. Test 12 (solid circles) was disturbed during sampling and had a vane shear strength (TV) of only 0.30 kg/cm^2 (610 psf), compared to the undisturbed specimen (Test 18) which had a vane shear strength (TV) of 0.51 kg/cm^2 (1040 psf). Thus sample disturbance can significantly lower the value of the maximum past pressure of saturated cohesive soil.
Stress history	In Fig. 9.9, the vertical effective stress σ'_{vo} has a numerical value that is close to the maximum past pressure σ'_{vm} for the undisturbed soil specimen (open circles). This means that the Orinoco clay is essentially normally consolidated. In Fig. 9.11, all of the consolidation test data for the Orinoco clay is summarized on one sheet. The vertical axis is depth (feet) and the horizontal axis is effective stress in kg/cm^2. In Fig. 9.11, the vertical effective stress σ'_{vo} versus depth for the Orinoco clay at borings E1 and F1 has been plotted. The maximum past pressure σ'_{vm} for each consolidation test has also been plotted with possible ranges in values based on the procedure shown in Fig. 9.10. The open symbols represent consolidation tests performed on undisturbed specimens, while the solid symbols represent consolidation tests performed on disturbed specimens. In general, the maximum past pressure data points σ'_{vm} are close to the vertical effective stress σ'_{vo} and for this offshore deposit of clay, it could be concluded that the clay is essentially normally consolidated (OCR = 1) to very slightly overconsolidated. This information on the stress history of saturated clay deposits is very important and it is often added to the subsoil profile. For example, note in Figs. 2.19 to 2.21 that the maximum past pressures (i.e., preconsolidation pressures) obtained from the results of oedometer tests on the cohesive soils have been included on these subsoil profiles.

TABLE 9.12 Laboratory Consolidation Test (*Continued*)

Topic (1)	Discussion (2)
Recompression index C_r and compression index C_c	In addition to using the consolidation curve to determine the maximum past pressure σ'_{vm} of a cohesive soil, the consolidation curve obtained from the oedometer test can also be used to estimate the primary consolidation settlement s_c due to a loading of the cohesive soil, such as from the construction of a building. Note in Fig. 9.10 that the consolidation curve can often be approximated as two straight line segments. The recompression index C_r represents the reloading of the saturated cohesive soil, and the compression index C_c represents the loading of the saturated cohesive soil beyond the maximum past pressure σ'_{vm}. The compression index is often referred to as the slope of the virgin consolidation curve because the *in situ* saturated cohesive soil has never experienced this loading condition. The reason for the relatively steep virgin consolidation curve, compared to the recompression curve, is that there is a tendency for the soil structure to break down and contract once the pressure exceeds the maximum past pressure σ'_{vm}. The values of the recompression index C_r and the compression index C_c are simply the slope of the lines shown in Fig. 9.10 and can be calculated as a change in void ratio (Δe) divided by the corresponding change in the effective pressures ($\log \sigma'_{vc2} - \log \sigma'_{vc1}$), or $C_r = \Delta e / \log (\sigma'_{vc2}/\sigma'_{vc1})$ for the recompression curve and $C_c = \Delta e / \log (\sigma'_{vc2}/\sigma'_{vc1})$ for the virgin consolidation curve. An easier method to obtain C_r and C_c is to determine Δe over one logcycle; for example, if $\sigma'_{vc2} = 100$ and $\sigma'_{vc1} = 10$, then the $\log (\sigma'_{vc2}/(\sigma'_{vc1}) = \log (100/10) = 1$. For one log cycle, the values of $C_r = \Delta e$ for the recompression curve and $C_c = \Delta e$ for the virgin consolidation curve.

TABLE 9.13 Calculating the Amount of Primary Consolidation Settlement

Topic (1)	Discussion (2)
Underconsolidated soil (OCR < 1)	On the basis of the results of the laboratory testing of undisturbed specimens (Table 9.12), the primary consolidation settlement s_c due to an increase in load $\Delta\sigma_v$ can be determined from the following equations: For underconsolidated soil (OCR < 1) $$s_c = C_c \frac{H_o}{1 + e_o} \log \frac{\sigma'_{vo} + \Delta\sigma'_v + \Delta\sigma_v}{\sigma'_{vo}}$$
Normally consolidated soil (OCR = 1)	For normally consolidated soil (OCR = 1) $$s_c = C_c \frac{H_o}{1 + e_o} \log \frac{\sigma'_{vo} + \Delta\sigma_v}{\sigma'_{vo}}$$
Overconsolidated soil (OCR > 1)	For overconsolidated soil (OCR > 1) Case I: $\sigma'_{vo} + \Delta\sigma_v \leq \sigma'_{vm}$ $$s_c = C_r \frac{H_o}{1 + e_o} \log \frac{\sigma'_{vo} + \Delta\sigma_v}{\sigma'_{vo}}$$ Case II: $\sigma'_{vo} + \Delta\sigma_v > \sigma'_{vm}$ $$s_c = C_r \frac{H_o}{1 + e_o} \log \frac{\sigma'_{vm}}{\sigma'_{vo}} + C_c \frac{H_o}{1 + e_o} \log \frac{\sigma'_{vo} + \Delta\sigma_v}{\sigma'_{vm}}$$
Definition of symbols	s_c = settlement due to primary consolidation caused by an increase in load (m or ft) C_c = compression index, from the virgin consolidation curve (dimensionless) C_r = recompression index, obtained from the recompression portion of the consolidation curve (dimensionless) H_o = initial thickness of the *in situ* saturated cohesive soil layer (m or ft) e_o = initial void ratio of the *in situ* saturated cohesive soil layer (dimensionless) σ'_{vo} = initial vertical effective stress of the in situ soil (kPa or psf) $\Delta\sigma'_v$ = for an underconsolidated soil, this represents the increase in vertical effective stress that will occur as the cohesive soil consolidates under its own weight (kPa or psf) $\Delta\sigma_v$ = increase in load, typically due to the construction of a building or the construction of a fill layer at ground surface (kPa or psf). The value of $\Delta\sigma_v$ (also known as σ_z) can be obtained from stress distribution theory as discussed in Sec. 6.3. Note that a drop in the groundwater table or a reduction in pore water pressure can also result in an increase in load on the cohesive soil. σ'_{vm} = maximum past pressure (kPa or psf), also known as the preconsolidation pressure σ'_p. It is obtained from the consolidation curve using the Casagrande (1936) construction technique (see Fig. 9.10).

TABLE 9.13 Calculating the Amount of Primary Consolidation Settlement (*Continued*)

Topic (1)	Discussion (2)
Two cases for overconsolidated soil	For overconsolidated soil, there are two possible cases (Cases I and II) that can be used to calculate the amount of settlement. Case I occurs when the existing vertical effective stress σ'_{vo} plus the increase in vertical stress $\Delta\sigma_v$ due to the proposed building weight does not exceed the maximum past pressure σ'_{vm}. For this first case, there will only be recompression of the cohesive soil. For Case II, the sum of the existing vertical effective stress σ'_{vo} plus the increase in vertical stress $\Delta\sigma_v$ due to the proposed building weight exceeds the maximum past pressure σ'_{vm}. For the second case, there will be virgin consolidation of the cohesive soil. Given the same cohesive soil and identical field conditions, the settlement due to the second case will be significantly more than the first case.
Modified recompression index $C_{r\epsilon}$ and modified compression index $C_{c\epsilon}$	If the consolidation data is plotted on a vertical strain ϵ_v versus consolidation stress σ'_{vc} plot such as shown in Fig. 9.9, then the recompression curve and virgin consolidation curve can also be approximated as straight lines. When using a vertical strain ϵ_v versus consolidation stress σ'_{vc} plot, the slope of the recompression curve is designated the *modified recompression index* $C_{r\epsilon}$ and the slope of the virgin consolidation curve is designated the *modified compression index* $C_{c\epsilon}$. These indices are related as follows: $$C_{r\epsilon} = \frac{C_r}{1 + e_o}$$ $$C_{c\epsilon} = \frac{C_c}{1 + e_o}$$ The values of $C_{r\epsilon}$ and $C_{c\epsilon}$ from the above equations can be substituted into the primary consolidation settlement equations s_c.
Dividing the soil stratum into several layers	The value of H_o in the primary consolidation settlement equations represents the initial thickness of the cohesive soil layer. Because σ'_{vo} and $\Delta\sigma_v$ can both change with depth, the cohesive soil layer may need to be broken into several horizontal layers in order to obtain an accurate value of the primary consolidation settlement s_c.

TABLE 9.14 Terzaghi One-Dimensional Consolidation Theory

Topic (1)	Discussion (2)
Terzaghi one-dimensional consolidation theory	The purpose of the Terzaghi theory of consolidation is to estimate the settlement versus time relationship after loading of the saturated cohesive soil. When such soil is loaded, there is an increase in the pore water pressure (known as excess pore water pressure u_e). As water flows out of the cohesive soil, this excess pore water pressure slowly dissipates. In order to determine how fast this consolidation process will take, Terzaghi (1925) developed the one-dimensional consolidation equation to describe the time-dependent settlement behavior of clays. The Terzaghi equation is a form of the diffusion equation from mathematical physics. There are other phenomena that can be described by the diffusion equation, such as the heat flow through solids.
Assumptions for Terzaghi one-dimensional consolidation theory	In order to derive Terzaghi's diffusion equation, the following assumptions are made (Terzaghi 1925, Taylor 1948, Holtz and Kovacs 1981): **1.** The clay is homogeneous and the degree of saturation S is 100% (saturated soil). **2.** Drainage is provided at the top and/or bottom of the compressible layer. **3.** Darcy's law ($v = ki$) is valid. **4.** The soil grains and pore water are incompressible. **5.** Both the compression and flow of water are one-dimensional. **6.** The load increment results in small strains so that the coefficient of permeability k and the coefficient of compressibility a_v remain constant. **7.** There is no secondary compression.
Derivation of Terzaghi one-dimensional consolidation theory	Terzaghi's one-dimensional consolidation theory is based on the continuity of flow out of a soil element, and is: $$c_v \frac{\partial^2 u_e}{\partial z^2} = \frac{\partial u_e}{\partial t}$$ where u_e = excess pore water pressure z = vertical height to the nearest drainage boundary t = time c_v = coefficient of consolidation, defined as: $$c_v = \frac{k(1 + e_o)}{\gamma_w a_v}$$ where k = coefficient of permeability (hydraulic conductivity) e_o = initial void ratio γ_w = unit weight of water a_v = coefficient of compressibility, commonly defined as: $$a_v = \frac{-\Delta e}{\Delta \sigma'_v}$$ where Δe is the change in void ratio corresponding to the change in vertical effective stress $\Delta \sigma'_v$.

TABLE 9.14 Terzaghi One-Dimensional Consolidation Theory (*Continued*)

Topic (1)	Discussion (2)
Solution to Terzaghi one-dimensional consolidation theory	The mathematical solution to Terzaghi's one-dimensional consolidation equation can be expressed in graphical form as shown in Fig. 9.12. The vertical axis is defined as $$Z = z/H$$ where z = vertical depth below the top of the clay layer and H = one-half the thickness of the clay layer (double drainage). The horizontal axis is the consolidation ratio U_z defined as: $$U_z = 1 - (u_e/u_o)$$ where u_e = excess pore water pressures in the cohesive soil and u_o = initial excess pore water pressure which is equal to the applied load on the saturated cohesive soil (i.e., $u_o = \Delta\sigma_v$).
Time factor T	In Fig. 9.12, the value of U_z can be determined for different time factors T, defined as (time factor equation): $$T = \frac{c_v t}{H_{dr}^2}$$ where T = time factor from Table 9.15 (dimensionless) c_v = coefficient of consolidation from laboratory testing (cm^2/sec or ft^2/day) H_{dr} = height of the drainage path (cm or ft). If water can drain out through the top and bottom of the cohesive soil layer (double drainage), then $H_{dr} = {}^1/2H$, where H = thickness of the cohesive soil layer. An example of double drainage would be if there are sand layers located on top and below the cohesive soil layer. If water can drain out of only the top or bottom of the cohesive soil layer (single drainage), then $H_{dr} = H$. An example of single drainage would be if the clay layer is underlain by dense shale that is essentially impervious. t = time since the application of the load (assumed to be applied instantaneously)
Average degree of consolidation settlement U_{avg}	As previously mentioned, the purpose of the Terzaghi theory of consolidation is to estimate the settlement versus time relationship after loading of the cohesive soil. The amount of settlement, otherwise known as the average degree of consolidation settlement (U_{avg}), is related to the time factor T as follows: $$U_{avg} = 100 - 100\ \text{(area within a } T \text{ curve)/(total area of Fig. 9.12)}$$ Using Fig. 9.12 and the above equation, the relationship between the average degree of consolidation settlement (U_{avg}) and the time factor T has been calculated and is summarized in Table 9.15. When the load is applied, the average degree of consolidation (U_{avg}) is zero (no settlement) and the time factor T is also zero. When one-half

TABLE 9.14 Terzaghi One-Dimensional Consolidation Theory (*Continued*)

Topic (1)	Discussion (2)
Average degree of consolidation settlement U_{avg} (*Continued*)	of the consolidation settlement has occurred (i.e., $U_{avg} = 50\%$), the time factor (T) = 0.197. The Terzaghi theory of consolidation predicts that the time t for complete consolidation ($U_{avg} = 100\%$, i.e., total settlement) is infinity, but as a practical matter, a time factor T equal to 1.0 is often assumed for $U_{avg} = 100\%$.
Four steps in using the Terzaghi one-dimensional consolidation theory	**1.** *Coefficient of consolidation.* The first step is to determine the coefficient of consolidation c_v of the cohesive soil. The usual procedure is to calculate c_v from the results of laboratory consolidation tests. During each incremental loading of the saturated cohesive soil in the oedometer apparatus, dial readings of vertical deformation can be recorded and plotted versus time (log scale), such as shown in Fig. 9.13. This vertical deformation versus time data was recorded from a laboratory consolidation test performed on a highly plastic soil (LL = 93, PI = 71, $e_o = 3.05$). The vertical deformation versus time data was recorded when the vertical load on the soil (σ'_{vc}) = 50 kPa. The laboratory data shown in Fig. 9.13 can be used to determine the coefficient of consolidation. The procedure is as follows: *a.* Determine end of primary consolidation. The end of primary consolidation (location of arrow) is estimated as the intersection of the two straight line segments. The value of d_{100} = the end of primary consolidation ($d_{100} = 2.0$ mm). *b.* Determine d_o. As an approximation, d_o can be assumed to be the dial reading at $t = 0.1$ minutes (i.e., the curve is extended back to $t = 0.1$ min), or $d_o = 0.1$ mm. *c.* Determine d_{50} and t_{50}. The value of d_{50} is equal to d_o plus d_{100} divided by 2 [i.e., $d_{50} = \frac{1}{2}(d_o + d_{100}) = \frac{1}{2}(0.1 + 2.0) = 1.05$]. Using the curve shown in Fig. 9.13, at a vertical deformation of 1.05 mm, find the corresponding time (t_{50}) = 150 min. The value of t_{50} (150 min) is the length of time it took for the saturated cohesive soil specimen to experience 50% of its primary consolidation (i.e., $U_{avg} = 50\%$) when subjected to a vertical pressure of 50 kPa. *d.* Calculate the coefficient of consolidation c_v. The time factor T equation is used to calculate the coefficient of consolidation c_v. From Table 9.15, the time factor T for an average degree of consolidation U_{avg} of 50% is 0.197. The height H of the specimen at $d_{50} = 9.0$ mm, and since the specimen has double drainage in the laboratory oedometer apparatus, $H_{dr} = 4.5$ mm. Inserting $t_{50} = 150$ min, $H_{dr} = 4.5$ mm, and $T = 0.197$ into the time factor T equation, the coefficient of consolidation $c_v = 0.027$ mm^2/min (4.4×10^{-6} cm^2/s). This very low coefficient of consolidation is due to the clay particles (montmorillonite) in the soil (i.e., PI = 71).

TABLE 9.14 Terzaghi One-Dimensional Consolidation Theory (*Continued*)

Topic (1)	Discussion (2)
Four steps in using the Terzaghi one-dimensional consolidation theory (*Continued*)	The coefficient of consolidation c_v for laboratory specimens under different sample heights, load increment ratios, and load duration has been determined for numerous cohesive soils (Taylor 1948, Leonards and Ramiah 1959). Ladd (1973) states that it can be assumed that variations in sample height, load increment ratio, and load duration will generally lead to insignificant difference in c_v provided that the resulting dial versus time reading have the characteristic shape shown in Fig. 9.13 (i.e., a Type 1 curve; Leonards and Altschaeffl 1964) when plotted on a log-time scale. This means that there must be appreciable primary consolidation during the loading increment. A factor that can significantly affect the coefficient of consolidation is sample disturbance. For example, in Fig. 9.9 the coefficient of consolidation is shown for both the disturbed Orinoco clay specimen (solid circles) and the undisturbed Orinoco clay specimen (open circles). Especially for that part of the consolidation test that involves recompression of the cohesive soil, sample disturbance can significantly reduce the values of the coefficient of consolidation. **2.** *Determine the drainage height H_{dr} of the in situ cohesive soil.* After the coefficient of consolidation c_v has been determined from laboratory testing of undisturbed soil specimens, the second step is to determine the drainage height of the *in situ* cohesive soil layer. As previously mentioned, if water can drain from the cohesive soil layer at both the top and bottom of the clay layer (double drainage), then $H_{dr} = \frac{1}{2}H$, where H = thickness of the cohesive soil layer. If water can drain from only the top or bottom of the cohesive soil layer (single drainage), then $H_{dr} = H$. **3.** *Determine the time factor T.* On the basis of an average degree of consolidation (U_{avg}), the time factor T can be obtained from Table 9.15. **4.** *Use the time factor equation T to determine the time t.* Once c_v, T, and H_{dr} are known, the time t corresponding to a certain amount of settlement U_{avg} can be calculated from the time factor equation. In summary, the four steps listed above basically consist of determining the coefficient of consolidation c_v from laboratory consolidation tests performed on undisturbed soil specimens and then this data is used to predict the time-settlement behavior of the *in situ* cohesive soil layer.
Limitations of the Terzaghi one-dimensional consolidation theory	Terzaghi's one-dimensional consolidation theory is one of the most widely taught and applied theories in geotechnical engineering. Some of the limitations of this theory are as follows (Duncan 1993): **1.** c_v is commonly assumed to be constant, which is often not the case in the field. **2.** The stress-strain behavior of the soil skeleton is assumed to be linear and elastic.

TABLE 9.14 Terzaghi One-Dimensional Consolidation Theory (*Continued*)

Topic (1)	Discussion (2)
Limitations of the Terzaghi one-dimensional consolidation theory (*Continued*)	**3.** The strains are assumed to be uniform. **4.** There are often permeable sand lenses within the cohesive soil that can significantly decrease the length of time for primary consolidation as predicted by assuming a constant c_v. **5.** The strain often decreases with depth because the stress increase caused by surface loads decreases with depth, or the clay compressibility decreases with depth, or both. **6.** A final factor is that the theory is based on the vertical flow of water from the cohesive soil, but many cases involve two- or three-dimensional loading which would allow the clay to drain partially from its sides. Because of all these factors, the Terzaghi theory of consolidation often overpredicts the time required to reach a certain average degree of consolidation. The theory should only be used as an approximation of the time-settlement behavior for loading of saturated cohesive soil. Field measurements (Chap. 23) are often essential in comparing the actual time-settlement behavior with the predicted behavior.

TABLE 9.15 Average Degree of Consolidation U_{avg} versus Time Factor T

Average degree of consolidation U_{avg}, % (1)	Time factor T (2)
0	0
5	0.002
10	0.008
15	0.017
20	0.031
25	0.049
30	0.071
35	0.092
40	0.126
45	0.159
50	0.197
55	0.238
60	0.286
65	0.340
70	0.403
75	0.477
80	0.567
85	0.683
90	0.848
95	1.13
100	∞

Assumptions: Terzaghi theory of consolidation, linear initial excess pore water pressures, and instantaneous loading.

Note: The above values are based on the following approximate equations by Casagrande (unpublished notes) and Taylor (1948):

For $U_{avg} < 60\%$, the time factor $T = \frac{1}{4}\pi(U_{avg}/100)^2$

For $U_{avg} \geq 60\%$, the time factor $T = 1.781 - 0.933 \log(100 - U_{avg})$

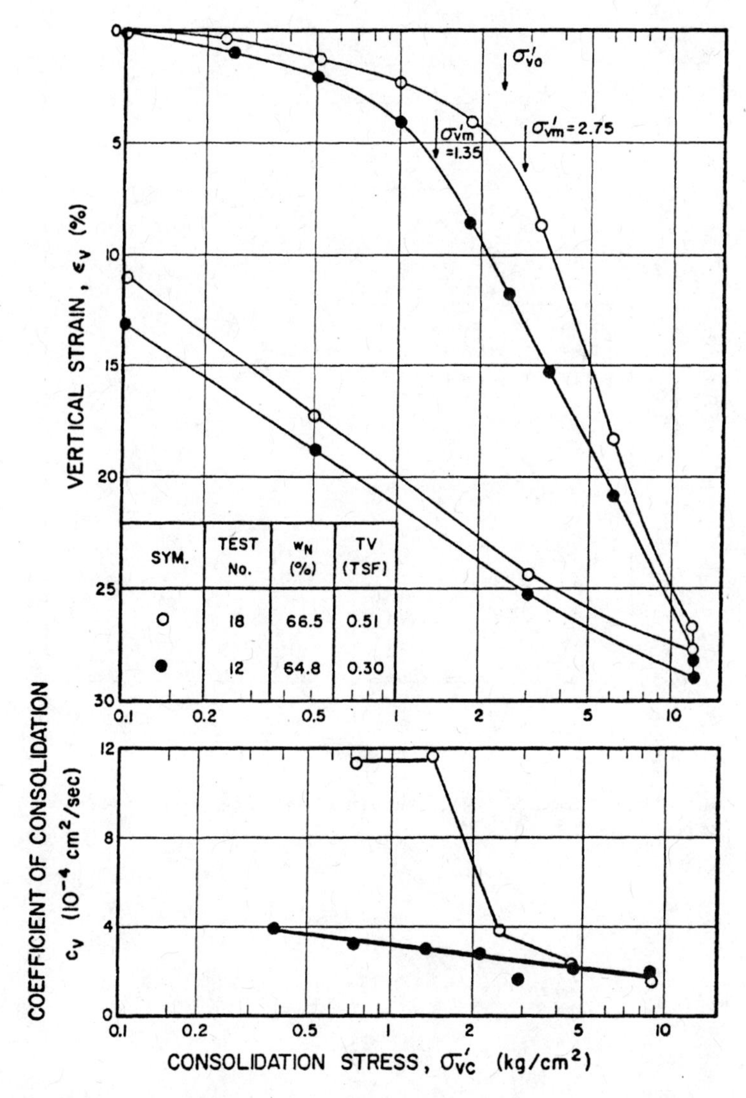

FIGURE 9.9 Consolidation test data for Orinoco clay at a depth of 40 m. See Table 9.12. (*From Ladd et al. 1980, Day 1980.*)

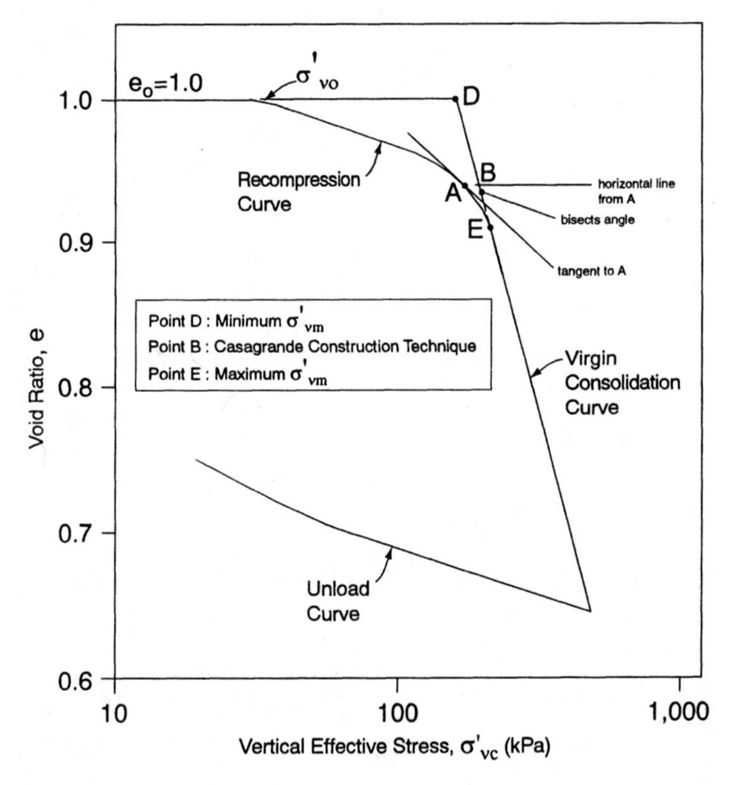

FIGURE 9.10 Casagrande (1936) technique for determining the maximum past pressure. See Table 9.12.

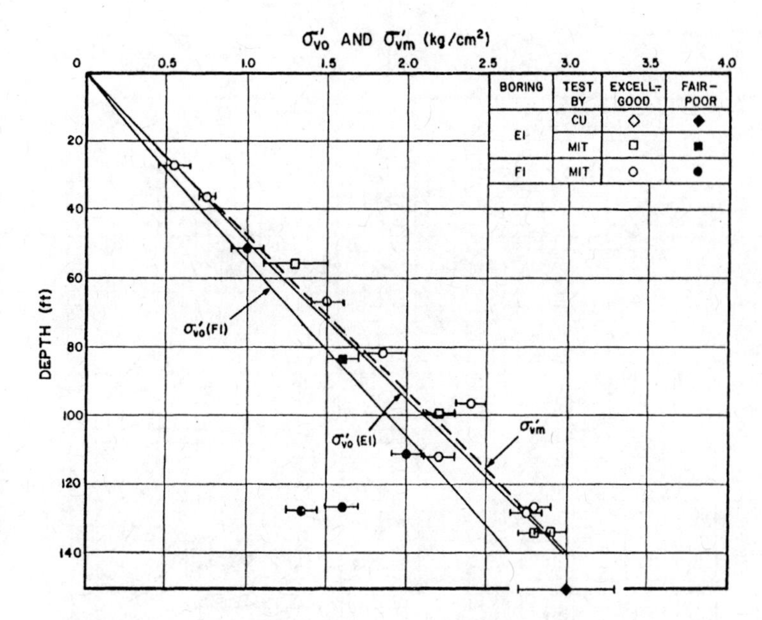

FIGURE 9.11 Stress history of Orinoco clay. See Table 9.12. (*From Ladd et al. 1980, Day 1980.*)

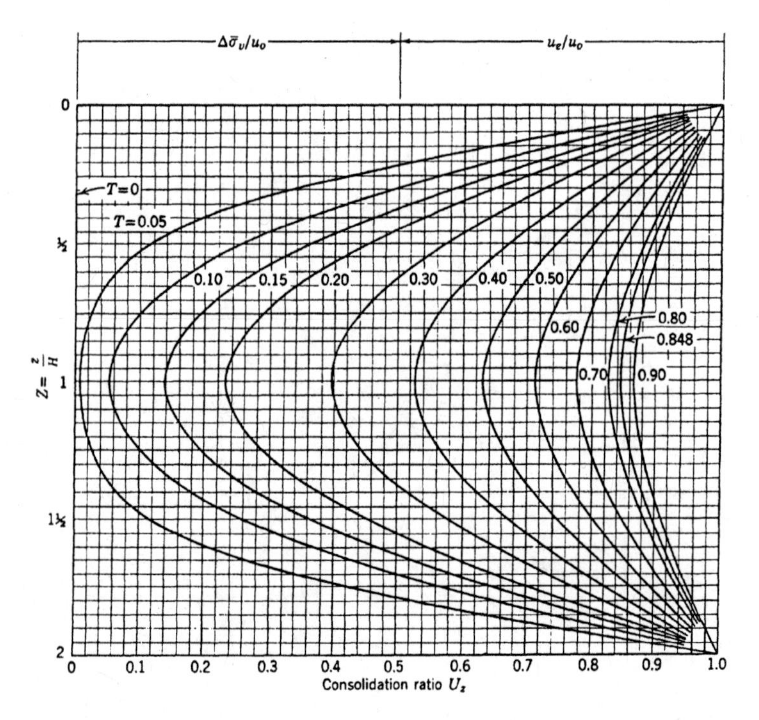

FIGURE 9.12 Consolidation ratio as a function of depth and time factor for uniform initial excess pore water pressure. See Table 9.14. (*From Lambe and Whitman 1969; reproduced with permission of John Wiley & Sons.*)

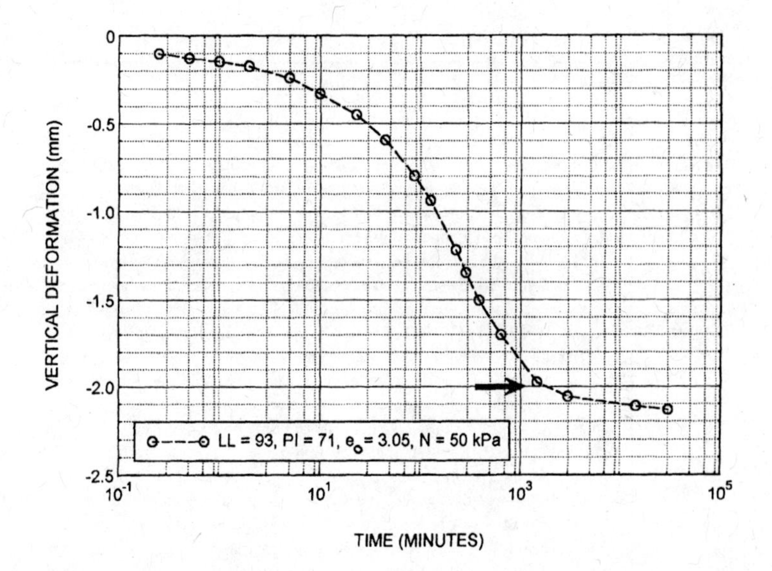

FIGURE 9.13 Data recorded from a consolidation test performed on saturated cohesive soil. At a vertical pressure N of 50 kPa, the vertical deformation as a function of time after loading was recorded and plotted. The arrow indicates the end of primary consolidation. See Table 9.14.

TABLE 9.16 Secondary Compression

Topic (1)	Discussion (2)
General discussion	The final component of settlement is that due to secondary compression, which is that part of the settlement that occurs after essentially all of the excess pore water pressures have dissipated (i.e., settlement that occurs at constant effective stress). The usual assumption is that secondary compression does not start until after primary consolidation is complete. In Fig. 9.13, the vertical deformation that occurs above the arrow is primary consolidation, while the vertical deformation that occurs below the arrow is secondary compression. The amount of secondary compression is often neglected because, as shown in Fig. 9.13, it is rather small compared to the primary consolidation settlement. However, secondary compression can constitute a major part of the total settlement for peat or other highly organic soil (Holtz and Kovacs 1981). Ladd (1973) describes secondary compression as a process where the particle contacts are still rather unstable at the end of primary consolidation and the particles will continue to move until they find a stable arrangement. This would explain why the rate of secondary compression often increases with compressibility. The more compressible the soil, the greater the tendency for a larger number of particles to be unstable at the end of primary consolidation (Ladd 1973). Because there are no excess pore water pressures during secondary compression, it is often described as *drained creep.*
Secondary compression equation	The amount of settlement s_s due to secondary compression can be calculated as follows: $$s_s = C_\alpha H_o \Delta \log t$$ where s_s = settlement due to secondary compression (occurs after the end of primary consolidation), m or ft C_α = secondary compression ratio (dimensionless parameter) which is defined as the slope of the secondary compression curve. For example, the secondary compression curve shown in Fig. 9.13 (portion of curve below the arrow) can be approximated as a straight line. If the vertical axis in Fig. 9.13 is converted to strain ϵ_v, then C_α equals the change in strain ($\Delta\epsilon_v$) divided by the change in time (log scale), or $C_\alpha = \Delta\epsilon_v / \Delta \log t$. Using one log cycle of time (i.e., $\Delta \log t = 1$), then the value of secondary compression $C_\alpha = \Delta\epsilon_v$. H_o = initial thickness of the *in situ* cohesive soil layer, m or ft $\Delta \log t$ = change in log of time from the end of primary consolidation to the end of the design life of the structure. For example, if primary consolidation is complete at a time = 10 years, and the design life of the structure is 100 years, then $\Delta \log t = \log 100 - \log 10 = 1$. As indicated in Table 9.11, secondary compression (aging) can create a slight overconsolidation of a cohesive soil, even though the effective stress has not changed. While the site history may indicate a normally consolidated soil, it is possible that a slight load can be applied to the cohesive soil without triggering virgin consolidation because of this secondary compression aging effect.

TABLE 9.16 Secondary Compression (*Continued*)

Topic (1)	Discussion (2)
Total settlement ρ_{max}	The final calculation for estimating the total settlement ρ_{max} of the *in situ* cohesive soil would be to add together the three components of settlement: $$\rho_{max} = s_i + s_c + s_s$$ where ρ_{max} = total settlement over the life of the structure s_i = immediate settlement (Table 9.9) s_c = primary consolidation settlement (Table 9.13) s_s = secondary compression settlement

TABLE 9.17 Settlement of Cohesionless Soil

Topic (1)	Discussion (2)
General discussion	A major difference between saturated cohesive soil and cohesionless soil is that the settlement of cohesionless soil is not time-dependent. Because of the generally high permeability of cohesionless soil, the settlement of a saturated cohesionless soil usually occurs as the load is applied during the construction of the building. Cohesionless soils typically do not have long-term settlement (such as secondary compression). Exceptions would be if the cohesionless soil is susceptible to collapse due to moisture infiltration (Table 9.5) or if it is subjected to seismic loading (Chap. 14). For cohesionless soil, there are many different methods that can be used to determine the settlement due to structural loading. Some of the more commonly used methods are presented in this table. All of these methods are used to determine the settlement caused by the weight of the structure (i.e., the first category of settlement; see Table 9.1).
Plate load tests	The plate load test can be used to directly estimate the settlement potential of the footing (Table 2.13). For settlement of sands caused by an applied surface loading, Table 2.13 presents an empirical equation from Terzaghi and Peck (1967) that relates the depth of penetration of the steel plate S_1 to the settlement of the actual footing (S, or ρ_{max} for that footing having the largest settlement). As previously mentioned in Table 2.13, the stress exerted by the steel plate corresponding to S_1 must be the same bearing stress exerted by the actual footing. The calculated settlement from the plate load settlement equation (see Table 2.13) can significantly underestimate the actual settlement at the site in cases where there is settlement due to secondary influences (such as collapsible soil) or in cases where there is a deep looser or softer layer that is not affected by the small plate, but is loaded by the much larger footing.
Laboratory oedometer tests	As discussed in the previous section, the main method of determining the settlement of cohesive soil is to test undisturbed soil specimens in the oedometer apparatus. If undisturbed soil specimens of the cohesionless soil can be obtained, such as those from test pits or Shelby tube samplers, then the compression curve of cohesionless soil could be obtained by using the oedometer apparatus. Usually the compression curve for cohesionless soil does not have recompression and virgin compression lines such as shown in Fig. 9.10. As an alternative to using the equations listed in Table 9.13, the following equation can be used to estimate the settlement (ρ_{max}) of the cohesionless soil: For data plotted on a void ratio e versus log σ'_{vc} plot: $$\rho_{max} = \Delta e H_o/(1 + e_o)$$ For data plotted on a vertical strain ϵ_v versus log σ'_{vc} plot: $$\rho_{max} = \Delta \epsilon_v H_o$$ where ρ_{max} = total settlement of the building (m or ft) Δe = change in void ratio (dimensionless) from the void ratio e versus log σ'_{vc} plot. Enter the plot at the vertical effective stress σ'_{vo} and upon intersecting the compression curve, determine the initial void ratio e_i. Also enter the plot at the final vertical effective stress (i.e., σ'_{vo} + $\Delta\sigma_v$), and upon intersecting the compression curve, determine the final void ratio e_f. And then $\Delta e = e_i - e_f$

TABLE 9.17 Settlement of Cohesionless Soil (*Continued*)

Topic (1)	Discussion (2)
Laboratory oedometer tests (*Continued*)	H_o = initial height of the *in situ* cohesionless soil layer (m or ft). Often several undisturbed soil specimens are obtained and tested in the laboratory and H_o represents the thickness of the *in situ* layer that is represented by an individual oedometer test. For the analysis of several soil layers, the above equations can be used for each layer and then the maximum settlement is the sum of the settlements calculated for each soil layer. e_o = initial void ratio of the cohesionless soil layer (dimensionless) $\Delta\epsilon_v$ = change in vertical strain (dimensionless) from the vertical strain ϵ_v versus log σ'_{vc} plot. Enter the plot at the vertical effective stress σ'_{vo} and upon intersecting the compression curve, determine the initial vertical strain ϵ_i. Also enter the plot at the final vertical effective stress (i.e., $\sigma'_{vo} + \Delta\sigma_v$) and upon intersecting the compression curve, determine the final vertical strain ϵ_f. And then $\Delta\epsilon = \epsilon_f - \epsilon_i$.
Empirical correlations	In many cases, undisturbed specimens of the cohesionless soil cannot be obtained. Thus a common approach is to use empirical correlations in order to determine the settlement of the cohesionless soil. For example, Fig. 9.14 shows a chart by Terzaghi and Peck (1967) that presents an empirical correlation between the measured *N*-value (obtained from the standard penetration test; see Table 2.11) and the allowable soil pressure (tsf) that will produce a settlement of the footing of 1 in. (2.5 cm). As an example of the use of Fig. 9.14, suppose a site contains a sand deposit and the proposed structure can be subjected to a maximum settlement ρ_{max} of 1.0 in. (2.5 cm). If the measured *N*-value from the standard penetration test = 10 and the width of the proposed footings = 5 ft (1.5 m), then the allowable soil pressure = 1 tsf (100 kPa). For measured *N*-values other than those for which the curves are drawn in Fig. 9.14, the allowable soil pressure can be obtained by linear interpolation between curves. According to Terzaghi and Peck (1967), if all of the footings are proportioned in accordance with the allowable soil pressure corresponding to Fig. 9.14, then the maximum settlement ρ_{max} of the foundation should not exceed 1 in. (2.5 cm) and the maximum differential settlement Δ should not exceed ¾ in. (2 cm). Figure 9.14 was developed for the groundwater table located at a depth equal to or greater than a depth of 2*B* below the bottom of the footing. For conditions of a high groundwater table close to the bottom of the shallow foundation, the values obtained from Fig. 9.14 should be reduced by 50%.
Elastic method	Similar to determining the immediate settlement for cohesive soil (Table 9.9), the theory of elasticity can be used to estimate the settlement of cohesionless soil. The settlement would be determined from the equation listed in Table 9.9, using the drained modulus E_s of the cohesionless soil. The drained modulus is often obtained from a drained triaxial compression test performed on an undisturbed

TABLE 9.17 Settlement of Cohesionless Soil (*Continued*)

Topic (1)	Discussion (2)
Elastic method (*Continued*)	soil specimen where the stress-strain curve is plotted. The initial tangent modulus to the stress-strain curve is often assumed to be the value of E_s. The drained modulus E_s of the cohesionless soil can also be obtained from empirical correlations with the standard penetration test (SPT) or the cone penetration test (CPT). Approximate correlations are as follows (NAVFAC DM-7.1, 1982):

Cohesionless soil type	E_s/N
Nonplastic silty sands or silt-sand mixtures	4
Clean, fine to medium sands	7
Coarse sands and sand with a little gravel	10
Sandy gravel and gravel	12

Topic (1)	Discussion (2)
Elastic method (*Continued*)	The units for E_s/N are tsf (1 tsf ≅ 100 kPa). For example, if a clean medium sand has a measured N-value = 10, then E_s = 70 tsf (7 MPa). For the cone penetration test (CPT), an approximate correlation (NAVFAC DM-7.1, 1982) is $E_s = 2q_c$, where q_c = cone resistance in kg/cm^2 (1 kg/cm^2 ≅ 1 tsf ≅ 100 kPa). In many cases, the soil modulus E_s will vary with depth. Schmertmann's method (Schmertmann 1970, Schmertmann et al. 1978) is ideally suited to determining the settlement of footings when the soil modulus E_s varies with depth. For complex foundation problems, computer programs have been developed to estimate the settlement based on the finite-element analysis. For example, the SIGMA/W (Geo-Slope 1993) computer program can determine the settlement, assuming the soil is an elastic or an inelastic material.
Subgrade modulus	This method includes the width and depth of the foundation, the elevation of the groundwater table, and the subgrade modulus in the determination of the settlement of cohesionless (nonplastic) soil. The subgrade modulus (also known as the modulus of subgrade reaction) has been discussed in Table 2.13. Figure 9.15 presents a correlation between the relative density (see Table 3.16) and the subgrade modulus K_v. Assuming the modulus of elasticity increases linearly with depth, the settlement ρ_{max} caused by a uniform vertical footing pressure q can be estimated from the following equations (NAVFAC DM-7.1, 1982): For shallow foundations with $D \leq B$ For $B \leq 20$ ft: $\rho_{max} = \dfrac{4qB^2}{K_v(B+1)^2}$ For $B \geq 40$ ft: $\rho_{max} = \dfrac{2qB^2}{K_v(B+1)^2}$

TABLE 9.17 Settlement of Cohesionless Soil (*Continued*)

Topic (1)	Discussion (2)
Subgrade modulus (*Continued*)	Interpolate for intermediate values of *B*. For deep foundations ($D \geq 5\ B$) having $B \leq 20$ ft $$\rho_{max} = \frac{2qB^2}{K_v(B+1)^2}$$ where ρ_{max} = total settlement in feet q = vertical footing pressure (tons per square foot) B = footing width (ft) D = depth of the footing below ground surface (ft) K_v = subgrade modulus (tcf) from Fig. 9.15. The values of K_v from Fig. 9.15 apply to dry or moist cohesionless (granular) soil with the groundwater table at a depth of at least 1.5*B* below the base of the footing. If a groundwater table is at the base of the footing, use $\frac{1}{2}K_v$ in computing the settlement. For continuous footings, multiply the settlement calculated above by a factor of 2. The above equations may underestimate the settlement in cases of large footings where soil deformation properties vary significantly with depth or where the thickness of granular soil is only a fraction of the width of the loaded area.

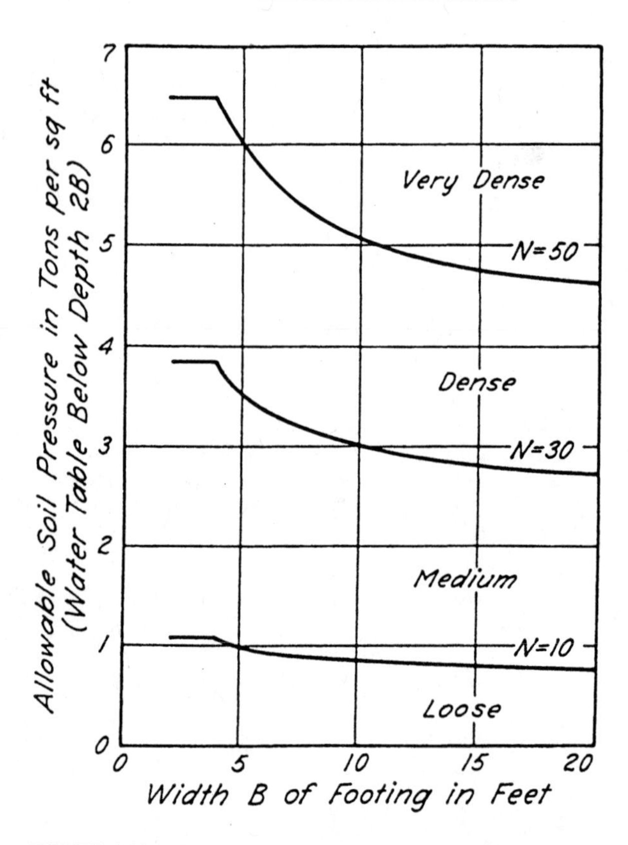

FIGURE 9.14 Allowable soil bearing pressures for footings on sand based on the standard penetration test. See Table 9.17. (*From Terzaghi and Peck, 1967; reprinted with permission of John Wiley & Sons.*)

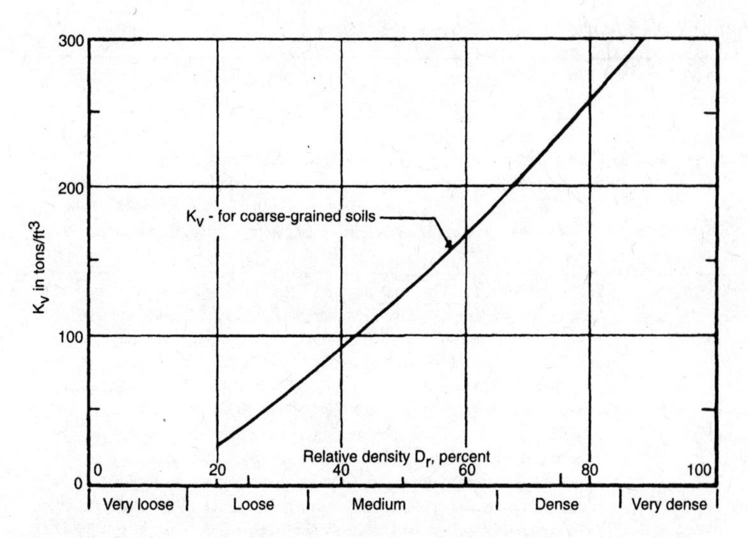

FIGURE 9.15 Correlation between relative density D_r and subgrade modulus K_v for cohesionless soil. See Table 9.17. (*Adapted from NAVFAV DM-7.1, 1982.*)

TABLE 9.18 Other Common Causes of Settlement

Topic (1)	Discussion (2)
General discussion	There are many other causes of the settlement of structures. The types of settlement discussed in this table are within the second basic category of settlement (see Table 9.1), settlement of the structure primarily due to secondary influences. An exception is the construction of a structure atop a landfill, where the weight of the structure could directly cause compression of the underlying loose material.
Limestone cavities or sinkholes	Settlement related to limestone cavities or sinkholes will usually be limited to areas having karst topography. Karst topography is a type of landform developed in a region of easily soluble limestone bedrock. It is characterized by vast numbers of depressions of all sizes, sometimes by great outcrops of limestone, sinks and other solution passages, an almost total lack of surface streams, and larger springs in the deeper valleys (Stokes and Varnes 1955). Identification techniques and foundations constructed on karst topography are discussed by Sowers (1997). Methods to investigate the presence of sinkholes are geophysical techniques and the cone penetration device (*Earth Manual* 1985, Foshee and Bixler 1994). A low cone penetration resistance could indicate the presence of raveling, which is a slow process where granular soil particles migrate into the underlying porous limestone. An advanced state of raveling will result in subsidence of the ground, commonly referred to as *sinkhole activity.* Figure 9.16 shows the settlement process starting with an initial condition of underground caverns (dark areas), followed by the development of surface depressions due to raveling of soil into the caverns, and the final condition of a collapsed sinkhole.
Undergound mines and tunnels	According to Gray (1988), damage to residential structures in the United States caused by the collapse of underground mines is estimated to be between $25 million and $35 million each year, with another $3 million to $4 million in damage to roads, utilities and services. There are approximately 2 million hectares of abandoned or inactive coal mines, with 200,000 of these hectares in populated urban areas (Dyni and Burnett 1993). It has been stated that ground subsidence associated with longwall mining can be predicted fairly well with respect to magnitude, time, and areal position (Lin et al. 1995). Once the amount of ground subsidence has been estimated, there are measures that can be taken to mitigate the effects of mine-related subsidence (National Coal Board 1975, Kratzsch 1983, Peng 1986, 1992). For example, in a study of different foundations subjected to mining induced subsidence, it was concluded that posttensioning of the foundation was most effective, because it prevented the footings from cracking (Lin et al. 1995). Like settlement in a canyon environment (Fig. 9.4), the collapse of underground mines and tunnels can produce tension and compression-type features within the buildings. Figure 9.17 (from Marino et al. 1988) shows that the location of the compression zone will be in the center of the subsided area. The tension zone is located along the perimeter of the subsided area. These tension and compression zones are similar to fill settlement in a canyon environment (Fig. 9.4).

TABLE 9.18 Other Common Causes of Settlement (*Continued*)

Topic (1)	Discussion (2)
Underground mines and tunnels (*Continued*)	Besides the collapse of underground mines and tunnels, there can be settlement of buildings constructed on spoil extracted from the mines. Mine operators often dispose of other debris, such as trees, scrap metal, and tires, within the mine spoil. In many cases, the mine spoil is dumped (no compaction) and can be susceptible to large amounts of settlement. For example, Cheeks (1996) describes an interesting case of a motel unknowingly built on spoil that had been used to fill in a strip-mining operation. The motel building experienced about 1 m (3 ft) of settlement within the monitoring period (5 years). The settlement and damage for this building actually started during construction and the motel owners could never place the building into service. A lesson from this case study was the importance of subsurface exploration. In many cases, the borings may encounter refusal on boulders generated from the mining operation and the geotechnical engineer may believe that the depth to solid rock is shallow. It is important when dealing with spoil generated from mining operations that the thickness of the spoil and the compression or collapse behavior of the material be adequately investigated. This may require use of rock coring or geophysical techniques to accurately define the limits and depth of the spoil pile.
Subsidence due to extraction of oil or groundwater	Large-scale pumping of water or oil from the ground can cause settlement of the ground surface over a large area. The pumping can cause a lowering of the groundwater table, which then increases the overburden pressure on underlying sediments. This can cause consolidation of soft clay deposits. In other cases, the removal of water or oil can lead to compression of the soil or porous rock structure, which then results in ground subsidence. Lambe and Whitman (1969) describe two famous cases of ground surface subsidence due to oil or groundwater extraction. The first is oil pumping from Long Beach, California, which affected a 65 km^2 (25 mi^2) area and caused 8 m (25 ft) of ground surface subsidence. Because of this ground surface subsidence, the Long Beach Naval Shipyard had to construct seawalls to keep the ocean from flooding the facilities. A second famous example is ground surface subsidence caused by pumping of water for domestic and industrial use in Mexico City. Rutledge (1944) shows that the underlying Mexico City clay, which contain a porous structure of microfossils and diatoms, has a very high void ratio (up to e = 14) and is very compressible. Ground surface subsidence in Mexico City has been 9 m (30 ft) since the beginning of the twentieth century. The theory of consolidation (Sec. 9.4) can often be used to estimate the ground surface subsidence caused by the pumping of shallow groundwater. In this case, the lowering of the groundwater table will increase the overburden pressure acting on the clay layer. For example, if pumping causes the groundwater table to be permanently lowered, then the increase in overburden stress ($\Delta\sigma_v$) will be equal to the difference between the final (σ'_{vf}) and initial (σ'_{vo}) vertical effective stress in the clay layer (i.e., $\Delta\sigma_v = \sigma'_{vf} - \sigma'_{vo}$). In other cases,

TABLE 9.18 Other Common Causes of Settlement (*Continued*)

Topic (1)	Discussion (2)
Subsidence due to extraction of oil or groundwater (*Continued*)	pumping of shallow groundwater may not cause a significant lowering of the groundwater table, but instead the pumping can cause a decrease in the pore water pressure in a permeable layer located below the clay layer. In this case, the reduction in pore water pressure will also lead to consolidation of the clay, and once again the increase in overburden stress ($\Delta\sigma_v$) will be equal to the difference between the final (σ'_{vf}) and initial (σ'_{vo}) vertical effective stress in the clay layer.
Landfills	Landfills can settle by compression of the loose waste products and decompos-ition of organic matter that was placed within the landfill. Landfills are a special case of settlement because they could have both basic types of settlement: settlement due directly to the weight of the structure, which causes an initial compression of the loose waste products, and settlement due to secondary effects, such as when the organic matter in the landfill slowly decomposes. There are many old abandoned municipal landfills throughout the United States. Other sites may contain buried debris or other waste products that have been dumped at the site. Settlement of structures could be a result of compression of the underlying loose waste products or the decomposition of any organic matter remaining in the landfill. During decomposition of the organic matter, there will also be the generation of methane (a flammable gas), which must be safely vented to the atmosphere.

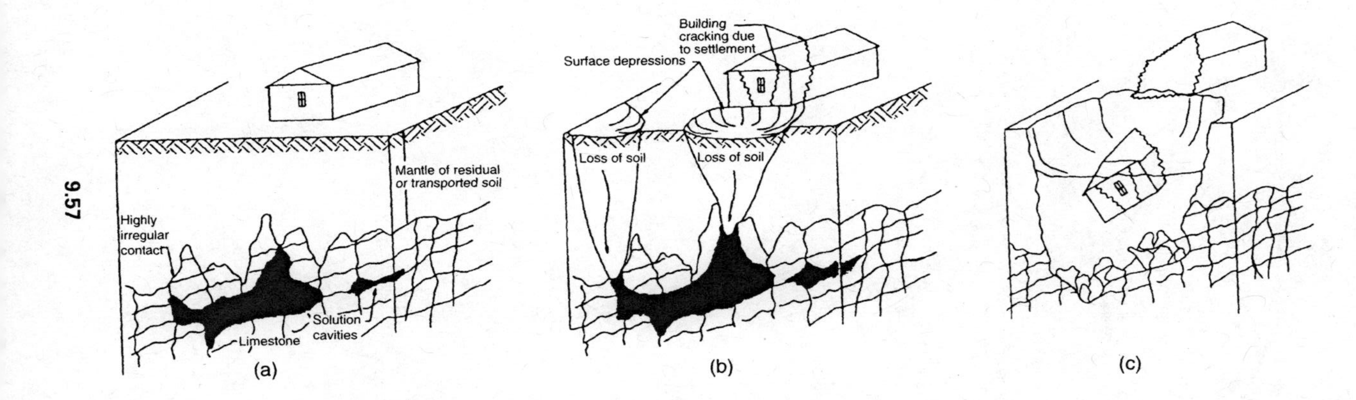

FIGURE 9.16 Development of sinkholes. (*a*) Initial condition; (*b*) subsidence or raveling sinkhole; (*c*) collapse sinkhole. See Table 9.18. (*From Rollings and Rollings 1996; reprinted with permission of McGraw-Hill Book Co.*)

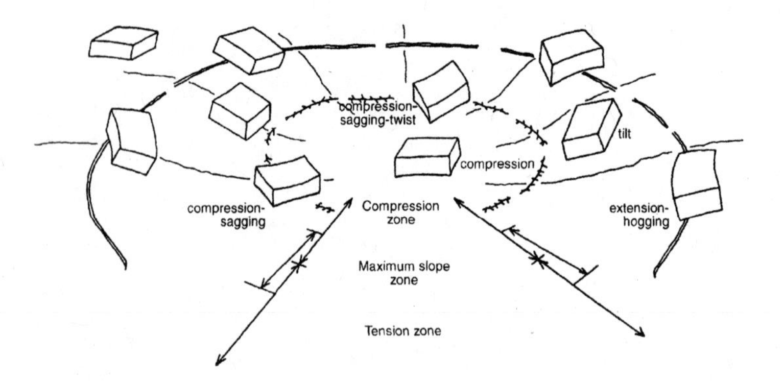

FIGURE 9.17 Location of tension and compression zones due to collapse of underground mines. See Table 9.18. (*From Marino et al. 1988; reprinted with permission from the American Society of Civil Engineers.*)

TABLE 9.19 Example Problems

Topic (1)	Problem and solution (2)
Building load	*Problem:* A proposed project will consist of an industrial building (20 m wide and 30 m long) that will be used by a furniture moving company to store household items. The structural engineer indicates that preliminary plans call for exterior tilt-up walls, interior isolated columns on spread footings, and a floor slab. Perimeter tilt-up wall dead loads are 60 kN per linear meter. Interior columns will be spaced 5 m on center in both directions and each column will support 800 kN of dead load. The interior floor slab dead load = 5 kPa. The structural engineer also indicates that it is likely that the industrial building will contain household items and this anticipated live load = 15 kPa. Assume the subsurface exploration has discovered the presence of a soft clay layer and, for the purposes of the settlement analysis, the weight of the industrial building can be assumed to be a uniform stress σ_o applied at ground surface. Determine the uniform stress σ_o to be used in the settlement analysis. *Solution:* Perimeter wall load = 60 [(20)(2)+(30)(2)] = 6000 kN Interior column loads = (15 columns)(800 kN) = 12,000 kN Equivalent pressure (σ_o) = (6000 + 12,000)/[(20)(30)] + 5 kPa + 15 kPa = 50 kPa
Primary consolidation	*Problem:* Use the same conditions as the previous problem. Also assume that the site has a level ground surface and a level groundwater table (with hydrostatic pore water pressures) located 5 m below the ground surface. Subsurface exploration has discovered that the site is underlain with sand, except for a uniform and continuous clay layer that is located at a depth of 10 to 12 m below ground surface. A consolidation test performed on an undisturbed specimen obtained from the center of the clay layer indicates a maximum past pressure σ'_{vm} = 150 kPa, a compression ratio C_c = 0.83, a void ratio e_o of 1.10, and buoyant unit weight γ_t = 7.9 kN/m^3. The total unit weight γ_t of the sand above the groundwater table = 18.9 kN/m^3, and γ_t of the sand below the groundwater table = 19.7 kN/m^3. Calculate the primary consolidation settlement s_c of the clay layer due to the building weight. *Solution:* The vertical effective stress σ'_{vo} at the center of the clay layer is equal to: σ'_{vo} = (5 m)(18.9 kN/m^3) + (5 m)(19.7 − 9.81 kN/m^3) + (1 m)(7.9 kN/m^3) = 152 kPa Since σ'_{vo} is approximately equal to σ'_{vm}, the clay layer is essentially normally consolidated (OCR = 1). Using the 2:1 approximation (Table 6.5) with B = 20 m, L = 30 m, z = 11 m, and σ_o = 50 kPa, then $\Delta\sigma_v$ = 24 kPa. From the normally consolidated equation (OCR = 1) from Table 9.13, with the following values: σ'_{vo} = 152 kPa, $\Delta\sigma_v$ = 24 kPa, H_o = 2 m, C_c = 0.83, and e_o = 1.10, the calculated primary consolidation settlement s_c = 0.050 m (2 in.).

TABLE 9.19 Example Problems (*Continued*)

Topic (1)	Problem and solution (2)
Terzaghi one-dimensional consolidation theory	*Problem:* Use the same conditions as the previous example. Assuming quick construction of the building and using the Terzaghi theory of consolidation, predict how long after construction it will take for 50% and 90% of the primary consolidation settlement to occur. On the basis of laboratory oedometer testing of an undisturbed clay specimen, where dial readings versus time were recorded at a vertical pressure of 200 kPa, the coefficient of consolidation c_v was calculated to be 1×10^{-4} cm^2/s. *Solution:* Since there is sand on both the top and bottom of the 2-m-thick clay layer, the clay layer has double drainage and $H_{dr} = 1$ m. For $U_{avg} = 50\%$, the time factor $T = 0.197$ and for $U_{avg} = 90\%$, the time factor $T = 0.848$ (Table 9.15). Using the time factor T equation in Table 9.14, with a value of $c_v = 1 \times 10^{-4}$ cm^2/s (0.32 m^2/year), it will take 0.62 years for 50% of the primary consolidation to occur (i.e., 0.62 years for 1 in of settlement) and it will take 2.7 years for 90% of the primary consolidation to occur (i.e., 2.7 years for 1.8 in. of settlement).

CHAPTER 10
BEARING CAPACITY ANALYSES

10.1 INTRODUCTION

A bearing capacity failure is defined as a foundation failure that occurs when the shear stresses in the soil exceed the shear strength of the soil. Bearing capacity failures of foundations can be grouped into three categories, as discussed in Table 10.1 and Figs. 10.1*a* to 10.1*c*. Compared to the number of structures damaged by settlement, there are far fewer structures that have bearing capacity failures (see Tables 10.2 and 10.3).

10.2 BEARING CAPACITY FOR SHALLOW FOUNDATIONS

As indicated in Table 1.2, common types of shallow foundation include spread footings for isolated columns, combined footings for supporting the load from more than one structural unit, strip footings for walls, and mats or raft foundations constructed at or near ground surface. Shallow footings often have an embedment that is less than the footing width. The Terzaghi bearing capacity equation is discussed in Table 10.4 and the bearing capacity factors are presented in Table 10.5 and Figs. 10.2 and 10.3. The bearing capacity analysis for shallow foundations on cohesionless (nonplastic) and cohesive (plastic) soils are discussed in Tables 10.6 and 10.7, respectively. Other design considerations, such as footings subjected to lateral loads and eccentric loads, are discussed in Table 10.8.

10.3 BEARING CAPACITY FOR DEEP FOUNDATIONS

Deep foundations are used when the upper soil stratum is too soft, weak, or compressible to support the foundation loads. Deep foundations are also used when there is a possibility of the undermining of the foundation. For example, bridge piers are often founded on deep foundations to prevent a loss of support due to flood conditions which could cause river bottom scour. Common types of deep foundations and design criteria are discussed in Tables 10.9 and 10.10. The bearing capacity analysis for deep foundations in cohesionless soil (Table 10.11 and Fig. 10.4) and cohesive soil (Table 10.12 and Figs. 10.5 and 10.6) are discussed in this section. Example problems are presented in Table 10.13.

TABLE 10.1 Three Categories of Bearing Capacity Failures

Topic (1)	Discussion (2)
General discussion	Bearing capacity failures of foundations can be grouped into three categories as indicated below (Vesic 1963, 1975).
General shear bearing capacity failure	As shown in Fig. 10.1*a*, a general shear failure involves total rupture of the underlying soil. There is a continuous shear failure of the soil (solid lines) from below the footing to the ground surface. When the load is plotted versus settlement of the footing, there is a distinct load at which the foundation fails (solid circle), and this is designated Q_{ult}. The value of Q_{ult} divided by the width B and length L of the footing is considered to be the ultimate bearing capacity (q_{ult}) of the footing. The ultimate bearing capacity has been defined as the bearing stress that causes a sudden catastrophic failure of the foundation (Lambe and Whitman 1969). Note in Fig. 10.1*a* that a general shear failure ruptures and pushes up the soil on both sides of the footing. For actual failures in the field, the soil is often pushed up on only one side of the footing with subsequent tilting of the structure. A general shear failure occurs for soils that are in a dense or hard state.
Punching shear bearing capacity failure	As shown in Fig. 10.1*b*, a punching shear failure does not develop the distinct shear surfaces associated with a general shear failure. For punching shear, the soil outside the loaded area remains relatively uninvolved and there is minimal movement of soil on both sides of the footing. The process of deformation of the footing involves compression of soil directly below the footing as well as the vertical shearing of soil around the footing perimeter. As shown in Fig. 10.1*b*, the load settlement curve does not have a dramatic break and for punching shear, the bearing capacity is often defined as the first major nonlinearity in the load-settlement curve (open circle). A punching shear failure occurs for soils that are in a loose or soft state.
Local shear bearing capacity failure	As shown in Fig. 10.1*c*, local shear failure involves rupture of the soil only immediately below the footing. There is soil bulging on both sides of the footing, but the bulging is not as significant as in general shear. Local shear failure can be considered as a transitional phase between general shear and punching shear. Because of the transitional nature of local shear failure, the bearing capacity could be defined as the first major nonlinearity in the load-settlement curve (open circle) or at the point where the settlement rapidly increases (solid circle). A local shear failure occurs for soils that have a medium density or firm state.
Summary	In summary, based on the density or consistency of the bearing soil, the type of bearing capacity failure would most likely be the following: • *General shear failure* (Fig. 10.1*a*). For cohesionless (nonplastic) soil in a dense to very dense state. For cohesive (plastic) soil having a very stiff to hard consistency. • *Local shear failure* (Fig. 10.1*c*). For cohesionless (nonplastic) soil in a medium density state. For cohesive (plastic) soil having a medium to stiff consistency. • *Punching shear failure* (Fig. 10.1*b*). For cohesionless (nonplastic) soil in a loose to very loose state. For cohesive (plastic) soil having a very soft to soft consistency.

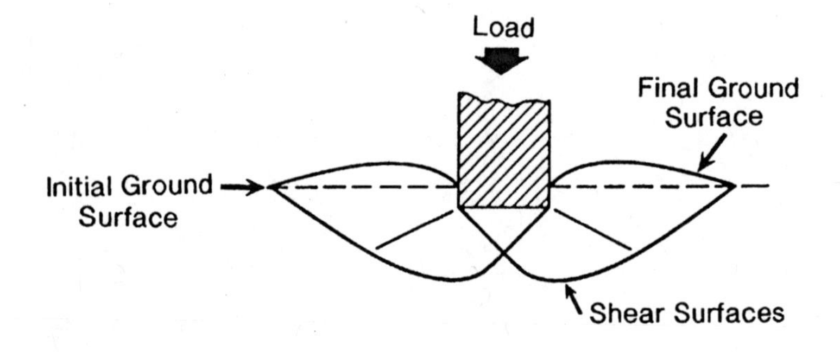

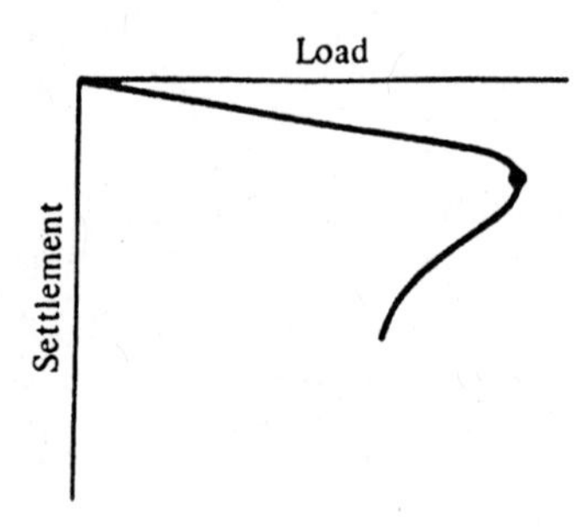

FIGURE 10.1a General shear foundation failure. See Table 10.1. (*After Vesic 1963.*)

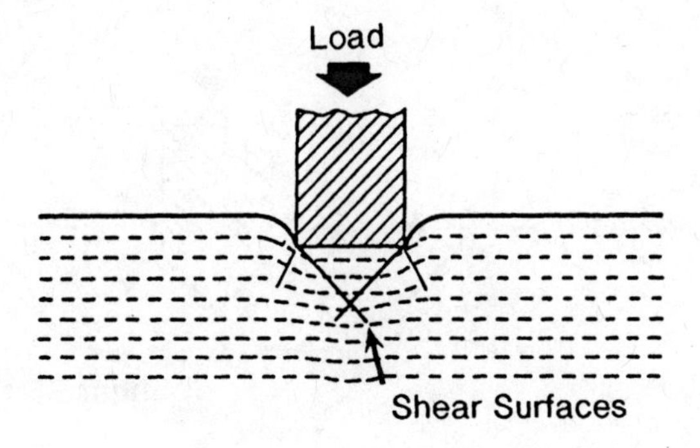

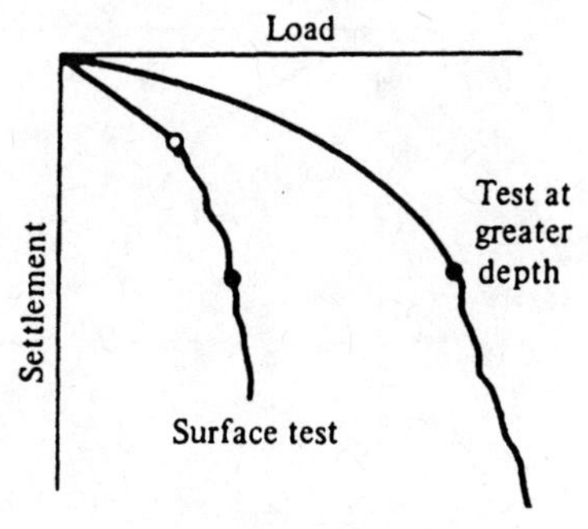

FIGURE 10.1b Punching shear foundation failure. See Table 10.1. (*After Vesic 1963.*)

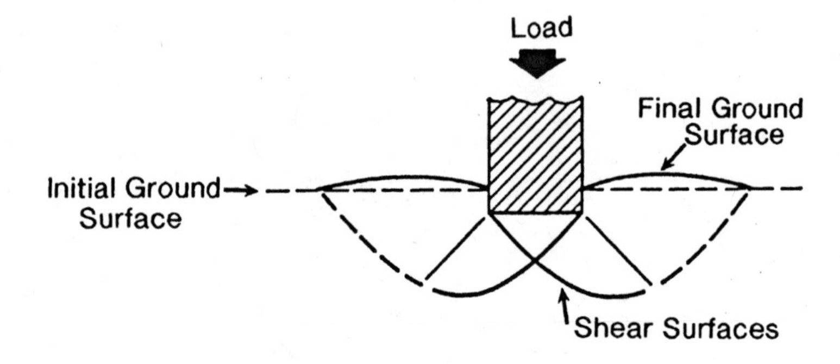

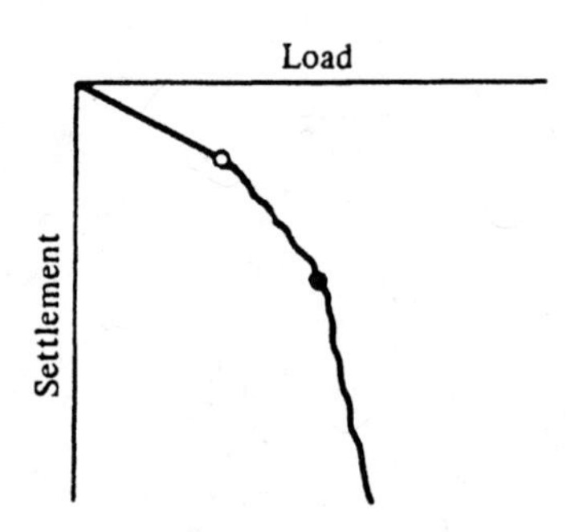

FIGURE 10.1c Local shear foundation failure. See Table 10.1. (*After Vesic 1963.*)

TABLE 10.2 Bearing Capacity Failures

Topic (1)	Discussion (2)
Frequency of bearing capacity failures	Compared to the number of structures damaged by settlement, there are far fewer structures that have bearing capacity failures. This is because of the following factors: **1.** *Settlement governs.* The foundation design is based on several requirements (see Chap. 18), and two of the main considerations are: (1) settlement due to the building loads must not exceed allowable values (Chap. 9) and (2) there must be an adequate factor of safety against a bearing capacity failure (Chap. 10). In most cases, settlement governs, and the foundation bearing pressures recommended by the geotechnical engineer are based on limiting the amount of settlement. **2.** *Extensive studies.* There have been extensive studies of bearing capacity failures which have led to the development of bearing capacity equations that are routinely used in practice to determine the ultimate bearing capacity of the foundation. **3.** *Factor of safety.* In order to determine the allowable bearing pressure q_{all}, the ultimate bearing capacity q_{ult} is divided by a factor of safety. The normal factor of safety used for bearing capacity analyses is 3. This is a high factor of safety as compared to other factors of safety, such as only 1.5 for slope stability analyses (Chap. 13). **4.** *Minimum footing sizes.* Building codes often require minimum footing sizes and embedment depths (for example, see the *Uniform Building Code* 1997, Table 18-I-C for minimum footing requirements for stud bearing walls). **5.** *Allowable bearing pressures.* In addition, building codes often have maximum allowable bearing pressures for different soil and rock conditions. Table 10.3 presents maximum allowable bearing pressures based on the *Uniform Building Code* 1977, Table 18-I-A. Especially in the case of dense or stiff soils, these allowable bearing pressures often have adequate factors of safety. For very loose cohesionless (nonplastic) soils and very soft cohesive (plastic) soil, the values presented in Table 10.3 may be too high. **6.** *Footing dimensions.* Usually the structural engineer will determine the size of the footings by dividing the maximum footing load (dead load plus live load) by the allowable bearing pressure. Typically the structural engineer uses values of dead and live loads that also contain factors of safety. For example, the live load may be from the local building code which specifies minimum live load requirements for specific building uses. Such building code values often contain a factor of safety, which is in addition to the factor of safety of 3 that was used to determine the allowable bearing pressure.
Depth of bearing capacity failures	Because the bearing capacity failure involves a shear failure of the underlying soil (Figs. 10.1*a* to *c*), the analysis will naturally include the shear strength of the soil (Chap. 7). As indicated in Figs. 10.1*a* to *c*,

TABLE 10.2 Bearing Capacity Failures (*Continued*)

Topic (1)	Discussion (2)
Depth of bearing capacity failures (*Continued*)	the depth of the bearing capacity failure is rather shallow. It is often assumed that the soil involved in the bearing capacity failure can extend to a depth equal to B (footing width) below the bottom of the footing. Thus, for bearing capacity analysis, this zone of soil should be evaluated for its shear strength properties.
Typical cause of bearing capacity failures	The documented cases of bearing capacity failures indicate that the main causes (separately or in combination) of the failure are: (1) there was an overestimation of the shear strength of the underlying soil, (2) the actual structural load at the time of the bearing capacity failure was greater than that assumed during the design phase, and (3) the site was altered, for example, by the construction of an adjacent excavation, which resulted in a reduction in support and a bearing capacity failure.

TABLE 10.3 Allowable Bearing Pressures

Material type (1)	Allowable bearing pressure* (2)	Maximum allowable bearing pressure† (3)
Massive crystalline bedrock	4000 psf (200 kPa)	12,000 psf (600 kPa)
Sedimentary and foliated rock	2000 psf (100 kPa)	6000 psf (300 kPa)
Gravel and sandy gravel (GW, GP)	2000 psf (100 kPa)	6000 psf (300 kPa)
Nonplastic soil: sands, silts, and NP silt (GM, SW, SP, SM, MN)‡	1500 psf (75 kPa)	4500 psf (220 kPa)
Plastic soil: silts and clays (ML, MI, MH, SC, CL, CI, CH)‡	1000 psf (50 kPa)	3000 psf (150 kPa)§

*Minimum footing width and embedment depth equals 1 ft (0.3 m).

†An increase of 20% of the allowable bearing pressure is allowed for each additional foot (0.3 m) of width or depth up to the maximum allowable bearing pressures listed in column 3. An exception is plastic soil.

‡Group symbols from the ISBP.

§No increase in the allowable bearing pressure is allowed for an increase in width of the footing.

For dense or stiff soils, allowable bearing values are generally conservative. For very loose or very soft soils, allowable bearing values may be too high.

Source: Data from the *Uniform Building Code* (1997).

TABLE 10.4 Terzaghi Bearing Capacity Equation

Topic (1)	Discussion (2)
General discussion	As indicated in Table 1.2, shallow foundations include spread footings for isolated columns, combined footings for supporting the load from more than one structural unit, strip footings for walls, and mats or raft foundations constructed at or near ground surface. Shallow footings often have an embedment that is less than the footing width.
Terzaghi bearing capacity equation	The most commonly used bearing capacity equation is that equation developed by Terzaghi (1943). For a uniform vertical loading of a strip footing, Terzaghi (1943) assumed a general shear failure (Fig. 10.1*a*) in order to develop the following bearing capacity equation: $$q_{ult} = \frac{Q_{ult}}{BL} = cN_c + \tfrac{1}{2}\gamma_t BN_\gamma + \gamma_t D_f N_q$$ where q_{ult} = ultimate bearing capacity for a strip footing (kPa or psf) Q_{ult} = vertical load causing a general shear failure of the soil (Fig. 10.1*a*) B = width of the strip footing (m or ft) L = length of the strip footing (m or ft) γ_t = total unit weight of the soil (kN/m^3 or pcf) D_f = distance from the ground surface to the bottom of the strip footing (m or ft) c = cohesion of the soil underlying the strip footing (kPa or psf), and N_c, N_γ, N_q = dimensionless bearing capacity factors (see Table 10.5).
The three terms in the Terzaghi bearing capacity equation	As indicated in the above equation, there are three terms that are added together to obtain the ultimate bearing capacity for strip footings. These terms represent the following: cN_c: The first term accounts for the cohesive shear strength of the soil located below the strip footing. If the soil below the footing is cohesionless (i.e., $c = 0$), then this term is zero. $\tfrac{1}{2}\gamma_t\, B\, N_\gamma$: The second term accounts for the frictional shear strength of the soil located below the strip footing. The friction angle ϕ is not included in this term, but is accounted for by the bearing capacity factor N_γ. Note that γ_t represents the total unit weight of the soil located below the footing. $\gamma_t\, D_f N_q$: This third term accounts for the soil located above the bottom of the footing. The value of γ_t times D_f represents a surcharge pressure that helps to increase the bearing capacity of the footing. If the footing were constructed at ground surface (i.e., $D_f = 0$), then this term would equal zero. This third term indicates that the deeper the footing, the greater the ultimate bearing capacity of the footing. In this term, γ_t represents the total unit weight of the soil located above the bottom of the footing. The total unit weight above and below the footing bottom may be different, in which case different values are used in the second and third terms in the Terzaghi bearing capacity equation.

TABLE 10.4 Terzaghi Bearing Capacity Equation (*Continued*)

Topic (1)	Discussion (2)
Bearing capacity factors	Table 10.5 and Figs. 10.2 and 10.3 present bearing capacity factors N_c, N_γ, and N_q. There are many other charts, graphs, and figures that present bearing capacity factors by other engineers and researchers based on varying assumptions (Vesic 1975, Myslivec and Kysela 1978). Some of these bearing capacity factors are even listed to an accuracy of five significant figures (i.e., Table 4.4.7.1A, Bearing Capacity Factors, AASHTO 1996). These bearing capacity factors imply an accuracy which simply does not exist because the bearing capacity equation is only an approximation of the actual bearing failure. As indicated in Table 10.5, the bearing capacity factors are directly related to the friction angle ϕ of the soil. A dense cohesionless soil would tend to have a high friction angle and high bearing capacity factors, resulting in a large ultimate bearing capacity. On the other hand, a loose cohesionless soil would tend to have a lower friction angle and lower ultimate bearing capacity. Thus a major disadvantage of building code values (such as Table 10.3) is that they consider only the material type, and not the density condition of the soil which influences the friction angle ϕ and bearing capacity factors. Building code values tend to underestimate the allowable bearing pressure for dense or stiff soil, but may overestimate the allowable bearing pressure for very loose or very soft soil. Note in Table 10.5 that the bearing capacity factors rapidly increase at high friction angles ϕ. These bearing capacity factors should be used with caution, because natural soils are not homogeneous and the natural variability of such soil will result in weaker layers that will be exploited during a bearing capacity failure. Terzaghi (1943) originally developed the bearing capacity equation for a general shear bearing capacity failure. This type of bearing capacity failure is shown in Fig. 10.1*a* and will develop for dense or stiff soil. For loose or soft soil, there will be a punching shear failure as shown in Fig. 10.1*b*. The bearing capacity factors presented in Table 10.5 have been empirically adjusted and automatically incorporate allowances for local shear and punching shear.
Allowable bearing pressure	In order to calculate the allowable bearing pressure q_{all}, which is used to determine the size of the footings, the following equation is used: $$q_{\text{all}} = q_{\text{ult}}/F$$ where q_{all} = allowable bearing pressure (kPa or psf), q_{ult} = ultimate bearing capacity from the Terzaghi bearing capacity equation, and F = factor of safety. The commonly used factor of safety $F = 3$. Table 10.3 presents the allowable bearing pressures q_{all} from the *Uniform Building Code* (1997).
Spread and combined footings	The Terzaghi bearing capacity equation was developed for strip footings. For other types of footings and loading conditions, corrections need to be applied to the bearing capacity equation. Many different types of corrections have been proposed (e.g., Meyerhof 1951, 1953, 1965). One commonly used form of the bearing capacity equation for spread (square footings) and combined footings (rectangular footings) subjected to uniform vertical loading is as follows (NAVFAC DM-7.2, 1982):

TABLE 10.4 Terzaghi Bearing Capacity Equation (*Continued*)

Topic (1)	Discussion (2)
Spread and combined footings (*Continued*)	$q_{ult} = \frac{Q_{ult}}{BL} = cN_c\,[1 + 0.3(B/L)] + 0.4\,\gamma_t BN_\gamma + \gamma_t D_f N_q$ This equation is similar to the Terzaghi equation and the terms have the same definitions. For the strip footing, the shear strength is based on a plane strain condition (soil is confined along the long axis of the footing). It has been stated that the friction angle ϕ is about 10% higher in the plane strain condition as compared to the friction angle ϕ measured in the triaxial apparatus (Meyerhof 1961, Perloff and Baron 1976). Ladd et al. (1977) indicated that the friction angle ϕ in plane strain is larger than ϕ in triaxial shear by 4° to 9° for dense sands. A difference in friction angle of 4° to 9° has a significant impact on the bearing capacity factors (see Table 10.5). The plane strain effect is usually ignored in practice and results in an additional factor of safety.

TABLE 10.5 Bearing Capacity Factors (N_c, N_γ, and N_q) That Automatically Incorporate Allowances for Local Shear and Punching Shear Failure

Friction angle (ϕ or ϕ'), degrees (1)	N_c (2)	N_γ (3)	N_q (4)
0	5	0	1
10	8	1	3
20	15	4	5
30	30	8	10
31	33	10	12
32	35	15	18
33	39	20	21
34	42	28	28
35	46	35	34
36	51	48	45
37	56	65	55
38	61	80	67
39	68	100	80
40	75	120	93
41	84	140	110

Note: At high friction angles, bearing capacity factors increase rapidly, and these values should be used with caution.
Source: Based on Terzaghi 1943, and Peck, Hansen, and Thornburn 1974.

TABLE 10.6 Bearing Capacity for Shallow Foundations on Cohesionless Soil

Topic (1)	Discussion (2)
General discussion	As discussed in Chap. 4, cohesionless soil is nonplastic and includes gravels, sands, and nonplastic silt such as rock flour (see Table 4.5). A cohesionless soil develops its shear strength as a result of the frictional and interlocking resistance between the soil particles. For cohesionless soil, $c = 0$ and the first term in the Terzaghi bearing capacity equation (Table 10.4) is equal to zero.
Location of groundwater table	For cohesionless soil, the location of the groundwater table can affect the ultimate bearing capacity. As mentioned in Table 7.1, the saturation of sand usually does not have much effect on the friction angle (i.e., $\phi' \cong \phi$). But a groundwater table creates a buoyant condition in the soil which results in less resistance of the soil and hence a lower ultimate bearing capacity. As previously mentioned, the depth of the bearing capacity failure is rather shallow (Figs. 10.1*a* to *c*) and it is often assumed that the soil involved in the bearing capacity failure can extend to a depth equal to B (footing width) below the bottom of the footing. Thus, for a groundwater table located in this zone, an adjustment to the ultimate bearing capacity is performed by adjusting the unit weight of the second term of the Terzaghi bearing capacity equation (Table 10.4) by using the following equation (Myslivec and Kysela 1978): $$\gamma_a = \gamma_b + \frac{h' - D_f}{B}(\gamma_t - \gamma_b)$$ where γ_a = adjusted unit weight that is used in place of γ_t for the second term in the Terzaghi bearing capacity equation (kN/m^3 or pcf) γ_b = buoyant unit weight of the soil (kN/m^3 or pcf) γ_t = total unit weight of the soil (kN/m^3 or pcf) h' = depth of the groundwater table below ground surface (m or ft) D_f = depth below ground surface to the bottom of the footing (m or ft) B = width of the footing (m or ft) This equation is valid for $0 \le (h' - D_f) \le B$. Note that no correction to the unit weight of the soil is required if the groundwater table is at a depth h' that is equal to or greater than $D_f + B$.
Calculation procedure	In order to determine the allowable bearing capacity for shallow foundations on cohesionless soil, the following steps would be performed: **1.** *Shear strength.* The first step would be to determine the effective friction angle ϕ' and unit weight γ_t of the underlying cohesionless soil deposit from the results of subsurface exploration and laboratory testing. For a given value of ϕ', the bearing capacity factors N_γ and N_q are obtained from Table 10.5. **2.** *Groundwater table.* The location of the groundwater table could impact the bearing capacity (use the above equation). **3.** *Ultimate bearing capacity.* The form of the ultimate bearing capacity equation (Table 10.4) depends on the type of footing (strip, square, or rectangular). Using $c = 0$ and inputting the values for B (footing

TABLE 10.6 Bearing Capacity for Shallow Foundations on Cohesionless Soil (*Continued*)

Topic (1)	Discussion (2)
Calculation procedure (*Continued*)	width), γ_t (unit weight), D_f (depth to bottom of footing), and N_γ and N_q (bearing capacity factors), the ultimate bearing capacity is calculated. **4.** *Allowable bearing capacity.* The allowable bearing capacity is obtained by dividing the ultimate bearing capacity by a factor of safety (usually $F = 3$). **5.** *Settlement governs.* The allowable bearing capacity may have to be reduced because of local building code requirements or because a lower bearing pressure is required in order to reduce the amount of settlement.

TABLE 10.7 Bearing Capacity for Shallow Foundations on Cohesive Soil

Topic (1)	Discussion (2)
General discussion	The bearing capacity of cohesive (plastic) soil, such as silts and clays, is more complicated than the bearing capacity of cohesionless (nonplastic) soil. In the southwestern United States, surface deposits of clay are common and can have a hard and rocklike appearance when they become dried out during the summer. Instead of being susceptible to bearing capacity failure, desiccated clays can cause heave (upward movement) of lightly loaded foundations when the clays get wet during the rainy season. Expansion of clay is discussed in Chap. 12. Plastic saturated soils (silts and clays) usually have a lower shear strength than nonplastic cohesionless soil, and are more susceptible to bearing capacity failure. For saturated plastic soils, the bearing capacity usually has to be calculated for two different conditions: (1) total stress analysis (short-term condition) which uses the undrained shear strength of the plastic soil and (2) effective stress analysis (long-term condition) which uses the drained shear strength parameters (c' and ϕ') of the plastic soil.
Total stress analysis	This type of analysis uses the undrained shear strength of the plastic soil (Sec. 7.3.1). The undrained shear strength s_u could be determined from field tests, such as the vane shear test (VST), or in the laboratory from unconfined compression tests. If the undrained shear strength is approximately constant with depth, then $s_u = c$ and $\phi = 0$ for the Terzaghi bearing capacity equation (Table 10.4). If unconsolidated undrained triaxial compression tests (ASTM D 2850-95, 1998) are performed, then an envelope similar to Fig. 7.7 should be obtained for the saturated plastic soil and the $\phi = 0$ concept should be utilized (Sec. 7.3.1). For example, as shown in Fig. 7.7, the undrained shear strength parameters are $c = 2.1$ psi (14.5 kPa) and $\phi = 0$. Note in Table 10.5 that for $\phi = 0$, the bearing capacity factors are $N_c = 5$, $N_\gamma = 0$, and $N_q = 1$. For both of these cases of either an undrained shear strength s_u or shear strength from unconsolidated undrained triaxial compression tests on saturated cohesive soil ($\phi = 0$ concept), the Terzaghi bearing capacity equations reduce to the following: For strip footings: $$q_{\text{ult}} = 5c + \gamma_t D_f = 5s_u + \gamma_t D_f$$ For spread footings: $$q_{\text{ult}} = 5c[1 + 0.3(B/L)] + \gamma_t D_f = 5s_u[1 + 0.3(B/L)] + \gamma_t D_f$$ Because of the use of total stress parameters, the groundwater table does not have an effect on the above equations. In some cases, it may be appropriate to use total stress parameters c and ϕ in order to calculate the ultimate bearing capacity. For example, a structure (such as an oil tank or grain elevator) could be constructed, and then sufficient time elapses so that the saturated plastic soil consolidates under this load. If the oil tank or grain elevator is then quickly filled, the saturated plastic soil would be subjected to an undrained loading. This condition can be modeled by performing consolidated undrained triaxial compression tests (ASTM D 4767-95, 1998) in order to

TABLE 10.7 Bearing Capacity for Shallow Foundations on Cohesive Soil (*Continued*)

Topic (1)	Discussion (2)
Total stress analysis (*Continued*)	determine the total stress parameters c and ϕ. On the basis of the ϕ value, the bearing capacity factors would be obtained from Table 10.5, and then the ultimate bearing capacity would be calculated from Terzaghi bearing capacity equations (Table 10.4).
Effective stress analysis	This type of analysis uses the drained shear strength c' and ϕ' of the plastic soil. The drained shear strength could be obtained from triaxial compression tests with pore water pressure measurements performed on saturated specimens of the plastic soil. This analysis is termed a *long-term analysis* because the shear-induced pore water pressures (positive or negative) from the loading have dissipated and the hydrostatic pore water conditions now prevail in the field. Because an effective stress analysis is being performed, the location of the groundwater table must be considered in the analysis (use γ_a; see Table 10.6). The first step to perform the bearing capacity analysis would be to obtain the bearing capacity factors N_c, N_γ, and N_q from Table 10.5 using the value of ϕ'. An adjustment to the total unit weight may be required, depending on the location of the groundwater table. Then the Terzaghi bearing capacity equation would be utilized (with c' substituted for c) to obtain the ultimate bearing capacity, with a factor of safety of 3 applied in order to calculate the allowable bearing pressure.
Governing condition	Usually the total stress analysis will provide a lower allowable bearing capacity for soft or very soft saturated plastic soils. This is because the foundation load will consolidate the plastic soil, leading to an increase in the shear strength as time passes. For the long-term case (effective stress analysis), the shear strength of the plastic soil is higher with a resulting higher bearing capacity. Usually the effective stress analysis will provide a lower allowable bearing capacity for very stiff or hard saturated plastic soils. This is because such plastic soils are usually heavily overconsolidated and they tend to dilate (increase in volume) during undrained shear deformation. A portion of the undrained shear strength is due to the development of negative pore water pressures during shear deformation. As these negative pore water pressures dissipate with time, the shear strength of the heavily overconsolidated plastic soil decreases. For the long-term case (effective stress analysis), the shear strength will be lower resulting in a lower bearing capacity. Firm to stiff saturated plastic soils are intermediate conditions. The overconsolidation ratio (OCR) and the tendency of the saturated plastic soil to consolidate (gain shear strength) will determine whether the short-term condition (total stress parameters) or the long-term condition (effective stress parameters) provide the lower bearing capacity.
Calculation procedure	In order to determine the allowable bearing capacity for shallow foundations on cohesive soil, the following steps would be performed: **1.** *Total stress or effective stress analysis.* The first step would be to determine the shear strength parameters for the total stress s_u or effective

TABLE 10.7 Bearing Capacity for Shallow Foundations on Cohesive Soil (*Continued*)

Topic (1)	Discussion (2)
Calculation procedure (*Continued*)	stress (c' and ϕ') analysis and unit weight γ_t of the underlying cohesive soil deposit based on the results of subsurface exploration and laboratory testing. On the basis of the shear strength parameters, the bearing capacity factors are obtained from Table 10.5. **2.** *Groundwater table.* For an effective stress analysis, the location of the groundwater table could impact the bearing capacity. **3.** *Ultimate bearing capacity.* The form of the ultimate bearing capacity equation depends on the type of footing (strip, square, or rectangular) and the type of analysis (total stress or effective stress analysis). **4.** *Allowable bearing capacity.* The allowable bearing capacity is obtained by dividing the ultimate bearing capacity by a factor of safety (usually $F = 3$). **5.** *Settlement governs.* The allowable bearing capacity may have to be reduced because of local building code requirements or because a lower bearing pressure is required in order to reduce the amount of settlement.

TABLE 10.8 Other Design Considerations for Shallow Foundations

Topic (1)	Discussion (2)
General discussion	There are many other possible considerations in the evaluation of bearing capacity. Some of the more common items are discussed in this table.
Lateral loads	In addition to the vertical load acting of the footing, it may also be subjected to a lateral load. A common procedure is to treat lateral loads separately and resist the lateral loads by using the soil pressure acting on the sides of the footing (passive pressure) and by using the frictional resistance along the bottom of the footing. Foundations subjected to lateral loads will be further discussed in Sec. 16.2.
Moments and eccentric loads	It is always desirable to design and construct shallow footings so that the vertical load is applied at the center of gravity of the footing. For combined footings that carry more than one vertical load, the combined footing should be designed and constructed so that the vertical loads are symmetric. There may be design situations where the footing is subjected to a moment, such as where there is a fixed end connection between the building frame and the footing. This moment can be represented by a load P that is offset a certain distance (known as the eccentricity) from the center of gravity of the footing. For other projects, there may be property line constraints and the load must be offset a certain distance (eccentricity) from the center of gravity of the footing. There are many different methods to evaluate eccentrically loaded footings. Because an eccentrically loaded footing will create a higher bearing pressure under one side as compared to the opposite side, the best approach is to evaluate the actual pressure distribution beneath the footing. The usual procedure is to assume a rigid footing (hence linear pressure distribution) and use the section modulus ($\frac{1}{6}B^2$) in order to calculate the largest and lowest bearing pressure. For a footing having a width B, the largest (q') and lowest (q'') bearing pressures are as follows: $$q' = \frac{Q(B + 6e)}{B^2}$$ $$q'' = \frac{Q(B - 6e)}{B^2}$$ where q' = largest bearing pressure underneath the footing that is located along the same side of the footing as the eccentricity (kPa or psf) q'' = lowest bearing pressure underneath the footing that is located at the opposite side of the footing (kPa or psf) Q = load applied to the footing (kN per linear meter of footing length or pounds per linear foot of footing length) e = eccentricity of the load Q, i.e., the lateral distance from Q to the center of gravity of the footing (m or ft) B = width of the footing (m or ft) A usual requirement is that the load Q must be located within the middle third of the footing, and the above equations are only valid for this condition. The value of q' must not exceed the allowable bearing pressure q_{all}.

TABLE 10.8 Other Design Considerations for Shallow Foundations (*Continued*)

Topic (1)	Discussion (2)
Footings at the top of slopes	Although methods have been developed to determine the allowable bearing capacity of foundations at the top of slopes (e.g., NAVFAC DM-7.2, 1982, page 7.2-135), these methods should be used with caution when dealing with plastic (cohesive) soil. This is because the outer face of slopes composed of plastic soils may creep downslope, leading to a loss in support to the footing constructed at the top of such slopes. Structures constructed at the top of clayey slopes are discussed in Sec. 13.8.
Inclined base of footing	Charts have been developed to determine the bearing capacity factors for footings having inclined bottoms. However, it has been stated that inclined bases should never be constructed for footings (AASHTO 1996). Sometimes a sloping contact of underlying hard material will be encountered during excavation of the footing. Instead of using an inclined footing base along the sloping contact, the hard material should be excavated in order to construct a level footing that is entirely founded within the hard material.
Earthquake loading	The effect of earthquakes on foundations will be discussed in Chap. 14. When performing foundation design for earthquake loadings, it is common to use a larger allowable bearing pressure in the analysis. For example, a common recommendation by the geotechnical engineer is that the allowable bearing pressure q_{all} can be increased by a factor of $\frac{1}{3}$ for earthquake analyses.

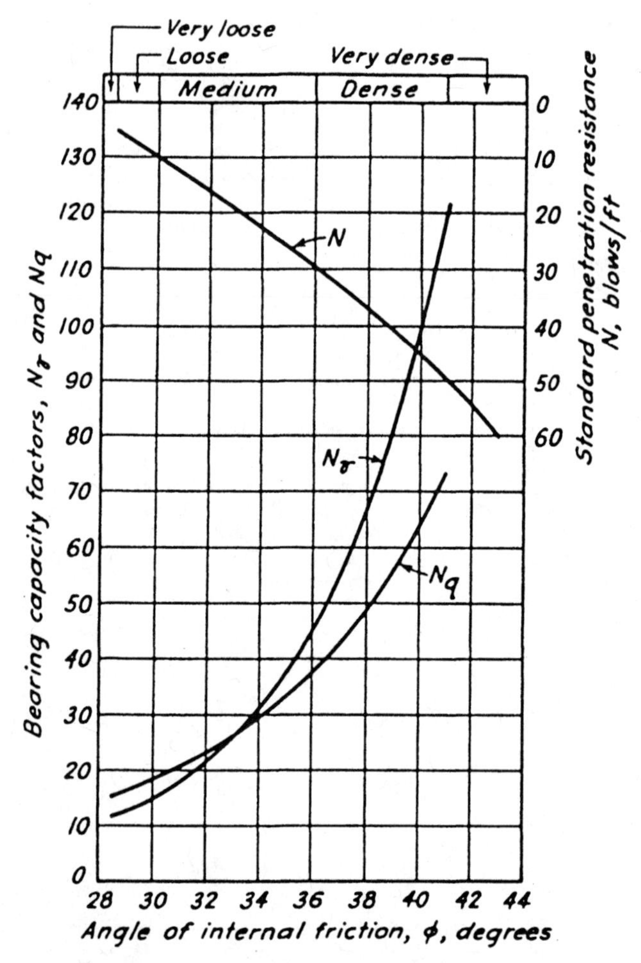

FIGURE 10.2 Bearing capacity factors N_γ and N_q, which automatically incorporate allowance for punching and local shear failure. See Table 10.4. (*From Peck et al. 1974; reproduced with permission of John Wiley & Sons.*)

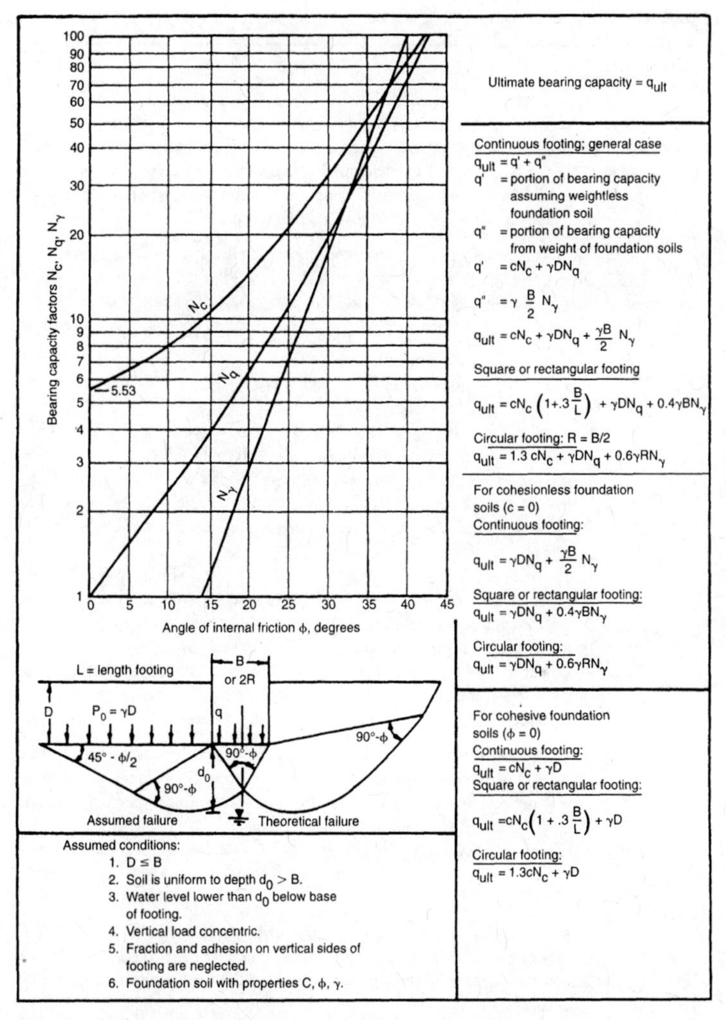

FIGURE 10.3 Bearing capacity factors N_γ, N_q, and N_c, which do not include allowance for punching and local shear failure. [*Note:* For local or punching shear of loose sands or soft clays, the value of ϕ to be used in this figure = $\tan^{-1}$ (0.67 tan ϕ) and the cohesion used in the bearing capacity equation = 0.67c]. See Table 10.4. (*Reproduced from NAVFAC DM-7.2, 1982.*)

TABLE 10.9 Types of Deep Foundations

Topic (1)	Discussion (2)
General discussion	Deep foundations are used when the upper soil stratum is too soft, weak, or compressible to support the foundation loads. Deep foundations are also used when there is a possibility of the undermining of the foundation. For example, bridge piers are often founded on deep foundations to prevent a loss of support due to flood conditions which could cause river bottom scour.
Types of deep foundations	The most common types of deep foundations are piles and piers that support individual footings or mat foundations (Table 1.2). Piles are defined as relatively long, slender, columnlike members often made of steel, concrete, or wood that are either driven into place or cast-in-place in predrilled holes. Common types of piles are as follows: • *Batter pile.* A pile driven in at an angle inclined to the vertical to provide high resistance to lateral loads. • *End-bearing pile.* A pile whose support capacity is derived principally from the resistance of the foundation material on which the pile tip rests. End-bearing piles are often used when a soft upper layer is underlain by a dense or hard strata. If the upper soft layer should settle, the pile could be subjected to down-drag forces, and the pile must be designed to resist these soil-induced forces. • *Friction pile.* A pile whose support capacity is derived principally from the resistance of the soil friction and/or adhesion mobilized along the side of the pile. Friction piles are often used in soft clays where the end-bearing resistance is small because of punching shear at the pile tip. • *Combined end-bearing and friction pile.* A pile that derives its support capacity from combined end-bearing resistance developed at the pile tip and frictional and/or adhesion resistance on the pile perimeter. A pier is defined as a deep foundation system, similar to a cast-in-place pile, that consists of a columnlike reinforced-concrete member. Piers are often of large enough diameter to enable downhole inspection. Piers are also commonly referred to as drilled shafts, bored piles, or drilled caissons. There are many other methods available for forming deep foundation elements. Examples include earth stabilization columns, such as (NAVFAC DM-7.2, 1982): • *Mixed-in-place piles.* A mixed-in-place soil-cement or soil-lime pile. • *Vibroreplacement stone columns.* Vibroflotation or other method is used to make a cylindrical, vertical hole which is filled with compacted gravel or crushed rock. • *Grouted stone columns.* This is similar to the above but includes filling voids with bentonite-cement or water-sand-bentonite cement mixtures. • *Concrete vibro columns.* Similar to stone columns, but concrete is used instead of gravel.

TABLE 10.10 Design Criteria for Deep Foundations

Topic (1)	Discussion (2)
General discussion	There are several different items that are used in the design and construction of piles. These items are individually discussed below.
Engineering analysis	On the basis of the results of subsurface exploration and laboratory testing, the bearing capacity of the deep foundation can be calculated in the same way as the previous discussion of shallow foundations. The next two tables (Tables 10.11 and 10.12) will describe the engineering analyses for deep foundations in cohesionless and cohesive soil.
Field load tests	Prior to the construction of the foundation, a pile or pier could be load tested in the field to determine its carrying capacity. Because of the uncertainties in the design of piles based on engineering analyses, pile load tests are common. The pile load test can often result in a more economical foundation then one based solely on engineering analyses. Pile load tests will be further discussed in Chap. 21.
Application of pile driving resistance	In the past, the pile capacity was estimated on the basis of the driving resistance during the installation of the pile. Pile driving equations, such as the *Engineering News formula* (Wellington 1888), were developed that related the pile capacity to the energy of the pile driving hammer and the average net penetration of the pile per blow of the pile hammer. But studies have shown that there is no satisfactory relationship between the pile capacity from pile driving equations and the pile capacity measured from load tests. Based on these studies, it has been concluded that use of pile driving equations is no longer justified (Terzaghi and Peck 1967). Often the pile driving resistance (i.e., blows per foot) is recorded as the pile is driven into place. When the anticipated bearing layer is encountered, the driving resistance (blows per foot) should substantially increase.
Specifications and experience	Other factors that should be considered in the deep foundation design include governing building code or agency requirements and local experience. Local experience is very important in the design and construction of pile foundations. For example, for the construction of a bridge pier, it was determined that 24-in.- (0.6-m-) diameter concrete-filled steel pipe piles would be required for the design loading conditions. Once driven into place and load-tested, the 24-in.- (0.6-m-) diameter pipe piles were unable to develop the necessary load capacity. On the basis of local experience advice, 12-in.- (0.3-m-) diameter steel pipe piles were installed, and when field tested, they developed a higher load capacity even though they have less surface area and less end-bearing area than the larger 24-in.- (0.6-m-) diameter pipe piles. The reason for the higher load capacity of the smaller driven piles could be that they caused less remolding and disturbance of the cohesive soil than the larger piles. This project demonstrated the importance of local experience on the selection of pile types and sizes.

TABLE 10.11 Bearing Capacity of Deep Foundations in Cohesionless Soil

Topic (1)	Discussion (2)
End-bearing pile or pier	For an end-bearing pile or pier, the bearing capacity equations can be used to determine the ultimate bearing capacity q_{ult}. As mentioned in Table 10.4, the original Terzaghi bearing capacity equation has been revised and used for square footings. Assuming a pile that has a square cross section (i.e., $B = L$) and since $c = 0$ (cohesionless soil), the bearing capacity equation for square footings (Table 10.4) reduces to: $$q_{ult} = \frac{Q_p}{B^2} = 0.4\gamma_t B N_\gamma + \gamma_t D_f N_q$$ In comparing the second and third terms in this equation, the value of B (width of pile) is much less than the embedment depth D_f of the pile. Therefore, the first term in the equation can be neglected. The value of $\gamma_t D_f$ in the equation is equivalent to the total vertical stress σ_v at the pile tip. For cohesionless soil (effective stress analysis), the groundwater table must be included in the analysis and the vertical effective stress σ'_v can be substituted for σ_v, and the equation reduces to: For end-bearing piles having a square cross section: $$q_{ult} = \frac{Q_p}{B^2} = \sigma'_v N_q$$ For end-bearing piles or piers having a circular cross section: $$q_{ult} = \frac{Q_p}{\pi r^2} = \sigma'_v N_q$$ where q_{ult} = the ultimate bearing capacity of the end-bearing pile or pier (kPa or psf) Q_p = point resistance force (kN or lb) B = width of the piles having a square cross section (m or ft) r = radius of the piles or piers having a round cross section (m or ft) σ'_v = vertical effective stress at the pile tip (kPa or psf) N_q = dimensionless bearing capacity factor For drilled piers or piles placed in predrilled holes, the value of N_q can be obtained from Table 10.5 on the basis of the friction angle ϕ of the cohesionless soil located at the pile tip. However, for driven piles, the values of N_q listed in Table 10.5 are generally too conservative. Figure 10.4 presents a chart prepared by Vesic (1967) that provides the bearing capacity factor N_q from several different sources. Note in Fig. 10.4 that at $\phi = 30°$, N_q varies from about 30 to 150, while at $\phi = 40°$, N_q varies from about 100 to 1000. This is a tremendous variation in N_q values and is related to the different approaches used by the various researchers, where in some cases the basis of the relationship shown in Fig. 10.4 is theoretical, while in other cases the relationship is based on analysis of field data such as pile load tests. There is a general belief that the bearing capacity factor N_q is higher for driven piles than for shallow foundations. One reason for a higher N_q value

TABLE 10.11 Bearing Capacity of Deep Foundations in Cohesionless Soil (*Continued*)

Topic (1)	Discussion (2)
End-bearing pile or pier (*Continued*)	is the effect of driving the pile, which displaces and densifies the cohesionless soil at the bottom of the pile. The densification could be due to both the physical process of displacing the soil and the driving vibrations. These actions would tend to increase the friction angle of the cohesionless soil in the vicinity of the driven pile. Large-diameter piles would tend to displace and densify more soil than smaller diameter piles. The ultimate end-bearing capacity Q_p of the pile is divided by a factor of safety (usually $F = 3$) to obtain the allowable end-bearing capacity (Q_{all}).
Friction piles	As the name implies, a friction pile develops its load-carrying capacity from the frictional resistance between the cohesionless soil and the pile perimeter. Piles subjected to vertical uplift forces would be designed as friction piles because there would be no end-bearing resistance as the pile is pulled from the ground. On the basis of a linear increase in frictional resistance with confining pressure, the average ultimate frictional capacity q_{ult} can be calculated as follows: For piles having a square cross section: $$q_{ult} = \frac{Q_s}{4BL} = \sigma'_h \tan \phi_w = \sigma'_v k \tan \phi_w$$ For piles or piers having a circular cross section: $$q_{ult} = \frac{Q_s}{2\pi r L} = \sigma'_h \tan \phi_w = \sigma'_v k \tan \phi_w$$ where q_{ult} = average ultimate frictional capacity for the pile or pier (kPa or psf) Q_s = ultimate skin friction resistance force (kN or lbs) B = width of the piles having a square cross section (m or ft) r = radius of the piles or piers having a round cross section (m or ft) L = length of the pile or pier (m or ft) σ'_h = average horizontal effective stress over the length of the pile or pier (kPa or psf) σ'_v = average vertical effective stress over the length of the pile or pier (kPa or psf) k = dimensionless parameter equal to σ'_h divided by σ'_v (i.e., similar to k_o; see Table 6.3). The equations for k_o in Table 6.3 can be used to estimate the value of k for sand deposits. Because of the densification of the cohesionless soil associated with driven displacement piles, values of k between 1 and 2 are often assumed. ϕ_w = friction angle between the cohesionless soil and the perimeter of the pile or pier (degrees). Commonly used friction angles are $\phi_w = \frac{3}{4}\phi$ for wood and concrete piles and $\phi_w = 20°$ for steel piles.

TABLE 10.11 Bearing Capacity of Deep Foundations in Cohesionless Soil (*Continued*)

Topic (1)	Discussion (2)
Friction piles (*Continued*)	Note in the above equations that the term $4BL$ (for square piles) and $2\pi rL$ (for circular piles) is the perimeter surface area of the pile or pier. In the above equations, the term σ'_h tan ϕ_w equals the shear strength τ_f between the pile or pier surface and the cohesionless soil. This term is identical to shear strength equation (equation in Table 7.2 with $c' = 0$), i.e., $\tau_f = \sigma'_n$ tan ϕ'. Thus the frictional resistance force Q_s in the above equations is equal to the perimeter surface area times the shear strength of the soil at the pile or pier surface. The ultimate skin friction resistance force Q_s is divided by a factor of safety (usually $F = 3$) to obtain the allowable skin friction resistance force Q_{all}.
Combined end-bearing and friction piles	For piles and piers subjected to vertical compressive loads and embedded in a deposit of cohesionless soil, they are usually treated in the design analysis as combined end-bearing and friction piles or piers. This is because the pile or pier can develop substantial load-carrying capacity from both end-bearing and frictional resistance. To calculate the ultimate pile or pier capacity for a condition of combined end-bearing and friction, the value of Q_p is added to the value of Q_s (see above equations). This value of Q_p plus Q_s is divided by a factor of safety (usually $F = 3$) to obtain the allowable combined end bearing and frictional resistance force Q_{all}.
Pile groups	The previous discussion has dealt with the load capacity of a single pile in cohesionless soil. Usually pile groups are used to support the foundation elements, such as a group of piles supporting a spread footing (pile cap) or a mat slab. In loose sand and gravel deposits, the load-carrying capacity of each pile in the group may be greater than a single pile because of the densification effect due to driving the piles. Because of this densification effect, the load capacity of the group is often taken as the load capacity of a single pile times the number of piles in the group. An exception would be a situation where a weak layer underlies the cohesionless soil. In this case, group action of the piles could cause them to punch through the cohesionless soil and into the weaker layer or cause excessive settlement of the weak layer located below the pile tips. In order to determine the settlement of the stratum underlying the pile group, the 2:1 approximation (see Table 6.5) can be used to determine the increase in vertical stress ($\Delta\sigma_v$) for those soil layers located below the pile tip. If the piles in the group are principally end bearing, then the 2:1 approximation starts at the tip of the piles (L = length of the pile group, B = width of the pile group, and z = depth below the tip of the piles; see Table 6.5). If the pile group develops its load-carrying capacity principally through side friction, then the 2:1 approximation starts at a depth of $\frac{2}{3}D$, where D = depth of the pile group.

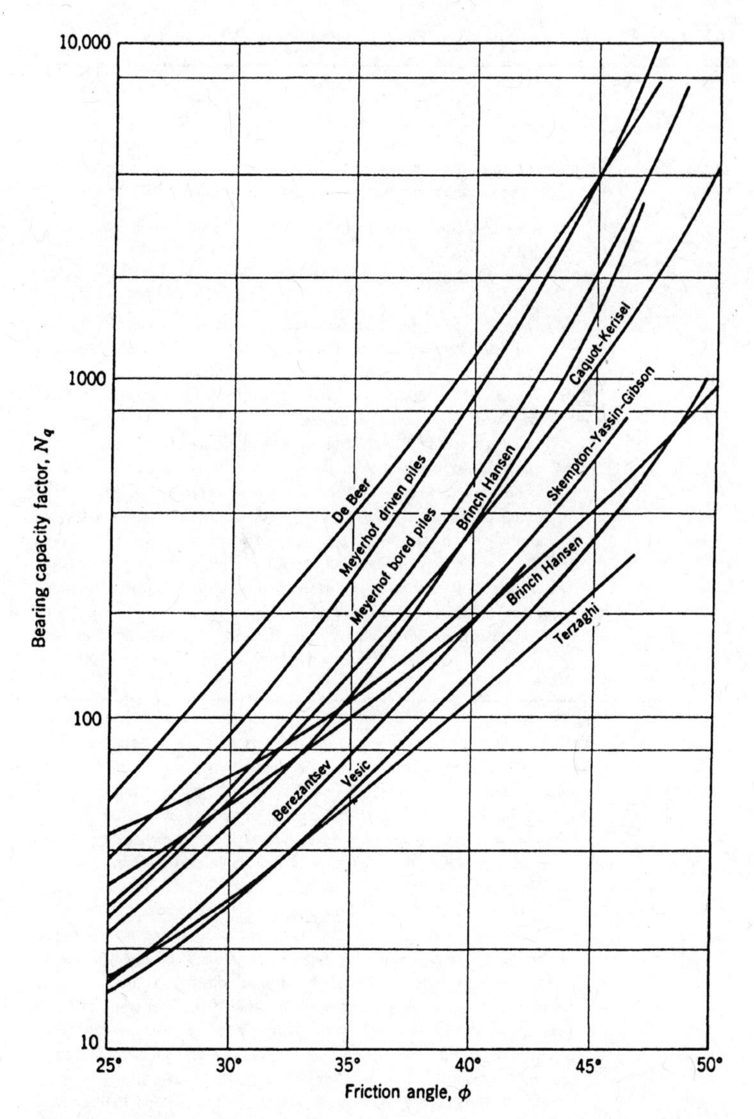

FIGURE 10.4 Bearing capacity factor N_q as recommended by various researchers for deep foundations. See Table 10.11. (*From Vesic 1967; reproduced from Lambe and Whitman 1969.*)

TABLE 10.12 Bearing Capacity of Deep Foundations in Cohesive Soil

Topic (1)	Discussion (2)
General discussion	The analysis of load-carrying capacity of piles and piers in cohesive soil is more complex than that for cohesionless soil. Some of the factors that may need to be considered in the analysis are as follows (AASHTO 1996): **1.** A lower load-carrying capacity of a pile in a pile group compared to that of a single pile. **2.** The settlement of the underlying cohesive soil due to the load of the pile group. **3.** The effects of driving piles on adjacent structures or slopes. The ground will often heave around piles driven into soft and saturated cohesive soil. **4.** The increase in load on the pile due to negative skin friction (i.e., down-drag loads) from consolidating soil. **5.** The effects of uplift loads from expansive and swelling clays. **6.** The reduction in shear strength of the cohesive soil due to construction techniques, such as the disturbance of sensitive clays or development of excess pore water pressures during the driving of the pile. There is often an increase in load-carrying capacity of a pile after it has been driven into a soft and saturated clay deposit. This increase in load-carrying capacity with time is known as *freeze* or *setup* and is caused primarily by the dissipation of excess pore water pressures. **7.** The influence of fluctuations in the elevation of the groundwater table on the load-carrying capacity when analyzed in terms of effective stresses.
Total stress analysis	The ultimate load capacity of a single pile or pier in cohesive soil is often determined by performing a total stress analysis. This is because the critical load on the pile, such as from wind or earthquake loads, is a short-term loading condition and thus the undrained shear strength of the cohesive soil will govern. The total stress analysis for a single pile or pier in cohesive soil typically is based on the undrained shear strength s_u of the cohesive soil or the value of cohesion c determined from unconsolidated undrained triaxial compression tests (i.e., $\phi = 0$ analysis; see Table 7.5). The ultimate load capacity of the pile or pier in cohesive soil would equal the sum of the ultimate end-bearing and ultimate side adhesion components. In order to determine the ultimate end-bearing capacity, the Terzaghi bearing capacity equation (Table 10.4) can be utilized with $B = L$, and $\phi = 0$, in which case $N_c = 5$, $N_\gamma = 0$, and $N_q = 1$ (Table 10.5). The term $\gamma_t D_f N_q = \gamma_t D_f$ (for $\phi = 0$) is the weight of overburden which is often assumed to be balanced by the pile weight, and thus this term is not included in the analysis. Note in Table 10.4 that the term $cN_c\,(1 + 0.3B/L) = c\,(5)(1.3) = 6.5c$ (or $N_c = 6.5$). However, N_c is commonly assumed to be equal to 9 for deep foundations (Mabsout et al. 1995). Thus the ultimate load capacity Q_{ult} of a single pile or pier in cohesive soil equals: Q_{ult} = end bearing + side adhesion = cN_c (area of tip) + c_A (surface area) or $$Q_{\text{ult}} = c9(\pi R^2) + c_A(2\pi R z) = 9\pi c R^2 + 2\pi c_A R z$$

TABLE 10.12 Bearing Capacity of Deep Foundations in Cohesive Soil (*Continued*)

Topic (1)	Discussion (2)
Total stress analysis (*Continued*)	where Q_{ult} = ultimate load capacity of the pile or pier (kN or kips) c = cohesion of the cohesive soil at the pile tip (kPa or psf). Because it is a total stress analysis, the undrained shear strength ($s_u = c$) is used, or the undrained shear strength is obtained from unconsolidated undrained triaxial compression tests (i.e., $\phi = 0$ analysis, c = peak point of Mohr circles, see Fig. 7.7). R = radius of the uniform pile or pier (m or ft). If the pier bottom is belled or a tapered pile is used, then R at the tip would be different from the radius of the shaft. z = embedment depth of the pile (m or ft) c_A = adhesion between the cohesive soil and pile or pier perimeter (kPa or psf). Figure 10.5 can be used to determine the value of the adhesion (c_A) for different types of piles and cohesive soil conditions. If the pile or pier is subjected to an uplift force, then the ultimate capacity T_{ult} for the pile or pier in tension is equal to: $$T_{ult} = 2\pi c_A R z$$ where c_A is the adhesion between the cohesive soil and the pile or pier perimeter (Fig. 10.5). In order to determine the allowable capacity of a pile or pier in cohesive soil, the values calculated from the above equations would be divided by a factor of safety. A commonly used factor of safety = 3.
Effective stress analysis	For long-term loading conditions of piles or piers, an effective stress analysis could be performed. In this case, the effective cohesion c' and effective friction angle ϕ' would be used in the analysis for end bearing. The location of the groundwater table would also have to be considered in the analysis. Along the pile perimeter, the ultimate resistance could be based on the effective shear strength between the pile or pier perimeter and the cohesive soil.
Pile groups	The bearing capacity of pile groups in cohesive soils is normally less than the sum of individual piles in the group, and this reduction in group capacity must be considered in the analysis. The *group efficiency* is defined as the ratio of the ultimate load capacity of each pile in the group to the ultimate load capacity of a single isolated pile. If the spacing between piles in the group are at a distance that is greater than about 7 times the pile diameter, then the group efficiency is equal to one (i.e., no reduction in pile capacity for group action). The group efficiency decreases as the piles become closer together in the pile group. For example, a 9 $\times$ 9 pile group with a pile spacing equal to 1.5 times the pile diameter has a group efficiency of only 0.3. Figure 10.6 can be used to determine the ultimate load capacity of a pile group in cohesive soil. The ultimate load capacity of the pile group is divided by a factor of safety (usually $F = 3$) to obtain the allowable group pile capacity Q_{all}.

TABLE 10.12 Bearing Capacity of Deep Foundations in Cohesive Soil (*Continued*)

Topic (1)	Discussion (2)
Pile groups (*Continued*)	Similar to pile groups in cohesionless soil, the settlement of the stratum underlying the pile group can be evaluated by using the 2:1 approximation (see Table 6.5) to calculate the increase in vertical stress ($\Delta\sigma_v$) for those soil layers located below the pile tip. If the piles in the group develop their load-carrying capacity principally by end bearing in cohesive soil, then the 2:1 approximation starts at the tip of the piles (L = length of the pile group, B = width of the pile group, and z = depth below the tip of the piles; see Table 6.5). If the pile group develops its load-carrying capacity principally through cohesive soil adhesion along the pile perimeter, then the 2:1 approximation starts at a depth of $\frac{2}{3}D$, where D = depth of the pile group.

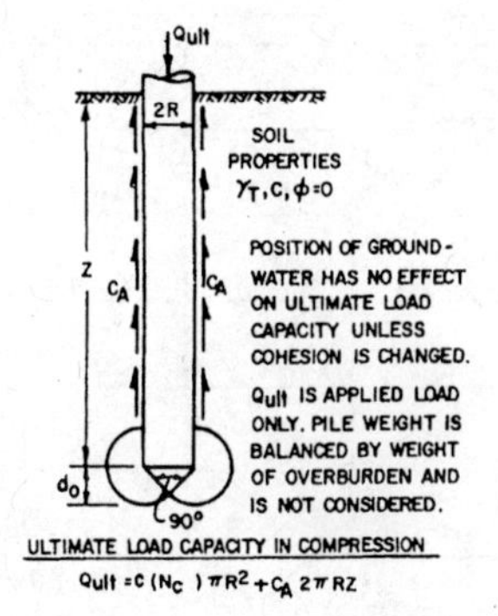

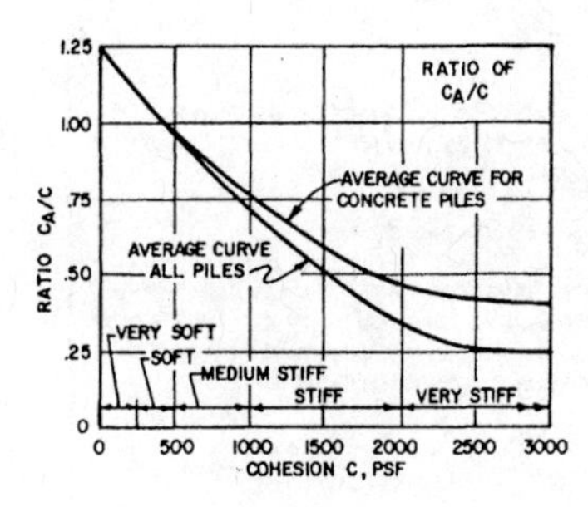

RECOMMENDED VALUES OF ADHESION

PILE TYPE	CONSISTENCY OF SOIL	COHESION, C PSF	ADHESION, C_A PSF
TIMBER AND CONCRETE	VERY SOFT	0 - 250	0 - 250
	SOFT	250 - 500	250 - 480
	MED. STIFF	500 - 1000	480 - 750
	STIFF	1000 - 2000	750 - 950
	VERY STIFF	2000 - 4000	950 - 1300
STEEL	VERY SOFT	0 - 250	0 - 250
	SOFT	250 - 500	250 - 460
	MED. STIFF	500 - 1000	460 - 700
	STIFF	1000 - 2000	700 - 720
	VERY STIFF	2000 - 4000	720 - 750

ULTIMATE LOAD CAPACITY IN TENSION

$T_{ult} = C_A\, 2\pi RZ$

FIGURE 10.5 Ultimate capacity for a single pile or pier in cohesive soil. See Table 10.12. (*Reproduced from NAVFAC DM-7.2, 1982.*)

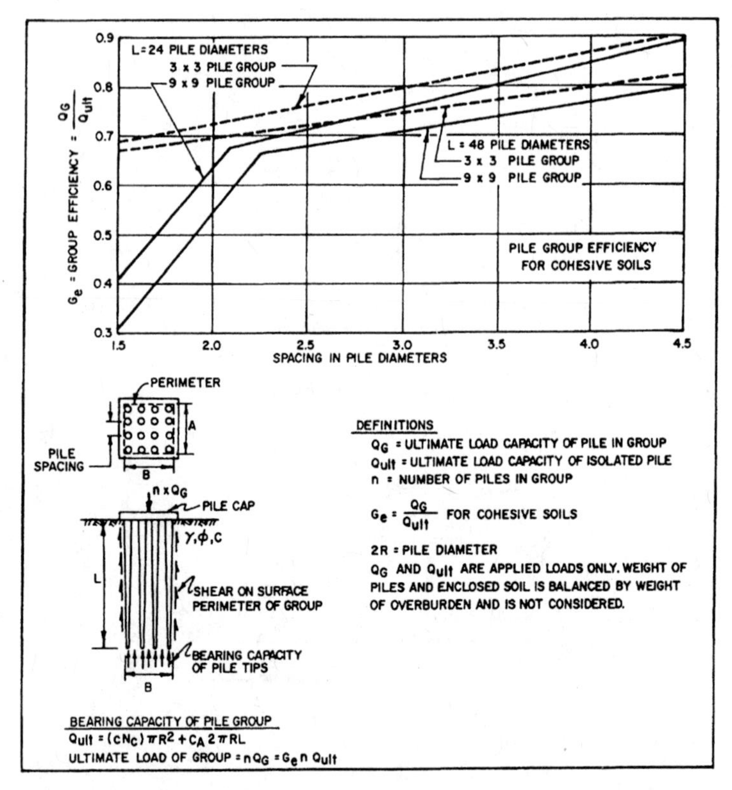

FIGURE 10.6 Ultimate capacity of a pile group in cohesive soil. See Table 10.12. (*Developed by Whitaker 1957; reproduced from NAVFAC DM-7.2, 1982.*)

TABLE 10.13 Example Problems

Topic (1)	Problem and solution (2)
Bearing capacity of a strip footing	*Problem:* A strip footing will be constructed on a nonplastic silty sand deposit which has the shear strength properties shown in Fig. 7.1 (i.e., $c = 0$ and $\phi' = 30°$) and a total unit weight of 19 kN/m^3 (120 pcf). Assume the strip footing will be 1.2 m (4 ft) wide and embedded 0.9 m (3 ft) below the ground surface. Using a factor of safety of 3 and assuming the groundwater table is well below the bottom of the footing, determine the allowable bearing pressure q_{all} for the nonplastic silty sand. *Solution:* $c = 0$, γ_t = 19 kN/m^3 (120 pcf), B = 1.2 m (4 ft), $N_\gamma = 8$ and N_q = 10 (Table 10.5 for $\phi' = 30°$), and D_f = 0.9 m (3 ft). Substituting these values into the Terzaghi bearing capacity equation (Table 10.4), the ultimate bearing capacity q_{ult} = 260 kPa (5500 psf). Using a factor of safety = 3, the allowable bearing pressure (q_{all}) = 90 kPa (1800 psf). This 1.2-m- (4-ft-) wide strip footing could carry a maximum vertical load Q_{all} of 110 kN per linear meter of wall (7.2 kips per linear foot of wall).
End-bearing capacity of piles	*Problem:* Assume a pile having a diameter of 0.3 m (1 ft) and a length of 6 m (20 ft) will be driven into a nonplastic silty sand deposit which has the shear strength properties shown in Fig. 7.1 (i.e., $c = 0$ and $\phi' = 30°$). Assume that the location of the groundwater table is located 3 m (10 ft) below the ground surface and the total unit weight above the groundwater table is 19 kN/m^3 (120 pcf) and the buoyant unit weight γ_b below the groundwater table is 9.9 kN/m^3 (63 pcf). Using the Terzaghi correlation shown in Fig. 10.4, calculate the allowable end-bearing capacity using a factor of safety of 3. *Solution:* The vertical effective stress (σ'_v) at the pile tip equals: σ'_v = (3 m) (19 kN/m^3) + (3 m) (9.9 kN/m^3) = 87 kPa or 1800 psf From Fig. 10.4, using the Terzaghi relationship, $N_q = 30$. Using the end-bearing equation in Table 10.11 (circular cross section), the ultimate end-bearing capacity (q_{ult}) = 2600 kPa (54,000 psf). Multiplying q_{ult} by the area of the pile tip (πr^2), the ultimate pile tip resistance (Q_p) = 180 kN (42 kips). For a factor of safety of 3, the allowable pile tip resistance = 60 kN (13 kips).
Friction pile	*Problem:* Assume the same conditions as the previous example and that the pile is made of concrete. For driven displacement piles and using $k = 1$, calculate the allowable frictional capacity of the pile. *Solution:* The easiest solution consists of dividing the pile into two sections. The first section is located above the groundwater table (z = 0 to 3 m) and the second section is that part of the pile below the groundwater table (z = 3 to 6 m). The average vertical stress will be at the midpoint of these two sections, or: σ_v (at z = 1.5 m) = (1.5 m) (19 kN/m^3) = 29 kPa or 600 psf σ'_v (at z = 4.5 m) = (3 m) (19 kN/m^3) + (1.5 m) (9.9 kN/m^3) = 72 kPa or 1500 psf For a concrete pile $\phi_w = \frac{3}{4}\phi = \frac{3}{4}(30°) = 22.5°$. Substitute values into the friction pile equation in Table 10.11: For z = 0 to 3 m and L = 3 m,

TABLE 10.13 Example Problems (*Continued*)

Topic (1)	Problem and solution (2)
Friction pile (*Continued*)	$Q_s = (2\pi rL)(\sigma'_v k \tan \phi_w) = (2\pi)(0.15\text{ m})(3\text{ m})(29\text{ kPa})(1)(\tan 22.5°) = 34$ kN or 7.6 kips For $z = 3$ m to 6 m and $L = 3$ m, $Q_s = (2\pi rL)(\sigma'_v k \tan \phi_w) = (2\pi)(0.15\text{ m})(3\text{ m})(72\text{ kPa})(1)(\tan 22.5°) = 84$ kN or 19 kips Adding together both values of Q_s, the total frictional resistance force = 34 kN + 84 kN = 118 kN or 26.6 kips. For a factor of safety of 3, the allowable frictional capacity of the pile is approximately equal to 40 kN (9 kips).
Combined	Combined end-bearing and frictional resistance = 60 kN + 40 kN = 100 kN.

CHAPTER 11
PAVEMENT AND PIPELINE DESIGN

11.1 PAVEMENT DESIGN

Because the thickness of the pavement design is governed by the shear strength of the soil supporting the road, it is usually the geotechnical engineer who tests the soil and determines the pavement design thickness. The transportation engineer often provides design data to the geotechnical engineer, such as the estimated traffic loading, required width of pavement, and design life of the pavement. Table 11.1 presents a general discussion of pavement types (i.e., flexible versus rigid) and the pavement section. Table 11.2 presents characteristics of compacted subgrade for roads and airfields.

There are many different methods that are used for the design of pavements. Some methods utilize the California Bearing Ratio (CBR; see Table 2.13) as a measure of the shear strength of the base and subgrade. Numerous charts have also been developed that relate the shear strength of the subgrade and the traffic loads to a recommended pavement thickness (e.g., Asphalt Institute 1984). When designing pavements, the geotechnical engineer should always check with the local transportation authority for design requirements as well as the local building department or governing agency for possible specifications on the type of method that must be used for the design.

11.1.1 California Method of Flexible Pavement Design

In California, the method used to design flexible pavement sections is based on the *R*-value, as discussed in Tables 11.3 and 11.4 and illustrated in Fig. 11.1.

11.2 PIPELINE DESIGN

The external load on a pipeline depends on many different factors. One important factor is the type of pipeline (rigid versus flexible). Another important factor is the placement conditions, i.e., whether the pipeline is constructed under an embankment, in a trench, or is pushed or jacked into place. Figure 11.2 illustrates the three placement conditions of trench, embankment, and tunnel (or pushed or jacked condition).

Other factors that affect the external load on a pipeline include the unit weight and thickness of overburden soil, the surface loads such as those applied by traffic, compaction procedures, and the presence of groundwater (i.e., buoyant conditions on an empty pipeline).

11.2.1 Rigid Pipeline and Flexible Pipeline Design

The discussion of pipeline loads is divided into two main categories: rigid pipelines (Table 11.5 and Fig. 11.3) and flexible pipelines (Table 11.6). An example of the specifications for the construction of a storm drain in a trench excavation is shown in Fig. 11.4.

TABLE 11.1 Pavement Types and Pavement Section

Topic (1)	Discussion (2)
General discussion	Besides shallow and deep foundations, other structures can be susceptible to bearing capacity failures. For example, unpaved roads and roads with an inadequate pavement section or weak subgrade can also be susceptible to bearing capacity failures caused by heavy wheel loads. The heavy wheel loads can cause a general bearing capacity failure or a punching-type shear failure. These bearing capacity failures are commonly known as rutting, and they develop when the unpaved road or weak pavement section is unable to support the heavy wheel load.
Pavement types	Pavements are usually classified as either *rigid* or *flexible* depending on how the surface loads are distributed. A rigid pavement consists of Portland cement concrete slabs, which tend to distribute the loads over a fairly wide area. A flexible pavement is defined as a pavement having a sufficiently low bending resistance, yet having the required stability to support the traffic loads, e.g., macadam, crushed stone, gravel, and asphalt (California Division of Highways 1973).
Flexible pavement section	A flexible pavement section can consist of the following: 1. *Asphalt concrete.* The uppermost layer (surface course) is typically "asphalt concrete" that distributes the vehicle load in a cone-shape area under the wheel and acts as the wearing surface. The ingredients in asphalt concrete are asphalt (the cementing agent), coarse and fine aggregates, mineral filler (i.e., fines such as limestone dust), and air. Asphalt concrete is usually hot-mixed in an asphalt plant and then hot-laid and compacted by smooth-wheeled rollers. Other common names for asphalt concrete are *blacktop, hot mix,* or simply *asphalt* (Atkins 1983). 2. *Base.* Although not always a requirement, in many cases there is a base material that supports the asphalt concrete. The base typically consists of aggregates that are well graded, hard, and resistant to degradation from traffic loads. The base material is compacted into a dense layer that has a high frictional resistance and good load distribution qualities. The base can be mixed with up to 6% portland cement to give it more strength, and this is termed a *cement-treated base* (CTB). 3. *Subbase.* In some cases, a subbase will be used to support the base and asphalt concrete layers. The subbase often consists of a lesser quality aggregate that is lower-priced than base material. 4. *Subgrade.* The subgrade supports the pavement section (i.e., the overlying subbase, base, and asphalt concrete). The subgrade could be native soil or rock, a compacted fill, or soil that has been strengthened by the addition of lime or other cementing agents. Instead of strengthening the subgrade, a geotextile could be placed on top of the subgrade in order to improve its load-carrying capacity. Table 11.2 lists soil type versus value as subgrade, potential frost action, drainage properties, CBR values, and subgrade modulus.

TABLE 11.12 Characteristics of Compacted Subgrade for Roads and Airfields

Major Divisions (1)	Subdivisions (2)	USCS symbol (3)	Name (4)	Value as subgrade (no frost action) (5)	Potential frost action (6)
Coarse-grained soils	Gravel and gravelly soils	GW	Well-graded gravels or gravel-sand mixtures, little or no fines	Excellent	None to very slight
		GP	Poorly graded gravels or gravelly sands, little or no fines	Good to excellent	None to very slight
		GM	Silty gravels, gravel-sand-silt mixtures	Good to excellent	Slight to medium
		GC	Clayey gravels, gravel-sand-clay mixtures	Good	Slight to medium
	Sand and sandy soils	SW	Well-graded sands or gravelly sands, little or no fines	Good	None to very slight
		SP	Poorly graded sands or gravelly sands, little or no fines	Fair to good	None to very slight
		SM	Silty sands, sand-silt mixtures	Fair to good	Slight to high
		SC	Clayey sands, sand-clay mixtures	Poor to fair	Slight to high
Fine-grained soils	Silts and clays with liquid limit less than 50	ML	Inorganic silts, rock flour, silts of low plasticity	Poor to fair	Medium to very high
		CL	Inorganic clays of low plasticity, gravelly clays, sandy clays, etc.	Poor to fair	Medium to high
		OL	Organic silts and organic clays of low plasticity	Poor	Medium to high
	Silts and clays with liquid limit greater than 50	MH	Inorganic silts, micaceous silts, silts of high plasticity	Poor	Medium to very high
		CH	Inorganic clays of high plasticity, fat clays, silty clays, etc.	Poor to fair	Medium
		OH	Organic silts and organic clays of high plasticity	Poor to very poor	Medium
Peat	Highly organic	PT	Peat and other highly organic soils	Not suitable	Slight

Note: The Unified Soil Classification System (see Table: 4.4) is used in this table.
Source: From U.S. Army (1960).

TABLE 11.12 Characteristics of Compacted Subgrade for Roads and Airfields (*Continued*)

Compressibility (7)	Drainage properties (8)	Compaction equipment (9)	Typical dry densities (10)		CBR (11)	Sub. mod*, pci (12)
			pcf	Mg/m³		
Almost none	Excellent	Crawler-type tractor, rubber-tired roller, steel-wheeled roller	125–140	2.00–2.24	40–80	300–500
Almost none	Excellent	Crawler-type tractor, rubber-tired roller, steel-wheeled roller	110–140	1.76–2.24	30–60	300–500
Very slight to slight	Fair to very poor	Rubber-tired roller, sheepsfoot roller	115–145	1.84–2.32	20–60	200–500
Slight	Poor to very poor	Rubber-tired roller, sheepsfoot roller	130–145	2.08–2.32	20–40	200–500
Almost none	Excellent	Crawler-type tractor, rubber-tired roller	110–130	1.76–2.08	20–40	200–400
Almost none	Excellent	Crawler-type tractor, rubber-tired roller	105–135	1.68–2.16	10–40	150–400
Very slight to medium	Fair to poor	Rubber-tired roller, sheepsfoot roller	100–135	1.60–2.16	10–40	100–400
Slight to medium	Poor to very poor	Rubber-tired roller, sheepsfoot roller	100–135	1.60–2.16	5–20	100–300
Slight to medium	Fair to poor	Rubber-tired roller, sheepsfoot roller	90–130	1.44–2.08	15 or less	100–200
Medium	Practically impervious	Rubber-tired roller, sheepsfoot roller	90–130	1.44–2.08	15 or less	50–150
Medium to high	Poor	Rubber-tired roller, sheepsfoot roller	90–105	1.44–1.68	5 or less	50–100
High	Fair to poor	Sheepsfoot roller, rubber-tired roller	80–105	1.28–1.68	10 or less	50–100
High	Practically impervious	Sheepsfoot roller, rubber-tired roller	90–115	1.44–1.84	15 or less	50–150
High	Practically impervious	Sheepsfoot roller, rubber-tired roller	80–110	1.28–1.76	5 or less	25–100
Very high	Fair to poor	Compaction not practical	—	—	—	—

*Subgrade Modulus.

TABLE 11.3 California Design Method for Flexible Pavements

Topic (1)	Discussion (2)
Design equation	The design of flexible pavement in California is usually based on the following equation (California Division of Highways 1973): $$T = \frac{0.0032\ \mathrm{TI}\ (100 - R)}{G_f}$$ where T = flexible pavement thickness for a 10-year design life (ft) TI = traffic index (dimensionless parameter) R = R-value (dimensionless parameter) G_f = gravel equivalent factor (dimensionless parameter) This flexible pavement design equation was developed from empirical relationships and the performance of pavements in actual service. Although the above equation is an empirical one, it does contain the major factors that influence pavement performance, such as the amount of traffic loads (TI) and the strength of the base and subgrade (R-values), and it includes a factor that relates the relative strengths of different types of materials in the pavement section (G_f). One limitation is that the equation does not have a factor to account for the expansiveness of the subgrade. This limitation will be further discussed in Chap. 12, "Expansive Soil." The terms in the equation are defined below.
Traffic index (TI)	The effect of traffic on the roadway over the design life of the pavement is expressed by the traffic index (TI). The higher the equivalent wheel loads (EWL) during the life of the pavement, the higher the traffic index. Traffic index values for different types of facilities are summarized in Table 11.4, and the design traffic index will often be provided to the geotechnical engineer by the transportation engineer. If the traffic index is 8 or more, the design should be based on the same criteria as used for state highways.
R-value	The R-value is the resistance of the subgrade or base and refers to the ability of the soil to resist lateral deformation when acted upon by a vertical load. In essence, the R-value is a relative measure of the shear strength of the soil. The R-value is determined by using standardized laboratory test procedures and equipment, such as specified by ASTM D 2844-94, "Standard Test Method for Resistance R-Value and Expansion Pressure of Compacted Soils" (1998). The R-value is determined on laboratory specimens compacted to anticipated field conditions. Prior to testing, the specimens are soaked in water to achieve saturation. By soaking the soil prior to testing, the R-value is supposed to represent the worst possible state that the soil might attain in the field. A crushed rock base can have an R-value of 75 to 85, while a clay subgrade will usually have an R-value of less than 10 (see Fig. 11.1).
Gravel equivalent factor G_f	The gravel equivalent G_f represents the strength factor of the materials in the pavement structural section. The gravel equivalent is an empirical factor developed through research and field experience and it relates the relative strength of a unit thickness of a particular material to an equivalent thickness of gravel. An aggregate base and subbase will have G_f = 1.1 and G_f = 1.0, respectively, while a class A cement-treated aggregate base has a G_f = 1.7. Asphalt concrete (AC) has the following gravel equivalent factors (California Division of Highways 1973): • For TI = 5 and below: asphalt concrete G_f = 2.50 • For TI = 6: asphalt concrete G_f = 2.32 • For TI = 7: asphalt concrete G_f = 2.14

TABLE 11.3 California Design Method for Flexible Pavements (*Continued*)

Topic (1)	Discussion (2)
Gravel equivalent factor G_f (*Continued*)	These gravel equivalent factors relate the relative strength of the asphalt concrete to the gravel base. Thus at a TI = 5 or less, 1 in. of asphalt concrete ($G_f = 2.50$) is considered to be equivalent to 2.5 in. of subbase ($G_f = 1.0$).
Minimum pavement thickness	In terms of the minimum design thickness of asphalt concrete and base, it has been stated (California Division of Highways 1973) that in general, it is difficult to obtain good results over an untreated base with less than 6.4 cm (2.5 in.) of asphalt concrete surfacing. Also, it is very difficult to properly place and compact layers of aggregate base or subbase when these layers measure less than 10 cm (4 in.). Therefore the geotechnical engineer should not recommend thickness values less than these minimum values (i.e., 2.5 in. of asphalt concrete and 4 in. of base or subbase).
Example problem	*Problem:* A pavement design (10-year life) for a secondary collector road is required. The design TI = 5, *R*-value of the base = 65, and *R*-value of the subgrade = 20. Determine the thickness of the base and asphalt concrete. *Solution:* The first step is to determine the required thickness of asphalt concrete using the *R*-value of the base (*R*-value of base = 65). Note that for asphalt concrete with a TI = 5, the gravel equivalent factor (G_f) = 2.50: $T = [0.0032\text{TI}(100 - R)]/G_f = [0.0032(5)(100 - 65)]/2.50 = 0.22 \text{ ft} = 2.7 \text{ in.}$ Usually ½-in. increments of asphalt concrete are specified, so use 3 in. of asphalt concrete. The second step is to determine the thickness of base material. Using the *R*-value of the subgrade (*R*-value = 20) and rearranging the design equation gives $TG_f = 0.0032\text{TI}(100 - R) = 0.0032(5)(100 - 20) = 1.28 \text{ ft}$ The thickness of the asphalt concrete has already been selected (3 in.). For this thickness of asphalt concrete: $TG_f = (3/12)(2.50) = 0.62 \text{ ft}$ Subtracting the asphalt concrete portion (0.62 ft) from the total required (1.28 ft) leaves a balance of 0.66 ft. Use this value to calculate the thickness of the base, recognizing that $G_f = 1.1$ for the base material: $TG_f = 0.66 \text{ ft}$ or $T = (0.66 \text{ ft})/G_f = 0.66/1.1 = 0.6 \text{ ft}$ Thus the base must be 0.6 ft (7.2 in.) thick. Usually it is best to use 1-in. increments of base thickness, and therefore the recommended thickness would be 8 in. The final recommended values are: • Thickness of asphalt concrete: 3 in. • Thickness of base material: 8 in. The third step is to check the recommended values. This is accomplished by multiplying the thickness of each layer by the gravel equivalent factors, or: Check: Asphalt concrete: $TG_f = (3/12)(2.50) = 0.62$ Base: $TG_f = (8/12)(1.1) = 0.73$ Total = $0.62 + 0.73 = 1.35$ Note that $1.35 > 1.28$, and thus the design checks.

TABLE11.4 Traffic Index (TI) for Various Types of Facilities

Type of facility (1)	Traffic index (2)
Minor residential streets and cul-de-sacs	4
Average residential streets	4.5
Residential collectors and minor or secondary collectors	5
Major or primary collectors providing for traffic movement between minor collectors and major arterials	6
Farm-to-market roads providing for the movement of traffic through agricultural areas to major arterials	5–7
Commercial roads (arterials serving areas which are primarily commercial in nature)	7–9
Connector roads (highways and arterials connecting two areas of relatively high population density)	7–9
Major city streets and thoroughfares and county highways	7–9

Note: If the traffic index (TI) is 8 or more, the design should be based on the same criteria as used for state highways
Source: California Division of Highways,(1973.)

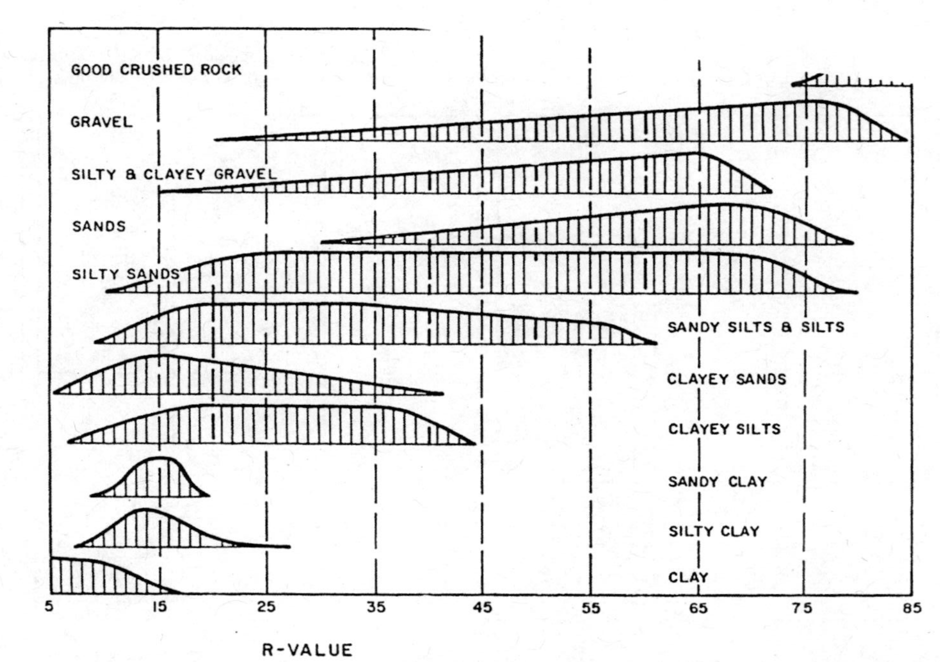

FIGURE 11.1 Typical R-values for different soil types. See Table 11.3. (*Reproduced from California Division of Highways 1973.*)

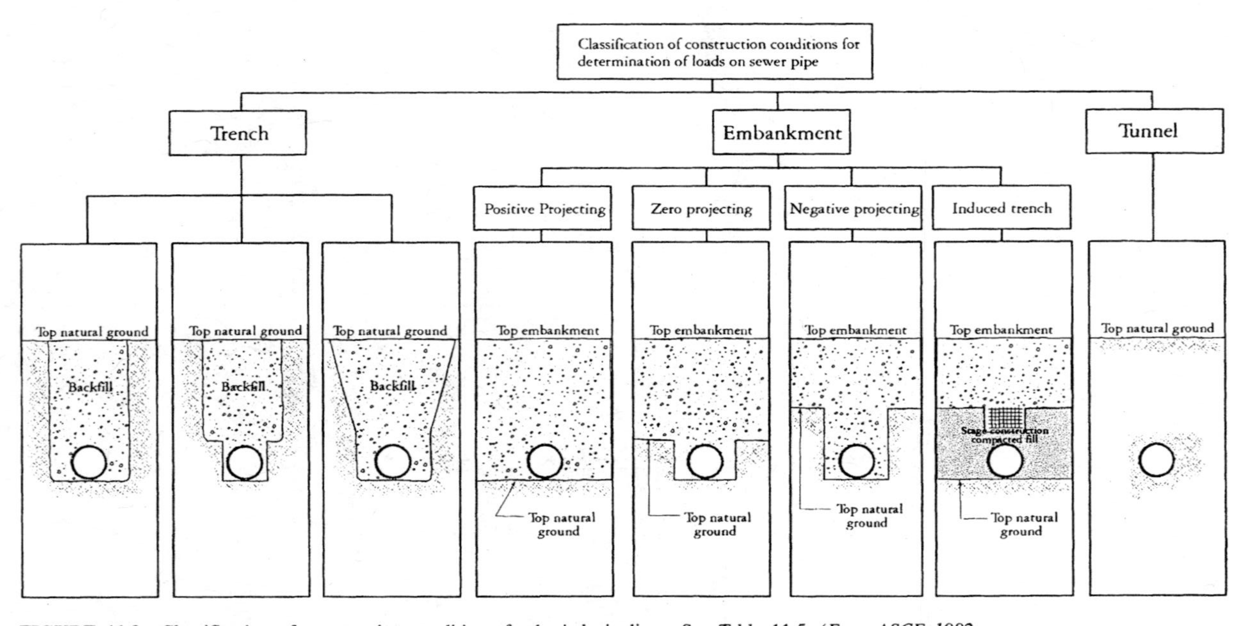

FIGURE 11.2 Classification of construction conditions for buried pipelines. See Table 11.5. (*From ASCE 1982, reprinted with permission of the American Society of Civil Engineers.*)

TABLE 11.5 Rigid Pipeline Design

Topic (1)	Discussion (2)
Types of rigid pipelines	Pipes made from precast concrete, cast-in-place concrete, and cast iron should be considered to be rigid pipelines.
Minimum design load	In general, the minimum vertical load W on a rigid pipeline is equal to the unit weight of soil γ_t times the height H of soil above the top of the pipeline times the diameter of the pipe D, or: $$W_{min} = \gamma_t HD$$ As an example, suppose the pipeline has a diameter D of 24 in. (2 ft), a depth of overburden H of 10 ft, and the backfill soil has a total unit weight γ_t of 125 pcf. Therefore, the minimum vertical load W_{min} acting on the pipeline equals the following: $$W_{min} = (125 \text{ pcf})(10 \text{ ft})(2 \text{ ft}) = 2500 \text{ lb per linear foot of pipe length}$$
Embankment condition	Different types of embankment conditions are shown in Fig. 11.2. In many cases, compaction of fill or placement conditions will impose vertical loads greater than the minimum values calculated above. Also, because the pipe is rigid, the arching affect of soil adjacent to the pipe will tend to transfer load to the rigid pipe. Figure 11.3*a* shows the recommendations for a pipeline to be constructed beneath a fill embankment. In Fig. 11.3*a*, W = vertical dead load on the pipeline, D = diameter of the pipeline, and B = width of the pipeline (i.e., $B = D$). Note that Fig. 11.3*a* was developed for an embankment fill having a total unit weight γ_t = 100 pcf, and an adjustment is required for conditions having different unit weights. As an example, use the same conditions as before ($B = D = 2$ ft, $H = 10$ ft, and γ_t = 125 pcf). Figure 11.3*a* is entered with H = 10 ft, the curve marked 24 in. (2 ft) is intersected, and the value of W read from the vertical axis is about 3800 lb. Therefore: $$W = (3800)(125/100) = 4750 \text{ lb per linear foot of pipeline length}$$ Note this value of 4750 lb is greater than the minimum dead load (2500 lb), and the above value (4750 lb) would be used for the embankment condition.
Trench condition	Different types of trench conditions are shown in Fig. 11.2. Figure 11.3*b* shows the recommendations for a pipeline to be constructed in a trench. Note that in Fig. 11.3*b* that the dimension B is *not* the diameter of the pipeline, but rather the width of the trench at the top of the pipeline. This is because studies have shown that if the pipeline is rigid, it will carry practically all the load on the plane defined by B (Marston 1930, ASCE 1982). Curves are shown for both sand and clay backfill in Fig. 11.3*b*. The procedure is to enter the chart with the H/B ratio, intersect the "Sands" or "Clays" curve, and then determine c_w. Once c_w is obtained, the vertical load W on the pipeline is calculated from the following equation: $$W = c_w \gamma_t B^2$$ As an example, use the same conditions as before (D = 2 ft, H = 10 ft, and γ_t = 125 pcf). Also assume that the trench width at the top of the pipeline

TABLE 11.5 Rigid Pipeline Design (*Continued*)

Topic (1)	Discussion (2)
Trench condition (*Continued*)	will be 4 ft (i.e., $B = 4$ ft) and the trench will be backfilled with sand. Figure 11.3*b* is entered with $H/B = 10/4 = 2.5$, the curve marked "Sands" is intersected, and the value of c_w of about 1.6 is obtained from the vertical axis. Therefore: $$W = c_w \gamma_t B^2 = (1.6)(125)(4)^2 = 3200 \text{ lb per linear foot}$$ Note this value of 3200 lb is greater than the minimum value (2500 lb) and thus 3200 lb would be used for the trench condition. It should be mentioned that, as the width of the trench increases, the values from this section may exceed the embankment values (Fig. 11.3*a*). If this occurs, the embankment condition (Fig. 11.3*a*) should be considered the governing loading condition.
Jacked or driven pipelines	The jacked or driven pipeline condition (i.e., tunnel condition) is shown in Fig. 11.2. Figure 11.3*c* shows the recommendations for a jacked or driven pipeline. Note in Figure 11.3*c* that the dimension *B* is equal to the diameter of the pipeline ($B = D$). The curves shown in Fig. 11.3*c* are for pipelines jacked or driven through sand, clay, or intermediate soils. The procedure is to enter the chart with the *H/B* ratio, intersect the appropriate curve, and then determine c_w. Once c_w is obtained, the vertical load *W* on the pipeline is calculated from the following equation: $$W = c_w \gamma_t B^2$$ As an example, use the same conditions as before ($D = B = 2$ ft, $H = 10$ ft, and $\gamma_t = 125$ pcf) and the pipeline will be jacked through a sand deposit. Figure 11.3*c* is entered with $H/B = 10/2 = 5$, the curve marked "Sand" is intersected, and the value of c_w of about 1.5 is obtained from the vertical axis. Therefore: $$W = c_w \gamma_t B^2 = (1.5)(125)(2)^2 = 750 \text{ lb per linear foot}$$ Note this value of 750 lb is less than the minimum load value (2500 lb) and thus the value of 2500 lb would be used for the jacked or driven pipe condition. Basic soil mechanics indicates that the long-term load for rigid pipelines will be at least equal to the overburden soil pressure (i.e., the minimum design load).
Factor of safety	A factor of safety should be applied to the design dead load *W* calculated above. The above values also only consider the vertical load *W* on the pipeline due to soil pressure. Other loads, such as traffic or seismic loads, may need to be included in the design of the pipeline. For pressurized pipes, rather than the exterior soil load *W*, the interior fluid pressure may govern the design.

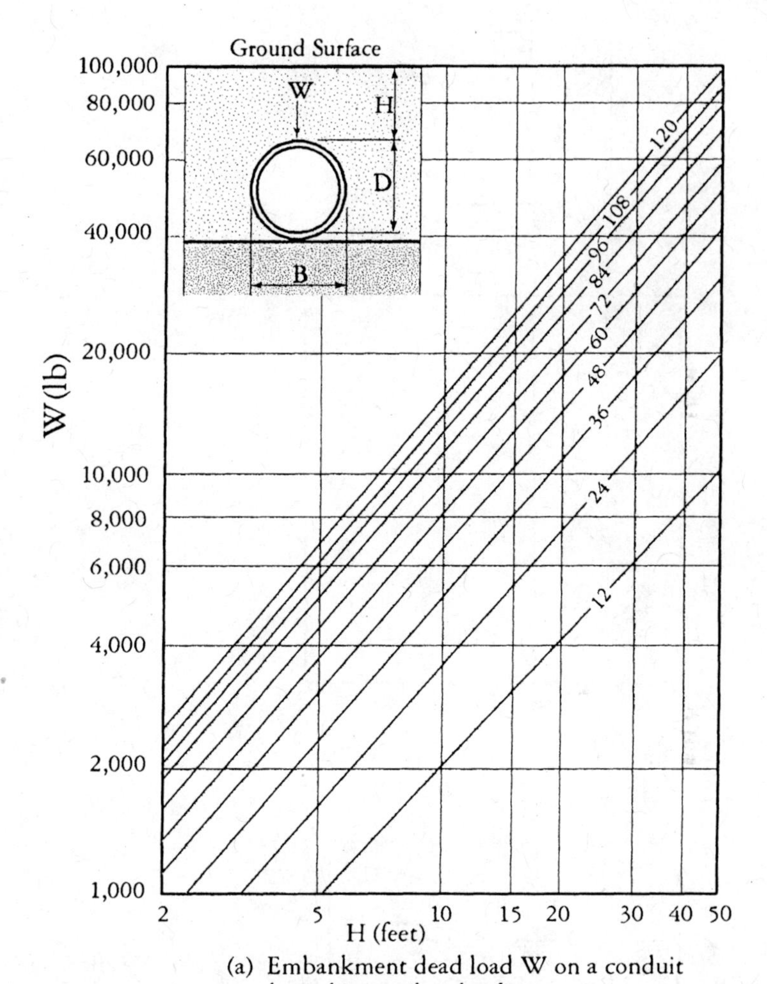

(a) Embankment dead load W on a conduit buried in a soil embankment.

FIGURE 11.3 Embankment load W and backfill coefficient C_w for rigid pipelines. See Table 11.5. (*Reproduced from NAVFAC DM-7.1, 1982.*)

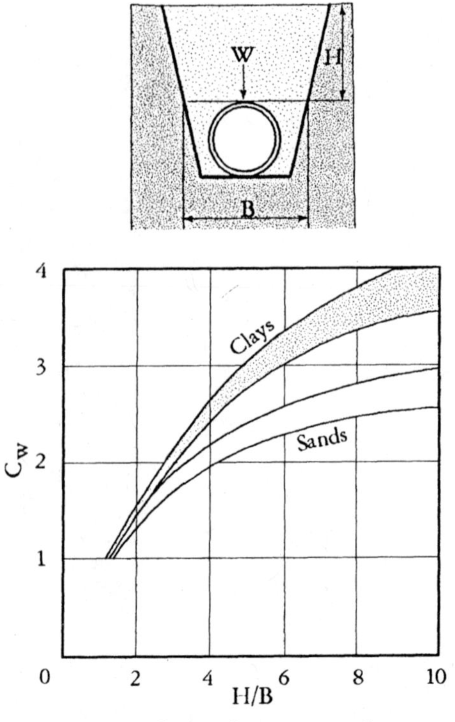

(b) C_w for conduit in a trench

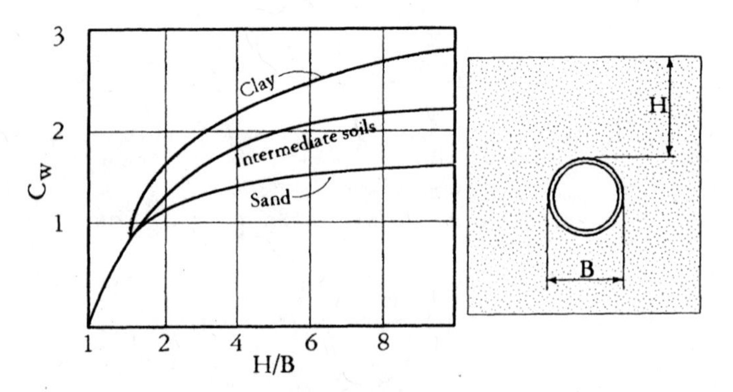

(c) C_w for jacked conduit

FIGURE 11.3 Embankment load W and backfill coefficient C_w for rigid pipelines. See Table 11.5. (*Reproduced from NAVFAC DM-7.1, 1982.*) (*Continued*)

TABLE 11.6 Flexible Pipeline Design

Topic (1)	Discussion (2)
General discussion	Flexible pipelines under embankments or in trenches derive their ability to support loads from their inherent strength plus the passive resistance of the soil as the pipe deflects and the sides of the flexible pipe move outward against the soil. Example of flexible pipes are ductile iron pipe, ABS pipe, polyvinyl chloride (PVC) pipe, and corrugated metal pipe (CMP). Proper compaction of the soil adjacent to the sides of the flexible pipe is essential in their long-term performance. Flexible pipes often fail by excessive deflection and by collapse, buckling, and cracking, rather than by rupture as in the case of rigid pipes.
Design of flexible pipelines	The design of flexible pipelines depends on the amount of deflection considered permissible, which in turn depends on the physical properties of the pipe material and the project use. Because the flexible pipe can deform, the dead load on the pipe (*W*) is usually less than that calculated for rigid pipes. Thus, as a conservative approach, the value of the design dead load *W* calculated from the rigid pipe section (Table 11.5) can be used for the flexible pipeline design. In order to complete the design of flexible pipelines, the designer will need to calculate the deflection of the pipeline. The deflection depends on the applied vertical dead load *W* as well as other factors such as the modulus of elasticity of the pipe, pipe diameter and thickness, modulus of soil resistance (E'; see ASCE 1982, Table 9-10), and bedding constant (K_b). Per ASCE (1982, Table 9-11), the values of the bedding constant (K_b) vary from 0.110 (bedding angle = 0°) to about 0.083 (bedding angle = 180°). The bedding angle may vary along the trench, and thus a conservative value of 0.10 is often recommended.

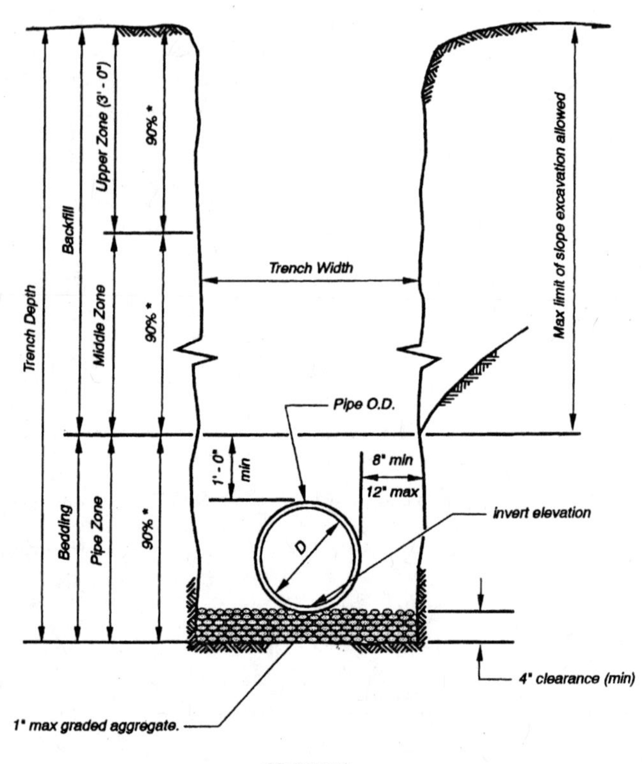

Notes:

1- For trenching on improved streets see Standard Drawing G-24 or G-25 for resurfacing details.
2- (*) indicated minimum relative compaction.

FIGURE 11.4 Pipe bedding and trench backfill for storm drains. (*From City of San Diego Standard Drawings 1986.*)

CHAPTER 12
EXPANSIVE SOIL

12.1 INTRODUCTION

Section 9.4 discusses the consolidation of clay, which is basically the compression of soft clays that have a high water content. Expansive clays are different in that the near-surface clay often varies in density and moisture condition from the wet season to the dry season. For example, near- or at-surface clays often dry out during periods of drought but then expand during the rainy season or when they get wet by irrigation water or water from leaky pipes. Table 12.1 lists the factors that govern the expansion behavior of soil, Table 12.2 presents typical soil properties versus expansion potential, and Figs. 12.1 and 12.2 present expansive soil classification charts.

12.2 LABORATORY TESTING

The geotechnical engineer can determine the presence of expansive soil by performing subsurface exploration and laboratory testing. Various types of laboratory tests for expansive soil are discussed in Tables 12.3 and 12.4. Figure 12.3 presents data from an expansion index test.

12.3 SWELLING OF DESICCATED CLAY

Desiccated clays are common in areas having near-surface clay deposits and periods of drought. Structures constructed on top of desiccated clay can be severely damaged by expansive soil heave (Jennings 1953). Table 12.5 and Fig. 12.4 discuss the depth of seasonal moisture change, and Table 12.6 and Fig. 12.5 list factors that affect the rate of swelling of desiccated clays.

12.4 TYPES OF EXPANSIVE SOIL MOVEMENT

Expansive soil movement can affect all types of civil engineering projects. Table 12.7 describes common types of expansive soil movement, and Fig. 12.6 shows the typical crack pattern due to center lift of a slab-on-grade foundation. The effect of vegetation on expansive soils is discussed in Table 12.8.

12.5 CALCULATING FOUNDATION HEAVE

Two commonly used methods for estimating the amount of foundation heave are presented in Table 12.9 and Figs. 12.7 and 12.8. The first method uses the results from swell tests on undisturbed soil specimens (see Table 12.4). The second approach (Fig. 12.8) is different in that the soil specimen is actually loaded in the oedometer apparatus, much like a consolidation test.

12.6 SOIL TREATMENT AND FOUNDATION DESIGN

There are many different methods that can be used to mitigate the effects of expansive soil. Table 12.10 presents commonly used soil treatment alternatives. Special foundations can also be constructed, and they are described in Tables 12.11 and 12.12 and Figs. 12.9 and 12.10.

12.7 PAVEMENTS AND FLATWORK ON EXPANSIVE SOIL

As mentioned in Table 12.1, it is the lightly loaded structures, such as pavements and flatwork, that are often impacted by expansive soil (see Table 12.13).

12.8 EXPANSIVE ROCK

There are several different mechanisms that can cause the expansion of rock. Some rock types, such as shale, slate, mudstone, siltstone, and claystone, can be especially susceptible to expansion. Some common mechanisms that can cause rock to expand are listed in Table 12.14.

TABLE 12.1 Factors That Govern the Expansion Behavior of Soil

Topic (1)	Discussion (2)
General discussion	Expansive soils are a worldwide problem, causing extensive damage to civil engineering structures. Jones and Holtz estimated in 1973 that the annual cost of damage in the United States due to expansive soil movement was $2.3 billion (Jones and Holtz 1973). A more up-to-date figure is about $9 billion in damage annually to buildings, roads, airports, pipelines and other facilities (Jones and Jones 1987). Although most states have expansive soil, Chen (1988) reported that certain areas of the United States, such as Colorado, Texas, Wyoming, and California, are more susceptible to damage from expansive soils than others. These areas have large surface deposits of clay and have climates characterized by alternating periods of rainfall and drought.
Expansive soil factors	There are many factors that govern the expansion behavior of soil. The primary factors are the availability of moisture, and the amount and type of the clay-size particles in the soil. For example, Seed et al. (1962) developed a classification chart based solely on the amount and type (activity) of clay-size particles (see Figure 12.1). Other factors affecting the expansion behavior include the type of soil (natural or fill), the condition of the soil in terms of dry density and moisture content, the magnitude of the surcharge pressure, the amount of nonexpansive material (gravel or cobble size particles), and the amount of aging (Holtz 1959, Ladd and Lambe 1961, Kassiff and Baker 1971, Day 1991b, 1992a). In general, expansion potential increases as the dry density increases and the moisture content decreases. Also, the expansion potential increases as the surcharge pressure decreases. Some of these factors are individually discussed below:
Amount of clay particles	As shown in Fig. 12.1, the more clay-size particles of a particular type a soil has, the more swell there will be (all other factors being the same). As mentioned in Table 4.1, clay-size particles attract water to their particle faces (double-layer effect). Thus the more clay-size particles in a dry soil, the greater the need for water to be drawn into the soil, and hence higher swell potential of the soil.
Type of clay-size particles	Also shown in Fig. 12.1, the type of clay-size particles significantly affects swell potential. Given the same dry weight, kaolinite clay particles (activity between 0.3 and 0.5) are much less expansive than sodium montmorillonite clay particles (activity between 4 and 7), Holtz and Kovacs (1981). This effect is also related to the attraction of water to the clay-size particle faces. Montmorillonite is a much smaller and more active clay mineral than kaolinite, and this results in much more attracted water per unit dry mass of clay particles. Once again, the need for more water results in a greater amount of water drawn into the dry soil and hence higher swell potential of the soil.

TABLE 12.1 Factors That Govern the Expansion Behavior of Soil (*Continued*)

Topic (1)	Discussion (2)
Type of clay-size particles (*Continued*)	Using such factors as the clay-particle content, Holtz and Gibbs (1956) developed a system to classify soils as having either a low, medium, high, or very high expansion potential. Table 12.2 lists typical soil properties versus the expansion potential (Holtz and Gibbs 1956, Holtz and Kovacs 1981, Meehan and Karp 1994, and *Uniform Building Code* 1997).
Surcharge pressure	The bottom three rows of Table 12.2 list typical values of percent swell versus expansion potential. Note in Table 12.2 the importance of surcharge pressure on percent swell. At a surcharge pressure of 31 kPa (650 psf) the percent swell is much less than at a surcharge pressure of 2.8 kPa (60 psf). For example, for "highly" expansive soil, the percent swell for a surcharge pressure of 2.8 kPa (60 psf) is typically 10 to 15%, while at a surcharge pressure of 31 kPa (650 psf), the percent swell is 4 to 6%. The effect of surcharge is important because it is usually the lightly loaded structures, such as concrete flatwork, pavements, slab-on-grade foundations, or concrete canal liners that are often impacted by expansive soil.
Figure 12.2	Figure 12.2 presents another expansive soil classification chart. In Fig. 12.2, the PI for the vertical axis = (PI) × (fraction of soil passing No. 40 sieve).

TABLE 12.2 Typical Soil Properties versus Expansion Potential

Expansion potential (1)	Very low (2)	Low (3)	Medium (4)	High (5)	Very high (6)
Expansion index	0–20	21–50	51–90	91–130	130+
Clay content (< 2 μm)	0–10%	10–15%	15–25%	25–35%	35–100%
Plasticity index	0–10	10–15	15–25	25–35	35+
% swell @ 2.8 kPa (60 psf)	0–3	3–5	5–10	10–15	15+
% swell @ 6.9 kPa (144 psf)	0–2	2–4	4–7	7–12	12+
% swell @ 31 kPa (650 psf)	0	0–1	1–4	4–6	6+

Notes:

1. Percent swell for specimens at moisture and density conditions according to HUD criteria.
2. The clay content (percent clay-size particles) is usually a less reliable indicator of expansion potential than the plasticity index (PI). This is because the type of clay mineral (i.e., kaolinite versus sodium montmorillonite) has such a large effect on expansion potential. When correlating the plasticity index and expansion potential, the PI for this table = (PI) × (fraction of soil passing No. 40 sieve).

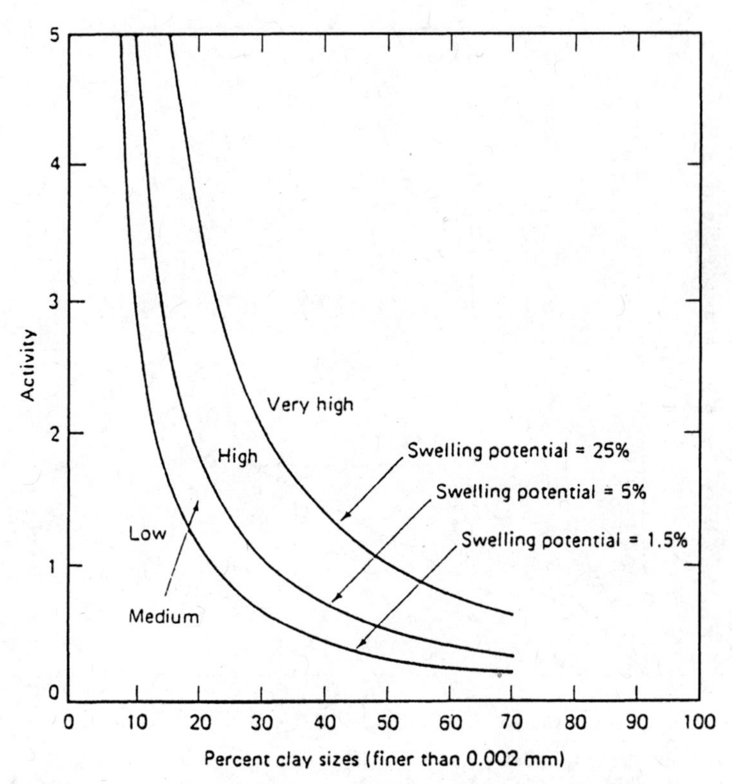

FIGURE 12.1 Classification chart for swelling potential. *Note:* Soil specimens were compacted with standard Proctor energy and tested with a normal stress of 6.9 kPa (1 psi); see Table 12.1. (*From Seed et al. 1962; reprinted with permission from the American Society of Civil Engineers.*)

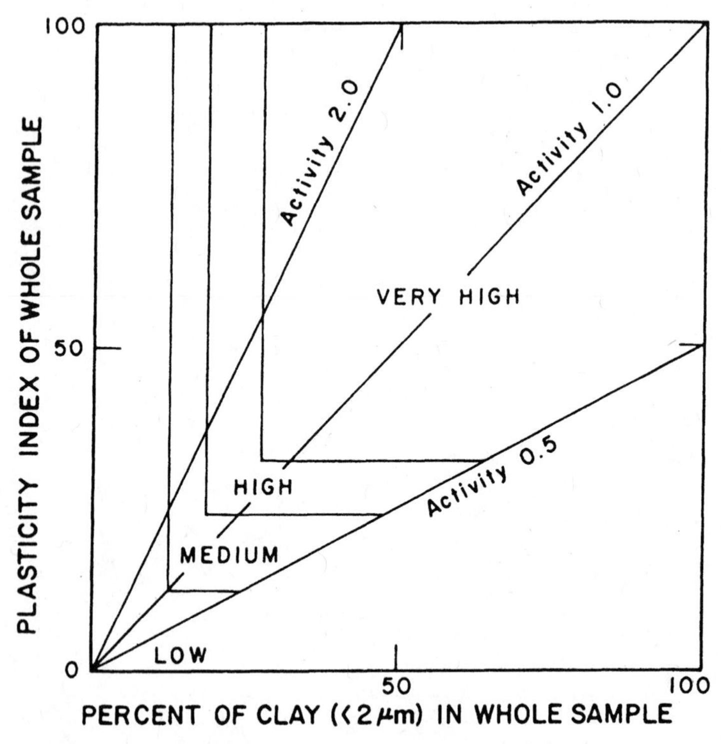

FIGURE 12.2 Classification chart for swelling potential. See Table 12.1. (*From NAVFAC DM-7.1, 1982.*)

TABLE 12.3 Laboratory Testing: Expansion Index Test

Topic (1)	Discussion (2)
General discussion	A common laboratory test used to determine the expansion potential of a soil is the expansion index test. The test provisions are stated in the *Uniform Building Code* (1997), under the title "Uniform Building Code Standard 18-2, Expansion Index Test," and in ASTM (1998), which has a nearly identical test specification (ASTM D 4829-95, 1998). The purpose of this laboratory test is to determine the expansion index, which is then used to classify the soil as having either a very low, low, medium, high, or very high expansion potential as shown in Table 12.2.
Laboratory testing procedures	The expansion index test is performed on soil passing the No. 4 sieve. The usual testing procedure is as follows: 1. If particles larger than the No. 4 sieve are possibly expansive (such as claystone and shale), these particles are to be broken down so that they pass the No. 4 sieve. 2. The soil is then compacted into a standard mold having an internal diameter of 10.2 cm (4 in.) using two layers with each layer receiving 15 blows from a 2.5-kg (5.5-lb) tamper having a 30.5-cm (12-in.) drop. 3. After compaction, the thickness of the compacted soil should be about 5.1 cm (2 in.). After removal from the compaction mold, the specimen is trimmed into a 10.2-cm (4-in.) diameter by 2.5-cm- (1.0-in.-) thick oedometer ring. 4. The degree of saturation of the test specimen is then determined by utilizing either the known water content of the compacted material or that determined on trimmings. 5. If the degree of saturation is between 49 and 51%, the specimen is ready for testing, or else the test specimen preparation procedure is repeated after adjustment of initial water content. 6. The test specimen is mounted in the oedometer apparatus with porous plates on the top and bottom of the specimen. 7. The specimen is then subjected to a vertical stress of 6.89 kPa (144 psf). 8. A dial gauge (part of the oedometer apparatus) is used to measure the vertical movement, an initial dial reading is taken, and the sample is submerged in distilled water. 9. After the specimen has swelled, the expansion index (EI) can be calculated as follows (Day 1993a): $\mathrm{EI} = 1000 \dfrac{(h_p - h_i)}{h_i} = 10\ (\%\ \text{primary swell})$ where h_p = height of the specimen at the end of primary swell and h_i = initial height of the specimen. From the expansion index, the expansion potential of the soil is determined as indicated in Table 12.2.

TABLE 12.3 Laboratory Testing: Expansion Index Test (*Continued*)

Topic (1)	Discussion (2)
End of primary swell h_p	In order to obtain the specimen height at the end of primary swell h_p, dial versus time readings can be recorded. The dial readings can be converted to percent swell and plotted versus time from the start of the test on a semilog plot. Figure 12.3 presents an expansion index test performed on a clay of low plasticity (CL). The shape of the swelling versus time curve (Fig. 12.3) is similar to the consolidation of a saturated clay (Fig. 9.13), except that swelling has been plotted as positive values and consolidation as negative values. From this similarity, the end of primary swell can be determined as the intersection of the straight line portions of the swell curve. The arrow in Fig. 12.3 indicates a primary swell of 6.2% (or EI = 62), indicating a medium expansion potential (Table 12.2).
Clayey gravels	In some cases, the expansion index test can yield misleading results for soils classified as clayey gravels (GC). The reason is that clayey gravels will have a significant fraction retained on the No. 4 sieve, but the expansion index test uses only those particles passing the No. 4 sieve. There is no correction in the test specification to account for nonexpansive (i.e., hard rock) gravel particles retained on the No. 4 sieve. One approach used in practice is to reduce the expansion index (EI) according to the percentage of particles by dry mass that pass the No. 4 sieve. This correction is as follows (Day 1993a): $$\text{EI (corrected)} = \frac{\text{(EI)(\% passing No. 4 sieve)}}{100}$$ Suppose a clayey gravel has 40% by dry mass passing the No. 4 sieve and for the particles passing the No. 4 sieve, the EI is 100. Then according to the above equation, the corrected EI would be 100 times 0.4, or 40 ("low" expansion potential). This procedure is not exact since it uses the dry mass, rather than the volume, of the oversize particles and does not consider the compaction energy (Day 1992b).
Coefficient of swell	As previously mentioned, there is a similarity of the shape of time versus deformation curves for consolidation and swell of clay in the oedometer apparatus. The rate of swell can be estimated from Terzaghi's diffusion equation, where the coefficient of consolidation (c_v; see time factor equation, Table 9.14) is replaced by the coefficient of swell c_s, or (Blight 1965): $$c_s = \frac{TH_{dr}^{\ 2}}{t}$$ where T = time factor (Table 9.15), H_{dr} = swell height, and t = time. The coefficient of swell c_s can be determined from laboratory time versus swell data such as shown in Fig. 12.3. For example, for the data shown in Fig. 12.3, the time for 50% primary swell (t_{50}) is about 6.5 minutes. The specimen height corresponding to 50% primary swell is 2.62 cm (1.03 in.) and, since water can be drawn into the soil from the top and bottom, the coefficient of swell c_s from the above equation is 9×10^{-4} cm^2/s.

TABLE 12.4 Laboratory Testing: Swell Tests on Undisturbed Specimens

Topic (1)	Discussion (2)
Test procedures	The most direct method of determining the amount of swelling is by performing a one-dimensional swell test by utilizing the oedometer apparatus (ASTM D 4546-96, 1998). The undisturbed soil specimen is placed in the oedometer and a vertical pressure (also referred to as surcharge pressure) is applied. Then the soil specimen is inundated with distilled water and the one-dimensional vertical swell is calculated as the increase in height of the soil specimen divided by the initial height, often expressed as a percentage. Such a test offers an easy and accurate method of determining the percent swell of the soil. After the soil has completed its swelling, the vertical pressure can be increased to determine the swelling pressure, which is defined as that pressure required to return the soil specimen to its original (initial) height (Chen 1988).
Limitations of the swell test	Concerning the limitations of the swell test, ASTM (1998) states: "Estimates of the swell of soil determined by this test method [ASTM D 4546-96] are often of key importance in design of floor slabs on grade and evaluation of their performance. However, when using these estimates, it is recognized that swell parameters determined from these test methods for the purpose of estimating *in situ* heave of foundations and compacted soils may not be representative of many field conditions because: 1. Lateral swell and lateral confining pressure are not simulated. 2. Rates of swell indicated by swell test are not always reliable indicators of field rates of heave due to fissures in the *in situ* soil mass and inadequate simulation of the actual availability of water to the soil. The actual availability of water to the foundation may be cyclic, intermittent, or depend on in-place situations, such as pervious soil-filled trenches and broken water and drain lines. 3. Secondary or long-term swell may be significant for some soils and should be added to primary swell. 4. Chemical content of the inundation water affects volume changes and swell pressure; that is, field water containing large concentrations of calcium ions will produce less swelling than field water containing large concentrations of sodium ions or even rain water. 5. Disturbance of naturally occurring soil samples greatly diminishes the meaningfulness of the results." Another major limitation of the swell test is that the water content of the tested soil may not be the same water content at the time of construction of the foundation. For example, if a drought occurs and the clay dries out, the percent expansion will be underestimated. On the other hand, if a rainy season occurs just prior to construction and the clay absorbs moisture, the percent swell will be overestimated. But even with all of these limitations, swell tests on undisturbed soil specimens are generally considered to be the most reliable method of predicting the future potential heave of the foundation.

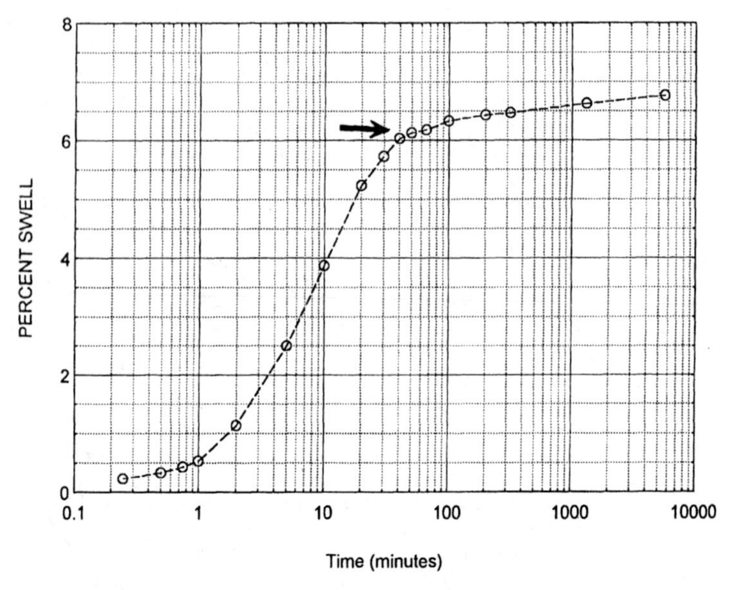

FIGURE 12.3 Percent swell versus time for an expansion index test of a clay of low plasticity. *Note:* The arrow indicates the end of primary swell. See Table 12.3.

TABLE 12.5 Depth of Seasonal Moisture Change

Topic (1)	Discussion (2)
General discussion	For desiccated clays, Chen (1988) states: "Very dry clays with natural moisture content below 15 percent usually indicate danger. Such clays will easily absorb moisture to as high as 35 percent with resultant damaging expansion to structures." There can also be desiccation and damage to final clay cover systems for landfills and site remediation projects, and for shallow clay landfill liners (Boardman and Daniel 1996).
Depth of seasonal moisture change	The geotechnical engineer can often visually identify desiccated clay because of the numerous ground surface cracks. Desiccated clay deposits also generally have a distinct water content profile, where the water content increases with depth. For example, Fig. 12.4 shows the water content versus depth for two clay deposits located in Irbid, Jordan (Al-Homoud et al. 1997). Soil deposit A has a liquid limit of 35 and a plasticity index of 22, while soil deposit B has a liquid limit of 79 and a plasticity index of 27 (Al-Homoud et al. 1995). Note in Fig. 12.4 that during the hot and dry summer, the water content of the soil is significantly lower than during the wet winter. During the summer, the lowest water contents are recorded near ground surface, and the water contents are below the shrinkage limit (SL). A near-surface water content below the shrinkage limit (SL) is indicative of severe desiccation of the clay. Below a depth of about 3.2 m (10 ft) for soil deposit A and a depth of about 4.5 m (15 ft) for soil deposit B, the water content is relatively unchanged between the summer and winter monitoring period, and this depth is commonly known as the "depth of seasonal moisture change." The depth of seasonal moisture change is also sometimes referred to as the "depth of the active zone." The depth of seasonal moisture change would depend on many factors such as the temperature and humidity, length of the drying season, presence of vegetation that can extract soil moisture, depth of the water table, and the nature of the soil in terms of clay content, clay mineralogy, and density. As shown in Fig. 12.4, soil deposit B has a greater variation in water content from the dry summer to wet winter and a greater depth of seasonal moisture change. This is probably because soil deposit B has a higher clay content than soil deposit A.
Likelihood of damage	A structure that is constructed during a hot and dry season when the near-surface water content of the desiccated clay is below the shrinkage limit, such as shown in Fig. 12.4, will have the greatest potential increase in water content and resulting heave of the foundation. The hydraulic conductivity (permeability) and rate of swell (Table 12.6) are important because they determine how fast the water will penetrate the soil.

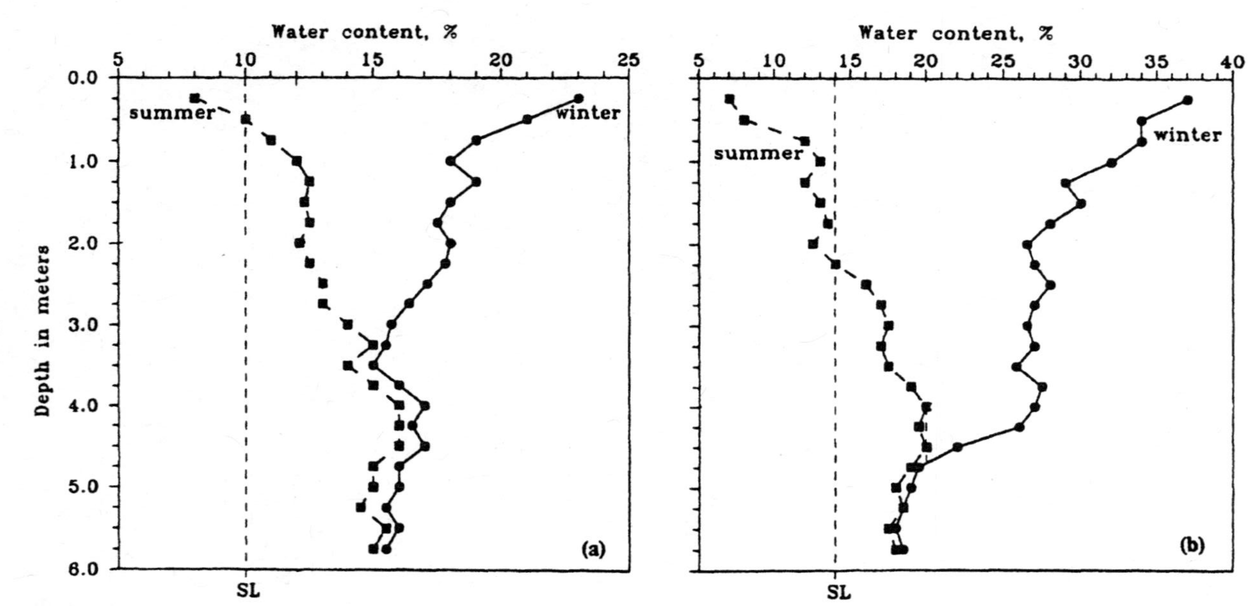

FIGURE 12.4 Water content versus depth for: (*a*) soil A; (*b*) soil B. See Table 12.5. (*From Al-Homoud et al. 1997; reprinted with permission from the American Society of Civil Engineers.*)

TABLE 12.6 Factors That Affect the Rate of Swelling

Topic (1)	Discussion (2)
Laboratory testing of desiccated clay	Figure 12.5 presents the results of a one-dimensional swell test (lower half of Fig. 12.5) and a falling head permeameter test (upper half of Fig. 12.5) performed on a specimen of desiccated Otay Mesa clay (Day 1997b). The clay particles in this soil are almost exclusively montmorillonite (Kennedy and Tan 1977, Cleveland 1960). At time zero, the desiccated clay specimen was inundated with distilled water. The data in Fig. 12.5 indicates three separate phases of swelling of the clay, as follows: 1. *Primary swell.* The first phase of swelling of the desiccated clay was primary swell. The primary swell occurs from time equals zero (start of wetting) to about 100 minutes. Figure 12.5 shows that during primary swell, there was a rapid decrease in the hydraulic conductivity of the clay. The rapid decrease in hydraulic conductivity was due to the closing of soil cracks as the clay swells. At the end of primary swell, the main soil cracks had probably closed and the hydraulic conductivity was about 7×10^{-7} cm/s. 2. *Secondary swell.* The second phase of swelling was secondary swell. The secondary swell occurred from a time of about 100 minutes to 20,000 minutes after wetting. Figure 12.5 shows that during secondary swell, the hydraulic conductivity continued to decrease as the clay continued to swell and the microcracks closed up. The lowest hydraulic conductivity of about 1.5×10^{-8} cm/s occurred at a time of about 5000 minutes, when most of the microcracks had probably sealed up. From a time of 5000 to 20,000 minutes after wetting, there was a slight increase in hydraulic conductivity. This was probably due to a combination of additional secondary swell, which increases the void ratio, and a reduction in entrapped air. 3. *Steady state.* The third phase started when the clay stopped swelling. This occurred at about 20,000 minutes after inundation with distilled water. No swell was recorded from a time of 20,000 minutes after wetting to the end of the test (50,000 minutes). As shown in Fig. 12.5, the hydraulic conductivity was constant once the clay had stopped swelling. From a time of 20,000 minutes after wetting to the end of the test (50,000 minutes), the hydraulic conductivity of the clay was constant at about 3×10^{-8} cm/s.
Desiccation cracks	The amount and distribution of desiccation cracks are probably the greatest factors in the rate of swelling. Clays will shrink until the shrinkage limit (usually a low water content) is reached. Even as the moisture content decreases below the shrinkage limit, there is probably still the development of additional microcracks as the clay dries. The more cracks in the clay, the greater the pathways for water to penetrate the soil, and the quicker the rate of swelling.

TABLE 12.6 Factors That Affect the Rate of Swelling (*Continued*)

Topic (1)	Discussion (2)
Increased suction at a lower water content	The second factor that governs the rate of swelling of a desiccated clay is suction pressure. It is well known that, as the water content decreases, the suction pressure of the clay increases (Fredlund and Rahardjo 1993). At low water contents, the water is drawn into the clay by the suction pressures. The combination of both shrinkage cracks and high suction pressures allows water to be quickly sucked into the clay, resulting in a higher rate of swell.
Slaking	Slaking also affects the rate of swell of desiccated clay. Slaking is defined as the breaking of dried clay when submerged in water, due either to compression of entrapped air by inwardly migrating capillary water or to the progressive swelling and sloughing off of the outer layers (Stokes and Varnes 1955). Slaking breaks apart the dried clay clods and allows water to quickly penetrate all portions of the desiccated clay. The process of slaking is quicker and more disruptive for clays having the most drying time and lowest initial water content.

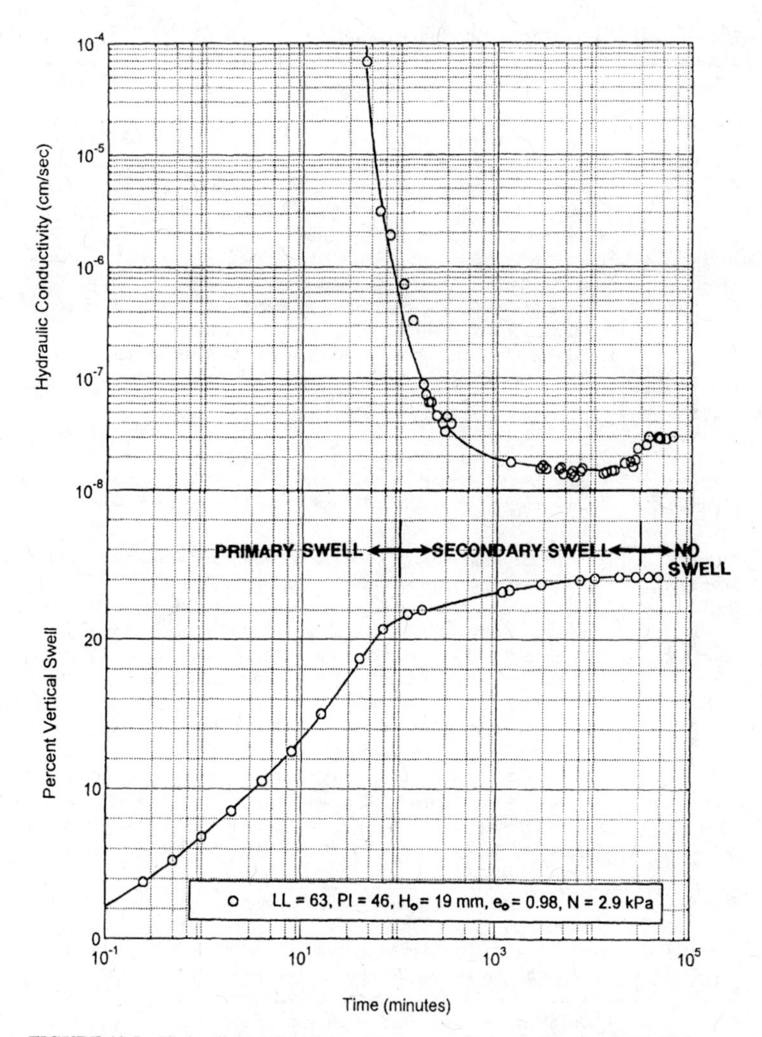

FIGURE 12.5 Hydraulic conductivity and percent swell versus time. See Table 12.6.

TABLE 12.7 Types of Expansive Soil Movement

Topic (1)	Discussion (2)
Lateral movement	Expansive soil movement can affect all types of civil engineering projects. For many retaining and basement walls, especially if the clay backfill is compacted below optimum moisture content, seepage of water into the clay backfill causes horizontal swelling pressures well in excess of at-rest values. Fourie (1989) measured the swell pressure of a compacted clay for zero lateral strain to be 420 kPa (8800 psf). Besides the swelling pressure induced by the expansive soil, there can also be groundwater or perched water pressure on the retaining or basement wall because of the poor drainage of clayey soils. Because of these detrimental effects of clay backfill, a common recommendation is to use only free-draining, granular material (clean sand or clean gravel) as retaining or basement wall backfill.
Vertical movement, general discussion	If a structure having a large area, such as a pavement or foundation, is constructed on top of a desiccated clay, there are usually two main types of expansive soil movement. The first is the cyclic heave and shrinkage around the perimeter of the structure. The second is the long-term progressive swell beneath the center of the structure.
Vertical movement, cyclic heave and shrinkage	Cyclic heave and shrinkage commonly affects the perimeter of the foundation, by uplifting the edge of the structure or shrinking away from it. For example, the perimeter of a pavement or slab-on-grade foundation will heave during the rainy season and then deform downward during the drought if the clay dries out. This causes cycles of up and down movement, causing cracking and damage to the structure. Field measurements of this up-and-down cyclic movement have been recorded by Johnson (1980). The amount of cyclic heave and shrinkage depends on the change in moisture content of the clays below the perimeter of the structure. The moisture change in turn depends on the severity of the drought and rainy seasons, the influence of drainage and irrigation, and the presence of live tree roots, which can extract moisture and cause clays to shrink. The cyclic heave and shrinkage around the perimeter of a structure is generally described as a seasonal or short-term condition.
Vertical movement, center lift	Two ways that moisture can accumulate underneath structures are by thermal osmosis and capillary action. It has been stated that water at a higher temperature than its surroundings will migrate in the soil toward the cooler area to equalize the thermal energy of the two areas (Chen 1988, Nelson and Miller 1992). This process has been termed *thermal osmosis* (Sowers 1979, Day 1996a). Especially during the summer months, the temperature under the center of a structure tends to be much cooler than at the exterior ground surface.

TABLE 12.7 Types of Expansive Soil Movement (*Continued*)

Topic (1)	Discussion (2)
Vertical movement, center lift (*Continued*)	Because of capillary action, moisture can move upward through soil, where it will evaporate at the ground surface. But when a structure is constructed, it acts as a ground surface barrier, reducing or preventing the evaporation of moisture. It is the effect of thermal osmosis and the evaporation barrier due to the structure that causes moisture to accumulate underneath the center of the structure. A moisture increase will result in swelling of expansive soils. The progressive heave of the center of the structure is generally described as a long-term condition, because the maximum value may not be reached until many years after construction. Figure 12.6 illustrates center lift beneath a house foundation, and Fig. 12.6 shows the typical crack pattern in the concrete slab-on-grade due to expansive soil center lift.

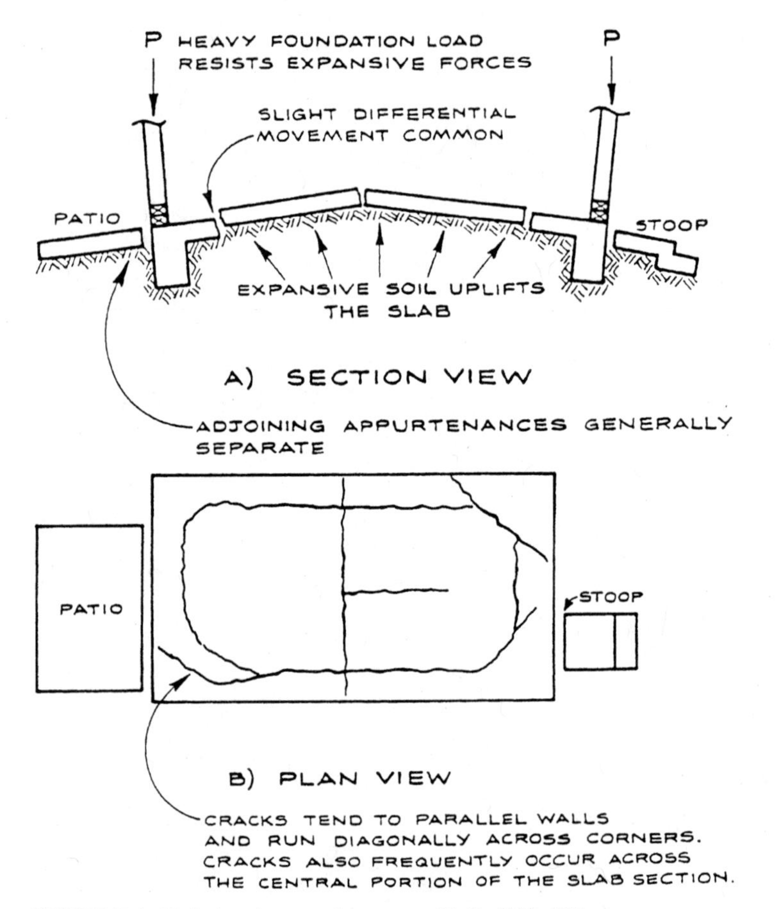

FIGURE 12.6 Typical crack pattern due to center lift. See Table 12.7.

TABLE 12.8 Effect of Vegetation on Expansive Soils

Topic (1)	Discussion (2)
General discussion	Foundation movement can be caused by the shrinkage of clay. For example, tree roots and rootlets can extract moisture from the ground, which can cause the near-surface clay to shrink and the foundation to deform downward. There are cases (Cheney and Burford 1975) where the opposite can also occur, where large trees have been removed and the clay has expanded as the soil moisture increases to its natural state.
Periods of drought	In the United States, Holtz (1984) states that large, broadleaf, deciduous trees located near the structure cause the greatest changes in moisture and the greatest resulting damage in both arid and humid areas. However, the most dramatic effects are during periods of drought, such as the severe drought in Britain from 1975 to 1976, when the amount of water used by the trees during transpiration greatly exceeded the amount of rainfall within the area containing tree roots. Biddle (1979, 1983) investigated 36 different trees, covering a range of tree species and clay types and concluded that poplars have much greater effects than other species and that the amount of soil movement will depend on the clay shrinkage characteristics. Ravina (1984) states that it is the nonuniform moisture changes and soil heterogeneities that cause the uneven soil movements that damage shallow foundations, structures, and pavements.
Levels of desiccation	Driscoll (1983) has suggested relationships between water content w of the soil and the liquid limit (LL), which indicate the level of desiccation, where $w = 0.5\text{LL}$ would indicate the onset of desiccation, and $w = 0.4\text{LL}$ would indicate when desiccation becomes significant. As mentioned in Table 12.5, a near-surface water content w below the shrinkage limit (SL) would be indicative of severe desiccation of the clay.
Planting of trees adjacent to structures	Tucker and Poor (1978) indicated that trees located at distances closer than their heights to structures caused significantly larger movements due to clay shrinkage than those trees located at greater distances. Hammer and Thompson (1966) stated that trees should not be planted a minimum of one-half their anticipated mature height from a shallow foundation, and slow-growing, shallow-rooted varieties of trees were preferred. However, Cutler and Richardson (1989) indicated that total crown volume (hence leaf area) is generally more important than absolute height in relation to water demand. Cutler and Richardson (1989) used case histories to determine the frequency of damage as a function of tree-trunk distance from the structure for different species. Biddle (1983) recommended that for very high shrinkage clay, the perimeter footings should be at least 1.5 m (5 ft) deep, and this would be sufficient to accommodate most tree-planting designs. When dealing with expansive clays, it is important for the geotechnical engineer to consider the possibility of damage due to clay shrinkage caused by the extraction of moisture by tree roots. Shallow foundations, such as slab-on-grade or raised wood floor foundations having shallow perimeter footings, are especially vulnerable to damage due to clay shrinkage.

TABLE 12.9 Calculating Foundation Heave

Topic (1)	Discussion (2)
General discussion	Figures 12.7 and 12.8 present two commonly used methods to calculate the total heave (also known as total swell) of a foundation on expansive soil. Both of these methods are based on the testing of undisturbed soil specimens in the oedometer apparatus. If the project will consist of compacted fill, then the soil specimens could be prepared by compacting the soil to anticipated field dry density and water content conditions.
Method based on swell tests (Fig. 12.7)	The method shown in Fig. 12.7 is based on testing undisturbed soil or compacted fill specimens by using the procedure outlined in Table 12.4. The soil specimens are loaded in the oedometer apparatus to a pressure that equals the ultimate value of the effective overburden pressure σ'_v plus the weight of the structure $\Delta\sigma_v$. The soil specimens are then submerged in distilled water and the percent swell is measured. The calculations used to determine the total swell are shown in Fig. 12.7.
Method based on swell curve (Fig. 12.8)	The approach shown in Fig. 12.8 is different in that the soil specimen is actually loaded in the oedometer apparatus much like a consolidation test. At a low seating pressure, the soil specimen is inundated with distilled water. The vertical pressure is increased in order to prevent the soil specimen from swelling. At a certain pressure, known as the swelling pressure P'_s, the soil specimen will begin to compress. As shown in Fig. 12.8 (void ratio versus log pressure plot), the soil specimen is loaded to a high pressure and then the soil specimen is unloaded and allowed to swell. Often the swell curve will be linear on a void ratio versus log of the vertical pressure plot, and the swell index C_s can be calculated in the same manner as the compression index C_c was calculated in Table 9.12. Note that the equation (Fig. 12.8) used to calculate the heave of the foundation is similar to the consolidation equations in Table 9.13, with h_i = the initial height of the *in situ* soil layer ($h_i = H_o$), the swelling index C_s used in place of the compression index C_c, the value of P_f = the final vertical effective stress in the soil after swelling has occurred, and P_o = the swelling pressure (i.e., $P_o = P'_s$). Also note that in Fig. 12.8 the swelling soil was divided into three layers and that the swell was calculated for each layer and then the total heave was calculated as the sum of the swell from the three layers.
Example problem	*Problem:* Use the data shown in Fig. 12.8. Assume that the final condition P_f will be a groundwater table at a depth of 0.5 m with hydrostatic pore water pressures below the groundwater table and zero pore water pressures above the groundwater table. Also assume the total unit weight γ_t of 18 kN/m^3 can be used for the clay above and below the groundwater table. A mat foundation will be constructed such that the bottom of the mat is located at a depth of 0.5 m and the mat foundation exerts a vertical pressure = 25 kPa on the soil at this level. If the mat foundation is large enough so that one-dimensional conditions exist beneath the center of the mat foundation, determine the total heave of the center of the mat foundation.

TABLE 12.9 Calculating Foundation Heave (*Continued*)

Topic (1)	Discussion (2)
Example problem (*Continued*)	*Solution:* The initial condition is a uniform P'_s condition = 200 kPa and the final condition is a groundwater table at a depth of 0.5 m. From Fig. 12.8, $$\Delta h = [(h_o C_s)/(1 + e_o)] \log (P_f/P_o)$$ Layer 2 $$P_f = \sigma'_v = 25 + (0.25)(18 - 9.81) = 27.0 \text{ kPa}$$ $$\Delta h = [(500)(0.1)/(1 + 1.0)] \log (27.0/200) = 21.7 \text{ mm}$$ Layer 3 $$P_f = \sigma'_v = 25 + (1.0)(18 - 9.81) = 33.2 \text{ kPa}$$ $$\Delta h = [(1000)(0.1)/(1 + 1.0)] \log (33.2/200) = 39.0 \text{ mm}$$ Total heave of center of mat foundation = 21.7 + 39.0 = 61 mm.

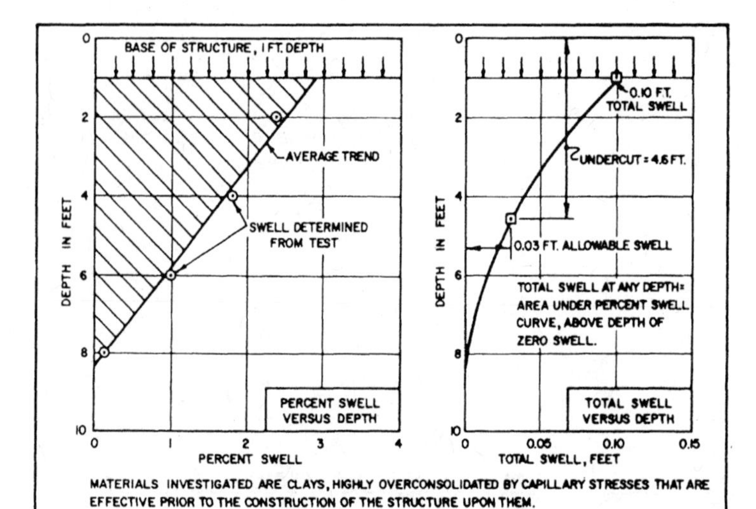

PROCEDURE FOR ESTIMATING TOTAL SWELL UNDER STRUCTURE LOAD.

1. OBTAIN REPRESENTATIVE UNDISTURBED SAMPLES OF THE SHALLOW CLAY STRATUM AT A TIME WHEN CAPILLARY STRESSES ARE EFFECTIVE; I.E., WHEN NOT FLOODED OR SUBJECTED TO HEAVY RAIN.
2. LOAD SPECIMENS (AT NATURAL WATER CONTENT) IN CONSOLIDOMETER UNDER A PRESSURE EQUAL TO THE ULTIMATE VALUE OF OVERBURDEN FOR HIGH GROUND WATER, PLUS WEIGHT OF STRUCTURE. ADD WATER TO SATURATE AND MEASURE SWELL.
3. COMPUTE FINAL SWELL IN TERMS OF PERCENT OF ORIGINAL SAMPLE HEIGHT AND PLOT SWELL VERSUS DEPTH, AS IN THE LEFT PANEL.
4. COMPUTE TOTAL SWELL WHICH IS EQUAL TO THE AREA UNDER THE PERCENT SWELL VERSUS DEPTH CURVE. FOR THE ABOVE EXAMPLE:

   TOTAL SWELL = 1/2 (8.2 − 1.0) × 2.8/100 = 0.10 FT.

PROCEDURE FOR ESTIMATING UNDERCUT NECESSARY TO REDUCE SWELL TO AN ALLOWABLE VALUE.

1. FROM PERCENT SWELL VERSUS DEPTH CURVE PLOT RELATIONSHIP OF TOTAL SWELL VERSUS DEPTH AT THE RIGHT. TOTAL SWELL AT ANY DEPTH EQUALS AREA UNDER THE CURVE AT LEFT, INTEGRATED UPWARD FROM THE DEPTH OF ZERO SWELL.
2. FOR A GIVEN ALLOWABLE VALUE OF SWELL, READ THE AMOUNT OF UNDERCUT NECESSARY FROM THE TOTAL SWELL VERSUS DEPTH CURVE. FOR EXAMPLE, FOR AN ALLOWABLE SWELL OF 0.03 FT, UNDERCUT REQUIRED = 4.6 FT. UNDERCUT CLAY IS REPLACED BY AN EQUAL THICKNESS OF NONSWELLING COMPACTED FILL.

FIGURE 12.7 Computation of foundation heave for expansive clays. See Table 12.9. (*From NAVFAC DM-7.1 1982.*)

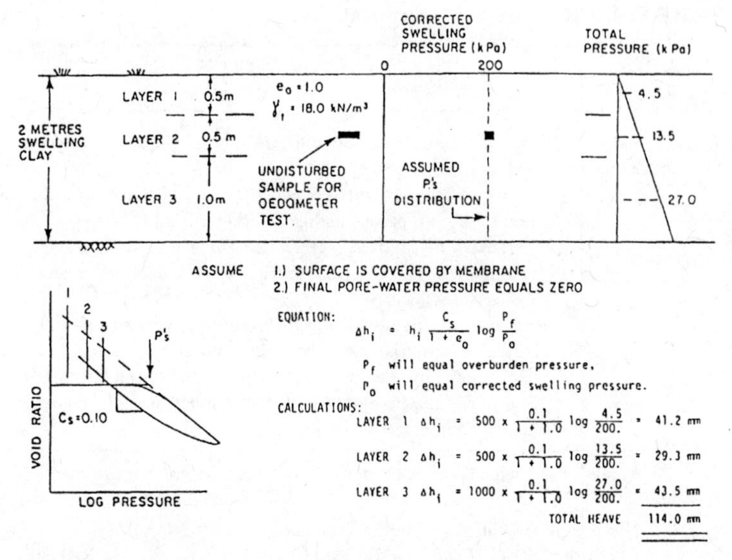

FIGURE 12.8 Computation of foundation heave for expansive clays. See Table 12.9. (*Method by Fredlund 1983; reproduced from Chen 1988. Reprinted with permission from Elsevier Science.*)

TABLE 12.10 Expansive Soil Treatment Alternatives

Method (1)	Salient points (2)
Removal and replacement	Nonexpansive, impermeable fill must be available and economical. Nonexpansive soils can be compacted at higher densities than expansive clay, producing high bearing capacities. If granular fill is used, special precautions must be taken to control drainage away from the fill, so water does not collect in the pervious material. Replacement can provide safe slab-on-grade construction. Expansive material may be subexcavated to a reasonable depth, then protected by a vertical and/or horizontal membrane. Sprayed asphalt membranes are effectively used in highway construction.
Remolding and compaction	Beneficial for soils having low potential for expansion, high dry density, low natural water content, and in a fractured condition. Soils having a high swell potential may be treated with hydrated lime, thoroughly broken up, and compacted—if they are lime reactive. If lime is not used, the bearing capacity of the remolded soil is usually lower since the soil is generally compacted wet of optimum at a moderate density. Quality control is essential. If the active zone is deep, drainage control is especially important. The specific moisture-density conditions should be maintained until construction begins and checked prior to construction.
Surcharge loading	If swell pressures are low and some deformation can be tolerated, a surcharge load may be effective. A program of soil testing is necessary to determine the depth of the active zone and the maximum swell pressures to be counteracted. Drainage control is important when using a surcharge. Moisture migration can be both vertical and horizontal.
Prewetting	Time periods up to as long as a year or more may be necessary to increase moisture contents in the active zone. Vertical sand drains drilled in a grid pattern can decrease the wetting time. Highly fissured, desiccated soils respond more favorably to prewetting.

TABLE 12.10 Expansive Soil Treatment Alternatives (*Continued*)

Method (1)	Salient points (2)
Prewetting (*Continued*)	Moisture contents should be increased to at least 2 to 3% above the plastic limit. Surfactants may increase the percolation rate. The time needed to produce the expected swelling may be significantly longer than the time to increase moisture contents. It is almost impossible to adequately prewet dense unfissured clays. Excess water left in the upper soil can cause swelling in deeper layers at a later date. Economics of prewetting can compare favorably to other methods, but funds must be available at an early date in the project. Lime treatment of the surface soil following prewetting can provide a working table for equipment and increase soil strength. Without lime treatment soil strength can be significantly reduced, and the wet surface may make equipment operation difficult. The surface should be protected against evaporation and surface slaking. Quality control improves performance.
Lime treatment	Sustained temperatures over 70°F for a minimum of 10 to 14 days is necessary for the soil to gain strength. Higher temperatures over a longer time produce higher strength gains. Organics, sulfates, and some iron compounds retard the poz zolanic reaction of lime. Gypsum and ammonium fertilizers can increase the soil's lime requirements. Calcareous and alkaline soil have good reactivity. Poorly drained soils have higher reactivities than well-drained soils. Usually 2 to 10% lime stabilizes reactive soil. Soil should be tested for lime reactivity and percentage of lime needed. The mixing depth is usually limited to 12 to 18 in., but large tractors with ripper blades have successfully allowed in-place mixing of 2 ft of soil. Lime can be applied dry or in a slurry, but excess water must be present. Some delay between application and final mixing improves workability and compaction.

TABLE 12.10 Expansive Soil Treatment Alternatives (*Continued*)

Method (1)	Salient points (2)
Lime treatment (*Continued*)	Quality control is especially important during pulverization, mixing, and compaction. Lime-treated soils should be protected from surface and groundwater. The lime can be leached out and the soil can lose strength if saturated. Dispersion of the lime from drill holes is generally ineffective unless the soil has an extensive network of fissures. Stress relief from drill holes may be a factor in reducing heave. Smaller-diameter drill holes provide less surface area to contact the slurry. Penetration of pressure-injected lime is limited by the slow diffusion rate of the lime, the amount of fracturing in the soil, and the small pore size of clay. Pressure injection of lime may be useful to treat layers deeper than possible with the mixed-in-place technique.
Cement treatment	Portland cement (4 to 6%) reduces the potential for volume change. Results are similar to lime, but shrinkage may be less with cement. Method of application is similar to mix-in-place lime treatment, but there is significantly less time delay between application and final placement. Portland cement may not be as effective as lime in treating highly plastic clays. Portland cement may be more effective in treating soils that are not lime reactive. Higher strength gains may result with cement. Cement-stabilized material may be prone to cracking and should be evaluated prior to use.
Salt treatment	There is no evidence that use of salts other than NaCl or $CaCl_2$ is economically justified. Salts may be leached easily. Lack of permanence of treatment may make salt treatment uneconomical. The relative humidity must be at least 30% before $CaCl_2$ can be used. Calcium and sodium chloride can reduce frost heave by lowering the freezing point of water. $CaCl_2$ may be useful to stabilize soils having a high sulfur content.
Fly ash	Fly ash can increase the pozzolanic reaction of silty soils. The gradation of granular soils can be improved.
Organic compounds	Spraying and injection are not very effective because of the low rate of diffusion in expansive soil.

TABLE 12.10 Expansive Soil Treatment Alternatives (*Continued*)

Method (1)	Salient points (2)
Organic compounds (*Continued*)	Many compounds are not water-soluble and react quickly and irreversibly. Organic compounds do not appear to be more effective than lime. None is as economical and effective as lime.
Horizontal barriers	Barrier should extend far enough from the roadway or foundation to prevent horizontal moisture movement into the foundation soils. Extreme care should be taken to securely attach barrier to foundation, seal the joints, and slope the barrier down and away from the structure. Barrier material must be durable and nondegradable. Seams and joints attaching the membrane to a structure should be carefully secured and made waterproof. Shrubbery and large plants should be planted well away from the barrier. Adequate slope should be provided to direct surface drainage away from the edges of the membranes.
Asphalt	When used in highway construction, a continuous membrane should be placed over subgrade and ditches. Remedial repair may be less complex than for concrete pavement. Strength of pavement is improved over untreated granular base. Can be effective when used in slab-on-grade construction.
Rigid barrier	Concrete sidewalks should be reinforced. A flexible joint should connect sidewalk and foundation. Barriers should be regularly inspected to check for cracks and leaks.
Vertical barrier	Placement should extend as deep as possible, but equipment limitations often restrict the depth. A minimum of half of the active zone should be used. Backfill material in the trench should be impervious. Types of barriers that have provided control of moisture content are capillary barrier (coarse limestone), lean concrete, asphalt and ground-up tires, polyethylene, and semihardening slurries. A trenching machine is more effective than a backhoe for digging the trench.
Membrane-encapsulated soil layers	Joints must be carefully sealed. Barrier material must be durable to withstand placement. Placement of the first layer of soil over the bottom barrier must be controlled to prevent barrier damage.
Compaction control	Clay is compacted at a water content wet of optimum, which reduces the swell potential (Gromko 1974, Holtz and Gibbs 1956). If the clay should dry out prior to placement of the structure, then the beneficial effect of compacting the clay wet of optimum will be destroyed, causing the soil to become significantly more expansive (Day 1994g).

Sources: Nelson and Miller (1992) and Day (1999).

TABLE 12.11 Foundation Alternatives for Expansive Soil

Topic (1)	Discussion (2)
Conventional slab-on-grade foundation	A conventional slab-on-grade consists of interior and exterior bearing wall footings and an interior slab. The concrete for the footings and slab are usually placed at the same time to create a monolithic foundation. For the conventional slab-on-grade foundation, the soils engineer rather than the structural engineer usually provides the construction details. The purpose of the design is to provide deepened perimeter footings that are below the primary zone of cyclic heave and shrinkage and to soak the subgrade soils (prior to construction) in order to reduce the long-term progressive swelling beneath the center of the slab. Table 12.12 presents an example of expansive soil foundation specifications for conventional slab-on-grade on expansive soils in southern California (Day 1994c). The design is empirical and depends on the expansion index (from the expansion index test). Table 12.12 indicates that both the depth of perimeter footings and the depth of presaturation should be increased as the soil becomes more expansive. This purpose of the conventional slab-on-grade design, which is based on the use of deepened footings and presoaking, is to reduce the effects of expansive soil forces rather than make the foundation strong enough to resist such forces. Because the design is empirical, damage can develop because: (1) the perimeter footings are not deep enough to resist the cyclic heave and shrinkage and (2) the presoaking beneath the slab is not deep enough or it was not done properly (the soil is not saturated), resulting in long-term progressive swelling beneath the center of the slab.
Posttensioned slab-on-grade	A second type of foundation for expansive soils is the posttensioned slab-on-grade. There can be many different types of posttensioned designs. In Texas and Louisiana, the early posttensioned foundations consisted of a uniform-thickness slab with stiffening beams in both directions, which became known as the "ribbed foundation." In California, a commonly used type of posttensioned slab consists of a uniform-thickness slab with an edge beam at the entire perimeter, but no or minimal interior stiffening beams or ribs. This type of posttensioned slab has been termed the "California slab" (Post-Tensioning Institute 1996). In "Design and Construction of Post-Tensioned Slabs-on-Ground," prepared by the Post-Tensioning Institute (1996), the design moments, shears, and differential deflections under the action of soil loading from changes in water contents of expansive soils are predicted by using equations developed from empirical data and a computer study of a plate on an elastic foundation. *The Uniform Building Code* (1997) presents nearly identical equations for the design of posttensioned slabs-on-grade. The idea for the

TABLE 12.11 Foundation Alternatives for Expansive Soil (*Continued*)

Topic (1)	Discussion (2)
Posttensioned slab-on-grade (*Continued*)	design of a posttensioned slab-on-grade in accordance with Post-Tensioning Institute is to construct a slab foundation that is strong enough and rigid enough to resist the expansive soil forces. To get the required rigidity to reduce foundation deflections, the stiffening beams (perimeter and interior footings) can be deepened. Although the differential movement to be expected for a given expansive soil is supplied by a soils engineer, the actual design of the foundation is usually by the structural engineer. The posttensioned slab-on-grade should be designed for two conditions: (1) center lift (also called center heave or doming) and (2) edge-lift (also called edge heave or dishing). These two conditions are illustrated in Fig. 12.9 (from Post-Tensioning Institute 1996). Center lift is the long-term progressive swelling beneath the center of the slab, or because the soil around the perimeter of the slab dries and shrinks (causing the perimeter to deform downward), or a combination of both. Edge lift is the cyclic heave beneath the perimeter of the foundation. In order to complete the design, the soils engineer usually provides the maximum anticipated vertical differential soil movement y_m and the horizontal distance of moisture variation from the slab perimeter e_m for both the center-lift and edge-lift conditions (see Fig. 12.9). Values of moisture variation from the slab perimeter e_m are usually obtained from the Post-tensioning Institute (1996), with typical values of e_m equal to 0.9 to 1.8 m (3 to 6 ft) for center lift and 0.6 to 1.8 m (2 to 6 ft) for edge lift. There are many design considerations for posttensioned slabs-on-grade. Three important considerations are as follows: 1. The design of the foundation is based on static values of y_m, but the actual movement is cyclic. To mitigate leaks from utilities, flexible utility lines that enter the slab should be used. Also, the posttensioned slabs-on-grade must be rigid enough so that cracks do not develop in interior wallboard because of cyclic movement. 2. The values of e_m are difficult to determine because they are dependent on soil and structural interaction. The structural parameters that govern e_m include the magnitude and distribution of dead loads, the rigidity of the foundation, and the depth of the perimeter footings. The soil parameters include the amount and specific limits of the heave and shrinkage. 3. The geotechnical engineer might test the expansive clays during the rainy season, but the foundation might be built during a drought. In this case, the values of y_m for center lift could be considerably underestimated. The geotechnical engineer must consider the worst-case conditions, which do not necessarily occur at the time of the subsurface exploration.

TABLE 12.11 Foundation Alternatives for Expansive Soil (*Continued*)

Topic (1)	Discussion (2)
Pier and grade beam support	A third common foundation type for expansive soil is pier and grade beam support, as shown in Fig. 12.10 (from Chen 1988). The basic principle is to construct the piers such that they are below the depth of seasonal moisture changes. The piers can be belled at the bottom to increase their uplift resistance. Grade beams and structural floor systems that are free of the ground are supported by the piers. Chen (1988) provides several examples of proper construction details for pier and grade beam foundations. The design considerations for the pier and grade beam support are as follows (Woodward et al. 1972, Jubenville and Hepworth 1981, Chen 1988): 1. *Sufficient pier length.* A model test for pier uplift indicated that the expansive soil uplift of a pier is similar to the extraction of a pile (i.e., uplift capacity) from the ground (Chen 1988). Based on this test data, the uplift force T_u on the pier due to swelling of soil in the active zone can be estimated as follows: $$T_u = c_A 2\pi R Z_a$$ where c_A = adhesion between the clay and pier, R = radius of the pier, and Z_a = depth of wetting, which usually corresponds to the depth of seasonal moisture change (the active zone). For a given type of pier and the undrained shear strength of the clay, the value for c_A can be obtained from Fig. 10.5. There are methods to reduce the uplift force T_u on the pier. For example, an air gap can be provided around the pier (Lambe and Whitman 1969, p. 7). Another option is to enlarge the pier hole and install an easily deformable material between the pier and expansive clay. The resisting force T_r for straight concrete piers can be calculated as follows: $$T_r = P + c_A 2\pi R Z_{\text{na}}$$ where P = dead load of the pier, which includes both the weight of the pier and the dead load applied to the top of the pier, and Z_{na} = portion of the pier below the active zone. Note that the total length of the pier = $Z_a + Z_{\text{na}}$. If the ends of the piers are belled, then there would be additional resisting forces. By equating the two equations, the value of Z_{na} can be obtained. By multiplying Z_{na} by an appropriate factor of safety (at least 1.5), the total depth of the pier can be determined. As an alternative to the above calculations, it has been stated that the depth of piles or piers should be 1.5 times the depth where the swelling pressure is equal to the overburden pressure (David and Komornik 1980).

TABLE 12.11 Foundation Alternatives for Expansive Soil (*Continued*)

Topic (1)	Discussion (2)
Pier and grade beam support (*Continued*)	2. *Pier diameter.* In regard to pier diameter, Chen (1988) states that to exert enough dead load pressure on the piers, it is necessary to use small-diameter piers in combination with long spans of the grade beams. Piers for expansive soil typically have a diameter of 0.3 m (1 ft). 3. *Pier reinforcement.* Because a large uplift force T_u can be developed on the pier as the clay swells in the active zone, steel reinforcement is required to prevent the piers from failing in tension. Reinforcement of the full length of the pier is essential to avoid tensile failures. 4. *Construction process.* It is important to use proper construction procedures when constructing the piers. For example, because the piers are often heavily loaded, the bottom of the piers must be cleaned of all loose debris or slough. After the pier hole is drilled and the concrete has been placed, excess concrete is often not removed from the top of the pier. This excess concrete has a mushroom shape, and because of its large area, the expansive soil can exert a substantial and unanticipated uplift force onto the top of the pier (Chen 1988). 5. *Void space below grade beams.* A common procedure is to use a void-forming material (such as cardboard) to create a void space below the grade beam. It is important to create a complete and open void below the grade beam so that when the soil expands, it will not come in contact with the grade beam and cause damaging uplift forces.

TABLE 12.12 Expansive Soil Foundation Recommendations for Conventional Slab-on-Grade

Expansive classification (1)	Depth of footing below adjacent grade (2)	Footing reinforcement (3)	Slab thickness and reinforcement conditions (4)	Presaturation below slabs (5)	Rock base below slabs (6)
None to low	0.5 m (18 in.) exterior 0.3 m (12 in.) interior	Exterior: 4 #4 bars, 2 top and 2 bottom Interior: 2 #4 bars, 1 top and 1 bottom	0.10 m (4 in.) nominal with #3 bars at 0.4 m (16 in.) on center, each way	to 0.3 m (12 in.)	Optional
Medium	0.6 m (24 in.) exterior 0.5 m (18 in.) interior	Exterior: 4 #5 bars, 2 top and 2 bottom Interior: 4 #4 bars, 2 top and 2 bottom	0.10 m (4 in.) net with #3 bars at 0.3 m (12 in.) on center, each way	to 0.5 m (18 in.)	0.10 m (4 in.)
High	0.8 m (30 in.) exterior 0.5 m (18 in.) interior	Exterior: 4 #5 bars, 2 top and 2 bottom Interior: 4 #5 bars, 2 top and 2 bottom	0.13 m (5 in.) net with #4 bars at 0.4 m (16 in.) on center, each way	to 0.6 m (24 in.)	0.15 m (6 in.)
Very high	0.9 m (36 in.) exterior 0.8 m (30 in.) interior	Exterior: 6 #5 bars, 3 top and 3 bottom Interior: 4 #5 bars, 2 top and 2 bottom	0.15 m (6 in.) nominal with #4 bars at 0.3 m (12 in.) on center, each way	to 0.8 m (30 in.)	0.20 m (8 in.)

Note: Design recommendations were empirically derived. See Table 12.11 for discussion of risk factors. Foundation requirements are often not adequate for high and very high expansive soil classifications, and a posttensioned slab designed in accordance with the Post-Tensioning Institute (1996) is often recommended.

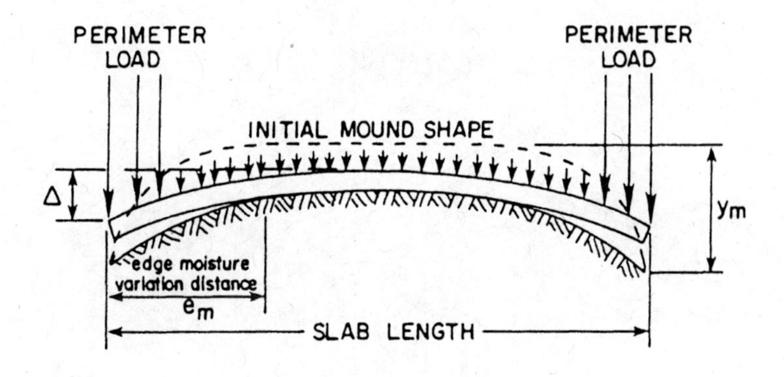

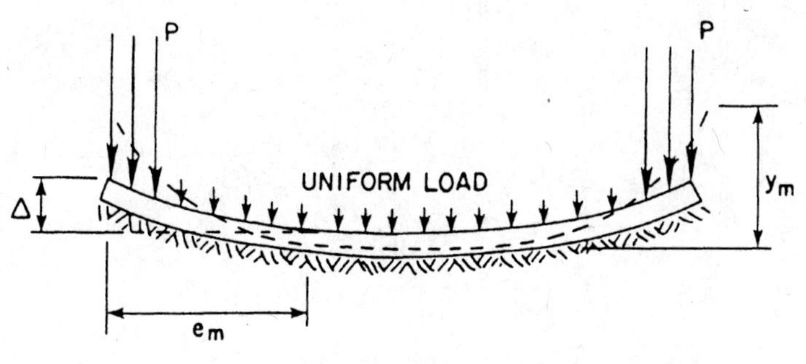

FIGURE 12.9 Depiction of center lift and edge lift for posttensioned slab-on-grade. See Table 12.11. (*Reproduced with permission from the Post-Tensioning Institute 1996.*)

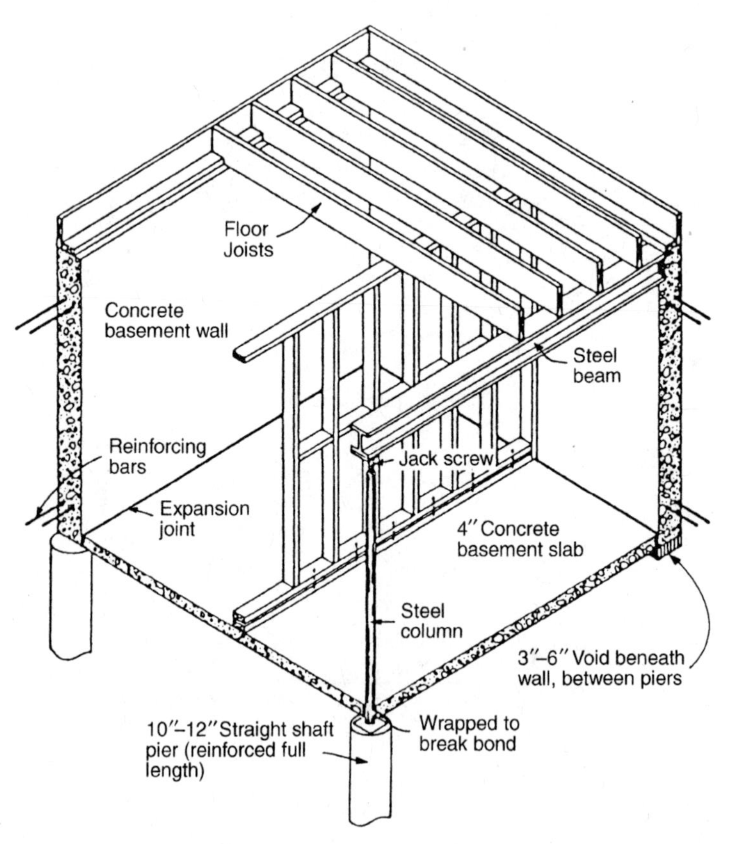

FIGURE 12.10 Typical detail of the grade beam and pier system. See Table 12.11. (*From Chen 1988. Reprinted with permission from Elsevier Science.*)

TABLE 12.13 Pavements and Flatwork on Expansive Soil

Topic (1)	Discussion (2)
Flexible pavements	Flexible pavements can be damaged by heave of compacted clay subgrade. This is often caused by water seeping through the joints in the concrete gutters or cracks in the AC pavement. The water then percolates downward through the base and is absorbed by the compacted clay subgrade, resulting in expansion of the clay. The asphalt pavement will have more heave relative to the curbs, resulting in cracks parallel to the curbs and gutters. Once cracked, the asphalt pavement allows for further infiltration of water, accelerating the heave process. There can be a variety of heave and crack patterns in pavements on expansive clay subgrade (Van der Merwe and Ahronovitz 1973, Day 1995c). Another effect of moisture infiltration into the compacted clay subgrade is a reduction in the undrained shear strength of the soil. Moisture causes a softening of the compacted clay. The process involves an increase in moisture content and a reduction in negative pore pressures, which results in a decrease in the undrained shear strength. This makes the subgrade weaker and more susceptible to pavement deterioration, such as alligator cracking and rutting. The flexible pavement design equation (Table 11.3) does not have a factor to account for the expansiveness of the subgrade. For example, it is not unusual for a medium expansive clay subgrade and a very highly expansive clay subgrade to both have an R-value of less than 10. As a result, for both the medium and very highly expansive clay subgrade, the thickness of asphalt concrete and base material are similar given the same traffic index. But because the percent swell is much greater for the very highly expansive soil, this pavement will have more uplift and cracking as moisture infiltrates into the clay subgrade. In some cases, project specifications require compaction of the subgrade to a high density, such as requiring the subgrade to be compacted to a minimum of 95% relative compaction (Modified Proctor). This can lead to problems, especially if the clays are wet, because it may be difficult to dry out and effectively compact the clays to a high density. Also, once the clays are dried out and densified, there is a much greater potential for additional expansion when water infiltrates the base. A better solution would be to keep the clay subgrade in a wet condition and increase the base thickness to compensate for the lower undrained shear strength of the subgrade.
Concrete pavements	The same expansive soil mechanisms that affect flexible pavement can also impact concrete pavements. The damage caused by expansive soil may be more severe for concrete pavements than asphalt pavements because they are usually more brittle.
Mitigation measures	Methods to mitigate the effects of clay expansion include compacting the clay subgrade wet of optimum or adding lime or cement to the clay subgrade (see Table 12.10). In compacting the clay subgrade wet of optimum, a geofabric can be used to prevent the aggregate base from being pushed down into and contaminated by the soft clay subgrade. Contractors often believe that for construction on expansive soil, the subgrade should be hard and dense. Thus they allow the clay to dry out and then fill in the desiccation cracks and tamp the soil to create a firm, level subgrade.

TABLE 12.13 Pavements and Flatwork on Expansive Soil (*Continued*)

Topic (1)	Discussion (2)
Mitigation measures (*Continued*)	This then creates a solid and rocklike subgrade of dense clay having a low moisture content. This method of subgrade preparation is probably the worst possible method for expansive soil. Once the flatwork is constructed, moisture inevitably accumulates beneath the flatwork, with resulting heave and extensive damage. As previously mentioned, the best approach is to get the subgrade in a wet condition prior to construction, so that when water does seep into the base after the construction is complete, there will not be damaging expansion of the clay subgrade. For construction atop expansive soil, asphalt paving is usually preferred because it is more flexible and can be more easily repaired than concrete. One construction method for concrete pavements is to use open graded gravel as base material over the subgrade material. Extra base can be added to compensate for some gravel being pushed into the clay subgrade, or a permeable geofabric can be placed atop the clay subgrade. Once the base (open graded gravel) is in place, the gravel is flooded with water. The water will seep down through the gravel and be absorbed by the underlying clay subgrade. Cylindrical holes can be excavated into the subgrade to facilitate the wetting of the subgrade. Periodically, soil samples can be taken from the subgrade to check the moisture condition and swell of the subgrade. After the subgrade has been sufficiently moistened, the pavement section can be constructed. Usually extra steel reinforcement and an increased thickness of concrete pavement is needed to compensate for the poor bearing conditions of the subgrade.
Flatwork—upward movement	Flatwork can be defined as appurtenant structures that surround a building or house such as concrete walkways, patios, driveways, and pool decks. It is the lightly loaded structures, such as pavements or lightly loaded foundations, that are commonly damaged by expansive soil. Because flatwork usually supports only its own weight, it can be especially susceptible to expansive soil–related damage. The expansive soil uplift of the flatwork tends to produce a distinct crack pattern, which has been termed a "spider" or "x" -type crack pattern. Besides the concrete itself, utility lines that are embedded in the concrete can also be damaged by the upward movement of the flatwork.
Walking of flatwork	Besides differential movement of the flatwork, there can also be progressive movement of flatwork away from the structure. This lateral movement is known as "walking." In many cases, there are appurtenant structures (such as patio shade cover) that are both attached to the house and also derive support from the flatwork. As the flatwork walks away from the house, these appurtenant structures are pulled laterally and frequently damaged. The results of a field experiment indicated that most walking occurs during the wet period (Day 1992c). The expansion of the clay causes the flatwork to move up and away from the structure. During the dry period, the flatwork does not return to its original position. Then during the next wet period, the expansion of the clay again causes an upward and outward movement of the flatwork. The cycles of wetting and drying cause a progressive movement of flatwork away from the building. An important factor in the amount of "walking" is the moisture condition of the clay prior to construction of the flatwork. If the clay is dry, then more initial upward and outward movement can occur during the first wet cycle.

TABLE 12.14 Expansive Rock

Topic (1)	Discussion (2)
Expansive rock	There are several different mechanisms that can cause the expansion of rock. Some rock types, such as shale, slate, mudstone, siltstone, and claystone, can be especially susceptible to expansion. Some common mechanisms that can cause rock to expand are listed below.
Rebound	For cut areas, where the overburden has been removed by erosion or during mass grading operations, the rock will rebound because of the release in overburden pressure. The rebound can cause the opening of cracks and joints. Usually rebound of rock occurs during the rock excavation, and because it is a relatively rapid process, it is not included in the engineering analyses.
Expansion due to physical factors	Rock, especially soft sedimentary and fractured rock, can expand as a result of the physical growth of plant roots or by the freezing of water in the fractures. Studies have also shown that the precipitation of gypsum in rock pores, cracks, and joints can cause rock expansion and disintegration. Such conditions occur in arid climates where subsurface moisture evaporates at ground surface, precipitating the minerals in the rock pores. Gypsum crystals have been observed to grow in rock fractures and are believed to exert the most force at their growing end (Hawkins and Pinches 1987). Gypsum growth has even been observed in massive sandstone, which resulted in significant heave of the rock (Hollingsworth and Grover 1992).
Expansion due to weathering of rock	Probably the most frequent cause of heave of rock and resulting damage to foundations is the weathering of rock. Weathering of rock can occur by physical and chemical methods. Typical types of chemical weathering include oxidation, hydration of clay minerals, and the chemical alteration of the silt-size particles to clay. Factors affecting oxidation include the presence of moisture and oxygen (aerobic conditions), biological activity, acidic environment, and temperature (Hollingsworth and Grover 1992). As indicated in Table 2.16, pyritic shale often expands upon exposure to air and moisture. Another example is bentonite, which is a rock that is composed of montmorillonite clay minerals. This type of rock will also rapidly weather and greatly expand when exposed to air and moisture.
Predicting expansion of rock	It is often difficult to predict the amount of heave that will occur because of expansive rock. One approach is similar to the testing of clay specimens, where a rock specimen is placed into an oedometer, subjected to the anticipated overburden pressure, submerged in distilled water, and then the expansion of the rock is measured (ASTM D 4546-96, 1998). However, this approach could considerably underestimate the expansion potential of the rock because in an intact (unweathered) state, it is much less expansive than in a fractured and weathered state. Another approach is to break apart the rock and then subject the fragments to wetting and drying cycles. The cycles of wetting and drying are often very effective in rapidly weathering expansive rock (Day 1994e). Once the rock fragments have sufficiently weathered, the index properties of the weathered material could be determined and then a foundation (such as a posttensioned slab) could be designed by the structural engineer.

CHAPTER 13
SLOPE STABILITY

13.1 TYPICAL TYPES OF SLOPE MOVEMENT

Slope movement can be divided into six basic categories, as described in Table 13.1. Table 13.2 presents a checklist for slope stability and landslide analysis. This table provides a comprehensive list of the factors that may need to be considered by the geotechnical engineer when designing slopes.

13.2 ALLOWABLE LATERAL MOVEMENT

Compared to the settlement of buildings, there is less work available on the allowable lateral movement. Table 13.3 and Fig. 13.1 present data on the allowable lateral movement of buildings.

13.3 ROCKFALL

A rockfall is defined as relatively free-falling rock or rocks that have detached themselves from a cliff, steep slope, cave, or arch (see Table 13.4 and Fig. 13.2).

13.4 SURFICIAL SLOPE STABILITY ANALYSIS

The surficial failure by definition is shallow with the failure surface usually at a depth of 1.2 m (4 ft) or less. In many cases, the failure surface is parallel to the slope face. This section presents a discussion of the surficial failure mechanism, the method of analysis, and various options that can be used to increase the factor of safety (Tables 13.5 to 13.8 and Figs. 13.3 to 13.6).

13.5 GROSS SLOPE STABILITY ANALYSIS

In contrast to the shallow (surficial) slope failure discussed in the previous section, a gross slope failure often involves shear displacement of the entire slope. Different terms, such as slides or slumps, are often used to identify gross slope instability. This section presents a discussion of the wedge method, the method of slices, important factors that may need to be considered in a slope stability analysis, and slope stability charts (Tables 13.9 to 13.13 and Figs. 13.7 to 13.13).

13.6 LANDSLIDE STABILITY ANALYSIS

Gross slope stability analysis could be referred to as landslide analysis. However, landslides in some cases may be so large that they involve several different slopes. This section presents a discussion of the classification of landslides (i.e., translation versus rotation), common nomenclature used to describe landslide features, method of analysis for landslides, and commonly used methods to stabilize slopes and landslides (Tables 13.14 to 13.17 and Figs. 13.14 to 13.16).

13.7 DEBRIS FLOW

A debris flow is commonly defined as soil with entrained water and air that moves readily as a fluid on low slopes. Debris flow is discussed in Table 13.18.

13.8 SLOPE SOFTENING AND CREEP

This section describes the process of slope softening and creep. Both of these processes cause lateral movement of slopes. As a practical matter, slope softening and creep need only be considered for plastic (cohesive) soil. See Table 13.19 and Fig. 13.17.

13.9 ANALYSIS FOR EARTH DAMS

An important consideration for the design of dams is slope stability, and therefore dams have been included in the chapter. The design and construction of dams are complex and will only be briefly discussed in this section (see Tables 13.20 and 13.21).

TABLE 13.1 Typical Types of Slope Movement

Type (1)	Discussion (2)
Rockfalls or topples	This is usually an extremely rapid movement that includes the free fall of rocks, movement of rocks by leaps and bounds down the slope face, and/or the rolling of rocks or fragments of rocks down the slope face (Varnes 1978). A rock topple is similar to a rockfall, except that there is a turning moment about the center of gravity of the rock, which results in an initial rotational-type movement and detachment from the slope face. Rockfalls are discussed in Sec. 13.3.
Surficial slope stability	Surficial slope instability involves shear displacement along a distinct failure (or slip) surface. As the name implies, surficial slope stability analysis deals with the outer face of the slope, generally up to 1.2 m (4 ft) deep. The typical surficial slope stability analysis assumes the failure surface is parallel to the slope face. Surficial slope stability analysis is discussed in Sec. 13.4.
Gross slope stability	In contrast to a shallow (surficial) stability analysis, gross slope stability involves an analysis of the entire slope. In the stability analysis of slopes, curved or circular slip surfaces are often assumed. Terms such as *fill slope stability analysis* and *earth* or *rock slump analysis* have been used to identify similar processes. Gross slope stability is discussed in Sec. 13.5.
Landslides	Gross slope stability could be referred to as landslide analysis. However, landslides in some cases may be so large that they involve several different slopes. Landslides are discussed in Sec. 13.6.
Debris flow	Debris flow is commonly defined as soil with entrained water and air that moves readily as a fluid on low slopes. Debris flow can include a wide variety of soil-particle sizes (including boulders) as well as logs, branches, tires, and automobiles. Other terms, such as *mud flow, debris slide, mud slide,* and *earth flow,* have been used to identify similar processes. While categorizing flows according to rate of movement or the percentage of clay particles may be important, the mechanisms of all these flows are essentially the same (Johnson and Rodine 1984). Debris flow are discussed in Sec. 13.7.
Creep	Creep is generally defined as an imperceptibly slow and more or less continuous downward and outward movement of slope-forming soil or rock (Stokes and Varnes 1955). Creep can affect both the near-surface (surficial) soil and deep-seated (gross) materials. The process of creep is frequently described as viscous shear that produces permanent deformations, but not failure as in landslide movement. Creep is discussed in Sec. 13.8.

TABLE 13.2 Checklist for the Study of Slope Stability and Landslides

Main topic (1)	Relevant items (2)
Topography	Contour map: Consider land form and anomalous patterns (jumbled, scarps, bulges). Surface drainage: Evaluate conditions such as continuous or intermittent drainage. Profiles of slope: Evaluate along with geology and the contour map. Topographic changes, such as the rate of change by time: Correlate with groundwater, weather, and vibrations.
Geology	Formations at site: Consider the sequence of formations, colluvium (bedrock contact and residual soil), formations with bad experience, and rock minerals susceptible to alteration. Structure: Evaluate three-dimensional geometry, stratification, folding, strike and dip of bedding or foliation (changes in strike and dip and relation to slope and slide), and strike and dip of joints with relation to slope. Also investigate faults, breccia, and shear zones with relation to slope and slide. Weathering: Consider the character (chemical, mechanical, and solution) and depth (uniform or variable).
Groundwater	Piezometric levels within slope, such as normal, perched levels, or artesian pressures with relation to formations and structure. Variations in piezometric levels due to weather, vibration, and history of slope changes. Other factors include response to rainfall, seasonal fluctuations, year-to-year changes, and effect of snowmelt. Ground surface indication of subsurface water, such as springs, seeps, damp areas, and vegetation differences. Effect of human activity on groundwater, such as groundwater utilization, groundwater flow restriction, impoundment, additions to groundwater, changes in ground cover, infiltration opportunity, and surface water changes. Groundwater chemistry, such as dissolved salts and gases and changes in radioactive gases.
Weather	Precipitation from rain or snow. Also consider hourly, daily, monthly, or annual rates. Temperature, such as hourly and daily means or extremes, cumulative degree-day deficit (freezing index), and sudden thaws. Barometric changes.
Vibration	Seismicity, such as seismic events, microseismic intensity, and microseismic changes. Human-induced from blasting, heavy machinery, or transportation (trucks, trains, etc.).
History of slope changes	Natural processes, such as long-term geologic changes, erosion, evidence of past movement, submergence, or emergence. Human activities, including cutting, filling, clearing, excavation, cultivation, paving, flooding, and sudden drawdown of reservoirs. Also consider changes caused by human activities, such as changes in surface water, groundwater, and vegetation cover.

TABLE 13.2 Checklist for the Study of Slope Stability and Landslides (*Continued*)

Main topic (1)	Relevant items (2)
History of slope changes (*Continued*)	Rate of movement from visual accounts, evidence in vegetation, evidence in topography, and photographs (oblique, aerial, stereoptical data, and spectral changes). Also consider instrumental data, such as vertical changes, horizontal changes, and internal strains and tilt, including time history. Correlate movements with groundwater, weather, vibration, and human activity.

Note: Adapted from Sowers and Royster 1978. Reprinted with permission from *Landslides: Analysis and Control, Special Report 176.* Copyright 1978 by the National Academy of Sciences. Courtesy of the National Academy Press, Washington, D.C.

TABLE 13.3 Allowable Lateral Movement of Buildings

Topic (1)	Discussion (2)
Horizontal strain	To evaluate the lateral movement of buildings, a useful parameter is the horizontal strain ε_h, defined as the change in length divided by the original length of the foundation. Figure 13.1 shows a correlation between horizontal strain ε_h and severity of damage (Boone 1996, originally from Boscardin and Cording 1989). It should be mentioned that in Fig. 13.1, Boscardin and Cording (1989) used a "distortion factor" in their calculation of angular distortion β for foundations subjected to settlement from mines, tunnels, and braced cuts. Because of this "distortion factor," the angular distortion β by Boscardin and Cording (1989) in Fig. 13.1 is different than the definition δ/L used in Chap. 9.
Foundation resistance	The ability of the foundation to resist lateral movement will depend on the tensile strength of the foundation. Those foundations that can not resist the tensile forces imposed by lateral movement will be the most severely damaged. Those foundations that have joints or planes of weakness will be most susceptible to damage from lateral movement. Buildings having a mat foundation or a post-tensioned slab would be less susceptible to damage because of the high tensile resistance of these foundations.

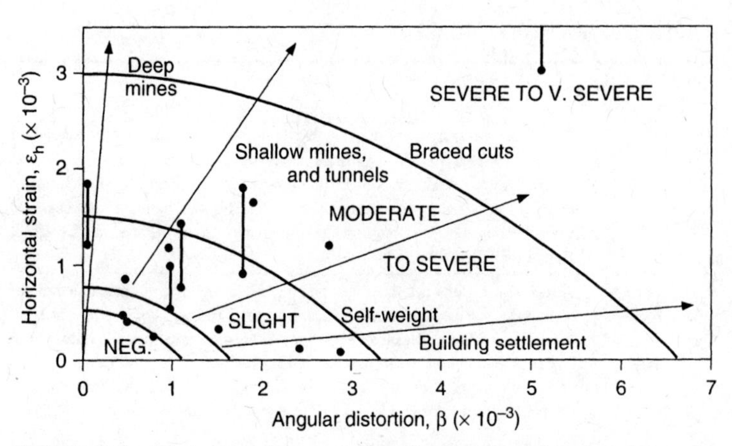

FIGURE 13.1 Relationship of damage to angular distortion and horizontal extension strain. See Table 13.3. (*From Boscardin and Cording 1989; reprinted with permission from the American Society of Civil Engineers.*)

TABLE 13.4 Rockfall

Topic (1)	Discussion (2)
Definition	A rockfall is defined as a relatively free-falling rock or rocks that have detached themselves from a cliff, steep slope, cave, arch, or tunnel (Stokes and Varnes 1955). The movement may be by the process of a vertical fall, by a series of bounces, or by rolling down the slope face. The free-fall nature of the rocks and the lack of movement along a well-defined slip surface differentiate a rockfall from a rockslide.
Basic factors governing a rockfall	A rock slope is characterized by a heterogeneous and discontinuous medium of solid rocks that are separated by discontinuities. The rocks composing a rockfall tend to detach themselves from these preexisting discontinuities in the slope. The sizes of the individual rocks in a rockfall are governed by the attitude, geometry, and spatial distribution of the rock discontinuities. The basic factors governing the potential for a rockfall include (Piteau and Peckover 1978): 1. The geometry of the slope. 2. The system of joints and other discontinuities and the relation of these systems to possible failure surfaces. 3. The shear strength of the joints and discontinuities. 4. Destabilizing forces such as water pressure in the joints, freezing water, or vibrations.
Remedial measures	The investigation and analysis of rockfall will often be jointly performed by the engineering geologist and geotechnical engineer. If the analysis indicates that rockfall are likely to impact the site, then remedial measures can be implemented. The main measures to prevent damage due to a rockfall are as follows: 1. *Alter the slope configuration.* Altering the slope can include such measures as removing the unstable or potentially unstable rocks, flattening the slope, or incorporating benches into the slope (Piteau and Peckover 1978). 2. *Retain the rocks on the slope face.* Measures to retain the rocks on the slope face include the use of anchoring systems (such as bolts, rods, or dowels), shotcrete applied to the rock slope face, or retaining walls. 3. *Intercept the falling rocks before they reach the structure.* Intercepting or deflecting falling rocks around the structure can be accomplished by using toe-of-slope ditches, wire mesh catch fences, and catch walls (Peckover 1975). Recommendations for the width and depth of toe-of-slope ditches have been presented by Ritchie (1963) and Piteau and Peckover (1978) and are shown in Fig. 13.2. 4. *Direct the falling rocks around the structure.* Walls, fences, or ditches can be constructed in order to deflect the falling rocks around the structure.

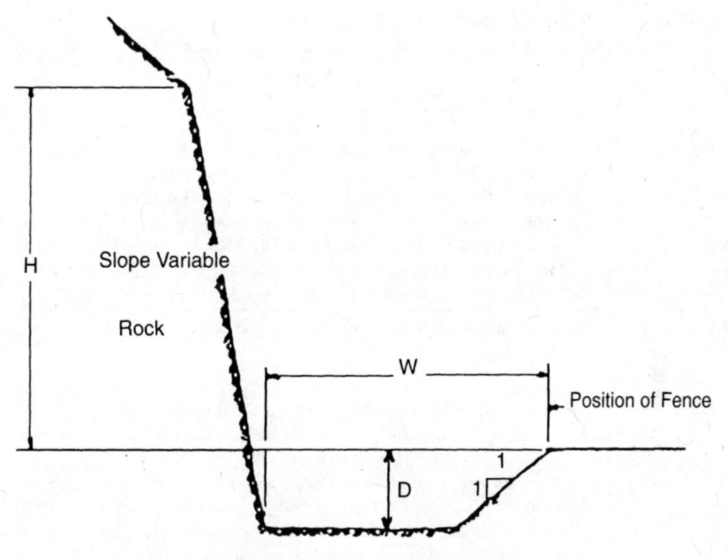

Rock Slope Angle	Rock Slope Height (m)	Fallout Area Width (m)	Ditch Depth (m)
Near vertical	5 to 10	3.7	1.0
	10 to 20	4.6	1.2
	>20	6.1	1.2
0.25 or 0.3:1	5 to 10	3.7	1.0
	10 to 20	4.6	1.2
	20 to 30	6.1	1.8*
	>30	7.6	1.8*
0.5:1	5 to 10	3.7	1.2
	10 to 20	4.6	1.8*
	20 to 30	6.1	1.8*
	>30	7.6	2.7*
0.75:1	0 to 10	3.7	1.0
	10 to 20	4.6	1.2
	>20	4.6	1.8*
1:1	0 to 10	3.7	1.0
	10 to 20	3.7	1.5*
	>20	4.6	1.8*

* May be 1.2 m if catch fence is used.

FIGURE 13.2 Design criteria for ditches at the base of rock slopes. See Table 13.4. (*From Piteau and Peckover 1978.*)

TABLE 13.5 Surficial Slope Stability: Failure Mechanism

Topic (1)	Discussion (2)
General discussion	Surficial failures of slopes are quite common throughout the United States (Day and Axten 1989, Wu et al. 1993). In southern California, surficial failures usually occur during the winter rainy season, after a prolonged rainfall or during a heavy rainstorm, and are estimated to account for a significant portion of the problems associated with slope movement on developed properties (Gill 1967). Figure 13.3 illustrates a typical surficial slope failure. The surficial failure by definition is shallow, with the failure surface usually at a depth of 1.2 m (4 ft) or less (Evans 1972). In many cases, the failure surface is parallel to the slope face.
Surficial failure mechanism for clay slopes	The common surficial failure mechanism for clay slopes in southern California is as follows: 1. During the hot and dry summer period, the slope face can become desiccated and shrunken. The extent and depth of the shrinkage cracks depend on many factors, such as the temperature and humidity, the plasticity of the clay, and the extraction of moisture by plant roots. 2. When the winter rains occur, water percolates into the fissures, causing the slope surface to swell and saturate with a corresponding reduction in shear strength. Initially, water percolates downward into the slope through desiccation cracks and in response to the suction pressures of the dried clay. 3. As the outer face of the slope swells and saturates, the permeability parallel to the slope face increases. With continued rainfall, seepage develops parallel to the slope face. 4. Because of a reduction in shear strength due to saturation and swell coupled with the condition of seepage parallel to the slope face, failure occurs. Studies by Pradel and Raad (1993) indicate that in southern California fill slopes made of clayey or silty soils are more prone to develop the conditions for surficial instability than slopes made of sand or gravel. This is probably because water tends to migrate downward in sand or gravel fill slopes, rather than parallel to the slope face.
Surficial failure mechanism for cut slopes in rock	For cut slopes in rock, the common cause of the surficial slope failure is due to weathering. Weathering breaks down the rock and reduces the effective shear strength of the material. The weathering process also opens up fissures and cracks which increase the permeability of the near surface rock and promotes seepage of water parallel to the slope face. This weathering process is illustrated in Fig. 13.4 (Ortigao et al. 1997). Failure will eventually occur when the material has weathered to such a point that the effective cohesion approaches zero. Surficial failures are most common for cut slopes in soft sedimentary rock, such as claystones or weakly cemented sandstones. The most common reason for the surficial failure is because a relatively steep slope (such as 1.5:1 or 1:1) is excavated into the sedimentary rock. Then with time, as vegetation is established on the slope face and the face of the cut slope weathers, the probability of surficial failures increases. The surficial failure usually develops during or after a period of heavy and prolonged rainfall.

TABLE 13.5 Surficial Slope Stability: Failure Mechanism (*Continued*)

Topic (1)	Discussion (2)
Surficial failure mechanism for natural slopes	A similar process to that described above can cause surficial failures in natural slopes. There is often an upper weathered zone of soil that slowly grades into solid rock with depth. The upper weathered zone of soil is often much more permeable because of its loose soil structure, and hence higher porosity, than the underlying dense rock. Water that seeps into the natural slope from rainstorms will tend to flow in the outer layer of the slope which has a much higher permeability than the deeper unweathered rock. Thus the seepage condition for natural slopes is often similar to that as shown in Fig. 13.4.
Surficial failure mechanism for earth dams	Surficial failures can also develop on the downstream face of earth dams. For example, Sherard et al. (1963) state: "Shallow slides, most of which follow heavy rainstorms, do not as a rule extend into the embankment in a direction normal to the slope more than 4 or 5 feet [1.2–1.5 m]. Some take place soon after construction, while others occur after many years of reservoir operation. . . . Shallow surface slips involving only the upper few inches of the embankment have sometimes occurred when the embankment slopes have been poorly compacted. This is a frequent difficulty in small, cheaply constructed dams, where the construction forces often do not make the determined effort necessary to prevent a loose condition in the outer slope. The outer few feet then soften during the first rainy season, and shallow slides result."
Typical damage due to surficial failures	For developed property, surficial failures can cause extensive damage to landscaping. Surficial failures can even carry large trees downslope. Besides the landscaping, there can be damage to the irrigation and drainage lines. The surficial failure can also damage appurtenant structures, such as fences, walls, and patios. A particularly dangerous condition occurs when the surficial failure mobilizes itself into a debris flow. In such cases, severe damage can occur to any structure located in the path of the debris flow. Surficial failures can be sudden and unexpected, without any warning of potential failure. Other surficial failures, especially in clays, may have semicircular ground cracks that are characteristic of imminent failure.
Effect of vegetation	A contributing factor in many surficial slope failures is the loss of vegetation due to fire. Roots can provide a large resistance to shearing. The shear resistance of root-permeated homogeneous and stratified soil has been studied by Waldron (1977) and Merfield (1992). The increase in shear strength due to plant roots is directly attributable to mechanical reinforcement of the soil and indirectly to removal of soil moisture by transpiration. Even grass roots can provide an increase in shear resistance of the soil equivalent to an effective cohesion of 3 to 5 kPa (60 to 100 psf), (Day 1993b). When the vegetation is damaged or destroyed by fire, the slope can be much more susceptible to surficial failure. In order to improve the surficial stability of the slope face, plants should consist of deep-rooted varieties requiring little watering. A landscape architect is often the best party to consult regarding actual types of plants and planting configuration. To retard the tendency of weathering of the slope face, irrigation

TABLE 13.5 Surficial Slope Stability: Failure Mechanism (*Continued*)

Topic (1)	Discussion (2)
Effect of vegetation (*Continued*)	should be planned to achieve uniform moisture conditions well below the saturation level. If automatic timing devices are utilized in conjunction with irrigation systems, provisions should be made for interrupting normal watering during and following periods of rainfall. Property owners and/or maintenance personnel should be made aware that improper slope maintenance, altering site drainage, overwatering, and burrowing animals can be detrimental to surficial slope stability. When installing landscape irrigation systems, it is often recommended that piping be anchored to the slope face instead of excavating trenches into the slope faces.

TABLE 13.6 Surficial Slope Stability: Method of Analysis

Topic (1)	Discussion (2)
Surficial stability equation	The factor of safety is used to determine the stability of a slope. A factor of safety of 1.0 indicates a failure condition, while a factor of safety greater than 1.0 indicates that the slope is stable. The higher the factor of safety, the higher the stability of a slope. To determine the factor of safety F for surficial stability, an effective stress slope stability analysis is performed and the following equation is used: $$F = \frac{\text{shear strength of soil}}{\text{shear stress in the soil}} = \frac{c' + \sigma'_n \tan \phi'}{\tau}$$ where σ'_n = effective normal stress on the assumed failure surface and τ = shear stress in the soil along the assumed failure surface. Figure 13.5 (from Lambe and Whitman 1969) shows the derivation of N' and T, assuming an infinite slope with seepage parallel to the slope face. If the failure surface is at a depth $= d$ and the width of the slip surface (C to D) is 1.0, then $a = \cos \alpha$ and $\sigma'_n = N'$ and $\tau = T$. Substituting these values into the above equation gives the surficial stability equation: $$F = \frac{c' + \gamma_b d \cos^2 \alpha \tan \phi'}{\gamma_t d \cos \alpha \sin \alpha}$$ Since an effective stress analysis is being performed with steady-state flow conditions, effective shear strength soil parameters (ϕ' = effective friction angle; c' = effective cohesion) must be used in the analysis. In the above equation, α = slope inclination, γ_b = buoyant unit weight of the soil, and γ_t = total unit weight of the soil. As mentioned, the parameter d = depth of the failure surface where the factor of safety is computed. The factor of safety for surficial stability is highly dependent on the effective cohesion value of the soil. Automatically assuming $c' = 0$ for the surficial stability equation may be overly conservative.
Effective normal pressure on the failure surface	Because of the shallow nature of surficial failures, the normal stress on the failure surface is usually low. Studies have shown that the effective shear strength envelope for soil can be nonlinear at low effective stresses. For example, Fig. 7.11 shows the effective stress failure envelope for compacted London clay. Note the nonlinear nature of the shear strength envelope. For an effective stress of about 100 to 300 kPa (2100 to 6300 psf), the failure envelope is relatively linear and has effective shear strength parameters of $\phi' = 16°$ and $c' = 25$ kPa (520 psf). But below about 100 kPa (2100 psf), the failure envelope is curved (see Fig. 7.11*b*) and the shear strength is less than the extrapolated line from high effective stresses. Because of the nonlinear nature of the shear strength envelope, the shear strength parameters (c' and ϕ') obtained at high normal stresses can overestimate the shear strength of the soil and should not be used in the surficial stability equation (Day 1994d).

TABLE 13.6 Surficial Slope Stability: Method of Analysis (*Continued*)

Topic (1)	Discussion (2)
Effective normal pressure on the failure surface (*Continued*)	As an example, suppose a fill slope having an inclination of 1.5:1 ($\alpha = 33.7°$) was constructed using compacted London clay having the shear strength envelope shown in Fig. 7.11. Table 13.7 lists the parameters used for the calculations and shows that using the linear portion of the shear strength envelope (Fig. 7.11*a*), the factor of safety = 2.5, but using the actual nonlinear shear strength envelope (Fig. 7.11*b*), the factor of safety = 0.98, indicating a failure condition.
Surficial stability analysis	In order to calculate the factor of safety for surficial stability of cut, natural, or fill slopes, the method of analysis should be as follows: 1. For cut slopes, determine if the rock is likely to weather. Local experience can be used in identifying sedimentary rocks, such as claystones or shales, that are known to quickly weather. 2. Obtain samples of the fill or weathered rock. Perform shear strength tests, such as drained direct shear tests, at confining pressures as low as possible to obtain the effective shear strength parameters ϕ' and c'. The factor of safety for surficial stability is very dependent on the value of effective cohesion c'. If the shear strength tests indicate a large effective cohesion intercept, then perform additional drained shear strength tests to verify this cohesion value. 3. Use the surficial stability equation to calculate the factor of safety. The parameters in this equation can be determined as follows: *a.* *Inclination* α. The slope inclination can be measured for natural slopes or based on the anticipated constructed condition for cut-and-fill slopes. *b.* *Total unit weight* γ_t. For fill slopes, the total unit weight γ_t can be based on anticipated unit weight conditions at the end of grading. For cut or natural slopes, the total unit weight γ_t can be obtained from the laboratory testing of undisturbed soil samples. Note that the total unit weight used in the surficial stability equation must be based on a saturated (S = 100%) condition. *c.* *Buoyant unit weight* γ_b. From the total unit weight for saturated soil, the buoyant unit weight γ_b can be calculated (see Table 5.3). *d.* *Depth of failure surface d*. In the surficial stability equation, the depth of the failure surface d is also the depth of seepage parallel to the slope face. In southern California, the value of d is typically assumed to equal 1.2 m (4 ft) for fill slopes. This may be overly conservative for cut or natural slopes, and different values of d may be appropriate given local rainfall and weathering conditions. 4. The acceptable minimum factor of safety for surficial stability is often 1.5. However, as previously mentioned, root reinforcement can significantly increase the surficial stability of a slope. It may be appropriate to accept a lower factor of safety in cases where deep rooting plants will be quickly established on the slope face.

TABLE 13.7 Surficial Slope Stability: Example

Shear Strength (1)	Factor of Safety (2)
Linear portion of shear strength envelope, $\phi' = 16°$ and $c' = 25$ kPa (520 psf), i.e., Fig. 7.11*a*	2.5
Nonlinear portion of shear strength envelope at low effective stress, i.e., Fig. 7.11*b*	0.98

Note: Both surficial stability analyses used the surficial stability equation (Table 13.6), with slope inclination (α) = 33.7°, total unit weight (γ_t) = 19.8 kN/m^3 (126 pcf), buoyant unit weight (γ_b) = 10.0 kN/m^3 (63.6 pcf), and depth of seepage and failure plane (d) = 1.2 m (4 ft).

TABLE 13.8 Surficial Slope Stability: Options to Increase the Factor of Safety

Topic (1)	Discussion (2)
Flatter slope inclination	If the factor of safety for surficial stability is deemed to be too low, there are many different methods that can be used to increase the factor of safety. For example, the surficial stability can be increased by building a flatter slope (i.e., decreasing the slope inclination). It has been observed that 1.5:1 (horizontal:vertical) or steeper slope inclinations are often most susceptible to surficial instability, while 2:1 or flatter slope inclinations have significantly fewer surficial failures.
Soil with a higher shear strength	The surficial stability could also be increased by facing the slope with soil that has a higher shear strength. This process can be performed during the grading of the site and is similar to the installation of a stabilization fill (see Standard Detail No. 3, App. B).
Gunite facing of slope	Another option is to face the slope with gunite. The gunite facing will not allow rainwater to infiltrate the slope, which should prevent surficial instability from developing.
Soil reinforcement	The outer zone of the slope could be constructed with layers of geogrid, as shown in Fig. 13.6. The typical construction steps for slopes reinforced with geogrid are as follows: 1. *Excavation of benches.* Benches are first cut into the hillside. Benches provide favorable (i.e., not out-of-slope) frictional contact between the new fill mass and the horizontal portion of the bench. The benches are also used for the placement of the drainage system. 2. *Installation of drains.* After the benches have been excavated, drains are installed. The vertical drains are often placed at a 3 m (10 ft) spacing and are used to intercept seepage that may be migrating through the ground. The horizontal drains collect the water from the vertical drains and dispose of the water off-site. 3. *Construction of the slope.* The slope is built by using layers of geogrid and compacted fill. The usual design specification is to compact the fill to a minimum of 90% of the Modified Proctor maximum dry density. 4. *Erosion control fabric.* At the end of the construction process, an erosion control fabric can be pinned to the face of the slope and then the slope face is planted. In the soil, the geogrid acts as soil reinforcement, providing reinforcement effects similar to those of plant roots. Note in Fig. 13.6 that the geogrid is tipped back into the slope in order to get the geogrid as perpendicular to the potential failure surface as possible. The main design requirements are the type and vertical spacing of the geogrid. The design will depend on such factors as the shear strength of the soil, the slope inclination, and the depth d of the potential failure surface used in the analysis. The design could be based on the surficial stability equation, where a geogrid resisting force is included in the numerator of the equation.

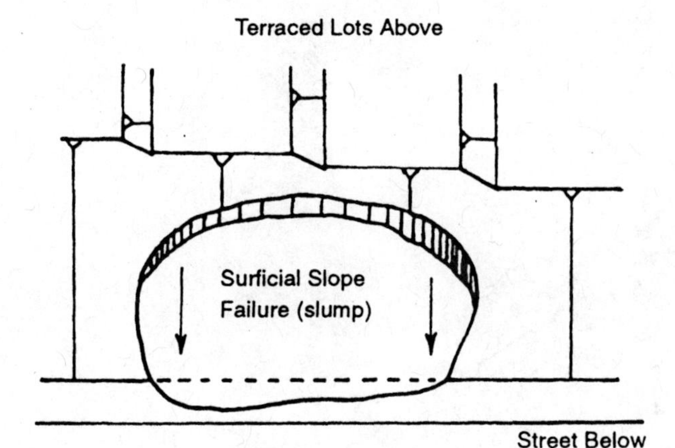

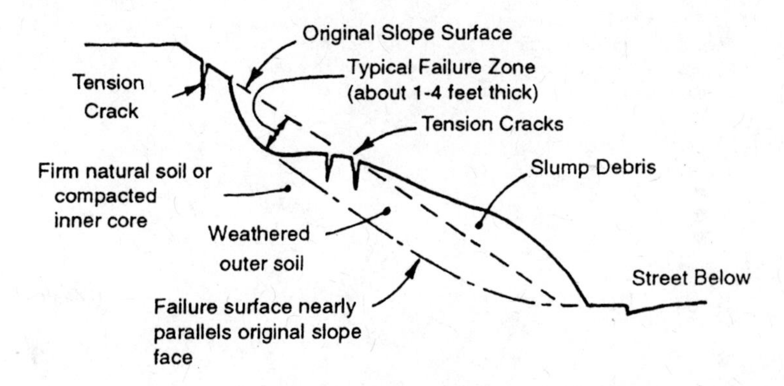

FIGURE 13.3 Illustration of typical surficial slope failure. (*a*) Plan view; (*b*) cross-sectional view. See Table 13.5.

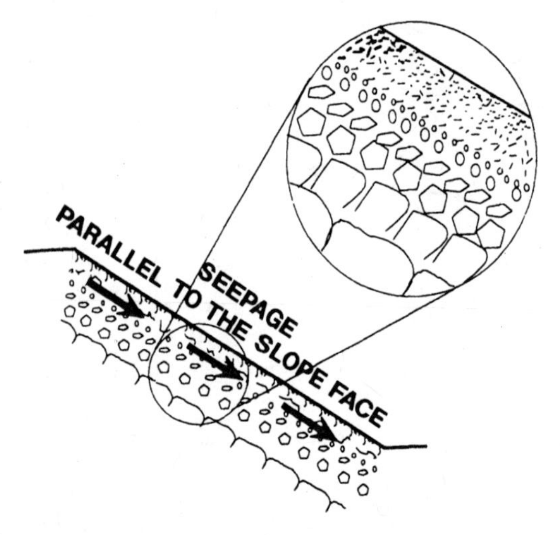

FIGURE 13.4 Illustration of near-surface weathering of claystone and zone of seepage parallel to slope face. See Table 13.5. (*Adapted from Ortigao et al. 1997; reprinted with permission from the American Society of Civil Engineers.*)

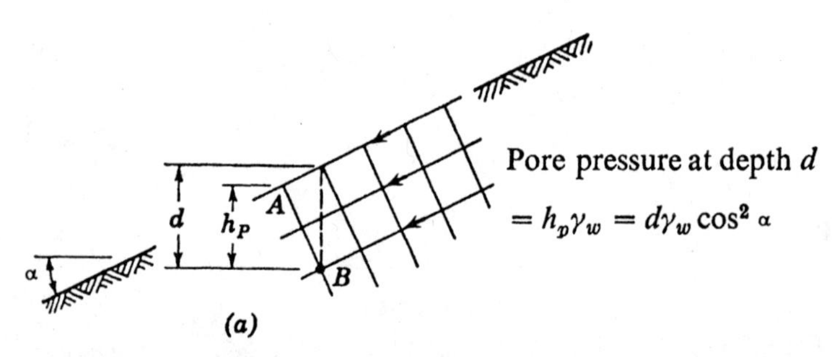

FIGURE 13.5 Analysis of infinite slope with seepage parallel to slope face. (*a*) Flow net. See Table 13.6. (*From Lambe and Whitman 1969; reproduced with permission from John Wiley & Sons.*)

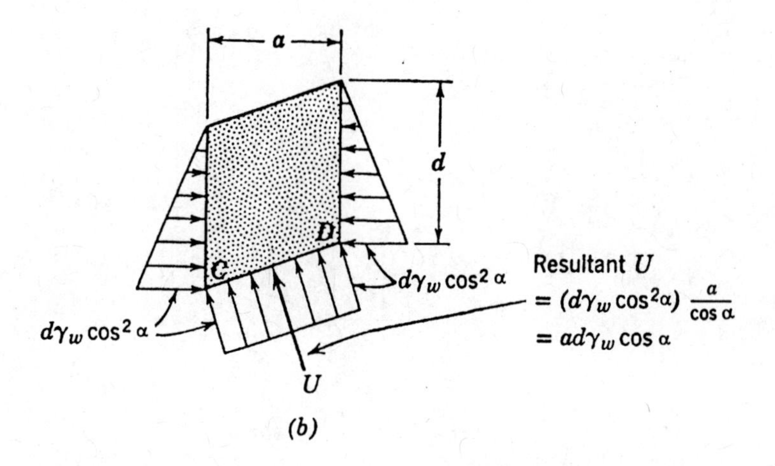

$$N' + U = \gamma_t ad \cos\alpha$$

$$\therefore \quad N' = \gamma_b ad \cos\alpha$$

$$T = \gamma_t ad \sin\alpha$$

$W = \gamma_t ad$

T

N'

$U = ad\gamma_w \cos\alpha$

(c)

FIGURE 13.5 Analysis of infinite slope with seepage parallel to slope face (*Continued*). (*b*) Boundary pore pressures. (*c*) Analysis of force equilibrium with moments balanced by side forces. See Table 13.6. (*From Lambe and Whitman 1969; reproduced with permission from John Wiley & Sons.*)

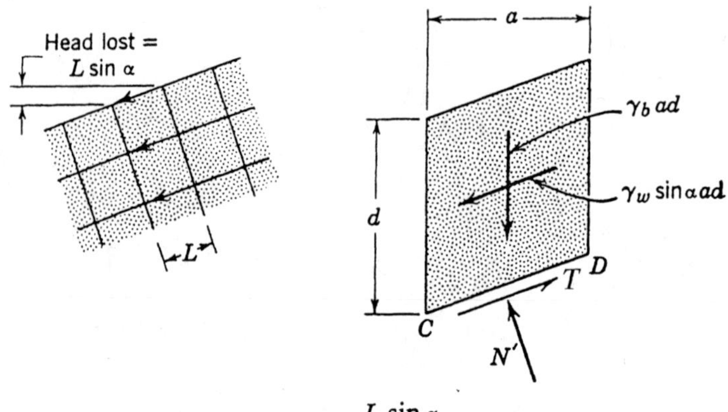

$$\text{Gradient } i = \frac{L \sin \alpha}{L} = \sin \alpha$$

Summing $\perp$ to CD:

$$N' = \gamma_b a d \cos \alpha$$

Summing $=$ to CD:

$$T = \gamma_b a d \sin \alpha + \gamma_w a d \sin \alpha = \gamma_t a d \sin \alpha$$

(d)

FIGURE 13.5 Analysis of infinite slope with seepage parallel to slope face (*Continued*). (*d*) Alternative analysis using seepage forces and buoyant unit weight. See Table 13.6. (*From Lambe and Whitman 1969; reproduced with permission from John Wiley & Sons.*)

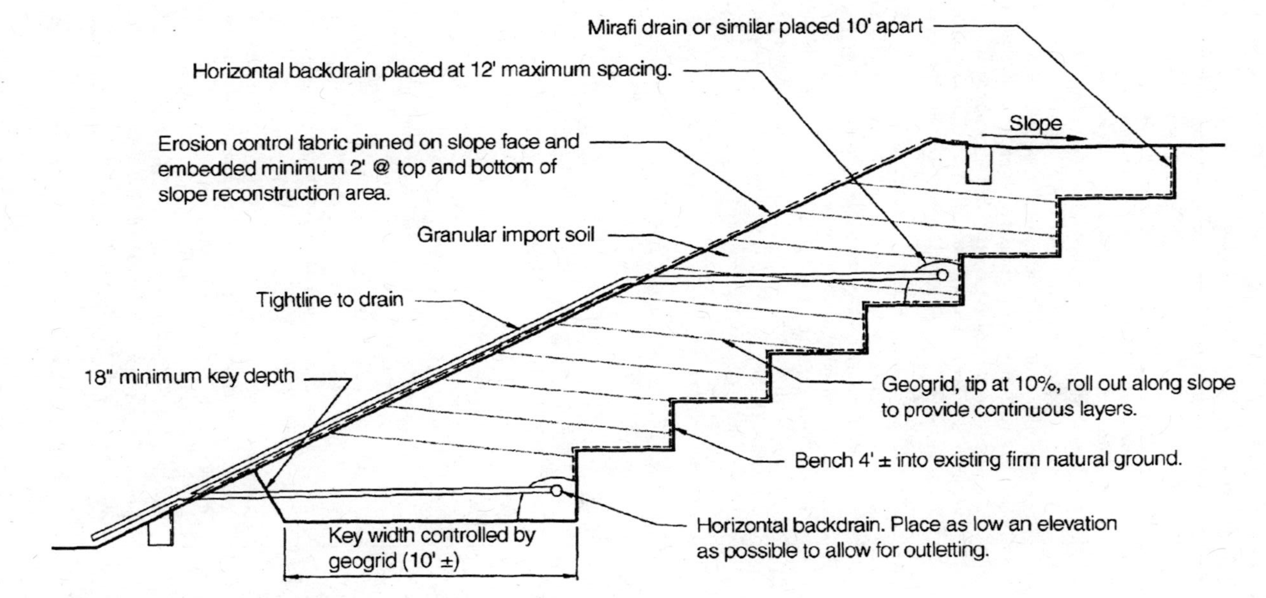

FIGURE 13.6 Construction of slope using geogrid. See Table 13.8.

TABLE 13.9 Gross Slope Stability: Wedge Method

Topic (1)	Discussion (2)
General discussion	As in surficial stability analysis, the objective of a gross slope stability analysis is to determine the factor of safety of an existing or proposed slope. For permanent slopes, the minimum factor of safety is 1.5. Gross slope stability analyses can be performed in terms of either a total stress analysis (short-term condition using undrained shear strength) or an effective stress analysis (long-term condition using the drained shear strength). For a total stress slope stability analysis, the total unit weight of the soil is utilized and the groundwater table is not considered in the analysis. For an effective stress analysis, the pore water pressures (based on the groundwater table, for example) must be included in the analysis.
Factor of safety for wedge method	The simplest type of gross slope stability analysis uses a free body-diagram such as illustrated in Fig. 13.7, where there is a planar slip surface inclined at an angle α to the horizontal. The wedge method is a two-dimensional analysis based on a unit length of slope. The assumption in this slope stability analysis is that there will be a wedge-type failure of the slope along a planar slip surface. The factor of safety F of the slope can be derived by summing forces parallel to the slip surface, and is: Total stress analysis: $F = \frac{\text{resisting force}}{\text{driving force}} = \frac{cL + N \tan \phi}{W \sin \alpha} = \frac{cL + W \cos \alpha \tan \phi}{W \sin \alpha}$ Effective stress analysis: $F = \frac{\text{resisting force}}{\text{driving force}} = \frac{c'L + N' \tan \phi'}{W \sin \alpha} = \frac{c'L + (W \cos \alpha - uL) \tan \phi'}{W \sin \alpha}$ where F = factor of safety for gross slope stability (dimensionless parameter) c, ϕ = undrained shear strength of the slide plane c', ϕ' = drained shear strength of the slide plane L = length of the slip surface (m or ft) N = normal force, i.e., the force acting perpendicular to the slip surface ($N = W \cos \alpha$) W = total weight of the failure wedge (i.e., $W = \gamma_t$ times the area of the wedge). Units are kN (or pounds) for a unit length of slope. u = average pore water pressure along the slip surface (kPa or psf) α = slip surface inclination (degrees) Because the wedge method is a two-dimensional analysis based on a unit length of slope (i.e., length = 1 m or 1 ft), the numerator and denominator of the above equations are in kN (or pounds). Similar to the surficial stability analysis, the resisting force in the above equations is equal to the shear strength (in terms of total stress or effective stress) of the soil along the slip surface. The driving force is caused by the pull of gravity and it is equal to the component of the weight of the wedge parallel to the slip surface.

TABLE 13.9 Gross Slope Stability: Wedge Method (*Continued*)

Topic (1)	Discussion (2)
Total stress analysis	The total stress analysis would be applicable to the quick construction of the slope or a condition where the slope is loaded very quickly. A total stress analysis could be performed by using the consolidated undrained shear strength c and ϕ or the undrained shear strength s_u of the slip surface material. In using the undrained shear strength, $s_u = c$ and $\phi = 0$ are substituted into the above total stress equation. *Example:* A slope has a height of 9.1 m (30 ft) and the slope face is inclined at a 2:1 (horizontal:vertical) ratio. Assume a wedge-type analysis where the slip surface is planar through the toe of the slope and is inclined at a 3:1 (horizontal:vertical) ratio. The total unit weight of the slope material γ_t = 18.1 kN/m^3 (115 pcf). Using the undrained shear strength parameters from Fig. 7.7 [i.e., c = 14.5 kPa (2.1 psi) and $\phi = 0$], calculate the factor of safety. *Solution:* The area of the wedge is first determined from simple geometry and is equal to 41.4 m^2 (450 ft^2). For a unit length of the slope, the total weight W of the wedge equals the area times total unit weight, or 750 kN per meter of slope length (52,000 lb per foot of slope length). For the total stress wedge analysis equation and the following values: c = 14.5 kPa (300 psf) $\phi = 0$ Length of slip surface (L) = 29 m (95 ft) Slope inclination (α) = 18° Total weight of wedge (W) = 750 kN/m (51,700 lb/ft) The factor of safety F of the slope = 1.80.
Effective stress analysis	The purpose of the effective stress slope stability analysis is to model the long-term condition of the slope, and effective stresses must be utilized along the slip surface. The term $W \cos \alpha - uL$ is equal to the effective normal force on the slip surface. Because the analysis uses the boundary pore water pressures and total unit weight of the soil, the weight W in the numerator and denominator of the effective stress equation is equal to the total weight of the wedge. *Example:* Same situation as the previous example except that the slip surface has the effective shear strength shown in Fig. 7.9 [i.e., c' = 3.4 kPa (70 psf), ϕ' = 29°]. Also assume that piezometers have been installed along the slip surface and the average measured steady-state pore water pressure (u) = 2.4 kPa (50 psf). Calculate the factor of safety of the failure wedge based on an effective stress analysis.

TABLE 13.9 Gross Slope Stability: Wedge Method (*Continued*)

Topic (1)	Discussion (2)
Effective stress analysis (*Continued*)	*Solution:* For the effective stress wedge analysis equation and the following values: $c' = 3.4$ kPa (70 psf) $\phi' = 29°$ Length of slip surface (L) = 29 m (95 ft) Slope inclination (α) = 18° Average pore water pressure acting on the slip surface (u) = 2.4 kPa (50 psf) Total weight of wedge (W) = 750 kN/m (51,700 lb/ft) The factor of safety F of the slope = 1.95.

TABLE 13.10 Gross Slope Stability: Method of Slices

Topic (1)	Discussion (2)
General discussion	The most commonly used method of gross slope stability analysis is the "method of slices," where the failure mass is subdivided into vertical slices and the factor of safety is calculated based on force equilibrium equations. A circular arc slip surface and rotational type of failure mode is often used for the method of slices, and for homogeneous soil, a circular arc slip surface provides a lower factor of safety than assuming a planar slip surface.
Example of analysis (Fig. 13.8)	Figure 13.8 shows an example of a slope stability analysis using a circular arc slip surface. The failure mass has then been divided into 30 vertical slices. The calculations are similar to the wedge-type analysis, except that the resisting and driving forces are calculated for each slice and then summed up in order to obtain the factor of safety of the slope. For the "Ordinary Method of Slices" (also known as the "Swedish Circle Method" or "Fellenius Method," Fellenius 1936), the equations used to calculate the factor of safety are identical to those presented in Table 13.9, with the resisting and driving forces calculated for each slice and then summed up in order to obtain the factor of safety.
Commonly used methods of slices	Commonly used methods of slices to obtain the factor of safety are listed in Table 13.11. The method of slices is not an exact method because there are more unknowns than equilibrium equations. This requires that an assumption be made concerning the interslice forces. Table 13.11 presents a summary of the assumptions for the various methods. For example, Figure 13.9 shows that for the Ordinary Method of Slices (Fellenius 1936), it is assumed that the resultant of the interslice forces are parallel to the average inclination of the slice (α). It has been determined that because of this interslice assumption for the Ordinary Method of Slices, this method provides a factor of safety that is too low for some situations (Whitman and Bailey 1967). As a result, the other methods listed in Table 13.11 are used more often than the Ordinary Method of Slices.
Computer programs	Because of the tedious nature of the calculations, computer programs are routinely used to perform the analysis. Duncan (1996) states that the nearly universal availability of computers and much improved understanding of the mechanics of slope stability analyses have brought about considerable change in the computational aspects of slope stability analysis. Analyses can be done much more thoroughly, and, from the point of view of mechanics, more accurately than was possible previously. However, problems can develop because of a lack of understanding of soil mechanics, soil strength, and the computer programs themselves, as well as the inability to analyze the results in order to avoid mistakes and misuse (Duncan 1996).
Total stress analysis	The slope stability analysis shown in Fig. 13.8 was performed by using the SLOPE/W (Geo-Slope 1991) computer program. The following data was inputted into the program: 1. Slope Cross Section: The slope has a height of 9.1 m (30 ft) and a 2:1 (horizontal:vertical) slope inclination.

TABLE 13.10 Gross Slope Stability: Method of Slices (*Continued*)

Topic (1)	Discussion (2)
Total stress analysis (*Continued*)	2. Type of Slope Stability Analysis: A total stress slope stability analysis was selected. 3. Shear Strength Parameters: The shear strength parameters from Fig. 7.7 were used in the slope stability analysis. 4. Total Unit Weight of Soil: The total unit weight (γ_t) = 18.1 kN/m^3 (115 pcf). 5. Critical Slip Surface: For this analysis, the computer program was requested to perform a trial and error search for the critical slip surface (i.e., the slip surface having the lowest factor of safety). Note the grid of points that has been produced above the slope. Each one of these points represents the center of rotation of a circular arc slip surface passing through the toe of the slope. The computer program has actually performed 121 slope stability analyses. In Fig. 13.8, the dot with the number 1.154 indicates the center of rotation of the circular arc slip surface with the lowest factor of safety (i.e., lowest factor of safety = 1.15). In Table 13.9 (total stress analysis), the factor of safety of this slope was determined based on the assumption of a planar failure surface through the toe. That analysis indicated a factor of safety of 1.80. The analysis with a circular arc slip surface (Fig. 13.8) indicates a factor of safety of 1.15. This demonstrates that for a homogeneous slope, a slope stability analysis using circular arc slip surfaces will generate a lower factor of safety than using planar slip surfaces.
Effective stress analysis	The slope stability program can also perform effective stress slope stability analyses. For this type of analysis, the effective shear strength parameters (c' and ϕ') are inputted into the computer program. The pore water pressures must also be inputted into the computer program. In most cases, the pore water pressures have a significant impact on slope stability and they are often very difficult to estimate. There are several different options that can be used concerning the pore water pressures, as follows 1. Zero Pore Water Pressure: A common assumption for slopes that have or will be constructed with drainage devices, such as the drainage systems shown in Standard Detail No. 4 (App. B), is to use a pore water pressure equal to zero. 2. Groundwater Table: A second option is to specify a groundwater table. For the soil above the groundwater table, it is common to assume zero pore water pressures. If the groundwater table is horizontal, then the pore water pressures below the groundwater table are typically assumed to be hydrostatic. For the condition of seepage through the slope (i.e., a sloping groundwater table), the computer program can develop a flow net in order to estimate the pore water pressures below the groundwater table (see Table 8.4).

TABLE 13.10 Gross Slope Stability: Method of Slices (*Continued*)

Topic (1)	Discussion (2)
Effective stress analysis (*Continued*)	3. Pore Water Pressure Ratio (r_u): The pore water pressure ratio is defined as: $$r_u = \frac{u}{\gamma_t h}$$ where u = pore water pressure, γ_t = total unit weight of the soil, and h = depth below the ground surface. If a value of $r_u = 0$ is selected, then the pore water pressures u are assumed to be equal to zero in the slope. Suppose an r_u value is used for the entire slope. In many cases the total unit weight is about equal to two times the unit weight of water (i.e., $\gamma_t = 2\gamma_w$), and thus a value of $r_u = 0.25$ is similar to the effect of a groundwater table at midheight of the slope. A value of $r_u = 0.5$ would be similar to the effect of a groundwater table corresponding to the ground surface. The pore water pressure ratio r_u can be used for existing slopes where the pore water pressures have been measured in the field, or for the design of proposed slopes where it is desirable to obtain a quick estimate of the effect of pore water pressures on the stability of the slope.

TABLE 13.11 Gross Slope Stability: Assumptions Concerning Interslice Forces for Different Method of Slices

Type of method of slices (1)	Assumption concerning interslice forces (2)	Reference (3)
Ordinary method of slices	Resultant of the interslice forces is parallel to the average inclination of the slice	Fellenius (1936)
Bishop simplified method	Resultant of the interslice forces is horizontal (no interslice shear forces)	Bishop (1955)
Janbu simplified method	Resultant of the interslice forces is horizontal (a correction factor is used to account for interslice shear forces)	Janbu (1968)
Janbu generalized method	Location of the interslice normal force is defined by an assumed line of thrust	Janbu (1957)
Spencer method	Resultant of the interslice forces is of constant slope throughout the sliding mass	Spencer (1967, 1968)
Morgenstern-Price method	Direction of the resultant of interslice forces is determined by using a selected function	Morgenstern and Price (1965)

Sources: Lambe and Whitman (1969); Geo-Slope (1991).

TABLE 13.12 Gross Slope Stability: Important Considerations

Topic (1)	Discussion (2)
General discussion	The object of the slope stability analysis is to accurately model the existing or design conditions of the slope. Some of the important factors that may need to be considered in a slope stability analysis are discussed in this table.
Different soil layers	If a proposed slope or existing slope contains layers of different soil types with different engineering properties, then these layers must be input into the slope stability computer program. Most slope stability computer programs have the capability to handle multiple layers.
Slip surfaces	In some cases, a composite-type slip surface may need to be included in the analysis. This option is discussed in Sec. 13.6, "Landslide Stability Analysis."
Tension cracks	It has been stated that tension cracks at the top of the slope can reduce the factor of safety of a slope by as much as 20% and are usually regarded as an early and important warning sign of impending failure in cohesive soil (Cernica 1995a). Slope stability programs often have the capability to model or input tension crack zones. The destabilizing effects of water in tension cracks or even the expansive forces caused by freezing water can also be modeled by some slope stability computer programs.
Surcharge loads	There may be surcharge loads (such as a building load) at the top of the slope or even on the slope face. Most slope stability computer programs have the capability of including surcharge loads. In some computer programs, other types of loads, such as those from tieback anchors, can also be included in the analysis.
Nonlinear shear strength envelope	In some cases, the shear strength envelope is nonlinear (for example, see Fig. 7.11). If the shear strength envelope is nonlinear, then a slope stability computer program that has the capability of using a nonlinear shear strength envelope in the analysis should be used.
Plane strain condition	Like strip footings, long uniform slopes will be in a plane strain condition. The friction angle ϕ is about 10% higher in the plane strain condition compared to the friction angle ϕ measured in the triaxial apparatus (Meyerhof 1961, Perloff and Baron 1976). Since plane strain shear strength tests are not performed in practice, there will be an additional factor of safety associated with the plane strain condition. For uniform fill slopes that have a low factor of safety, it is often observed that the "end" slopes (slopes that make a 90° turn) are the first to show indications of slope movement. This is because the end slope is not subjected to a plane strain condition and the shear strength is actually lower than in the center of a long, continuous slope.
Progressive failure	For the method of slices, the factor of safety is an average value of all the slices. Some slices, such as at the toe of the slope, may have a lower factor of safety, which is balanced by other slices that have a higher factor of safety. For those slices that have a low factor of safety, the shear stress and strain may exceed the peak shear strength. For some soils, such as stiff-fissured clays, there may be a significant drop in shear strength as the soil deforms beyond the peak values. This reduction in shear strength will then transfer the

TABLE 13.12 Gross Slope Stability: Important Considerations (*Continued*)

Topic (1)	Discussion (2)
Progressive failure (*Continued*)	load to an adjacent slice, which will cause it to experience the same condition. Thus the movement and reduction of shear strength will progress along the slip surface, eventually leading to failure of the slope. The progressive nature of the failure may even reduce the shear strength of the soil to its residual value ϕ'_r.
Other conditions	Slope stability analysis can also be adapted to perform earthquake analysis (i.e., pseudo-static analysis, Chap. 14). Slope stability analysis can be used for other types of engineering structures. For example, the stability of a retaining wall is often analyzed by considering a slip surface beneath the foundation of the wall.

TABLE 13.13 Gross Slope Stability: Stability Charts

Topic (1)	Discussion (2)
General discussion	Many different charts have been developed that can be used to determine the factor of safety of slopes having constant slope inclination and composed of a single soil where the soil properties are approximately constant. Stability charts are useful because they can provide a quick check on the factor of safety of a slope for different design conditions.
Taylor chart (Fig. 13.10)	The Taylor (1937) chart is presented in Fig. 13.10. This chart is used only for a total stress analysis of plastic soils. The steps in using this chart are as follows: 1. *Depth to firm base.* The parameter d is determined as the vertical distance below the toe of the slope to a firm base (D) divided by the height of the slope (H); i.e., $d = D/H$. For slope inclinations greater than 53°, the slope failure occurs through the toe of the slope and the value of d is not required. 2. *Stability number N_o.* The upper chart in Fig. 13.10 is entered with the slope inclination β and, upon intersecting the curve for a specific value of d, the stability number N_o is determined from the vertical axis. 3. *Factor of safety.* The factor of safety F is calculated from the following equation: $$F = \frac{N_o c}{\gamma_t H}$$ where N_o = stability number (step 2) γ_t = total unit weight of the soil (kN/m^3 or pcf) H = height of the slope (m or ft) c = undrained shear strength of the soil (kPa or psf) For those situations where the undrained shear strength s_u is known, then $c = s_u$ in the above equation. In those cases where $\phi = 0$, the value of cohesion is obtained from data such as shown in Fig. 7.7. Note that this chart can not be used for cases where the consolidated undrained shear strength (i.e., $\phi \neq 0$) has been obtained.
Stability charts by Cousins (Figs. 13.11 to 13.13)	Figures 13.11 through 13.13 present stability charts prepared by Cousins (1978). These three charts were developed for failure through the toe of the slope. The pore water pressure ratio r_u has been discussed in Table 13.10. Each chart has been developed for a different pore water pressure ratio r_u, as follows: • Figure 13.11: $r_u = 0$. For this chart, pore water pressure equals zero in the slope. • Figure 13.12: $r_u = 0.25$. This chart is similar to the effect of a groundwater table at midheight of the slope. • Figure 13.13: $r_u = 0.50$. This chart is similar to the effect of a groundwater table corresponding to the ground surface. Note that in each chart there are lines that are labeled with various D values. The value of D is defined as follows: $$D = \frac{\text{vertical distance between top of slope and lowest point on the slip surface}}{\text{height of the slope}}$$

TABLE 13.13 Gross Slope Stability: Stability Charts (*Continued*)

Topic (1)	Discussion (2)
Stability charts by Cousins (Figs. 13.11 to 13.13) (*Continued*)	The Cousins charts can be used only if the soil has a cohesion value. The steps to be used in the analysis are presented below: *A.* Effective stress analysis: 1. On the basis of the existing groundwater table or the anticipated groundwater level in the slope, select a value of r_u. Use Fig. 13.11, 13.12, or 13.13, depending on the value of r_u. 2. Calculate the value of $\lambda_{c\phi}$, which is defined as follows: $$\lambda_{c\phi} = \frac{\gamma_t H \tan \phi'}{c'}$$ where γ_t = total unit weight of the soil (kN/m^3 or pcf) H = height of the slope (m or ft) ϕ' = effective friction angle of the soil (degrees) c' = effective cohesion of the soil (kPa or psf) 3. Enter the chart along the horizontal axis at the value of the slope inclination α. Select the appropriate curve based on the value of $\lambda_c \phi$ and then determine the stability number N_F from the vertical axis. 4. Calculate the factor of safety F from the following equation: $$F = \frac{N_F c'}{\gamma_t H}$$ *B.* Total stress analysis: For a total stress analysis based on the undrained shear strength from consolidated undrained triaxial compression tests (i.e., ϕ and c), Fig. 13.11 would be used (r_u = 0) and the four steps outlined above would be performed using ϕ and c (in place of ϕ' and c').
Nonplastic (cohesionless) soil	For a constant slope inclination α with the slope consisting of a uniform nonplastic cohesionless soil (i.e., c' = 0) and no pore water pressures (u = 0), the equations in Table 13.9 reduce to: $$F = \frac{\tan \phi}{\tan \alpha}$$ Note that the above equation is independent of the total unit weight γ_t of the cohesionless soil and the height of the slope H. According to the above equation, if the slope inclination of a cohesionless soil is slowly increased until the slope just begins to fail, this maximum slope inclination will equal the friction angle of the cohesionless soil. This maximum slope inclination at which the soil is stable has been termed the *angle of repose.*
Example problem	*Problem:* Same problem as illustrated in Fig. 13.10, except calculate the factor of safety for failure through the toe of the slope. *Solution:* Enter chart (Fig. 13.10) at β = 35°, intersect solid line and N_o = 6.2. The factor of safety = [(6.2)(600)]/[(115)(25)] = 1.29. As a check, enter Fig. 13.11 with ϕ = 0, therefore $\lambda_{c\phi}$ = 0 and for α = 35°, N_F = 6.2 and F = [(6.2)(600)]/[(115)(25)] = 1.29.

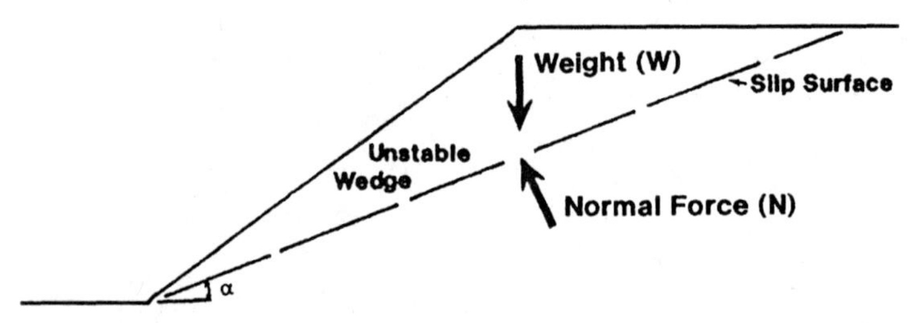

FIGURE 13.7 Wedge method. See Table 13.9.

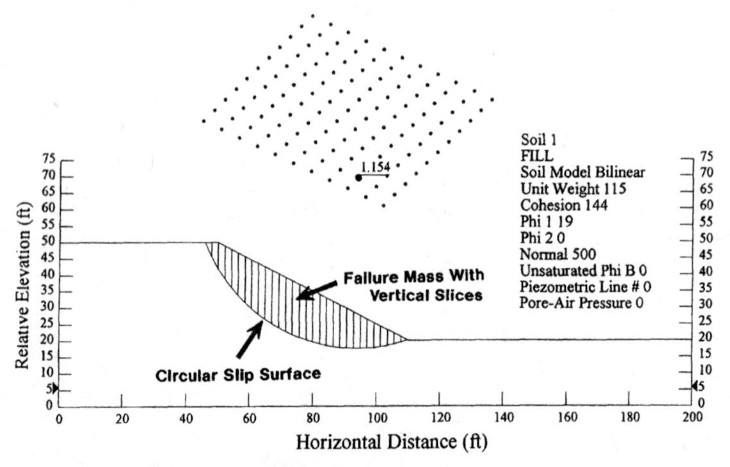

FIGURE 13.8 Example of a slope stability analysis using the SLOPE/W (Geo-Slope 1991) computer program. See Table 13.10.

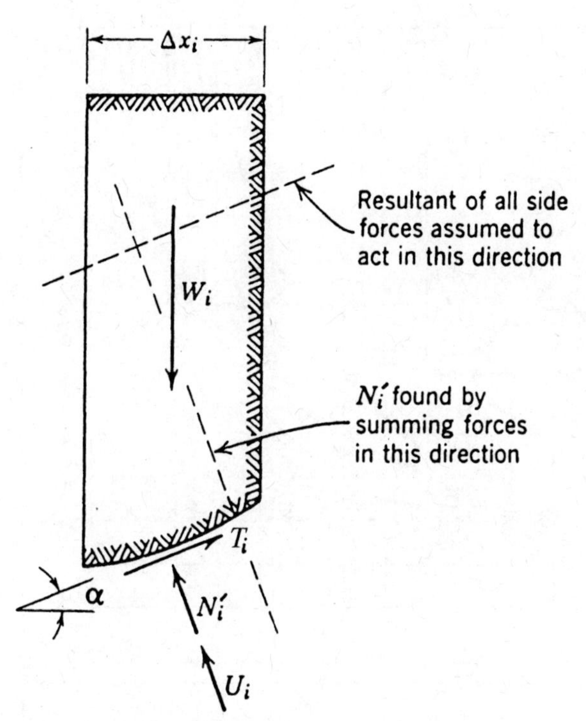

FIGURE 13.9 Forces acting on a vertical slice and the assumption concerning interslice forces for the ordinary method of slices. See Table 13.10. (*Adapted from Lambe and Whitman 1969.*)

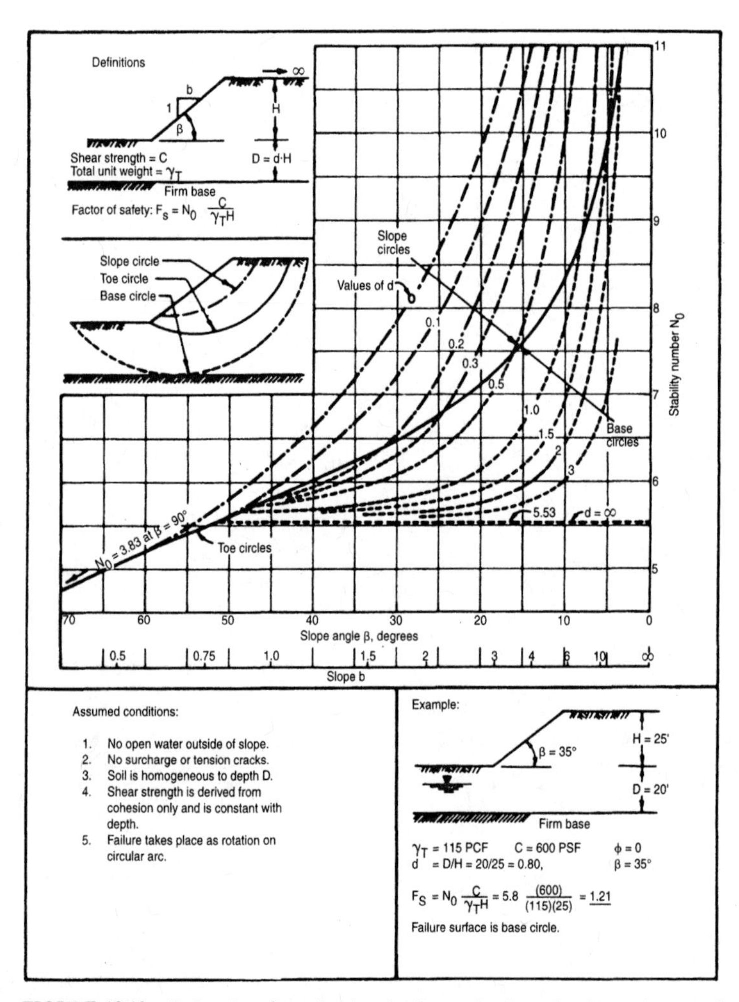

FIGURE 13.10 Taylor chart for estimating the factor of safety of a slope using a total stress analysis. See Table 13.13. (*Reproduced from NAVFAC DM-7.1, 1982.*)

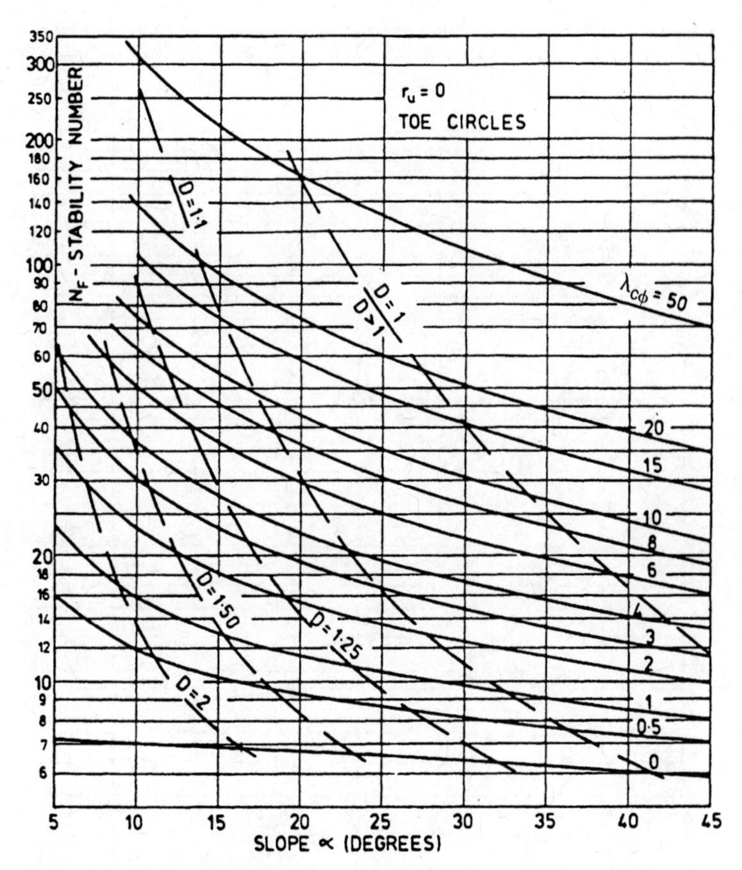

FIGURE 13.11 Cousins (1978) chart for failure analysis through the toe of the slope and zero pore water pressures in the slope ($r_u = 0$). See Table 13.13. (*Reprinted with permission of the American Society of Civil Engineers.*)

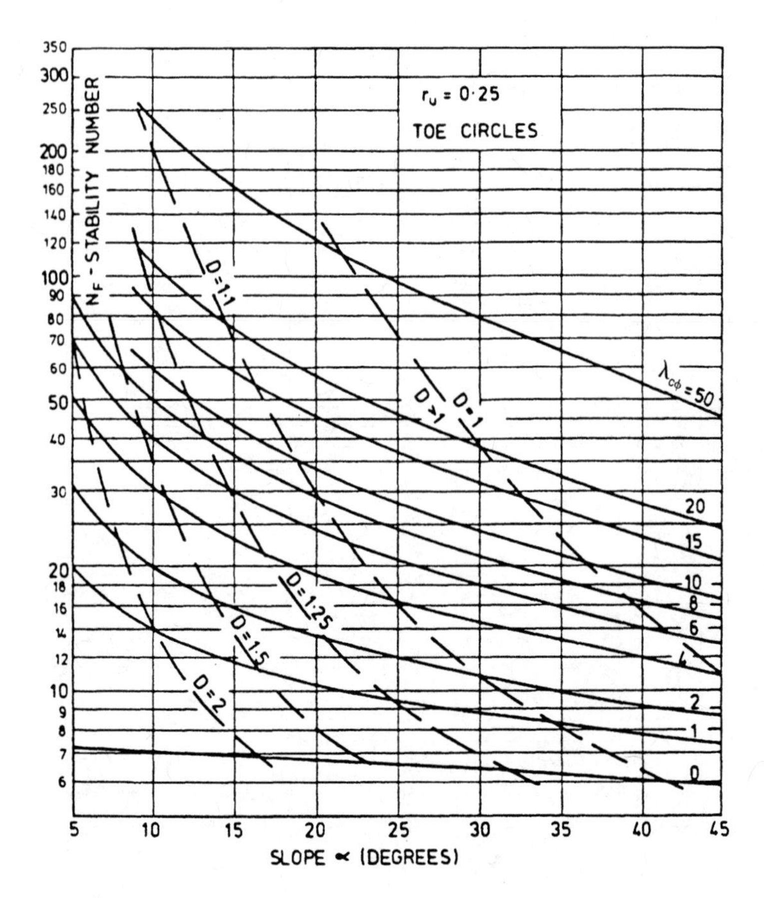

FIGURE 13.12 Cousins (1978) chart for failure analysis through the toe of the slope and a pore water pressure ratio (r_u = 0.25). See Table 13.13. (*Reprinted with permission of the American Society of Civil Engineers.*)

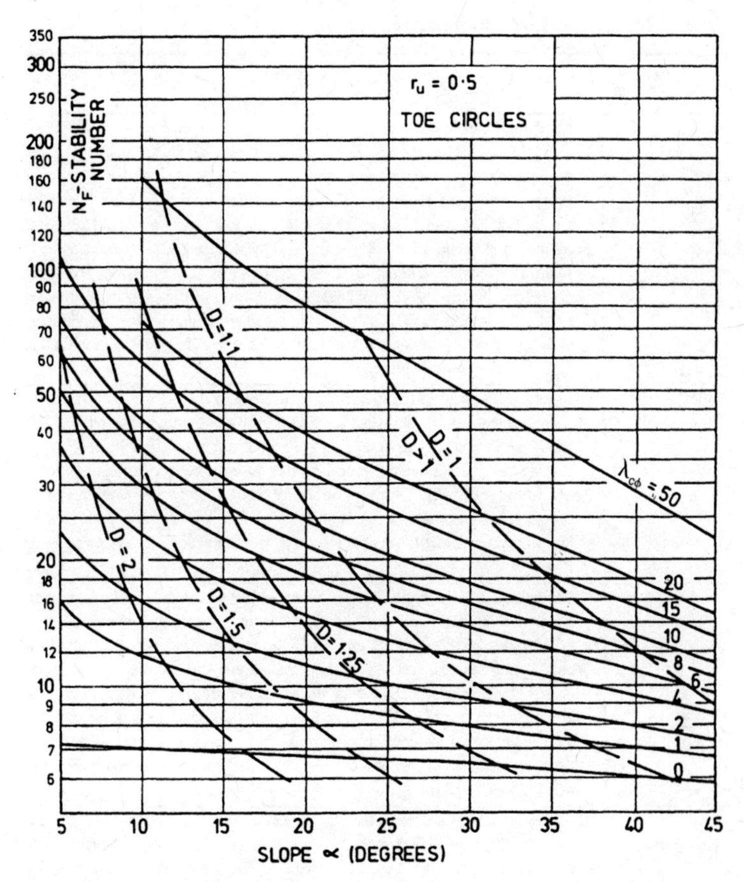

FIGURE 13.13 Cousins (1978) chart for failure analysis through the toe of the slope and a pore water pressure ratio (r_u = 0.50). See Table 13.13. (*Reprinted with permission of the American Society of Civil Engineers.*

TABLE 13.14 Landslide Classification

Topic (1)	Discussion (2)
General discussion	Landslides can be some of the most challenging projects worked on by geotechnical engineers and engineering geologists. Landslides can cause extensive damage to structures and may be very expensive to stabilize when they impact developed property. In terms of damage, the National Research Council (1985) states: "Landsliding in the United States causes at least $1 to $2 billion in economic losses and 25 to 50 deaths each year. Despite a growing geologic understanding of landslide processes and a rapidly developing engineering capability for landslide control, losses from landslides are continuing to increase. This is largely a consequence of residential and commercial development that continues to expand onto the steeply sloping terrain that is most prone to landsliding."
Landslide description	Figure 13.14 shows an example of a landslide and Table 13.15 presents common nomenclature used to describe landslide features (Varnes 1978). Landslides are described as mass movement of soil or rock that involves shear displacement along one or several rupture surfaces, which are either visible or may be reasonably inferred (Varnes 1978). As previously mentioned, it is the shear displacement along a distinct rupture surface that distinguishes landslides from other types of soil or rock movement such as falls, topples, or flows.
Landslide classification	Landslides are generally classified as either rotational or translational. Rotational landslides are due to forces that cause a turning movement about a point above the center of gravity of the failure mass, which results in a curved or circular surface of rupture. Translational landslides occur on a more or less planar or gently undulatory surface of rupture. Translational landslides are frequently controlled by weak layers, such as faults, joints, or bedding planes; examples include the variations in shear strength between layers of tilted bedded deposits or the contact between firm bedrock and weathered overlying material.
Active landslides	Active landslides are those that are either currently moving or that are only temporarily suspended, which means that they are not moving at present but have moved within the last cycle of seasons (Varnes 1978). Active landslides have fresh features, such as a main scarp, transverse ridges and cracks, and a distinct main body of movement. The fresh features of an active landslide enable the limits of movement to be easily recognized. Generally active landslides are not significantly modified by the processes of weathering or erosion.
Ancient landslides	Landslides that have long since stopped moving are typically modified by erosion and weathering, or may be covered with vegetation so that the evidence of movement is obscure. The main scarp and transverse cracks will have been eroded or filled in with debris. Such landslides are generally referred to as ancient or fossil landslides (Zaruba and Mencl 1969, Day 1995d). These landslides have commonly developed under different climatic conditions thousands or more years ago.

TABLE 13.14 Landslide Classification (*Continued*)

Topic (1)	Discussion (2)
Landslide triggering mechanisms	Many different conditions can trigger a landslide, for example, an increase in shear stress or a reduction in shear strength. The following factors contribute to an increase in shear stress: (1) removing lateral support, such as erosion of the toe of the landslide by streams or rivers; (2) applying a surcharge at the head of the landslide, such as the construction of a fill mass for a road; (3) applying a lateral pressure, such as the raising of the groundwater table; and (4) applying vibration forces, such as an earthquake or construction activities. Factors that result in a reduction in shear strength include: (1) natural weathering of soil or rock; (2) development of discontinuities, such as faults or bedding planes; and (3) an increase in moisture content or pore water pressure of the slide plane material.

TABLE 13.15 Landslide Nomenclature (See Fig. 13.14)

Terms (1)	Definitions (2)
Main scarp	A steep surface on the undisturbed ground around the periphery of the slide, caused by the movement of slide material away from the undisturbed ground. The projection of the scarp surface under the displaced material becomes the surface of rupture.
Minor scarp	A steep surface on the displaced material produced by differential movements within the sliding mass.
Head	The upper parts of the slide material along the contact between the displaced material and the main scarp.
Top	The highest point of contact between the displaced material and the main scarp.
Toe, surface of rupture	The intersection (sometimes buried) between the lower part of the surface of rupture and the original ground surface.
Toe	The margin of displaced material most distant from the main scarp.
Tip	The point on the toe most distant from the top of the slide.
Foot	That portion of the displaced material that lies downslope from the toe of the surface of rupture.
Main body	That part of the displaced material that overlies the surface of rupture between the main scarp and toe of the surface of rupture.
Flank	The side of the landslide.
Crown	The material that is still in place, practically undisplaced and adjacent to the highest parts of the main scarp.
Original ground surface	The slope that existed before the movement which is being considered took place. If this is the surface of an older landslide, that fact should be stated.
Left and right	Compass directions are preferable in describing a slide, but if right and left are used they refer to the slide as viewed from the crown.
Surface of separation	The surface separating displaced material from stable material but not known to have been a surface on which failure occurred.
Displaced material	The material that has moved away from its original position on the slope. It may be a deformed or undeformed state.
Zone of depletion	The area within which the displaced material lies below the original ground surface.
Zone of accumulation	The area within which the displaced material lies above the original ground surface.

Source: Varnes 1978.
Note: Reprinted with permission from *Landslides: Analysis and Control, Special Report 176.* Copyright 1978 by the National Academy of Sciences. Courtesy of the National Academy Press, Washington, D.C.

TABLE 13.16 Landslide Analysis

Topic (1)	Discussion (2)
Effective stress analysis	Except in cases where the landslide fails because of a sudden loading or unloading (such as during an earthquake, Chap. 14), the usual procedure is to perform an effective stress slope stability analysis.
Method of analysis	If an active or ancient landslide is discovered during the design stage of a new project, then extensive subsurface exploration would be required so that the engineering geologist could develop a cross section of the ancient landslide based on subsurface exploration. The drained residual friction angle (see Table 3.13) could be obtained from laboratory shear strength tests. The unit weight of the landslide could be estimated from laboratory testing of undisturbed specimens of the landslide materials. A major unknown in the effective stress slope stability analysis of the ancient landslide would be the groundwater table. It is not uncommon that the groundwater table will rise once the project has been completed. This is because there will be additional infiltration of water into the landslide mass from irrigation or leaky pipes. The approximate location of the future (long-term) groundwater table could be based on the local topography and presence of drainage facilities that are to be installed during development of the site. Once the cross section of the landslide, location of the slip surface, estimated location of the long-term groundwater table, residual shear strength parameters, and unit weight are determined, the factor of safety of the landslide mass could be calculated by using a slope stability computer program. The standard requirement is that a landslide must have a factor of safety of at least 1.5.
Example of landslide analysis	Figure 13.15 shows an example of an effective stress analysis for a landslide. The actual slope stability analysis was performed by using the SLOPE/W (Geo-Slope 1991) computer program. Although the method of slices was originally developed for circular slip surfaces, the analyses can be readily adapted to planar or composite slip surfaces. In order to calculate the factor of safety for the landslide based on an effective stress analysis, the following parameters were inputted into the SLOPE/W (Geo-Slope 1991) computer program: 1. *Landslide cross section.* The landslide cross section was developed by the engineering geologist. The different layers in Fig. 13.15 have been identified as ef (engineered fill), Q_{al} (alluvium), T_m (fractured formational rock), and slide plane. 2. *Failure mass.* The computer program can efficiently search for the critical failure surface that has the lowest factor of safety or the location of the slip surface can be specified. For this landslide, the location of the slip surface was determined from inclinometer monitoring. This location of the slip surface was then input into the computer program (i.e., the slip surface is "fully specified"). Note in Fig. 13.15 that, for the stability analysis of the inputted failure mass, the computer program has divided the failure mass into 64 vertical slices.

TABLE 13.16 Landslide Analysis (*Continued*)

Topic (1)	Discussion (2)
Example of landslide analysis (*Continued*)	3. *Groundwater table.* The location of the groundwater table was determined from piezometer readings and inputted into the slope stability program. The dashed line in Fig. 13.15 is the location of the groundwater table. 4. *Shear strength.* Based on the results of laboratory testing, a drained residual friction angle ϕ'_r of 12° was used for the slope stability analysis. For the other soil layers, the effective shear strength parameters c' and ϕ' were determined from laboratory shear strength tests, and the data is summarized in Fig. 13.15. 5. *Total unit weight.* By laboratory testing of soil and rock specimens, total unit weights of the various strata were determined and are listed in Fig. 13.15. Based on the input parameters, the factor of safety based on the Spencer method of slices as calculated by the SLOPE/W (Geo-Slope 1991) computer program is 0.995. This value is consistent with the actual failure (factor of safety = 1.0) of the landslide.

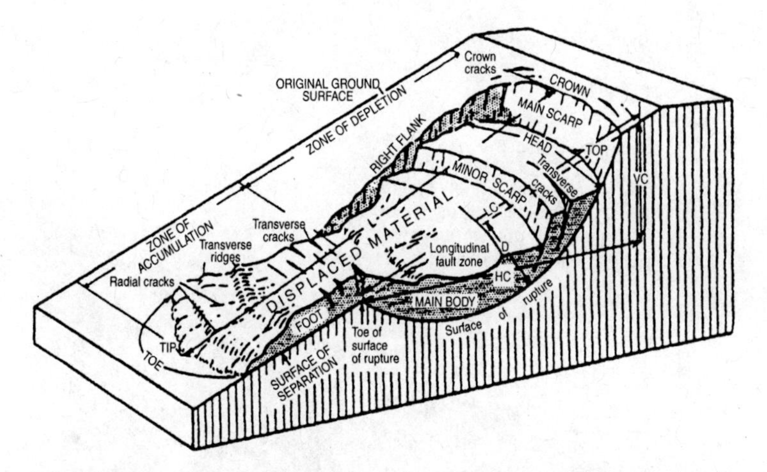

FIGURE 13.14 Landslide illustration. See Table 13.15. [*Reprinted with permission from Landslides: Analysis and Control, Special Report 176 (from Varnes 1978). Copyright 1978 by the National Academy of Sciences, courtesy of the National Academy Press, Washington, D.C.*]

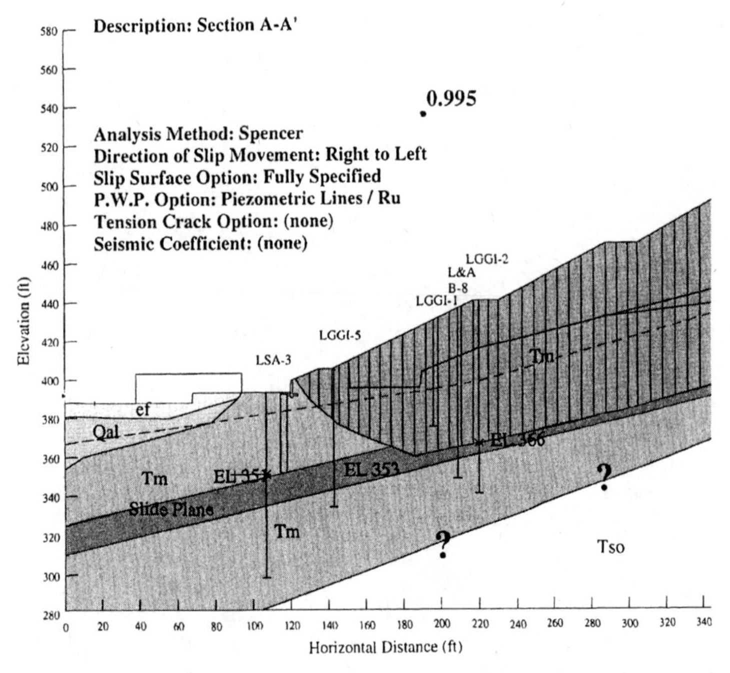

FIGURE 13.15 Slope stability analysis of a landslide. See Table 13.16.

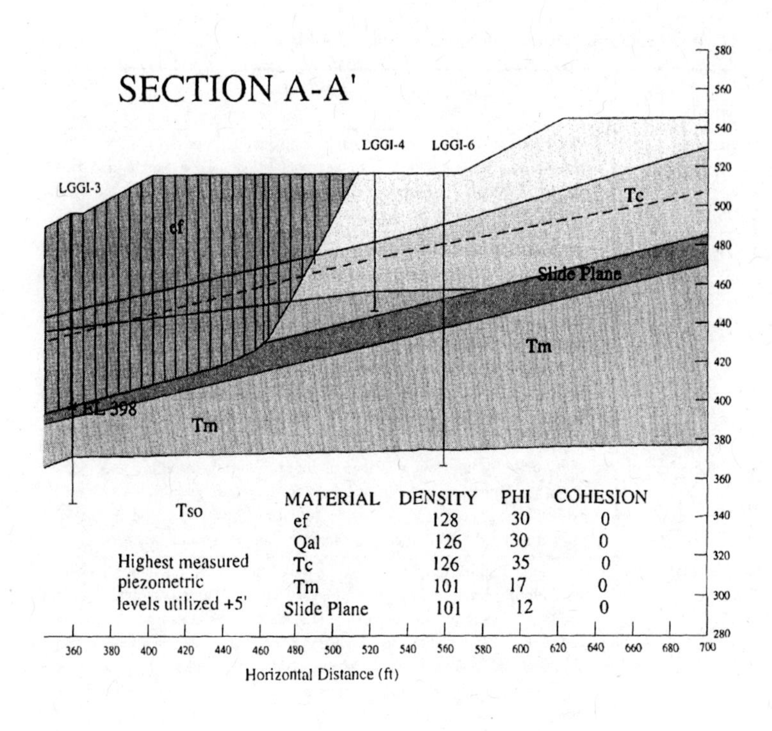
SECTION A-A'
LGGI-3
LGGI-4
LGGI-6
ef
Tc
Slide Plane
Tm
EL 398
Tm
Tso
Highest measured
piezometric
levels utilized +5'
MATERIAL DENSITY PHI COHESION
ef 128 30 0
Qal 126 30 0
Tc 126 35 0
Tm 101 17 0
Slide Plane 101 12 0
580
560
540
520
500
480
460
440
420
400
380
360
340
320
300
280
360 380 400 420 440 460 480 500 520 540 560 580 600 620 640 660 680 700
Horizontal Distance (ft)

TABLE 13.17 Landslide and Slope Stabilization Methods

Topic (1)	Discussion (2)
General discussion	If the stability analysis shows that the landslide or slope has too low of a factor of safety, there are three basic approaches that can be used to increase the factor of safety: (1) increase the resisting forces, (2) decrease the driving forces, or (3) rebuild the slope.
Increase the resisting forces	Methods to increase the resisting forces include the construction of a buttress at the toe of the slope (see Standard Detail No. 3, App. B), or the installation of piles or reinforced concrete pier walls which provide added resistance to the slope. The design and construction of reinforced pier walls to stabilize landslides will be further discussed in Chap. 16. Another technique is soil nailing, which is a practical and proven system used to stabilize slopes by reinforcing the slope with relatively short, fully bonded inclusions such as steel bars (Bruce and Jewell 1987).
Decreasing the driving forces	Methods to decrease the driving forces include the lowering of the groundwater table by improving surface drainage facilities, installing underground drains, and pumping groundwater from wells. Other methods to decrease the driving forces consist of removing soil from the head of the landslide or regrading the slope in order to decrease its height or slope inclination.
Rebuild the slope	The slope could also be rebuilt and strengthened by using geogrids or other soil reinforcement techniques (Rogers 1992). Other techniques could be employed during the grading of the site, such as the construction of a shear key. A shear key is defined as a deep and wide trench cut through the landslide mass and into intact material below the slide. The shear key is excavated from ground surface to below the basal rupture surface, and then backfilled with soil that provides a higher shear strength than the original rupture surface. By interrupting the original weak rupture surface with a higher-strength soil, the factor of safety of the landslide is increased. During construction, the shear key is also normally provided with a drainage system. It is generally recognized that during the excavation of a shear key, there is a risk of the failure of the landslide. To reduce this risk, a shear key is usually constructed in several sections with only a portion of the slide plane exposed at any given time. Figure 13.16 shows a cross section through an ancient landslide where a shear key was installed during grading of the site. At this project, the shear key was not successful in stabilizing the landslide and a portion of the landslide became reactivated (Day and Thoeny 1998).
Other measures	There are many other methods that can be used to increase the factor of safety of a slope, such as grouting, freezing, electroosmosis, and vacuum pumping.

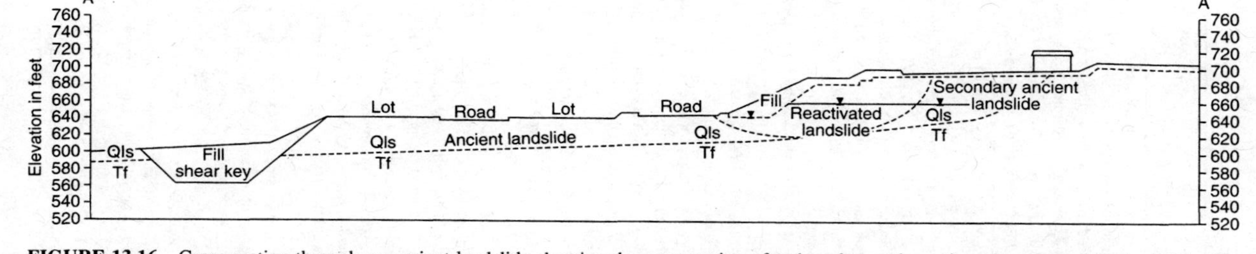

FIGURE 13.16 Cross section through an ancient landslide showing the construction of a shear key and reactivated portion of the landslide. See Table 13.17.

TABLE 13.18 Debris Flow

Topic (1)	Discussion (2)
General discussion	Debris flows cause a tremendous amount of damage and loss of life throughout the world. An example is the loss of 6000 lives from the devastating flows that occurred in Leyte, Philippines, on November 5, 1991, because of deforestation and torrential rains from tropical storm Thelma. As a result of continued population growth, deforestation, and poor land development practices, it is expected that debris flows will increase in frequency and devastation.
Definition of debris flow	A debris flow is commonly defined as soil with entrained water and air that moves readily as a fluid on low slopes. In many cases there is an initial surficial slope failure that transforms itself into a debris flow (Ellen and Fleming 1987, Anderson and Sitar 1995, 1996). Debris flow can include a wide variety of soil-particle sizes (including boulders) as well as logs, branches, tires, and automobiles. Other terms, such as *mud flow, debris slide, mud slide,* and *earth flow,* have been used to identify similar processes. While categorizing flows according to rate of movement or the percentage of clay particles may be important, the mechanisms of all these flows are essentially the same (Johnson and Rodine 1984).
Three segments of a debris flow	There are generally three segments of a debris flow: the source area, main track, and depositional area (Baldwin et al. 1987). The source area is the region where a soil mass becomes detached and transforms itself into a debris flow. The main track is the path over which the debris flow descends the slope and increases in velocity, depending on the slope steepness, obstructions, channel configuration, and the viscosity of the flowing mass. When the debris flow encounters a marked decrease in slope gradient and deposition begins, this is called the *depositional area.*
Prediction of a debris flow	It can be very difficult to predict the potential for a debris flow at a particular site. The historical method is one means of predicting debris-flow activity in a particular area. For example, as Johnson and Rodine (1984) indicated, many alluvial fans in southern California contain previous debris-flow deposits, which in the future will likely again experience debris flow. However, using the historical method for predicting debris flow is not always reliable. For example, the residences of Los Altos Hills experienced an unexpected debris flow mobilization from a road fill after several days of intense rainfall (Johnson and Hampton 1969). Using the historical method to predict debris flow is not always reliable, because the area can be changed, especially by society's activities. Johnson and Rodine (1984) stated that a single parameter should not be used to predict either the potential or actual initiation of a debris flow. Several parameters appear to be of prime importance. Two such parameters, which numerous investigators have studied, are rainfall amount and rainfall intensity. For example, Neary and Swift (1987) stated that hourly rainfall intensity of 90 to 100 mm/h (3.5 to 4 in./h) was the key to triggering debris flows in the southern Appalachians. Other important factors include the type and thickness of soil in the source area, the steepness and length of the slope in the source area, the destruction of vegetation by fire or logging, and other society-induced factors such as the cutting of roads.

TABLE 13.18 Debris Flow *(Continued)*

Topic (1)	Discussion (2)
Mitigation measures	The engineering geologist is usually the best individual to investigate the possibility of a debris flow impacting the site. On the basis of this analysis, measures can be taken to prevent a debris flow from damaging the site. For example, grading could be performed to create a raised building pad so that the structure is elevated above the main debris flow path. Other measures include the construction of retention basins, deflection walls, or channels to control or direct the debris flow away from the structures.

TABLE 13.19 Slope Softening and Creep

Topic (1)	Discussion (2)
General discussion	This table describes the process of slope softening and creep. Both of these processes cause lateral movement of the slopes. As a practical matter, slope softening and creep need only be considered for plastic (cohesive) soil. Slopes composed of cohesionless soil, such as sands and gravels (Table 4.5), usually do not need to be evaluated for slope softening or creep.
Slope softening	In many urban areas, there is a tendency toward small lot sizes because of the high cost of the land, with most of the lot occupied by building structures. A common soil engineering recommendation is to have the bottom edge of the footing at least 1.5 m (5 ft) horizontally from the face of any slope, irrespective of the slope height or soil type. This recommendation results in many buildings being constructed near the top of fill slopes. Fill in slope areas is generally placed and compacted at near-optimum moisture content, which is often well below saturation. After construction of the slope, additional moisture is introduced into the fill by irrigation, rainfall, groundwater sources, and leaking water pipes. At optimum moisture content, a compacted clay fill can have a high shear strength because of negative pore water pressures. As water infiltrates the clay, the slope softens as the pore spaces fill with water and the pore water pressures tend toward zero. If a groundwater table then develops, the pore water pressures will become positive. The elimination of negative pore water pressure results in a decrease in effective stress and deformation of the slope in order to mobilize the needed shear stress to maintain stability. This process of moisture infiltration into a compacted clay slope which results in slope deformation has been termed *slope softening* (Day and Axten 1990). The moisture migration into a cohesive fill slope that leads to slope deformation has also been termed *lateral fill extension.* Some indications of slope softening include the rear patio pulling away from the structure, pool decking pulling away from the coping, tilting of improvements near the top of the slope, stair-step cracking in walls perpendicular to the slope, and downward deformation of that part of the building near the top of the slope. In addition to the slope movement caused by the slope softening process, there can be additional movement due to the process of creep.
Slope creep	Slope creep is generally defined as an imperceptibly slow and more or less continuous downward and outward movement of slope-forming soil or rock (Stokes and Varnes 1955). Creep can affect both the near-surface soil or deep-seated materials. The process of creep is frequently described as viscous shear that produces permanent deformations, but not failure as in landslide movement. Typically the amount of movement is governed by the following factors: shear strength of the clay, slope angle, slope height, elapsed time, moisture conditions, and thickness of the active creep zone (Lytton and Dyke 1980).

TABLE 13.19 Slope Softening and Creep (*Continued*)

Topic (1)	Discussion (2)
Slope creep (*Continued*)	The process of creep is often divided into three different stages: primary or transient, steady state or secondary, and tertiary which could lead to failure of the slope. These three stages of creep are illustrated in Fig. 13.17. Because of its relatively short duration, the primary (transient) creep is often ignored in slope deformation analysis. The secondary phase of creep often produces a relatively constant rate of strain, and depending on the slope conditions, it could continue for a considerable number of years. If the tertiary phase of creep is reached, the strain rate accelerates, and the slope could ultimately be subjected to a shear failure.
Method of analysis	The engineering geologist will often be involved with the studies of creep of natural slopes. In many cases, subsurface exploration such as test pits or trenches excavated into the slope face will reveal the depth of active creep, which can be identified by the lateral offset of rock strata or soil layers. The zone of a compacted clay slope subjected to slope softening and creep is often difficult to estimate because it depends on many different factors. In general, the higher the plasticity index of the clay, the larger the zone of slope softening and creep. Also, the lower the factor of safety of the slope face, the more active the creep zone. In some cases grading options such as facing the slope with a cohesionless soil stabilization fill (see Standard Detail No. 3, App. B) can be used to mitigate the effects of slope creep. For projects where clay slopes can not be avoided, the depth of slope softening and creep will be at least as deep as the depth of seasonal moisture changes in the clay (see Fig. 12.4). Structures should not be founded on the slope face or near the top of slope that corresponds to the zone of clay having seasonal moisture changes. Also, this zone should not be relied upon for support, such as passive pressure support for retaining wall footings (Chap. 16). The clay slope could creep at a depth that is deeper than the depth of seasonal moisture changes. One design approach is to test the clay in the laboratory and model its slope softening and creep behavior. For example, a clay specimen could be prepared at anticipated field conditions and then placed in the triaxial apparatus, sealed in a rubber membrane, and then subjected to the anticipated horizontal and vertical total stresses based on the slope configuration (stress path method). The measurement of the deformation upon saturation and deformation measurements versus time during the loading could be used to estimate the depth of the creep zone and the amount of anticipated slope movement.

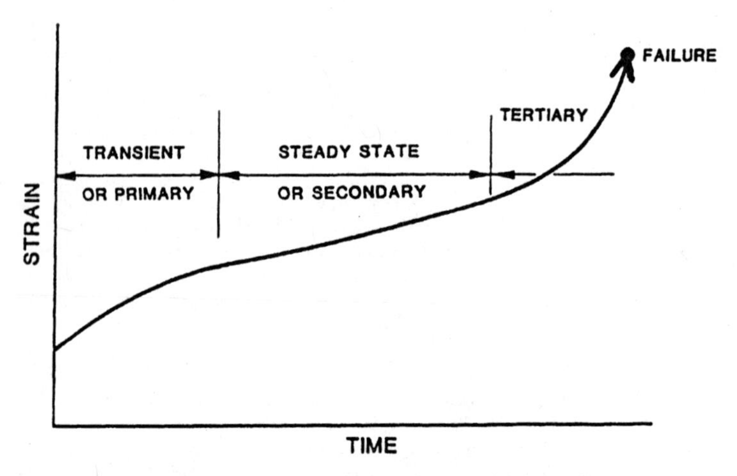

FIGURE 13.17 Three stages of creep. See Table 13.19. (*After Price 1966.*)

TABLE 13.20 Classification of Dams

Topic (1)	Discussion (2)
General discussion	A dam is often grouped into one of the three following categories: (1) large dams, (2) small dams, and (3) landslide dams.
Large dams	A dam failure has the potential to cause more damage and death than the failure of any other type of civil engineering structure. The worst type of failure is when the reservoir behind a large dam is full and the dam suddenly ruptures, which causes a massive flood wave to surge downstream. When this type of sudden dam failure occurs without warning, the toll can be especially high. For example, the sudden collapse in 1928 of the St. Francis Dam in California killed about 450 people; it was California's second most destructive disaster, exceeded in loss of life and property only by the San Francisco earthquake of 1906. As in many dam failures, most of the dam was washed away, and the exact cause of the failure is unknown. The consensus of forensic geologists and engineers is that the failure was caused by adverse geologic conditions at the site. Three distinct geologic conditions could have led to the disaster: (1) slipping of the rock beneath the easterly side of the dam along weak geologic planes; (2) slumping of rocks on the westerly side of the dam as a result of water saturation; or (3) seepage of water under pressure along a fault beneath the dam (*Committee Report for the State* 1928, Association of Engineering Geologists 1978, Schlager 1994).
Small dams	Small dams have been classified as those dams having a height of less than 12 m (40 ft) or those that impound a volume of water less than 1 million m^3 (1000 acre-ft) (Corns 1974). Sowers (1974) states that although failures of large dams are generally more spectacular, failures of small dams occur far more frequently. Reasons for the higher frequency of failures include: (1) lack of appropriate design, (2) owners believe that the consequences of failure of a small dam will be minimal, (3) small dam owners frequently have no previous experience with dams, and (4) smaller dams are often not maintained (Sowers 1974). On the basis of his experience, Griffin (1974) lists some of the common deficiencies with small dams: 1. Subsurface or geological investigations were minimal or nonexistent. 2. No provision was made for future maintenance, and maintenance was totally ignored for many years. 3. Slopes of many of the structures were constructed too steeply for routine maintenance. 4. Construction supervision varied from inadequate to nonexistent. 5. Hydrological design was deficient in that the top elevation of the dam was set an arbitrary distance above the pool with no flood routings being made. These hydrological deficiencies are now amplified by development within the watershed. 6. Inadequate purchase of land to protect the investment in the project.

TABLE 13.20 Classification of Dams (*Continued*)

Topic (1)	Discussion (2)
Landslide dams	Landslide dams usually develop when there is a blockage of a valley by landslides, debris flow, or rock fall. Concerning landslide dams, Schuster (1986) states: "Landslide dams have proved to be both interesting natural phenomena and significant hazards in many areas of the world. A few of these blockages attain heights and volumes that rival or exceed the world's largest man-made dams. Because landslide dams are natural phenomena and thus are not subject to engineering design (although engineering methods can be utilized to alter their geometries or add physical control measures), they are vulnerable to catastrophic failure by overtopping and breaching. Some of the world's largest and most catastrophic floods have occurred because of failure of these natural dams." According to Schuster and Costa (1986), most landslide dams are short-lived. In their study of 63 landslide dams, 22% failed in less than one day after formation, and half failed within 10 days. According to Schuster and Costa (1986), overtopping was by far the most frequent cause of landslide-dam failure.

TABLE 13.21 Common Causes of Earth Dam Failure

Topic (1)	Discussion (2)
Overtopping	According to Middlebrooks' (1953) comprehensive study of earth dams, the most common cause of catastrophic failure of an earth dam is overtopping, where water flows over the top of the dam. This generally happens during heavy or record-breaking rainfall, which causes so much water to enter the reservoir that the spillway can not handle the flow or the spillway becomes clogged. Once the earth dam is overtopped, the erosive action of the water can quickly cut through the shells and core of the dam.
Piping	The second most common cause of earth dam failure is piping (Middlebrooks 1953). Piping is defined as the progressive erosion of the dam at areas of concentrated leakage. As water seeps through the earth dam, seepage forces are generated that exert a viscous drag force on the soil particles. If the forces resisting erosion are less than the seepage forces, the soil particles are washed away and the process of piping commences. The forces resisting erosion include the cohesion of the soil, interlocking of individual soil particles, confining pressure from the overlying soil, and the action of any filters (Sherard et al. 1963). There can be many different reasons for the development of piping in an earth dam. Some of the more common reasons are as follows (Sherard et al. 1963): • Poor construction control, which can result in inadequately compacted or pervious layers in the embankment. • Inferior compaction adjacent to concrete outlet pipes or other structures. • Poor compaction and bond between the embankment and the foundation or abutments. • Leakage through cracks that developed when portions of the dam were subjected to tensile strains caused by differential settlement of the dam. • Cracking in outlet pipes, which is often caused by foundation settlement, spreading of the base of the dam, or deterioration of the pipe itself. • Leakage through the natural foundation soils under the dam. Leakage of the natural soils under the dam can be a result of the natural variation of the foundation material. Any seepage erupting on the downstream side of the dam is likely to cause "sand boils," which are circular mounds of soil deposited as the water exits the ground surface. Sand boils, if unobserved or unattended, can lead to complete failure by piping (Sherard et al. 1963). Clean (uncemented) sand and nonplastic silt are probably the most susceptible to piping with clay being the most resistant. There can be exceptions, such as dispersive clays, which Perry (1987) defines as follows: "Dispersive clays are a particular type of soil in which the clay fraction erodes in the presence of water by a process of deflocculation. This occurs when the interparticle forces of repulsion exceed those of attraction so that the clay particles go into suspension and, if the water is flowing such as in a crack in an earth embankment, the detached particles are carried away and piping occurs."

TABLE 13.21 Common Causes of Earth Dam Failure (*Continued*)

Topic (1)	Discussion (2)
Piping (*Continued*)	According to Sherard (1972), one of the largest known areas of dispersive clays in the United States is north-central Mississippi. The cause of failure for several earth dams has been attributed to the piping of dispersive clays (Bourdeaux and Imaizumi 1977, Stapledon and Casinader 1977, Sherard et al. 1972). The "pinhole test" is a laboratory test that can be used to classify the clay as being either highly dispersive, moderately dispersive, slightly dispersive, or nondispersive. The test method consists of evaluating the erodibility of clay soils by causing water to flow through a small hole punched through the specimen (ASTM D 4647-93, 1998).
Slope instability	Another common cause of dam failures is slope instability. There could be a gross slope failure of the upstream or downstream faces of the dam or sliding of the dam foundation. The development of these failures is similar to the development of gross slope failures and landslides discussed in Secs. 13.5 and 13.6, and slope stability analyses can be used to design the upstream and downstream faces of the dam. Design analyses can be grouped into three general categories (Sherard et al. 1963): 1. Stability during construction, usually involving a failure through the natural ground underlying the dam. A total and/or effective stress slope stability analysis could be performed, depending on the type of soil beneath the dam and the rate of construction. 2. Stability analyses of the downstream slope during reservoir operation. Usually an effective stress analysis is performed, assuming a full reservoir condition. A flow net can be used to estimate the groundwater level and pore water pressures within the earth dam. 3. Stability analyses of the upstream slope after reservoir drawdown. Like item 1 above, the type of slope stability analysis (total stress versus effective stress analysis) would depend on the soil type and assumed rate of drawdown. Besides the three most important design considerations (overtopping, piping, and slope instability), there can be many other analyses required for the design of earth dams (see Sherard et al. 1963). The checklist provided in Table 13.2 may also prove useful for the design of dams. Depending on the size and location of the dam, the design and construction may have to meet various state, local, and federal regulations.

CHAPTER 14
EARTHQUAKE ANALYSES

14.1 INTRODUCTION

This chapter presents a brief discussion of earthquakes. A structure could be damaged by several different earthquake effects. For example, a common cause of damage is fault rupture, which can cause extensive damage to civil engineering structures (see Table 14.1).

14.2 LIQUEFACTION

Another common cause of damage is liquefaction, which is defined in Table 14.2. The main factors that govern the liquefaction process are listed in Table 14.3 and the engineering analyses for liquefaction are summarized in Table 14.4 and Figs. 14.1 to 14.4.

14.3 EARTHQUAKE-INDUCED SETTLEMENT AND SLOPE MOVEMENT

In addition to damage caused by fault rupture and liquefaction, the earthquake could also cause settlement of loose sand and slope movement. Engineering analyses for earthquake-induced settlement and slope movement are summarized in Table 14.5 and Figure 14.5.

TABLE 14.1 Ground Rupture

Topic (1)	Discussion (2)
General discussion	A structure could be damaged by the seismic energy of the earthquake or actual rupture of the ground due to fault movement. Other problems caused by the ground shaking from earthquakes include the settlement of loose soil, slope movement or failure, and liquefaction of loose granular soil below the groundwater table. Earthquakes throughout the world cause a considerable amount of death and destruction. Much as diseases will attack the weak and infirm, earthquakes damage those structures that have inherent weaknesses or age-related deterioration. Those buildings that are nonreinforced, poorly constructed, weakened from age or rot, or underlain by soft or unstable soil are most susceptible to damage. Besides buildings, earthquakes can damage other structures, such as earth dams. Sherard et al. (1963) indicate that in the majority of dams shaken by severe earthquakes, two primary types of damage have developed: (1) longitudinal cracks at the top of the dam and (2) crest settlement. In the case of the Sheffield Dam, a complete failure did occur, probably by liquefaction of the very loose and saturated lower portion of the embankment (Sherard et al. 1963).
Ground rupture	Surface fault rupture caused by the earthquake is important because it has caused severe damage to buildings, bridges, dams, tunnels, canals, and underground utilities (Lawson et al. 1908, Ambraseys 1960, Duke 1960, California Department of Water Resources 1967, Bonilla 1970, Steinbrugge 1970). Fault displacement is defined as the relative movement of the two sides of a fault, measured in a specific direction (Bonilla 1970). Examples of very large surface fault rupture are the 11 m (35 ft) of vertical displacement in the Assam earthquake of 1897 (Oldham 1899) and the 9 m (29 ft) of horizontal movement during the Gobi-Altai earthquake of 1957 (Florensov and Solonenko 1965). Most structures would be unable to accommodate such huge displacement. The length of the fault rupture can also be quite significant. For example, the estimated length of surface faulting in the 1964 Alaskan earthquake varied from 600 to 720 km (Savage and Hastie 1966, Housner 1970). Besides fault rupture, there can also be ground rupture away from the main trace of the fault. These ground cracks could be caused by many different factors, such as movement of subsidiary faults, auxiliary movement that branches off from the main fault trace, or ground rupture caused by the differential or lateral movement of underlying soil deposits.
Mitigation measures	Since most structures will be unable to resist the shear movement associated with surface faulting and ground rupture, one design approach is to simply restrict construction in the active fault shear zone. The best individual to determine the location and width of the active fault shear zone is the engineering geologist. Seismic study maps, such as the *State of California Special Studies Zones Maps* (1982), which were developed as part of the Alquist-Priolo Special Studies Zones Act, delineate the approximate location of active fault zones that require special geologic studies. These maps also indicate the approximate locations of historic fault offsets, which are indicated by year of earthquake-associated event, as well as the locations of ongoing fault displacement due to fault creep. There are many other geologic references, such as the cross section shown in Fig. 2.2, that can be used to identify active shear fault zones. Trenches can be excavated across the fault zone to more accurately identify the width of the active fault shear zone. Critical structures, such as essential transportation routes (see Table 1.6) that must cross the active shear fault zones, will need special designs to resist the earthquake forces induced by ground rupture.

TABLE 14.2 Definition of Liquefaction

Topic (1)	Discussion (2)
Definition of liquefaction	The typical subsurface condition that is susceptible to liquefaction is a loose or very loose sand that has been newly deposited or placed, with a groundwater table near ground surface. During an earthquake, the ground shaking causes the loose sand to contract, resulting in an increase in pore water pressure. Because the seismic shaking occurs so quickly, the cohesionless soil is subjected to an undrained loading (total stress analysis). The increase in pore water pressure causes an upward flow of water to the ground surface, where it emerges in the form of mud spouts or sand boils. The development of high pore water pressures due to the ground shaking (i.e., the effective stress becomes zero) and the upward flow of water may turn the sand into a liquefied condition, a process that has been termed *liquefaction.* Structures on top of a loose sand deposit that has liquefied during an earthquake will sink or fall over, and buried tanks will float to the surface when the loose sand liquefies (Seed 1970).
Lateral movement of slopes	Liquefaction can also cause lateral movement of slopes and create flow slides (Ishihara 1993). Seed (1970) states: "If liquefaction occurs in or under a sloping soil mass, the entire mass will flow or translate laterally to the unsupported side in a phenomena termed a flow slide. Such slides also develop in loose, saturated, cohesionless materials during earthquakes and are reported at Chile (1960), Alaska (1964), and Niigata (1964)." There can also be liquefaction of seams of loose saturated sands within a slope. This can cause the entire slope to move laterally along the liquefied layer at the base. These types of gross slope failures caused by liquefied seams of soil caused extensive damage during the 1964 Alaskan earthquake (Shannon and Wilson, Inc. 1964, Hansen 1965). It has been observed that slope movement of this type typically results in little damage to structures located on the main intact slide mass, but buildings located in the graben area are subjected to large differential settlements and are often completely destroyed (Seed 1970).
Mitigation measures	Depending on the estimation of ground movement due to seismic shaking, there are many different options that can be used to reduce foundation displacement during the earthquake. For example, prior to construction of the project, the engineering properties of the soil or slope could be improved. Loose sands could be densified to reduce liquefaction susceptibility. The potential lateral deformation of slopes could be decreased by installing shear keys or other slope stabilization devices. For other projects, deep foundations can be installed that derive support from soil that is located below the depth of possible liquefaction. The type of foundation can be very important in the performance of the structure during the earthquake. Foundations such as mats or posttensioned slabs may enable the building to remain intact, even with substantial movements. To resist damage during the earthquake, the foundation should be monolithic with no gaps in the footings or planes of weakness due to free-floating slabs.

TABLE 14.3 Main Factors That Govern the Liquefaction Process

Topic (1)	Discussion (2)
Earthquake intensity and duration	There are many factors that govern the liquefaction process. The most important factors are the earthquake intensity and duration. It is the earthquake-induced shear strains and subsequent contraction of the soil particles that lead to the development of excess pore water pressures and ultimately liquefaction. The potential for liquefaction increases as the earthquake intensity and duration increase. Sites located near the epicenter of major earthquakes will be subjected to the largest intensity and duration of ground shaking (i.e., higher number of applications of cyclic shear strain). Besides earthquakes, other conditions can cause liquefaction, such as subsurface blasting.
Groundwater table	The condition most conducive to liquefaction is a near-surface groundwater table. Unsaturated soil located above the groundwater table will not liquefy.
Soil type	The soil types susceptible to liquefaction are nonplastic (cohesionless) soils. Seed et al. (1983) state that, on the basis of both laboratory testing and field performance, the great majority of clayey soils will not liquefy during earthquakes. An approximate listing of cohesionless soils from least to most resistant to liquefaction are clean sands, nonplastic silty sands, nonplastic silt, and gravels. There could be numerous exceptions to this sequence. For example, Ishihara (1985, 1993) describes the case of tailings derived from the mining industry that were essentially composed of ground-up rocks and were classified as rock flour. Ishihara (1985, 1993) states that the cohesionless rock flour behaved as if it were a clean sand. These tailings were shown to exhibit as low a resistance to liquefaction as clean sand.
Soil relative density D_r	Cohesionless soils in a very loose relative density state are susceptible to liquefaction while the same soil in a very dense relative density state will not liquefy. Very loose nonplastic soils will contract during the seismic shaking, which will cause the development of excess pore water pressures. Very dense soils will dilate during seismic shaking and are not susceptible to liquefaction. See Table 3.16 for the definition of relative density.
Particle size gradation	Poorly graded nonplastic soils tend to form more unstable particle arrangements and are more susceptible to liquefaction than well-graded soils.
Placement conditions	Hydraulic fills (fill placed under water) are more susceptible to liquefaction because of the loose and segregated soil structure created by the soil particles falling through water.
Drainage conditions	If the excess pore water pressure can quickly dissipate, the soil may not liquefy. Thus gravel drains or gravel layers can reduce the liquefaction potential of adjacent soil.
Confining pressures	The greater the confining pressure, the less susceptible the soil is to liquefaction. Conditions that can create a higher confining pressure are a deeper groundwater table, soil that is located at a deeper depth below ground surface, and a surcharge pressure applied at ground surface. Case studies have shown that the possible zone of liquefaction usually extends from the ground surface to a maximum depth of about 15 m (50 ft). Deeper soils generally do not liquefy because of the higher confining pressures.
Aging	Newly deposited soils tend to be more susceptible to liquefaction than old deposits of soil. Older soil deposits may already have been subjected to seismic shaking, or the soil particles may have deformed or been compressed into more stable arrangements.

TABLE 14.4 Liquefaction Analyses

Topic (1)	Discussion (2)
Method of analysis	The most common type of analysis to determine the liquefaction potential is to use the standard penetration test (SPT) or the cone penetration test (CPT) (Seed et al. 1985, Stark and Olson 1995). The analysis is based on the simplified method proposed by Seed and Idriss (1971). The method of analysis is outlined below.
Seismic shear stress ratio (SSR) caused by the earthquake	The first step in the liquefaction analysis is to determine the seismic shear stress ratio (SSR). The seismic shear stress ratio (SSR) induced by the earthquake at any point in the ground is estimated as follows (Seed and Idriss 1971): $$\mathrm{SSR} = 0.65 r_d (a_{max}/g)(\sigma_{vo}/\sigma'_{vo})$$ where SSR = seismic shear stress ratio (dimensionless parameter) a_{max} = peak acceleration measured or estimated at the ground surface of the site (m/s^2) g = acceleration of gravity (9.81 m/s^2). Usually the engineering geologist will determine the peak acceleration at the ground surface at the site from fault, seismicity, and attenuation studies. Typically the engineering geologist provides a peak ground acceleration in the form of a_{max}/g = a constant. For example, the engineering geologist may determine that the peak ground surface acceleration at a site is $a_{max}/g = 0.1$, in which case the value of 0.1 (dimensionless) is substituted into the above equation in place of a_{max}/g. σ_{vo} = total vertical stress at a particular depth where the liquefaction analysis is being performed (kPa). In order to calculate the total vertical stress, the total unit weight γ_t of the soil layers must be known. σ'_{vo} = vertical effective stress at that same depth in the soil deposit where σ_{vo} was calculated (kPa). In order to calculate the vertical effective stress, the location of the groundwater table must be known. r_d = depth reduction factor, which can be estimated in the upper 10 m of soil as (Kayen et al. 1992) $r_d = 1 - 0.012\,z$, where z = depth in meters below the ground surface where the liquefaction analysis is being performed (i.e., the same depth used to calculate σ_{vo} and σ'_{vo})
Seismic shear stress ratio (SSR) that will cause liquefaction of the soil, based on SPT tests	The second step is to determine the seismic shear stress ratio (SSR) that will cause liquefaction of the *in situ* soil. Figure 14.1 presents a chart that can be used to determine the seismic shear stress ratio (SSR) that will cause liquefaction of the *in situ* soil. In order to use this chart, the results of the standard penetration test (SPT) must be expressed in terms of the SPT $(N_1)_{60}$ value. In liquefaction analysis, the SPT N_{60} value (see Table 2.11 for equation used to calculate the SPT N_{60} value) is corrected for the overburden pressure. When a correction is applied to the SPT N_{60} value to account for the effect of overburden pressure,

TABLE 14.4 Liquefaction Analyses (*Continued*)

Topic (1)	Discussion (2)
Seismic shear stress ratio (SSR) that will cause liquefaction of the soil, based on SPT tests (*Continued*)	these values are referred to as SPT $(N_1)_{60}$ values. The procedure consists of multiplying the N_{60} value by a correction C_N in order to calculate the SPT $(N_1)_{60}$ value, or: $$(N_1)_{60} = C_N N_{60} = (100/\sigma'_{vo})^{0.5} N_{60}$$ where $(N_1)_{60}$ = standard penetration N-value corrected for both field testing procedures and overburden pressure C_N = correction factor to account for the overburden pressure. As indicated in the above equation, C_N is approximately equal to $(100/\sigma'_{vo})^{0.5}$, where σ'_{vo} is the vertical effective stress in kPa. N_6 = standard penetration N-value corrected for testing procedures. Note that N_{60} is calculated by using the equation in Table 2.11. Once the corrected SPT $(N_1)_{60}$ has been calculated, Fig. 14.1 can be used to determine the seismic shear stress ratio (SSR) that will cause liquefaction of the *in situ* soil. Note that Fig. 14.1 is for a projected earthquake of 7.5 magnitude. The figure also has different curves that are to be used, depending on the percent fines in the soil. For a given $(N_1)_{60}$ value, soils with more fines have a higher seismic shear stress ratio (SSR) that will cause liquefaction of the *in situ* soil. Figure 14.2 presents a chart for clean sands (5% or less fines) and different magnitude earthquakes. The magnitude 7.5 curve in Fig. 14.2 is similar to the magnitude 7.5 curve for 5% or less fines in Fig. 14.1.
Seismic shear stress ratio (SSR) that will cause liquefaction of the soil, based on CPT tests	As an alternative to the second step, the cone penetration test (CPT) can be used instead of the standard penetration test (SPT). The procedure consists of multiplying the cone penetration tip resistance q_c (see Table 2.13) by a correction factor C_q to account for the overburden pressure, in order to calculate the corrected CPT tip resistance q_{c1}, or: $$q_{c1} = C_q q_c = \frac{1.8\, q_c}{0.8 + (\sigma'_{vo}/100)}$$ where q_{c1} = corrected CPT tip resistance (corrected for the overburden pressure) C_q = correction factor to account for the overburden pressure. As indicated in the above equation, C_q is approximately equal to $1.8/[0.8 + (\sigma'_{vo}/100)]$, where σ'_{vo} is the vertical effective stress in kPa q_c = cone penetration tip resistance (see Table 2.13). Once the corrected CPT tip resistance q_{c1} has been calculated, Fig. 14.3 can be used to determine the seismic shear stress ratio (SSR) that will cause liquefaction of the in situ soil. Note that Fig. 14.3 is for a projected earthquake of 7.5 magnitude. The figure also has different curves that are to be used depending on the percent fines in the soil (i.e., F.C. = percent fines in the soil). For a given q_{c1} value, soils with more fines have a higher seismic shear stress ratio (SSR) that will cause liquefaction of the in situ soil. Figure 14.4 presents a chart that can be used to assess the liquefaction of clean gravels (5% or less fines) and silty gravels.

TABLE 14.4 Liquefaction Analyses (*Continued*)

Topic (1)	Discussion (2)
Compare seismic shear stress ratios	The third step in the liquefaction analysis is to compare the seismic shear stress ratio (SSR) values. If the SSR induced by the earthquake is greater than the SSR value obtained from Fig. 14.1, 14.2, 14.3, or 14.4, then liquefaction could occur during the earthquake, and vice versa.
Analysis accuracy	In the above liquefaction analysis, there are many different equations and corrections that are applied to the seismic shear stress ratio (SSR). For example, there are four different corrections (i.e., E_m, C_b, C_r, and σ'_{vo}) that are applied to the SPT N-value in order to calculate the $(N_1)_{60}$ value. All of these different equations and various corrections may provide the engineer with a sense of high accuracy, when in fact, the entire analysis is only a gross approximation. The analysis should be treated as such and engineering experience and judgment are essential in the final determination of whether or not a site has liquefaction potential.
Example problem	*Problem:* It is planned to construct a building on a cohesionless soil deposit (fines $< 5\%$). There is a nearby major active fault and the engineering geologist has determined that the peak ground acceleration $(a_{max}) = 0.4g$. Assume the site conditions consist of a level ground surface with the groundwater table located 1.5 m below ground surface and the standard penetration test was performed at a depth of 3 m. Also assume that at a depth of 3 m, $\sigma'_{vo} = 43$ kPa, $\sigma_{vo} = 58$ kPa, and $N_{60} = 5$. If the earthquake magnitude $(M) = 7.5$, will the saturated clean sand located at a depth of 3 m below ground surface liquefy during the anticipated earthquake? *Solution:* Since $z = 3$ m, $r_d = 0.96$. Use the following values: $r_d = 0.96$ $a_{max}/g = 0.4$ $(\sigma_{vo}/\sigma'_{vo}) = (58/43) = 1.35$ The first step is to insert the above values into the equation by Seed and Idriss (1971), and find the seismic shear stress ratio (SSR) caused by the earthquake = 0.34. The second step is to determine the seismic shear stress ratio (SSR) that will cause liquefaction of the *in situ* soil. Correcting the N_{60} value for the effective overburden pressure (i.e., $\sigma'_{vo} = 43$ kPa) gives a value of $(N_1)_{60} = 8$. Entering Fig. 14.1 with $(N_1)_{60} = 8$ and intersecting the curve labeled less than 5% fines, find the seismic shear stress ratio (SSR) that will cause liquefaction of the *in situ* soil at a depth of 3 m = 0.08. The third step is to compare the SSR caused by the earthquake (SSR = 0.34) with the SSR that will cause liquefaction of the *in situ* soil (SSR = 0.08). From a comparison of the SSR values, it is probable that during the earthquake the *in situ* sand located at a depth of 3 m below ground surface will liquefy.

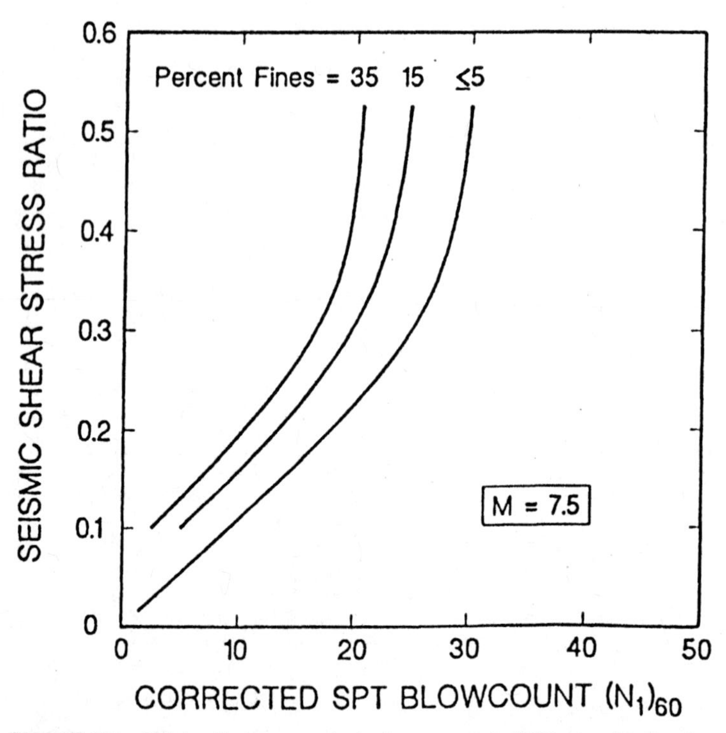

FIGURE 14.1 Relationship between seismic shear stress ratio (SSR) triggering liquefaction and $(N_1)_{60}$ values for clean and silty sands for $M = 7.5$ earthquakes. See Table 14.4. (*After Seed and DeAlba 1986, from Stark and Olson 1995; reprinted with permission of the American Society of Civil Engineers.*)

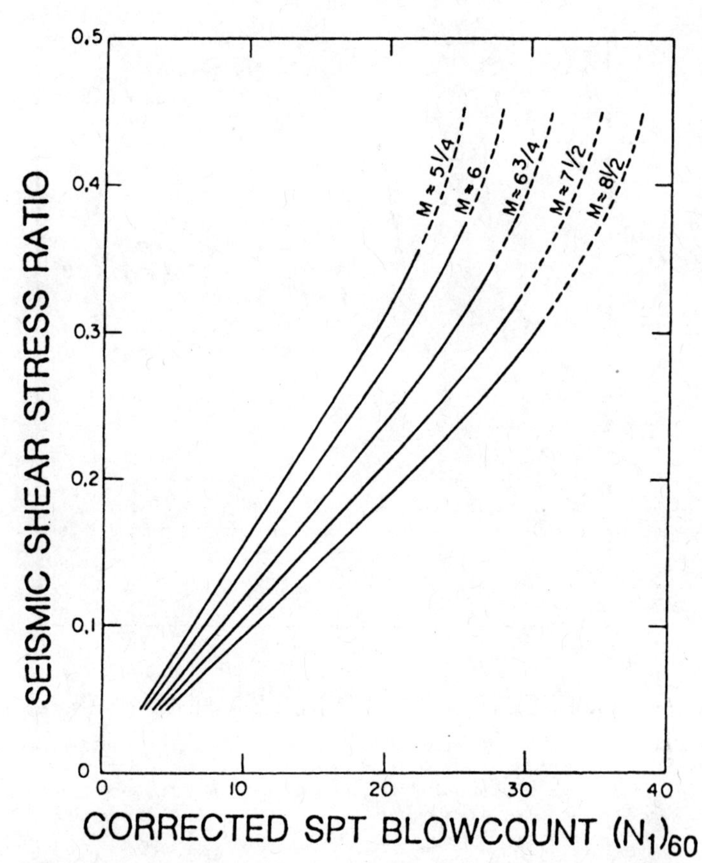

FIGURE 14.2 Relationship between seismic shear stress ratio (SSR) triggering liquefaction and $(N_1)_{60}$ values for clean sand for different magnitude earthquakes. See Table 14.4. (*After Seed et al. 1983; reprinted with permission of the American Society of Civil Engineers.*)

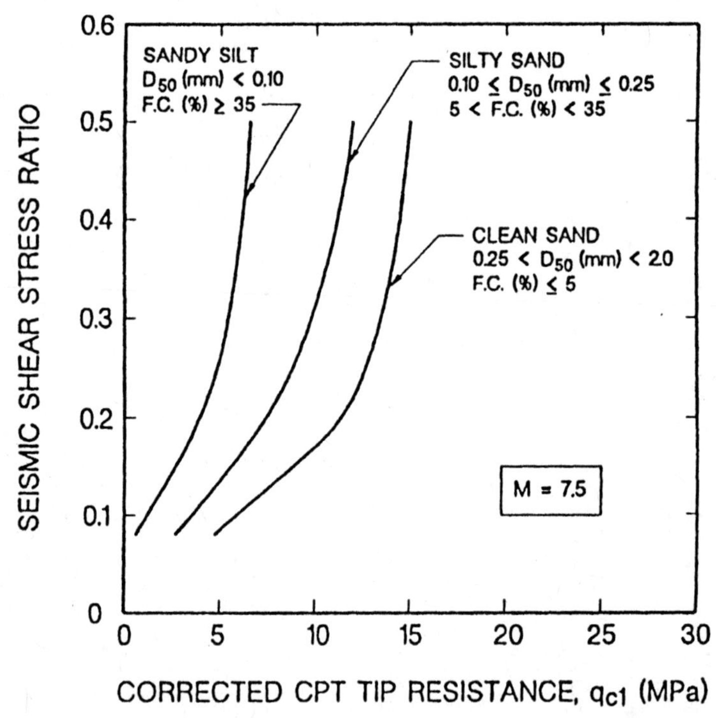

FIGURE 14.3 Relationship between seismic shear stress ratio (SSR) triggering liquefaction and corrected CPT tip resistance values for clean sand, silty sand, and sandy silt for $M = 7.5$ earthquakes. See Table 14.4. (*From Stark and Olson 1995; reprinted with permission of the American Society of Civil Engineers.*)

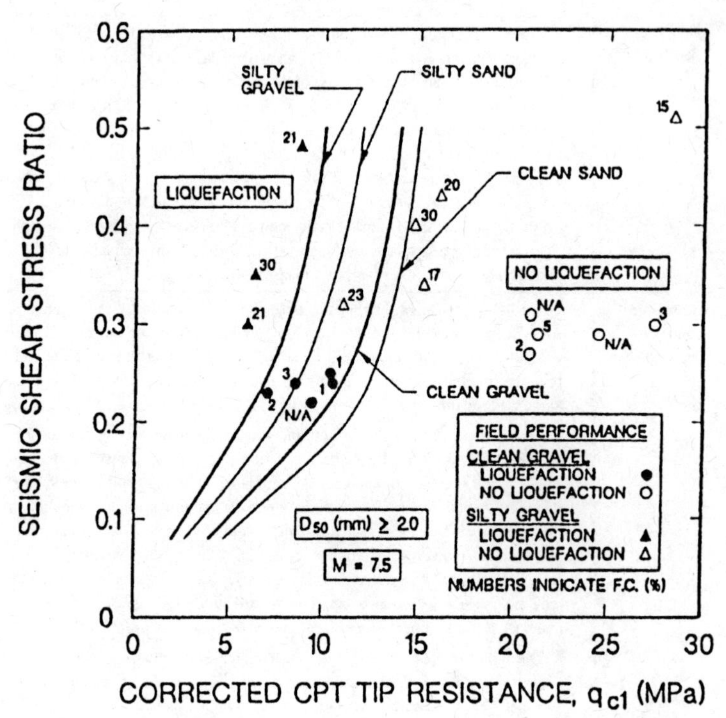

FIGURE 14.4 Relationship between seismic shear stress ratio (SSR) triggering liquefaction and corrected CPT tip resistance values for clean gravel and silty gravel for $M = 7.5$ earthquakes. See Table 14.4. (*From Stark and Olson 1995; reprinted with permission of the American Society of Civil Engineers.*)

TABLE 14.5 Earthquake-Induced Settlement and Slope Movement

Topic (1)	Discussion (2)
General discussion	Besides liquefaction of loose saturated sands, other soil conditions can result in slope movement or settlement during an earthquake. For example, Grantz et al. (1964) described an interesting case of ground vibrations from the 1964 Alaskan earthquake that caused 0.8 m (2.6 ft) of alluvium settlement. Other loose soils, such as cohesionless sand and gravel, will also be susceptible to settlement due to the ground vibrations from earthquakes. Slopes having a low factor of safety can experience large horizontal movement during an earthquake. Types of slopes most susceptible to movement during earthquakes include those slopes composed of soil that lose shear strength with strain (such as sensitive soil) or ancient landslides that can become reactivated by seismic forces (Day and Poland 1996).
Estimating seismic-induced ground movement	Often the geotechnical engineer will be required to estimate the amount of foundation displacement caused by earthquake-induced soil movement. For example, the *Uniform Building Code* (1997), which is the building code required for construction in California, states (code provision submitted by the author, adopted in May 1994): "The potential for soil liquefaction and soil strength loss during earthquakes shall be evaluated during the geotechnical investigation. The geotechnical report shall assess potential consequences of any liquefaction and soil strength loss, including estimation of differential settlement, lateral movement or reduction in foundation soil-bearing capacity, and discuss mitigating measures. Such measures shall be given consideration in the design of the building and may include, but are not limited to, ground stabilization, selection of appropriate foundation type and depths, selection of appropriate structural systems to accommodate anticipated displacement or any combination of these measures." The intent of this building code requirement is to obtain an approximate estimate of the foundation displacement caused by the earthquake-induced soil movement. In terms of accuracy of the calculations used to determine the earthquake-induced soil movement, Tokimatsu and Seed (1984) conclude: "It should be recognized that, even under static loading conditions, the error associated with the estimation of settlement is on the order of ±25 to 50%. It is therefore reasonable to expect less accuracy in predicting settlements for the more complicated conditions associated with earthquake loading....In the application of the methods, it is essential to check that the final results are reasonable in light of available experience."
Earthquake-induced settlement	Estimating the settlement of a structure caused by an earthquake is usually very difficult. For example, if the upper layer of a sand deposit supports a heavy structure and this sand layer liquefies during an earthquake, the building will literally sink into the liquefied soil. In this instance, there would be a large amount of foundation settlement and extensive damage to the building. Figure 14.5 shows one method that can be used to estimate ground surface settlement of saturated clean sands. As in Figs. 14.1 and 14.2, the SPT $(N_1)_{60}$ value is first calculated. Also, the seismic shear stress ratio (SSR) induced by the earthquake is calculated (i.e., use the Seed and Idriss 1971 equation; see Table 14.4). By entering the chart with the SPT $(N_1)_{60}$ value and the seismic shear stress ratio (SSR) induced by the earthquake, the volumetric strain (%) can be estimated. Note that Fig. 14.5 can be used only to estimate the volumetric strain (%) of the soil and it does not include the amount of settlement of the building due to its weight, which can cause the building to sink into the soil.

TABLE 14.5 Earthquake-Induced Settlement and Slope Movement (*Continued*)

Topic (1)	Discussion (2)
Earthquake-induced settlement (example)	*Example*: In the example problem in Table 14.4, the value of SPT $(N_1)_{60} = 8$ and the earthquake-induced seismic shear stress ratio (SSR) = 0.34. Entering Fig. 14.5 with these two values, find that the volumetric strain (%) is about 2.8%. If the *in situ* sand layer is 1 m thick, then the estimated settlement of this 1-m-thick layer during this earthquake would be 2.8 cm (i.e., 1 m × 0.028 = 2.8 cm). Tokimatsu and Seed (1984) also present a simplified method of determining the settlement of *in situ* dry loose sands during seismic shaking.
Pseudo-static approach	The vast majority of foundation and earthwork designs are based on the pseudo-static approach (Coduto 1994). This method ignores the cyclic nature of earthquakes and treats them as if they apply an additional static force upon the slope, retaining wall, or foundation element. For example, as will be discussed in Chap. 16, a common approach for the design of retaining walls is to subject the retaining wall to an additional static force P_E that represents the seismic energy of the earthquake.
Earthquake slope stability analysis (pseudo-static approach)	For slope stability analysis, the pseudo-static approach is to apply a lateral force acting through the center of the sliding mass, where the lateral force = $(W)(a_{max}/g)$ and W = weight of the failure wedge or sliding mass of soil. Many local jurisdictions require specific values of a_{max}/g that must be used in the pseudo-static analysis (typically a_{max}/g = 0.1 to 0.2 in southern California). Most slope stability computer programs (Sec. 13.5) have the ability to perform pseudo-static slope stability analyses, and the only additional data that needs to be inputted is the value of a_{max}/g. In southern California, an acceptable minimum factor of safety of the slope is 1.1 for a pseudo-static slope stability analysis. The major unknown in the pseudo-static method is the shear strength of the soil. In most cases, the dynamic shear resistance of the soil is assumed to be equal to the static shear strength (i.e., the shear strength of the soil that exists prior to the earthquake). This pseudo-static approach can be substantially in error for the following conditions (Coduto 1994): 1. *Sensitive clay*. These soils will lose a portion of their shear strength during cyclic loading. The higher the sensitivity (see Table 4.9), the greater the loss of shear strength for a given shear strain. 2. *Saturated loose to medium sands*. Even if these cohesionless soils do not liquefy, they can still develop large excess pore water pressures during the earthquake and lose some or a significant portion of their shear strength. 3. *Immediate settlement*. There can be significant settlement for foundations on soft saturated clays because of undrained plastic flow when the foundations are overloaded during the seismic shaking. 4. *Uplift loads*. The foundation may be subjected to uplift loads during an earthquake. The foundation is often more susceptible to failure during an uplift load, as was demonstrated by uplift failures of deep foundations in Mexico City during the 1985 earthquake. Coduto (1994) suggests that in most situations, it is probably best to use dynamic shear strengths equal to or less than the static shear strengths. Lower shear strengths would be used for sensitive clays or soils likely to experience a reduction in shear strength, such as loose sands that may not liquefy, but will lose shear strength during an earthquake.

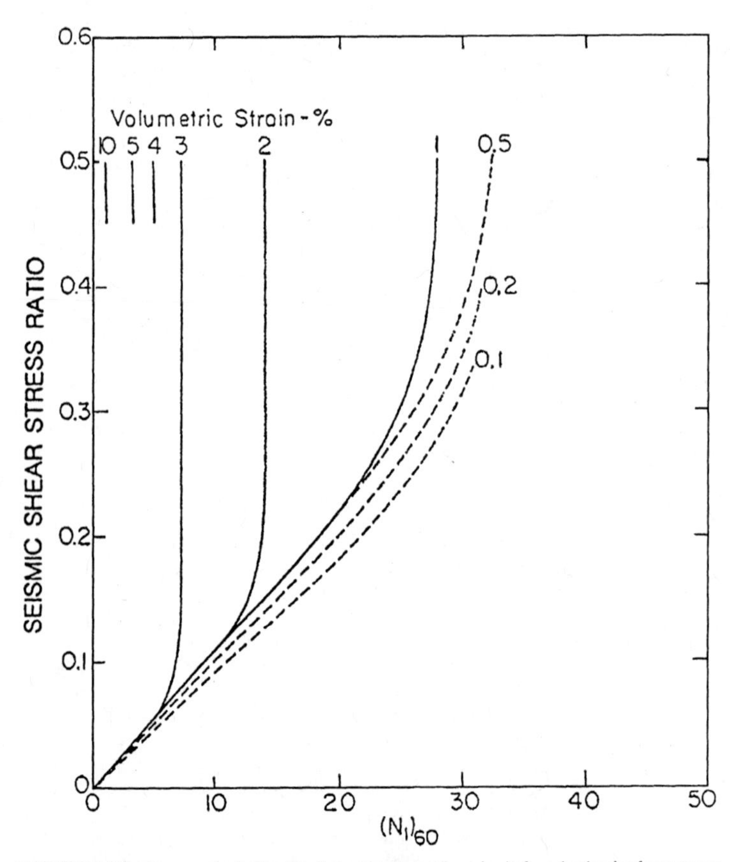

FIGURE 14.5 Proposed relationship between the earthquake-induced seismic shear stress ratio $(N_1)_{60}$ and volumetric strain for saturated clean sand. See Table 14.5. (*From Tokimatsu and Seed 1984.*)

CHAPTER 15
EROSION ANALYSES

15.1 INTRODUCTION

Underground erosion, such as the piping of soil from or underneath an earth dam, is discussed in Sec. 13.9. The purpose of this chapter is to focus solely on ground surface erosion. Table 15.1 presents factors that affect erosion, Table 15.2 lists the five levels of erosion, and Table 15.3 presents examples of landforms that are especially susceptible to erosion.

15.2 UNIVERSAL SOIL LOSS EQUATION

The purpose of the universal soil loss equation is to determine the soil loss caused by surface erosion (see Tables 15.4 to 15.7 and Fig. 15.1).

15.3 DESIGN AND CONSTRUCTION

The basic principles of erosion control are discussed in this section, including methods that can be used to reduce erosion during construction, and the design and construction of systems that can be used to trap sediment on site (Table 15.8 and Figs. 15.2 to 15.4).

TABLE 15.1 Factors That Affect Erosion

Topic (1)	Discussion (2)
Process of erosion	The process of slope erosion can begin as slopewash. The term *slopewash* refers to one form of erosion in which water moves as a thin and relatively uniform film (Rice 1988). With time, the flow of water may concentrate into a series of slightly deeper pathways (called *rills*) and this is known as *rillwash.* With continued erosion, gullies could emerge and ultimately there could be the formation of channeled stream flow through the slope.
Factors that affect erosion	The factors that affect sheet and rill erosion have been discussed by Smith and Wischmeier (1957). They indicate that there are two principal processes of sheet erosion: raindrop impact and transportation of soil particles by flowing water. According to Smith and Wischmeier (1957), the amount of soil loss is governed by six factors: length of slope, slope gradient, ground cover, soil type, management, and rainfall. Soil type would include such factors as type and degree of cementation of particles.
Evaluation of erosion potential	Based on the results of subsurface exploration and laboratory testing, the geotechnical engineer is often the best individual to evaluate the erosion potential of the on-site soils. For example, local building departments often require that the geotechnical engineer provide recommendations on "special planting and irrigation measures, slope coverings and other erosion control measures which may be apparent from the preparation of the geotechnical report" (*Orange County Grading Manual* 1993).

TABLE 15.2 Level of Erosion

Level of erosion (1)	Classification (2)	Description (3)
1	Very slight	Minor erosion; at the base of the slope, minor accumulation of debris
2	Slight	Erosion consists of rills, which may be up to about 8 cm deep; some debris at the base of the slope
3	Appreciable	Rills up to about 0.3 m deep; debris at the base of the slope
4	Severe	Rills from about 0.3 to 1 m deep and gullies are beginning to form; considerable debris at the base of the slope
5	Very severe	Deep erosion channels, consisting of rills and gullies; development of pipes causing underground erosion; very large accumulation of debris at the base of the slope

TABLE 15.3 Erosion-Prone Landforms

Type of landform (1)	Discussion (2)
Loss of vegetation during construction	It has been stated that although agriculture produces the largest percentage of the total sediment load, construction activities cause the most concentrated form of erosion (Goldman et al. 1986). The disturbance of the land surface during construction can increase the soil erosion by a factor of 2 to 40,000 times the preconstruction erosion rate (Wolman and Schick 1967, Virginia Soil and Water Conservation Commission 1980). The most damaging effect of construction is the removal of ground cover, which normally means vegetation. Ground cover can also include ground surface stabilization treatments, such as open-graded gravel, mulch, and jute netting. According to Goldman et al. (1986), vegetation performs the following functions: • Shields the soil surface from the impact of falling rain • Slows the velocity of runoff • Holds soil particles in place • Maintains the soil's capacity to absorb water
Dispersive clays	As mentioned in Sec. 13.9, clean (uncemented) sand and nonplastic silt are most susceptible to erosion with clay being the most resistant with certain exceptions, such as dispersive clays. McElroy (1987) describes unique erosion features caused by dispersive clays. For example, the erosion of fill or cut slopes can take the form of vertical or near vertical tunnels called "jugs." McElroy (1987) states that jugs may develop as a result of small drying cracks, rodent holes, openings created by decaying roots, animal and vehicle tracks, or other small surface depressions that permit rainfall or surface water to collect. The jugs frequently consist of a small hole (often less than 25 mm) on the surface of the cut or fill slope, and this hole can extend to a depth of 3 m (10 ft) with a bottom diameter of 1 m (3 ft). When these jugs collapse, the slope can erode to form severe rills and gullies.
Sea cliffs	Sea cliffs can be more susceptible to erosion because of a lack of vegetation caused by salt accumulation in the soil from sea spray. A reduction in vegetation will provide less rain impact protection and less root reinforcement to hold the soil in place. Another factor that causes sea cliff erosion that is unique to the ocean environment is toe erosion due to ocean wave impact and scour.
Badlands	Badlands have been defined as follows (Stokes and Varnes 1955): "An area, large or small, characterized by extremely intricate and sharp erosional sculpture. Badlands usually develop in areas of soft sedimentary rocks such as shale, but may also occur in decomposed igneous rocks, loess, etc. The divides are sharp and the slopes are scored by intricate systems of ravines and furrows. Fantastic erosional forms are commonly developed through the unequal erosion of hard and soft layers. Vegetation is scanty or lacking, and there is a notable lack of coarse detritus. Badlands occur chiefly in arid or semiarid climates where the rainfall is concentrated in sudden heavy showers. They may, however, occur in humid regions where vegetation has been destroyed, or where soil and coarse detritus are lacking."

TABLE 15.4 Universal Soil Loss Equation

Topic (1)	Discussion (2)
Universal soil loss equation	The universal soil loss equation calculates the soil loss A by multiplying together five different factors, as follows (Goldman et al. 1986): $$A = RKSCP$$ where A = soil loss, tons per acre per year [tons/(acre·year)] R = rainfall erosion index [100(ft·tons/acre)·(in./h)] K = soil erodibility (tons/acre per unit of R) S = combined slope length and steepness factor (dimensionless) C = vegetative cover factor (dimensionless) P = erosion control practice factor (dimensionless) Note that the universal soil loss equation is an empirical equation and the units on the right side of the equation are tons per acre per year. In order to determine the amount of soil loss (in tons) per year, the value of A would be multiplied by the area (in acres) under consideration. To convert A from tons per acre to SI units (tonnes per hectare), multiply A by 2.24. The individual terms in the universal soil loss equation are discussed below.
Rainfall erosion index R	The rainfall erosion index combines both a rainfall energy component and a rainfall intensity component into one number. There is a two-step process in calculating the rainfall erosion index R. First, an average rainfall year is selected and the total energy E times the maximum 30-minute rainfall intensity I_{30} is calculated for a given storm. Second, the E times I_{30} values for all the major storms in an area are added up for this average rainfall year. Figure 15.1 presents rainfall erosion index R values for the continental United States. Those values west of 104° are only rough approximations because the rainfall erosion index values can vary significantly between adjacent areas because of the differing topography. In Fig. 15.1, the highest values are R = 550 for the general area of New Orleans, which receives numerous storms of high intensity in an average year. The lower values in the west are a result of the much lower amount of rainfall in an average year. The rainfall erosion index R also does not consider erosion due to snowmelt runoff. As previously mentioned, the universal soil loss equation determines the soil loss (in tons) per acre for an average year. It is possible to determine A for a shorter time period. For example, suppose grading work will be performed during January and the average rainfall in January is 2 inches. If the annual average rainfall is 20 inches, then the value of A for only January would be 10% (i.e., 2/20 = 10%) of the value calculated from the universal soil loss equation.
Soil erodibility factor K	The soil erodibility factor K is a measure of the ability of rainfall and runoff to detach and transport the soil particles. There are two main factors that govern a soils susceptibility to erosion: particle size and soil plasticity. Values of the soil erodibility factor K range from a low of 0.02 to a high of 0.7. Table 15.5 presents recommended values of the soil erodibility factor K. Note that nonplastic silt (MN) has the highest susceptibility to erosion (K = 0.7). This is because of the small size of the silt-size particles, and without any soil plasticity to hold the soil particles together, they are easily eroded. At the other extreme is a highly plastic clay (CH), where the plasticity is so great

TABLE 15.4 Universal Soil Loss Equation (*Continued*)

Topic (1)	Discussion (2)
Soil erodibility factor K (*Continued*)	(PI > 30) that rainwater is unable to detach the clay particles ($K = 0.02$). Other soils with relatively low soil erodibility factors K are gravels, which are large-size particles that resist erosion because of their weight. In Table 15.5, ranges of the soil erodibility factor K are provided. Lower values would be used if the soil contains a higher gravel content or has a higher plasticity index, and vice versa. Note that some soils, such as a silt of low plasticity (ML), have a wide range in values for the soil erodibility factor (K). For silt of low plasticity (ML), the value of K increases as the plasticity index decreases. As the PI approaches zero, the soil becomes nonplastic and the value of K approaches 0.7 (value for nonplastic silt). Likewise, for clayey sands (SC), the soil erodibility factor K is very low when the clayey sand has a high plasticity index ($K = 0.1$), but as the plasticity index approaches zero, the value of K increases to the nonplastic sand value ($K = 0.7$). There are other factors that may need to be considered in the selection of the soil erodibility factor. Some of these factors are as follows: • *Organic matter.* A small amount of organic matter in the soil tends to decrease the soil erodibility factor K. This is probably because of the absorptive and possibly binding capability of organic matter. • *Cementing agents.* In some cases there may be cementing agents that bind the soil particles together. If these cementing agents have a low solubility in water, then the soil erodibility factor K listed in Table 15.5 would be too high. • *Dispersive clays.* Dispersive clays may have a high plasticity index, but as previously discussed, they can be susceptible to erosion because of the process of deflocculation of the clay size particles (Perry 1987). Values listed in Table 15.5 would be too low for dispersive clays.
Combined slope length and steepness factor S	This factor combines the effects of both the slope length and the slope gradient. The combined slope length and slope gradient factor S is the ratio of soil loss per unit area on a site to the corresponding loss from a 73-ft-long experimental plot with a 9% slope (Goldman et al. 1986). The following equation (Wischmeier and Smith 1965, 1978) can be used to calculate the combined slope length and steepness factor S: $$S = \frac{65s^2L'}{s^2 + 10{,}000} + \frac{4.6\,sL'}{(s^2 + 10{,}000)^{0.5}} + 0.065L'$$ where s = slope steepness expressed as a percentage and L' = length factor, defined as follows: $$L' = (L/73)^m$$ where L = length of the slope in feet and m has one of the following values: 0.2 for $s < 1\%$ 0.3 for $1\% \le s < 3\%$ 0.4 for $3\% \le s < 5\%$ 0.5 for $s \ge 5\%$

TABLE 15.4 Universal Soil Loss Equation (*Continued*)

Topic (1)	Discussion (2)
Combined slope length and steepness factor S (*Continued*)	For the above equations, the slope steepness s must be inserted as a percentage. For example, a 2:1 (horizontal:vertical) slope has a slope steepness of 50%, and thus $s = 50$. Note that by plugging the values from the experimental plot ($s = 9\%$ and $L = 73$ ft) into the above equations, the value of the combined slope length and steepness factor s is equal to 1.0.
Combined slope length and steepness factor S (example problems)	*Example 1:* Suppose that a building pad will be constructed at a slope gradient of 100:1 (horizontal:vertical) and the slope length L of the graded pad is equal to 100 ft. For this example, $s = 1$, $m = 0.3$, and $L = 100$ ft, and thus $L' = 1.1$. Substituting $L' = 1.1$ and $s = 1$ into the main equation, the value of the combined slope length and steepness factor S is 0.13. This value is less than 1 because the slope gradient ($s = 1\%$) is much less than the slope gradient used for the experimental plot ($s = 9\%$). *Example 2:* Suppose a slope will be constructed such that it has a height of 70 ft and will have a slope gradient of 1:1 (horizontal:vertical). In this case, with a slope height of 70 ft, the length of the slope surface L will be 100 ft (i.e., $L = H/\sin\alpha$, where H = height of slope and α = slope inclination). For a 1:1 slope, the slope gradient s equals 100. For $L = 100$ and $m = 0.5$, the value of $L' = 1.2$. Substituting $s = 100$ and $L' = 1.2$ into the main equation, the value of the combined slope length and steepness factor S is 43. This value is greater than 1 because the slope gradient ($s = 100\%$) is much greater than the slope gradient used for the experimental plot ($s = 9\%$). In these two examples, the values of the combined slope length and steepness factor S are 0.13 for the first example and 43 for the second example. The second example has an S value that is 330 times greater than the first example. As these two examples illustrate, the S value can have a wider variation than any other factor in the universal soil loss equation.
Vegetative cover factor C	The vegetative cover factor accounts for the type and extent of vegetation at the site. Table 15.6 (from Goldman et al. 1986) presents C values for different ground cover conditions. Note in Table 15.6 that bare soil, which is the typical condition for grading or construction projects, has a C value of 1.0. For a site where the native vegetation is undisturbed, the C value is only 0.01. Thus the erosion for a bare site will be 100 times greater than the same site having native vegetation in an undisturbed state. For sites where there is little native vegetation, such as badlands, a C value of 1.0 should be used.
Erosion control practice factor P	The final term in the universal soil loss equation is the erosion control practice factor P. It is included in the equation to account for the management of the site. For example, Table 15.7 (from Goldman et al. 1986) presents typical P values for the construction of a site. Higher P values apply to sites where the graded surface is smooth and uniform, while lower P values are applicable for surfaces that are rough and irregular. As indicated in Table 15.7, changing the graded surface condition does not have much effect on the P value and of the five factors in the universal soil loss equation, the P value has the least effect on soil loss at a construction site.

TABLE 15.4 Universal Soil Loss Equation (*Continued*)

Topic (1)	Discussion (2)
Limitations of the universal soil loss equation	In using the universal soil loss equation, the final result should be considered only an estimate of the soil loss at a site. This is because of the many limitations of the universal soil loss equation, such as (Goldman et al. 1986): • *Empirical equation.* The equation is empirical, and it obtains the soil loss by simply multiplying together five different factors that have been observed to have a significant impact on the erosion process. • *Average conditions.* The equation is based on average rainfall conditions. Unusual weather patterns, such as El Niño, which caused much heavier rainfall in California during the winter of 1997–1998, can produce significantly more soil loss. • *Sheet and rill erosion.* The equation was developed to predict the soil loss for sheet and rill erosion. Concentrated flow of water, such as in gullies, will result in a much larger volume of eroded soil than predicted by the universal soil loss equation. • *Sediment deposition.* The universal soil loss equation predicts soil loss and not soil deposition. Gravel size particles and coarse sand may be eroded only to the toe of a slope and may not need to be considered in the analysis of sediment storage facilities.
Example problem	*Problem:* It is proposed to construct an earth dam, 800 ft long and 40 ft high, in the northwest corner of South Carolina. The downstream face of the dam will be constructed at a 2.5:1 (horizontal:vertical) slope ratio. The soil type is red clay fill (CI), having LL = 38 and PI = 14. The downstream slope face will be trackwalked up and down the slope. Determine the sediment loss in 1 year for a bare downstream face of the dam. *Solution:* $R = 300$ (Fig. 15.1*b*) $K = 0.1$ (from Table 15.5, for a CI soil with the PI closer to the lower limit) $S = 13$ (slope height = 40 ft, thus slope length L = 108 ft; slope steepness s = 40) $C = 1.0$ (from Table 15.6, for a bare face) $P = 0.9$ (from Table 15.7, for a trackwalked condition) Therefore: $A = RKSCP = (300)(0.1)(13)(1.0)(0.9) = 350$ tons/acre·year) Area = (800 ft)(108 ft)/(43,560 ft^2/acre) = 2 acres Total soil loss per year = (350 tons/acre)(2 acres) = 700 tons (1.4 million lb)

TABLE 15.5 Universal Soil Loss Equation: Recommended Values of the Soil Erodibility Factor ***K***

Major divisions (1)	Subdivisions (2)	ISBP symbol (3)	Typical names (4)	Laboratory classification criteria (5)	Soil erodibility factor K (6)
Nonplastic soils	Gravels (greater fraction of total sample is retained on No. 4 sieve)	GW	Well-graded gravels, sandy gravels, silty-sandy gravels	$C_u \geq 4$ and $1 \leq C_c \leq 3$	0.05 to 0.1
		GP	Poorly graded gravels, gravel-sand-silt mixtures	Does not meet above criteria, % passing No. 200 sieve $< 15\%$	0.1 to 0.15
		GM	Poorly graded, nonplastic silty gravels	Does not meet above criteria, % passing No. 200 sieve $\geq 15\%$	0.15 to 0.2
	Sands (greater fraction of total sample is between No. 4 and No. 200 sieves)	SW	Well-graded sands and gravelly sands	$C_u \geq 6$ and $1 \leq C_c \leq 3$	0.5 to 0.6
		SP	Poorly graded sands or sand-gravel-silt mixtures	Does not meet above criteria, % passing No. 200 sieve $< 15\%$	0.6 to 0.7
		SM	Poorly graded, nonplastic silty sands	Does not meet above criteria, % passing No. 200 sieve $\geq 15\%$	0.6 to 0.7
	NP silt	MN	Nonplastic silts, rock flour	Greater fraction of the total sample passes the No. 200 sieve.	0.7
Plastic soils	Plastic silts (minus No. 40 fraction plots below A-line)	GM*	Plastic silty gravels, gravel-silt mixtures	$\geq 50\%$ retained on the No. 200 sieve with the greater fraction of gravel size	0.1 to 0.2
		SM*	Plastic silty sands, sand-silt mixtures	$\geq 50\%$ retained on the No. 200 sieve with the greater fraction of sand size	0.3 to 0.7
		ML MI MH	Plastic silts, sandy silts, and clayey silts	Silt of low plasticity (ML) PI ≤ 10 Intermediate plasticity (MI) $10 <$ PI ≤ 30 Silt of high plasticity (MH) PI > 30	0.3 to 0.7 0.2 to 0.3 0.1 to 0.2
	Clays (minus No. 40 fraction plots on or above A-line)	GC*	Clayey gravels, gravel-clay mixtures	$\geq 50\%$ retained on the No. 200 sieve with the greater fraction of gravel size	0.05 to 0.2
		SC*	Clayey sands, sand-clay mixtures	$\geq 50\%$ retained on the No. 200 sieve with the greater fraction of sand size	0.1 to 0.7
		CL CI CH	Clay, sandy clays, and silty clays	For clay of low plasticity (CL) PI ≤ 10 Intermediate plasticity (CI) $10 <$ PI ≤ 30 For clay of high plasticity (CH) PI > 30	0.1 to 0.6 0.02 to 0.1 0.02

*High, intermediate, or low plasticity based on the ISBP classification system

TABLE 15.6 Universal Soil Loss Equation: Vegetative Cover Factor *C*

Type of cover (1)	*C* factor (2)
No vegetative cover (bare ground)	1.0
Native vegetation (undisturbed)	0.01
Temporary seeding:	
90% cover, annual grasses, no mulch	0.1
Wood fiber mulch, 3/4 ton/acre, with seed*	0.5
Excelsior mat, jute*	0.3
Straw mulch:*	
1.5 tons/acre, tacked down	0.2
4 tons/acre, tacked down	0.05

*For slopes up to 2:1 (horizontal:vertical).

Sources: This table was developed by Goldman et al. 1986 and based on the work by Kay (1983), U.S. Department of Agriculture (1977), and Wischmeier and Smith (1965).

TABLE 15.7 Universal Soil Loss Equation: Erosion Control Practice Factor *P*

Surface condition (1)	*P* value (2)
Compacted and smooth	1.3
Trackwalked along contour*	1.2
Trackwalked up and down slope†	0.9
Punched straw	0.9
Rough, irregular cut	0.9
Loose to 12 in. (30 cm) depth	0.8

*Tread marks oriented up and down slope.

†Tread marks oriented parallel to contours.

Sources: This table was developed by Goldman et al. 1986 and based on the work by the U.S. Department of Agriculture (1977).

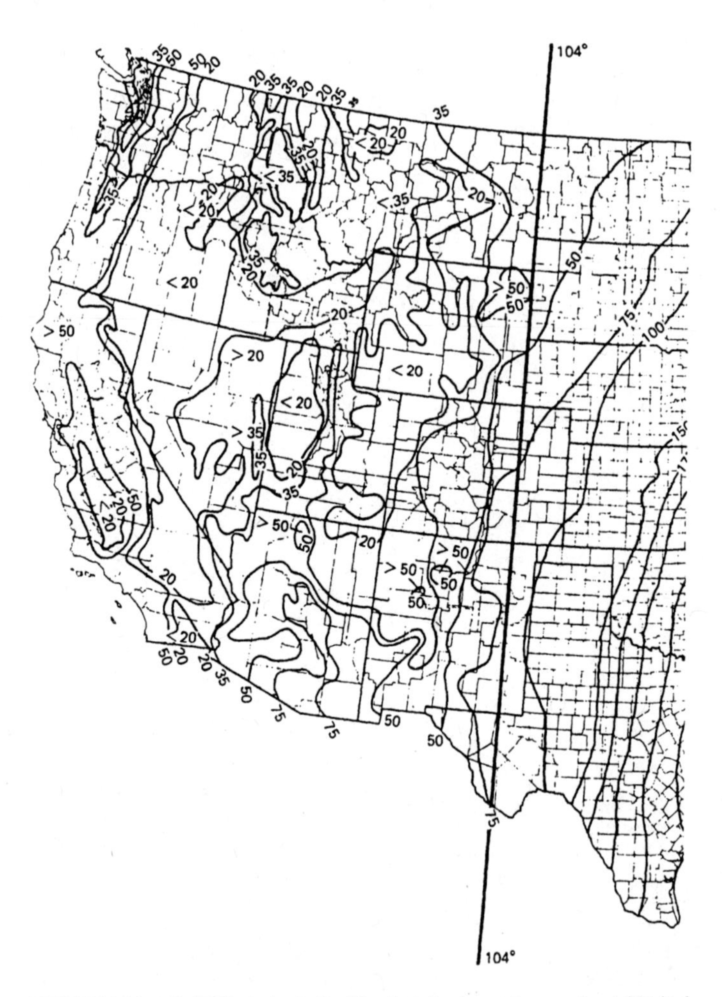

FIGURE 15.1a Rainfall erosion index (R) values for the western continental United States. The units of R are 100 (ft·tons/acre)·(in./h). See Table 15.4. (*Reproduced from Goldman et al. 1986; reprinted with permission from McGraw-Hill, Inc.*)

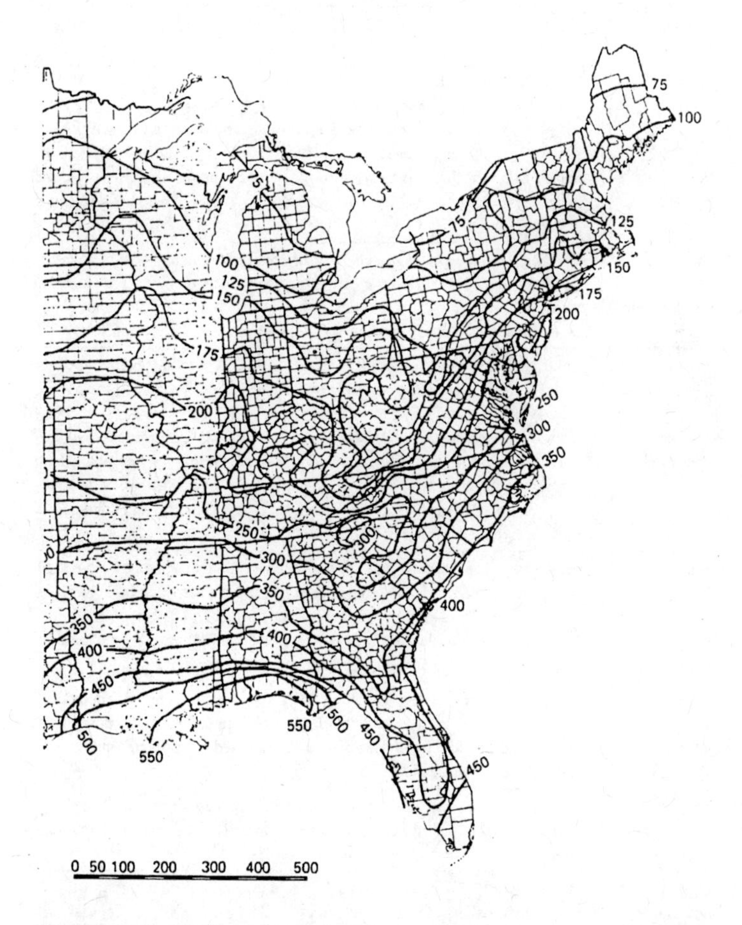

FIGURE 15.1b Rainfall erosion index (R) values for the eastern continental United States. The units of R are 100 (ft·tons/acre)·(in./h). See Table 15.4. (*Reproduced from Goldman et al. 1986; reprinted with permission from McGraw-Hill, Inc.*)

TABLE 15.8 Design and Construction for Erosion Control

Topic (1)	Discussion (2)
General discussion	The advantage of the universal soil loss equation is that it can provide an estimate of the soil loss for a given construction project. Obviously, for the example problem in Table 15.4 (construction of an earth dam), a total soil loss of 1.4 million lb would be unacceptable. In terms of design and construction, the basic principles that can be used to reduce the amount of soil erosion and avoid sediment transport onto adjacent property are listed below (Goldman et al. 1986).
Fit the development to the terrain	The best method to mitigate the risk of creating erosion and sedimentation problems is to disturb as little of the land surface as possible. For example, the development could be tailored to the natural contours of the land, which would then require little grading with less ground surface erosion.
Time the grading and construction to minimize soil exposure	Rainfall intensity is a key factor in slope erosion. If possible, it would be best to not perform the grading and construction activities during the time of year subjected to intense rainstorms. For example, in many parts of the country, there is a dry season when the amount of erosion will be much less than during the wet season. In California, the dry season is during the summer and the wet season is during the winter. The best situation would be to plan the grading operation during the summer months and have the grading completed in time to have the slopes seeded prior to the rainy season.
Retain existing vegetation whenever possible	As previously mentioned, the vegetative cover factor C can vary from 0.01 for native vegetation (undisturbed state) to 1.0 for a bare slope (Table 15.6). Thus there will be 100 times more soil loss for the denuded slope condition. Whenever possible, the existing trees and other natural vegetation should be incorporated into the site development plan.
Vegetate and mulch denuded areas	The areas that have been denuded by the site grading or construction process should be planted as soon as possible. In many cases, it may be possible to perform the site grading in phases, with the denuded areas planted at the completion of each phase of grading.
Divert runoff away from denuded areas	Runoff should always be diverted away from denuded areas and not allowed to flow over the tops of slopes. If surface runoff is not diverted, there could be the formation of rills and eventually gullies.
Minimize length and steepness of slopes	Slope length and slope steepness are critical factors in the amount of soil loss caused by erosion. This is because the slope length and slope steepness primarily determine the velocity of runoff. The energy (hence erosive potential) of flowing water increases as the square of the velocity. Thus long continuous slopes allow for the surface runoff to gain momentum, which can result in the runoff becoming concentrated in gullies. The combined length and steepness factor S can be significantly decreased by installing gunite drainage ditches (see Fig. 15.2) on the face of the slope. As an example of the use of drainage ditches, suppose that for the example of the earth dam (Table 15.4) that a midslope drainage ditch will be installed. In this case, the slope length will be cut in half (L = 54 ft). The combined length andsteepness factor S is 9.2. Therefore the amount of soil loss would be reduced by about 30% by simply constructing a midslope drainage ditch.

TABLE 15.8 Design and Construction for Erosion Control (*Continued*)

Topic (1)	Discussion (2)
Keep runoff velocities low	Soil erosion can be reduced by keeping the runoff velocities low. For example, as indicated in Table 15.7, the erosion control practice factor P can be reduced by constructing a rough and irregular slope face rather than a smooth slope surface.
Prepare drainage facilities	The grading and construction process will increase the amount of runoff from the project. The project plans should include proper drainage ways and outlets to handle this increased runoff.
Trap sediment on site	There are many different devices that can be used to trap sediment on site. The most basic are devices that filter the sediment from the water, such as silt fences. Examples of reliable devices that can be used to trap sediment on site are shown in Figs. 15.3 and 15.4. Figure 15.3 shows the typical features of a sediment trap and Fig. 15.4 shows an elevated sediment trap connected to a storm drain line. In a sediment basin (Fig. 15.3), loose and porous soil will accumulate. If fill is placed atop the basin sediment and a structure is built on top of the fill, then the structure could be subjected to substantial settlement and distress as the loose and porous sediment consolidates or collapses. It is important to remove and recompact all sediment located in the basin prior to placing a structural fill at the basin area.
Inspect and maintain control measures	Inspection and maintenance of erosion control systems are essential. Without cleaning and maintenance, the systems shown in Figs. 15.3 and 15.4 will fill with sediment and no longer trap the eroded soil.

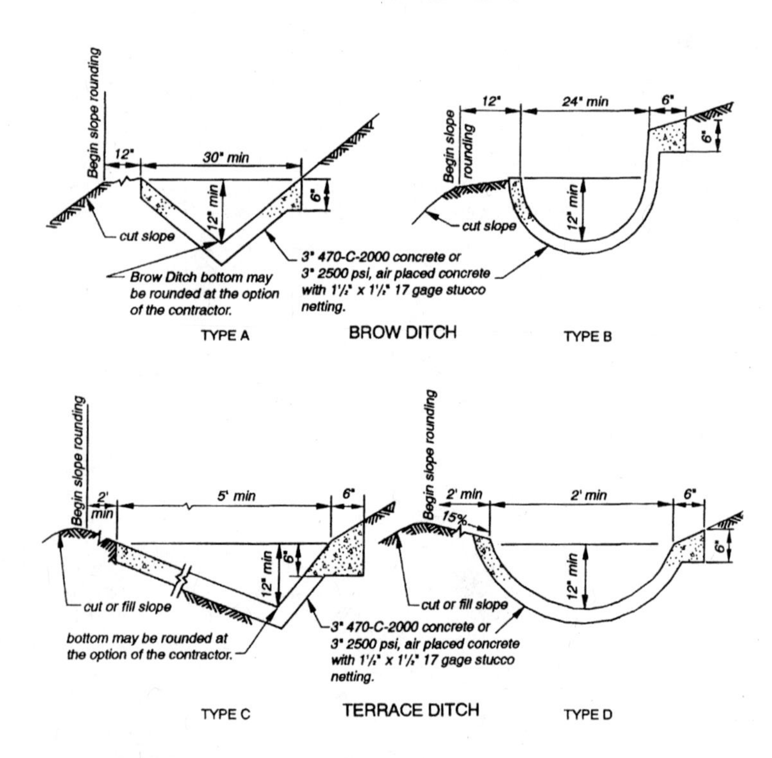

FIGURE 15.2 Drainage ditches. See Table 15.8. (*From City of San Diego Standard Drawings 1986.*)

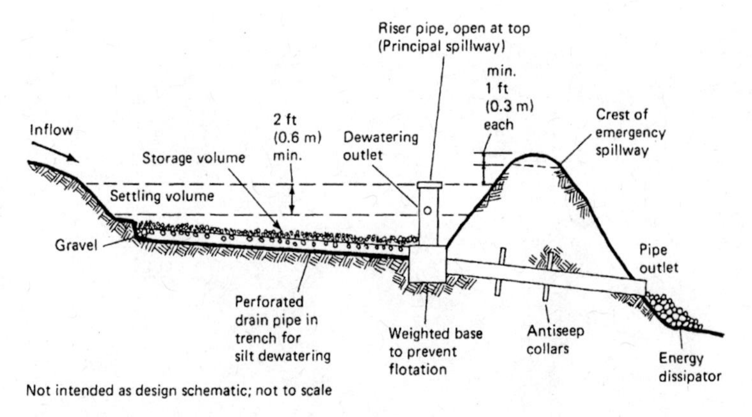

FIGURE 15.3 Typical features of a sediment basin. See Table 15.8. (*From Goldman et al. 1986; reprinted with permission from McGraw-Hill, Inc.*)

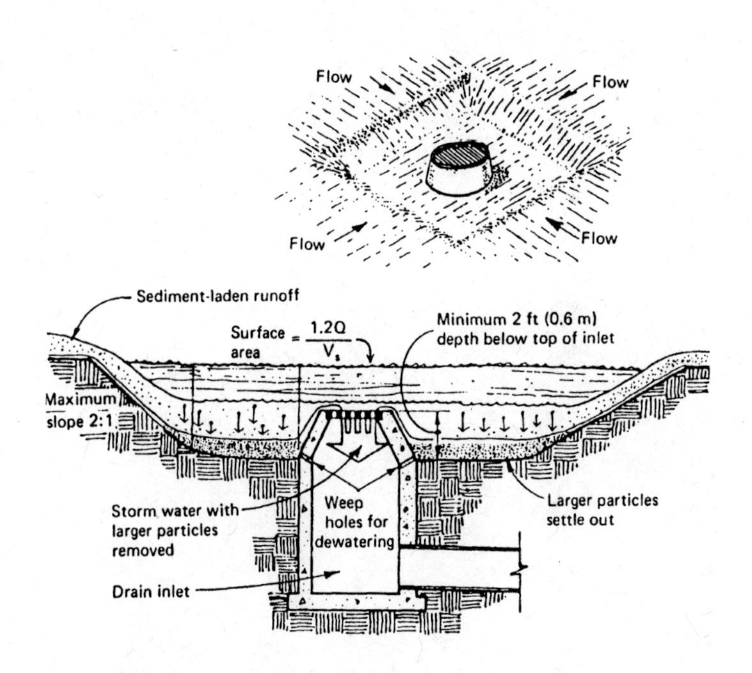

FIGURE 15.4 Elevated sediment trap connected to a storm drain line. See Table 15.8. (*From Goldman et al. 1986; based on the Virginia Erosion and Sediment Control Handbook, 1980; reprinted with permission from McGraw-Hill, Inc.*)

CHAPTER 16
RETAINING WALLS

16.1 INTRODUCTION

A retaining wall is defined as a structure whose primary purpose is to provide lateral support for soil or rock. In some cases, the retaining wall may also support vertical loads. Examples include basement walls and certain types of bridge abutments (see Fig. 16.1). Table 16.1 lists and describes various types of retaining walls and backfill conditions.

16.2 RETAINING WALL ANALYSES

Tables 16.2 and 16.3 and Fig. 16.2 present retaining wall analyses neglecting the friction between the wall and backfill soil. The analysis is more complicated and generally less conservative when the friction between the wall and backfill soil is included in the analysis (see Table 16.4 and Fig. 16.3). An example of retaining wall analysis is presented in Fig. 16.4.

16.3 DESIGN AND CONSTRUCTION OF RETAINING WALLS

The previous section dealt with simple retaining walls, such as those shown in Figs. 16.1 and 16.4. This section provides an additional discussion of the design and construction of retaining walls (see Table 16.5 and Fig. 16.5).

16.4 RESTRAINED RETAINING WALLS

In order for the active wedge (Fig. 16.2b) to be developed, there must be sufficient movement of the retaining wall (Table 16.3). There are many cases where movement of the retaining wall is restricted. Examples include massive bridge abutments, rigid basement walls, and retaining walls that are anchored in nonyielding rock. These cases are often described as *restrained retaining walls.*

In order to determine the earth pressure acting on a restrained retaining wall, the coefficient of earth pressure at rest (k_o) is substituted for k_A in Table 16.2. The value of k_o can be estimated from the equations in Table 6.3. A common value of k_o that is used for restrained retaining walls is 0.5. Restrained retaining walls are especially susceptible to higher earth pressures induced by heavy compaction equipment, and extra care must be taken during the compaction of backfill for restrained retaining walls.

16.5 MECHANICALLY STABILIZED EARTH RETAINING WALLS

Mechanically stabilized earth retaining walls (also known as MSE retaining walls) are typically composed of strip- or grid-type (geosynthetic) reinforcement (Koerner 1990). Because they are often more economical to construct than conventional concrete retaining walls, mechanically stabilized earth retaining walls have become very popular in the past decade. The design and construction of MSE retaining walls are discussed in Table 16.6 and Figs. 16.6 and 16.7.

16.6 SHEET PILE WALLS

Sheet pile retaining walls are widely used for waterfront construction and consist of interlocking members that are driven into place. Individual sheet piles come in different sizes and shapes, such as those shown in Fig. 16.8. Sheet piles have an interlocking joint (see Fig. 16.8) that enables the individual segments to be connected together to form a solid wall. This section presents a discussion of anchored sheet pile walls, cantilevered sheet pile walls, cofferdams, and important design considerations (Tables 16.7 to 16.10 and Figs. 16.9 to 16.11).

16.7 TEMPORARY RETAINING WALLS

Temporary retaining walls are often used during construction, for example, for the support of the sides of an excavation that is made to construct the building foundation. Examples include braced retaining walls, steel I-beam and wood lagging retaining walls, and temporary shoring for utility trenches (see Table 16.11 and Figs. 16.12 to 16.14).

16.8 PIER WALLS

Contiguous drilled piers are sometimes constructed to retain large open cuts more than 30 m (100 ft) deep (Abramson et al. 1996). When used to stabilize slopes, the drilled piers must pass through the slip surface and be embedded into a bearing stratum that can provide passive resistance to the destabilizing force transmitted by the unstable slope (see Tables 16.12 to 16.15 and Figs. 16.15 and 16.16).

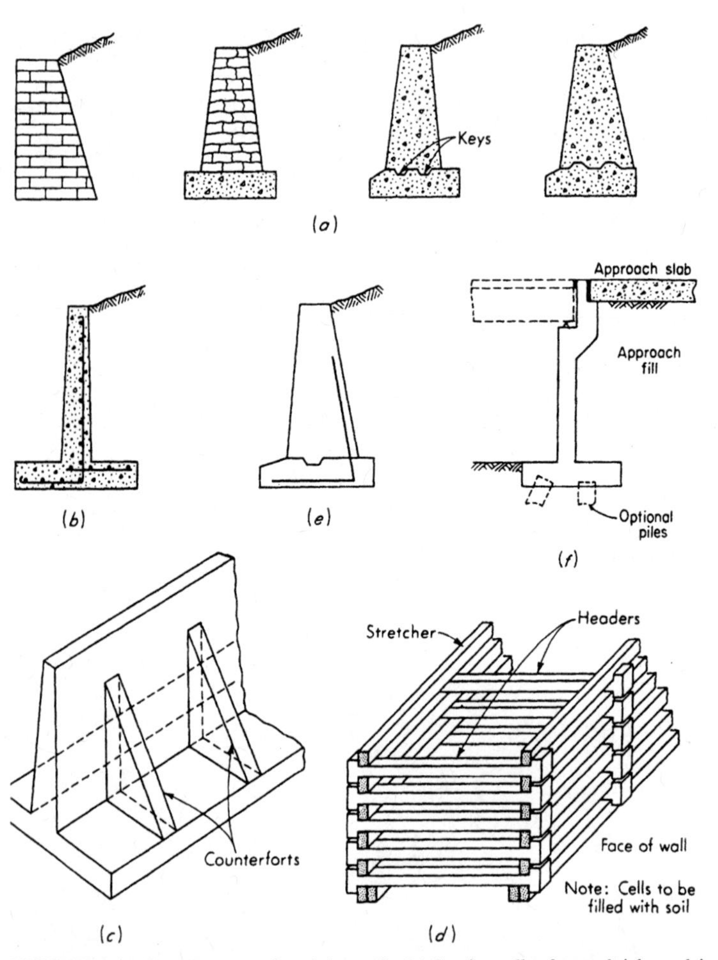

FIGURE 16.1 Common types of retaining walls. (*a*) Gravity walls of stone, brick, or plain concrete. Weight provides overturning and sliding stability. (*b*) Cantilevered wall. (*c*) Counterfort, or buttressed, wall. If backfill covers the counterforts, the wall is termed a counterfort retaining wall. (*d*) Crib wall. (*e*) Semigravity wall (often steel reinforcement is used). (*f*) Bridge abutment. See Table 16.1. (*Reproduced from Bowles 1982 with permission of McGraw-Hill, Inc.*)

TABLE 16.1 Types of Retaining Walls and Backfill Conditions

Topic (1)	Discussion (2)
Types of retaining walls	As shown in Fig. 16.1, some of the more common types of retaining walls are gravity walls, counterfort walls, cantilevered walls, and crib walls (Cernica 1995a). Gravity retaining walls are routinely built of plain concrete or stone, and the wall depends primarily on its massive weight to resist failure from overturning and sliding. Counterfort walls consist of a footing, a wall stem, and intermittent vertical ribs (called *counterforts*) which tie the footing and wall stem together. Crib walls consist of interlocking concrete members that form cells, which are then filled with compacted soil. Although mechanically stabilized earth retaining walls have become more popular in the past decade, cantilever retaining walls are still probably the most common type of retaining structure. There are many different types of cantilevered walls, with the common features being a footing that supports the vertical wall stem. Typical cantilevered walls are T-shaped, L-shaped, or reverse L–shaped (Cernica 1995a).
Backfill material	Clean granular material (no silt or clay) is the standard recommendation for backfill material. There are several reasons for this recommendation: 1. *Predictable behavior.* Import granular backfill generally has a more predictable behavior in terms of earth pressure exerted on the wall. Also, expansive soil–related forces (Chap. 12) will not be generated by clean granular soil. 2. *Drainage system.* To prevent the buildup of hydrostatic water pressure on the retaining wall, a drainage system is often constructed at the heel of the wall. The drainage system will be more effective if highly permeable soil, such as clean granular soil, is used as backfill. 3. *Frost action.* In cold climates, frost action has caused many retaining walls to move so much that they have become unusable. If freezing temperatures prevail, the backfill soil can be susceptible to frost action, where ice lenses will form parallel to the wall and cause horizontal movements of up to 0.6 to 0.9 m (2 to 3 ft) in a single season (Sowers and Sowers 1970). Backfill soil consisting of clean granular soil and the installation of a drainage system at the heel of the wall will help to protect the wall from frost action.
Plane strain condition	Movement of retaining walls (i.e., active condition) involves the shear failure of the wall backfill, and the analysis will naturally include the shear strength of the backfill soil (Chap. 7). As in the analysis of strip footings (Chap. 10) and slope stability (Chap. 13), for most field situations involving retaining structures, the backfill soil is in a plane strain condition (i.e., the soil is confined along the long axis of the wall). As previously mentioned, the friction angle ϕ is about 10% higher in the plane strain condition compared to the friction angle ϕ measured in the triaxial apparatus. In practice, plane strain shear strength tests are not performed, which often results in an additional factor of safety for retaining wall analyses.

TABLE 16.2 Retaining Wall Analyses Neglecting Friction between the Wall and Backfill Soil

Topic (1)	Discussion (2)
General discussion	Figure 16.2*a* shows a reverse L–shaped cantilever retaining wall. This type of simple retaining wall will be used to introduce the basic types of retaining wall design analyses. The pressure exerted on the back side of the wall is the active earth pressure. The footing is supported by the vertical bearing pressure of the soil or rock. Lateral movement of the wall is resisted by passive earth pressure and slide friction between the footing and bearing material.
Active earth pressure	As shown in Fig. 16.2*a*, the active earth pressure is often assumed to be horizontal by neglecting the friction developed between the vertical wall stem and the backfill. This friction force has a stabilizing effect on the wall, and therefore it is usually a safe assumption to ignore friction. However, if the wall should settle more than the backfill, for example, because of high vertical loads imposed on the top of the wall, then a negative skin friction can develop between the wall and backfill which has a destabilizing effect on the wall. In the evaluation of the active earth pressure, it is common for the soil engineer to recommend clean granular soil as backfill material. In order to calculate the active earth pressure resultant force P_A in kN per linear meter of wall or pounds per linear foot of wall, the following equation is used for clean granular backfill: $$P_A = \tfrac{1}{2} k_A \gamma_t H^2$$ where k_A = active earth pressure coefficient, γ_t = total unit weight of the backfill, and H = height over which the active earth pressure acts as defined in Fig. 16.2*a*. The active earth pressure coefficient k_A is equal to: $$k_A = \tan^2 (45° - \tfrac{1}{2}\phi)$$ where ϕ = friction angle of the clean granular backfill. This equation is known as the *active Rankine state,* after the British engineer Rankine who in 1857 obtained this relationship.
Equivalent fluid pressure	In the active earth pressure resultant force equation, the product of k_A times γ_t is referred to as the *equivalent fluid pressure* (even though the product is actually a unit weight). In the design analysis, the soil engineer usually assumes a total unit weight γ_t of 18.9 kN/m^3 (120 pcf) and a friction angle ϕ of 30° for the granular backfill. Using the above equation and ϕ = 30°, the active earth pressure coefficient k_A is 0.333. Multiplying 0.333 times the total unit weight γ_t of backfill results in an equivalent fluid pressure of 6.3 kN/m^3 (40 pcf). This is a common recommendation for equivalent fluid pressure from soil engineers. It is valid for the conditions of clean granular backfill, a level ground surface behind the wall, a backdrain system, and no surcharge loads. Note that this recommended value of equivalent fluid pressure of 6.3 kN/m^3 (40 pcf) does not include a factor of safety and is the actual pressure that would be exerted on a smooth wall when the friction angle ϕ of the granular

TABLE 16.2 Retaining Wall Analyses Neglecting Friction between the Wall and Backfill Soil (*Continued*)

Topic (1)	Discussion (2)
Equivalent fluid pressure (*Continued*)	backfill equals 30°. In designing the vertical wall stem in terms of wall thickness and size and location of steel reinforcement, a factor of safety F can be applied to the active earth pressure resultant force equation. A factor of safety may be prudent because higher wall pressures will most likely be generated during compaction of the backfill or when translation of the footing is restricted (Goh 1993).
Important details for the active earth pressure	Additional important details concerning the active earth pressure are as follows: 1. *Sufficient movement.* There must be sufficient movement of the retaining wall in order to develop the active earth pressure of the backfill. For example, Table 16.3 (from NAVFAC DM 7.2, 1982) indicates the amount of wall rotation that must occur for different backfill soils in order to reach the active earth pressure state. 2. *Triangular distribution.* As shown in Fig. 16.2*a,* the active earth pressure is a triangular distribution and thus the active earth pressure resultant force P_A is located at a distance equal to $\frac{1}{3}H$ above the base of the wall. 3. *Surcharge pressure.* If there is a uniform surcharge pressure Q acting on the entire ground surface behind the wall, then there would be an additional horizontal pressure exerted upon the retaining wall equal to the product of k_A times Q. Thus the resultant force P_Q, in kN per linear meter of wall or pounds per linear foot of wall, acting on the retaining wall due to the surcharge Q is equal to: $$P_Q = QHk_A$$ where Q = uniform vertical surcharge (kPa or psf) acting on the entire ground surface behind the retaining wall, k_A = active earth pressure coefficient, and H = height of the retaining wall. Because this pressure acting on the retaining wall is uniform, the resultant force P_Q is located at midheight of the retaining wall. 4. *Active wedge.* The *active wedge* is defined as that zone of soil involved in the development of the active earth pressures upon the wall. This active wedge must move laterally in order to develop the active earth pressures. It is important that building footings or other load-carrying members are not supported by the active wedge, or else they will be subjected to lateral movement. Figure 16.2*b* shows an illustration of the active wedge of soil behind the retaining wall. As indicated in Fig. 16.2*b,* the active wedge is inclined at an angle of 45° + ϕ/2 from the horizontal.
Passive earth pressure	As shown in Fig. 16.2*a,* the passive earth pressure is developed along the front side of the footing. Passive pressure is developed when the wall footing moves laterally into the soil and a passive wedge is developed as shown in Fig. 16.2*b.* In order to calculate the passive resultant force P_p, the following equation is used, assuming that there is cohesionless soil in front of the wall footing:

TABLE 16.2 Retaining Wall Analyses Neglecting Friction between the Wall and Backfill Soil (*Continued*)

Topic (1)	Discussion (2)
Passive earth pressure (*Continued*)	$$P_p = \tfrac{1}{2}k_p\gamma_t D^2$$ where P_p = passive resultant force in kN per linear meter of wall or pounds per linear foot of wall, k_p = passive earth pressure coefficient, γ_t = total unit weight of the soil located in front of the wall footing, and D = depth of the wall footing (vertical distance from the ground surface in front of the retaining wall to the bottom of the footing). The passive earth pressure coefficient k_p is equal to: $$k_p = \tan^2(45° + \tfrac{1}{2}\phi)$$ where ϕ = friction angle of the soil in front of the wall footing. This equation is known as the *passive Rankine state*. In order to develop passive pressure, the wall footing must move laterally into the soil. As indicated in Table 16.3 (from NAVFAC DM 7.2, 1982), the wall translation to reach the passive state is at least twice that required to reach the active earth pressure state. Usually it is desirable to limit the amount of wall translation by applying a reduction factor to the passive pressure. A commonly used reduction factor is 2.0 (Lambe and Whitman 1969). The soil engineer routinely reduces the passive pressure by ½ (reduction factor = 2.0) and then refers to the value as the allowable passive pressure. To limit wall translation, the structural engineer should use the allowable passive pressure for design of the retaining wall. The passive pressure may also be limited by building codes. For example, the allowable passive soil pressure, in terms of equivalent fluid pressure, is 16 to 32 kN/m^3 (100 to 200 pcf) according to the *Uniform Building Code* (1997). If the soil in front of the retaining wall is a plastic (clayey) soil, then usually the long-term effective stress analysis will govern. For the effective stress analysis, the effective cohesion c' and effective friction angle ϕ' can be determined from laboratory tests. As a conservative approach, often the effective cohesion c' is ignored and the effective friction angle ϕ' is used to determine the allowable passive resistance.
Footing bearing pressure	In order to calculate the footing bearing pressure, the first step is to sum the vertical loads, such as the wall and footing weights. The vertical loads can be represented by a single resultant vertical force, per linear meter or foot of wall, that is offset by a distance from the toe of the footing. This can then be converted to a pressure distribution as shown in Fig. 16.2*a*. The largest bearing pressure is routinely at the toe of the footing (point *A*, Fig. 16.2*a*). The largest bearing pressure should not exceed the allowable bearing pressure (Chap. 10), which is usually provided by the soil engineer or by local building code specifications.
Sliding analyses	The factor of safety *F* for sliding of the retaining wall is often defined as the resisting forces divided by the driving force. The forces are per linear meter or foot of wall, or:

TABLE 16.2 Retaining Wall Analyses Neglecting Friction between the Wall and Backfill Soil (*Continued*)

Topic (1)	Discussion (2)
Sliding analyses (*Continued*)	$$F = \frac{W \tan \delta + P_p}{P_A}$$ where δ = friction angle between the bottom of the concrete foundation and bearing soil, W = weight of the wall and footing, P_p = allowable passive resultant force (P_p divided by a reduction factor), and P_A = active earth pressure resultant force. The typical recommendation for minimum factor of safety for sliding is 1.5 to 2.0 (Cernica 1995a). In some situations, there may be adhesion between the bottom of the footing and the bearing soil. This adhesion is often neglected because the wall is designed for active pressures, which typically develop when there is translation of the footing. Translation of the footing will break the adhesive forces between the bottom of the footing and the bearing soil, and therefore adhesion is often neglected for the factor of safety of sliding.
Overturning analyses	The factor of safety F for overturning of the retaining wall is calculated by taking moments about the toe of the footing (point A, Fig. 16.2*a*): $$F = \frac{a}{\frac{1}{3} P_A H}$$ where W = weight of the wall and footing, a = lateral distance from W to the toe of the footing, and P_A = active earth pressure resultant force. Note in this equation that the moment due to passive pressure is neglected. The reason is that, with a rotation-type failure mode, the wall may not move enough laterally to induce passive earth pressures. The typical recommendations for minimum factor of safety for overturning is 1.5 to 2.0 (Cernica 1995a).

TABLE 16.3 Magnitudes of Wall Rotation to Reach Active and Passive States

soil type and condition (1)	Rotation (Y/H) for active state (2)	Rotation (Y/H) for passive state (3)
Dense cohesionless	0.0005	0.002
Loose cohesionless	0.002	0.006
Stiff cohesi ve	0.01	0.02
Soft cohesi ve	0.02	0.04

Note: Y = wall displacement and H = height of the wall.
Source: From NAVFAC DM-7.2, 1982.

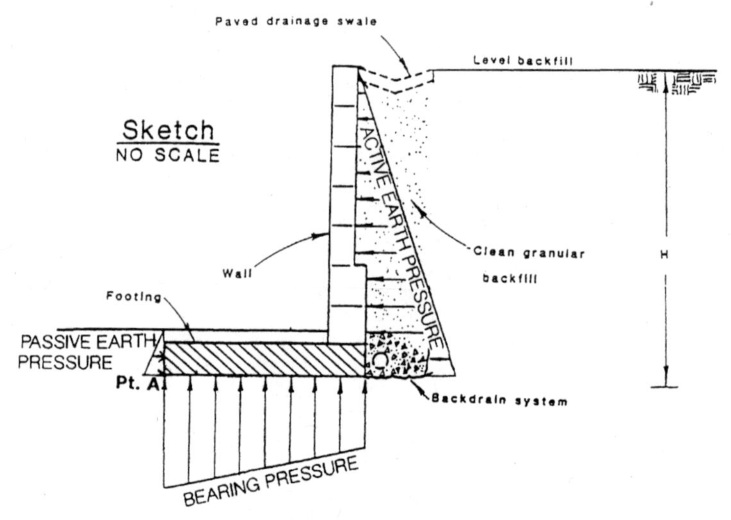

FIGURE 16.2a Retaining wall design pressures. See Table 16.2.

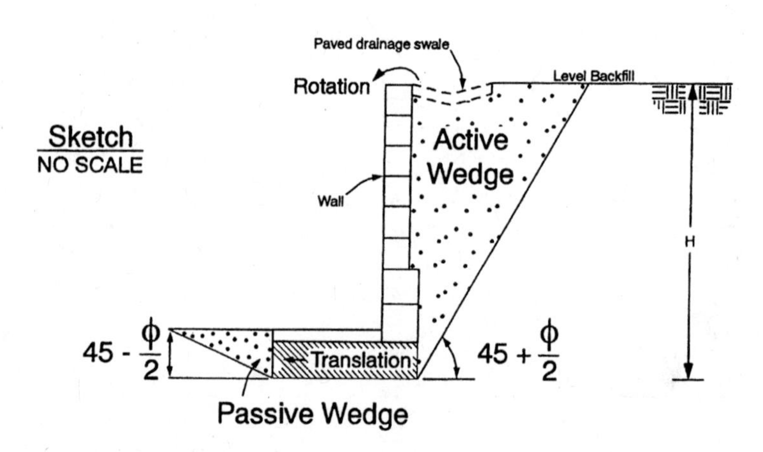

Note: For active and passive wedge development there must be movement of the retaining wall as illustrated above.

FIGURE 16.2b Active wedge behind retaining wall. See Table 16.2.

TABLE 16.4 Retaining Wall Analyses That Include the Friction between the Wall and Backfill Soil

Topic (1)	Discussion (2)
General discussion	In some cases, the geotechnical engineer may want to include the friction between the soil and the rear side of the retaining wall. For this situation, the design analysis is more complicated, as discussed below.
Active earth pressure	A common equation that is used to calculate the active earth pressure coefficient k_A for the case of wall friction is the Coulomb equation, which is shown in Fig. 16.3 (from NAVFAC DM-7.2, 1982). The Coulomb equation can also be used if the back face of the wall is sloping or if there is a sloping backfill behind the wall. Once the active earth pressure coefficient k_A is calculated, the active earth pressure resultant force P_A can be calculated by using the equation from Table 16.2. Figure 16.4 (from Lambe and Whitman 1969) presents an example of a proposed concrete retaining wall that will have a height of 20 ft (6.1 m) and a base width of 7 ft (2.1 m). The wall will be backfilled with sand that has a total unit weight γ_t of 110 pcf (17.3 kN/m^3), friction angle ϕ of 30°, and an assumed wall friction ϕ_w of 30°. Although $\phi_w = 30°$ will be used for this example problem, more typical values of wall friction are $\phi_w = \frac{3}{4}\,\phi$ for the wall friction between granular soil and wood or concrete walls and $\phi_w = 20°$ for the wall friction between granular soil and steel walls such as sheet pile walls. For the example problem shown in Fig. 16.4, the value of the active earth pressure coefficient k_A can be calculated by using Coulomb's equation (Fig. 16.3) and inserting the following values: • Slope inclination: $\beta = 0$ (no slope inclination) • Back face of the retaining wall: $\theta = 0$ (vertical back face of the wall) • Friction between the back face of the wall and the soil backfill: $\delta = \phi_w = 30°$ • Friction angle of backfill sand: $\phi = 30°$ Inputting the above values into Coulomb's equation (Fig. 16.3) gives the value of the active earth pressure coefficient $(k_A) = 0.297$. By using the active earth pressure resultant force equation (Table 16.2) with $k_A = 0.297$, total unit weight $(\gamma_t) = 110$ pcf (17.3 kN/m^3), and the height of the retaining wall $(H) = 20$ ft (6.1 m, see Fig. 16.4*a*), find the active earth pressure resultant force $(P_A) = 6540$ lb per linear foot of wall (95.4 kN per linear meter of wall). As indicated in Fig. 16.4*a*, the active earth pressure resultant force ($P_A = 6540$ lb/ft) is inclined at an angle of 30° because of the wall friction assumptions. The vertical ($P_v = 3270$ lb/ft) and horizontal ($P_H = 5660$ lb/ft) components of P_A are also shown in Fig. 16.4*a*. Note in Fig. 16.3 that, even with wall friction, the active earth pressure is still a triangular distribution acting upon the retaining wall, and thus the location of the active earth pressure resultant force (P_A) is at a distance of $\frac{1}{3}H$ above the base of the wall, or 6.7 ft (2.0 m).

TABLE 16.4 Retaining Wall Analyses That Include the Friction between the Wall and Backfill Soil (*Continued*)

Topic (1)	Discussion (2)
Passive earth pressure	As shown in Fig. 16.4*a*, the passive earth pressure is developed by the soil located at the front of the retaining wall. Usually wall friction is ignored for the passive earth pressure calculations. For the example problem shown in Fig. 16.4, the passive resultant force P_p was calculated by using the equations from Table 16.2, which neglected wall friction and the slight slope of the front of the retaining wall (see Fig. 16.4*c* for passive earth pressure calculations).
Footing bearing pressure	The procedure for the calculation of the footing bearing pressure is as follows: 1. *Calculate N.* As indicated in Fig. 16.4*b*, the first step is to calculate N (15,270 lb/ft), which equals the sum of the weight of the wall, footing, and vertical component of the active earth pressure resultant force (i.e., $N = W + P_A \sin \phi$). 2. *Determine $\bar{x}$.* The value of $\bar{x}$ (2.66 ft) is calculated as shown in Fig. 16.4*b*. The moments are determined about the toe of the retaining wall. Then $\bar{x}$ equals the difference in the opposing moments divided by N. 3. *Determine average bearing pressure.* The average bearing pressure (2180 psf) is calculated in Fig. 16.4*c* as N divided by the width of the footing (7 ft). 4. *Calculate moment about the centerline of the footing.* The moment about the centerline of the footing is calculated as N times the eccentricity (0.84 ft). 5. *Obtain the section modulus.* The section modulus of the footing is calculated as shown in Fig. 16.4*c*. 6. *Determine the portion of the bearing stress due to the moment.* The portion of the bearing stress due to the moment (σ_{mom}) is determined as the moment divided by the section modulus. 7. *Calculate the maximum bearing stress.* The maximum bearing stress is then calculated as the sum of the average stress (σ_{avg} = 2180 psf) plus the bearing stress due to the moment (σ_{mom} = 1570 psf). As indicated in Fig. 16.4*c*, the maximum bearing stress is 3750 psf (180 kPa). This maximum bearing stress must be less than the allowable bearing pressure (Chap. 10). It is also a standard requirement that the resultant normal force N be located within the middle third of the footing, as illustrated in Fig. 16.4*b*.
Sliding analysis	The factor of safety F for sliding of the retaining wall is often defined as the resisting forces divided by the driving force. The forces are per linear meter or foot of wall, or: $$F = \frac{N \tan \delta + P_p}{P_H}$$ where $\delta = \phi_{cv}$ = friction angle between the bottom of the concrete foundation and bearing soil, N = sum of the weight of the wall, footing, and vertical component of the active earth pressure resultant force (i.e., $N = W + P_A \sin \phi_w$), P_p = allowable passive resultant force (P_p from Table 16.2 divided by a reduction factor), and P_H = horizontal component of the active earth pressure resultant force (i.e., $P_H = P_A \cos \phi_w$).

TABLE 16.4 Retaining Wall Analyses That Include the Friction between the Wall and Backfill Soil (*Continued*)

Topic (1)	Discussion (2)
Sliding analysis (*Continued*)	There are variations of the above equation that are used in practice. For example, as illustrated in Fig. 16.4*c*, the value of P_p is subtracted from P_H in the denominator, instead of P_p being used in the numerator. For the example problem shown in Fig. 16.4, the factor of safety for sliding (F) = 1.79 when passive pressure is included and F = 1.55 when passive pressure is excluded. As previously mentioned, the typical recommendation for minimum factor of safety for sliding is 1.5 to 2.0 (Cernica 1995a).
Overturning analysis	The factor of safety (F) for overturning of the retaining wall is calculated by taking moments about the toe of the footing (point A, Fig. 16.12*b*), and is: $$F = \frac{Wa}{\frac{1}{3}P_H H - P_v e}$$ where a = lateral distance from the resultant weight of the wall and footing (W) to the toe of the footing P_H = horizontal component of the active earth pressure resultant force P_v = vertical component of the active earth pressure resultant force e = lateral distance from the location of P_v to the toe of the wall In Fig. 16.4*b*, the factor of safety (ratio) for overturning is calculated to be 3.73. As previously mentioned, the typical recommendation for minimum factor of safety for overturning is 1.5 to 2.0 (Cernica 1995a).

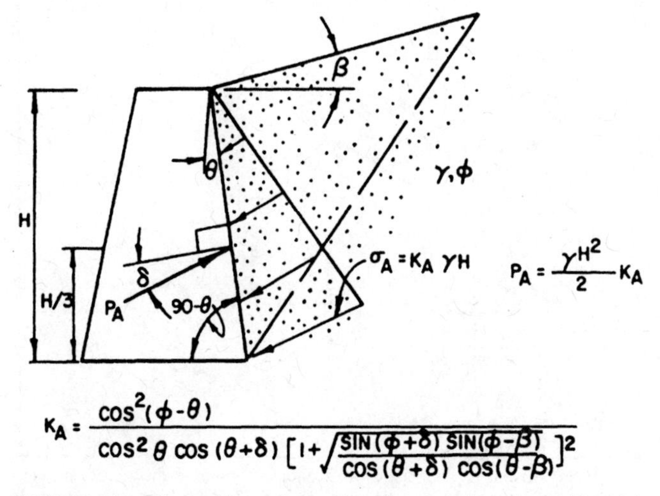

FIGURE 16.3 Coulomb's earth pressure (k_A) equation. See Table 16.4. (*Reproduced from NAVFAC DM-7.2, 1982.*)

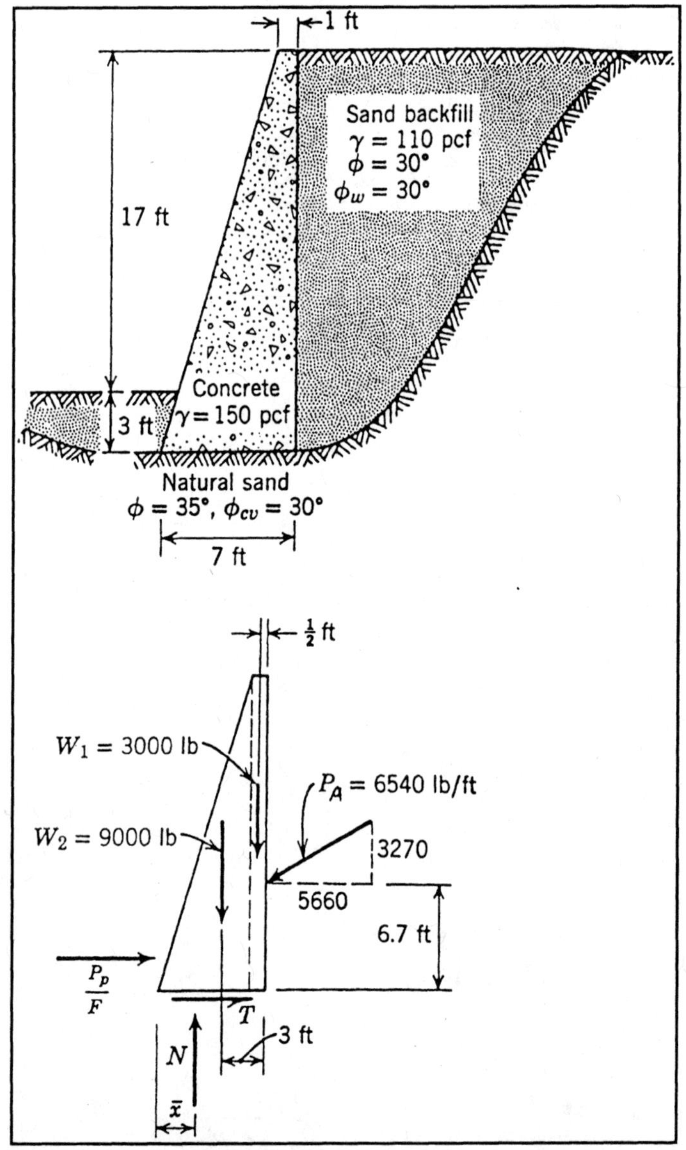

FIGURE 16.4a Example problem: Cross section of proposed retaining wall and resultant forces acting on the retaining wall. See Table 16.4. (*From Lambe and Whitman 1969; reproduced with permission of John Wiley & Sons.*)

Find. Adequacy of wall.
Solution. The first step is to determine the active thrust;
The next step is to compute the weights:

$$W_1 = (1)(20)(150) = 3000 \text{ lb/ft}$$

$$W_2 = \tfrac{1}{2}(6)(20)(150) = 9000 \text{ lb/ft}$$

Next N and $\bar{x}$ are computed:

$$N = 9000 + 3000 + 3270 = 15{,}270 \text{ lb/ft}$$

$$\text{Overturning moment} = 5660(6.67) - 3270(7) = 37{,}800 - 22{,}900 = 14{,}900$$

$$\text{Moment of weight} = (6.5)(3000) + (4)(9000) = 19{,}500 + 36{,}000 = 55{,}500$$

$$\text{Ratio} = 3.73 \qquad \underline{\underline{\text{OK}}}$$

$$\bar{x} = \frac{55{,}500 - 14{,}900}{15{,}270} = \frac{40{,}600}{15{,}270} = 2.66 \text{ ft} \qquad \underline{\underline{\text{OK}}}$$

The location of N

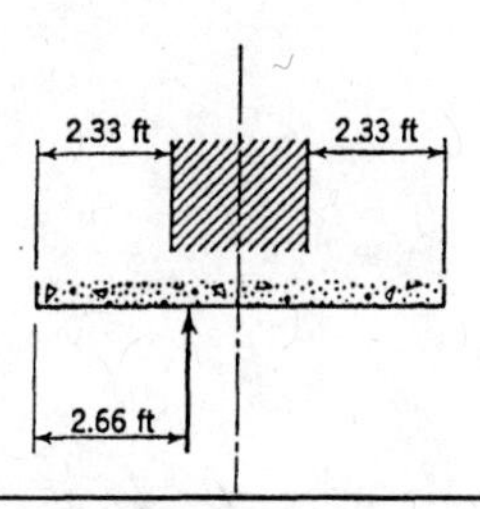

FIGURE 16.4b Example problem (*Continued*): Calculation of the factor of safety of overturning and the location of the resultant force N. See Table 16.4. (*From Lambe and Whitman 1969; reproduced with permission of John Wiley & Sons.*)

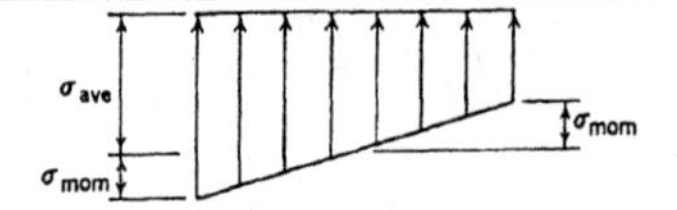

Next the bearing stress is computed. The average bearing stress is 15,270/7 = 2180 psf. Assuming that the bearing stress is distributed linearly, the maximum stress can be found

$$\sigma_{\text{mom}} = \frac{M}{S}$$

where

$$M = \text{moment about } \mathbb{C} = 15{,}270(3.5 - 2.66) = 12{,}820 \text{ lb-ft/ft}$$

$$S = \text{section modulus} = \tfrac{1}{6}B^2 = \tfrac{1}{6}(7)^2 = 8.17 \text{ ft}^2$$

where B is width of base

$$\sigma_{\text{mom}} = \frac{12{,}820}{8.17} = 1570 \text{ psf}$$

$$\text{Maximum stress} = 2180 + 1570 = 3750 \text{ psf}$$

Finally, the resistance to horizontal sliding is checked. Assuming passive resistance without wall friction,

$$K_p = 3$$

$$P_p = \tfrac{1}{2}(110)(3^2)(3) = 1500 \text{ lb/ft}$$

With reduction factor of 2,

$$\frac{P_p}{F} = 750 \text{ lb/ft}$$

$$T = 5660 - 750 = 4910 \text{ lb/ft}$$

$$N \tan 30^\circ = 8810 \text{ lb/ft}$$

$$\frac{N \tan \phi_{cv}}{T} = 1.79 < 2 \qquad \text{not OK}$$

Ignoring passive resistance

$$T = 5660 \text{ lb/ft}$$

$$\frac{N \tan \phi_{cv}}{T} = 1.55 > 1.5 \qquad \text{OK}$$

FIGURE 16.4c Example problem (*Continued*): Calculation of the maximum bearing stress and the factor of safety of sliding. See Table 16.4. (*From Lambe and Whitman 1969; reproduced with permission of John Wiley & Sons.*)

TABLE 16.5 Design and Construction of Retaining Walls

Topic (1)	Discussion (2)
Gravity, semigravity, T-shaped cantilevered, and counterfort walls	Figure 16.5 (from NAVFAC DM-7.2, 1982) shows several examples of different types of retaining walls. Figure 16.5*a* shows gravity and semigravity retaining walls, Fig. 16.5*b* shows cantilever and counterfort retaining walls, and the design analyses are shown in Fig. 16.5*c*. Although the equations in Figure 16.5*c* include an adhesion value c_a, as previously mentioned, the adhesion is often neglected. This is because active pressures develop when there is translation (movement) of the footing which would tend to break the adhesive resistance.
Retaining walls at the top of slopes	Retaining walls are sometimes constructed at the top of slopes. In this case, there is a descending ground surface in front of the retaining wall and the standard k_p equation (Table 16.2) can not be used to determine the passive earth pressure coefficient. For either a descending slope $(-\beta)$ or ascending slope $(+\beta)$ in front of the retaining wall, the following equation can be used to determine the passive earth pressure coefficient k_p: $$k_p = \frac{\cos^2 \phi}{[1 - (\sin^2 \phi + \sin \phi \cos \phi \tan \beta)^{0.5}]^2}$$ where ϕ = friction angle of the soil in front of the retaining wall and β = slope inclination measured from a horizontal plane, where a descending slope in front of the retaining wall has a negative β value. Although not readily apparent, if $\beta = 0$, the above equation will give exactly the same values of k_p as the equation from Table 16.2. For example, substituting $\beta = 0$ and $\phi = 30°$ into the above equation gives k_p = 3, which is exactly the value obtained from the equation in Table 16.2. Suppose that a retaining wall is constructed at the top of a 2:1 (horizontal:vertical) slope. In this case, the slope inclination is $\beta = -26.6°$. Assuming $\phi = 30°$ for the soil composing the slope, then according to the above equation, the value of $k_p = 1.12$. Thus for the case of a 2:1 descending slope in front of the retaining wall, the passive resistance will be significantly reduced (k_p decreases from 3 to 1.12). Many retaining walls tilt or deform downslope because the condition of a sloping ground surface in front of the retaining wall significantly reduces the passive resistance for the wall foundation. There can be additional factors that contribute to the failure of retaining walls constructed at the top of slopes. In some cases, there may be a low factor of safety of the entire slope which is exacerbated by the construction of a top of slope wall. In other cases, where the soil in front of the retaining wall is clayey, there can be slope creep which causes a soil gap to open up at the front of the retaining wall footing. In these cases of a descending clayey slope in front of the retaining wall, the depth of creep of the outer face of the slope must be estimated (see Table 13.19) and then the passive pressure must be utilized only below the depth of surface slope creep.

TABLE 16.5 Design and Construction of Retaining Walls (*Continued*)

Topic (1)	Discussion (2)
Earthquake analysis	It is difficult to accurately predict the additional lateral forces that will be generated on a retaining wall during an earthquake. Some factors affecting the magnitude of earthquake forces on the wall are the size and duration of the earthquake, the distance from the earthquake epicenter to the site, and the mass of soil retained by the wall. Many retaining walls are designed for only the active earth pressure and then fail when additional forces are generated by the earthquake. A simple approach, based on the work of Seed and Whitman (1970), is to include in the design analysis an additional horizontal force P_E to account for the additional loads imposed on the retaining wall by the earthquake, as follows: $$P_E = \tfrac{3}{8}(a_{max}/g)\gamma_t H^2$$ where a_{max} = peak acceleration at the ground surface γ_t = total unit weight of the backfill soil H = height of the retaining wall g = earth's gravity (9.81 m/s^2) As an example of the use of the above equation, assume that the maximum horizontal ground acceleration for the retaining wall shown in Fig. 16.4 will be 0.20g. Then substituting a_{max} = 0.2 g, γ_t = 110 pcf, and H = 20 ft into the above equation, find the additional horizontal force acting on the retaining wall due to the earthquake (P_E) = 3300 lb per linear foot of wall (48.1 kN per meter of wall). The location of this earthquake-induced force can be assumed to act at a distance of 0.6H above the base of the wall. Because P_E is a short-term loading that may never occur during the life of the retaining wall, it is common to allow a one-third increase in the bearing pressure and passive resistance for earthquake analysis. Also, for the analysis of sliding and overturning of the retaining wall, it is common to accept a lower factor of safety (1.1 to 1.2) under the combined static and earthquake loads.
Construction of the retaining wall	There are many construction factors that can result in excessive lateral movement, bearing capacity failures, sliding failures, or failure by overturning of the retaining wall. Common causes can include inadequate subsurface exploration or laboratory testing, incorrect design, improper construction, or unanticipated loadings. Typical construction-related problems are discussed below.
Clay backfill	A frequent cause of failure is that the wall was backfilled with clay. As previously mentioned, clean granular sand or gravel is usually recommended as backfill material. This is because of the undesirable effects of using clay or silt as a backfill material (see Table 12.7). When clay is used as backfill material, the clay backfill can exert swelling pressures on the wall (Fourie 1989, Marsh and Walsh 1996). The highest swelling pressures develop when water infiltrates a backfill consisting of a clay that was compacted to a high dry density at a low moisture content. The type of clay particles that will exert the highest swelling pressures is montmorillonite. Because the clay backfill is not free-draining, there could also be additional hydrostatic forces or ice-related forces that substantially increase the thrust on the wall.

TABLE 16.5 Design and Construction of Retaining Walls (*Continued*)

Topic (1)	Discussion (2)
Inferior backfill soil	To reduce construction costs, soil available on-site is sometimes used for backfill. This soil may not have the properties, such as being a clean granular soil with a high shear strength, that were assumed during the design stage. Using on-site available soil, instead of importing granular material, is probably the most common reason for retaining wall failures.
Compaction-induced pressures	As previously mentioned, one reason for applying a factor of safety F to the active earth pressure resultant force P_A is that larger wall pressures will typically be generated during compaction of the backfill. By using heavy compaction equipment in close proximity to the wall, excessive pressures can be developed that damage the wall. The best compaction equipment, in terms of exerting the least compaction-induced pressures on the wall, is small vibrator plate (hand-operated) compactors such as models VPG 160B and BP 19/75 (Duncan et al. 1991). The vibrator plates effectively densify the granular backfill, but do not induce high lateral loads because of their light weight. Besides hand-operated compactors, other types of relatively lightweight equipment (such as a bobcat) can be used to compact the backfill.
Failure of the backcut	There could also be the failure of the backcut for the retaining wall. A vertical backcut is often used when the retaining wall is less than 1.5 m (5 ft) high. In other cases, the backcut is usually sloped. The backcut can fail if it is excavated too steeply and does not have an adequate factor of safety.
Retaining wall deformation	Retaining wall movement is gradual and the wall deforms by intermittently tilting or moving laterally. It is possible that a failure can occur suddenly, such as when there is a slope-type failure beneath the wall or when the foundation of the wall fails because of inadequate bearing capacity. These rapid failure conditions could develop if the wall is supported by soft clay and there is an undrained shear failure beneath the foundation.

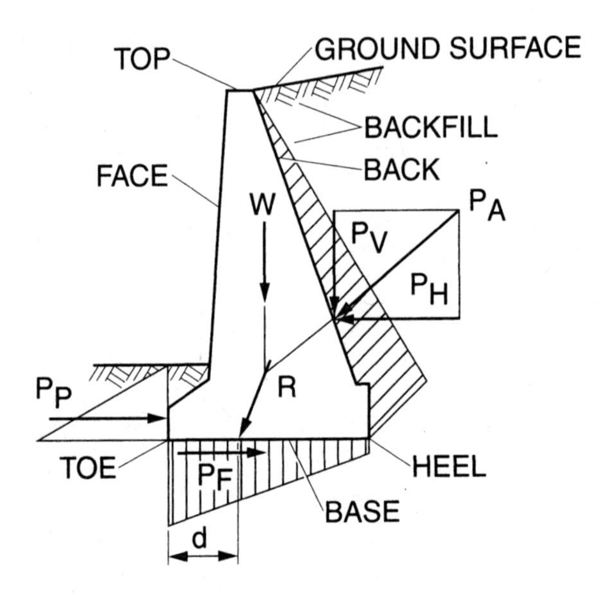

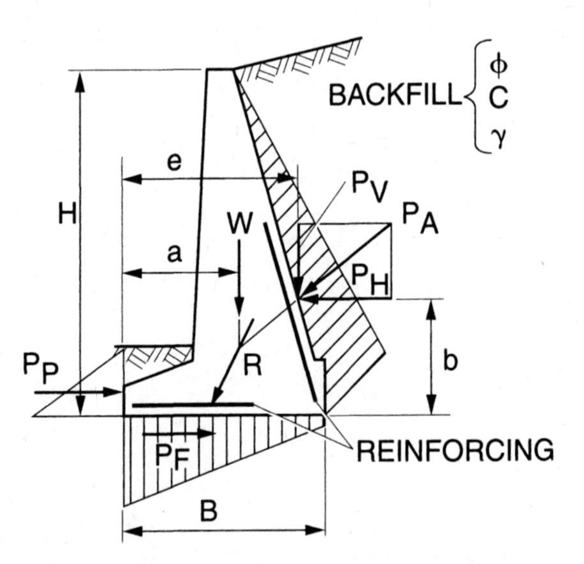

FIGURE 16.5a Gravity and semigravity retaining walls. See Table 16.5. (*From NAVFAC DM-7.2, 1982.*)

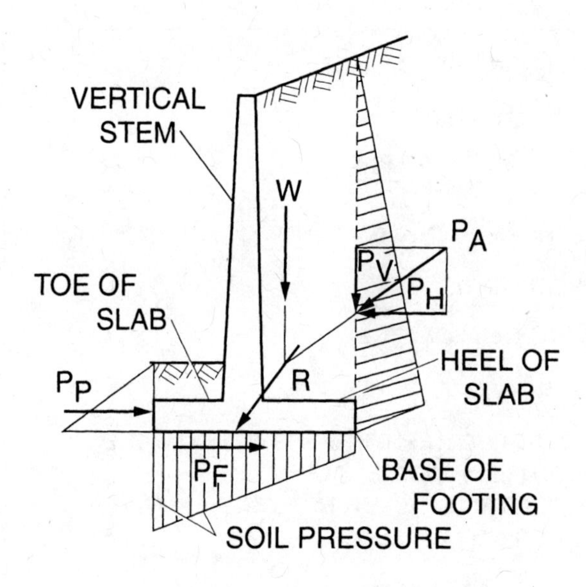

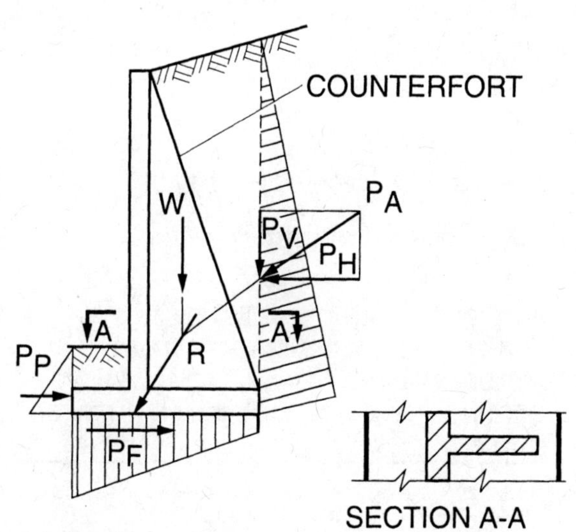

FIGURE 16.5b Cantilever and counterfort retaining walls. See Table 16.5. (*From NAVFAC DM-7.2, 1982.*)

LOCATION OF RESULTANT

MOMENTS ABOUT TOE:

$$d = \frac{Wa + P_V e - P_H b}{W + P_V}$$

ASSUMING $P_P = 0$

OVERTURNING

MOMENTS ABOUT TOE:

$$F = \frac{Wa}{P_H b - P_V e} \geqq 1.5$$

IGNORE OVERTURNING IF R IS WITHIN MIDDLE THIRD (SOIL), MIDDLE HALF (ROCK).
CHECK R AT DIFFERENT HORIZONTAL PLANES FOR GRAVITY WALLS.

RESISTANCE AGAINST SLIDING

$$F = \frac{(W + P_V)\ \mathrm{TAN}\ \delta + C_a B}{P_H} \geqq 1.5$$

$$F = \frac{(W+P_V)\ \mathrm{TAN}\ \delta + C_a B + P_P}{P_H} \geqq 2.0$$

$$P_F = (W + P_V)\ \mathrm{TAN}\ \delta + C_a B$$

C_a = ADHESION BETWEEN SOIL AND BASE

TAN δ = FRICTION FACTOR BETWEEN SOIL AND BASE

W = INCLUDES WEIGHT OF WALL AND SOIL IN FRONT FOR GRAVITY AND SEMIGRAVITY WALLS.
INCLUDES WEIGHT OF WALL AND SOIL ABOVE FOOTING, FOR CANTILEVER AND COUNTERFORT WALLS.

FIGURE 16.5c Design analysis for retaining walls shown in Figs. 16.5*a* and 16.5*b*. See Table 16.5. (*From NAVFAC DM-7.2, 1982.*)

TABLE 16.6 Mechanically Stabilized Earth Retaining Walls

Topic (1)	Discussion (2)
Construction of MSE retaining walls	The first step in the construction of a mechanically stabilized earth (MSE) retaining wall is to install a drainage system. A key can be excavated into the natural ground and then a drainage system can be constructed at the back of the key. The drainage system often consists of a perforated drain pipe, surrounded by open graded gravel, which is in turn surrounded by the geofabric (see "Geofabric Alternative," Standard Detail No. 1, App. B). A mechanically stabilized earth retaining wall is composed of three elements: (1) wall facing material; (2) soil reinforcement, such as strip- or grid-type reinforcement; and (3) compacted fill between the soil reinforcement. The soil reinforcement is often attached to the wall facing elements. The compacted soil typically consists of a granular (permeable) soil in order to prevent the buildup of pore water pressure behind or in the mechanically stabilized earth zone. The soil reinforcement and compacted fill derive frictional resistance and interlocking resistance between each other. When the mechanically stabilized soil mass is subjected to shear stress, the soil tends to transfer the shear stress to the stronger geogrid. Also, if high stress concentrations develop in the mechanically stabilized soil mass, along a potential slip surface, for example, then the geogrid tends to redistribute stresses away from areas of high stress. Most often lightweight equipment is used to compact the fill that is placed on top of each layer of soil reinforcement. Heavy compaction equipment could push the wall facing elements out of alignment.
Design analyses for MSE retaining walls	The design analysis for a mechanically stabilized earth retaining wall is more complex than for a cantilevered retaining wall. For a mechanically stabilized retaining wall, both the internal and external stability must be checked. 1. *External stability.* The analysis for the external stability is similar to a gravity retaining wall. For example, Figs. 16.6 and 16.7 (adapted from AASHTO 1996) present the design analysis for external stability for a level backfill condition and a sloping backfill condition. In both Figs. 16.6 and 16.7, the zone of mechanically stabilized earth mass is treated in a similar fashion as a massive gravity retaining wall. The following analyses must be performed: *a.* *Allowable bearing pressure.* The bearing pressure due to the reinforced soil mass must not exceed the allowable bearing pressure. *b.* *Factor of safety of sliding.* The reinforced soil mass must have an adequate factor of safety for sliding (F = 1.5 to 2). *c.* *Factor of safety of overturning.* The reinforced soil mass must have an adequate factor of safety (F = 2) for overturning about point O. *d.* *Resultant of vertical forces N.* The resultant of the vertical forces N must be within the middle one-third of the base of the reinforced soil mass. *e.* *Stability of reinforced soil mass.* The stability of the entire reinforced soil mass (i.e., shear failure below the bottom of the wall) would have to be checked.

TABLE 16.6 Mechanically Stabilized Earth Retaining Walls (*Continued*)

Topic (1)	Discussion (2)
Design analyses for MSE retaining walls (*Continued*)	Note in Fig. 16.6 that two forces (P_1 and P_2) are shown acting on the reinforced soil mass. The first force (P_1) is determined from the standard active earth pressure resultant equation (i.e., Table 16.2). The second force (P_2) is a result of a uniform surcharge Q applied to the entire ground surface behind the mechanically stabilized earth retaining wall. If the wall does not have a surcharge, then P_2 is equal to zero. Figure 16.7 presents the active earth pressure force for an inclined slope behind the retaining wall. Note in Fig. 16.7 that the friction δ of the soil along the back side of the reinforced soil mass has been included in the analysis. The value of k_A would be obtained from Coulomb's earth pressure equation (Fig. 16.3). As a conservative approach, the friction angle δ can be assumed to be equal to zero and then $P_H = P_A$. Note in both Figs. 16.6 and 16.7 that the minimum width of the reinforced soil mass must be at least $\frac{7}{10}$ the height of the reinforced soil mass. 2. *Internal stability.* In terms of the internal stability, there is good agreement between the predicted (from the method of slices) and measured slip surface in a mechanically stabilized zone consisting of geotextile (Zornberg et al. 1998). To check the stability of the mechanically stabilized zone, a slope stability analysis can be performed where the soil reinforcement is modeled as horizontal forces equivalent to the allowable tensile resistance of the geogrid. In addition to calculating the factor of safety, the pull-out resistance of the reinforcement along the slip surface should also be checked.
Movement of MSE retaining walls	The analysis of mechanically stabilized earth retaining walls is based on active earth pressures. It is assumed that the wall will move enough to develop the active wedge. As in concrete retaining walls, it is important that building footings or other load-carrying members are not supported by the mechanically stabilized earth retaining wall and the active wedge, or else they will be subjected to lateral movement.
Example problem	*Problem:* Using the mechanically stabilized earth retaining wall shown in Fig. 16.6, assume $H = 20$ ft, the width of the mechanically stabilized earth retaining wall = 14 ft, the depth of embedment at the front of the mechanically stabilized zone = 3 ft, and the soil behind the mechanically stabilized zone is a clean sand with a friction angle $\phi = 30°$ and a total unit weight of $\gamma_t = 110$ pcf. Also assume that there is sand in front of the wall with these same properties, and there is a level backfill with no surcharge pressures (i.e., $P_2 = 0$). Further assume that the mechanically stabilized zone will have a total unit weight $\gamma_t = 120$ pcf, there will be no shear stress (i.e., $\delta = 0°$) along the vertical back side of the mechanically stabilized zone, and $\delta = 23°$ along the bottom of the mechanically stabilized zone. Determine the factor of safety for sliding and the factor of safety for overturning. *Solution:* $k_A = 0.333$, $k_p = 3.0$, $P_A = 7330$ lb/ft, $P_p = 740$ lb/ft (reduction factor = 2.0), and $W = 33{,}600$ lb. Factor of safety for sliding = [(33,600)(tan 23°) + 740]/7,330 = 2.05 Factor of safety for overturning = [(33,600)(14/2)]/[(7330)(20/3)] = 4.81

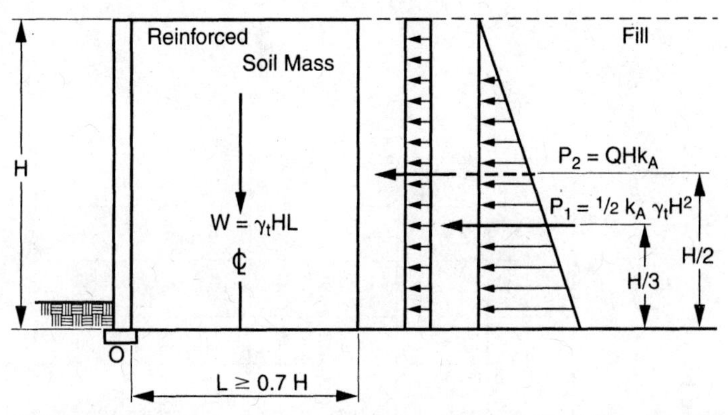

FIGURE 16.6 Design analysis for mechanically stabilized earth retaining wall having horizontal backfill. See Table 16.6. (*Adapted from AASHTO 1996.*)

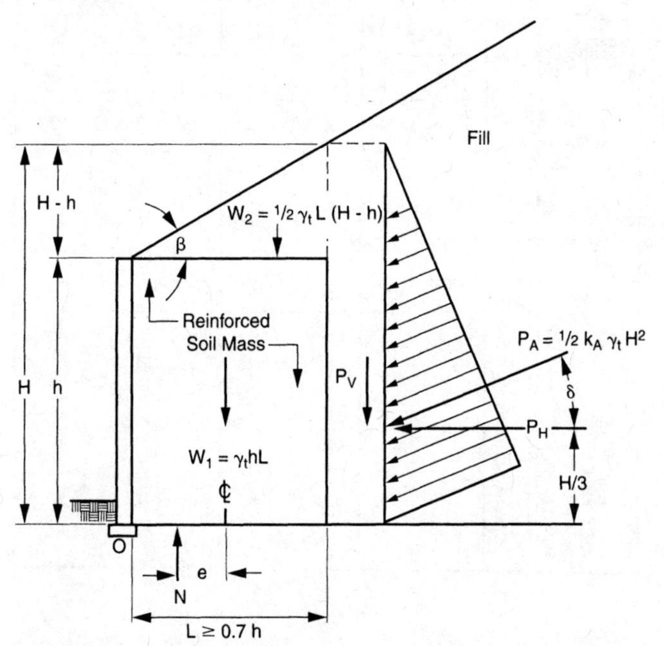

FIGURE 16.7 Design analysis for mechanically stabilized earth retaining wall having sloping backfill. See Table 16.6. (*Adapted from AASHTO 1996.*)

Suggested Allowable Design Stresses—Sheet Piling

Steel Brand or Grade	Minimum Yield Point, psi	Allowable Design Stress, psi*
USS-EX-TEN 55 (ASTM A572 GR 55)	55,000	35,000
USS EX-TEN 50 (ASTM A572 GR 50)	50,000	32,000
USS MARINER STEEL	50,000	32,000
USS EX-TEN 45 (ASTM A572 GR 45)	45,000	29,000
Regular Carbon Grade (ASTM A 328)	38,500	25,000

Steel Sheet Piling Sections

Profile		Section Index	District Rolled	Driving Distance per Pile	Weight		Web Thickness	Section Modulus		Area	Moment of Inertia	
					Per Foot	Per Square Foot of Wall		Per Pile	Per Foot of Wall	Per Pile	Per Pile	Per Foot of Wall
				In.	Lbs.	Lbs.	In.	In.³	In.³	In.²	In.⁴	In.⁴
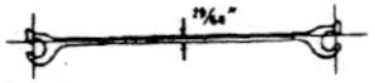 29/64"	Interlock with Each Other	PSX 32	H.	16½	44.0	32.0	29/64	3.3	2.4	12.94	5.1	3.7
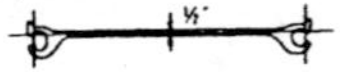 ½"		PS32*	H.S.	15	40.0	32.0	½	2.4	1.9	11.77	3.6	2.9
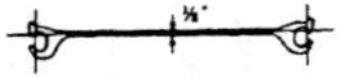 ⅜"		PS28	H.S.	15	35.0	28.0	⅜	2.4	1.9	10.30	3.5	2.8
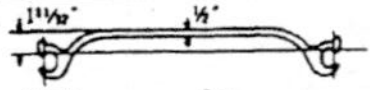 1 11/32"; ½"	Interlock with Each Other	PSA 28*	H.	16	37.3	28.0	½	3.3	2.5	10.98	6.0	4.5
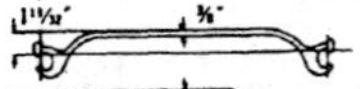 1 11/32"; ⅜"		PSA 23	H.S.	16	30.7	23.0	⅜	3.2	2.4	8.99	5.5	4.1
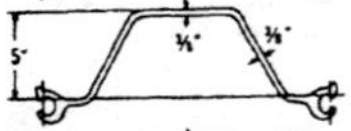 5"; ⅜"; ⅜"		PDA 27	H.S.	16	36.0	27.0	⅜	14.3	10.7	10.59	53.0	39.8
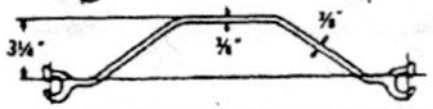 3½"; ⅜"; ⅜"		PMA 22	H.S.	19⅝	36.0	22.0	⅜	8.8	5.4	10.59	22.4	13.7

Section	Interlock	Section										
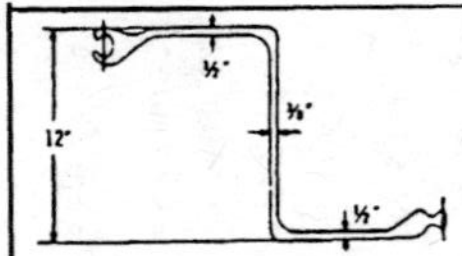 12"; ½"; ½"; ½"	Interlock with Each Other and with PSA23 or PSA28	PZ38	H.	18	57.0	38.0	⅜	70.2	46.8	16.77	421.2	280.8
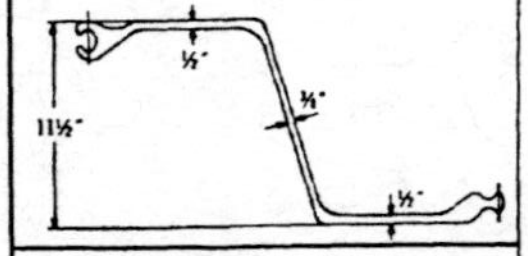 11½"; ½"; ½"; ½"		PZ32	H.	21	56.0	32.0	⅜	67.0	38.3	16.47	385.7	220.4
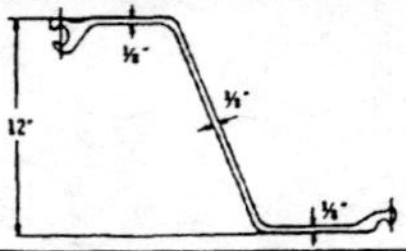 12"; ⅜"; ⅜"; ⅜"		PZ27	H.	18	40.5	27.0	⅜	45.3	30.2	11.91	276.3	184.2
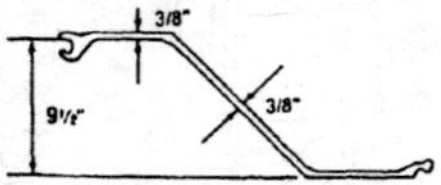 9½"; 3/8"; 3/8"		PZ 22	H.	22	40.3	22.0	⅜	34.8	19.0	11.9	167	91.1

FIGURE 16.8 Steel sheet piling sections. See Table 16.7. (*From USS Steel Sheet Piling Design Manual 1984.*)

TABLE 16.7 Anchored Sheet Pile Walls

Topic (1)	Discussion (2)
General discussion	There are many different types of design methods that are used for sheet pile walls. Figure 16.9 (from NAVFAC DM-7.2, 1982) shows the most common type of design method. In Fig. 16.9, the term *H* represents the unsupported face of the sheet pile wall. As indicated in Fig. 16.9, this sheet pile wall is being used as a waterfront retaining structure and the level of water in front of the wall is at the same elevation as the groundwater table elevation behind the wall. For highly permeable soil, such as clean sand and gravel, this often occurs because the water can quickly flow underneath the wall in order to equalize the water levels. In Fig. 16.9, the term *D* represents that portion of the sheet pile wall that is anchored in soil. Also shown in Fig. 16.9 is a force designated as A_p. This represents a restraining force on the sheet pile wall due to the construction of a tieback, such as a rod that has a grouted end or is attached to an anchor block. Tieback anchors are often used in sheet pile wall construction in order to reduce the bending moments in the sheet pile. When tieback anchors are used, the sheet pile wall is typically referred to as an *anchored bulkhead,* while if no tiebacks are utilized, the wall is called a *cantilevered sheet pile wall.*
Earth pressures acting on the sheet pile wall	Sheet pile walls tend to be relatively flexible. Thus, as indicated in Fig. 16.9, the design is based on active and passive earth pressures. 1. *Active earth pressure.* The soil behind the wall is assumed to exert an active earth pressure on the sheet pile wall. At the groundwater table (point *A*), the active earth pressure is equal to: Active earth pressure at point *A* (kPa or psf) $= k_A \gamma_t d_1$ where k_A = active earth pressure coefficient from Table 16.2 (dimensionless parameter). The friction between the sheet pile wall and the soil is usually neglected in the design analysis. γ_t = total unit weight of the soil above the groundwater table (kN/m^3 or pcf) d_1 = depth from the ground surface to the groundwater table (m or ft) For sheet pile wall analyses, a unit length (1 m or 1 ft) of sheet pile wall is assumed. At point *B* in Fig. 16.9, the active earth pressure equals: Active earth pressure at point *B* (kPa or psf) $= k_A\gamma_t d_1 + k_A\gamma_b d_2$ where γ_b = buoyant unit weight of the soil below the groundwater table d_2 = depth from the groundwater table to the bottom of the sheet pile wall. For a sheet pile wall having assumed values of *H* and *D* (see Fig. 16.9), and using the calculated values of active earth pressure at points *A* and *B*, the active earth pressure resultant force P_A, in kN per linear meter of wall or pounds per linear foot of wall, can be calculated. 2. *Passive earth pressure.* The soil in front of the wall is assumed to exert a passive earth pressure on the sheet pile wall. The passive earth pressure at point *C* in Fig. 16.9 is equal to:

TABLE 16.7 Anchored Sheet Pile Walls (*Continued*)

Topic (1)	Discussion (2)
Earth pressures acting on the sheet pile wall (*Continued*)	Passive earth pressure at point *C* (kPa or psf) $= k_p \gamma_b D$ where the passive earth pressure coefficient k_p can be calculated from the equation in Table 16.2. As in the analysis of cantilever retaining walls, if it is desirable to limit the amount of sheet pile wall translation, then a reduction factor can be applied to the passive pressure. Once the allowable passive pressure is known at point *C*, the passive resultant force P_p can be readily calculated. As an alternative solution for the passive pressure, the equation for P_p in Table 16.2 can be used with the buoyant unit weight γ_b substituted for the total unit weight γ_t and the depth *D* as shown in Fig. 16.9. Note that a water pressure has not been included in the analysis. This is because the water level is the same on both sides of the wall and water pressure cancels out, and thus should not be included in the analysis.
Design analysis	The design of sheet pile walls requires the following analyses: (1) evaluation of the earth pressures that act on the wall, such as shown in Fig. 16.9; (2) determination of the required depth *D* of piling penetration; (3) calculation of the maximum bending moment (M_{max}) which is used to determine the maximum stress in the sheet pile; and (4) selection of the appropriate piling type, size, and construction details. A typical design process is to assume a depth *D* (Fig. 16.9) and then calculate the factor of safety for toe failure (i.e., toe kick-out) by the summation of moments at the tieback anchor (point *D*). The factor of safety is defined as the moment due to the passive force divided by the moment due to the active force. Values of acceptable factor of safety for toe failure are 2 to 3. An alternative solution is to first select the factor of safety and then develop the active and passive resultant forces and moment arms in terms of *D*. By solving the equation, the value of *D* for a specific factor of safety can be directly calculated. Once the depth *D* of the sheet pile wall is known, the anchor pull A_p must be calculated. The anchor pull is determined by the summation of forces in the horizontal direction: $A_p = P_A - P_p/F$ where P_A and P_p are the resultant active and passive forces (see Fig. 16.9) and *F* is the factor of safety that was obtained from the toe failure analysis. Based on the earth pressure diagram (Fig. 16.9) and the calculated value of A_p, elementary structural mechanics can be used to determine the maximum moment in the sheet pile wall. The maximum moment divided by the section modulus can then be compared with the allowable design stresses listed in Fig. 16.8.

TABLE 16.8 Cantilevered Sheet Pile Walls

Topic (1)	Discussion (2)
General discussion	In the case of cantilevered sheet pile walls, the sheet piling is driven to an adequate depth to become fixed as a vertical cantilever in resisting the lateral active earth pressure. Cantilevered walls usually undergo large lateral deflections and are readily affected by scour and erosion in front of the wall. Because of these factors, penetration depths can be quite high, resulting in excess stresses and severe yield. According to the *USS Steel Sheet Piling Design Manual* (1984), cantilevered walls using steel sheet piling are restricted to a maximum height H of approximately 4.6 m (15 ft).
Design chart	Design charts have been developed for the analysis of cantilevered sheet pile walls. For example, Fig. 16.10 is a design chart reproduced from the *USS Steel Sheet Piling Design Manual* (1984). The value of α is based on the location of the water level as indicated in Fig. 16.10. The chart is entered with the ratio of the passive earth pressure coefficient k_p divided by the active earth pressure coefficient k_A. By intersecting either the depth ratio curves or moment ratio curves, the depth of the sheet pile wall D or maximum moment M_{max} can be calculated. Note that the depth D from Fig. 16.10 corresponds to a factor of safety of 1.0. The *USS Steel Sheet Piling Design Manual* (1984) states that increasing the depth of embedment D by 20 to 40% will provide a factor of safety of toe failure of approximately 1.5 to 2.0.
Example problem	*Problem:* Assume a cantilevered sheet pile wall with $k_A = 0.2$, $k_p = 8.65$, $\alpha = 1.0$, $H = 13.8$ ft (4.2 m), and $\gamma_b = 57$ pcf (9 kN/m^3). Determine the total length and the maximum moment in the sheet pile wall. *Solution:* 1. *Total length* ($H + D$). Entering Fig. 16.10 at $k_p/k_A = 43$, and intersecting the depth ratio curve of $\alpha = 1$, find $D/H = 0.6$. Therefore, for $H = 13.8$ ft (4.2 m), the value of D for a factor of safety of 1 is equal to 8.4 ft (2.5 m). For a 30% increase in embedment depth, the required embedment depth $D = 11$ ft (3.3 m). Thus, the total length of the sheet pile wall is 25 ft (7.6 m). 2. *Maximum moment* (M_{max}). Entering Fig. 16.10 at $k_p/k_A = 43$, and intersecting the moment ratio curve of $\alpha = 1$, find $M_{max}/\gamma_b k_A H^3) = 0.53$. Thus the maximum moment in the sheet pile wall = 15,900 ft-lb/ft (70.7 kN·m/m).

TABLE 16.9 Cofferdams

Topic (1)	Discussion (2)
General discussion	Steel sheet piling is widely used for the construction of a cofferdam, which is defined as a temporary structure designed to support the sides of an excavation and to exclude water from the excavation. Tomlinson (1986) presents an in-depth discussion of the different types of cofferdams and construction techniques. For example, the first step in the construction of a bridge foundation could be the installation of a large circular cofferdam. Then the river water would be pumped out from inside the cofferdam and the river bottom would be excavated to the desired bearing stratum for the bridge foundation.
Loads on a cofferdam	In general, the following loads could be exerted on the cofferdam: 1. Hydrostatic groundwater pressures located outside the cofferdam. If the cofferdam is constructed in a river or other body of water, then the hydrostatic water pressure should be based on the anticipated high water level. 2. Hydrostatic pressure of water inside the cofferdam if the water level outside falls below the interior level during initial pumping of water from the cofferdam. This condition could cause bursting of the cofferdam by interlock tension. 3. Earth pressure outside the cofferdam. 4. Other loads, such as surcharge loads, wave pressures, and earthquake loading.
Design conditions	Under the above loading conditions, the cofferdam must have an adequate factor of safety for the following circumstances: 1. Heave of the soil located at the bottom of the cofferdam. 2. Sliding of the cofferdam along its base. 3. Overturning and tilting of the cofferdam. 4. The inward or outward yielding of the cofferdam (i.e., maximum moment and shear in the sheet piling and interlock tension). 5. Piping of soil, which could undermine the cofferdam or lead to sudden flooding of the interior of the cofferdam.
Design earth pressures	An important consideration in the design of cofferdams is the ability of the sheet pile wall to deflect. If struts are used that brace the opposite sides of the excavation (i.e., braced excavation), the deflection of the sheet pile wall will be restricted. The next section will discuss the earth pressures exerted on braced excavations. If the sheet pile wall is able to deform—if raking braces are used, for example—then the deflections of the wall during construction will permit mobilization of active pressures in accordance with the previous discussion. The *USS Steel Sheet Piling Design Manual* (1984) and NAVFAC DM-7.2 (1982) present a further discussion of the design of cofferdams.

TABLE 16.10 Important Design Considerations for Sheet Pile Walls and Cofferdams

Topic (1)	Discussion (2)
General discussion	Some other important design considerations for sheet pile walls are included in this table.
Soil layers	The active and passive earth pressures should be adjusted for soil layers having different engineering properties.
Penetration depth	The penetration depth D of the sheet pile wall should be increased by at least an additional 20% to allow for the possibility of dredging and scour. Deeper penetration depths may be required, depending on the results of a scour analysis.
Surcharge loads	The ground surface behind the sheet pile wall is often subjected to surcharge loads. The equation for P_Q in Table 16.2 can be used to determine the active earth pressure resultant force due to a uniform surcharge pressure applied to the ground surface behind the wall. Note that for this equation, the entire height of the sheet pile wall (i.e., $H + D$; see Fig. 16.9) must be used in place of H. Typical surcharge pressures exerted on sheet pile walls are caused by railroads, highways, dock loading facilities and merchandise, ore piles, and cranes.
Unbalanced hydrostatic and seepage forces	The previous analyses have assumed that the water level on both sides of the sheet pile wall are at the same elevation. Depending on factors such as the watertightness of the sheet pile wall and the backfill permeability, it is possible that the groundwater level could be higher than the water level in front of the wall, in which case the wall would be subjected to water pressures. This condition could develop when there is a receding tide or a heavy rainstorm that causes a high groundwater table. Figure 16.11 shows a method of analysis for unbalanced hydrostatic and seepage forces. In Fig. 16.11, the water level behind the sheet pile wall is above the water level in front of the sheet pile wall by a vertical distance defined as H_w. Because of this difference in groundwater levels, water will flow underneath the sheet pile wall. In Fig. 16.11, the lines with the arrows are the flow lines. The procedure is as follows: 1. The first step in the analysis is to assume no groundwater flow and then determine the active earth pressure resultant force (P_A from the method outlined in Table 16.7), the hydrostatic water pressure resultant force behind the sheet pile wall ($P_{wA} = \frac{1}{2}\gamma_w H^2$), the passive earth pressure resultant force ($P_p = \frac{1}{2} k_p \gamma_b D^2$), and the hydrostatic water pressure resultant force in front of the sheet pile wall [$P_{wp} = \frac{1}{2}\gamma_w (H - H_w)^2$]. 2. The second step in the analysis is to determine $\Delta P_p = PH\gamma_w H_w$ and $\Delta P_A = AD\gamma_w H_w$, which are adjustment factors to account for the flow of water underneath the sheet pile wall. 3. As indicated in Fig. 16.11, the final step is to subtract the values of ΔP_p and ΔP_A from the values obtained from step 1.

TABLE 16.10 Important Design Considerations for Sheet Pile Walls and Cofferdams (*Continued*)

Topic (1)	Discussion (2)
Other loading conditions	The sheet pile wall may have to be designed to resist the lateral loads due to ice thrust, wave forces, ship impact, mooring pull, and earthquake forces. If granular soil behind or in front of the sheet pile wall is in a loose state, it could be susceptible to liquefaction during an earthquake.
Factors increasing the stability	A factor usually not considered in the design analysis is the densification of loose sand during driving of the sheet piles. However, since sheet piles are relatively thin, the densification effect would be less for a sheet pile than for a comparable round pile, because the sheet pile displaces less soil. Another beneficial effect in the design analysis is that many sheet pile walls used for waterfront construction are relatively long and hence the soil is in a plane strain condition. As previously discussed, the plane strain shear strength is higher than the shear strength determined from conventional triaxial shear strength tests.
Cohesive soil	For cohesive soil, the long-term condition often governs the design. In this case, an effective stress analysis can be performed (c' and ϕ') using the estimated location of the groundwater table. For convenience, the effective cohesion is neglected and the analysis is performed by using only ϕ'. Therefore, the long-term effective stress condition for cohesive soil is analyzed as described in the preceding subsections for sheet pile walls having granular soil. Since ϕ' for a cohesive soil is usually less than ϕ' for granular soil, the active earth pressure will be higher and the passive resistance lower for cohesive soil.
Example problem	*Problem:* Using the sheet pile wall diagram in Fig. 16.9, assume that the soil behind and in front of the sheet wall is uniform sand with a friction angle $\phi' = 33°$, buoyant unit weight $\gamma_b = 64$ pcf, and, above the groundwater table, the total unit weight $\gamma_t = 120$ pcf. Also assume that the sheet pile wall has $H = 30$ ft, $D = 20$ ft; the water level in front of the wall is at the same elevation as the groundwater table, which is located 5 ft below the ground surface; and the tieback anchor is located 4 ft below the ground surface. In the analysis, neglect wall friction and calculate the factor of safety for toe kick-out and the anchor pull force A_p, assuming that the anchors will be spaced 10 ft on center. *Solution:* $k_A = 0.295$, $k_p = 3.39$, $P_A = 27{,}500$ lb/ft, and $P_p = 43{,}400$ lb/ft (no reduction factor). Location of active earth pressure resultant force = 28.3 ft and the location of the passive earth pressure resultant force = 39.3 ft below the anchor. Factor of safety for toe kick-out = [(43,400)(39.3)]/[(27,500)(28.3)] = 2.19 Anchor pull force = 27,500 − (43,400/2.19) = 7680 lb = 7.68 kips For a 10-ft spacing, $A_p = (7.68)(10) = 76.8$ kips

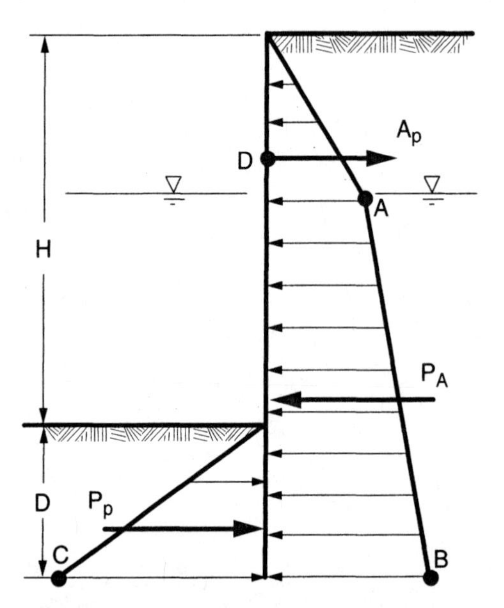

FIGURE 16.9 Earth pressure diagram for design of anchored sheet pile wall. See Table 16.7. *(Reproduced from NAVFAC DM-7.2, 1982.)*

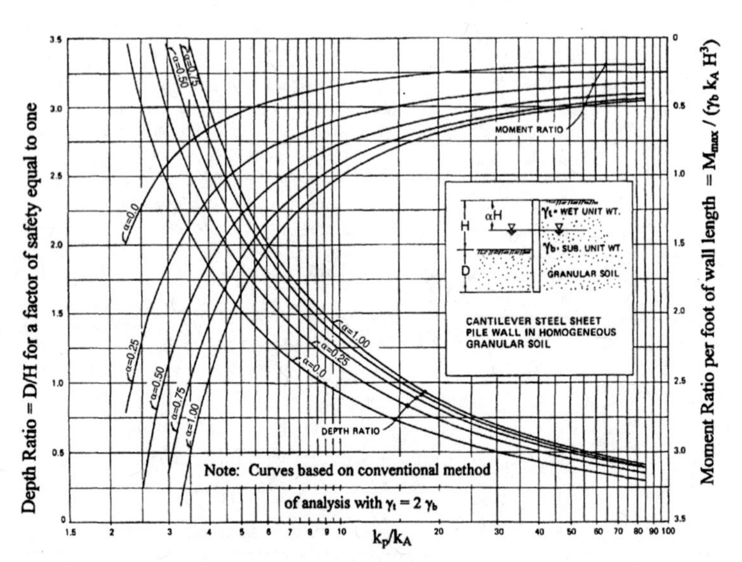

FIGURE 16.10 Design chart for cantilevered sheet pile wall. See Table 16.8. *(From USS Sheet Piling Design Manual 1984.)*

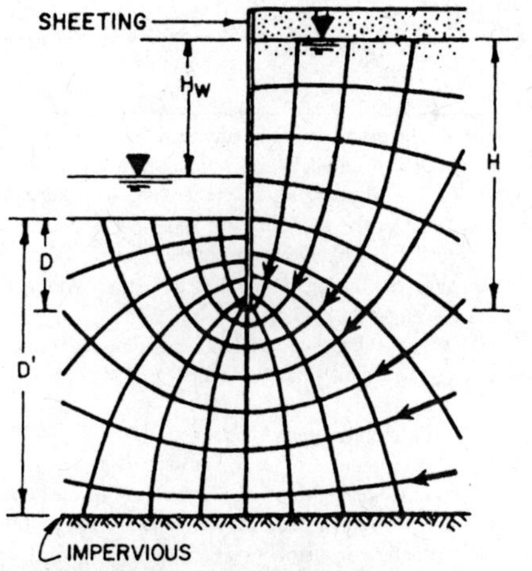

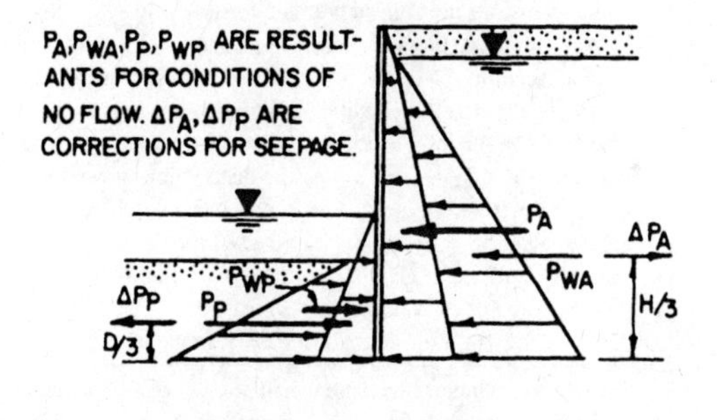

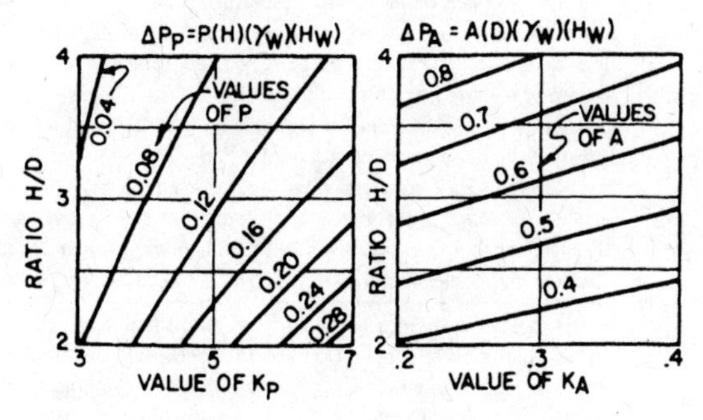

FIGURE 16.11 Effect of seepage beneath sheet pile wall. See Table 16.10. (*From NAVFAC DM-7.2, based on work by Richart and Schmertmann.*)

TABLE 16.11 Temporary Retaining Walls

Topic (1)	Discussion (2)
Braced retaining walls	For some projects, the temporary retaining wall may be constructed of sheeting (such as sheet piles) that is supported by horizontal braces, also known as *struts*. Near or at the top of the temporary retaining wall, the struts restrict movement of the retaining wall and prevent the development of the active wedge. Because of this inability of the retaining wall to deform at the top, earth pressures near the top of the wall are in excess of the active (k_A) pressures (see Fig. 16.12). At the bottom of the wall, the soil is usually able to deform into the excavation, which results in a reduction in earth pressure, and the earth pressures at the bottom of the excavation tend to be constant or even decrease, as shown in Fig. 16.12. The earth pressure distributions shown in Fig. 16.12 were developed from actual measurements of the forces in struts during the construction of braced excavations (Terzaghi and Peck 1967). In Fig. 16.12, case *a* shows the earth pressure distribution for braced excavations in sand and cases *b* and *c* show the earth pressure distribution for clays. In Fig. 16.12, the distance H represents the depth of the excavation (i.e., the height of the exposed wall surface). The earth pressure distribution is applied over the exposed height H of the wall surface with the earth pressures transferred from the wall sheeting to the struts (the struts are labeled with the forces F_1, F_2, etc.). Any surcharge pressures, such as surcharge pressures on the ground surface adjacent to the excavation, must be added to the pressure distributions shown in Fig. 16.12. In addition, if the sand deposit has a groundwater table that is above the level of the bottom of the excavation, then water pressures must be added to the case *a* pressure distribution shown in Fig. 16.12. Because the excavations are temporary (i.e., short-term condition), the undrained shear strength ($s_u = c$) is used for the analysis of the earth pressure distributions for clay. The earth pressure distributions for clay (i.e., cases *b* and *c*) are not valid for permanent walls or for walls where the groundwater table is above the bottom of the excavation.
Steel I-beam and wood lagging retaining wall	This type of temporary retaining wall consists of steel I beams that are either driven into place or installed in predrilled holes with the bottom portion cemented into place. During the main excavation of the interior area, wood lagging is installed between the steel I beams. The earth pressure exerted on the wood lagging is transferred to the steel I-beam flanges. In some cases, the steel I beams will be braced with struts and the earth pressure distributions would be similar to those as shown in Fig. 16.12. For cantilevered steel I-beam and wood lagging retaining walls, there will be sufficient movement of the top of the wall to develop the active earth pressure. Tieback anchors, such as shown in Fig. 16.13, are often installed to reduce the bending moments of the steel I beams for deep excavations.
Utility trench shoring	A trench is generally defined as a narrow excavation made below the ground surface that may or may not be shored. Each year, numerous workers are killed or injured in trench or excavation cave-ins. These accidents typically are due to the fact that the trench was not shored or the shoring system was

TABLE 16.11 Temporary Retaining Walls (*Continued*)

Topic (1)	Discussion (2)
Utility trench shoring (*Continued*)	inadequate (Thompson and Tanenbaum 1977). In the United States, fatalities in trench cave-ins represent a substantial portion of the construction fatalities. For example, the Ohio Bureau of Workers' Compensation indicated a total of 271 cave-ins during the period of 1986–1990. The cause of death in many cases is suffocation. Other workers die from the force of the cave-in, which causes a crushing injury to the chest. On large construction projects or deep excavations for buildings, money is generally available to perform an exploration program that defines both the soil and groundwater conditions over the full extent of the project. This results in an engineered shoring design that is commonly adjusted to satisfy the varying site conditions. However, utility trenches are smaller projects with typically less money for subsurface exploration. Also, utility trenches frequently extend over longer distances than large projects or deep building excavations. In many situations, it is the contractor who determines which shoring system to install on the basis of experience and the observed soil and groundwater conditions. When trench walls are loose or sandy, the danger is apparent to the contractor and shoring is installed. Many problems develop when the trench looks hard and compact, and the contractor will take chances and install minimal or no shoring, on the assumption that it will be stable for a few hours. A common form of shoring for unstable soil is close sheeting. Figure 16.14 presents a diagram of close sheeting ("Trench" 1984). The main components of the shoring system are continuous sheeting from the top to the bottom of the trench, which is held in place by stringers and cross-braces (struts). In many cases, the cross-braces are hydraulic jacks rather than wood members. Another common type of shoring uses a sliding trench metal shield. Many injuries occur when it is time to backfill the trench. If the trench has remained open for a long time, there can be much more pressure on the shoring system than when it was installed. A worker who is in the trench and knocks out a cross-brace will release this built-up pressure and perhaps cause failure of other members or collapse. According to Petersen (1963), the following is the correct procedure to backfill a trench. 1. Backfill and compact the trench to a point just below the bottom cross-brace. 2. Enter the trench and remove only the bottom cross-brace. 3. Backfill and compact to the next level of cross-braces. 4. Remove these cross-braces and continue the procedure until only the uprights or sheeting remain in the ground. 5. Pull out the uprights or sheeting using either a backhoe or front-end loader. When designing trench shoring, the engineer could use the earth pressures from braced excavations (e.g., Fig. 16.12). Using structural engineering principles, this earth pressure diagram could be used to determine the maximum moments and maximum shear in the sheeting and the compressive force in the cross-braces. From this analysis, and with an appropriate factor of safety, the size of individual members could then be selected.

TABLE 16.11 Temporary Retaining Walls (*Continued*)

Topic (1)	Discussion (2)
Utility trench shoring (*Continued*)	After obtaining the size of the individual shoring members, the design engineer should check the requirements versus the minimum state or local requirements. For example, the California Occupational Safety and Health Administration (Cal/OSHA) has developed a system to design shoring with different material types (see *Standards* 1991). In designing the shoring system, other important factors to be considered by the design engineer include the following: 1. The possibility of a temporary increase in the depth, length, or width of the trench. 2. The excavation of a trench below the groundwater table. 3. The presence of surcharge loads from adjacent soil or material stockpiles, buildings, and other loads. 4. The presence of vibrations from passing traffic, earthquakes, jackhammers, and other sources. 5. The possibility that the trench will be left open a long time. 6. The possibility of a change in climate, such as frequent rainstorms or thawing of frozen ground. 7. The possibility that surface-water flow may not be diverted away from the top of the trench.

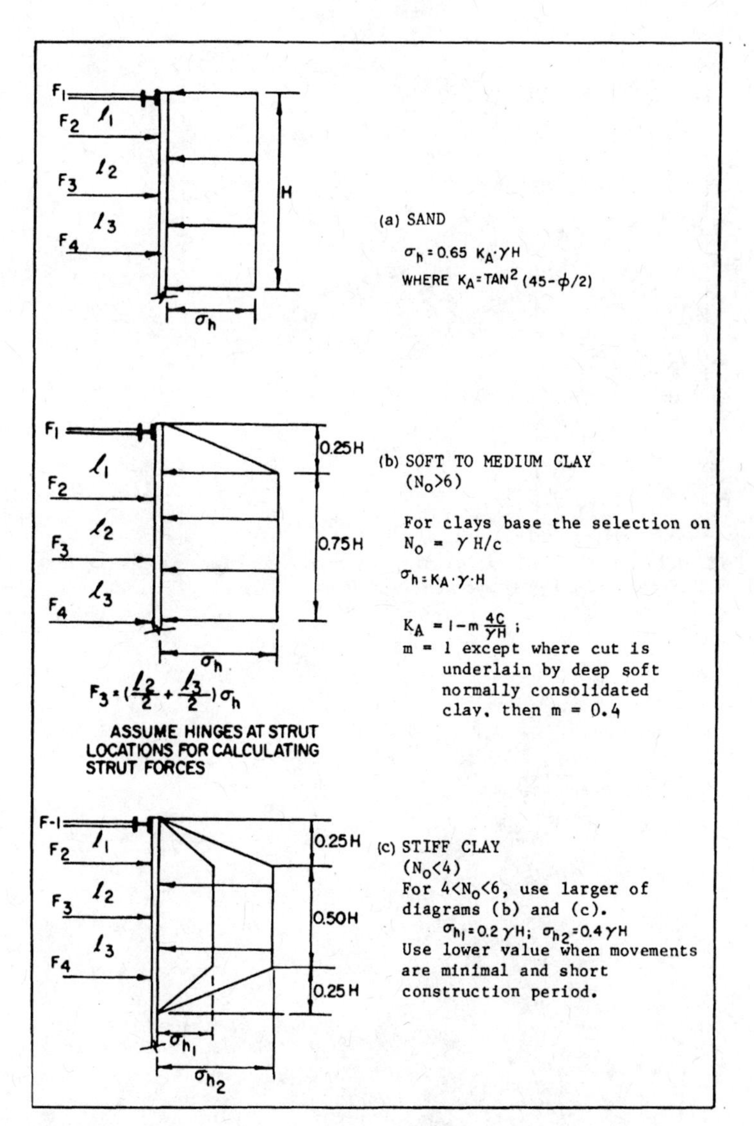

FIGURE 16.12 Earth pressure distribution on temporary braced walls. See Table 16.11. *(By Terzaghi and Peck 1967; reproduced from NAVFAC DM-7.2, 1982.)*

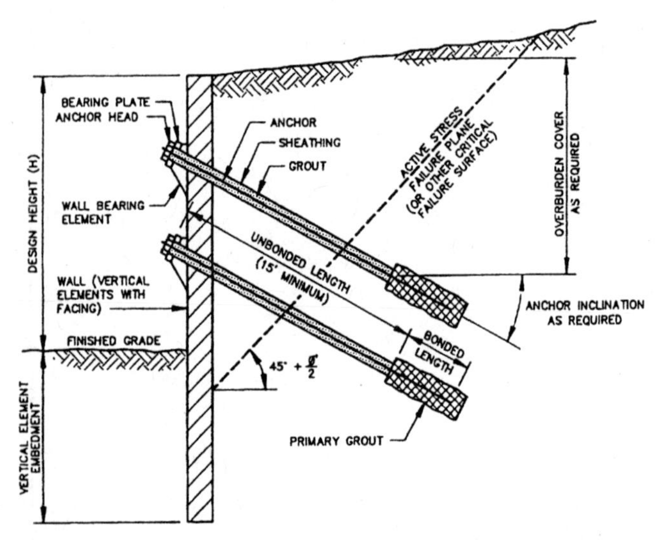

FIGURE 16.13 Cross section showing tieback anchors for retaining walls. See Table 16.11. (*Reproduced with permission from AASHTO 1996.*)

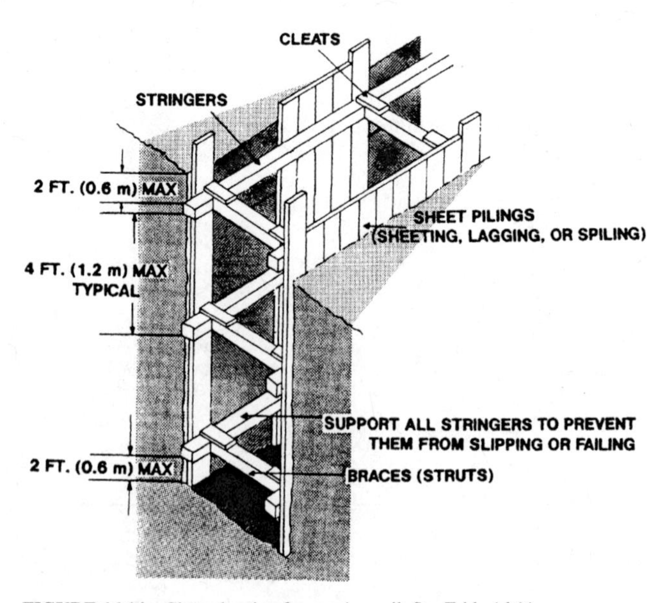

FIGURE 16.14 Close sheeting for running soil. See Table 16.11.

TABLE 16.12 Pier Walls

Topic (1)	Discussion (2)
General discussion	The terms *piers, shafts,* and *caissons* are sometimes used interchangeably by engineers. The common feature is that a cylindrical hole is drilled into the ground and then the hole is filled with concrete. The hole may be cased with a metal shell (a casing) in order to keep the hole from collapsing and sometimes to facilitate the cleaning of the bottom of the hole. The lower part of the hole may be belled out to develop a larger end-bearing area, thereby increasing the vertical load-carrying capacity of the drilled pier for a given allowable end-bearing pressure (Cernica 1995b). There are many different uses for drilled piers. Drilled piers are frequently used as a restraining system to stabilize slopes (Zaruba and Mencl 1969). Because soil arching can transfer lateral loads to the drilled piers, the piers are typically spaced a distance of 2 or 3 pier diameters apart. To stabilize slopes, the resistance required of the pier wall may be very large. This can result in deep pier walls, having large pier diameters and substantial reinforcement to resist the overturning moment (Abramson et al. 1996). The construction cost can be reduced when the pier wall is combined with tiebacks. A tieback is normally constructed at the top of the pier by drilling a hole into the slope, installing a tieback into the hole, and then filling the anchoring portion of the hole with high-strength grout. The purpose of the tieback is to transfer a portion of the destabilizing force to a zone behind the slip surface. The tiebacks consist of high-tensile-strength steel cables, tendons, or rods.
Advantages and disadvantages of piers	The advantages and disadvantages of piers are as follows (Reese 1978): Advantages: 1. Drilled piers can be successfully constructed in soils where it might be difficult to install other types of deep foundations. 2. Soil movements during construction due to heave or vibration are minimized. 3. Soil is exposed during the construction operation, and inspection can reveal whether the soil is consistent with that predicted by the subsurface exploration. 4. The size of the drilled pier can be readily adjusted during the construction operation so that variations in subsurface conditions can be accommodated. 5. Drilled piers can be built rapidly, compared to some other types of deep foundations. 6. Construction materials are readily available and construction equipment is generally available in all parts of the United States as well as many other parts of the world. 7. Construction noise is tolerable compared to other types of construction of deep elements. Disadvantages: 1. An excellent subsurface investigation must be carried so that designs are made properly and the appropriate construction procedure is selected.

TABLE 16.12 Pier Walls (*Continued*)

Topic (1)	Discussion (2)
Advantages and disadvantages of piers (*Continued*)	2. Drilled piers of good quality are critically dependent on the construction techniques that are employed. 3. The appropriate inspection of the construction requires a considerable amount of knowledge and experience. It is normally not possible to investigate the complete pier to see whether or not a good construction job has been obtained. 4. The shear strength of the supporting soil in general is reduced by the construction operation. 5. Drilled piers of small diameter can not be successfully constructed. 6. Failures of a drilled pier can be expensive because each pier is usually designed to resist a load of large magnitude.

TABLE 16.13 Pier Walls for the Stabilization of Slopes

Topic (1)	Discussion (2)
Design of pier walls to stabilize slopes (Method A)	One approach for the design of pier walls is to use slope stability analysis (Method A). The factor of safety of the slope, when stabilized by the pier wall, is first selected. Depending on such factors as the size of the slope failure, the proximity of critical facilities, and the nature of the pier wall stabilization (temporary versus permanent), a factor of safety of 1.2 to 1.5 is routinely selected. The next step is to determine the lateral design force P_L that each pier must resist in order to increase the factor of safety of the slope up to the selected value. Figure 16.15 shows an unstable slope having a planar slip surface inclined at an angle α to the horizontal (wedge method; see Table 13.9). The most common approach is to perform an effective stress analysis (long-term condition). For an effective stress analysis, the shear strength of the slip surface can be defined by the effective friction angle ϕ' and effective cohesion c'. The factor of safety of the slope with the pier wall is derived by summing forces parallel to the slip surface: $$F = \frac{\text{resisting forces}}{\text{driving force}} = \frac{c'L + (W\cos\alpha - uL)\tan\phi' + P_i}{W\sin\alpha}$$ where L = length of the slip surface W = total weight of the failure wedge u = average pore water pressure along the slip surface P_i = required pier wall force that is inclined at an angle α as shown in Fig. 16.15. The elements of the above equation are determined as follows: α and L: The angle of inclination α and the length of the slip surface L are based on the geometry of the failure wedge. W: Samples of the failure wedge material can be used to obtain the total unit weight. Knowing the total unit weight, the total weight W of the failure wedge (Fig. 16.15) can be calculated. ϕ' and c': The shear strength parameters (ϕ' and c') can be determined from laboratory shear testing of slip surface specimens. u: By installing piezometers in the slope, the pore water pressure u can be measured. F: As previously mentioned, a factor of safety F of 1.2 to 1.5 is routinely selected for the slope stabilized with a pier wall. The only unknown in the above equation is P_i (inclined pier wall force, Fig. 16.15). The following equation can be used to calculate the lateral design force P_L for each pier having an on-center spacing of S: $$P_L = SP_i\cos\alpha$$

TABLE 16.13 Pier Walls for the Stabilization of Slopes (*Continued*)

Topic (1)	Discussion (2)
Design of pier walls to stabilize slopes (Method A) (*Continued*)	The location of the lateral design force P_L is ordinarily assumed to be at a distance of ⅓ H above the slip surface, where H is defined in Fig. 16.15. The lateral design force P_L would be resisted by passive pressure exerted on that portion of the pier below the slip surface. The above analysis to determine the lateral design force P_L for each pier was based on a wedge type of slope failure, as illustrated in Fig. 16.15. If the slip surface is a circular arc, then the method of slices can be used to calculate the lateral design force P_L. Circular arc slip surfaces may develop for slope failures in uniform soil.
Design of pier walls to stabilize slopes (Method B)	A second approach for the design of pier walls is to use earth pressure theory (Method B), as illustrated in Fig. 16.16. The destabilizing pressure acting on the pier wall is assumed to be the active earth pressure. The active pressure is applied in a tributary fashion as if a continuous wall exists. The active pressure can be applied over the entire length Z_T of the pier. The resistance to movement of the pier wall is from passive pressure. To account for arching, the passive pressure can be applied over 2 pier diameters (for pier spacing of $2D$ or greater). Note that the passive pressure begins at a depth of Z_1, which is defined as the depth to adequate lateral bearing material, determined from subsurface investigation. To calculate the lateral design force P_L for each pier, the following equation is used: $$P_L = 0.5 k_A \gamma_t Z_T^2 S$$ where k_A = active earth pressure coefficient and γ_t = total unit weight. Because the active earth pressure is the actual driving pressure exerted on the pier wall, a factor of safety F can be included in the above equation. The factor of safety F could be the same value (1.2 to 1.5) used in Method A (slope stability method). As mentioned in Table 16.2, the value of k_A times γ_t is ordinarily referred to as the *equivalent fluid pressure.* The value of k_A can be calculated for a given shear strength of unstable slope material. In Fig. 16.16, the value of 60 psf/ft is equal to the equivalent fluid pressure recommended for the design of a pier wall to stabilize fill and landslide debris which moved during the Northridge earthquake (Day and Poland 1996). Earth pressure theory (Method B) can be used to calculate the lateral design force P_L when the slope deformation varies with depth (such as creep of clayey slopes), rather than failure on a distinct slip surface (Method A). Because soil arching can transfer lateral loads to the drilled piers, the piers are typically spaced a distance of 2 or 3 diameters apart. Table 16.14 presents recommendations for maximum spacing of piers versus soil or rock type. This table is based on the performance of existing pier walls. In general, the pier spacing should be decreased as the rock becomes more fractured or the soil becomes more plastic.

TABLE 16.13 Pier Walls for the Stabilization of Slopes (*Continued*)

Topic (1)	Discussion (2)
Construction of pier walls to stabilize slopes	General construction details for drilled piers are presented by Reese et al. (1981, 1985). A major difficulty with the excavation of piers is access for drilling equipment. Because of the normal large size of the pier hole, great depth of excavation, and high resistance of the lateral bearing strata, a truck-mounted drill rig is required. This can make access a serious limitation for the construction of pier walls. If there is not enough room for the drill rig or if the topography is too steep, the option of using a pier wall for slope stabilization may have to be abandoned. Vertical loads can be imposed on the pier by the tieback anchor or the slope movement. These vertical loads must be resisted by end-bearing resistance and/or skin friction in the bearing strata. Loose soil or debris is frequently knocked into the hole during the construction of the pier. This generally occurs when the prefabricated steel cage is lowered into the pier hole. When the steel cage strikes the side wall of the pier, soil is inevitably knocked to the bottom of the pier hole. Once the steel cage is in place, it is nearly impossible to remove the accumulation of loose soil at the bottom of the pier. One solution to this problem is to design the piers to resist vertical loads only through skin friction between the concrete and the bearing strata. In this case, end-bearing resistance is neglected. Since lateral loads usually govern the depth of embedment, it may be economical to simply neglect end bearing and design the piers to resist vertical loads through skin friction. A small-diameter inclined hole is ordinarily used for the tieback anchor. This means that it will not be possible to visually observe the installation conditions of the tieback. Common problems are tiebacks not centered in the hole and inadequately grouted tieback zones which are unable to generate the required frictional resistance. Because of the numerous difficulties in construction of tieback anchors, field testing of the anchors is essential in order to be sure of an acceptable capacity. Table 16.15 presents test procedures and acceptance criteria for tieback anchors.

TABLE 16.14 Allowable Pier Spacing for Slope Stabilization

Material type (1)	Largest on-center pier spacing (2)
Intact rock	No limit
Fractured rock	$4D$
Clean sand or gravel	$3D$
Clayey sand or silt	$2D$
Highly plastic clay	$1.5D$

Note: D = diameter of pier.

TABLE 16.15 Test Procedures and Acceptance Criteria for Each Tieback

A. Stages and Observation Periods for 24-Hour Test (Performance Test)

Load level	Recommended period of observations (in minutes)
Seating or alignment load = 0.1DL	None
0.25DL	10
0.50DL	10
0.75DL	10
1.00DL	10
1.25DL	10
1.50DL	10
1.75DL	30
2.00DL	480

DL = design load

After each stage of loading, the load should be reduced to seating or alignment load and replaced after 2 minutes. Deflection should be measured immediately after application of each load level and at 5-minute intervals thereafter. Measurements should also be taken immediately prior to any unloading and reloading steps.

B. Load stages and observation periods for proof test

Load Level	Recommended period of observations (in minutes)
Seating or alignment load = 0.1DL	None
0.50DL	10
1.00DL	10
1.25DL	10
1.50DL	30

DL = design load

After each stage of loading, the load should be reduced to seating or alignment load and replaced after 2 minutes. Deflection should be measured before and after application of each loading increment.

Acceptance criteria:

1. Total deflection during the long term testing should not exceed 12 inches.
2. Creep deflection at 200% design load should not exceed 0.1 in. during the 4-hour period.
3. Total deflection during short-term testing should not exceed 12 inches.
4. Creep deflection at 150% design load should not exceed 0.1 in. during the 15-minute period.
5. All the tiebacks should be locked-off at 120% of the design load.
6. Following final tensioning, the nongrouted portion of the anchor boring should be grouted.

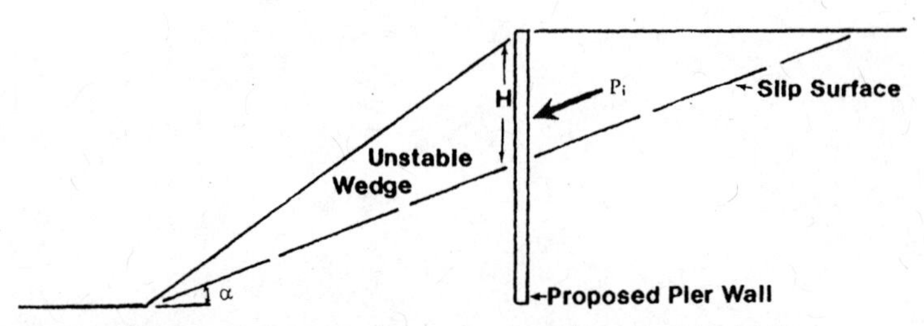

FIGURE 16.15 Design of pier wall for wedge slope failure. See Table 16.13.

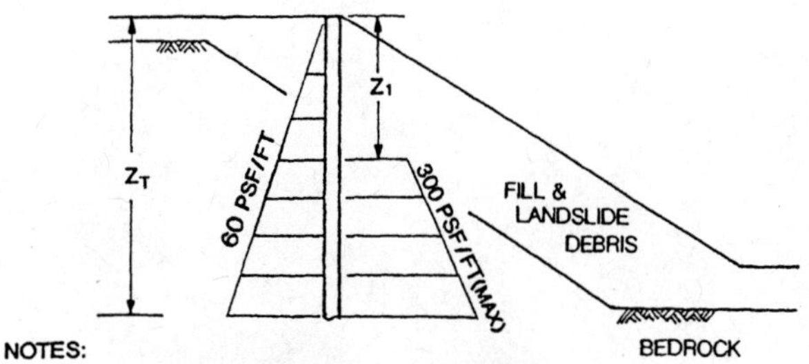

NOTES:

1) MAXIMUM PIER SPACING = 3 DIAMETERS
2) APPLY ACTIVE PRESSURE IN TRIBUTARY FASHION AS IF A WALL EXISTS. APPLY PASSIVE RESISTANCE OVER 2 PIER DIAMETERS
3) FOR TYPICAL CANTILEVER DESIGN, STRUCTURAL ENGINEER SHOULD ASSUME ABOUT 0.5 TO 1% ROTATION OF WALL IS REQUIRED TO MOBILIZE MAXIMUM PASSIVE RESISTANCE. TOTAL DEFLECTION SHOULD BE CALCULATED BY ADDING THIS ROTATION TO THE STRUCTURAL DEFLECTION FROM BENDING STRESS.

FIGURE 16.16 Design of pier walls using earth pressure theory. See Table 16.13.

CHAPTER 17
DETERIORATION

17.1 INTRODUCTION

All man-made and natural materials are susceptible to deterioration. This topic is so broad that it is not possible to cover every type of geotechnical and foundation element susceptible to deterioration. Instead, this chapter will discuss three of the more common types of deterioration: sulfate attack of concrete, pavement distress, and frost-related damage. In regard to deterioration, the National Science Foundation (NSF 1992) states: "The infrastructure deteriorates with time, due to aging of the materials, excessive use, overloading, climatic conditions, lack of sufficient maintenance, and difficulties encountered in proper inspection methods. All of these factors contribute to the obsolescence of the structural system as a whole. As a result, repair, retrofit, rehabilitation, and replacement become necessary actions to be taken to insure the safety of the public."

17.2 SULFATE ATTACK OF CONCRETE

There are four principal causes of concrete deterioration: (1) sulfate attack, (2) alkali-silica reaction (ASR), (3) corrosion of reinforcing steel, and (4) freeze-thaw cycles (Collepardi 1999). Only two of the principal causes of concrete deterioration are covered in this book, sulfate attack (this section) and deterioration due to freeze-thaw cycles (see Sec. 17.4). The main mechanisms of sulfate attack of concrete and measures to mitigate potential damage are discussed in Tables 17.1 to 17.4.

17.3 PAVEMENT DETERIORATION

There can be many types of pavement deterioration or failure, including rutting, alligator cracking, bleeding, block cracking, raveling, corrugation, and potholes or depressions. Descriptions and photographs of these types of pavement deterioration are presented by ASTM (e.g., ASTM D 5340-93 and ASTM E 1778-96a, 1997). Pavement deterioration is further discussed in Table 12.13 (expansive soil) and Table 17.5.

17.4 FROST

There have been extensive studies on the detrimental effects of frost (Casagrande 1931, Kaplar 1970, Yong and Warkentin 1975, and Reed et al. 1979). Two common types of damage related to frost are: (1) freezing of water in cracks and (2) formation of ice lenses (see Table 17.6 and Figs. 17.1 and 17.2). Potential frost action of compacted subgrade soil is presented in column 6 of Table 11.2.

Permafrost is often defined as perennially frozen soil. Another frequently used definition of permafrost is that portion of the ground that remains below freezing temperatures for 2 or more years. The bottom of permafrost lies at depths ranging from a few feet to over a thousand feet. The *active layer* is defined as the upper few inches to several feet of ground that is frozen in winter but thawed in summer. Permafrost often requires special design and construction measures, which are beyond the scope of this book.

TABLE 17.1 Mechanisms of Sulfate Attack of Concrete

Topic (1)	Discussion (2)
General discussion	Sulfate attack of concrete is defined as a chemical and/or physical reaction between sulfates (usually in the soil or groundwater) and concrete or mortar, primarily with hydrated calcium aluminate in the cement-paste matrix, often causing deterioration (*ACI* 1990). Sulfate attack of concrete occurs throughout the world, especially in arid areas, such as the southwestern United States. In arid regions, the salts can be drawn up into the concrete and then deposited on the concrete surface as the groundwater evaporates. Sulfate attack of concrete can cause a physical loss of concrete, unusual cracking, and discoloration of concrete. Typically the geotechnical engineer obtains the representative soil or groundwater samples to be tested for sulfate content. The geotechnical engineer can analyze the soil samples or groundwater in house, but a more common situation is to send the samples to a chemical laboratory for testing. There are different methods to determine the soluble sulfate content in soil or groundwater. One method is to precipitate out and then weigh the sulfate compounds. A faster and easier method is to add barium chloride to the solution and then compare the turbidity (relative cloudiness) of barium sulfate with known concentration standards. There has been considerable research, testing, and chemical analysis of sulfate attack. Two different mechanisms of sulfate attack have been discovered: chemical reactions and the physical growth of crystals.
Chemical reactions	The chemical reactions involving sulfate attack of concrete are complex. Studies (Lea 1971, Mehta 1976) have discovered two main chemical reactions. The first is a chemical reaction of sulfate and calcium hydroxide (which was generated during the hydration of the cement) to form calcium sulfate, commonly known as gypsum. The second is a chemical reaction of gypsum and hydrated calcium aluminate to form calcium sulfoaluminate, commonly called ettringite (*ACI* 1990). As with many chemical reactions, the final product of ettringite causes an increase in volume of the concrete. Hurst (1968) indicates that the chemical reactions produce a compound of twice the volume of the original tricalcium aluminate compound. Concrete has a low tensile strength, and thus the increase in volume fractures the concrete, allowing for more sulfates to penetrate the concrete, resulting in accelerated deterioration.
Physical growth of crystals	The physical reaction of sulfate has been studied by Tuthill (1966) and Reading (1975). They conclude that there can be crystallization of the sulfate salts in the pores of the concrete. The growth of crystals exerts expansive forces within the concrete, causing flaking and spalling of the outer concrete surface. Besides sulfate, the concrete, if porous enough, can be disintegrated by the expansive force exerted by the crystallization of almost any salt in its pores (Tuthill 1966 and Reading 1975). Damage due to crystallization of salt is commonly observed in areas where water is migrating through the concrete and then evaporating at the concrete surface. Examples include the surfaces of concrete dams, basement and retaining walls that lack proper waterproofing, and concrete structures that are partially immersed in salt-bearing water (such as seawater) or soils.

TABLE 17.2 Factors Affecting the Sulfate Resistance of Concrete

Topic (1)	Discussion (2)
General discussion	The geotechnical engineer should be aware of the factors that increase the sulfate resistance of concrete. In general, the degree of sulfate attack of concrete will depend on the type of cement used, quality of the concrete, soluble sulfate concentration that is in contact with the concrete, and the surface preparation of the concrete (Mather 1968).
Type of cement	There is a correlation between the sulfate resistance of cement and its tricalcium aluminate content. As previously discussed, it is the chemical reaction of hydrated calcium aluminate and gypsum that forms ettringite. Therefore, limiting the tricalcium aluminate content of cement reduces the potential for the formation of ettringite. It has been stated that the tricalcium aluminate content of the cement is the greatest single factor that influences the resistance of concrete to sulfate attack, where in general, the lower the tricalcium aluminate content, the greater the sulfate resistance (Bellport 1968). As indicated in Table 17.3, of the types of portland cements, the most resistant cement is type V, in which the tricalcium aluminate content must be less than 5%. The ACI (*ACI* 1990), Portland Cement Association (*Design* 1994), and *Uniform Building Code* (Table 19-A-4, 1997) have essentially the same requirements (see Table 17.3) for normal-weight concrete subjected to sulfate attack. Depending on the percentage of soluble sulfate (SO_4) in the soil or groundwater, a certain cement type is required as indicated in Table 17.3.
Quality of concrete	In general, the more impermeable the concrete, the more difficult for the waterborne sulfate to penetrate the concrete surface. To have a low permeability, the concrete must be dense, have a high cement content, and a low water-cement ratio. Using a low water-cement ratio decreases the permeability of mature concrete (*Design* 1994). A low water-cement ratio is a requirement of ACI (1990) for concrete subjected to soluble sulfate in the soil or groundwater. For example, the water-cement ratio must be equal to or less than 0.45 for concrete exposed to severe or very severe sulfate conditions. There are many other conditions that can affect the quality of the concrete. For example, a lack of proper consolidation of the concrete can result in excessive voids. Another condition is the corrosion of reinforcement, which may crack the concrete and increase its permeability. Cracking of concrete may also occur when structural members are subjected to bending stresses. For example the tensile stress due to a bending moment in a footing may cause the development of microcracks, which increase the permeability of the concrete.
Concentration of sulfates	As previously mentioned, it is often the geotechnical engineer who obtains the soil or water specimens that will be tested for soluble sulfate concentration. In some cases, after construction is complete, the sulfate may become concentrated on crack faces. For example, water evaporating through cracks in concrete flatwork will deposit the sulfate on the crack faces. This concentration of sulfate may cause accelerated deterioration of the concrete.
Surface preparation of concrete	An important factor in concrete resistance is the surface preparation, such as the amount of curing of the concrete. Curing results in a stronger and more impermeable concrete (*Design* 1994), which is better able to resist the effects of salt intrusion.

TABLE 17.3 Typical Requirements for Concrete Foundations Exposed to Soluble Sulfate

Classification of sulfate exposure (1)	Soluble sulfate (SO_4) in the soil, %, based on dry weight (2)	Dissolved sulfate (SO_4) in the groundwater, parts per million (3)	Required portland cement type (4)	Maximum water-cement ratio (5)
Negligible	0.0 to 0.1	0 to 150	Any type	No requirement
Moderate	0.1 to 0.2	150 to 1500	Type II	0.50
Severe	0.2 to 2.0	1500 to 10,000	Type V	0.45
Very severe	Over 2.0	Over 10,000	Type V with pozzolan	0.45

Sources: ACI (*ACI* 1990), Portland Cement Association (*Design* 1994), and *Uniform Building Code* (Table 19-A-4, 1997).

TABLE 17.4 Determining Sulfate Concentration at the Site

Topic (1)	Discussion (2)
Obtaining soil and groundwater specimens	The soil or groundwater samples should be obtained after the site has been graded and the location of the proposed building is known. For the planned construction of shallow foundations, near-surface soil samples should be obtained for the sulfate testing. For the planned construction of deep foundations that consist of concrete piles or piers, soil and groundwater samples for sulfate testing should be taken at various depths that encompass the entire length of the concrete foundation elements. The geotechnical engineer should select representative soil samples for testing. If the site contains clean sand or gravel, then these types of soil often have low sulfate contents because of their low capillary rise and high permeability, which enables any sulfate to be washed from the soil. But clays can have a higher sulfate content because of their high capillary rise, which enables them to draw up sulfate-bearing groundwater, and their low permeability, which prevents the sulfate from being washed from the soil.
Natural variations in sulfate concentration	Sulfate concentrations in the soil and groundwater can vary throughout the year. For example, the highest concentration of sulfates in the soil and groundwater will tend to occur at the end of the dry season or after a long dry spell. Likewise, the lowest concentration of sulfates will occur at the end of the rainy season, when the sulfate has been diluted or partially flushed out of the soil. The geotechnical engineer should recognize that sampling of soil or groundwater at the end of a heavy rainfall season will most likely be unrepresentative of the most severe condition.
Soil containing gypsum	Soil that contains gypsum should always be tested for sulfate content. This is because soluble sulfate (SO_4) is released as gypsum ($CaSO_4 \cdot 2\ H_2O$) weathers. Gypsum is an evaporate and can rapidly weather upon exposure to air and water.
Corrosive environments	There are many environments that could lead to the chemical attack of concrete. For example, if the site had been previously used as a farm, there may be fertilizers or animal wastes in the soil that will have a detrimental effect on concrete. A corrosive environment, such as acid mine water drainage will also lead to the deterioration of concrete. Another corrosive environment is seawater, which has both a moderate soluble sulfate content and a high salt content which can attack concrete through the process of chemical reactions involving sulfate attack and deterioration of concrete by the physical growth of salt crystals.
Design recommendations	Once the soluble sulfate content has been determined, the geotechnical or foundation engineer can then recommend measures, such as applying the requirements listed in Table 17.3, to mitigate the effects of sulfate on the concrete. In order to reduce the potential for a concentration of sulfate on crack surfaces, a capillary break can be installed below the concrete. A common type of capillary break consists of the installation of open graded gravel directly below the bottom of the concrete.

TABLE 17.5 Pavement Deterioration

Topic (1)	Discussion (2)
Water trapped in pavements or base	The most common cause of premature deterioration of pavements is water trapped in slow-draining pavements and base. Table 11.2 presents the drainage properties of different compacted soils. It has been stated that groundwater in pavements accelerates the damage rates by hundreds of times over the damage rates of pavements with no groundwater. This premature failure of thousands of miles of pavements and billions of dollars in losses a year could be avoided by good pavement drainage practices (Cedergren 1989). The key element in a good drainage system is a layer of highly permeable material (such as open graded gravel) protected by filters or geofabric so that the permeable material will not become clogged by the intrusion of soil fines. A drainage system, to remove the water from the base, is also required. There are several ways that groundwater can enter the base material. In areas having a high groundwater table or artesian condition, water can be forced upward into the base material. Water can also flow downward through pavement cracks or joints. There can also be the development of a perched groundwater condition, where water moves laterally through the base from adjacent planter areas, medians, or shoulders. There can also be the flow of groundwater in permeable utility trenches, which can also lead to premature deterioration of the pavements. One construction method to prevent the flow of groundwater through utility trenches is to use a grout (such as a cement slurry) in the utility line or storm drain bedding zone. The remainder of the trench could then be backfilled and compacted with on-site native soil. This should provide the trench with a permeability equal to or less than the surrounding native soil.
Heavy traffic loads	Besides expansive soil and groundwater, there can be other factors that contribute to pavement deterioration. Probably the most common causes of premature pavement deterioration or failure are heavier-than-expected traffic loads or an unanticipated higher volume of traffic.
Mitigation of pavement deterioration	Factors that should be considered during the design and construction phases of the pavement in order to prevent deterioration include (NAVFAC DM-21.3, 1978): • Characteristics, strength, and in-place density of the subgrade, base, and asphalt or concrete surface. • Seasonal fluctuations of groundwater and effectiveness of pavement drainage. • Frost susceptibility of the pavement section and the effect of freeze-thaw conditions on the subgrade. • Presence of weak or compressible layers in the subgrade. • Variability of the subgrade, which may cause differential surface movements.

TABLE 17.5 Pavement Deterioration (*Continued*)

Topic (1)	Discussion (2)
Mitigation of pavement deterioration (*Continued*)	Other important factors in preventing deterioration of pavements include: • *Thickness of pavement section.* The construction of the pavement surface and base course must be at least as thick as calculated during the design phase. • *Strength parameters.* If the *R*-value or California bearing ratio (CBR) values were assumed during the design phase, then the geotechnical engineer should check these values prior to construction at the site. • *Proper construction.* It is important to have proper compaction of the subgrade and base, and to use high-quality aggregate that does not degrade during compaction and from traffic loads. The pavement may need a drainage system to prevent the buildup of water pressures in the subgrade and base materials.

TABLE 17.6 Frost

Topic (1)	Discussion (2)
Freezing of water in cracks	There is about a 10% increase in volume of water when it freezes, and this volumetric expansion of water upon freezing can cause deterioration or damage to many different types of materials. Common examples include rock slopes and concrete, as discussed below. *Rock slopes.* The expansive forces of freezing water result in a deterioration of the rock mass, additional fractures, and added driving (destabilizing) forces. Feld and Carper (1997) describe several rock slope failures caused by freezing water, such as the February 1957 failure where 900 Mg (1000 tons) of rock fell out of the slope along the New York State Thruway, closing all three southbound lanes north of Yonkers. *Concrete.* Durability is defined by the American Concrete Institute as the ability to resist weathering, chemical attack, abrasion, or any other type of deterioration (ACI 1982). Durability is affected by strength, but also by density, permeability, air entrainment, dimensional stability, characteristics and proportions of constituent materials, and construction quality (Feld and Carper 1997). Durability is harmed by freezing and thawing, sulfate attack, corrosion of reinforcing steel, and reactions between the various constituents of the cements and aggregates. Damage to concrete caused by freezing could occur during the original placement of the concrete or after it has hardened. To prevent damage during placement, it is important that the fresh concrete not be allowed to freeze. Air-entraining admixtures can be added to the concrete mixture to help protect the hardened concrete from freeze-thaw deterioration.
Formation of ice lenses	Frost penetration and the formation of ice lenses in the soil can damage shallow foundations and pavements. The frost penetration will cause heave of the structure if moisture is available to form ice lenses in the underlying soil. The spring thaw will then melt the ice, resulting in settlement of the foundation or a weakened subgrade that will make the pavement surface susceptible to deterioration or failure. Damage to highways in the United States and Canada because of frost action is estimated to amount to millions of dollars annually (Holtz and Kovacs 1981). It is well known that silty soils are more likely to form ice lenses because of their high capillarity and sufficient permeability that enables them to draw up moisture to the ice lenses. Figure 17.1 shows a diagram illustrating three different conditions for the formation of ice lenses. The diagram on the left shows a condition where there is no source of water and the formation of ice lenses and heave will be small. Likewise, the diagram on the right shows a capillary break, such as open graded gravel, above the groundwater table and the formation of ice lenses and heave will also be small. The diagram in the middle shows a high groundwater table and soil having sufficient capillary rise, which is the condition most favorable to capillary rise and resulting heave. Feld and Carper (1997) describe several interesting cases of damage due to frost action. At Fredonia, New York, the frost from a deep-freeze storage facility froze the soil and heaved the foundations upward 100 mm (4 in.). A system of electrical wire heating was installed to maintain soil volume stability.

TABLE 17.6 Frost (*Continued*)

Topic (1)	Discussion (2)
Formation of ice lenses (*Continued*)	Another case involved an extremely cold winter in Chicago, where frost penetrated below an underground garage and broke a buried sprinkler line. This caused an ice buildup which heaved the structure above the street level and sheared off several supporting columns. As these cases show, it is important that the foundation be constructed below the depth of frost action. Figure 17.2 (from Bowles 1982) shows the approximate depth of frost penetration, based on a survey of selected cities. In addition, there may be local building requirements on the minimum depth of the foundation to prevent damage caused by the formation of ice lenses. There have also been studies to determine the annual maximum frost depths for various site conditions and 50-year or 100-year return periods (e.g., DeGaetano et al. 1997).

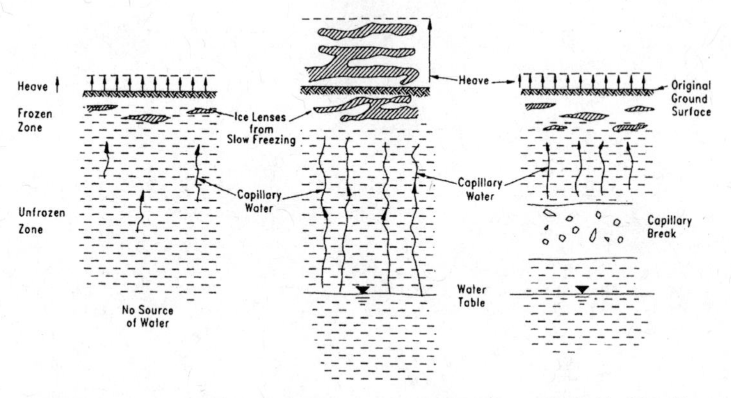

FIGURE 17.1 Formation of ice lenses and ground surface heave. See Table 17.6. (*From Rollings and Rollings 1996; reprinted with permission of McGraw-Hill, Inc.*)

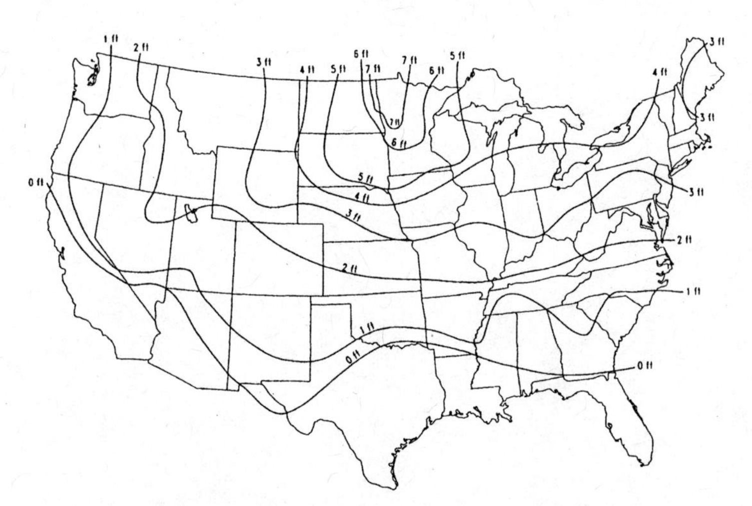

FIGURE 17.2 Approximate depth of frost penetration in the continental United States. See Table 17.6. (*From Bowles 1982; reprinted with permission of McGraw-Hill, Inc.*)

CHAPTER 18
FOUNDATIONS

18.1 INTRODUCTION

This final chapter of Part 1 deals with the selection of the type of foundation. Common types of foundations are listed in Table 1.2, and typical factors that govern the selection of a particular type of foundation are presented in Table 18.1.

18.2 SHALLOW FOUNDATIONS

A shallow foundation is often selected when the structural load will not cause excessive settlement of the underlying soil layers. In general, shallow foundations are more economical to construct than deep foundations. Common types of shallow foundations are listed in Table 1.2 and described in Table 18.2 and Figs. 18.1 and 18.2.

18.3 DEEP FOUNDATIONS

Different types of deep foundations are listed in Table 1.2 and described in Tables 18.3 and 18.4 and Figs. 18.3 to 18.5. Example problems concerning the selection of an appropriate type of foundation are presented in Table 18.5. As an option to using deep foundations, the soil at the site could be improved by performing grading or other site improvement techniques (see Chap. 19).

TABLE 18.1 Selection of Foundation Type

Topic (1)	Discussion (2)
Selection of the foundation type	On the basis of an analysis of the factors listed below, a specific type of foundation (i.e., shallow versus deep, see Table 1.2) would be recommended by the geotechnical engineer.
Adequate depth	The foundation must have an adequate depth to prevent frost damage (see Table 17.6). For such foundations as bridge piers, the depth of the foundation must be sufficient to prevent undermining by scour.
Bearing capacity failure	The foundation must be safe against a bearing capacity failure (see Chap. 10).
Settlement	The foundation must not settle to such an extent that it damages the structure (see Chap. 9).
Quality	The foundation must be of adequate quality so that it is not subjected to deterioration, for example, from sulfate attack (see Chap. 17).
Adequate strength	The foundation must be designed with sufficient strength that it does not fracture or break apart under the applied superstructure loads. The foundation must also be properly constructed in conformance with the design specifications.
Adverse soil changes	The foundation must be able to resist long-term adverse soil changes. An example is expansive soil (Chap. 12), which could expand or shrink, causing movement of the foundation and damage to the structure.
Seismic forces	The foundation must be able to support the structure during an earthquake without excessive settlement or lateral movement (Chap. 14).
Required specifications	The foundation may also have to meet special requirements or specifications required by the local building department or governing agency.

TABLE 18.2 Common Types of Shallow Foundations

Topic (1)	Discussion (2)
Spread footings	Spread footings are often square in plan view, are of uniform reinforced-concrete thickness, and are used to support a single load directly in the center of the footing.
Strip footings	Strip footings, also known as wall footings, are often used to support load bearing walls. They are usually long reinforced-concrete members of uniform width and shallow depth.
Combined footings	Reinforced-concrete combined footings are often rectangular or trapezoidal in plan view and carry more than one column load (see Fig. 18.1).
Other types of footings	Figure 18.1 shows other types of footings, such as the cantilever (also known as strap) footing, an octagonal footing, and an eccentric loaded footing with the resultant coincident with area so that the soil pressure is uniform.
Mat foundation	If a mat foundation is constructed at or near ground surface, then it is considered to be a shallow foundation. Figure 18.2 shows different types of mat foundations. For economic considerations, mat foundations are often constructed for the following reasons (NAVFAC DM-7.2, 1982): 1. *Large individual footings.* A mat foundation is often constructed when the sum of individual footing areas exceeds about one-half of the total foundation area. 2. *Cavities or compressible lenses.* A mat foundation can be used when the subsurface exploration indicates that there will be unequal settlement caused by small cavities or compressible lenses below the foundation. A mat foundation would tend to span over the small cavities or weak lenses and create a more uniform settlement condition. 3. *Shallow settlements.* A mat foundation can be recommended when shallow settlements predominate and the mat foundation would minimize differential settlements. 4. *Unequal distribution of loads.* For some structures, there can be a large difference in building loads acting on different areas of the foundation. Conventional spread footings could be subjected to excessive differential settlement, but a mat foundation would tend to distribute the unequal building loads and reduce the differential settlements. 5. *Hydrostatic uplift.* When the foundation will be subjected to hydrostatic uplift due to a high groundwater table, a mat foundation could be used to resist the uplift forces.
Conventional slab-on-grade	A continuous reinforced-concrete foundation consisting of bearing wall footings and a slab-on-grade. Concrete reinforcement often consists of steel rebar in the footings and wire mesh in the concrete slab.
Posttensioned slab-on-grade	Posttensioned slabs-on-grade are common in southern California and other parts of the United States. They are an economical foundation type when there is no ground freezing or the depth of frost penetration is low. The most common uses of posttensioned slab-on-grade are to resist expansive soil forces or when the projected differential settlement exceeds the tolerable value for a conventional (lightly reinforced) slab-on-grade. For example,

TABLE 18.2 Common Types of Shallow Foundations (*Continued*)

Topic (1)	Discussion (2)
Posttensioned slab-on-grade (*Continued*)	posttensioned slabs-on-grade are frequently recommended if the projected differential settlement is expected to exceed 2 cm (0.75 in.). Installation and field inspection procedures for posttensioned slab-on-grade have been prepared by the Post-Tensioning Institute (1996). Posttensioned slab-on-grade consists of concrete with embedded steel tendons that are encased in thick plastic sheaths. The plastic sheath prevents the tendon from coming in contact with the concrete and permits the tendon to slide within the hardened concrete during the tensioning operations. Usually tendons have a dead end (anchoring plate) in the perimeter (edge) beam and a stressing end at the opposite perimeter beam to enable the tendons to be stressed from one end. However, the Post-Tensioning Institute (1996) does recommend that the tendons in excess of 30 m (100 ft) be stressed from both ends. The Post-Tensioning Institute (1996) also provides typical anchorage details for the tendons. Because posttensioned slabs-on-grade perform better (i.e., less shrinkage-related concrete cracking) than conventional slabs-on-grade, they are more popular even for situations where low levels of settlement are expected.
Raised wood floor	Perimeter footings that support wood beams and a floor system. Interior support is provided by pad or strip footings. There is a crawl space below the wood floor.
Shallow foundation alternatives	If the expected settlement for a proposed shallow foundation is too large, then other options for foundation support or soil stabilization must be evaluated. Commonly used alternatives include deep foundations, grading options, or other site improvement techniques. Deep foundations are discussed in this chapter and grading and other site improvement techniques are discussed in Chap. 19.

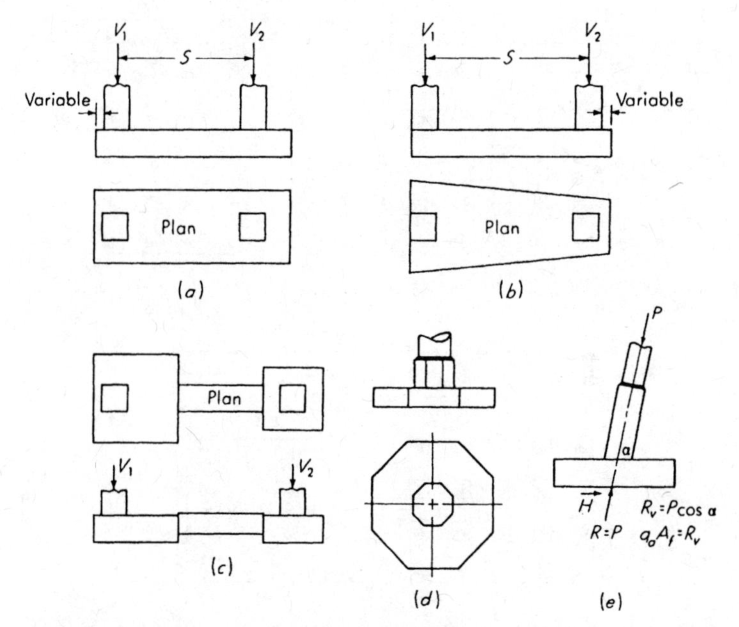

FIGURE 18.1 Examples of shallow foundations. (*a*) Combined footing; (*b*) combined trapezoidal footing; (*c*) cantilever or strap footing; (*d*) octagonal footing; (*e*) eccentric loaded footing with resultant coincident with area so soil pressure is uniform. See Table 18.2. (*Reproduced from Bowles 1982 with permission of McGraw-Hill, Inc.*)

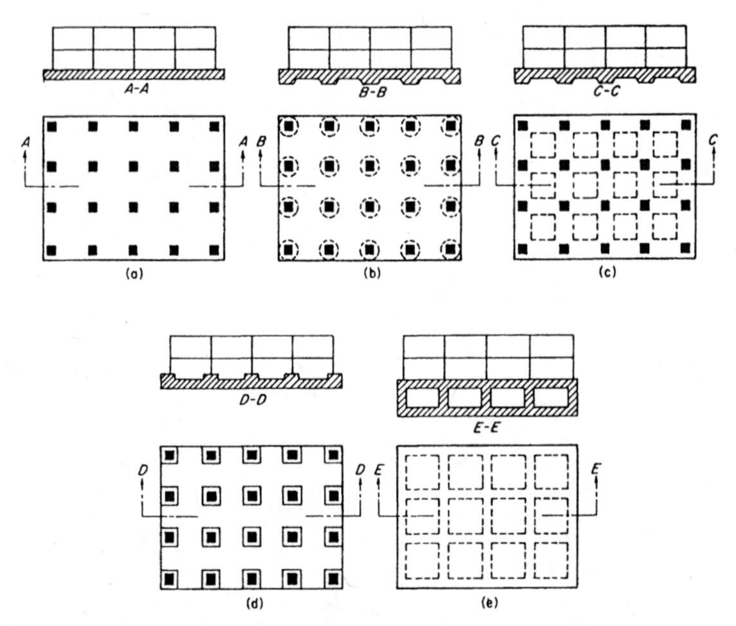

FIGURE 18.2 Examples of mat foundations. (*a*) Flat plate; (*b*) plate thickened under columns; (*c*) beam-and-slab; (*d*) plate with pedestals; (*e*) basement walls as part of mat. See Table 18.2. (*Reproduced from Bowles 1982 with permission of McGraw-Hill, Inc.*)

TABLE 18.3 Common Types of Deep Foundations

Topic (1)	Discussion (2)
Pile foundations	Probably the most common type of deep foundation is the pile foundation. Piles can consist of wood (timber), steel H sections, precast concrete, cast-in-place concrete, pressure-injected concrete, concrete-filled steel pipe piles, and composite-type piles (also see Table 18.4). Piles are either driven into place or installed in predrilled holes. Piles that are driven into place are generally considered to be low displacement or high displacement, depending on the amount of soil that must be pushed out of the way as the pile is driven. Examples of low-displacement piles are steel H sections and open-ended steel pipe piles that do not form a soil plug at the end. Examples of high-displacement piles are solid section piles, such as round timber piles or square precast concrete piles, and steel pipe piles with a closed end. Various types of piles are as follows: • *Batter pile.* A pile driven in at an angle inclined to the vertical to provide high resistance to lateral loads. • *End-bearing pile.* A pile whose support capacity is derived principally from the resistance of the foundation material on which the pile tip rests. End-bearing piles are often used when a soft upper layer is underlain by a dense or hard stratum. If the upper soft layer should settle, the pile could be subjected to downdrag forces, and the pile must be designed to resist these soil-induced forces. • *Friction pile.* A pile whose support capacity is derived principally from the resistance of the soil friction and/or adhesion mobilized along the side of the pile. Friction piles are often used in soft clays where the end-bearing resistance is small because of punching shear at the pile tip. A pile that resists upward loads (i.e., tension forces) would also be considered to be a friction pile. • *Combined end-bearing and friction pile.* A pile that derives its support capacity from combined end-bearing resistance developed at the pile tip and frictional and/or adhesion resistance on the pile perimeter. Piles are usually driven into specific arrangements and are used to support reinforced concrete pile caps or a mat foundation. For example, the building load from a steel column may be supported by a concrete pile cap that is in turn supported by four piles located near the corners of the concrete pile cap.
Cast-in-place piles	A cast-in-place pile is formed by making a hole in the ground and then filling the hole with concrete. As shown in Fig. 18.3, in its simplest form, the case-in-place pile consists of an uncased hole that is filled with concrete. If the soil tends to cave into the hole, then a shell-type pile can be installed (see Fig. 18.3). This consists of driving a steel shell or casing into the ground. The casing may be driven with a mandrel which is then removed and the casing is filled with concrete. In other cases, the casing can be driven into place and then slowly removed as the hole is filled with concrete. Table 18.4 also describes various cast-in-place concrete piles.

TABLE 18.3 Common Types of Deep Foundations (*Continued*)

Topic (1)	Discussion (2)
Concrete-filled steel pipe piles	Another option is a concrete-filled steel pipe pile. In this case, the steel pipe pile is driven into place. The pipe pile can be driven with either an open or closed end. If the end is open, the soil within the pipe pile is removed (by jetting) prior to placement of the steel reinforcement and concrete. Table 18.4 provides additional details on concrete-filled steel pipe piles.
Prestressed-concrete piles	Figure 18.4 presents details on typical prestressed-concrete piles that are delivered to the job site and then driven into place.
Other types of piles	Table 18.4 and Fig. 18.5 provide additional details on various types of piles.
Piers	A pier is defined as a deep foundation system, similar to a cast-in-place pile, that consists of a columnlike reinforced-concrete member. Piers are often of large enough diameter to enable downhole inspection. Piers are also commonly referred to as *drilled shafts, bored piles,* or *drilled caissons.*
Caissons	Large piers are sometimes referred to as *caissons.* A caisson can also be a watertight underground structure within which work is carried on.
Mat or raft foundation	If a mat or raft foundation is constructed below ground surface, or if the mat or raft is supported by piles or piers, then it should be considered to be a deep foundation system.
Floating foundation	A floating foundation is a special type of deep foundation where the weight of the structure is balanced by the removal of soil and construction of an underground basement.

TABLE 18.4 Typical Pile Characteristics and Uses

Pile type	Timber	Steel	Cast-in-place concrete piles (shells driven without mandrel)	Cast-in-place concrete piles (shells withdrawn)
Maximum length	35 m	Practically unlimited	45 m	36 m
Optimum length	9–20 m	12–50 m	9–25 m	8–12 m
Applicable material specifications	ASTM-D25 for piles; P1-54 for quality of creosote; C1-60 for creosote treatment (standards of American Wood Preservers Assoc.)	ASTM-A36 for structural sections ASTM-A1 for rail sections	ACI	ACI†
Recommended maximum stresses	Measured at midpoint of length: 4–6 MPa for cedar, western hem lock, Norway pine, spruce, and depending on code 5–8 MPa for southern pine. Douglas fir, oak cypress, hickory	f_s = 65 to 140 MPa $f_s = 0.35–0.5 f_y$	$0.33 f'_c$; $0.4 f'_c$ if shell gage ≤ 14; shell stress $= 0.35 f_y$ if thickness of shell ≥ 3 mm	$0.25–0.33f'_c$
Maximum load for usual conditions	270 kN	Maximum allowable stress × cross section	900 kN	1300 kN
Optimum-load range	130–225 kN	350–1050 kN	450–700 kN	350–900 kN
Disadvantages	Difficult to splice Vulnerable to damage in hard driving Vulnerable to decay unless treat ed, when piles are intermittently submerged	Vulnerable to corrosion HP section may be damaged or deflected by major obstructions	Hard to splice after concreting Considerable displacement	Concrete should be placed in dry hole More than average dependence on quality of workmanship

TABLE 18.4 Typical Pile Characteristics and Uses (*Continued*)

Pile type	Timber	Steel	Cast-in-place concrete piles (shells driven without mandrel)	Cast-in-place concrete piles (shells withdrawn)
Advantages	Comparatively low initial cost Permanently submerged piles are resistant to decay Easy to handle	Easy to splice High capacity Small displacement Able to penetrate through light obstructions	Can be redriven Shell not easily damaged	Initial economy
Remarks	Best suited for friction pile in granular material	Best suited for end bearing on rock Reduce allowable capacity for corrosive locations	Best suited for friction piles of medium length	Allowable load on pedestal pile is controlled by bearing capacity of stratum immediately below pile
Typical illustrations	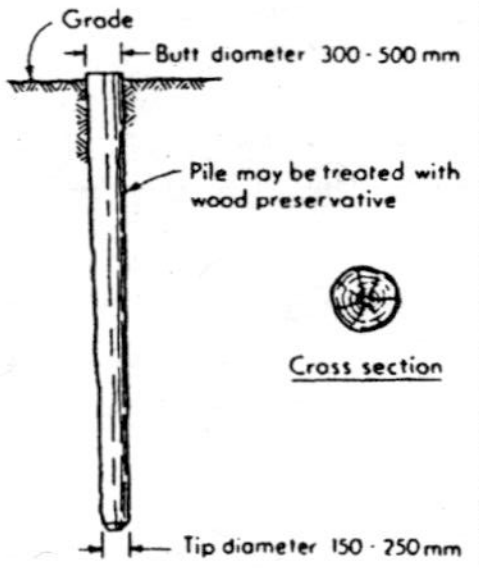 Grade Butt diameter 300 - 500 mm Pile may be treated with wood preservative Cross section Tip diameter 150 - 250 mm	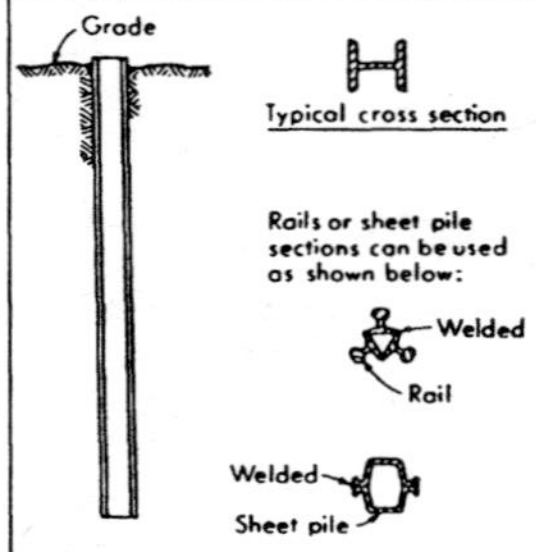 Grade Typical cross section Rails or sheet pile sections can be used as shown below: Welded Rail Welded Sheet pile	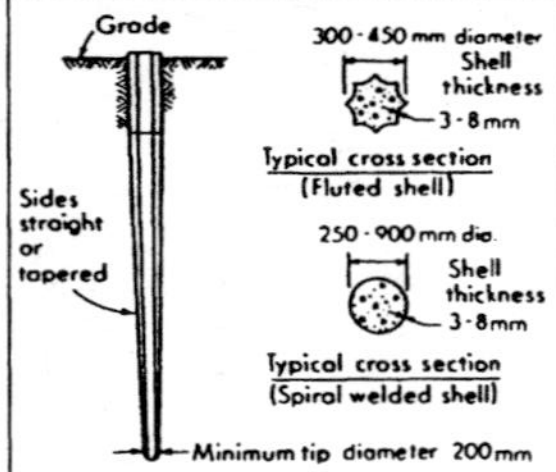 Grade Sides straight or tapered 300 - 450 mm diameter Shell thickness 3 - 8 mm Typical cross section (Fluted shell) 250 - 900 mm dia. Shell thickness 3 - 8 mm Typical cross section (Spiral welded shell) Minimum tip diameter 200 mm	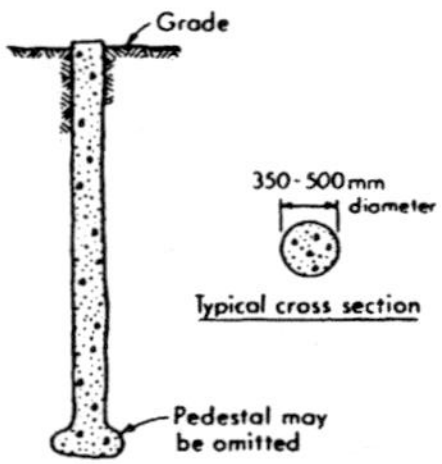 Grade 350 - 500 mm diameter Typical cross section Pedestal may be omitted

Notes: Stresses given for steel piles and shells are for noncorrosive locations. For corrosive locations estimate possible reduction in steel cross section or provide protection from corrosion.

Pile type	Concrete filled steel pipe piles	Composite piles	Precast concrete (including prestressed)	Cast in place (thin shell driven with mandrels)	Auger placed pressure-injected concrete (grout) piles
Maximum length	Practically unlimited	55 m	30 m for precast 60 m for prestressed	30 m for straight sections 12 m for tapered sections	9–25 m
Optimum length	12–36 m	18–36 m	12–15 m for precast 18–30 m for prestressed	12–18 m for straight 5–12 m for tapered	12–18 m
Applicable material specifications	ASTM A36 for core ASTM A252 for pipe ACI Code 318 for concrete	ACI Code 318 for concrete ASTM A36 for structural section ASTM A252 for steel pipe ASTM D25 for timber	ASTM A15 reinforcing steel ASTM A82 cold-drawn wire ACI Code 318 for concrete	ACI	See ACI†
Recommended maximum stresses	$0.40 f_y$ reinforcement < 205 MPa $0.50 f_y$ for core < 175 MPa $0.33 f'_c$ for concrete	Same as concrete in other piles Same as steel in other piles Same as timber piles for wood composite	$0.33 f'_c$ unless local building code is less; $0.4 f_y$ for reinforced unless prestressed	$0.33 f'_c$; $f_s = 0.4 f_y$ if shell gauge is ≤ 14; use $f_s = 0.35 f_y$ if shell thickness ≥ 3 mm	$0.225–0.40 f'_c$
Maximum load for usual conditions	1800 kN without cores 18,000 kN for large sections with steel cores	1800 kN	8500 kN for prestressed 900 kN for precast	675 kN	700 kN
Optimum-load range	700–1100 kN without cores 4500–14,000 kN with cores	250–725 kN	350–3500 kN	250–550 kN	350–550 kN
Disadvantages	High initial cost Displacement for closed-end pipe	Difficult to attain good joint between two materials	Difficult to handle unless prestressed High initial cost Considerable displacement Prestressed difficult to splice	Difficult to splice after concreting Redriving not recommended Thin shell vulnerable during driving Considerable displacement	Dependence on workmanship Not suitable in compressible soil

TABLE 18.4 Typical Pile Characteristics and Uses (*Continued*)

Pile type	Concrete filled steel pipe piles	Composite piles	Precast concrete (including prestressed)	Cast in place (thin shell driven with mandrels)	Auger placed pressure-injected concrete (grout) pile
Advantages	Best control during installation No displacement for open-end installation Open-end pipe best against obstructions High load capacities Easy to splice	Considerable length can be provided at comparatively low cost	High load capacities Corrosion resistance can be attained Hard driving possible	Initial economy Taped sections provide higher bearing resistance in granular stratum	Freedom from noise and vibration Economy High skin friction No splicing
Remarks	Provides high bending resistance where unsupported length is loaded laterally	The weakest of any material used shall govern allowable stresses and capacity	Cylinder piles in particular are suited for bending resistance	Best suited for medium-load friction piles in granular materials	Patented method
Typical illustrations	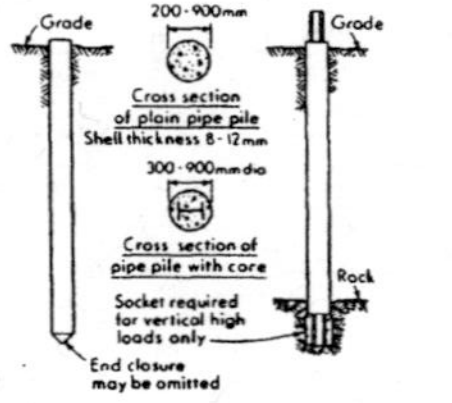 Grade 200-900mm Cross section of plain pipe pile Shell thickness 8-12mm 300-900mm dia Cross section of pipe pile with core Socket required for vertical high loads only End closure may be omitted Grade Rock	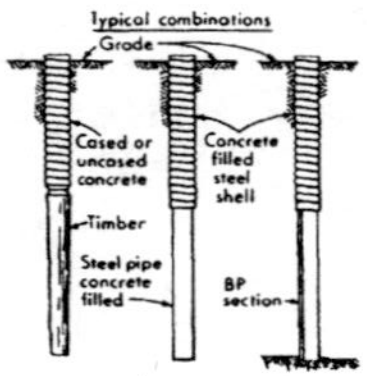 Typical combinations Grade Cased or uncased concrete Timber Steel pipe concrete filled Concrete filled steel shell BP section	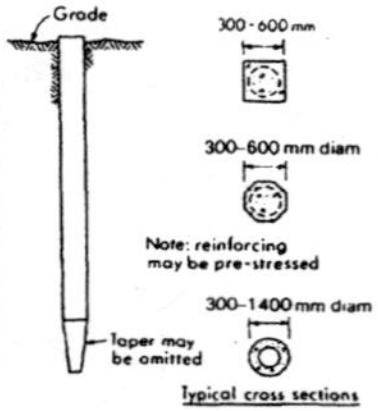 Grade 300-600 mm 300-600 mm diam Note: reinforcing may be pre-stressed 300-1400 mm diam Taper may be omitted Typical cross sections	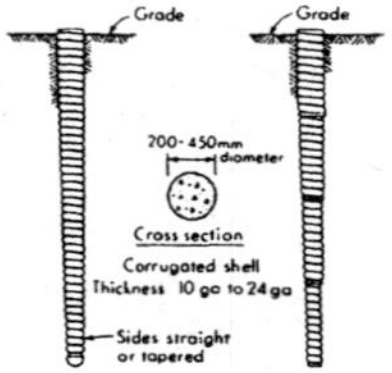 Grade Grade 200-450mm diameter Cross section Corrugated shell Thickness 10 ga to 24 ga Sides straight or tapered	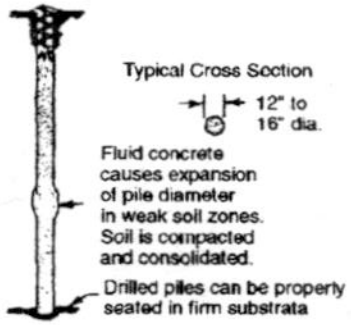 Typical Cross Section 12" to 16" dia. Fluid concrete causes expansion of pile diameter in weak soil zones. Soil is compacted and consolidated. Drilled piles can be properly seated in firm substrata

†ACI Committee 543: "Recommendations for Design, Manufacture, and Installation of Concrete Piles," *JACI,* August 1973, October 1974.

Sources: NAVFAC DM-7.2, 1982 and Bowles (1982).

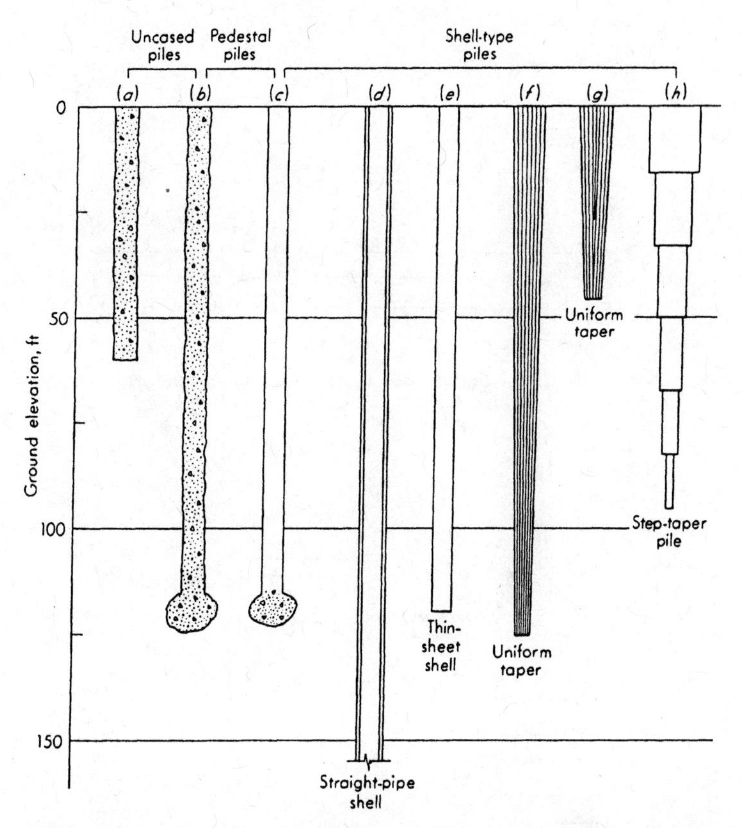

FIGURE 18.3 Common types of cast-in-place concrete piles. (*a*) Uncased pile; (*b*) Franki uncased-pedestal pile; (*c*) Franki cased-pedestal pile; (*d*) welded or seamless pipe pile; (*e*) cased pile with a thin sheet shell; (*f*) monotube pile; (*g*) uniform tapered pile; (*h*) step-tapered pile. See Table 18.3. (*Reproduced from Bowles 1982 with permission of McGraw-Hill, Inc.*)

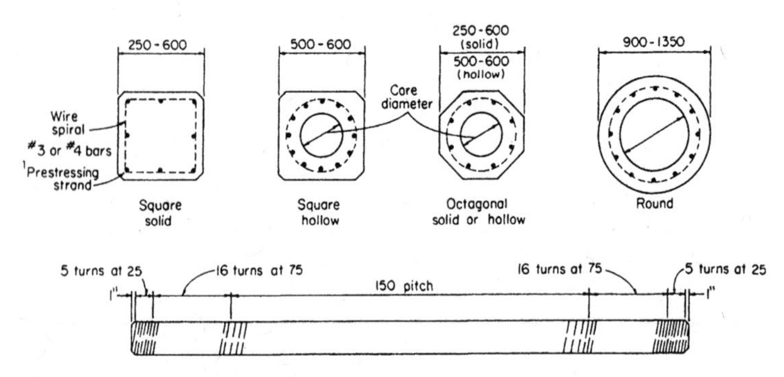

FIGURE 18.4 Typical prestressed concrete piles; dimensions in millimeters. See Table 18.3. (*Reproduced from Bowles 1982 with permission of McGraw-Hill, Inc.*)

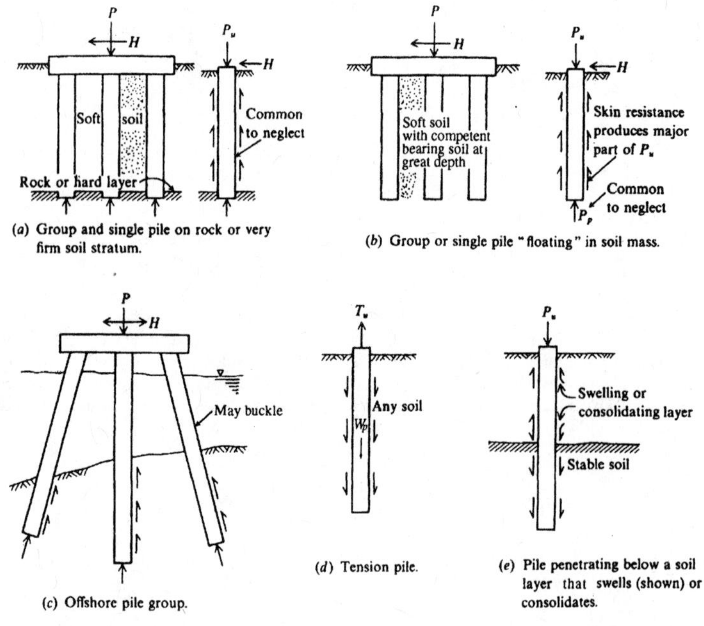

FIGURE 18.5 Typical pile configurations. See Table 18.3. (*Reproduced from Bowles 1982 with permission of McGraw-Hill, Inc.*)

TABLE 18.5 Example Problems

Topic (1)	Example Problem (2)
Shallow foundation	*Problem:* Assume that the subsoil conditions at a site are as shown in Fig. 2.20. It is proposed to construct single-family detached housing at this site. What type of foundation should be used? *Solution:* Because the upper 10 ft of the site consists of overconsolidated clay, it would be desirable to use a shallow foundation system based on the assumption of light building loads.
Floating foundation	*Problem:* Assume that the subsoil conditions at a site are as shown in Fig. 2.19. It is proposed to construct a two-story research facility that will also contain an underground basement to be used as the laboratory. On the basis of these requirements, what type of foundation should be installed for the research facility? *Solution:* Assuming the weight of soil excavated for the basement is approximately equal to the weight of the two-story structure, a floating foundation would be desirable.
Deep foundation	*Problem:* Assume that the subsoil conditions at a site are as shown in Fig. 2.18. It is proposed to build a state highway bridge (essential facility) at this location. It is anticipated that during major flood conditions, there could be scour that could remove sediments down to an elevation of −10 m. What type of bridge foundation should be installed for the highway bridge? *Solution:* A deep foundation system consisting of piles or piers embedded in the sandstone.
Deep foundation	*Problem:* Assume the subsoil conditions at a site are as shown in Fig. 2.21. The structural engineer is recommending that driven, high displacement piles be installed to a depth of 50 ft in the clay deposit. Are these type of piles appropriate? *Solution:* No; because of the very high sensitivity of the clay, high displacement piles will remold the clay and result in a loss of shear strength. The preferred option is to install low displacement piles or use predrilled, cast-in-place concrete piles.

P · A · R · T · 2

CONSTRUCTION

CHAPTER 19
GRADING AND OTHER SITE IMPROVEMENT METHODS

19.1 INTRODUCTION

Since most building sites start out as raw land, the first step in site construction work usually involves the grading of the site. Grading is defined as any operation consisting of excavation, filling, or combination thereof. The glossary (App. A, Glossary 4) presents a list of common construction and grading terms and their definitions. Most projects involve grading, and it is an important part of geotechnical engineering.

19.2 GRADING SPECIFICATIONS

It is important to prepare a set of grading specifications for the project. These specifications are often used to develop the grading plans, which are basically a series of maps that indicate the type and extent of grading work to be performed at the site. Appendix B presents an example of grading specifications. Included at the end of the grading specifications in App. B are Standard Details Nos. 1 through 9, which illustrate the proper grading procedure for various activities. Often the grading specifications will be included as an appendix in the preliminary or feasibility report prepared by the geotechnical engineer and engineering geologist. The typical steps in the grading of a project and key grading activities are presented in Tables 19.1 and 19.2.

19.3 COMPACTION FUNDAMENTALS

An important part of the grading of the site often includes the compaction of fill. Compaction is defined as the densification of a fill by mechanical means. This physical process of getting the soil into a dense state can increase the shear strength, decrease the compressibility, and decrease the permeability of the soil. Compaction fundamentals are presented in Tables 19.3 to 19.7 and Figs. 19.1 and 19.3. Laboratory testing compaction fundamentals have been presented in Table 3.15 (Modified Proctor and Standard Proctor Laboratory Compaction Tests) and Fig. 3.8 (Laboratory Compaction Curve).

19.4 SITE IMPROVEMENT METHODS

If the expected settlement for a proposed structure is too large, then different foundation support or soil stabilization options must be evaluated. As discussed in Chap. 18, one alternative is a deep foundation system that can transfer structural loads to adequate bearing material in order to bypass a compressible soil layer. Another option is to construct a floating foundation, which is a special type of deep foundation where the weight of the structure is balanced by the removal of soil and construction of an underground basement. Other alternatives include site improvement methods, such as the following (see Table 19.8).

19.4.1 Soil Replacement

As indicated in Table 19.8, there are basically two types of soil replacement methods: (1) removal and replacement and (2) displacement. The first method is the more common approach. It consists of the removal of the compressible soil layer and replacement with structural fill during the grading operations. Usually the remove-and-replace grading option is economical only if the compressible soil layer is near the ground surface and the groundwater table is below the compressible soil layer or the groundwater table can be economically lowered.

19.4.2 Water Removal

Table 19.8 lists several different types of water removal site improvement techniques. If the site contains an underlying compressible cohesive soil layer, the site can be surcharged with a fill layer placed at ground surface. Vertical drains (such as wick drains or sand drains) can be installed in the compressible soil layer to reduce the drainage path and speed up the consolidation process. Once the compressible cohesive soil layer has had sufficient consolidation, the fill surcharge layer is removed and the building is constructed.

19.4.3 Site Strengthening

There are many different methods that can be used to strengthen the on-site soil (see Table 19.8). For example, deep vibratory techniques, such as illustrated in Fig. 19.4, are often used to increase the density of loose sand deposits.

19.4.4 Grouting

In order to stabilize the ground, fluid grout can be injected into the ground to fill in joints, fractures, or underground voids (Graf 1969, Mitchell 1970). For the releveling of existing structures, one option is mudjacking, which has been defined as a process whereby a water and soil-cement or soil-lime cement grout is pumped beneath the slab, under pressure, to produce a lifting force that literally floats the slab to the desired position (Brown 1992).

Another commonly used site improvement technique is compaction grouting, which consists of intruding a mass of very thick consistency grout into the soil, which both displaces and compacts the loose soil (Brown and Warner 1973, Warner 1978, 1982). Compaction grouting has proved successful in increasing the density of poorly compacted fill, alluvium, and compressible or collapsible soil. The advantages of compaction grouting are less expense and disturbance to the structure than foundation underpinning, and it can be used to relevel the structure. The disadvantages of compaction grouting are that it is difficult to analyze the results, it is usually ineffective near slopes or for near-surface soils because of the lack of confining pressure, and there is the danger of filling underground pipes with grout (Brown and Warner 1973).

19.4.5 Thermal

As indicated in Table 19.8, the thermal site improvement method consists of either heating or freezing the soil in order to improve its shear strength and reduce its permeability.

19.4.6 Summary

Figure 19.5 presents a summary of site improvement methods as a function of soil grain size.

19.5 OBSERVATIONAL METHOD

The final section of Chap. 19 describes the observational method, which is a valuable method that can be utilized during the construction of the project (see Table 19.9).

TABLE 19.1 Typical Steps in a Grading Operation

Topic (1)	Discussion (2)
Easements	The first step in the grading operation is to determine the location of any on-site utilities and easements. The on-site utilities and easements often need protection so that they are not damaged during the grading operation.
Clearing, brushing, and grubbing	Clearing, brushing, and grubbing are defined as the removal of vegetation (grass, brush, trees, and similar plant types) by mechanical means. This debris is often temporarily stockpiled at the site. It is important that this debris be removed from the site and not accidentally placed within the structural fill mass.
Cleanouts	This grading process deals with the removal of unsuitable bearing material at the site, such as loose or porous alluvium, colluvium, and uncompacted fill.
Benching (hillside areas)	Benching is defined as the excavation of relatively level steps into earth material on which fill is to be placed. Standard Details Nos. 1 through 6 (App. B) show benching operations required for various grading operations.
Canyon subdrain	A subdrain is defined as a pipe and gravel or similar drainage system placed in the alignment of canyons or former drainage channels. After placement of the subdrain, structural fill is placed on top of the subdrain. See App. B, Standard Detail No. 1, for details on the construction of a canyon subdrain.
Scarifying and recompaction	In flat areas that have not been benched, scarifying and recompaction of the ground surface is performed by compaction equipment in order to get a good bond between the in-place material and compacted fill.
Cut-and-fill rough grading operations	Rough grading operations involve the cutting of earth materials from high areas and compaction of fill in low areas, in conformance with grading plans. Other activities could be performed during rough grading operations, such as: • *Ripping or blasting of rock.* Large rock fragments can be removed from the site or disposed of in windrows. Standard Detail No. 7 (App. B) illustrates the construction of a windrow. • *Cut-fill transition.* Standard Detail No. 6 (App. B) illustrates a cut-fill transition. It is the location in a building pad where on one side the pad has been cut down exposing natural or rock material, while on the other side, fill has been placed. Standard Detail No. 6 (App. B) presents one method to deal with cut-fill transitions. • *Slope stabilization.* Examples of slope stabilization using earth materials include stabilization fill (Standard Detail No. 3, App. B), buttress fill (Standard Detail No. 3, App. B), drainage buttress, and shear keys. Such devices should be equipped with backdrain systems. • *Fill slopes.* In creating a fill slope, it is often difficult to compact the outer edge of the fill mass. Because there is no confining pressure, the soil deforms downslope without increasing in density. To deal with this situation, the slope can be overbuilt and then cut back to the compacted core. The second best alternative is to use conventional construction procedures such as backrolling techniques, for example, using a bulldozer to track-walk the slope. • *Revision of grading operations.* Every grading job is different and there could be a change in grading operations based on field conditions.

TABLE 19.1 Typical Steps in a Grading Operation (*Continued*)

Topic (1)	Discussion (2)
Protection of adjacent property	If the grading operations will be adjacent to developed property, the geotechnical engineer and engineering geologist should evaluate the possibility of damage to off-site facilities. Many lawsuits are initiated when off-site property is damaged during the grading operations.
Fine grading	Fine grading is also known as precise grading. At the completion of the rough grading operations, fine grading is performed to obtain the finish elevations in accordance with the precise grading plan.
Slope protection and erosion control	Although usually not the responsibility of the grading contractor, on completion of the fine grading, slope protection and permanent erosion control devices are installed.
Trench excavations	Utility trenches are excavated in the proposed road alignments and building pads for the installation of the on-site utilities. The excavation and compaction of utility trenches is often part of the grading process. Once the utility lines are installed, scarifying and recompaction of the road subgrade is performed and base material is placed and compacted.
Footing and foundation excavations	Although usually not part of the grading operation, the footing and foundation elements can be excavated at the completion of grading in accordance with the foundation plans.

TABLE 19.2 Key Grading Activities

Topic (1)	Discussion (2)
General discussion	The geotechnical engineer and engineering geologist should be actively involved with the grading process. Typical activities include observing the grading operation as well as performing compaction tests during the placement of fill. Usually the compaction tests are performed by an experienced technician. Some of the key grading activities that should be attended or observed by the geotechnical engineer and engineering geologist are presented in this table.
Pregrading meeting	A pregrading meeting should be arranged between the geotechnical engineer, engineering geologist, grading contractor, client, and local building inspector. The purpose of the pregrading meeting is to review the grading plans and specifications, discuss the construction schedule such as the dates when key activities will be started and finished, and evaluate potential problems that may develop during the grading.
Initial observation	Usually the pregrading meeting is held at the site. This will give the geotechnical engineer and engineering geologist an opportunity to view the site and note any changes since the subsurface exploration, such as recently dumped debris and erosion.
Clearing, grubbing, and canyon cleanout	A site visit should be planned as a final check that all brush and unsuitable material has been removed from the site. This site visit should be performed before fill placement operations are started.
Adjacent property	Critical grading operations that may adversely affect adjacent property should be observed by the geotechnical engineer or engineering geologist. Monitoring devices (see Chap. 23) may need to be installed in order to monitor the performance of adjacent property during the grading operations.
Rough grading operations	During rough grading of the site, the geotechnical engineer and engineering geologist should observe key grading activities, such as ripping, benching, keying of slopes, construction of earth structures (buttresses, fill stabilization, etc.), treatment of cut-fill transitions, and installation of subsurface drainage systems. The compaction process must also be reviewed and field density tests taken to assess the quality of the fill.
Final grade observation	The final condition of the site should be observed at the completion of grading.

TABLE 19.3 Compaction Fundamentals: Basic Factors That Affect Compaction

Topic (1)	Discussion (2)
Soil type	Nonplastic (i.e., cohesionless) soil, such as sands and gravels, can be effectively compacted by using a vibrating or shaking type of compaction operation. Plastic (i.e., cohesive) soil, such as silts and clays, are more difficult to compact and require a kneading or manipulation type of compaction operation. If the soil contains oversize particles, such as coarse gravel and cobbles, then they tend to interfere with the compaction process and reduce the effectiveness of compaction for the finer soil particles. Typical values of dry density for different types of compacted soil are listed in column 10 of Table 11.2.
Material gradation	Those soils that have a well-graded grain size distribution can generally be compacted into a denser state than a poorly graded soil that is composed of soil particles of about the same size. For example, a well-graded decomposed granite (DG) can have a maximum dry density of 2.2 Mg/m^3 (137 pcf), while a poorly graded sand can have a maximum dry density of only 1.6 Mg/m^3 (100 pcf, Modified Proctor).
Water content	The water content is an important parameter in the compaction of soil. Water tends to lubricate the soil particles, thus helping them slide into dense arrangements. However, too much water and the soil becomes saturated and often difficult to compact. There is an optimum water content at which the soil can be compacted into its densest state for a given compaction energy. Typical optimum moisture contents (Modified Proctor) for different soil types are as follows: Clay of high plasticity (CH): Optimum moisture content ≥ 18% Clay of low plasticity (CL): Optimum moisture content = 12 to 18% Well-graded sand (SW): Optimum moisture content = 10% Well-graded gravel (GW): Optimum moisture content = 7% Some soils may be relatively insensitive to compaction water content. For example, open-graded gravels and clean coarse sands are so permeable that water simply drains out of the soil or is forced out of the soil during the compaction process. These types of soil can often be placed in a dry state and then vibrated into dense particle arrangements.
Compaction effort or energy	The compactive effort is a measure of the mechanical energy applied to the soil. Usually the greater the amount of compaction energy applied to a soil, the denser the soil will become. There are exceptions, such as pumping soils that can not be densified by an increased compaction effort. Compactors are designed to use one or a combination of the following types of compaction effort: 1. Static weight or pressure 2. Kneading action or manipulation 3. Impact or a sharp blow 4. Vibration or shaking

TABLE 19.4 Compaction Fundamentals: Field Compaction

<table>
<tr><th>Topic
(1)</th><th>Discussion
(2)</th></tr>
<tr><td>Common types of compaction equipment</td><td>For mass-grading operations, usually heavy construction equipment is used to excavate and compact fill. The three most common types of equipment used for grading operations are the bulldozer, scraper, and truck (Nichols and Day 1999):
1. Bulldozer. The bulldozer is used to clear the land of debris and vegetation (clearing, brushing, and grubbing), excavate soil from the borrow area, cut haul roads, spread out dumped fill, and compact the soil.
2. Scraper. The scraper is used to excavate (scrape up) soil from the borrow area, transport it to the site, and dump it at the site, and the rubber-tires of the scraper can be used to compact the soil. Push-pull scrapers can be used in tandem to provide additional energy to excavate hard soil or soft rock.
3. Truck. If the borrow area is quite a distance from the site, then dump trucks may be required to transport the borrow soil to the site. Dump trucks are also needed to transport soil on public roads or to import select material. Especially in the arid climate of the southwestern United States, soil taken from the borrow area may be in a dry state and water will need to be added to the soil to bring it up to optimum moisture content. Water trucks are typically used for this operation.
The Caterpillar Performance Handbook (1997), which is available at Caterpillar dealerships, is a valuable reference because it not only lists rippability versus types of equipment (see Figs. 2.14 to 2.17), but also indicates types and models of compaction equipment, equipment sizes and dimensions, and performance specifications.</td></tr>
<tr><td>Method of compaction</td><td>There are generally five different methods that are used to compact soil, as follows:
1. Static weight or pressure. This type of compaction equipment applies a static or relatively uniform pressure to the soil. Examples include the compaction by the rubber tires of a scraper, from the tracks of a bulldozer, and by using smooth drum rollers.
2. Kneading action or manipulation. The sheepsfoot roller, which has round- or rectangular-shaped protrusions or "feet," is ideally suited to applying a kneading action to the soil. This has proven to be effective in compacting silts and clays.
3. Impact or a sharp blow. There are compaction devices, such as the high-speed tamping foot and the Caterpillar tamping foot, that compact the soil by imparting an impact or sharp blow to the soil.
4. Vibration or shaking. Nonplastic sands and gravels can be effectively compacted by vibration or shaking. An example is the smooth drum vibratory soil compactor.
5. Chopper wheels. This type of compaction equipment has been specially developed for the compaction of waste products at municipal landfills.
Column 9 of Table 11.2 presents a summary of different types of compaction equipment best suited to compact different types of soil.</td></tr>
<tr><td>Cut and fill</td><td>A common objective of the grading operations is to balance the volume of cut and fill. This means that just enough earth material is cut from the high areas to fill in the low areas. A balanced cut and fill operation means that no soil needs to be imported or exported from the site, leading to a reduced cost of the grading operation.</td></tr>
</table>

TABLE 19.5 Compaction Fundamentals: Relative Compaction

Topic (1)	Discussion (2)
Definition	The most common method of assessing the quality of the field compaction is to calculate the relative compaction (RC) of the fill, defined as: $$RC = \frac{100\rho_d}{\rho_{d\,max}}$$ where $\rho_{d\,max}$ = laboratory maximum dry density and ρ_d = field dry density. In California, typical mass grading specifications require a minimum relative compaction of 90%. For example, the common fill compaction specifications adopted by city and county agencies are from the *Uniform Building Code* (1997), which states: "All fills shall be compacted to a minimum of 90 percent of maximum density." In some cases, such as the compaction of roadway base or the lower portions of deep fill, a higher compaction standard of a minimum relative compaction of 95% is specified. As discussed in Table 3.15, the maximum dry density $\rho_{d\,max}$ is obtained from the laboratory compaction curve (see Fig. 3.8).
Field density tests	In order to determine the field dry density ρ_d for the above equation, a field density test must be performed. Field density tests can be classified as either destructive or nondestructive tests (Holtz and Kovacs 1981), as follows: 1. *Sand cone test.* Probably the most common destructive method of determining the field dry density is through the use of the sand cone apparatus (ASTM D 1556-96, 1998). The test procedure consists of excavating a hole in the ground, filling the hole with sand using the sand cone apparatus, and then determining the volume of the hole from the amount of sand required to fill the hole. From the wet mass of soil removed from the hole divided by the volume of the hole, the wet density of the fill can be calculated. The water content *w* of the soil extracted from the hole can be determined, and thus the dry density ρ_d can then be calculated. 2. *Drive cylinder test.* Another type of destructive test for determining the field dry density is the drive cylinder test (ASTM D 2937-94, 1998). This method involves the driving of a steel cylinder of known volume into the soil. From the mass of soil within the cylinder, the wet density can be calculated. Once the water content *w* of the soil is obtained, the dry density ρ_d of the fill can be calculated. 3. *Nuclear method.* Probably the most common type of nondestructive field test is the nuclear method (ASTM D 2922-96, 1998). In this method, the wet density is determined by the attenuation of gamma radiation. The nuclear method can give inaccurate results (density too high) where oversize particles, such as coarse gravel and cobbles, are present. Likewise, if there is a large void in the source-detector path, then unusually low density values may be recorded.

TABLE 19.5 Compaction Fundamentals: Relative Compaction (*Continued*)

Topic (1)	Discussion (2)
Number of field density tests	There are guidelines on the number of field density tests for different types of grading projects, as follows (NAVFAC DM-7.2, 1982): • One test for every 380 m^3 (500 yd^3) of material placed for embankment construction. • One test for every 380 to 750 m^3 (500 to 1000 yd^3) of material for canal or reservoir linings or other relatively thin fill sections. • One test for every 75 to 150 m^3 (100 to 200 yd^3) of backfill in trenches or around structures, depending on total quantity of material involved. • At least one test for every full shift of compaction operations on mass earthwork. • One test whenever there is a definite suspicion of a change in the quality of moisture control or effectiveness of compaction. There are many different guidelines concerning the number of field density tests for specific grading activities. For example, the grading specifications presented in App. B indicate that field density tests for mass graded structural fill should be taken at about every 2 vertical feet or 750 m^3 (1000 yd^3) of fill placed and that the actual test interval may vary as field conditions dictate.
Role of technicians	It is rare for the licensed geotechnical engineer to perform field density testing on a daily basis because of the repetitive and time-consuming nature of such work. For large mass grading operations, it is common to have technicians performing the field density testing. The technician will have to be able to perform the field density tests, classify different soil types (based on visual and tactile methods), and insist on remedial measures when compaction falls below the specifications. As previously mentioned, for mass grading projects, the number of field density tests per volume of compacted fill is often very low (i.e., such as only one field density test per 1000 yd^3 of fill). It is important that the field technician perform the density tests on areas where compaction is suspect. For example, the technician should not perform field compaction tests in the haul road area, because this path receives continuous traffic and will usually be in a dense compacted state. Likewise, testing in the wheel paths of the compaction equipment will yield high values. Often the field technician uses a metal rod to probe for possible poorly compacted fill zones. Field density tests would then be performed in these areas of possible poor compaction.
Pumping	Pumping is a form of bearing capacity failure that occurs during compaction of fill. A commonly used definition of pumping is the softening and squeezing of wet to saturated plastic soil (such as a soft clay) from underneath the compaction equipment. Continual passes of the compaction equipment can cause a decrease in the undrained shear strength of the wet clay and the pumping may progressively worsen.

TABLE 19.5 Compaction Fundamentals: Relative Compaction (*Continued*)

Topic (1)	Discussion (2)
Pumping (*Continued*)	Pumping is dependent on the penetration resistance of the compacted clay. Figure 19.1 (from Turnbull and Foster 1956) presents data on the California bearing ratio (CBR) of compacted clay and shows that the penetration resistance approaches zero (i.e., the clay can exhibit pumping) when the clay has a water content that is wet of optimum. Also note in the lower part of Fig. 19.1 that the laboratory maximum dry density increases and the optimum moisture content decreases as the compaction energy increases (more blows per layer). There are many different methods to stabilize pumping soil. The most commonly used method is to simply allow the plastic soil to dry out. Other methods include adding a chemical agent (such as lime) to the clay or placing a geotextile on top of the pumping clay to stabilize its surface (Winterkorn and Fang 1975). Another common procedure to stabilize pumping clay is to add gravel to the clay. The typical procedure is to dump angular gravel at ground surface and then work it in from the surface. The angular gravel produces a granular skeleton, which then increases both the undrained shear strength and penetration resistance of the mixture (Day 1996b).

TABLE 19.6 Compaction Fundamentals: Types of Fill

Topic (1)	Discussion (2)
Types of fill	Fill is defined as a deposit of earth material placed by artificial means. There are different types of fill, as follows (Monahan 1986, Greenfield and Shen 1992): 1. *Engineered (or structural) fill.* This refers to a fill in which the geotechnical engineer has, during grading, made sufficient tests to enable the conclusion that the fill has been placed in substantial compliance with the recommendations of the geotechnical engineer and the governing agency requirements. Standard Detail No. 5 in App. B shows typical canyon fill placement specifications. Structural fills are used to support all types of structures. 2. *Hydraulic fill.* This refers to a fill placed by transporting soils through a pipe using large quantities of water. These fills are generally loose because they have little or no mechanical compaction during construction. 3. *Dumped or uncontrolled fill.* This refers to fill that was not documented with compaction testing as it was placed or fill that may have been compacted but there is no documentation of testing or the amount of effort that was used to perform the compaction. Dumped or uncontrolled fill should not be used to support structures.
Selection of laboratory maximum dry density for structural fill	In placing structural fill, it is usually not economical or possible from a time standpoint to perform a laboratory maximum dry density test for every field density test. NAVFAC DM-7.2 (1982) recommends a laboratory maximum dry density test for every 10 to 20 field density tests, depending on the variability of the materials. For mass grading operations in southern California, it is common to have a much higher ratio, such as one laboratory maximum dry density test for every 60 to 70 field density tests. The typical situation when placing structural fill is that the field technician will have a family of compaction curves corresponding to different soils. It is then the technician's responsibility to select the appropriate laboratory maximum dry density corresponding to the field-tested soil. The technician must use experience and judgment to match up the soil types from the laboratory maximum dry density tests with the field soil. When given several different laboratory maximum dry density tests on the same general soil type, it is common for the technician to select a laboratory maximum dry density that will provide a passing result. For example, if the field dry density is low, a low maximum dry density is selected. If the field dry density is high, a high maximum dry density is selected. The result is fill that may not meet project specifications because of the uncertainty in matching the laboratory maximum dry density with the fill soil type. The purpose of the remainder of this table to discuss types of fill and the selection of the appropriate laboratory maximum dry density for each fill type. Structural fill can generally be divided into four basic types, as follows: select (processed) import, uniform borrow, mixed borrow, and borrow having oversize particles. These four types of structural fill are individually discussed below.

TABLE 19.6 Compaction Fundamentals: Types of Fill (*Continued*)

Topic (1)	Discussion (2)
Structural fill derived from select import	Select import refers to a processed material. The material may be derived from several different sources, then screened and mixed to provide a material of specified gradation. For example, a common select import is granular base material, which may have to meet specifications for gradation, wear resistance, and shear strength (*Standard Specifications for Public Works Construction* 1997). Other uses for select import include backfill for retaining walls and utilities, and even for mass graded fill. Table 19.7 presents different methods that can be used to produce a select import material. The main characteristics of select import are a well-graded granular soil, which has a high laboratory maximum dry density, typically in the range of 2.0 to 2.2 Mg/m^3 (125 to 135 pcf). As a processed material, the particle size gradation for each batch of fill should be similar. Usually an import material will have all laboratory maximum dry density values within 0.05 Mg/m^3 (3 pcf) and a standard deviation of 0.02 Mg/m^3 (1 pcf) or less. Since results of the laboratory maximum dry density are within a narrow band, a field technician could use either the average or highest value of the laboratory maximum dry density without much effect on the relative compaction.
Structural fill derived from uniform borrow	Uniform borrow typically refers to a natural material that will consistently have the same soil classification and similar grain size distribution. An example of a possible uniform borrow could be a natural deposit of beach sand. Other uniform borrow could be formational rock, such as deposits of sandstone or siltstone. Figure 19.2 presents an example of a uniform borrow material. The fill was derived from a formational rock, classified as a weakly cemented shale. When used as fill, the material is classified as a silty clay, having a liquid limit between 41 and 50. The laboratory maximum dry density varies from 1.84 to 1.97 Mg/m^3 (115 to 123 pcf), with an average value of 1.92 Mg/m^3 (120 pcf) and a standard deviation of 0.035 Mg/m^3 (2.2 pcf). As the name implies, the main characteristic of the material is its uniformity. Usually a uniform borrow material has consistently the same soil classification, with all laboratory maximum dry density values within 0.13 Mg/m^3 (8 pcf) and a standard deviation of 0.05 Mg/m^3 (3 pcf) or less. For uniform borrow soil, such as soil having the data shown in Fig. 19.2, it is often not possible to match a specific laboratory maximum dry density to the soil in the field. This is because the grain size distribution curves and the plasticity characteristics are too similar to distinguish the soil by visual or tactile methods. Unless the maximum dry density happens to be performed on exactly the same soil tested in the field, it is not possible to accurately match the laboratory maximum dry density to the soil tested in the field. The maximum error in relative compaction for a uniform borrow is about 6%, or a difference in relative compaction from 84 to 90%. In order to have an error on the safe side, one of the higher maximum dry density values should be selected. For example, the following value of laboratory maximum dry density $\rho_{d\,\max}$ could be used:

TABLE 19.6 Compaction Fundamentals: Types of Fill (*Continued*)

Topic (1)	Discussion (2)
Structural fill derived from uniform borrow (*Continued*)	$$\rho_{d\,\max} = \rho_a + s$$ where ρ_a = average value of laboratory maximum dry density and s = standard deviation. When the above equation is used, the maximum error on the unsafe side for uniform borrow is about 1%, or a difference in relative compaction from 89 to 90%.
Structural fill derived from mixed borrow	Mixed borrow contains material of different classifications. For example, mixed borrow could be a deposit of alluvium that contains alternating layers of sand, silt, and clay. Mixed borrow could also be formational rock that contains thin alternating layers of sandstone and claystone. The main characteristics of mixed borrow are that each load of fill could have soils with significantly different grain size distributions and soil classifications. The fill contains many different soil types, all jumbled up and mixed together. One method to deal with mixed borrow material is to thoroughly mix each load of import and then perform a laboratory maximum dry density test on that batch of import soil. Another option is to thoroughly mix each batch of mixed borrow material and then perform a "one-point Proctor test" (Holtz and Kovacs 1981). This method consists of performing a single-point maximum dry density test (in the field) on soil that has a water content that is dry of optimum moisture content. The maximum dry density is then estimated based on the observation that compaction curves on similar soil types have the same basic shape. This procedure is illustrated in Fig. 19.3, where point *A* represents the one-point Proctor test performed on the soil that has a water content dry of optimum and then the compaction curve is drawn. Note in Fig. 19.3 that the laboratory maximum dry density (point *B*) is obtained by using the line of optimums, which is a line drawn through the peak point of the compaction curves.
Structural fill derived from borrow with oversize particles	The last basic type of structural fill is borrow having oversize particles, which are typically defined as those particles retained on the ¾ in. (19 mm) U.S. standard sieve, i.e., coarse gravel and cobble size particles. The soil matrix is defined as those soil particles that pass the ¾ in. (19 mm) U.S. standard sieve. When a field density test (such as a sand cone test) is performed, the soil excavated for the test can be sieved on the ¾ in. (19 mm) sieve in order to determine the mass of oversize particles. The elimination method (Day 1989) can then be used to mathematically eliminate the volume of oversize particles in order to calculate the dry density of the matrix material (ρ_{dm}), by using the following equation: $$\rho_{dm} = \frac{M_{ds}}{V - M_o/(G\rho_w)}$$ where M_{ds} = dry mass of the matrix soil V = total volume of the excavated hole M_o = mass of oversize particles

TABLE 19.6 Compaction Fundamentals: Types of Fill (*Continued*)

Topic (1)	Discussion (2)
Structural fill derived from borrow with oversize particles (*Continued*)	G = bulk specific gravity of the oversize particles ρ_w = density of water The relative compaction is calculated by dividing the dry density of the matrix material ρ_{dm} by the laboratory maximum dry density, where the laboratory compaction test is performed on the matrix material. By using the elimination method to calculate the relative compaction of the matrix material, the compaction state of the matrix soil is controlled. This is desirable because it is the matrix soil (not the oversize particles) that usually govern the compressibility, shear strength, and permeability of the soil mass. If the matrix soil can be considered to be a uniform borrow material, then the procedure for selecting the laboratory maximum dry density in the field is the same as previously discussed for uniform borrow material. If the matrix material is a mixed borrow material, then the procedure for selecting the laboratory maximum dry density in the field is the same as previously discussed for a mixed borrow material. Other methods have been developed to deal with fill containing oversize particles (Saxena et al. 1984, Houston and Walsh 1993).

TABLE 19.7 Methods Used to Produce a Select Import Material

Method (1)	Description (2)	Effect (3)
Screening	Material processed over vibrating screens (can be combined with spray washing on screens).	Divide by particle size.
Crushing	Material run through a crusher.	Produces angular shape.
Log washers	Material is run through an inclined unit with dual rotating shafts mounted with paddles. Continuous flow of water carries fine material out of low end of the unit while cleaned aggregate is discharged at the upper end.	Removes deleterious material (e.g., clay) present in the aggregate or removes coating on aggregates.
Sand classifying unit	Continuous flow of water containing sand is fed into horizontal unit. Coarse sand settles first; finer sands later; finer contaminants are carried out of far end by the flow of the water.	Divides sand into fractions based on particle size.
Screw classifier	Water and sand are fed into the low end of an inclined unit having a rotating screw auger. Sand is moved up the unit and out of the water by the screws. Waste water at the low end carries off fines and lightweight contaminants.	Removes lightweight material and fine contaminants.
Rotary scrubber	Water and aggregate are fed into a revolving, inclined drum equipped with lifting angles. The aggregate tumbles upon itself as it proceeds through the scrubber.	Capable of removing large quantities of soluble contaminants.
Jig benefaction	Mechanical or air pulses agitate water, allowing material to sink to the bottom of the unit and form layers of different density.	Separates aggregate on the basis of specific gravity.
Heavy separator	Aggregate fed into medium of given specific gravity. Denser particles sink; lighter particles float or are suspended in medium.	Precise separation on basis of specific gravity of medium.

Note: See Table 19.6 for further discussion of select import.
Source: Rollings and Rollings (1996).

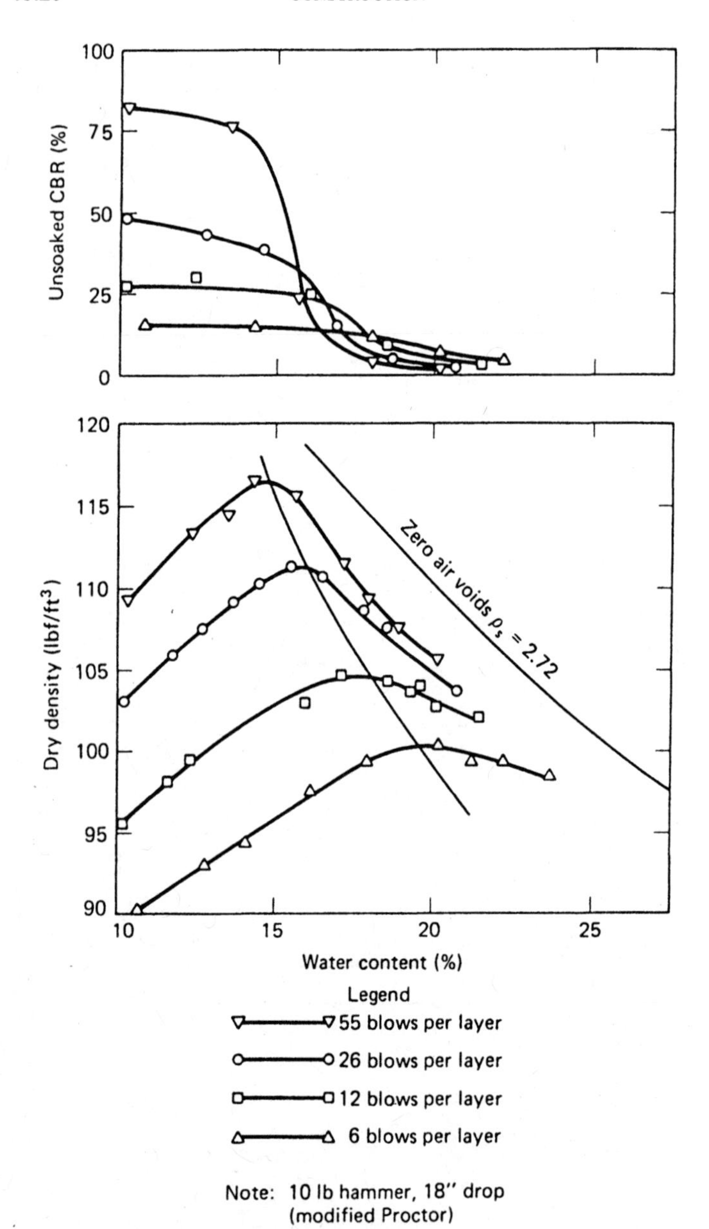

FIGURE 19.1 California bearing ratio (CBR) versus water content for a compacted clay. See Table 19.5. (*From Turnbull and Foster 1956; reprinted with permission from the American Society of Civil Engineers.*)

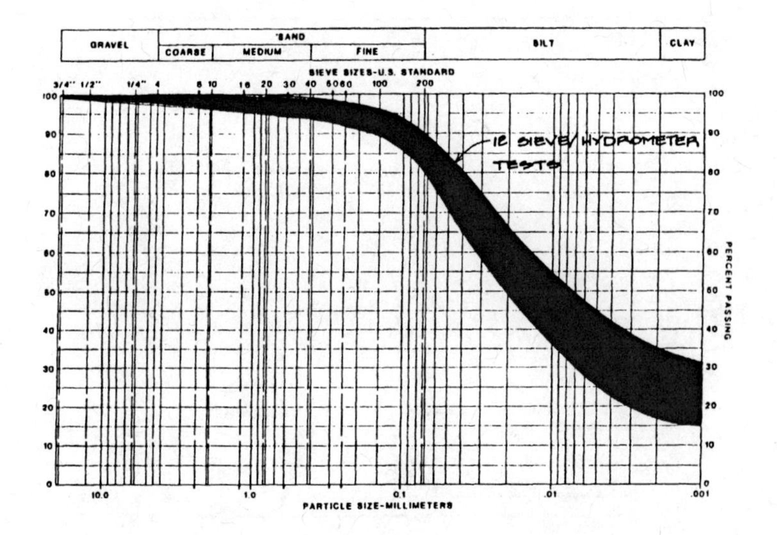

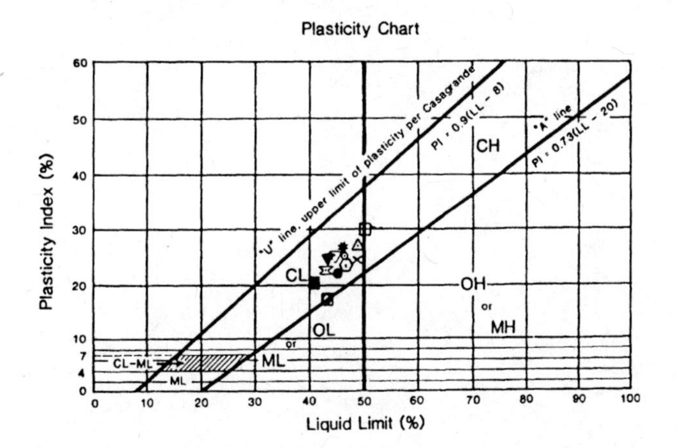

FIGURE 19.2 Example of a uniform borrow soil. See Table 19.6.

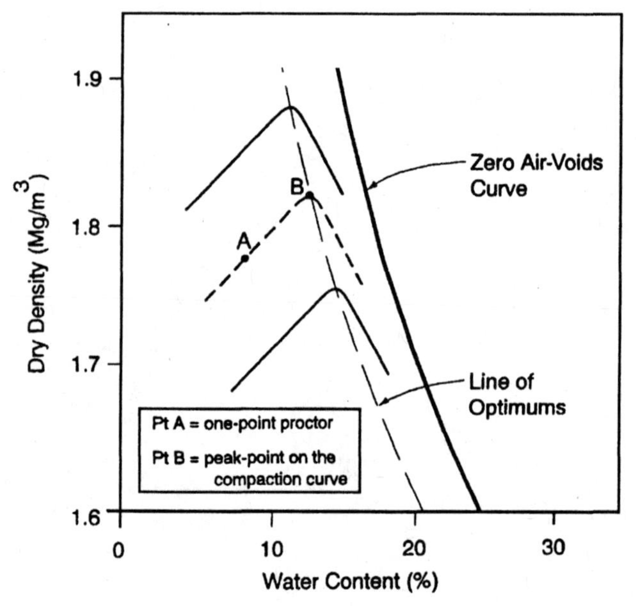

FIGURE 19.3 Procedure to estimate the laboratory maximum dry density based on the one-point Proctor test. See Table 19.6.

TABLE 19.8 Site Improvement Methods

Method (1)	Technique (2)	Principles (3)	Suitable Soils (4)	Remarks (5)
Soil replacement methods	Remove and replace	Excavate weak or undesirable material and replace with better soils	Any	Limited depth and area where cost-effective; generally ≤ 30 ft
	Displacement	Overload weak soils so that they shear and are displaced by stronger fill	Very soft	Problems with mud-waves and trapped compressible soil under the embankment; highly dependent on specific site
Water removal methods	Trenching	Allows water drainage	Soft, fine-grained soils and hydraulic fills	Effective depth up to 10 ft; speed dependent on soil and trench spacing; resulting desiccated crust can improve site mobility
	Precompression	Loads applied prior to construction to allow soil consolidation	Normally consolidated fine-grained soil, organic soil, fills	Generally economical; long time may be needed to obtain consolidation; effective depth limited only by ability to achieve needed stresses
	Precompression with vertical drains	Shortens drainage path to speed consolidation	Same as above	More costly; effective depth usually limited to ≤ 100 ft
	Electroosmosis	Electric current causes water to flow to cathode	Normally consolidated silts and silty clays	Expensive; relatively fast; usable in confined area; not usable in conductive soils; best for small areas

TABLE 19.8 Site Improvement Methods (*Continued*)

Method (1)	Technique (2)	Principles (3)	Suitable Soils (4)	Remarks (5)
Site strengthening methods	Dynamic compaction	Large impact loads applied by repeated dropping of a 5- to 35-ton weight; larger weights have been used	Cohesionless best; possible use for soils with fines; cohesive soils below groundwater table give poorest results	Simple and rapid; usable above and below the groundwater table; effective depths up to 60 ft; moderate cost; potential vibration damage to adjacent structures
	Vibrocompaction	Vibrating equipment densifies soils	Cohesionless soils with < 20 percent fines	Can be effective up to 100-ft depth; can achieve good density and uniformity; grid spacing of holes critical, relatively expensive
	Vibroreplacement	Jetting and vibration used to penetrate and remove soil; compacted granular fill then placed in hole to form support columns surrounded by undisturbed soil	Soft cohesive soils ($s_u = 15$ to 50 kPa)	Relatively expensive
	Vibrodisplacement	Similar to vibroreplacement except soil is displaced laterally rather than removed from the hole	Stiffer cohesive soils ($s_u = 30$ to 60 kPa)	Relatively expensive

TABLE 19.8 Site Improvement Methods (*Continued*)

Method (1)	Technique (2)	Principles (3)	Suitable Soils (4)	Remarks (5)
Grouting	Injection of grout	Fill soil voids with cementing agents to strengthen and reduce permeability	Wide spectrum of coarse- and fine-grained soils	Expensive; more expensive grouts needed for finer-grained soils may use pressure injection, soil fracturing, or compaction techniques
	Deep mixing	Jetting or augers used to physically mix stabilizer and soil	Wide spectrum of coarse- and fine-grained soils	Jetting poor for highly cohesive clays and some gravelly soils; deep mixing best for soft soils up to 165 ft deep
Thermal	Heat	Heat used to achieve irreversible strength gain and reduced water susceptibility	Cohesive soils	High energy requirements; cost limits practicality
	Freezing	Moisture in soil frozen to hold particles together and increase shear strength and reduce permeability	All soils below the groundwater table; cohesive soils above the groundwater table	Expensive; highly effective for excavations and tunneling; high groundwater flows troublesome; slow process
Geosynthetics	Geogrids, geotextiles, geonets, and geomembranes	Use geosynthetic materials for filters, erosion control, water barriers, drains, or soil reinforcing (see Chap. 22)	Effective filters for all soils; reinforcement often used for soft soils	Widely used to accomplish a variety of tasks; commonly used in conjunction with other methods (e.g., strip drain with surcharge or to build a construction platform for site access)

Note: See the text at the beginning of Chap. 19 for a further discussion of these methods.
Source: Rollings and Rollings (1996).

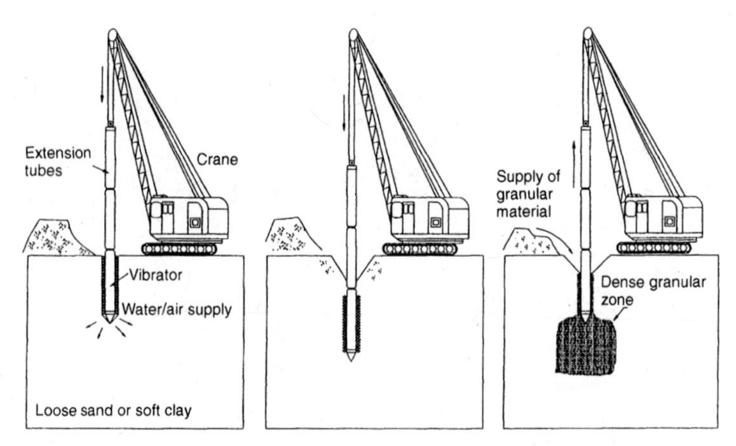

FIGURE 19.4 Equipment used for deep vibratory techniques. See Table 19.8. (*From Rollings and Rollings 1996; reprinted with permission of McGraw-Hill, Inc.*)

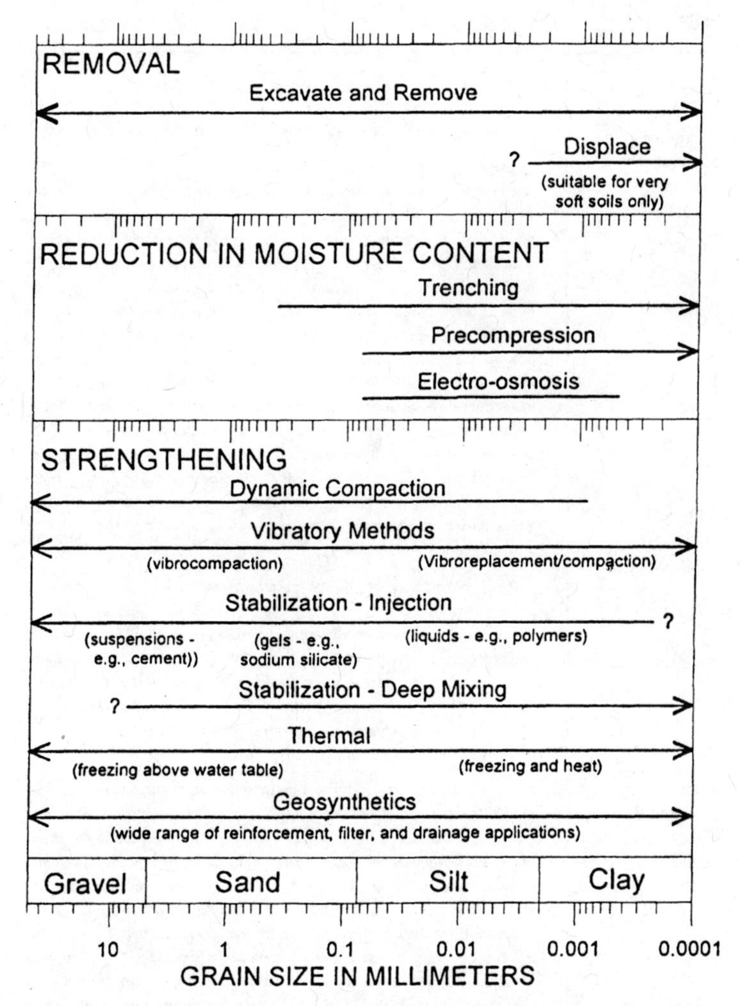

FIGURE 19.5 Site improvement methods as a function of soil grain size. See Table 19.8. (*From Rollings and Rollings 1996; reprinted with permission of McGraw-Hill, Inc.*)

TABLE 19.9 Observational Method

Topic (1)	Discussion (2)
Description of the observational method	Concerning the observational method, Terzaghi and Peck (1969) state: "Design on the basis of the most unfavorable assumptions is inevitably uneconomical, but no other procedure provides the designer in advance of construction with the assurance that the soil-supported structure will not develop unanticipated defects. However, if the project permits modifications of the design during construction, important savings can be made by designing on the basis of the most probable rather than the most unfavorable possibilities. The gaps in the available information are filled by observations during construction, and the design is modified in accordance with the findings. This basis of design may be called the observational procedure....In order to use the observational procedure in earthwork engineering, two requirements must be satisfied. First of all, the presence and general characteristics of the weak zones must be disclosed by the results of the subsoil exploration in advance of construction. Secondly, special provisions must be made to secure quantitative information concerning the undesirable characteristics of these zones during construction before it is too late to modify the design in accordance with the findings." As mentioned above, the observational method is used during the construction of the project. It is a valuable technique because it allows the geotechnical engineer, depending on observations and testing during construction, to revise the design and provide a more economical foundation or earth structure. The method is often used during the installation of deep foundations, where field performance testing or observations are essential in confirming that the foundation is bearing on the appropriate stratum.
Example of the misuse of the observational method	In some cases, the observational method can be misunderstood or misused. For example, at one project the geotechnical engineer discovered the presence of a shallow groundwater table and indicated in the feasibility report that the best approach would be to use the observational method, where the available information on the groundwater table would be supplemented by observations during construction. When the excavation for the underground garage was made, extensive groundwater control was required and an expensive dewatering system was installed. But the cost of the expensive dewatering system had not been anticipated by the client. The client was very upset at the high cost of the dewatering system because a simple design change of using the first floor of the building as the garage (i.e., above-grade garage) and adding an extra floor to the building would have been much less expensive than dealing with the groundwater. As the client stated: "If I had known that the below-grade garage was going to cost this much, I would never have attempted to construct it."

TABLE 19.9 Observational Method (*Continued*)

Topic (1)	Discussion (2)
Advantages and disadvantages	As the above case illustrates, the observational method must not be used in place of a plan of action, but rather to make the plan more economical on the basis of observed subsurface conditions during construction. The main advantage of the observational method is the savings due to modification of the design during construction. However, the client must understand that the savings associated with the observational method may be offset by construction delays as the project is redesigned.

CHAPTER 20
GROUNDWATER AND PERCOLATION TESTS

20.1 INTRODUCTION

This chapter deals with groundwater control and percolation tests for sewage disposal for rural areas. The groundwater table (also known as the *phreatic surface*) is the top surface of underground water, the location of which is often determined from piezometers, such as an open standpipe. A perched groundwater table refers to groundwater occurring in an upper zone separated from the main body of groundwater by underlying unsaturated rock or soil.

Groundwater and moisture migration can affect all types of civil engineering projects. Probably more failures in geotechnical and foundation engineering are either directly or indirectly related to groundwater and moisture migration than any other factor. Table 20.1 presents a brief summary of the importance of groundwater as discussed in prior chapters.

20.2 GROUNDWATER

Groundwater can cause or contribute to failure because of excess saturation, seepage pressures, or uplift forces. It has been stated that uncontrolled saturation and seepage causes many billions of dollars a year in damage (Cedergren 1989). Other geotechnical and foundation problems caused by groundwater are as follows (Harr 1962, Collins and Johnson 1988, Cedergren 1989):

- Piping failures of dams, levees, and reservoirs
- Seepage pressures that cause or contribute to slope failures and landslides
- Deterioration and failure of roads due to the presence of groundwater in the base or subgrade
- Highway and other fill foundation failures caused by perched groundwater
- Earth embankment and foundation failures caused by excess pore water pressures
- Retaining wall failures caused by hydrostatic water pressures
- Canal linings, drydocks, and basement or spillway slabs uplifted by groundwater pressures
- Soil liquefaction, caused by earthquake shocks, resulting from presence of loose granular soil that is below the groundwater table
- Transportation of contaminants by the groundwater

Proper drainage design and construction of drainage facilities can mitigate many of these groundwater problems. For example, for canyon and drainage channels where fill is to be placed, a canyon subdrain system such as shown in Standard Detail No. 1 (App. B) should be installed to prevent the buildup of groundwater in the canyon fill. The drain consists of a perforated pipe (perforations on the underside of the pipe) and open graded gravel around the pipe, with the gravel wrapped in a geofabric that prevents the gravel and pipe from being clogged with soil particles.

20.2.1 Groundwater Control

As indicated in Table 20.2, there are many different methods of groundwater control. A commonly used method of groundwater control for excavations is wellpoint systems with suction pumps. The purpose of this method is to lower the groundwater table by installing a system of perimeter wells. As illustrated in Fig. 20.1, the system consists of closely spaced wellpoints installed around the excavation. A wellpoint is a small-diameter pipe having perforations at the bottom end. A pump is used to extract water from the pipe which then lowers the groundwater table as illustrated in Fig. 20.1. It is important to consider the possible damage to adjacent structures caused by the lowering of the groundwater table at the site. For example, a lowering of the groundwater table could lead to consolidation of soft clay layers or rotting of wood piling.

20.2.2 Pavements

Probably the most common engineering facilities damaged by groundwater are pavements. Table 11.2 presents the drainage properties of different compacted soils. It has been stated that groundwater in pavements accelerates the damage rates by hundreds of times over the damage rates of pavements with no groundwater. This premature failure of thousands of miles of pavements and billions of dollars in losses a year could be avoided by good pavement drainage practices (Cedergren 1989). The key element in a good drainage system is a layer of highly permeable material (such as open graded gravel) protected by filters or geofabric so that the permeable material will not become clogged by the intrusion of soil fines. A drainage system, to remove the water from the base, is also required.

There are several ways that groundwater can enter the base material. In areas having a high groundwater table or artesian condition, water can be forced upward into the base material. Water can also flow downward through pavement cracks or joints. There can also be the development of a perched groundwater condition, where water moves laterally through the base from adjacent planter areas, medians, or shoulders.

A common problem is the flow of groundwater in permeable utility trenches. One construction method to prevent the flow of groundwater through utility trenches is to use a grout (such as a cement slurry) in the utility line or storm drain bedding zone. The remainder of the trench could then be backfilled and compacted with on-site native soil. This should provide the trench with a permeability equal to or less than the surrounding native soil.

20.2.3 Slopes

Groundwater can affect slopes in many different ways. Table 20.3 presents common examples and the influence of groundwater on slope failures. The main destabilizing factors of groundwater on slope stability are as follows (Cedergren 1989):

1. Reducing or eliminating cohesive strength
2. Producing pore water pressures which reduce effective stresses, thereby lowering shear strength
3. Causing horizontally inclined seepage forces, which increase the driving forces and reduce the factor of safety of the slope
4. Providing for the lubrication of slip surfaces
5. Trapping of groundwater in soil pores during earthquakes or other severe shocks, which leads to liquefaction failures

There are many different construction methods to mitigate the effects of groundwater on slopes. During construction of slopes, built-in drainage systems can be installed. For example, Standard Detail No. 4 in App. B shows a drainage detail for the construction of a stabilization fill slope. For existing slopes, drainage devices such as trenches or galleries, relief wells, or horizontal drains can be installed. Another common slope stabilization method is the construction of a drainage buttress at the toe of a slope. In its simplest form, a drainage buttress can consist of cobbles or crushed rock placed at the toe of a slope. The objective of the drainage buttress is to be as heavy as possible to stabilize the toe of the slope and also to have a high permeability so that seepage is not trapped in the underlying soil.

There can also be other indirect effects of groundwater on slopes. For example, when the groundwater evaporates at the toe of the slope, salt deposits can form. Both the high groundwater table and the surface salt deposits can kill or stunt the growth of plants and trees. Deposits of salts due to evaporation of groundwater are known as *evaporites.* In arid or semiarid regions where moisture is evaporating at the ground surface, they can form on or just beneath the ground surface. The three most common evaporites are gypsum, anhydrite, and sodium chloride (rock salt).

20.3 MOISTURE MIGRATION THROUGH FLOOR SLABS AND BASEMENT WALLS

Moisture migration into buildings is one of the major challenges faced by engineers, architects, and contractors. Many times, the project architect provides waterproofing recommendations. But in other instances, the geotechnical engineer may need to provide recommendations. Moisture migration through floor slabs and basement walls is discussed in Tables 20.4 and 20.5 and Figs. 20.3 and 20.4.

20.4 PERCOLATION TESTS FOR SEWAGE DISPOSAL SYSTEM

The geotechnical engineer is often involved with the design and construction of subsurface sewage disposal systems for buildings in rural areas. This is because the design and construction involve the determination of key geotechnical parameters, such as the location of groundwater and the permeability of the soil. The permeability of the soil is assessed by performing percolation tests.

The procedure for performing percolation tests varies in different localities. Because the specifications vary so much from place to place, the geotechnical engineer should always obtain the governing applicable standards. Appendix C (from the County of San Diego 1991) presents an example of the percolation test procedure for the design of subsurface sewage disposal systems for single-family residences in rural areas. The glossary (App. A,

Glossary 5) presents a list of common percolation and sewage disposal terms and their definitions. The field performance of percolation tests and design and construction of sewage disposal systems for rural areas is discussed in Table 20.6 and Figs. 20.5 to 20.7.

20.5 SURFACE DRAINAGE AND PIPE BREAKS

The final section of Chap. 20 presents a brief discussion of the infiltration of water into the soil due to poor surface drainage or from pipe leaks (see Table 20.7).

TABLE 20.1 Brief Summary of Groundwater Discussed in Prior Chapters

Topic (1)	Discussion (2)
Subsurface exploration	One of the main purposes of subsurface exploration is to determine the location of the groundwater table. The groundwater table (▼ symbol) is often shown on the soil profiles (see Figs. 2.18 to 2.20).
Laboratory testing	The rate of flow through soil is dependent on its permeability. The constant-head permeameter (Fig. 3.4) or the falling-head permeameter (Fig. 3.5) can be used to determine the coefficient of permeability (also known as the hydraulic conductivity). Equations to calculate the coefficient of permeability in the laboratory are based on Darcy's law ($v = ki$).
Pore water pressure	The pore water pressure u is calculated by using the equation in Table 6.1 for soil below the groundwater table and the equation in Table 6.2 for soil above the groundwater table that is saturated because of capillary rise.
Shear strength	The shear strength of saturated soil can be influenced by the water that fills the soil pores. For example, normally consolidated plastic soils have an increase in pore water pressure because the soil structure wants to contract during undrained shear, while heavily overconsolidated plastic soils have a decrease in pore water pressure because the soil structure wants to dilate during undrained shear (Chap. 7).
Permeability and seepage	The basic principles of permeability and seepage have been presented in Chap. 8. The primary method to estimate the seepage quantities and the distribution of pore water pressure is the two-dimensional flow net. Flow nets have numerous applications, such as the study of earth dams; determining the effect of cutoff walls, consequences of dewatering, drawdown due to pumping of groundwater from wells, and effect of groundwater on foundations; and in the study of slope stability.
Collapsible soil	As discussed in Table 9.5, collapsible soil is broadly defined as a soil with an unstable particle structure, often of low density and low moisture content, that collapses when there is moisture migration into the soil.
Consolidation	The increase in vertical pressure due to the weight of a structure constructed on top of a soft saturated clay or organic soil will initially be carried by excess pore water pressure u_e. As water slowly drains from the clay, the excess pore water pressure slowly dissipates. Drainage of water from the clay layer causes the soil particles to compress together (i.e., consolidate, see Table 9.10).
Subsidence due to extraction of groundwater	Pumping can cause a lowering of the groundwater table, which then increases the overburden pressure on underlying sediments. This can cause consolidation of soft clay deposits or compression of the loose soil or porous rock structure, which then results in ground surface subsidence (Table 9.18).
Bearing capacity	The bearing capacity of soil with a high groundwater table is lower than that of the same soil having a groundwater table at a great depth (Chap. 10). Groundwater can also exert buoyancy effects on submerged foundation elements and hydrostatic uplift forces on underground basements, storage tanks, etc.

TABLE 20.1 Brief Summary of Groundwater Discussed in Prior Chapters (*Continued*)

Topic (1)	Discussion (2)
Expansive soil	Moisture migration from the clay, such as that due to surface drying, creates a desiccated clay deposit (Chap. 12). If a structure is then constructed at ground surface, moisture will migrate by capillarity and thermoosmosis to the clay located underneath the structure, resulting in an increase in water content of the clay and heave of the center of the structure (center lift).
Surficial slope stability	The development of surficial instability is often due to saturation and groundwater seepage in the outer face of the slope. The surficial stability equation (Table 13.6) was derived by assuming seepage of groundwater parallel to the slope surface to a depth *d*.
Debris flow	Debris flow is commonly defined as soil with entrained water and air that moves readily as a fluid on low slopes. As discussed in Table 13.18, in many cases there is an initial surficial slope failure that transforms itself into a debris flow.
Slope softening	As water infiltrates a compacted clay slope, the pore spaces fill with water and the pore water pressure tends to zero. If a groundwater table then develops, the pore water pressures will become positive. The elimination of negative pore water pressure results in a decrease in effective stress and deformation of the slope (i.e., slope softening) in order to mobilize the needed shear stress to maintain stability (Table 13.19).
Dams	Design analyses of dams can be grouped into three general categories: (1) stability during construction, usually involving a failure through the natural ground that often has a groundwater table; (2) stability of the downstream slope during reservoir operation, where there is a steady-state groundwater flow through the dam; and (3) stability analysis of the upstream slope after rapid drawdown, which causes seepage of groundwater down and through the slope surface (Table 13.21).
Frost	Near-surface silty soils can form ice lenses because of their high capillarity and sufficient permeability that enables them to draw up moisture from the groundwater table. The development of ice lenses can cause heave of the structure. The spring thaw then will melt the ice, resulting in settlement of the foundation or a weakened subgrade that will make the pavement surface susceptible to deterioration or failure (Table 17.6).

TABLE 20.2 Methods of Groundwater Control

Method (1)	Soils suitable for treatment (2)	Uses (3)	Comments (4)
Sump pumping	Clean gravels and coarse sands	Open shallow excavations.	Simplest pumping equipment. Fines easily removed from the ground. Encourages instability of formation. See Fig. 20.2.
Wellpoint system with suction pump	Sandy gravels down to fine sands (with proper control can also be used in silty sands)	Open excavations including utility trench excavations.	Quick and easy to install in suitable soils. Suction lift is limited to about 18 ft (5.5 m). If greater lift is needed, multistage installation is necessary. See Fig. 20.1.
Deep wells with electric submersible pumps	Gravels to silty fine sands, and water-bearing rocks	Deep excavation in, through, or above water-bearing formations.	No limitation on depth of drawdown. Wells can be designed to draw water from several layers throughout its depth. Wells can be sited clear of working area.
Jetting system	Sands, silty sand, and sandy silts	Deep excavations in confined space where multistage wellpoints can not be used.	Jetting system uses high pressure water to create vacuum as well as to lift the water. No limitation on depth of drawdown.
Sheet piling cutoff wall	All types of soil (except boulder beds)	Practically unrestricted use.	Tongue-and-groove wood sheeting utilized for shallow excavations in soft and medium soils. Steel sheet piling for other cases. Well-understood method and can be rapidly installed. Steel sheet piling can be incorporated into permanent works or recovered. Interlock leakage can be reduced by filling interlock with bentonite, cement, grout, or similar materials.
Slurry trench cutoff wall	Silts, sands, gravels, and cobbles	Practically unrestricted use. Extensive curtain walls around open excavations.	Rapidly installed. Can be keyed into impermeable strata such as clays or soft shales. May be impractical to key into hard or irregular bedrock surfaces, or into open gravels.

TABLE 20.2 Methods of Groundwater Control (*Continued*)

Method (1)	Soils suitable for treatment (2)	Uses (3)	Comments (4)
Freezing: ammonium and brine refrigerant	All types of saturated soils and rock	Formation of ice in void spaces stops groundwater flow.	Treatment is effective from a working surface outward. Better for large applications of long duration. Treatment takes longer time to develop.
Freezing: liquid nitrogen refrigerant	All types of saturated soils and rock	Formation of ice in voids spaces stops groundwater flow.	Better for small applications of short duration where quick freezing is required. Liquid nitrogen is expensive and requires strict site control. Some ground heave could occur.
Diaphragm structural walls: structural concrete	All soil types including those containing boulders	Deep basements, underground construction, and shafts.	Can be designed to form a part of the permanent foundation. Particularly efficient for circular excavations. Can be keyed into rock. Minimum vibration and noise. Can be used in restricted space. Also can be installed very close to the existing foundation.
Diaphragm structural walls: bored piles or mixed-in-place piles	All soil types, but penetration through boulders may be difficult and costly	Deep basements, underground construction, and shafts.	A type of diaphragm wall that is rapidly installed. It can be keyed into impermeable strata such as clays or soft shales.

Sources: NAVFAC DM-7.2 (1982), based on the work by Cashman and Harris (1970).

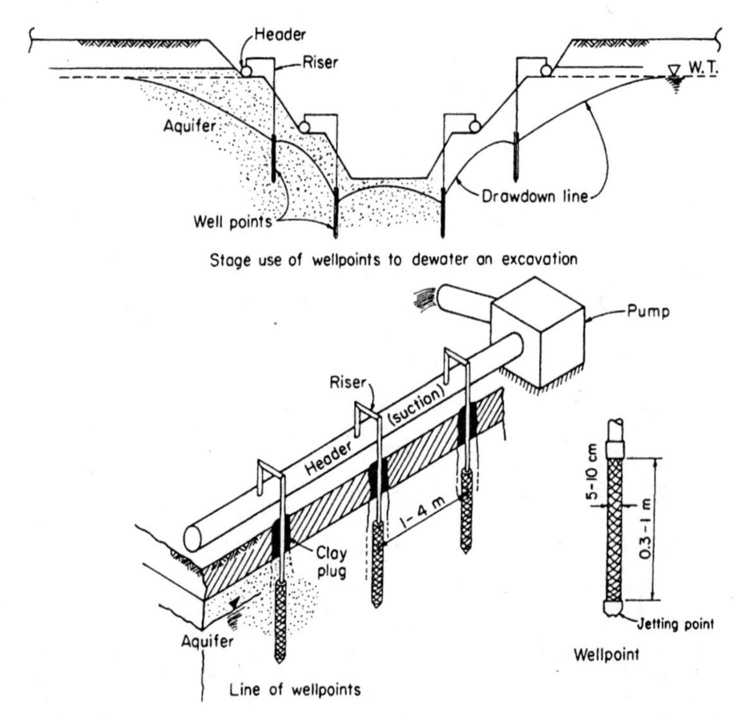

FIGURE 20.1 Groundwater control: wellpoint system with suction pump. See Table 20.2. (*From Bowles 1982; reprinted with permission of McGraw-Hill, Inc.*)

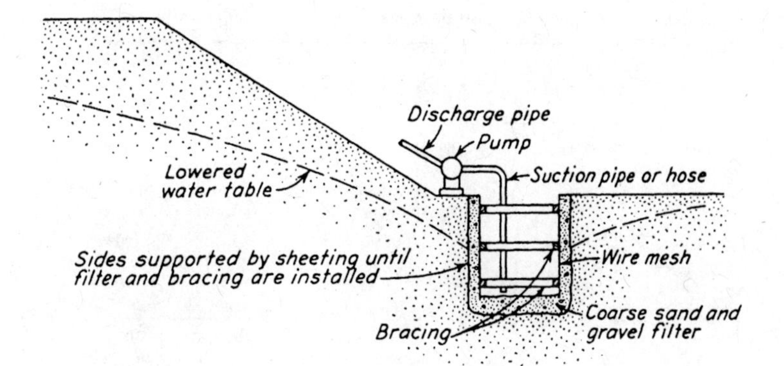

FIGURE 20.2 Groundwater control: example of a sump being used to lower the groundwater table. See Table 20.2. (*From Peck, Hanson, and Thornburn 1974; reprinted with permission of John Wiley & Sons.*)

TABLE 20.3 Common Groundwater Conditions Causing Slope Failures

Kind of slope (1)	Conditions leading to failure (2)	Type of failure and its consequences (3)
Natural earth slopes above developed land areas (homes, industrial)	Earthquake shocks, heavy rains, snow, freezing and thawing, undercutting at toe, mining excavations	Mud flows, avalanches, landslides; destroying property, burying villages, damming rivers
Natural earth slopes within developed land areas	Undercutting of slopes, heaping fill on unstable slopes, leaky sewers and water lines, lawn sprinkling	Usually slow creep type of failure; breaking water mains, sewers, destroying buildings, roads
Reservoir slopes	Increased soil and rock saturation, raised water table, increased buoyancy, rapid drawdown	Rapid or slow landslides; damaging highways, railways, blocking spillways, leading to overtopping of dams, causing flood damage with serious loss of life
Highway or railway cut or fill slopes	Excessive rain, snow, freezing, thawing, heaping fill on unstable slopes, undercutting, trapping groundwater	Cut slope failures blocking roadways, foundation slipouts removing roadbeds or tracks; property damage, some loss of life
Earth dams and levees, reservoir ridges	High seepage levels, earthquake shocks; poor drainage	Sudden slumps leading to total failure and floods downstream; much loss of life, property damage
Excavations	High groundwater level, insufficient groundwater control, breakdown of dewatering systems	Slope failures or heave of bottoms of excavations; largely delays in construction, equipment loss, property damage

Source: Cedergren 1989.

TABLE 20.4 Moisture Migration Through Floor Slabs

Topic (1)	Discussion (2)
Moisture migration	Moisture can migrate into the structure through the foundation, exterior walls, and through the roof. Four ways that moisture can penetrate a concrete floor slab are by water vapor, hydrostatic pressure, leakage, and capillary action (WFCA 1984). Water vapor acts in accordance with the physical laws of gases, where water vapor will travel from one area to another area whenever there is a difference in vapor pressure between the two areas. Hydrostatic pressure is the buildup of water pressure beneath the floor slab, which can force large quantities of water through slab cracks or joints. Leakage refers to water traveling from a higher to a lower elevation due solely to the force of gravity, and such water can surround or flood the area below the slab. Capillary action is different from leakage in that water can travel from a lower to a higher elevation. The controlling factor in the height of capillary rise in soils is the pore size (Holtz and Kovacs 1981). Open graded gravel has large pore spaces and hence very low capillary rise. This is why open graded gravel is frequently placed below the floor slab to act as a capillary break (Butt 1992).
Damage caused by moisture migration	Moisture that travels through the concrete floor slab can damage such floor coverings as carpet, hardwood, and vinyl. When a concrete floor slab has floor coverings, the moisture can collect at the top of the slab, where it weakens the floor-covering adhesive. Hardwood floors can be severely affected by moisture migration through slabs because they can warp or swell from the moisture. Moisture that penetrates the floor slab can also cause musty odors or mildew growth in the space above the slab. Some people are allergic to mold and mildew spores. They can develop health problems from the continuous exposure to such allergens. In most cases, the moisture that passes through the concrete slab contains dissolved salts. As the water evaporates at the slab surface, the salts form a white crystalline deposit, commonly called *efflorescence.* The salt can build up underneath the floor covering, where it attacks the adhesive as well as the flooring material itself. Oliver (1988) believes that it is the shrinkage cracks in concrete that provide the major pathways for rising dampness. Sealing of slab cracks may be necessary in situations of rising dampness affecting sensitive floor coverings.
Construction details	The geotechnical engineer often provides the design and construction specifications that are used to prevent both water vapor and capillary rise through floor slabs. An example of below-slab recommendations (WFCA 1984) is as follows: "Over the subgrade, place 4 inches (10 cm) to 8 inches (20 cm) of washed and graded gravel. Place a leveling bed of 1 to 2 inches (2.5 to 5 cm) of sand over the gravel to prevent moisture barrier puncture. Place a moisture barrier over the sand leveling bed and seal the joints to prevent moisture penetration. Place a 2 inch (5 cm) sand layer over the moisture barrier."

TABLE 20.4 Moisture Migration Through Floor Slabs (*Continued*)

Topic (1)	Discussion (2)
Construction details (*Continued*)	The gravel layer should consist of open graded gravel. This means that the gravel should not contain any fines and that all the soil particles are retained on the gravel size sieves. This will provide for large void spaces between the gravel particles. Large void spaces in the gravel will help prevent capillary rise of water through the gravel (Day 1992d). In addition to a gravel layer, the installation of a moisture barrier (such as visqueen) will further reduce the moisture migration through concrete (Brewer 1965).
Construction specifications	In some areas, there may be local building requirements for the construction of moisture barriers. For example, Fig. 20.3 presents the County of San Diego requirements for below slab moisture barriers in construction on clays. Note in this figure that 4 in. (10 cm) of open graded gravel or rock is required below the concrete slabs. A 6-mil visqueen moisture barrier is required below the open graded gravel. These recommendations are similar to the example listed above, except that the sand layers have been omitted. The purpose of the sand layer is to prevent puncture of the visqueen moisture barrier. In using specifications similar to Fig. 20.3, it is best to use rounded gravel or a thicker visqueen barrier so that the visqueen is not punctured during placement and densification of the gravel.
Groundwater collection system (sump)	The below-slab design and construction details presented above will not be able to resist a high groundwater table or artesian groundwater pressure. In these cases, a more extensive waterproofing system will need to be installed. One common approach is to provide a below-foundation waterproofing system and a sump equipped with a pump to catch and dispose of any water that seeps through the waterproofing system. A sump is defined as a small pit excavated in the ground or through the basement floor to serve as a collection basin for surface runoff or groundwater. A sump pump is used to periodically drain the pit when it fills with water. When the water level reaches a certain level in the collector box, the submerged pump will be activated, and the collection box will be emptied of water.

TABLE 20.5 Moisture Migration through Basement Walls

Topic (1)	Discussion (2)
Moisture migration	As in concrete floor slabs, water can penetrate basement walls by hydrostatic pressure, capillary action, and water vapor. If a groundwater table exists behind the basement wall, then the wall will be subjected to hydrostatic pressure, which can force large quantities of water through wall cracks or joints. A subdrain is usually placed behind the basement wall to prevent the buildup of hydrostatic pressure. Such drains will be more effective if the wall is backfilled with granular (permeable) soil rather than clay. Another way for moisture to penetrate basement walls is by capillary action in the soil or the wall itself. By capillary action, water can travel from a lower to higher elevation in the soil or wall. Capillary rise in walls is related to the porosity of the wall and the fine cracks in both the masonry and, especially, the mortar. To prevent moisture migration through basement walls, an internal or surface waterproofing agent is used. Chemicals can be added to cement mixes to act as internal waterproofs. More common are the exterior applied waterproof membranes. Oliver (1988) lists and describes various types of surface-applied waterproof membranes. A third way that moisture can penetrate through basement walls is by water vapor. As in concrete floor slabs, water vapor can penetrate a basement wall whenever there is a difference in vapor pressure between the two areas.
Contributing factors in moisture migration through basement walls	The effects of dampness, freezing, and salt deposition are major contributors to the weathering and deterioration of basement walls. Some of the other common deficiencies that contribute to moisture migration through basement walls are as follows (Diaz et al. 1994, Day 1994f): 1. The wall is poorly constructed (for example, joints are not constructed to be watertight), or poor-quality concrete that is highly porous or shrinks excessively is used. 2. There is no waterproofing membrane or there is a lack of waterproofing on the basement-wall exterior. 3. There is improper installation, such as a lack of bond between the membrane and the wall, or deterioration with time of the waterproofing membrane. 4. There is no drain, the drain is clogged, or there is improper installation of the drain behind the basement wall. Clay, rather than granular backfill, is used. 5. There is settlement of the wall, which causes cracking or opening of joints in the basement wall. 6. There is no protection board over the waterproofing membrane. During compaction of the backfill, the waterproofing membrane is damaged. 7. The waterproofing membrane is compromised. This happens, for example, when a hole is drilled through the basement wall. 8. There is poor surface drainage, or downspouts empty near the basement wall.

TABLE 20.5 Moisture Migration through Basement Walls (*Continued*)

Topic (1)	Discussion (2)
Damage caused by moisture migration	Moisture that travels through basement walls can damage wall coverings such as wood paneling. Moisture traveling through the basement walls can also cause musty odors or mildew growth in the basement areas. If the wall should freeze, then the expansion of freezing water in any cracks or joints may cause deterioration of the wall. As in concrete floor slabs, the moisture that is passing through the basement wall will usually contain dissolved salts. The penetrating water may often contain salts originating from the ground or mineral salts naturally present in the wall materials. As the water evaporates at the interior wall surface, the salts form white crystalline deposits (efflorescence) on the basement walls. The salt crystals can accumulate in cracks or wall pores, where they can cause erosion, flaking, or ultimate deterioration of the contaminated surface. This is because the process of crystallization often involves swelling and considerable forces are generated. Another problem occurs when the penetrating water contains dissolved sulfates, because they can accumulate and thus increase their concentration on the exposed wall surface, resulting in chemical deterioration of the concrete (ACI 1990).
Construction details	The main structural design and construction details to prevent moisture migration through basement walls are a drainage system at the base of the wall to prevent the buildup of hydrostatic water pressure and a waterproofing system applied to the exterior wall surface. A typical drainage system for basement walls is shown in Fig. 20.4. A perforated drain is installed at the bottom of the basement wall footing. Open graded gravel, wrapped in a geofabric, is used to convey water down to the perforated drain. The drain outlet should be tied to a storm drain system. The waterproofing system frequently consists of a high-strength membrane, a primer for wall preparation, a liquid membrane for difficult-to-reach areas, and a mastic to seal holes in the wall. The primer is used to prepare the concrete wall surface for the initial application of the membrane and to provide long-term adhesion of the membrane. A protection board is commonly placed on top of the waterproofing membrane to protect it from damage during compaction of the backfill soil. Self-adhering waterproofing systems have been developed to make the membrane easier and quicker to install.

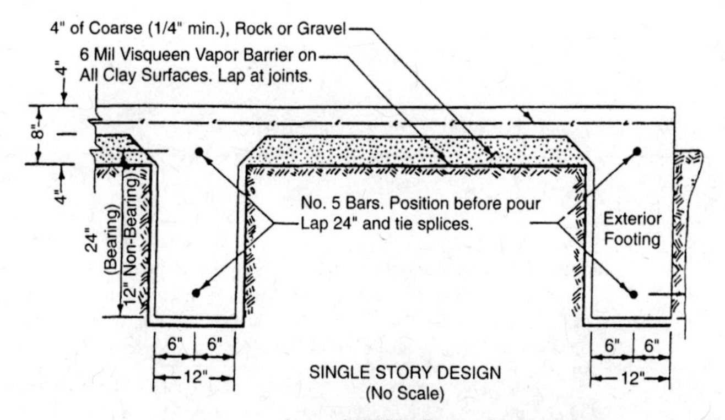

Layer of 1/4" Min. Rock or Gravel (4" deep)
Provides a capillarity breaker type of material. Moisture will not rise through the material as it would in the original clay or in a less coarse material, such as fine sand or fine granite.

Visqueen Moisture Barrier (6 mil)
Properly lapped it provides a vapor seal atop and along the vertical sides of any exposed clay soil to seal off the moisture within the soil so as to reduce variations in the volume of the soil due to changes in moisture and to also reduce the risk of moisture migrating into the area between the under side of the slab and the top of the clay soil.

FIGURE 20.3 Moisture barrier specifications of the County of San Diego (1983). See Table 20.4.

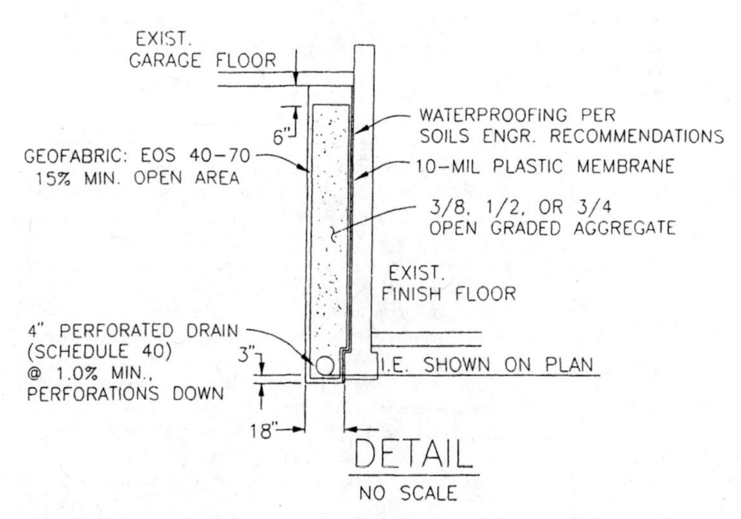

FIGURE 20.4 Typical waterproofing and drainage system for a basement wall. See Table 20.5.

TABLE 20.6 Percolation Tests and Sewage Disposal System

Topic (1)	Discussion (2)
Location of sewage disposal system	The first step in the analysis is to determine the approximate location of the sewage disposal system. Figure 20.5 shows an example of the proposed construction of a single family dwelling in a rural area. The area delineated as "Prop. 23.5′ × 50′ disposal field" in Fig. 20.5 shows the location of the proposed subsurface sewage disposal system.
Septic tank	Sewage from the residence first flows into the septic tank. Figure 20.6 presents a diagram of a two-compartment septic tank. As indicated in Table C.1 (App. C), the septic tank size is based on the number of bedrooms (hence number of occupants) in the house, with a 1000-gallon septic tank required for a 2- to 3-bedroom house, a 1200-gallon septic tank required for a 4-bedroom house, and a 1500-gallon septic tank required for a 5- to 6-bedroom house.
Leach field	After settling of solids and partial decomposition in the septic tank, the effluent drains through a tight line pipe to the leach field. Figure 20.7 shows a diagram of a leach field. The leach field is constructed by first excavating trenches, then placing perforated pipes that are surrounded by open graded gravel in the trench, and finally the trench is backfilled. In the leach field, the effluent seeps from the perforated pipes and into the ground.
Percolation test	As indicated in App. C, the purpose of the percolation test is to "determine the area necessary to properly treat and maintain sewage underground, to size the disposal system with adequate infiltration surface on the basis of expected hydraulic conductivity of the soil and the rate of loading, and to provide for a system with long term expectation of satisfactory performance." The steps in the percolation test as indicated in App. C are as follows: 1. A cylindrical hole is excavated into the ground. The hole should have a minimum diameter of 6 in. (15 cm) and a maximum diameter of 10 in. (25 cm). The depth of the hole must be at least 14 in. (36 cm) and should correspond to the approximate depth that the leach lines will be installed. Often the hole can be excavated by using a hand auger. 2. A 2-in.- (5-cm)-thick layer of open graded gravel is placed in the bottom of the hole. Clear water is then added to the hole and the depth of water is maintained at 12 to 14 in. (30 to 36 cm) in order to allow the soil to presoak (see App. C for details on presoaking). Presoaking is important because it allows desiccation cracks in clayey soil to close up and the suction pressures to decrease. If the test were performed without presoaking a clayey soil, a high and unrepresentative percolation rate could be obtained as the water quickly flows through the desiccation cracks. 3. After presoaking is complete, the percolation test is performed. Depending on the results of the presoaking, there are three procedures (Cases 1 through 3) that are used to perform the percolation test (see App. C). In general, the percolation test consists of having an initial height of 6 in. (15 cm) of water above the open graded gravel, and the drop of this water level over a time period of 30 minutes is recorded.

TABLE 20.6 Percolation Tests and Sewage Disposal System *(Continued)*

Topic (1)	Discussion (2)
Average percolation rate	The percolation rate is calculated as 30 minutes divided by the drop in elevation (inches) of the water level over the 30-minute time period. The units of the percolation rate are minutes per inch (mpi). As an example of the calculation of the percolation rate, suppose the water level in the hole drops ¾ in. in the 30-minute time period. Then the percolation rate would be 30 minutes divided by 0.75 in., or a percolation rate of 40 mpi. The average percolation rate is calculated as the sum of the percolation rate at each test hole divided by the number of test holes.
Design parameters	From the average percolation rate, two design parameters are obtained: 1. *Primary disposal trench length.* For an average percolation rate, the total primary disposal trench length (which contains the perforated pipe for the leach field) can be determined by using Table C.1 (App. C). 2. *Lot size.* As indicated in Table C.2 (App. C), the average percolation rate determines the minimum required lot size. For example, if the average percolation rate is 61 to 90 mpi, then the minimum lot size is 3 acres. If the average percolation rate is 91 to 120 mpi, then the minimum lot size is 5 acres.
Site factors	There are many site factors that may need to be considered in the design of the sewage disposal system. For example, site factors that could impact the design include a high groundwater table, placement of the leach field on a steep slope, nitrate impacts on vegetation growth, and a confined soil layer having a high horizontal permeability which could transport the effluent onto adjacent property.
Advantages and disadvantages of percolation tests	The percolation test is a straightforward and simple test that is widely used in the United States to design sewage disposal systems for rural properties. Besides its simplicity, the percolation test has an advantage in that it is a field test that measures the actual performance of the *in situ* soil in terms of water infiltration. Disadvantages are that the design is empirical and there are many site factors besides soil percolation that can impact a sewage disposal system. Another disadvantage is that there are many factors that can lead to the failure of the septic system. For example, a common cause of failure is the clogging of the leach field, often caused by a mat of impermeable vegetative growth (i.e., algae, etc.) that develops at the bottom and sides of the leach field trenches.

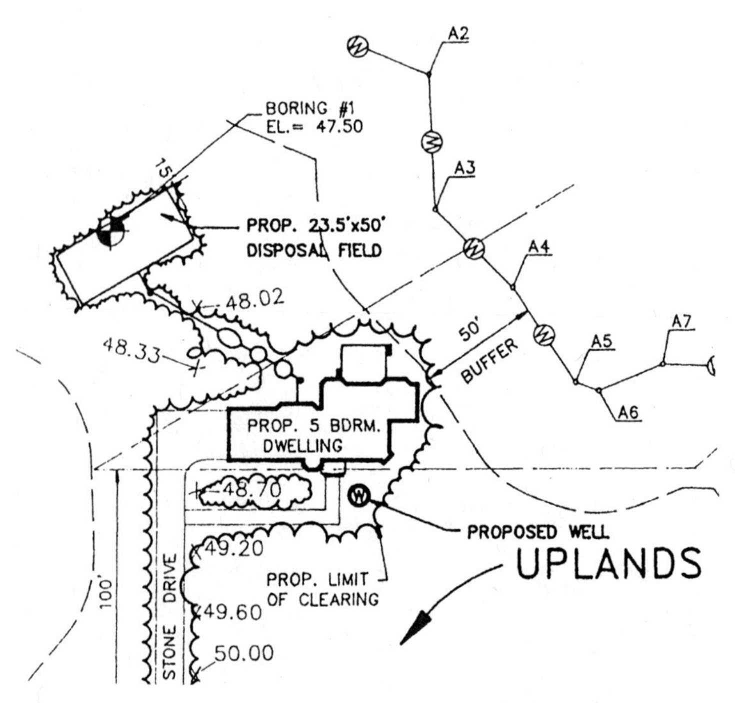

FIGURE 20.5 Plot plan showing the location of the proposed sewage disposal system for a single-family residence. See Table 20.6.

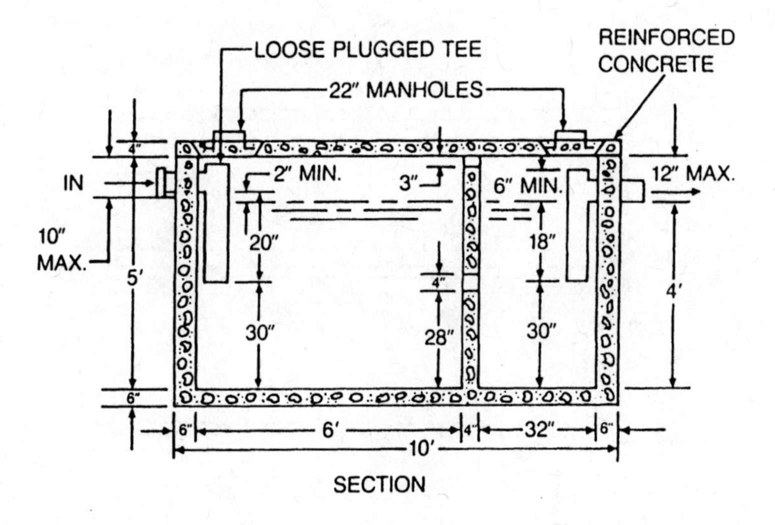

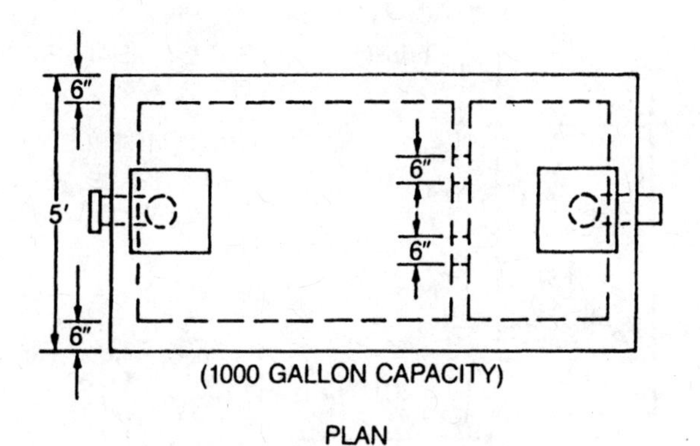

FIGURE 20.6 Example of a two-compartment septic tank. See Table 20.6. (*Reproduced from the County of San Diego, undated.*)

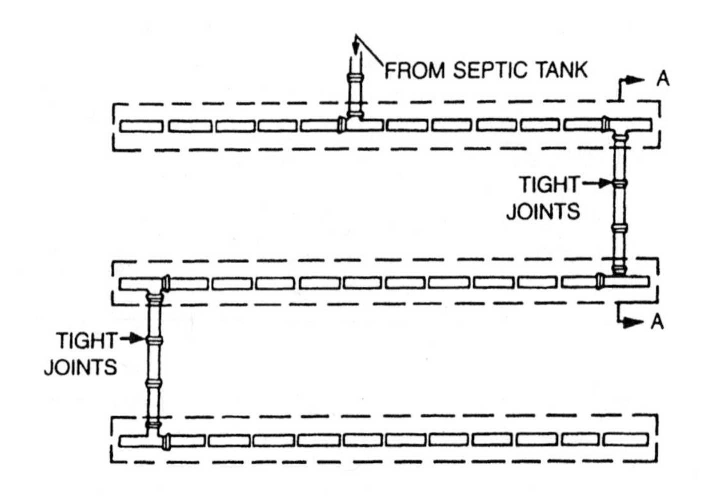

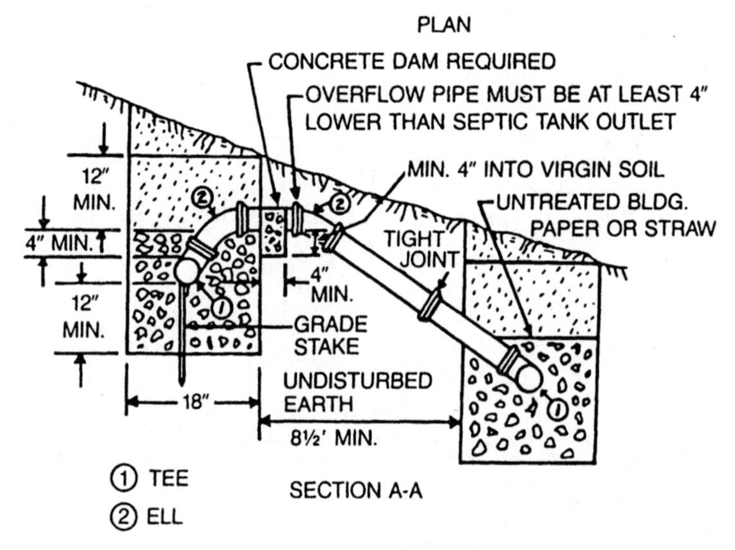

FIGURE 20.7 Example of a leach line disposal system. See Table 20.6. (*Reproduced from the County of San Diego, undated.*)

TABLE 20.7 Surface Drainage and Pipe Breaks

Topic (1)	Discussion (2)
Surface drainage	Inadequate surface drainage can be an important factor in triggering soil problems. For example, water ponding adjacent to a foundation can contribute to expansive soil edge lift, the infiltration of ponding water can raise the groundwater table, and inadequate surface drainage can contribute to moisture intrusion problems through floor slabs or basement walls. In order to prevent ponding of water at the site, it is important to have proper drainage gradients around the structure. In addition, the surface runoff must be transferred to suitable disposal systems, such as storm drain lines. Standard Detail No. 9 (App. B) shows two alternatives for lot surface drainage. Figure 15.2 shows typical construction specifications for drainage ditches that are used to drain surface runoff from slopes.
Pressurized pipe breaks	Pipes can be classified as either pressurized or nonpressurized. Common pressurized pipes are the water lines that provide potable water to the building occupants. An example of a nonpressurized pipe is a sewer line, which may be only intermittently filled with effluent. A pressurized pipe break can introduce large volumes of water into the ground. This water can trigger collapse of loose soil, heave of expansive soil, or a raising of the groundwater table, which may lead to slope instability. The large volume of water from a broken pressurized pipe can also erode and transport soil particles, leading to the development of voids below the structure.
Nonpressurized pipe break	Structural damage can also develop because of a nonpressurized break. At one building, there was substantial damage to the front bearing wall caused by the sudden subsidence of the ground surface. Subsurface exploration discovered that there was a sewer line that ran underneath the bearing wall. The top of the sewer line was broken and soil had filtered down into the sewer line. Periodic cleaning of the sewer line probably helped to enlarge the void that was developing above the sewer line. Eventually the void collapsed, causing the ground surface subsidence and damage to the overlying bearing wall.
Pipe clogs	Besides pipe breaks, there can also be damage because of pipe clogs. There are many different ways that a pipe can become clogged. For example, the pipe can become clogged with debris or the overburden pressure can crush the pipe.
Construction measures	The geotechnical engineer should determine if the site contains soil conditions, such as expansive soil or collapsible soil, that may be especially susceptible to pipe leaks or breaks. If this is the case, special pipes, connections, or shutoff valves that are triggered by a reduction in water pressure could be installed during construction.

CHAPTER 21
EXCAVATION, UNDERPINNING, AND FIELD LOAD TESTS

21.1 INTRODUCTION

In addition to grading and other site improvement techniques (Chap. 19) and groundwater control (Chap. 20), there are many other construction activities that require the services of a geotechnical engineer. Examples of these types of services are presented in this chapter.

21.2 EXCAVATIONS

There are many different types of excavations performed during the construction of a project. For example, soil may be excavated from the cut or borrow area and then used as fill (see Chap. 19). Another example is the excavation of a shear key or buttress that will be used to stabilize a slope or landslide. Other examples of excavations are discussed in this section.

An important aspect of the excavation may be groundwater control, which is discussed in Table 20.2.

21.2.1 Footing Excavations

This type of service involves measuring the dimensions of geotechnical elements (such as the depth and width of footings) to make sure that they conform to the requirements of the construction plans. This service is often performed at the same time as the field observation to confirm bearing conditions. Table 21.1 presents a discussion of footing excavation requirements.

21.2.2 Excavation of Piers

In excavation of piers, as in excavation of footings, the geotechnical engineer may be required to confirm embedment depths and bearing conditions for piers. Figure 21.1 presents typical steps in the construction of a drilled pier.

21.2.3 Open Excavations

An open excavation is defined as an excavation that has stable and unsupported side slopes. Table 21.2 presents a discussion of the general factors that control the excavation stability and Table 21.3 lists factors that control the stability of excavation slopes in some problem soils.

21.2.4 Braced Excavations

A braced excavation is defined as an excavation where the sides are supported by retaining structures. Figure 21.2 shows common types of retaining systems and braced excavations. Table 21.4 lists the design considerations for braced excavations and Table 21.5 indicates factors that are involved in the choice of a support system for a deep excavation.

21.2.5 Tunnels

Only a brief introduction to tunnels is provided in this book. Tables 21.6 and 21.7 present common terminology that is used to describe materials exposed in tunnels.

Of all engineering structures, tunnel excavations are most vulnerable to small troublesome and large catastrophic failures (Feld and Carper 1997). One reason is because the tunnel excavation allows for the sudden release of large confining pressures resulting in rock strains which are not easily predicted. Another reason is the unpredictability of subsurface conditions, such as the possibility of encountering groundwater which can flood the tunnel. Because of the problems with tunnel excavations, tunnel coring or boring machines (for soft and hard rock) that can provide some protection during excavation of the heading have been developed. Other measures to help stabilize tunnel rock faces include anchoring systems (such as bolts, rods, or dowels) and shotcrete. Chemical grouts have also been used to fill in rock discontinuities and joints in order to stabilize the tunnel crown and walls.

21.3 UNDERPINNING

There are many different situations where a structure may need to be underpinned. Common reasons for underpinning and various types of underpinning are discussed in Tables 21.8 and 21.9 and Figs. 21.3 to 21.6.

21.4 FIELD LOAD TESTS

There are numerous types of field load or performance tests. For example, load tests are common for pile foundations and are used to determine their load-carrying capacity. This process involves driving or installing the pile to the desired depth. Then the load is applied to the pile and the deformation behavior of the pile is measured. ASTM standards for the field testing of piles have been developed. Typical pile load tests and the applicable ASTM standard are listed below:

- Static Axial Compressive Load Test (ASTM D 1143-94, 1998).
- Static Axial Compressive Load Test for Piles in Permafrost (ASTM D 5780-95, 1998)
- Static Axial Tension Load Test (ASTM D 3689-90, 1998)
- Lateral Load Test (ASTM D 3966-90, 1998)
- Dynamic Testing of Piles (ASTM D 4945-89, 1998)

Figure 21.7 shows a commonly used method of analysis for static axially compressive or tension load testing of piles. Besides load tests, there can be all types of performance tests during construction that will need to be observed by the geotechnical engineer. An example is the field testing of tieback anchors as discussed in Table 16.15.

TABLE 21.1 Footing Excavations

Type (1)	Discussion (2)
Dimensions of footing	The geotechnical engineer will often be required to confirm the dimensions of the footings according to the building plans. The depth of the footing should always be referenced from the final grade, which may be different from the grade at the time of the footing observation.
Bearing conditions	The bearing conditions exposed in the footing should be checked with the conditions used during the design of the project. If the bearing soil or rock is substantially weaker than that assumed during the design phase, the footings may need to be deepened.
Groundwater conditions	The presence of groundwater can impact bearing conditions. For example, groundwater in a footing excavation may cause the side of the hole to cave in or loose slough to accumulate at the bottom of the footing. Especially for end-bearing footings, the groundwater table may need to be lowered in order to clean out any loose debris at the bottom of the footing.
Local building department requirements	In many cases, field observations to confirm bearing conditions and check foundation dimensions will be required by the local building department. In addition, a letter indicating the outcome of the observations must be prepared by the engineer to satisfy the local building department. Building departments often refer to these types of reports as "final inspection reports." The local building department often considers these reports to be so important that they may not issue a certificate of occupancy until the reports have been submitted and accepted. An example of such a report for the construction of a foundation is as follows: *Footing observations.* The footings at the site have been observed and are generally in conformance with the approved building plans. Additionally, the footings have been approved for installing steel reinforcement and the soil conditions are substantially in conformance with those observed during the subsurface exploration. Furthermore, the footing excavations extend to the proper depth and bearing strata. Care should be taken to keep all loose soil and debris out of the footing excavations prior to placement of concrete.

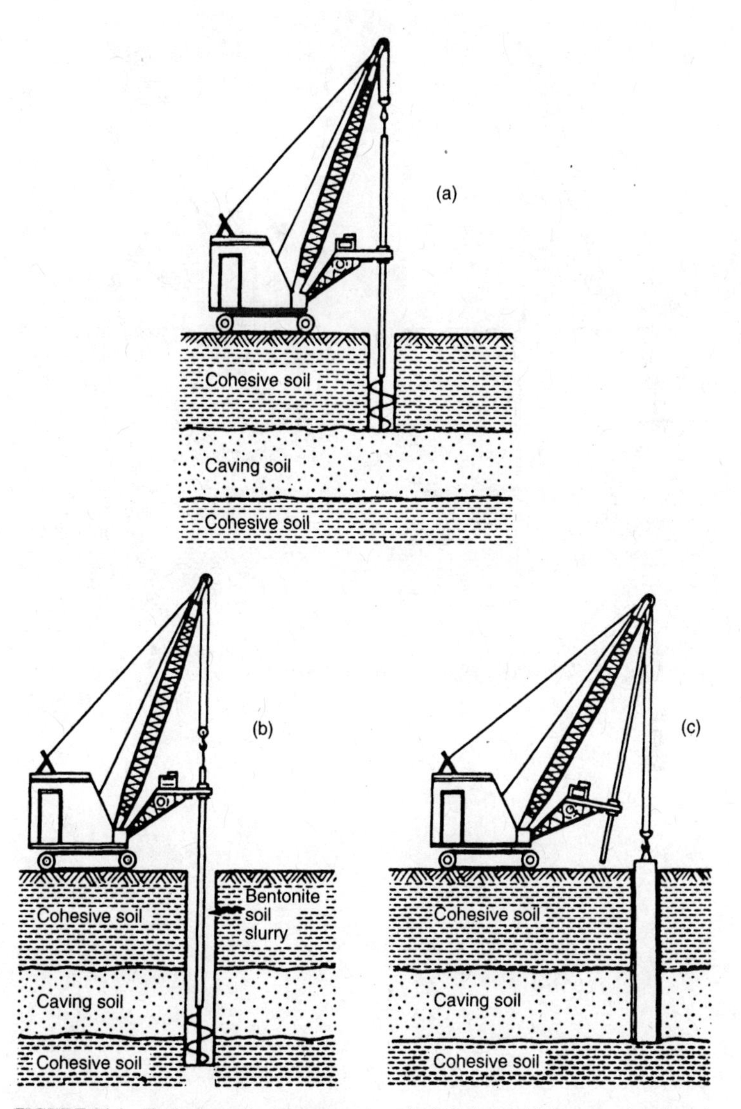

FIGURE 21.1 Typical steps in the construction of a drilled pier. (*a*) Dry augering through self-supporting cohesive soil; (*b*) augering through water-bearing cohesionless soil with aid of slurry; (*c*) setting the casing.

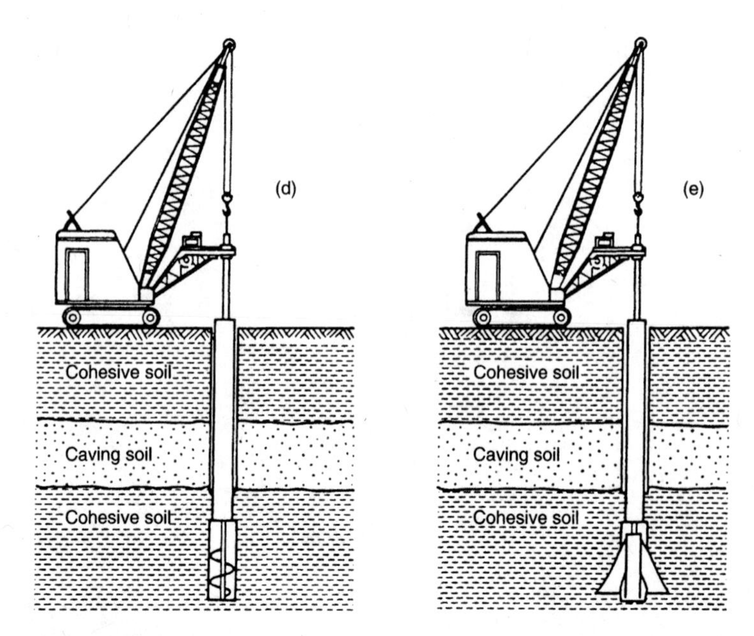

FIGURE 21.1 (*Continued*) (*d*) Dry augering into cohesive soil after sealing; (*e*) forming a bell. (*After O'Neill and Reese 1970; reproduced from Peck, Hanson, and Thornburn 1974.*)

TABLE 21.2 General Factors That Control the Stability of the Excavation Slopes

Construction activity (1)	Objectives (2)	Comments (3)
Dewatering (also see Chap. 20)	In order to prevent boiling, softening, or heave of the excavation bottom, reduce lateral pressures on sheeting, reduce seepage pressures on face of open cut, and eliminate piping of fines through sheeting.	Investigate soil compressibility and effect of dewatering on settlement of nearby structures; consider recharging or slurry wall cutoff. Examine for presence of lower aquifer and need to dewater. Install piezometers if needed. Consider effects of dewatering in cavity-laden limestone. Dewater in advance of excavation.
Excavation and grading (also see Chap. 19)	Utility trenches, basement excavations, and site grading.	Analyze safe slopes (see Chap. 13) or bracing requirements, and effects of stress reduction on overconsolidated, soft, or swelling soils and shales. Consider horizontal and vertical movements in adjacent areas due to excavation and effect on nearby structures. Keep equipment and stockpiles a safe distance from the top of the excavation.
Excavation and wall construction	To support vertical excavation walls, and to stabilize trenching in limited space.	See Chap. 16 for retaining wall design. Reduce earth movements and bracing stresses, where necessary, by installing lagging on front flange of soldier pile. Consider effect of vibrations due to driving sheet piles or soldier piles. Consider dewatering requirements as well as wall stability in calculating sheeting depth. Movement monitoring may be warranted.
Blasting	To remove or to facilitate the removal of rock in the excavation.	Consider the effect of vibrations on settlement or damage to adjacent areas. Design and monitor or require the contractor to design and monitor blasting in critical areas, and require a preconstruction survey of nearby structures.

TABLE 21.2 General Factors That Control the Stability of the Excavation Slopes
(*Continued*)

Construction activity (1)	Objectives (2)	Comments (3)
Anchor or strut installation	To obtain support system stiffness and interaction.	Major excavations require careful installation and monitoring: case anchor holes in collapsible soil, measure stress in ties and struts, etc.

Sources: NAVFAC DM-7.2, 1982, Clough and Davidson 1977, and Departments of the Army and the Air Force 1979.

TABLE 21.3 Factors That Control the Stability of Excavation Slopes in Some Problem Soils

Topic (1)	Discussion (2)
General discussion	The depth and slope of an excavation and groundwater conditions control the overall stability and movements of open excavations. Factors that control the stability of the excavation for different material types are as follows: 1. *Rock.* For rock, stability is controlled by depths and slopes of excavation, particular joint patterns, *in situ* stresses, and groundwater conditions. 2. *Granular soils.* For granular soils, instability usually does not extend significantly below the bottom of the excavation, provided that seepage forces are controlled. 3. *Cohesive soils.* For cohesive soils, stability typically involves side slopes but may also include the materials well below the bottom of the excavation. Instability of the bottom of the excavation, often referred to as *bottom heave,* is affected by soil type and strength, depth of cut, side slope and/or berm geometry, groundwater conditions, and construction procedures.
Stiff-fissured clays and shales	Field shear resistance may be less than suggested by laboratory testing. Slope failures may occur progressively and shear strengths are reduced to the residual value compatible with relatively large deformations. Some case histories suggest that the long-term performance is controlled by the drained residual friction angle. The most reliable design would involve the use of local experience and recorded observations.
Loess and other collapsible soil	Such soils have a strong potential for collapse and erosion of relatively dry materials upon wetting. Slopes in loess are frequently more stable when cut vertically to prevent water infiltration. Benches at intervals can be used to reduce effective slope angles. Evaluate potential for collapse as described in Tables 9.5 to 9.7.
Residual soil	Depending on the weathering profile from the parent rock, residual soil can have a significant local variation in properties. Guidance based on recorded observations provides a prudent base for design.
Sensitive clay	Very sensitive and quick clays have a considerable loss of strength upon remolding, which could be generated by natural or man-made disturbance. Minimize disturbance and use total stress analysis based on undrained shear strength from unconfined compression tests or field vane tests.
Talus	Talus is characterized by loose aggregation of rock that accumulates at the foot of rock cliffs. Stable slopes are commonly between 1.25:1 to 1.75:1 (horizontal:vertical). Instability is often associated with abundance of water, mostly when snow is melting.
Loose sands	Loose sands may settle under blasting vibrations, or liquefy, settle, and lose shear strength if saturated. Such soils are also prone to erosion and piping.

TABLE 21.3 Factors That Control the Stability of Excavation Slopes in Some Problem Soils (*Continued*)

Topic (1)	Discussion (2)
Engineering evaluation	Methods described in Chap. 13 (slope stability analyses) may be used to evaluate the stability of open excavations in soils where the behavior of such soils can be reasonably determined by field investigations, laboratory testing, and engineering analysis. As described above, in certain geologic formations stability is controlled by construction procedures, side effects during and after excavation, and inherent geologic planes of weaknesses.

Sources: NAVFAC DM-7.2, 1982 and Clough and Davidson 1977.

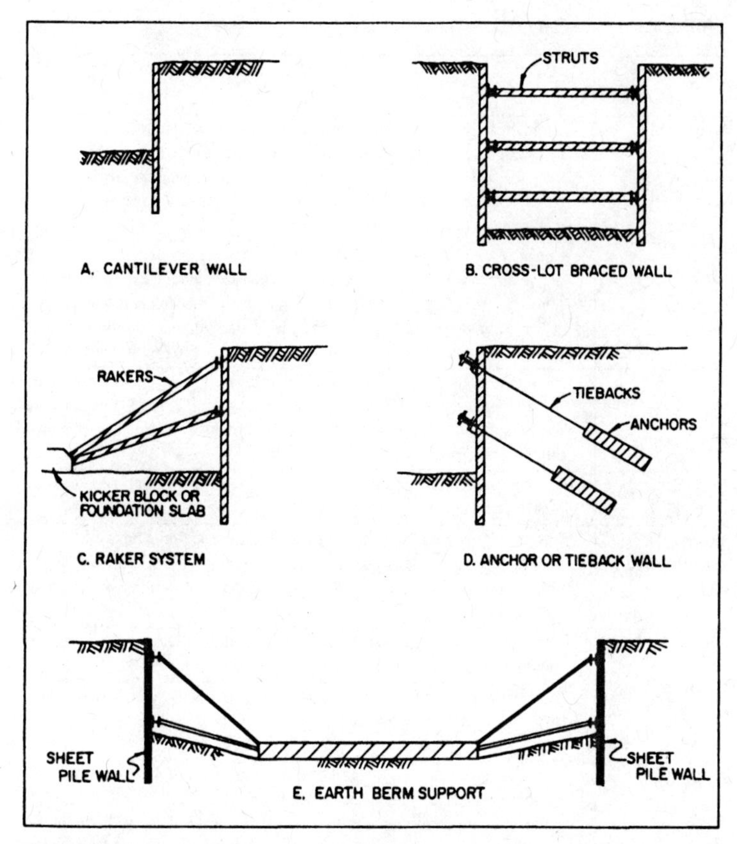

FIGURE 21.2 Common types of retaining systems and braced excavations. Also see Tables 21.4 and 21.5. (*Reproduced from NAVFAC DM-7.2, 1982.*)

TABLE 21.4 Design Considerations for Braced Excavations

Design factor (1)	Comments (2)
Water loads	Often greater than earth loads on an impervious wall. Recommend piezometers during construction to monitor water levels. Should also consider possible lower water pressures as a result of seepage of water through or under the wall. Dewatering can be used to reduce the water loads. Seepage under the wall reduces the passive resistance.
Stability	Consider the possible instability in any berm or exposed slope. The sliding potential beneath the wall or behind the tiebacks should also be evaluated. For weak soils, deep-seated bearing failure due to the weight of the supported soil should be checked. Also include in stability analysis the weight of surcharge or weight of other facilities in close proximity to the excavation.
Piping	Piping due to a high groundwater table causes a loss of ground, especially for silty and fine sands. Difficulties occur because of flow of water beneath the wall, through bad joints in the wall, or through unsealed sheet pile handling holes. Dewatering may be required.
Movements	Movements can be minimized through the use of a stiff wall supported by preloaded tiebacks or a braced system.
Dewatering and recharge	Dewatering reduces the loads on the wall system and minimizes the possible loss of ground due to piping. Dewatering may cause settlements, and in order to minimize settlements, there may be the need to recharge outside of the wall system.
Surcharge	Construction materials are usually stored near the wall systems. Allowances should always be made for surcharge loads on the wall system.
Prestressing of tieback anchors	In order to minimize soil and wall movements, it is useful to remove slack by prestressing tieback anchors.
Construction sequence	The amount of wall movement depends on the depth of the excavation. The amount of load on the tiebacks depends on the amount of wall movement which occurs before they are installed. Movements of the wall should be checked at every major construction stage. Upper struts should be installed as early as possible.
Temperature	Struts may be subjected to load fluctuations due to temperature differences. This may be important for long struts.
Frost penetration	In cold climates, frost penetration can cause significant loading on the wall system. Design of the upper portion of the wall system should be conservative. Anchors may have to be heated. Freezing temperatures also can cause blockage of flow of water and thus unexpected buildup of water pressure.
Earthquakes	Seismic loads may be induced during an earthquake.
Factors of safety	The following are suggested minimum factors of safety F for overall stability. Note that these values are suggested guidelines only. Design factors of safety depend on project requirements.

TABLE 21.4 Design Considerations for Braced Excavations (*Continued*)

Design factor (1)	Comments (2)		
Factors of safety (*Continued*)	Earth berms:	Permanent, $F = 2.0$	Temporary, $F = 1.5$
	Cut slopes:	Permanent, $F = 1.5$	Temporary, $F = 1.3$
	General stability:	Permanent, $F = 1.5$	Temporary, $F = 1.3$
	Bottom heave:	Permanent, $F = 2.0$	Temporary, $F = 1.5$

Source: NAVFAC DM-7.2, 1982.

TABLE 21.5 Factors That Are Involved in the Choice of a Support System for a Deep Excavation

Requirements (1)	Type of support system (2)	Comments (3)
Open excavation area	Tiebacks or rakers. For shallow excavation, use cantilever walls.	Consider design items listed in Table 21.4.
Low initial cost	Soldier pile or sheet pile walls. Consider combined soil slope and wall.	Consider design items listed in Table 21.4.
Use as part of permanent structure	Diaphragm or pier walls.	Diaphragm wall is the most common type of permanent wall.
Subsurface conditions of deep, soft clay	Struts or rakers that support a diaphragm or pier wall.	Tieback capacity not adequate in soft clays.
Subsurface conditions of dense, gravelly sands or clay	Soldier pile, diaphragm wall, or pier wall.	Sheet piles may lose interlock on hard driving.
Subsurface conditions of overconsolidated clays	Struts, long tiebacks, or combination of tiebacks and struts.	High *in situ* lateral stresses are relieved in overconsolidated soil. Lateral movements may be large and extend deep into the soil.
Avoid dewatering	Use diaphragm walls or possibly sheet pile walls in soft subsoils.	Soldier pile wall is too pervious for this application.
Minimize lateral movements of wall	Use high preloads on stiff strutted or tieback walls.	Analyze the stability of the bottom of the excavation.
Wide excavation (greater than 65 ft wide)	Use tiebacks or rackers.	Tiebacks are preferable except in very soft clay soils.
Narrow excavation (less than 65 ft wide)	Use cross-excavation struts.	Struts are more economical, but tiebacks still may be preferred in order to keep the excavation open.

Note: Deep excavation is defined as an excavation that is more than 20 ft (6 m) below ground surface.

Source: NAVFAC DM-7.2, 1982.

TABLE 21.6 Classification of Rock for Tunnel Excavations

Rock type (1)	Description (2)
Intact rock	Intact rock contains neither joints nor hairline cracks. Hence, if it breaks, it breaks across sound rock. On account of the disruption of the rock due to blasting, spalls may drop off the roof several hours or days after blasting. This is known as a spalling condition. Hard, intact rock may also be susceptible to the popping condition (i.e., rock burst) involving the spontaneous and violent detachment of rock slabs from sides or roof.
Stratified rock	Stratified rock consists of individual strata with little or no resistance against separation along the boundaries between strata. The strata may or may not be weakened by transverse joints. In such rock, the spalling condition is quite common.
Moderately jointed rock	Moderately jointed rock contains joints and hairline cracks, but the blocks between the joints are locally grown together or so intimately interlocked that vertical walls do not require lateral support. In rocks of this type, both the spalling and the popping condition may be encountered.
Blocky and seamy rock	Blocky and seamy rock consists of chemically intact or almost intact rock fragments which are entirely separated from each other and imperfectly interlocked. In such rock, vertical wall support may be required.
Crushed but chemically intact rock	Crushed but chemically intact rock has the character of a crusher run. If most or all of the fragments are as small as fine sand-size particles and no recementation has taken place, then the crushed rock below the groundwater table exhibits the properties of water-bearing sand.
Squeezing rock	Squeezing rock slowly advances into the tunnel without a perceptible volume increase. The prerequisite for squeeze is a high percentage of microscopic and submicroscopic particles of micaceous minerals or of clay minerals having a low swelling capacity.
Swelling rock	Swelling rock advances into the tunnel chiefly on account of expansion. The volume increase is in response to an increase in water content of the rock. The capacity to swell seems to be limited to those rocks which contain montmorillonite.

Note: The terms listed above for the classification of rock for tunnels are from the *Engineering Geology Field Manual* (1987). Also see Sandström 1963a, 1963b; Szechy 1973; Wahlstrom 1973; Attewell et al. 1986; Sinha 1989, 1991; and Mahtab and Grasso 1992.

TABLE 21.7 Classification of Soil for Tunnel Excavations

Soil type (1)	Description (2)
Hard	For hard soil, the tunnel heading may be advanced without roof support.
Firm	For firm soil, the roof section of a tunnel can be left unsupported for several days without inducing a perceptible movement of the ground.
Raveling	For raveling soil, chunks or flakes of soil begin to drop out of the roof at some point during the ground-movement period.
Slow raveling	The time required to excavate 5 ft (1.5 m) of tunnel and install a rib set and lagging in a small tunnel is about 6 hours. Therefore, if the stand-up time of a raveling ground is more than 6 hours, then the soil would be classified as slow raveling.
Fast raveling	In contrast to slow-raveling ground, for fast-raveling ground, chunks or flakes of soil drop out of the tunnel roof within 6 hours of being exposed by the tunnel excavation.
Squeezing	For squeezing soil, the ground slowly advances into the tunnel without any signs of fracturing. Yet the loss of ground caused by squeeze and the resulting settlement of the ground surface can be substantial.
Swelling	For swelling soil, the ground slowly advances into the tunnel partly or chiefly because of an increase in volume of the ground. The volume increase is in response to an increase in water content of the soil. In every other respect, swelling ground in a tunnel behaves like a stiff nonsqueezing, or slowly squeezing, nonswelling clay.
Running	The removal of lateral support on any surface rising at an angle of more than 34° (to the horizontal) is immediately followed by a running movement of the soil particles. This movement does not stop until the slope of the moving soil becomes roughly equal to 34°. If running ground has a trace of cohesion, then the run is preceded by a brief period of progressive raveling.
Very soft squeezing	For very soft squeezing soil, the ground advances rapidly into the tunnel in the form of a plastic flow.
Flowing	Soil supporting a tunnel excavation can not be classified as flowing ground unless water flows or seeps through it toward the tunnel. For this reason, a flowing condition is encountered only in free air tunnels below the groundwater table or under compressed air when the pressure is not high enough in the tunnel to stabilize the excavation. A second prerequisite for flowing is low cohesion of the soil. Therefore, conditions for flowing ground occur only in inorganic silt, fine silty sand, clean sand or gravel, or sand and gravel with some clay binder. Organic silt may behave either as a flowing or as a very soft, squeezing ground.

Source: Reproduced from *Engineering Geology Field Manual* 1987, originally from Terzaghi.

TABLE 21.8 Common Reasons for Underpinning

Topic (1)	Discussion (2)
Common reasons for underpinning	Underpinning is commonly defined as the replacement or strengthening of the foundation of an existing structure. Common reasons for underpinning are listed below (Tomlinson 1986): • To support a structure that is sinking or tilting due to ground subsidence or instability of the superstructure. • As a safeguard against possible settlement of a structure when excavating close to and below its foundation level. • To support a structure while making alterations to its foundation or main supporting members. • To enable the foundations to be deepened for structural reasons, for example to construct a basement beneath a building. • To increase the width of a foundation to permit heavier loads to be carried, for example, when increasing the story height of a building. • To enable a building to be moved bodily to a new site.
Unique projects	Each underpinning project is unique and requires highly skilled personnel, and therefore it should be attempted only by experienced firms. Because each job is different, individual consideration of the most economical and safest scheme is required for each project. Common methods of underpinning include the construction of continuous strip foundations, piers, and piles. To facilitate the underpinning process, the ground can be temporarily stabilized by freezing the ground or by injecting grout or chemicals into the soil (Tomlinson 1986, Prentis and White 1950, and Thorburn and Hutchison 1985).
Shoring	The excavation below the foundation of a structure usually necessitates the temporary support of the structure by the use of shoring. Figure 21.3 presents two methods of shoring, as follows: 1. *Notched-wall method.* For the notched-wall method, notches are cut into the structure and it is supported by inclined timber shoring that is supported by foundation pads. 2. *Needle beam method.* A more commonly used shoring method is the construction of a needle beam (see Fig. 21.3). This method consists of cutting the wall or columns at the base, inserting the needle beam, and then supporting the ends of the needle beam on a foundation pad equipped with screw jacks. Once the wall or column load has been transferred to the needle beam, an underpinning pit can be constructed as shown in Fig. 21.3.

TABLE 21.9 Various Types of Underpinning

Topic (1)	Discussion (2)
Introduction	This table presents a discussion of underpinning to support a structure that is sinking or tilting because of ground subsidence or instability of the superstructure. The most expensive and rigorous method of underpinning would be to entirely remove the existing foundation and install a new foundation. This method of repair is usually reserved for cases involving a large magnitude of soil movement.
Underpinning with a deep foundation system	A commonly used underpinning option is to strengthen or remove the existing foundation and then underpin it with a deep foundation system. For example, underpinning with a deep foundation system could consist of the construction of a mat foundation supported by piers, which are embedded in a firm bearing material. For a condition of soil settlement, the piers will usually be subjected to downdrag loads from the settling soil. The piers are usually at least 0.6 m (2 ft) in diameter to enable downhole logging to confirm end-bearing conditions. The piers can either be built within the building, or constructed outside the building with grade beams used to transfer loads to the piers. Given the height of a drill rig, it is usually difficult to drill within the building (unless it is raised). The advantages of constructing the piers outside the building are that the height restriction is no longer a concern and a large, powerful drill rig can be used to quickly and economically drill the holes for the piers. Besides piers, the foundation can also be underpinned with piles or screw or earth anchors (Brown 1992). Greenfield and Shen (1992) present a list of the advantages and disadvantages of pier and pile underpinning installations.
Underpinning with a shallow foundation system	Another type of underpinning consists of the construction of a shallow foundation system. This is usually a less expensive and rigorous method of underpinning than the deep foundation underpinning system. Underpinning by a shallow foundation is often used when the amount of soil movement is less or the depth of problem soil is shallower than the conditions requiring a deep foundation system. Figure 21.4 shows an example of a shallow foundation underpinning system that consists of a continuous strip footing. The construction of the footing starts with the excavation of slots in order to install the hydraulic jacks. The hydraulic jacks are used to temporarily support the foundation until the entire footing is exposed. Steel reinforcement is then tied to the existing foundation by dowels. The final step is to fill the excavation with concrete. The jacks are left in place during the placement of the concrete.
Repair of existing foundation cracks—strip replacement method	In addition to the underpinning, cracks in the foundation (if not removed and replaced) will often need repair. Figure 21.5 shows the strip replacement method, which is one type of repair for concrete slab-on-grade cracks. The construction of the strip replacement starts by cutting out the area containing the concrete crack with a saw. Figure 21.5 indicates that a distance of 0.3 m (1 ft) on both sides of the concrete crack be saw-cut. This is to provide enough working space to install reinforcement and the dowels. After the new reinforcement (No. 3 bars) and dowels are installed, the area is filled with a new portion of concrete.

TABLE 21.9 Various Types of Underpinning (*Continued*)

Topic (1)	Discussion (2)
Repair of existing foundation cracks—patching of existing cracks	Another option for the repair of existing foundation cracks is to patch the existing concrete cracks. The objective is to return the concrete slab to a satisfactory appearance and provide structural strength at the cracked areas. It has been stated (Transportation Research Board 1977) that a patching material must meet the following requirements: 1. Be at least as durable as the surrounding concrete. 2. Require a minimum of site preparation. 3. Be tolerant of a wide range of temperature and moisture conditions. 4. Be noninjurious to the concrete through chemical incompatibility. 5. Preferably be similar in color and surface texture to the surrounding concrete. Figure 21.6 shows a typical detail for concrete crack repair. If there is differential movement at the crack, then the concrete may require grinding or chipping to provide a smooth transition across the crack. The material commonly used to fill the concrete crack is epoxy. Epoxy compounds consist of a resin, a curing agent or hardener, and modifiers that make them suitable for specific uses. The typical range (3400 to 35,000 kPa, 500 to 5000 psi) in tensile strength of epoxy is similar to its range in compressive strength (Schutz 1984). Performance specifications for epoxy have been developed (e.g., ASTM C 881-90, "Standard Specification for Epoxy-Resin-Base Bonding Systems for Concrete," 1997). In order for the epoxy to be effective, it is important that the crack faces be free of contaminants (such as dirt) that could prevent bonding. In many cases, the epoxy is injected under pressure so that it can penetrate the full depth of the concrete crack.
Alternative methods to foundation underpinning	The previous discussion has dealt with the underpinning of the foundation in order to resist soil movement or bypass the problem soil. There are many other types of underpinning and soil treatment alternatives (Brown 1990, 1992; Greenfield and Shen 1992; Lawton 1996). In some cases, the magnitude of soil movement may be so large that the only alternative is to demolish the structure. For example, movement of the Portuguese Bend landslide in Palos Verdes, California, has destroyed about 160 homes. But a few homeowners refuse to abandon their homes as they slowly slide downslope. Some owners have underpinned their house foundations with steel beams which are supported by hydraulic jacks that are periodically used to relevel the house. Other owners have tried bizarre underpinning methods, such as supporting the house on huge steel drums. For expansive soil, underpinning alternatives can include horizontal or vertical moisture barriers to reduce the cyclic wetting and drying around the perimeter of the structure (Nadjer and Werno 1973, Snethen 1979, Williams 1965). Drainage improvements and the repair of leaky water lines are also performed in conjunction with the construction of the moisture barriers. Other expansive soil stabilization options include chemical injection (such as a lime slurry) into the soil below the structure. The goal of such mitigation measures is to induce a chemical-mineralogical change of the clay particles which will reduce the soil's tendency to swell.

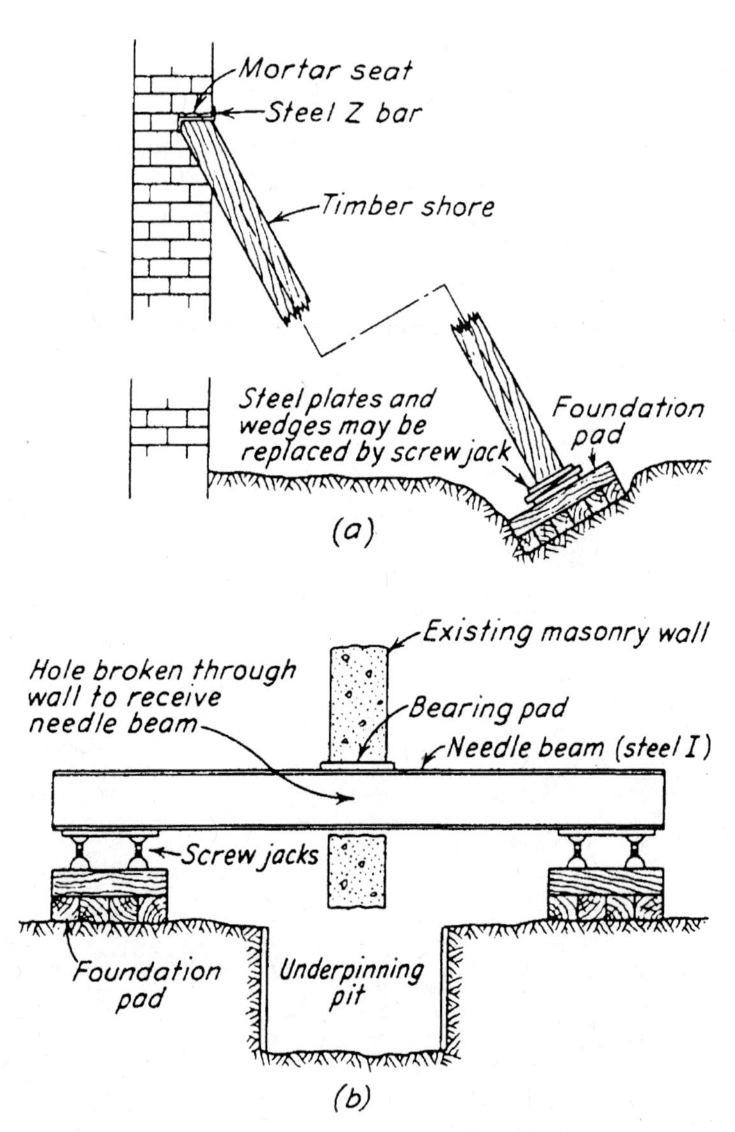

FIGURE 21.3 Methods of shoring. See Table 21.8. (*a*) Notched wall; (*b*) needle beam. (*From Peck, Hanson, and Thornburn 1974; reprinted with permission of John Wiley & Sons.*)

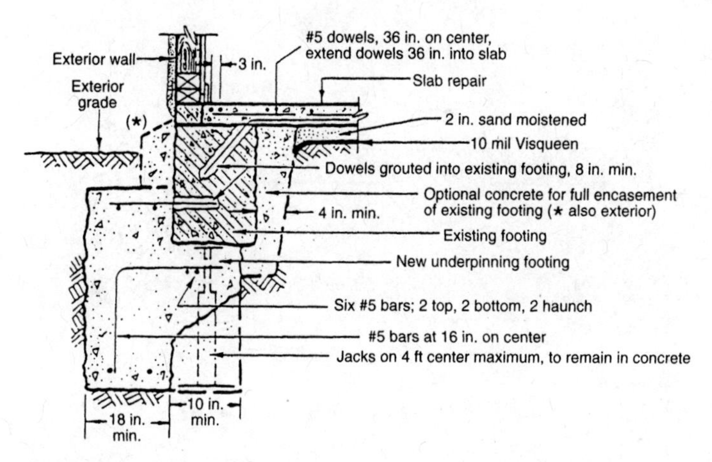

FIGURE 21.4 Example of underpinning with a shallow foundation system. See Table 21.9.

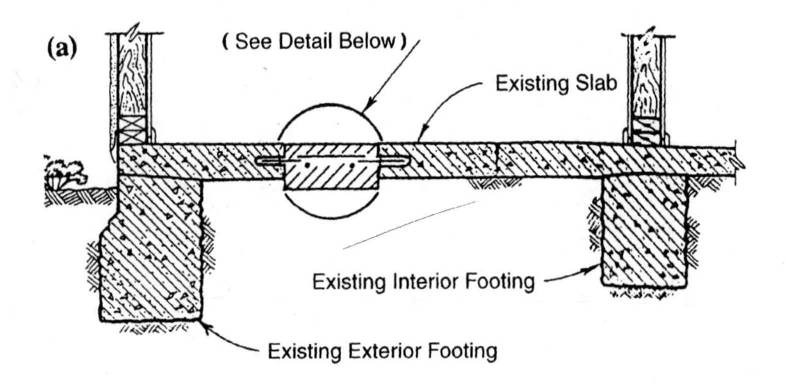

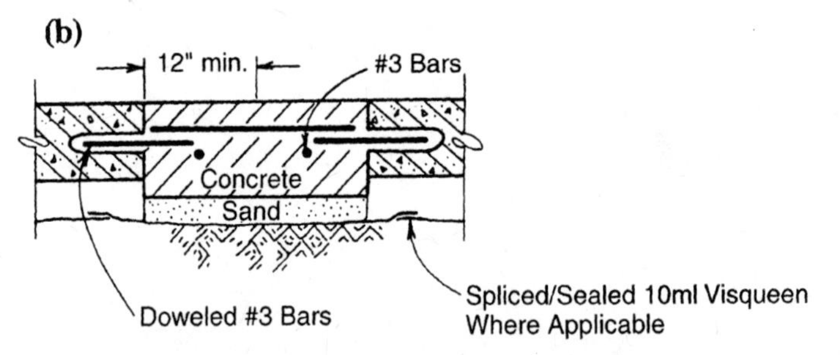

Saw-Cut 12" Each Side of Crack, Dowel #3 Bars 6"(min.) Into Existing Slab. Provide 5" Concrete Section with #3 Bars, 12" O.C. Both Ways. Underlay with 2" Moist Sand. Where Visqueen Exists, Splice/Seal in a Replacement Section

FIGURE 21.5 Concrete crack repair. (*a*) Strip replacement of floor cracks; (*b*) strip replacement detail. See Table 21.9.

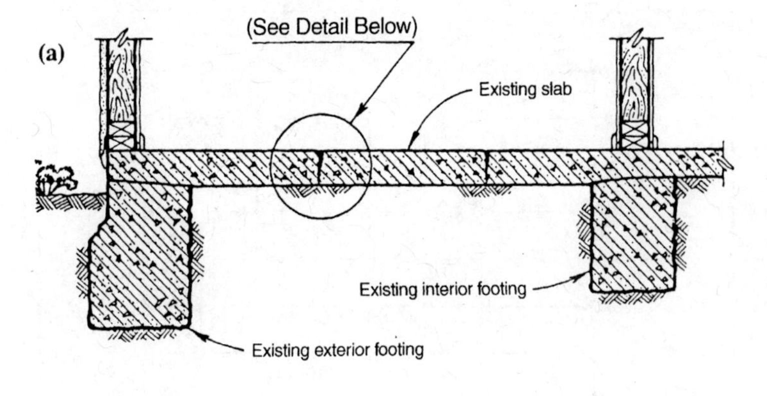

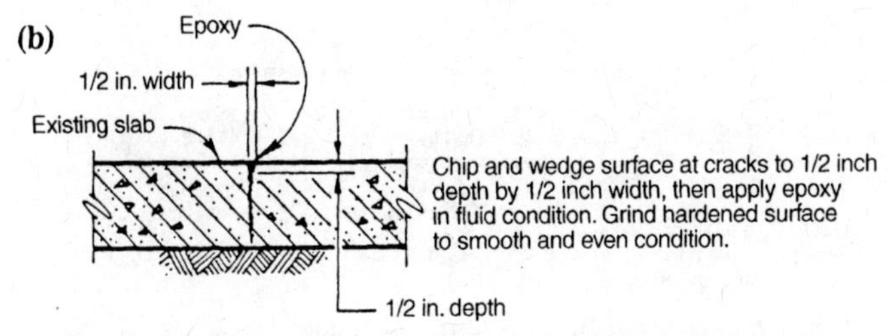

FIGURE 21.6 Concrete crack repair. (*a*) Epoxy repair of floor cracks; (*b*) detail of crack repair with epoxy. See Table 21.9.

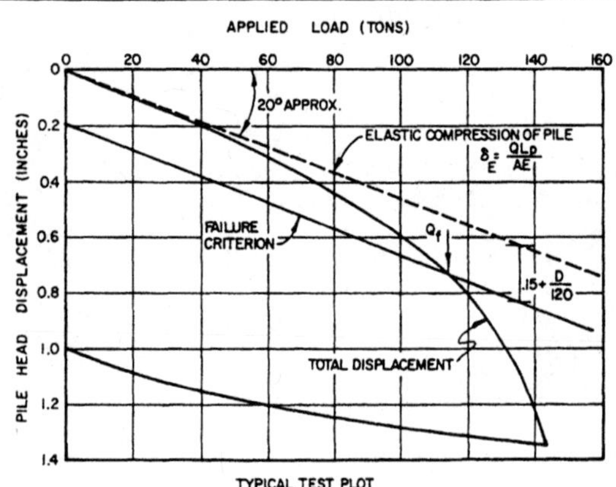

1. Calculate elastic compression of pile (δ_E) when considered as a free column by:

$$\delta_E = \frac{QL_p}{AE}$$

Q = test load, lbs
L_p = pile length, in. (for end-bearing pile)
A = cross-sectional area of pile material, sq in
E = Young's Modulus for pile material, psi

2. Determine scales of plot such that slope of pile elastic compression line is approximately 20°.

3. Plot pile head total displacment vs. applied load.

4. Failure load is defined as that load which produces a displacement of the pile head equal to:

$$S_f = \delta_E + (.15 + \frac{D}{120})$$

S_f = displacement at failure, in.
D = pile diameter, in.

5. Plot failure criterion as described in (4), represented as a straight line, parallel to line of pile elastic compression. Intersection of failure criterion with observed load deflection curve defines failure load, Q_f.

6. Where observed load displacement curve does not intersect failure criterion, the maximum test load should be taken as the failure load.

7. Apply factor of safety of at least 2.0 to failure load to determine allowable load.

FIGURE 21.7 Method of analysis for static axially compressive or tension load testing of piles. See Sec. 21.4. (*Reproduced from NAVFAC DM-7.2, 1982.*)

CHAPTER 22
GEOSYNTHETICS

22.1 INTRODUCTION

A geosynthetic is defined as a planar product manufactured from polymeric material and typically placed in soil to form an integral part of a drainage, reinforcement, or stabilization system. Common types of geosynthetics used during construction are discussed in this chapter and summarized in Table 22.1.

22.2 GEOGRIDS

Figure 22.1 shows a photograph of a geogrid, which contains relatively high strength polymer grids consisting of longitudinal and transverse ribs connected at their intersections. Geogrids have a large and open structure and are principally used as soil reinforcement (see Table 22.2).

22.3 GEOTEXTILES

Geotextiles are the most widely used type of geosynthetic. Geotextiles are often referred to as *fabric*. For example, common construction terminology for geotextiles includes *geofabric, filter fabric, construction fabric, synthetic fabric,* and *road-reinforcing fabric*. As shown in Figs. 22.2 and 22.3, geotextiles are usually categorized as being either woven or nonwoven depending on the type of manufacturing process. Table 22.3 presents a further discussion of geotextiles.

22.4 GEOMEMBRANES

Common construction terminology for geomembranes includes *liners, membranes, visqueen, plastic sheets,* and *impermeable sheets.* Geomembranes are used as barriers to reduce water or vapor migration through soil (see Fig. 22.4 and Table 22.4).

22.5 GEONETS AND GEOCOMPOSITES

Geonets are three-dimensional netlike polymeric materials used for drainage (flow of water within the geosynthetic). Figure 22.5 shows a photograph of a geonet. Geonets are usually used in conjunction with a geotextile and/or geomembrane, hence geonets are technically a geocomposite.

Depending on the particular project requirements, different types of geosynthetics can be combined to form a geocomposite. For example, a geocomposite consisting of a geotextile and a geomembrane provides a barrier that has increased tensile strength and resistance to punching and tearing. Figure 22.6 shows a photograph of a geocomposite consisting of a textured geomembrane, geonet, and geotextile (filter fabric).

22.6 GEOSYNTHETIC CLAY LINERS

Geosynthetic clay liners are frequently used as liners for municipal landfills. The geosynthetic clay liner typically consists of dry bentonite sandwiched between two geosynthetics. When moisture infiltrates the geosynthetic clay liner, the bentonite swells and creates a soil layer having a very low hydraulic conductivity, transforming it into an effective barrier to moisture migration.

TABLE 22.1 Geosynthetic Types and Applications

Geosynthetic type (1)	Applications (2)
Geogrids	1. *Soil reinforcement.* Used for subgrade stabilization, slope reinforcement, erosion control (reinforcement), and mechanically stabilized earth retaining walls. Also used to strengthen the junction between the top of soft clays and overlying embankments. 2. *Asphalt overlays.* Used in asphalt overlays to reduce reflective cracking.
Geotextiles	1. *Soil reinforcement.* Used for subgrade stabilization, slope reinforcement, and mechanically stabilized earth retaining walls. Also used to strengthen the junction between the top of soft clays and overlying embankments. 2. *Sediment control.* Used as silt fences to trap sediment on-site. 3. *Erosion control.* Installed along channels and under riprap, and used for shore and beach protection. 4. *Asphalt overlays.* Used in asphalt overlays to reduce reflective cracking. 5. *Separation.* Used between two dissimilar materials, such as an open graded base and a clay subgrade, in order to prevent contamination. 6. *Filtration and drainage.* Used in place of a graded filter where the flow of water occurs across (perpendicular to) the plane of the geotextile. For drainage applications, the water flows within the geotextile.
Geomembranes	*Barriers.* Used as barriers to reduce water or vapor migration through soil.
Geonets and geocomposites	1. *Drainage.* Geonets are three-dimensional netlike polymeric materials used for drainage (flow of water within the geosynthetic). Geonets are usually used in conjunction with a geotextile and/or geomembrane, hence geonets are technically a geocomposite. 2. *Dual applications.* Depending on the particular project requirements, different types of geosynthetics can be combined together to form a geocomposite. For example, a geocomposite consisting of a geotextile and a geomembrane provides for a barrier that has increased tensile strength and resistance to punching and tearing.
Geosynthetic clay liners	*Barriers.* Used as barriers to reduce water or vapor migration through soil. Commonly used as a waste containment liner and as a cover system liner.

Sources: Rollings and Rollings 1996, Fluet 1988, and Richardson and Koerner 1990.

FIGURE 22.1 Photograph of a geogrid. See Table 22.2. (*From Rollings and Rollings 1996; reprinted with permission of McGraw-Hill, Inc.*)

TABLE 22.2 Geogrids

Topic (1)	Discussion (2)
General discussion	Figure 22.1 shows a photograph of a geogrid, which contains relatively high-strength polymer grids consisting of longitudinal and transverse ribs connected at their intersections. Geogrids have a large and open structure and are principally used as soil reinforcement. The openings (i.e., apertures) are usually 0.5 to 4 inches (1.3 to 10 cm) in length and/or width. Geogrid can be either biaxial or uniaxial depending on the size of the apertures and shape of the interconnecting ribs.
Common usage	As indicated in Table 22.1, the most common usage of geogrids is as soil reinforcement. For example, Fig. 13.6 shows the use of geogrid to reinforce a slope face. Compacted soil tends to be strong in compression but weak in tension. The geogrid is just the opposite, strong in tension but weak in compression. Thus, layers of compacted soil and geogrid tend to compliment each other and produce a soil mass having both high compressive and tensile strength. The open structure of the geogrid (see Fig. 22.1) allows the compacted soil to bond in the open geogrid spaces. Geogrids provide soil reinforcement by transferring local tensile stresses in the soil to the geogrid. Because geogrids are continuous, they also tend to transfer and redistribute stresses away from areas of high stress concentrations (such as beneath a wheel load). Similar to other geosynthetics, geogrids are transported to the site in rolls 3 ft (1 m) to 10 ft (3 m) wide. It is generally not feasible to connect the ends of the geogrid, and it is typically overlapped at joints. Typical design methods for using geogrids are summarized by Koerner (1990).
Limitations	Some of the limitations of geogrid are as follows: 1. *Ultraviolet light.* Even geogrids produced of carbon black (i.e., ultraviolet stabilized geogrids) can degrade when exposed to long-term ultraviolet light. It is important to protect the geogrid from sunlight and cover the geogrid with fill as soon as possible. 2. *Nonuniform tensile strength.* Geogrids often have different tensile strengths in different directions as a result of the manufacturing process. For example, a Tensar SS-2 (BX1200) biaxial geogrid has an ultimate tensile strength of 2100 lb/ft in the main direction and only 1170 lb/ft in the minor (perpendicular) direction. It is essential that the engineer always check the manufacturer's specifications and determine the tensile strengths in the main and minor directions. 3. *Creep.* Polymer material can be susceptible to creep. Thus, it is important to use an allowable tensile strength that does allow for creep of the geosynthetic. Often, this allowable tensile design strength is much less than the ultimate strength of the geogrid. For example, for a Tensar SS-2 (BX1200) biaxial geogrid, the manufacturer's recommended tensile strength is about 300 lb/ft, which is only one-seventh the ultimate tensile strength (2100 lb/ft). The engineer should never apply an arbitrary factor of safety to the ultimate tensile strength, but rather obtain the allowable geogrid tensile design strength from the manufacturer.

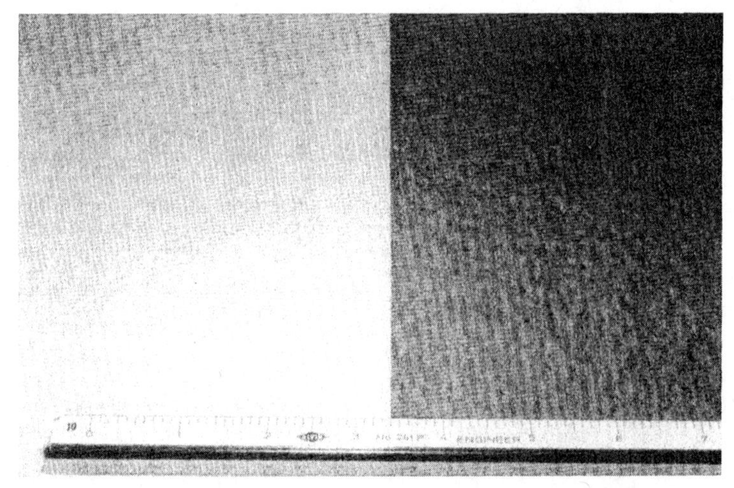

FIGURE 22.2 Photograph of nonwoven geotextiles. The geotextile on the left has no ultraviolet protection, while the geotextile on the right has ultraviolet protection. See Table 22.3. (*From Rollings and Rollings 1996; reprinted with permission of McGraw-Hill, Inc.*)

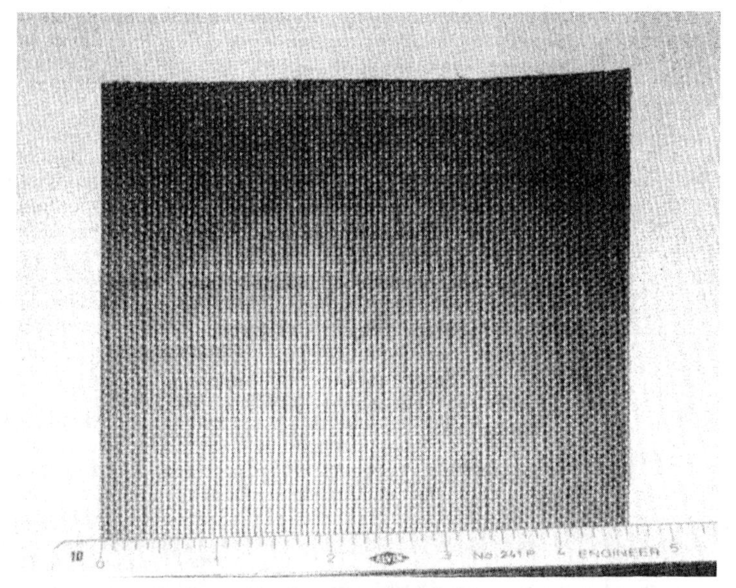

FIGURE 22.3 Photograph of a woven geotextile. See Table 22.3. (*From Rollings and Rollings 1996; reprinted with permission of McGraw-Hill, Inc.*)

TABLE 22.3 Geotextiles

Topic (1)	Discussion (2)
General discussion	Geotextiles are the most widely used type of geosynthetic. Geotextiles are often referred to as *fabric*. For example, common construction terminology for geotextiles includes *geofabric, filter fabric, construction fabric, synthetic fabric,* and *road-reinforcing fabric.* As shown in Figs. 22.2 and 22.3, geotextiles are usually categorized as being either woven or nonwoven, depending on the type of manufacturing process.
Common usage	As indicated in Table 22.1, geotextiles are used for more applications than any other type of geosynthetic. Probably the most common usage of geotextiles is for filtration (flow of water through the geotextile). For example, Standard Detail No. 1 (App. B) shows a geotextile that will be used to isolate the soil from the open graded gravel surrounding the drainage pipe. For filtration, the geotextile should be at least 10 times more permeable than the soil. In addition, the geotextile must always be placed between a less permeable (i.e., the soil) and a more permeable (i.e., the open graded gravel) material. An inappropriate use of a geotextile would be to place it around the drainage pipe (see Standard Detail No. 1, App. B), because then it would have more permeable material on both sides of the geotextile and it would tend to restrict flow. Two important design properties for geotextiles used as filtration devices are that they have an adequate flow capacity and also a proper soil retention capability. These two items are discussed below. 1. *Flow capacity.* Although specifications have been developed that limit the open area of the filtration geotextile to 10% or even 5%, it is best to have a larger open area to develop an adequate flow capacity. As indicated in Standard Detail No. 1 (App. B), the recommended minimum percentage of open area is 15%. 2. *Soil retention capability.* The equivalent opening size (EOS), also known as the apparent opening size (AOS), determines the soil retention capability. The EOS is often expressed in terms of opening size (mm) or equivalent sieve size (e.g., EOS = 40–70 indicates openings equivalent to the No. 40 to No. 70 sieves). Obviously, if the geotextile openings are larger than the largest soil particle diameter, then all of the soil particles will migrate through the geotextile and clog the drainage system. A common recommendation is that the required EOS be less than or equal to D_{85} (grain size corresponding to 85% passing).
Limitations	Some of the limitations of geotextile are as follows: 1. *Ultraviolet light.* Geotextile that has no ultraviolet light protection can rapidly deteriorate. For example, certain polypropylene geotextiles lost 100% of their strength after only 8 weeks of exposure (Raumann 1982, Koerner 1990). 2. *Sealing of geotextile.* When used for filtration, an impermeable soil layer can develop adjacent to the geotextile if it has too low an open area or too small an EOS.

TABLE 22.3 Geotextiles (*Continued*)

Topic (1)	Discussion (2)
Limitations (*Continued*)	3. *Construction problems.* Some of the more common problems related to construction with geotextiles are as follows (Richardson and Wyant 1987): (*a*) fill placement or compaction techniques damage the geotextile; (*b*) installation loads are greater than design loads, leading to failure during construction; (*c*) construction environment leads to a significant reduction in assumed fabric properties, causing failure of the completed project; (*d*) field seaming or overlap of the geotextile fails to fully develop desired fabric mechanical properties; and (*e*) instabilities during various construction phases may render a design inadequate even though the final product would have been stable.

FIGURE 22.4 Photograph of a geomembrane that has a rough surface texture for added friction. See Table 22.4. (*From Rollings and Rollings 1996; reprinted with permission of McGraw-Hill, Inc.*)

TABLE 22.4 Geomembranes

Topic (1)	Discussion (2)
General discussion	Common construction terminology for geomembranes includes *liners, membranes, visqueen, plastic sheets,* and *impermeable sheets.* Geomembranes are used as barriers to reduce water or vapor migration through soil (see Fig. 22.4). The surface of the geomembrane can be textured in order to provide more frictional resistance between the soil and geomembrane surface.
Common usage	As indicated in Table 22.1, geomembranes are used almost exclusively as barriers to reduce water or vapor migration through soil. For example, Fig. 20.3 shows design specifications for a moisture barrier, which includes a 6-mil visqueen vapor barrier (i.e., a geomembrane). In the United States, 1 mil is one-thousandth of an inch. Another common usage for geomembranes is for the lining and capping systems in municipal landfills. For liners in municipal landfills, the thickness of the geomembrane is usually at least 80 mils.
Limitations	Some of the limitations of geomembranes are as follows: 1. *Puncture resistance.* The geomembrane must be thick enough so that it is not punctured during installation and subsequent usage. 2. *Slide resistance.* Slope failures have developed in municipal liners because of the smooth and low frictional resistance between the geomembrane and overlying or underlying soil. Textured geomembranes (such as shown in Fig. 22.4) have been developed to increase the frictional resistance of the geomembrane surface. 3. *Sealing of seams.* A common cause of leakage through geomembranes is inadequate sealing of seams. The following are different methods commonly used to seal geomembrane seams (Rollings and Rollings 1996): *a. Extrusion welding.* Suitable for all polyethylenes. A ribbon of molten polymer is extruded over the edge (filet weld) or between the geomembrane sheets (flat weld). This melts the adjacent surfaces, which are then fused together upon cooling. *b. Thermal fusion.* Suitable for thermoplastics. Adjacent surfaces are melted and then pressed together. Commercial equipment is available that uses a heated wedge (most common) or hot air to melt the materials. Also, ultrasonic energy can be used for melting rather than heat. *c. Solvent-based systems.* Suitable for materials that are compatible with the solvent. A solvent is used with pressure to join adjacent surfaces. Heating may be used to accelerate the curing. The solvent may contain some of the geomembrane polymer already dissolved in the solvent liquid (bodied solvent) or an adhesive to improve the seam quality. *d. Contact adhesive.* Primarily suitable for thermosets. Solution is brushed onto surfaces to be joined, and pressure is applied to ensure good contact. Upon curing, the adhesive bonds the surfaces together.

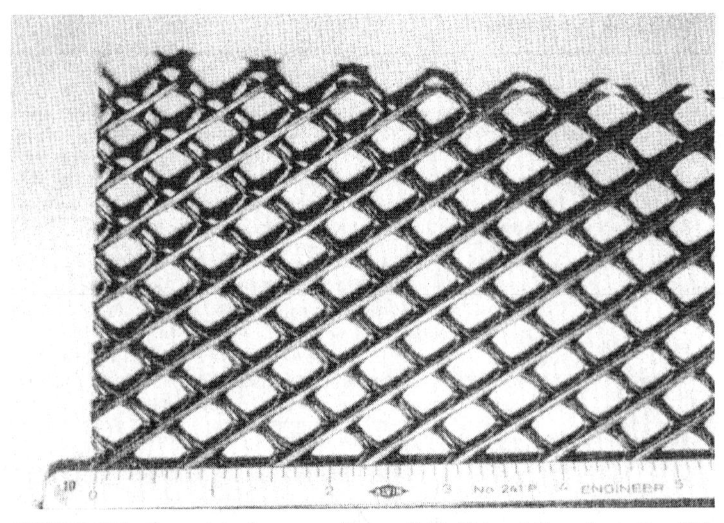

FIGURE 22.5 Photograph of a geonet. Also see Table 22.1 and discussion in Sec. 22.5. (*From Rollings and Rollings 1996; reprinted with permission of McGraw-Hill, Inc.*)

FIGURE 22.6 Photograph of a geocomposite. The geocomposite consists of a geonet having a textured geomembrane on top and a filter fabric (geotextile) on the bottom. Also see Table 22.1 and discussion in Sec. 22.5. (*From Rollings and Rollings 1996; reprinted with permission of McGraw-Hill, Inc.*)

CHAPTER 23
INSTRUMENTATION

23.1 INTRODUCTION

A common type of construction service performed by geotechnical engineers is the installation of monitoring devices. There are many types of monitoring devices used by geotechnical engineers. The usual purpose of the monitoring devices is to measure the performance of the structure as it is being built. Monitoring devices could also be installed to monitor existing adjacent structures, groundwater conditions, or slopes that may be impacted by the new construction.

Monitoring devices are especially important in urban areas where there are often adjacent structures that could be damaged by the construction activities. Frequent causes of damage to an adjacent structure include the lowering of the groundwater table or lateral movement of temporary underground shoring systems. Monitoring devices are essential for adjacent historic structures, which tend to be brittle and easily damaged, and can be very expensive to repair.

23.2 COMMONLY USED MONITORING DEVICES

Some of the more common geotechnical monitoring devices are described in Table 23.1 and shown in Figs. 23.1 to 23.9.

TABLE 23.1 Commonly Used Geotechnical Monitoring Devices

Monitoring device (1)	Discussion (2)
Inclinometer	The horizontal movement preceding or during the movement of slopes can be investigated by successive surveys of the shape and position of flexible vertical casings installed in the ground (Terzaghi and Peck 1967). The surveys are performed by lowering an inclinometer probe into the flexible vertical casing. The inclinometer probe is capable of measuring its deviation from the vertical. An initial survey (base reading) is performed and then successive surveys are compared to the base reading to obtain the horizontal movement of the slope. Figure 23.1 (from Slope Indicator Company 1998) shows a sketch of the inclinometer probe in the casing and the calculations used to obtain the lateral deformation. Inclinometers are often installed to monitor the performance of earth dams and during the excavation and grading of slopes where lateral movement might affect off-site structures. Inclinometers are also routinely installed to monitor the lateral ground movement due to the excavation of building basements and underground tunnels.
Piezometer	Piezometers are installed in order to monitor pore water pressures in the ground. There are many different types of piezometers, such as observation wells (Fig. 23.2), standpipe (Casagrande) piezometers (Fig. 23.3), and pneumatic piezometers (Fig. 23.4). The observation well is the simplest type of piezometer, and it consists of a standpipe having a filter tip that is installed in a vertical borehole. The hole is filled with sand or gravel and a bentonite or cement plug is installed at ground surface (see Fig. 23.2). This simple type of piezometer can be used to determine the depth of the groundwater table (i.e., the level of water in the pipe) and obtain groundwater samples. As indicated in Fig. 23.3, the Casagrande piezometer has only the filter tip surrounded by sand or gravel and the remainder of the borehole is filled with a bentonite or grout seal. This type of piezometer can be used to determine the piezometric water level, which may not correspond to the groundwater level. The pneumatic piezometer (Fig. 23.4) uses a pressure cell to measure the pore water pressure and the borehole backfill conditions are similar to the Casagrande piezometer. It is standard procedure to install piezometers when an urban project requires dewatering in order to make excavations below the groundwater table. Piezometers are also used to monitor the performance of earth dams and dissipation of excess pore water pressure associated with the consolidation of soft clay deposits.
Settlement monuments or cells	Settlement monuments or settlement cells can be used to monitor settlement or heave. Figure 23.5 (from Slope Indicator Company 1998) shows a diagram of the installation of a pneumatic settlement cell and plate. More advanced equipment includes settlement systems installed in borings that can not only measure total settlement, but also the incremental settlement at different depths. Settlement monuments or cells are often installed to measure the deformation of the foundation or embankments during construction or to monitor the movement of existing structures that are located adjacent to the area of construction.

TABLE 23.1 Commonly Used Geotechnical Monitoring Devices (*Continued*)

Monitoring device (1)	Discussion (2)
Pressure and load cells	A total pressure cell measures the sum of the effective stress and pore water pressure. The total pressure cell can be manufactured from two circular plates of stainless steel. The edges of the plates are welded together to form a sealed cavity, which is filled with fluid. Then a pressure transducer is attached to the cell. The total pressure acting on the sensitive surface is transmitted to the fluid inside the cell and measured by the pressure transducer (Slope Indicator Company 1998). Total pressure cells are often used to monitor total pressure exerted on a structure to verify design assumptions and to monitor the magnitude, distribution, and orientation of stresses. For example, as shown in Fig. 23.6, total pressure cells are commonly installed during the construction of earth dams to monitor the stresses within the dam core. During construction of the earth dam, the total pressure cells are often installed in arrays with each cell being placed in a different orientation and then covered with compacted fill (see Fig. 23.6). For the monitoring of earth pressure on retaining walls, the total pressure cell is typically placed into a recess so that the sensitive side is flush with the retaining wall surface. Load cells are similar in principle to total pressure cells. They can be used for many different types of geotechnical engineering projects. For example, Fig. 23.7 (from Slope Indicator Company 1998) shows a center-hole load cell which is designed to measure loads in tiebacks. This center-hole load cell can also be used to measure loads in rock bolts and cables. As shown in Fig. 23.7, for best results, the load cell is centered on the tieback bar and bearing plates are placed above and below the cell. The bearing plates must be able to distribute the load without bending or yielding.
Crack monitoring devices	For construction in congested urban areas, it is essential to monitor the performance of adjacent buildings, especially if they already have existing cracking. This can often be the case in historic districts of cities, where old buildings may be in a weakened or cracked state. Monitoring of existing cracks in adjacent buildings should be performed where there is pile driving or blasting at the construction site. The blasting of rock could be for the construction of an underground basement or for the construction of road cuts. People are often upset by the noise and vibrations from pile driving and blasting and will claim damage due to the vibrations from these construction activities. By monitoring the width of existing cracks, the geotechnical engineer will be able to evaluate these claims of damage. A simple method to measure the widening of cracks in concrete or brickwork is to install crack pins on both sides of the crack. By periodically measuring the distance between the pins, the amount of opening or closing of the crack can be determined.

TABLE 23.1 Commonly Used Geotechnical Monitoring Devices (*Continued*)

Monitoring device (1)	Discussion (2)
Crack monitoring devices (*Continued*)	Other crack monitoring devices are commercially available. For example, Fig. 23.8 shows two types of crack monitoring devices. For the Avongard crack monitoring device, there are two installation procedures: (1) the ends of the device are anchored by the use of bolts or screws or (2) the ends of the device are anchored with epoxy adhesive. The center of the Avongard crack monitoring device is held together with clear tape, which is cut once the ends of the monitoring device have been securely fastened with bolts, screws, or epoxy adhesive.
Other monitoring devices	There are many other types of monitoring devices that can be used by the geotechnical engineer. Some commercially available devices include soil strainmeters (see Fig. 23.9), borehole and tape extensometers, beam sensors and tiltmeters, and strain gauges.

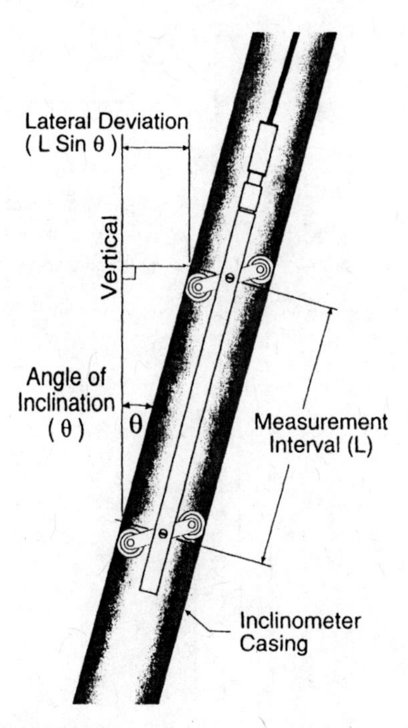

FIGURE 23.1 Inclinometer probe in a casing. See Table 23.1. (*Reprinted with permission from the Slope Indicator Company.*)

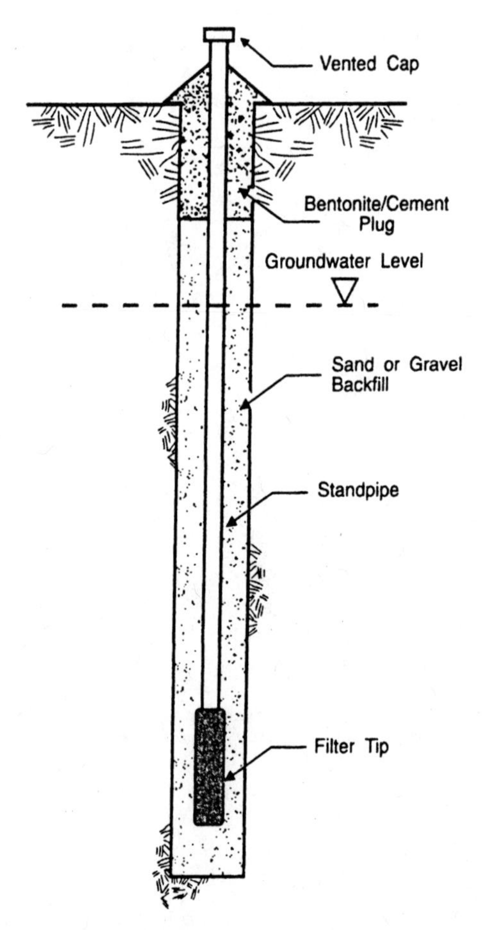

FIGURE 23.2 Observation well. See Table 23.1. (*Reprinted with permission from the Slope Indicator Company.*)

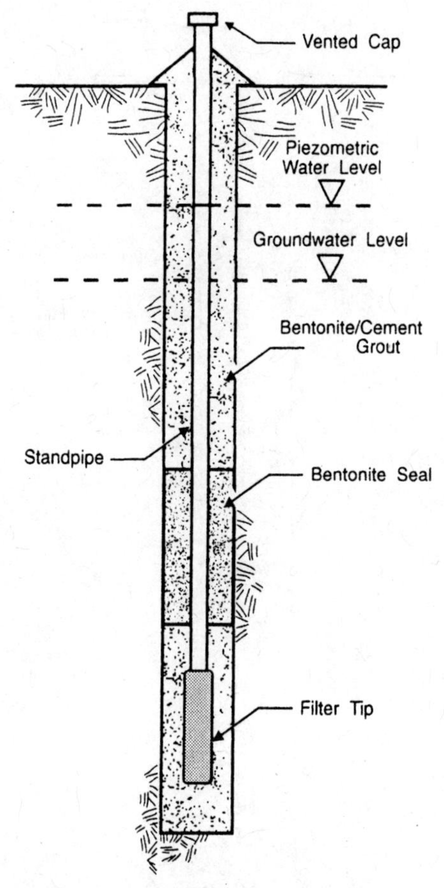

FIGURE 23.3 Standpipe (Casagrande) piezometer. See Table 23.1. (*Reprinted with permission from the Slope Indicator Company.*)

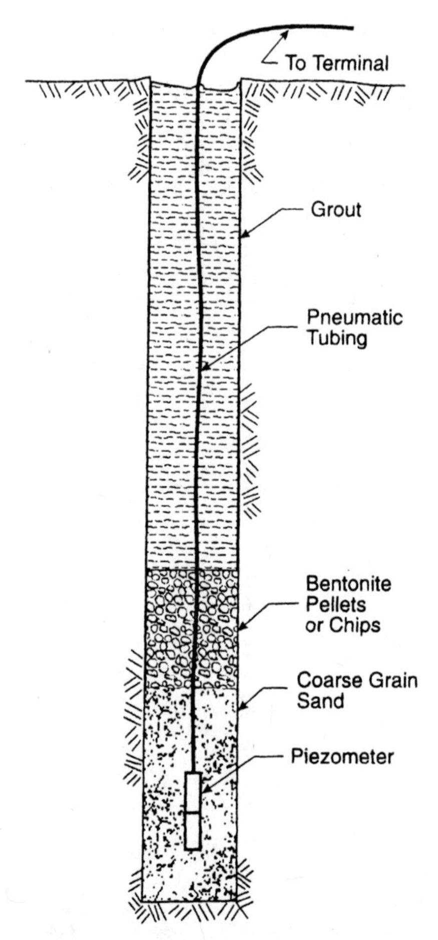

FIGURE 23.4 Pneumatic piezometer installed in a borehole. See Table 23.1. (*Reprinted with permission from the Slope Indicator Company.*)

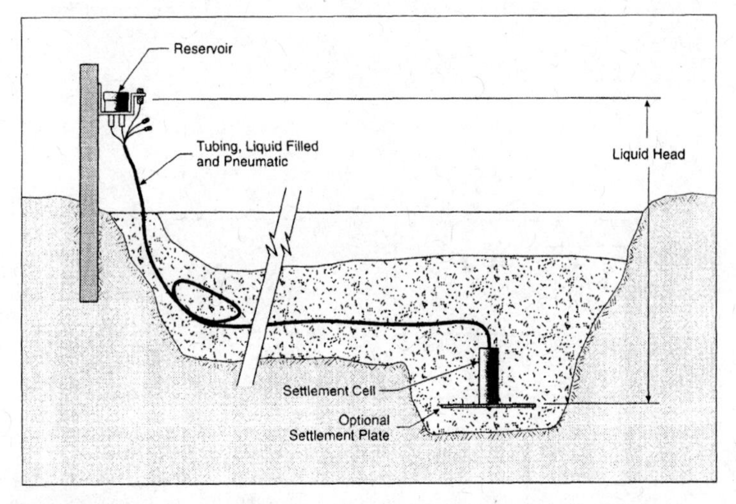

FIGURE 23.5 Pneumatic settlement cell installation. See Table 23.1. (*Reprinted with permission from the Slope Indicator Company.*)

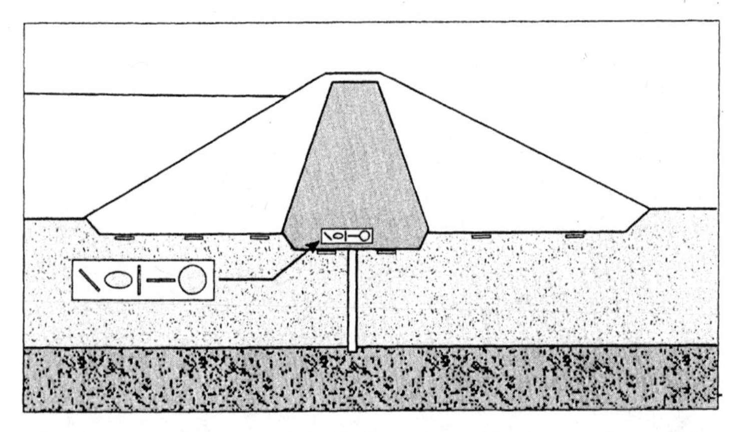

FIGURE 23.6 Total pressure cells installed at the base of the clay core of an earth dam. See Table 23.1. *(Reprinted with permission from the Slope Indicator Company.)*

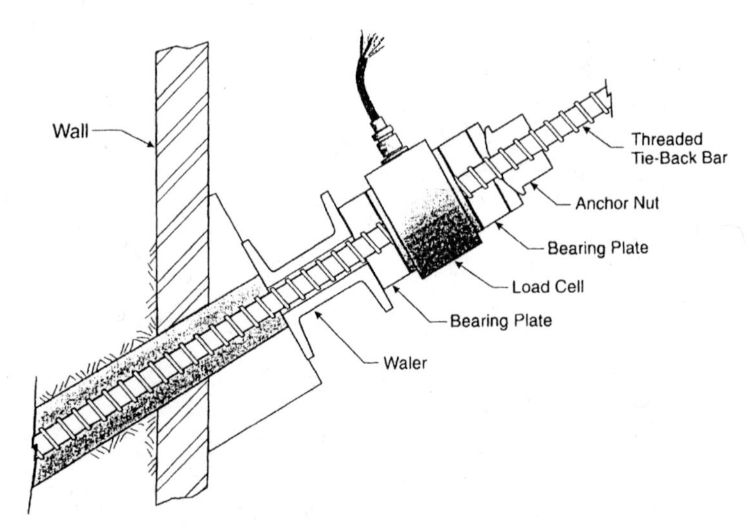

FIGURE 23.7 Center-hole load cell. See Table 23.1. *(Reprinted with permission from the Slope Indicator Company.)*

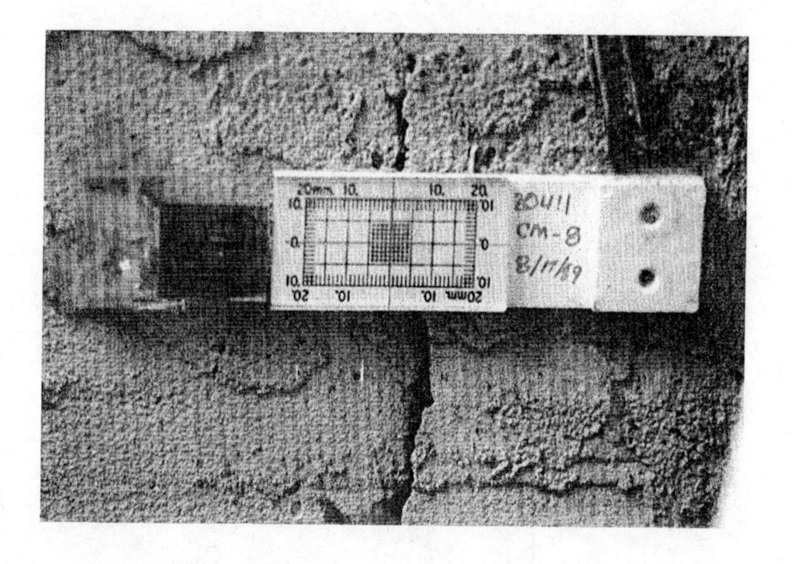

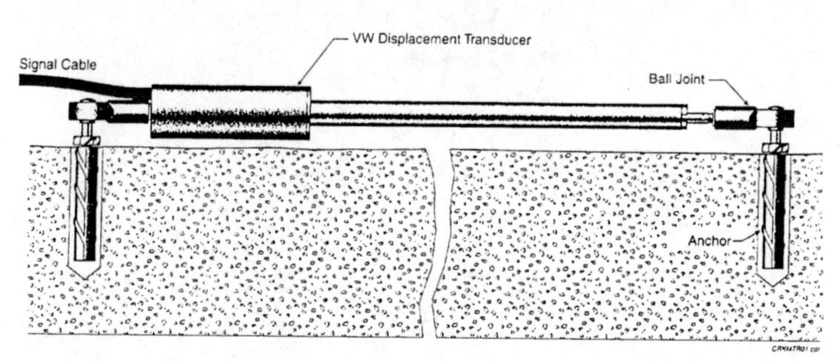

FIGURE 23.8 Crack monitoring devices. The upper photograph shows the Avongard crack monitoring device. The lower diagram shows the VW Crackmeter. See Table 23.1. (*Lower diagram reprinted with permission from the Slope Indicator Company.*)

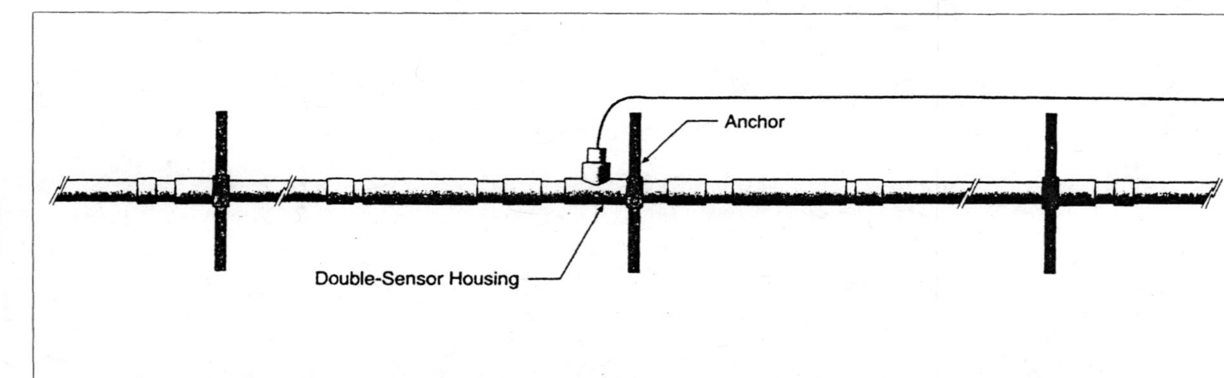

FIGURE 23.9 Double-sensor version of soil strainmeter. See Table 23.1. (*Reprinted with permission from the Slope Indicator Company.*)

APPENDIX A
GLOSSARY

INTRODUCTION

The following is a list of commonly used geotechnical engineering and engineering geology terms and definitions. The glossary has been divided into five main categories:

1. Engineering Geology and Subsurface Exploration Terminology
2. Laboratory Testing Terminology
3. Terminology for Engineering Analysis and Computations
4. Construction and Grading Terminology
5. Terminology for Sewage Disposal and Percolation Tests

References

References used for the Glossary include:

- *Glossary of Selected Geologic Terms with Special Reference to Their Use in Engineering* (1955)
- *Soil Mechanics in Engineering Practice* (1967)
- *Soil Mechanics* (1969)
- *Essentials of Soil Mechanics and Foundations* (1977)
- *An Introduction to Geotechnical Engineering* (1981)

- NAVFAC DM-7.1 (1982), NAVFAC DM-7.2 (1982), and NAVFAC DM-7.3 (1983)
- *Thickness Design—Asphalt Pavements for Highways and Streets* (1984)
- *Orange County Grading and Excavation Code* (1993)
- *Foundation Design, Principles and Practices* (1994)
- *Standard Specifications for Highway Bridges* (1996)
- *Caterpillar Performance Manual* (1997)
- *Uniform Building Code* (1997)
- American Society for Testing and Materials (ASTM D 653-97 and ASTM D 4439-97, 1998)
- *Sewage Disposal for Individual Homes, Small Industries and Institutions* (undated)

Basic Terms

civil engineer A professional engineer who is registered to practice in the field of civil works.

civil engineering The application of the knowledge of the forces of nature, principles of mechanics, and properties of materials for the evaluation, design, and construction of civil works for the beneficial uses of mankind.

engineering geologist A geologist who is experienced and knowledgeable in the field of engineering geology.

engineering geology The application of geologic knowledge and principles in the investigation and evaluation of naturally occurring rock and soil for use in the design of civil works.

geologist An individual educated and trained in the field of geology.

geotechnical engineer A licensed individual who performs an engineering evaluation of earth materials including soil, rock, groundwater, and man-made materials and their interaction with earth retention systems, structural foundations, and other civil engineering works.

geotechnical engineering A subdiscipline of civil engineering. Geotechnical engineering requires a knowledge of engineering laws, formulas, construction techniques, and the performance of civil engineering works influenced by earth materials. Geotechnical engineering encompasses many of the engineering aspects of soil mechanics, rock mechanics, foundation engineering, geology, geophysics, hydrology, and related sciences.

rock mechanics The application of the knowledge of the mechanical behavior of rock to engineering problems dealing with rock. Rock mechanics overlaps with engineering geology.

soil mechanics The application of the laws and principles of mechanics and hydraulics to engineering problems dealing with soil as an engineering material.

soils engineer Synonymous with geotechnical engineer (see **geotechnical engineer**).

soils engineering Synonymous with geotechnical engineering (see **geotechnical engineering**).

GLOSSARY 1

ENGINEERING GEOLOGY AND SUBSURFACE EXPLORATION TERMINOLOGY

adobe Sun-dried bricks composed of mud and straw. Adobe is commonly used for construction in the southwestern United States and in Mexico.

aeolian (or eolian) Particles of soil that have been deposited by the wind. Aeolian deposits include dune sands and loess.

alluvium Detrital deposits resulting from the flow of water, including sediments deposited in riverbeds, canyons, flood plains, lakes, fans at the foot of slopes, and estuaries.

aquiclude A relatively impervious rock or soil stratum that will not transmit groundwater fast enough to furnish an appreciable supply of water to a well or spring.

aquifer A relatively pervious rock or soil stratum that will transmit groundwater fast enough to furnish an appreciable supply of water to a well or spring.

artesian water Groundwater that is under pressure and is confined by impervious material. If the trapped pressurized water is released, by drilling a well, for example, the water will rise above the groundwater table and may even rise above the ground surface.

ash Fine fragments of rock, between 4 mm and 0.25 mm in size, that originated as airborne debris from explosive volcanic eruptions.

badlands An area, large or small, characterized by extremely intricate and sharp erosional sculpture. Badlands occur chiefly in arid or semiarid climates where the rainfall is concentrated in sudden heavy showers. They may, however, occur in humid regions where vegetation has been destroyed, or where soil and coarse detritus are lacking.

bedding The arrangement of rock in layers, strata, or beds.

bedrock A more or less solid, relatively undisturbed rock in place either at the surface or beneath deposits of soil.

bentonite A soil or formational material that has a high concentration of the clay mineral montmorillonite. Bentonite is usually characterized by high swelling upon wetting. The term bentonite also refers to manufactured products that have a high concentration of montmorillonite, for example, bentonite pellets.

bit A device that is attached to the end of the drill stem and is used as a cutting tool to bore into soil and rock.

bog A peat-covered area with a high groundwater table. The surface is often covered with moss, and it tends to be nutrient-poor and acidic.

boring A method of investigating subsurface conditions by drilling a hole into the earth materials. Usually soil and rock samples are extracted from the boring. Field tests, such as the standard penetration test (SPT) and the vane shear test (VST), can also be performed in the boring.

boring log A written record of the materials penetrated during the subsurface exploration. See Table 21.4 for other types of information that should be recorded on the boring log.

boulder A large detached rock fragment with an average dimension greater than 300 mm (12 in.).

California bearing ratio (CBR) The CBR can be determined for soil in the field or soil compacted in the laboratory. See Sec. 3.6 for specific details on the procedure for determining the CBR. The CBR is frequently used for the design of roads and airfields.

casing A steel pipe that is temporarily inserted into a boring or drilled shaft in order to prevent the adjacent soil from caving in.

cobble A rock fragment, usually rounded or semirounded, with an average dimension between 75 and 300 mm (3 and 12 in.).

cohesionless soil A soil, such as a clean gravel or sand, that when unconfined, falls apart in either a wet or dry state.

cohesive soil A soil, such as a silt or clay, that when unconfined, has considerable shear strength when dried, and will not fall apart in a saturated state. Cohesive soil is also known as *plastic soil,* or soil that has a plasticity index.

colluvium Generally loose deposits usually found near the base of slopes and brought there chiefly by gravity through slow continuous downhill creep.

cone penetration test (CPT) A field test used to identify and determine the *in situ* properties of soil deposits and soft rock.

electric cone A cone penetrometer that uses electric-force transducers built into the apparatus for measuring cone resistance and friction resistance.

mechanical cone A cone penetrometer that uses a set of inner rods to operate a telescoping penetrometer tip and to transmit the resistance force to the surface for measurement.

mechanical-friction cone A cone penetrometer with the additional capability of measuring the local side friction component of penetration resistance.

piezocone A cone penetrometer with the additional capability of measuring pore water pressure generated during the penetration of the cone.

core drilling Also known as *diamond drilling,* the process of cutting out cylindrical rock samples in the field.

core recovery (RQD) The RQD (Rock Quality Designation) is computed by summing the lengths of all pieces of the rock core (NX size) equal to or longer than 10 cm (4 in.) and dividing by the total length of the core run. The RQD is multiplied by 100 to express it as a percentage.

deposition The geologic process of laying down or accumulating natural material into beds, veins, or irregular masses. Deposition includes mechanical settling (such as sedimentation in lakes), precipitation (such as the evaporation of surface water to form halite), and the accumulation of dead plants (as in a peat bog).

detritus Any material worn or broken down from rocks by mechanical means.

diatomaceous earth Diatomaceous earth usually consists of fine, white, siliceous powder, composed mainly of diatoms and their remains.

erosion The wearing away of the ground surface as a result of the movement of wind, water, and/or ice.

fault A fracture in the earth's crust along which movement has occurred. A fault is considered active if movement has occurred within the last 11,000 years (Holocene geologic time).

fines Refers to the silt- and clay-size particles in the soil.

fold A bent or flexed layer (or layers) of rock. Examples of folded rock include anticlines and synclines. Usually folds are created by the massive compression of rock layers.

fracture A visible break in a rock mass. Examples include joints, faults, and fissures.

geophysical techniques Various methods of determining subsurface soil and rock conditions without performing subsurface exploration. A common geophysical technique is to induce a shock wave into the earth and then measure the seismic velocity of the wave's travel through the earth material. The seismic velocity has been correlated with the rippability of the earth material.

groundwater table The top surface of underground water (also known as *phreatic surface*), the location of which is often determined from piezometers, such as an open standpipe. A perched groundwater table refers to groundwater occurring in an upper zone separated from the main body of groundwater by underlying unsaturated rock or soil.

horizon One of the layers of a soil profile that can be distinguished by its texture, color, and structure.

***A* horizon** The uppermost layer of a soil profile which often contains remnants of organic life. Inorganic colloids and soluble materials are often leached from this horizon.

***B* horizon** The layer of a soil profile in which material leached from the overlying *A* horizon is accumulated.

***C* horizon** Undisturbed parent material from which the overlying soil profile has been developed.

inclinometer The horizontal movement preceding or during the movement of slopes can be investigated by successive surveys of the shape and position of flexible vertical casings installed in the ground. The surveys are performed by lowering an inclinometer probe into the flexible vertical casing.

in situ Used in reference to the original in-place (or *in situ*) condition of the soil or rock.

Iowa borehole shear test (BST) A field test where the device is lowered into an uncased borehole and then expanded against the sidewalls. The force required to pull the device toward ground surface is measured and, much as in a direct shear test, the shear strength properties of the *in situ* soil can then be determined.

karst topography A type of landform developed in a region of easily soluble limestone. It is characterized by vast numbers of depressions of all sizes, sometimes by great outcrops of limestone ledges, sinks and other solution passages, an almost total lack of surface streams, and large springs in the deeper valleys.

kelly A heavy tube or pipe, usually square or rectangular in cross section, that is used to provide a downward load in excavating an auger borehole.

landslide Mass movement of soil or rock that involves shear displacement along one or several rupture surfaces, which are either visible or may be reasonably inferred.

landslide debris Material, generally porous and of low density, produced from instability of natural or man-made slopes.

leaching The removal of soluble materials in soil or rock caused by percolating or moving groundwater.

loess A wind-deposited silt often having a high porosity and low density, which is often susceptible to collapse of its soil structure upon wetting.

mineral An inorganic substance that has a definite chemical composition and distinctive physical properties. Most minerals are crystalline solids.

overburden The soil that overlies bedrock. In other cases, it refers to all material overlying a point of interest in the ground, such as the overburden pressure exerted on a clay layer.

peat A naturally occurring highly organic deposit derived primarily from plant materials.

penetration resistance See **standard penetration test.**

percussion drilling A drilling process in which a borehole is advanced by applying a series of impacts to the drill rods and attached bit.

permafrost Perennially frozen soil. Also defined as ground that remains below freezing temperatures for 2 or more years. The bottom of permafrost lies at depths ranging from a few feet to over 1000 feet. The *active layer* is defined as the upper few inches to several feet of ground that is frozen in winter but thawed in summer.

piezometer A device installed for measuring the pore water pressure (or pressure head) at a specific point within the soil mass.

pit (or test pit) An excavation made for the purpose of observing subsurface conditions, performing field tests, and obtaining soil samples. A pit also refers to an excavation in the surface of the earth from which ore is extracted, such as an open-pit mine.

pressuremeter test (PMT) A field test that involves the expansion of a cylindrical probe within an uncased borehole.

refusal During subsurface exploration, refusal means an inability to excavate any deeper with the boring equipment. Refusal could be a result of many different factors, such as hard rock, boulders, or a layer of cobbles.

residual soil Soil derived by in-place weathering of the underlying material.

rock Rock is a relatively solid mass that has permanent and strong bonds between the minerals. Rock can be classified as sedimentary, igneous, or metamorphic.

rotary drilling A drilling process in which a borehole is advanced by rotation of a drill bit under constant pressure without impact.

rubble Rough stones of irregular shape and size that are naturally or artificially broken from larger masses of rock. Rubble is often created during quarrying, stone cutting, and blasting.

screw plate compressometer (SPC) test A field test that involves a plate that is screwed down to the desired depth, and then as pressure is applied, the settlement of the plate is measured.

seep A small area where water oozes from the soil or rock.

slaking The crumbling and disintegration of earth materials when exposed to air or moisture. Slaking can also refer to the breaking up of dried clay when submerged in water, a result either of compression of entrapped air by inwardly migrating water or of the progressive swelling and sloughing off of the outer layers.

slickensides Surfaces within a soil mass which have been smoothed and striated by shear movements on these surfaces.

slope wash Soil and/or rock material that has been transported down a slope by mass wasting assisted by runoff water not confined by channels (also see **colluvium**).

soil Sediments or other accumulations of mineral particles produced by the physical and chemical disintegration of rocks. Inorganic soil does not contain organic matter, while organic soil contains organic matter.

soil sampler A device used to obtain soil samples during subsurface exploration. On the basis of the inside clearance ratio and the area ratio, soil samples can be either disturbed or undisturbed.

standard penetration test (SPT) A field test that consists of driving a thick-walled sampler (I.D. = 1.5 in., O.D. = 2 in.) into the soil by using a 140-lb hammer falling 30 inches. The number of blows to drive the sampler 18 inches is recorded. The N-value (penetration resistance) is defined as the number of blows required to drive the sampler from a depth interval of 6 to 18 inches.

strike and dip Characteristics of a planar structure, such as a shear surface, fault, bed, etc. The strike is the compass direction of a level line drawn on the planar structure. The dip angle is measured between the planar structure and a horizontal surface.

subgrade modulus This value is often obtained from field plate load tests and is used in the design of pavements and airfields. (Also known as the modulus of subgrade reaction.)

subsoil profile Developed from subsurface exploration, a cross section of the ground that shows the soil and rock layers. A summary of field and laboratory tests could also be added to the subsoil profile.

till Material created directly by glaciers, without transportation or sorting by water. Till often contains particles in a wide range of sizes, including boulders, gravel, sand, and clay.

topsoil The fertile upper zone of soil which contains organic matter and is usually darker in color and loose.

vane shear test (VST) An *in situ* field test that consists of inserting a four-bladed vane into the borehole and then pushing the vane into the clay deposit located at the bottom of the borehole. Once inserted into the clay, the maximum torque required to rotate the vane and shear the clay is measured. From the dimensions of the vane and the maximum torque, the undrained shear strength s_u of the clay can be calculated.

varved silt or varved clay A lake deposit with alternating thin layers of sand and silt (varved silt) or sand and clay (varved clay). It is formed by the process of sedimentation from the summer to winter months. The sand is deposited during the summer and the silt or clay is deposited in the winter when the lake surface is covered with ice and the water is tranquil.

wetland Land which has a groundwater table at or near the ground surface, or land that is periodically under water, and supports various types of vegetation that are adapted to a wet environment.

GLOSSARY 2

LABORATORY TESTING TERMINOLOGY

absorption Defined as the mass of water in the aggregate divided by the dry mass of the aggregate. Absorption is used in soil mechanics for the study of oversize particles or in concrete mix design.

activity of clay The ratio of plasticity index to percent dry mass of the total sample that is smaller than 0.002 mm in grain size. This property is related to the types of clay minerals in the soil.

Atterberg limits Water contents corresponding to different behavior conditions of plastic soil.

liquid limit The water content corresponding to the behavior change between the liquid and plastic state of a soil. The liquid limit is arbitrarily defined as the water content at which a pat of soil, cut by a groove of standard dimensions, will flow together for a distance of 12.7 mm (0.5 in.) under the impact of 25 blows in a standard liquid limit device.

plastic limit The water content corresponding to the behavior change between the plastic and semisolid state of a soil. The plastic limit is arbitrarily defined as the water content at which the soil will just begin to crumble when rolled into a thread approximately 3.2 mm ($1/8$ in.) in diameter.

shrinkage limit The water content corresponding to the behavior change between the semisolid and the solid state of a soil. The shrinkage limit is also defined as the water content at which any further reduction in water content will not result in a decrease in volume of the soil mass.

capillarity Also known as *capillary action*, the rise of water through a soil due to the fluid property known as *surface tension*. Because of capillarity, the pore water pressures are less than atmospheric because of the surface tension of pore water acting on the meniscus formed in void spaces between the soil particles. The height of capillary rise is inversely proportional to the pore size of the soil.

clay-size particles Clay-size particles are finer than 0.002 mm. Most clay particles are flat or platelike in shape, and as such they have a large surface area. The most common clay minerals belong to the kaolin, montmorillonite, and illite groups.

coarse-grained soil According to the Unified Soil Classification System, coarse-grained soils have more than 50% soil particles (by dry mass) retained on the No. 200 U.S. standard sieve.

coefficient of consolidation A coefficient used in the theory of consolidation. It is obtained from laboratory consolidation tests and is used to predict the time-settlement behavior of field loading of fine-grained soil.

cohesion There are two types of cohesion: (1) cohesion in terms of total stress and (2) cohesion in terms of effective stress. For total cohesion *c,* the soil particles are predominantly held together by capillary tension. For effective stress cohesion c', there must be actual bonding or attraction forces between the soil particles.

colloidal soil particles Generally refers to clay-size particles (finer than 0.002 mm) where the surface activity of the particle has an appreciable influence on the properties of the soil.

compaction (laboratory)

compaction curve A curve showing the relationship between the dry density and the water content of a soil for a given compaction energy.

compaction test A laboratory compaction procedure whereby a soil at a known water content is compacted into a mold of specific dimensions. The procedure is repeated for various water contents to establish the compaction curve. The most common testing procedures (compaction energy, number of soil layers in the mold, etc.) are the Modified Proctor (ASTM D 1557) or Standard Proctor (ASTM D 698).

compression index For a consolidation test, the slope of the linear portion of the vertical pressure versus void ratio curve on a semilog plot.

consistency of clay Generally refers to the firmness of a cohesive soil. For example, a cohesive soil can have a consistency that varies from very soft to hard.

consolidation test A laboratory test used to measure the consolidation properties of cohesive soil. The specimen is laterally confined in a ring and is compressed between porous plates (oedometer apparatus).

contraction (during shear) During the shearing of soil, the tendency of loose soil to decrease in volume (or contract).

density Density is defined as mass per unit volume. In the International System of Units (SI), typical units for the density of soil are Mg/m^3.

deviator stress Difference between the major and minor principal stresses in a triaxial test.

dilation (during shear) During the shearing of soil, the tendency of dense soil to increase in volume (or dilate).

double layer A grossly simplified interpretation of the positively charged water layer, together with the negatively charged surface of the particle itself. Two reasons for the attraction of water to the clay particle are: (1) dipolar structure of water molecule which causes it to be electrostatically attracted to the surface of the clay particle and (2) the clay particles attract cations which contribute to the attraction of water by the hydration process. The *absorbed water layer* consists of water molecules that are tightly held to the clay particle face, by the process of hydrogen bonding, for example.

fabric (of soil) Definitions vary, but in general the fabric of soil often refers only to the geometric arrangement of the soil particles. In contrast, the soil structure refers to both the geometric arrangement of soil particles and the interparticle forces which may act between them.

fine-grained soil According to the Unified Soil Classification System, a fine-grained soil contains more than 50 percent (by dry mass) of particles finer than the No. 200 sieve.

flocculation The process of fines, when in suspension in water, attracting each other to form a larger particle. In the hydrometer test, a dispersing agent is added to prevent flocculation of fines.

friction angle In terms of effective shear stress, the soil friction is usually considered to be a result of the interlocking of the soil or rock grains and the resistance to sliding between the grains. A relative measure of a soil's frictional shear strength is the friction angle.

gravel-size fragments Rock fragments and soil particles that will pass the 3-in. (76-mm) sieve and be retained on a No. 4 (4.75-mm) U.S. standard sieve.

hydraulic conductivity (or coefficient of permeability) For laminar flow of water in soil, both terms are synonymous and indicate a measure of the soil's ability to allow water to flow through its soil pores. The hydraulic conductivity is often measured in a constant-head or falling-head permeameter.

laboratory maximum dry density The peak point of the compaction curve (see **compaction**).

moisture content (or water content) Moisture content and water content are synonymous. The definition of moisture content is the ratio of the mass of water in the soil divided by the dry mass of the soil, usually expressed as a percentage.

optimum moisture content The moisture content, determined from a laboratory compaction test, at which the maximum dry density of a soil is obtained by using a specific compaction energy.

overconsolidation ratio (OCR) The ratio of the preconsolidation vertical effective stress to the current vertical effective stress.

peak shear strength The maximum shear strength along a shear failure surface.

permeability The ability of water (or other fluid) to flow through a soil by traveling through the void spaces. A high permeability indicates flow occurs rapidly, and vice versa. A measure of the soil's permeability is the hydraulic conductivity, also known as the coefficient of permeability.

plasticity Term applied to silt and clay, to indicate the soil's ability to be rolled and molded without breaking apart. A measure of a soil's plasticity is the plasticity index.

plasticity index The plasticity index is defined as the liquid limit minus the plastic limit, often expressed as a whole number (also see **Atterberg limits**).

sand equivalent (SE) A measure of the amount of silt or clay contamination in fine aggregate as determined by ASTM D 2419 test procedures.

sand-size particles Soil particles that will pass the No. 4 (4.75-mm) sieve and be retained on the No. 200 (0.075-mm) U.S. standard sieve.

shear strength The maximum shear stress that a soil or rock can sustain. Shear strength of soil is based on total stresses (i.e., undrained shear strength) or effective stresses (i.e., effective shear strength).

shear strength in terms of total stress Shear strength of soil based on total stresses. The undrained shear strength of soil could be expressed in terms of the undrained shear strength s_u, or by using the failure envelope that is defined by total cohesion c and total friction angle ϕ.

effective shear strength Shear strength of soil based on effective stresses. The effective shear strength of soil could be expressed in terms of the failure envelope, that is defined by effective cohesion c' and effective friction angle ϕ'.

shear strength tests (laboratory) There are many types of shear strength tests that can be performed in the laboratory. The objective is to obtain the shear strength of the soil. Laboratory tests can generally be divided into two categories:

shear strength tests based on effective stress The purpose of these laboratory tests is to obtain the effective shear strength of the soil based on the failure envelope in terms of effective stress. An example is a direct shear test where the saturated, submerged, and consolidated soil specimen is sheared slowly enough that excess pore water pressures do not develop (this test is known as a consolidated-drained test).

shear strength tests based on total stress The purpose of these laboratory tests are to obtain the undrained shear strength of the soil or the failure envelope in terms of total stresses. An example is the unconfined compression test, which is also known as an unconsolidated-undrained test.

sieve Laboratory equipment consisting of a pan with a screen at the bottom. U.S. standard sieves are used to separate particles of a soil sample into their various sizes.

silt-size particles That portion of a soil that is finer than the No. 200 sieve (0.075 mm) and coarser than 0.002 mm. Silt and clay size particles are considered to be fines.

soil structure Definitions vary, but in general, the soil structure refers to both the geometric arrangement of the soil particles and the interparticle forces that may act between them. Common soil structures are as follows:

cluster structure Soil grains that consist of densely packed silt or clay size particles.

dispersed structure The clay-size particles are oriented parallel to each other.

flocculated (or cardhouse) structure The clay-size particles are oriented in edge-to-face arrangements.

honeycomb structure Loosely arranged bundles of soil particles, having a structure that resembles a honeycomb.

single-grained structure An arrangement composed of individual soil particles. This is a common structure of sands.

skeleton structure An arrangement where coarser soil grains form a skeleton with the void spaces partly filled by a relatively loose arrangement of soil fines.

specific gravity The specific gravity of soil or oversize particles can be determined in the laboratory. Specific gravity is generally defined as the ratio of the density of the soil particles divided by the density of water.

tensile test For a geosynthetic, a laboratory test in which the geosynthetic is stretched in one direction to determine the force-elongation characteristics, breaking force, and the breaking elongation.

texture (of soil) The degree of fineness of the soil, such as smooth, gritty, or sharp, when the soil is rubbed between the fingers.

thixotropy The property of a remolded clay that enables it to stiffen (gain shear strength) in a relatively short time.

triaxial test A laboratory test in which a cylindrical specimen of soil or rock encased in an impervious membrane is subjected to a confining pressure and then loaded axially to failure.

unconfined compressive strength The vertical stress which causes the shear failure of a cylindrical specimen of a plastic soil or rock in a simple compression test. For the simple compression test, the undrained shear strength s_u of the plastic soil is defined as one-half the unconfined compressive strength.

unit weight Unit weight is defined as weight per unit volume. In the International System of Units (SI), unit weight has units of kN/m^3. In the United States Customary System, unit weight has units of pcf (pounds-force per cubic foot).

water content (or moisture content) See **moisture content.**

zero air voids curve On the laboratory compaction curve, the zero air voids curve is often included. It is the relationship between water content and dry density for a condition of saturation (S = 100%) for a specified specific gravity.

GLOSSARY 3

TERMINOLOGY FOR ENGINEERING ANALYSIS AND COMPUTATIONS

adhesion Shearing resistance between two different materials. For example, for piles driven into clay deposits, there is adhesion between the surface of the pile and the surrounding clay.

allowable bearing pressure Allowable bearing pressure is the maximum pressure that can be imposed by a foundation onto soil or rock supporting the foundation. It is derived from experience and general usage, and provides an adequate factor of safety against shear failure and excessive settlement.

anisotropic soil A soil mass having different properties in different directions at any given point, referring primarily to stress-strain or permeability characteristics.

arching The transfer of stress from an unconfined area to a less-yielding or restrained structure. Arching is important in the design of pile or pier walls that have open gaps between the members.

bearing capacity

allowable bearing capacity The maximum allowable bearing pressure for the design of foundations.

ultimate bearing capacity The bearing pressure that causes failure of the soil or rock supporting the foundation.

bearing capacity failure A foundation failure that occurs when the shear stresses in the adjacent soil exceed the shear strength.

bell The enlarged portion of the bottom of a drilled shaft foundation. A bell is used to increase the end-bearing resistance. Not all drilled shafts have bells.

collapsible formations Examples of collapsible formations include limestone formations and deep mining of coal beds. Limestone can form underground caves and caverns which can gradually enlarge, resulting in a collapse of the ground surface and the formation of a sinkhole. Sites that are underlain by coal or salt mines could also experience ground surface settlement when the underground mine collapses.

collapsible soil Collapsible soil can be broadly classified as soil that is susceptible to a large and sudden reduction in volume upon wetting. Collapsible soil usually has a low dry density and low moisture content. Such soil can withstand a large applied vertical stress with a small compression, but then experience much larger settlements after wetting, with no increase in vertical pressure. Collapsible soil can include fill compacted dry of optimum and natural collapsible soil, such as alluvium, colluvium, or loess.

compressibility A decrease in volume that occurs in the soil mass when it is subjected to an increase in loading. Some highly compressible soils are loose sands, organic clays, sensitive clays, highly plastic and soft clays, uncompacted fills, municipal landfills, and permafrost soils.

consolidation The consolidation of a saturated clay deposit is generally divided into three separate categories:

initial or immediate settlement The initial settlement of the structure caused by undrained shear deformations, or in some cases contained plastic flow, due to two- or three-dimensional loading.

primary consolidation The compression of clays under load that occurs as excess pore water pressures slowly dissipate with time.

secondary compression The final component of settlement, which is that part of the settlement that occurs after essentially all of the excess pore water pressures have dissipated.

creep An imperceptibly slow and more or less continuous movement of slope-forming soil or rock debris.

critical height and critical slope Critical height refers to the maximum height at which a vertical excavation or slope will stand unsupported. Critical slope refers to the maximum angle at which a sloped bank of soil or rock of given height will stand unsupported.

crown Generally, the highest point. For tunnels, the crown is the arched roof. For landslides, the crown is the area above the main scarp of the landslide.

dead load Structural loads due to the weight of beams, columns, floors, roofs, and other fixed members. Does not include nonstructural items such as furniture, snow, occupants, or inventory.

debris flow An initial shear failure of a soil mass which then transforms itself into a fluid mass that can move rapidly over the ground surface.

depth of seasonal moisture change Also known as the active zone, the layer of expansive soil subjected to shrinkage during the dry season and swelling during the wet season. This zone extends from ground surface to the depth of significant moisture fluctuation.

desiccation The process of shrinkage of clays. The process involves a reduction in volume of the grain skeleton and subsequent cracking of the clay caused by the development of capillary stresses in the pore water as the soil dries.

design load All forces and moments that are used to proportion a foundation. The design load includes the deadweight of a structure, and in some cases, can include live loads. Considerable judgment and experience are required to determine the design load that is to be used to proportion a foundation.

downdrag Force induced on a deep foundation resulting from downward movement of adjacent soil relative to a foundation element. Also referred to as negative skin friction.

earth pressure Usually used in reference to the lateral pressure imposed by a soil mass against an earth-supporting structure such as a retaining wall or basement wall.

active earth pressure (k_A) Horizontal pressure for a condition where the retaining wall has yielded sufficiently to allow the backfill to mobilize its shear strength.

at-rest earth pressure (k_o) Horizontal pressure for a condition where the retaining wall has not yielded or compressed into the soil. This would also be applicable to a soil mass in its natural state.

passive earth pressure (k_p) Horizontal pressure for a condition such as a retaining wall footing that has moved into and compressed the soil sufficiently to develop its maximum lateral resistance.

effective stress The total stress minus the pore water pressure.

equipotential line A line connecting points of equal total head.

equivalent fluid pressure Horizontal pressures of soil, or soil and water in combination, that increase linearly with depth and are equivalent to those that would be produced by a soil of a given density. Equivalent fluid pressure is often used in the design of retaining walls.

exit gradient The hydraulic gradient near the toe of a dam or the bottom of an excavation through which groundwater seepage is exiting the ground surface.

finite element A soil and structure profile subdivided into regular geometrical shapes for the purpose of numerical stress analysis.

flow line The path of travel traced by moving groundwater as it flows through a soil mass.

flow net A graphical representation used to study the flow of groundwater through a soil. A flow net is composed of flow lines and equipotential lines.

head From Bernoulli's energy equation, the total head is defined as the sum of the velocity head, pressure head, and elevation head. Head has units of length. For seepage problems in soil, the velocity head is usually small enough to be neglected and thus for laminar flow in soil, the total head h is equal to the sum of the pressure head h_p and elevation head h_e.

heave The upward movement of foundations or other structures caused by frost or expansive soil and rock. Frost heave refers to the development of ice layers or lenses within the soil that causes the ground surface to heave upward. Heave due to expansive soil and rock is caused by an increase in water content of clays or rocks, such as shale or slate.

homogeneous soil Soil that exhibits essentially the same physical properties at every point throughout the soil mass.

hydraulic gradient Difference in total head at two points divided by the distance between them. Hydraulic gradient is used in seepage analyses.

isotropic soil A soil mass having essentially the same properties in all directions at any given point, referring primarily to stress-strain or permeability characteristics.

laminar flow Groundwater seepage in which the total head loss is proportional to the velocity.

liquefaction The sudden, large decrease of shear strength of a cohesionless soil caused by collapse of the soil structure, produced by shock or earthquake-induced shear strains, associated with a sudden but temporary increase of pore water pressures. Liquefaction causes the cohesionless soil to behave as a fluid.

live load Structural load due to nonstructural members, such as furniture, occupants, inventory, and snow.

Mohr circle A graphical representation of the stresses acting on the various planes at a given point in the soil.

normally consolidated The condition that exists if a soil deposit has never been subjected to an effective stress greater than the existing overburden pressure and if the deposit is completely consolidated under the existing overburden pressure.

overconsolidated The condition that exists if a soil deposit has been subjected to an effective stress greater than the existing overburden pressure.

piping The movement of soil particles as a result of unbalanced seepage forces produced by percolating water, leading to the development of ground surface boils or underground erosion voids and channels.

plastic equilibrium The state of stress of a soil mass that has been loaded and deformed to such an extent that its ultimate shearing resistance is mobilized at one or more points.

pore water pressure The water pressure that exists in the soil void spaces.

> **excess pore water pressure** The increment of pore water pressures greater than hydrostatic values, produced by consolidation stress in compressible materials or by shear strain.
>
> **hydrostatic pore water pressure** Pore water pressure or groundwater pressures exerted under conditions of no flow where the magnitude of pore pressures increase linearly with depth below the groundwater table.

porosity The ratio, usually expressed as a percentage, of the volume of voids divided by the total volume of the soil or rock.

preconsolidation pressure The greatest vertical effective stress to which a soil, such as a clay layer, has been subjected.

pressure (or stress) The load divided by the area over which it acts.

principal planes The three mutually perpendicular planes through a point in the soil mass on which the shearing stress is zero. For soil mechanics, compressive stresses are positive.

major principal plane The plane normal to the direction of the major principal stress (highest stress in the soil).

intermediate principal plane The plane normal to the direction of the intermediate principal stress.

minor principal plane The plane normal to the direction of the minor principal stress (lowest stress in the soil).

principal stresses The stresses that occur on the principal planes. Also see **Mohr circle.**

progressive failure Formation and development of localized stresses that lead to fracturing of the soil, which spreads and eventually forms a continuous rupture surface and a failure condition. Stiff fissured clay slopes are especially susceptible to progressive failure.

quick clay A clay that has a sensitivity greater than 16. Upon remolding, such clays can exhibit a fluid (or quick) condition.

quick condition A condition in which groundwater is flowing upward with a sufficient hydraulic gradient to produce a zero effective stress condition in the sand deposit. Such a deposit is called quicksand.

relative density Term applied to a sand deposit to indicate its relative density state, defined as the ratio of (1) the difference between the void ratio in the loosest state and the *in situ* void ratio to (2) the difference between the void ratios in the loosest and in the densest states.

saturation (degree of) The volume of water in the void space divided by the total volume of voids. It is usually expressed as a percentage. A completely dry soil has a degree of saturation of 0 percent and a saturated soil has a degree of saturation of 100 percent.

seepage The infiltration or percolation of water through soil and rock.

seepage analysis An analysis to determine the quantity of groundwater flowing through a soil deposit. For example, by using a flow net, the quantity of groundwater flowing through or underneath an earth dam can be determined.

seepage force The frictional drag of water flowing through the soil voids.

seepage velocity The velocity of flow of water in the soil, while the superficial velocity is the velocity of flow into or out of the soil.

sensitivity The ratio of the undrained shear strength of the undisturbed plastic soil to the remolded shear strength of the same plastic soil.

settlement The permanent downward vertical movement experienced by structures as the underlying soil consolidates, compresses, or collapses due to the structural load or secondary influences.

differential settlement The difference in settlement between two foundation elements or between two points on a single foundation.

total settlement The absolute vertical movement of the foundation.

shear failure A failure in a soil or rock mass caused by shearing strain along one or more slip (rupture) surfaces.

general shear failure Failure in which the shear strength of the soil or rock is mobilized along the entire slip surface.

local shear failure Failure in which the shear strength of the soil or rock is mobilized only locally along the slip surface.

progressive shear failure See **progressive failure.**

punching shear failure Shear failure where the foundation pushes (or punches) into the soil because of the compression of soil directly below the footing as well as vertical shearing around the footing perimeter.

shear plane (or slip surface) A plane along which failure of soil or rock occurs by shearing.

shear stress Stress that acts parallel to the surface element.

slope stability analyses

gross slope stability The stability of slope material below a plane approximately 0.9 to 1.2 m (3 to 4 ft) deep measured from and perpendicular to the slope face.

surficial slope stability The stability of the outer 0.9 to 1.2 m (3 to 4 ft) of slope material measured from and perpendicular to the slope face. See App. B, Standard Detail No. 8 for typical repair of surficial failures.

strain The change in shape of soil when it is acted upon by stress.

normal strain A measure of compressive or tensile deformation, defined as the change in length divided by the initial length. In geotechnical engineering, strain is positive when it results in compression of the soil.

shear strain A measure of the shear deformation of soil.

subsidence Settlement of the ground surface over a very large area, such as that caused by the extraction of oil from the ground or the pumping of groundwater from wells.

swell Increase in soil volume, typically referring to volumetric expansion of clay due to an increase in water content.

time factor (T) A dimensionless factor, used in the Terzaghi theory of consolidation or swelling of cohesive soil.

total stress The effective stress plus the pore water pressure. The vertical total stress for uniform soil and a level ground surface can be calculated by multiplying the total unit weight of the soil by the depth below ground surface.

underconsolidation The condition that exists if a soil deposit is not fully consolidated under the existing overburden pressure and excess pore water pressures exist within the soil. Underconsolidation occurs in areas where a cohesive soil is being deposited very rapidly and not enough time has elapsed for the soil to consolidate under its own weight.

void ratio The volume of voids divided by the volume of soil solids.

GLOSSARY 4

CONSTRUCTION AND GRADING TERMINOLOGY

aggregate A granular material used for a pavement base, wall backfill, etc.

coarse aggregate Gravel or crushed rock that is retained on the No. 4 sieve (4.75 mm).

fine aggregate Often refers to sand (passes the No. 4 sieve and is retained on the No. 200 U.S. standard sieve).

open graded aggregate Generally refers to a gravel that does not contain any soil particles finer than those retained by the No. 4 sieve.

apparent opening size For a geotextile, a property that indicates the approximate largest particle that would effectively pass through the geotextile.

approval A written engineering or geologic opinion by the responsible engineer, geologist of record, or responsible principal of the engineering company concerning the process and completion of the work unless it specifically refers to the building official.

approved plans The current grading plans that bear the stamp of approval of the building official.

approved testing agency A facility whose testing operations are controlled and monitored by a registered civil engineer and which is equipped to perform and certify the tests as required by the local building code or building official.

as-graded (or as-built) The surface conditions at the completion of grading.

asphalt A dark brown to black cementitious material in which the main ingredient is bitumen (high-molecular-weight hydrocarbons) that occurs in nature or is obtained from petroleum processing.

asphalt concrete (AC) A mixture of asphalt and aggregate that is compacted into a dense pavement surface. Asphalt concrete is often prepared in a batch plant.

backdrain Generally a pipe-and-gravel or similar drainage system placed behind earth-retaining structures such as buttresses, stabilization fills, and retaining walls. See App. B, Standard Detail No. 4.

backfill Soil material placed behind or on top of an area that has been excavated. For example, backfill is placed behind retaining walls and in utility trench excavations.

base course or base A layer of specified or selected material of planned thickness constructed on the subgrade or subbase for the purpose of providing support to the overlying concrete or asphalt concrete surface of roads and airfields.

bench A relatively level step excavated into earth material on which fill is to be placed.

berm A raised bank or path of soil. For example, a berm is often constructed at the top of slopes to prevent water from flowing over the top of the slope.

borrow Earth material acquired from an off-site location for use in grading on a site.

brooming The crushing or separation of wood fibers at the butt (top of the pile) of a timber pile while it is being driven.

building official The city engineer, director of the local building department, or a duly delegated representative.

bulking The increase in volume of soil or rock caused by its excavation. For example, rock or dense soil will increase in volume upon excavation or by being dumped into a truck for transportation.

buttress fill A fill mass, the configuration of which is designed by engineering calculations to stabilize a slope exhibiting adverse geologic features. A buttress is generally specified by minimum key width and depth and by maximum backcut angle. A buttress normally contains a backdrainage system. See App. B, Standard Detail No. 3.

caisson Sometimes large-diameter piers are referred to as caissons. Another definition is a large structural chamber utilized to keep soil and water from entering into a deep excavation or construction area. Caissons may be installed by being sunk in place or by excavating the bottom of the unit as it slowly sinks to the desired depth.

Cat Slang for Caterpillar grading or construction equipment.

clearing, brushing, and grubbing The removal of vegetation (grass, brush, trees, and similar plant types) by mechanical means.

clogging For a geotextile, a decrease in permeability due to soil particles that have either logged in the geotextile openings or have built up a restrictive layer on the surface of the geotextile.

compaction The densification of a fill by mechanical means.

compaction equipment Compaction equipment can be grouped generally into five different types or classifications: sheepsfoot, vibratory, pneumatic, high-speed tamping foot, and chopper wheels (for municipal landfill). Combinations of these types are also available.

compaction production Rate of compaction expressed in compacted cubic meters (m^3) or compacted cubic yards (yd^3) per hour.

concrete A mixture of aggregates (sand and gravel) and paste (portland cement and water). The paste binds the aggregates together into a rocklike mass as the paste hardens because of the chemical reactions between the cement and the water.

contractor A person or company under contract or otherwise retained by the client to perform demolition, grading, and other site improvements.

cut-fill transition The location in a building pad where on one side the pad has been cut down exposing natural or rock material, while on the other side, fill has been placed. See App. B, Standard Detail No. 6 for one method to deal with cut-fill transition lots.

dam A structure built to impound water or other fluid products such as tailing waste and wastewater effluent.

homogeneous earth dam An earth dam whose embankment is formed of one soil type without a systematic zoning of fill materials.

zoned earth dam An earth dam embankment zoned by the systematic distribution of soil types according to their strength and permeability characteristics, usually with a central impervious core and shells of coarser materials.

debris All products of clearing, grubbing, and demolition, or contaminated soil materials, that are unsuitable for reuse as compacted fill and/or any other material so designated by the geotechnical engineer or building official.

dewatering The process used to remove water from a construction site, such as pumping from wells, in order to lower the groundwater table during a foundation excavation.

dozer Slang for bulldozer construction equipment.

drainage The removal of surface water from the site. See App. B, Standard Detail No. 9, for typical lot drainage specifications.

drawdown The lowering of the groundwater table that occurs in the vicinity of a well that is in the process of being pumped.

earth material Any rock, natural soil, or fill, or any combination thereof.

electroosmosis A method of dewatering, applicable for silts and clays, in which an electric field is established in the soil mass to cause the movement by electroosmotic forces of pore water to wellpoint cathodes.

erosion control devices (temporary) Devices which are removable and can rarely be salvaged for subsequent reuse. In most cases they will last no longer than one rainy season. They include sandbags, gravel bags, plastic sheeting (visqueen), silt fencing, straw bales, and similar items.

erosion control system A combination of desilting facilities and erosion protection, including effective planting to protect adjacent private property, watercourses, public facilities, and receiving waters from any abnormal deposition of sediment or dust.

excavation The mechanical removal of earth material.

fill A deposit of earth material placed by artificial means. An engineered (or structural) fill refers to a fill in which the geotechnical engineer has, during grading, made sufficient tests to conclude that the fill has been placed in substantial compliance with the recommendations of the geotechnical engineer and the governing agency requirements. See App. B, Standard Detail No. 5, for typical canyon fill placement specifications.

hydraulic fill A fill placed by transporting soils through a pipe using large quantities of water. These fills are generally loose because they have little or no mechanical compaction during construction.

footing A structural member typically installed at a shallow depth that is used to transmit structural loads to the soil or rock stratum. Common types of footings include combined footings, spread (or pad) footings, and strip (or wall) footings.

forms Usually made of wood, forms are used during the placement of concrete. Forms confine and support the fluid concrete as it hardens.

foundation That part of the structure that supports the weight of the structure and transmits the load to underlying soil or rock.

deep foundation A foundation that derives its support by transferring loads to soil or rock at some depth below the structure.

shallow foundation A foundation that derives its support by transferring load directly to soil or rock at a shallow depth.

freeze Also known as setup, an increase in the load capacity of a pile after it has been driven. Freeze is caused primarily by the dissipation of excess pore water pressures.

geosynthetic A planar product manufactured from polymeric material and typically placed in soil to form an integral part of a drainage, reinforcement, or stabilization system. Types include geotextiles, geogrids, geonets, and geomembranes.

geotextile A permeable geosynthetic composed solely of textiles.

grade The vertical location of the ground surface.

existing grade The ground surface prior to grading.

finished grade The final grade of the site which conforms to the approved plan.

lowest adjacent grade Adjacent to the structure, the lowest point of elevation of the finished surface of the ground, paving, or sidewalk.

natural grade The ground surface unaltered by artificial means.

rough grade The stage at which the grade approximately conforms to the approved plan.

grading Any operation consisting of excavation, filling, or combination thereof.

grading contractor A contractor licensed and regulated who specializes in grading work or is otherwise licensed to do grading work.

grading permit An official document or certificate issued by the building official authorizing grading activity as specified by approved plans and specifications.

grouting The process of injecting grout into soil or rock formations to change their physical characteristics. Common examples include grouting to decrease the permeability of a soil or rock stratum, and compaction grouting to densify loose soil or fill.

hillside site A site that entails cut and/or fill grading of a slope which may be adversely affected by drainage and/or stability conditions within or outside the site, or which may cause an adverse affect on adjacent property.

jetting The use of a water jet to facilitate the installation of a pile. It can also refer to the fluid placement of soil, such as jetting in the soil for a utility trench.

key A designed compacted fill placed in a trench excavated in earth material beneath the toe of a proposed fill slope.

keyway An excavated trench into competent earth material beneath the toe of a proposed fill slope.

lift During compaction operations, a lift is a layer of soil that is dumped by the construction equipment and then subsequently compacted as structural fill.

necking A reduction in cross-sectional area of a drilled shaft as a result of the inward movement of the adjacent soils.

owner Any person, agency, firm, or corporation having a legal or equitable interest in a given real property.

permanent erosion control devices Improvements which remain throughout the life of the development. They include terrace drains, downdrains, slope landscaping, channels, storm drains, etc.

permit An official document or certificate issued by the building official authorizing performance of a specified activity.

pier A deep foundation system, similar to a cast-in-place pile, that consists of columnlike reinforced-concrete members. Piers are often of large enough diameter to enable downhole inspection. Piers are also commonly referred to as drilled shafts, bored piles, or drilled caissons.

pile A deep foundation system, consisting of relatively long, slender, columnlike members that are often driven into the ground.

> **batter pile** A pile driven in at an angle inclined to the vertical to provide higher resistance to lateral loads.
>
> **combination end-bearing and friction pile** A pile that derives its capacity from combined end-bearing resistance developed at the pile tip and frictional and/or adhesion resistance on the pile perimeter.

end-bearing pile A pile whose support capacity is derived principally from the resistance of the foundation material on which the pile tip rests.

friction pile A pile whose support capacity is derived principally from the resistance of the soil friction and/or adhesion mobilized along the side of the embedded pile.

pozzolan For concrete mix design, a siliceous or siliceous and aluminous material that will chemically react with calcium hydroxide within the cement paste to form compounds having cementitious properties.

precise grading permit A permit that is issued on the basis of approved plans that show the precise structure location, finish elevations, and all on-site improvements.

relative compaction The degree of compaction (expressed as a percentage) defined as the field dry density divided by the laboratory maximum dry density.

ripping or rippability The characteristic of rock or dense and rocky soils by which they can be excavated without blasting. Ripping is accomplished by using equipment such as a Caterpillar ripper, ripper-scarifier, tractor-ripper, or impact ripper. Ripper performance has been correlated with the seismic wave velocity of the soil or rock (see *Caterpillar Performance Handbook* 1997).

riprap Rocks that are generally less than 1800 kg (2 tons) in mass that are placed on the ground surface, on slopes or at the toe of slopes, or on top of structures to prevent erosion by wave action or strong currents.

running soil or running ground In tunneling or trench excavations, a granular material that tends to flow or "run" into the excavation.

sand boil The ejection of sand at ground surface, usually in a cone shape, caused by underground piping.

shear key Similar to a buttress; however, it is generally constructed by excavating a slot within a natural slope in order to stabilize the upper portion of the slope without grading encroachment into the lower portion of the slope. A shear key is also often used to increase the factor of safety of an ancient landslide.

shotcrete Mortar or concrete pumped through a hose and projected at high velocity onto a surface. Shotcrete can be applied by a "wet" or "dry" mix method.

shrinkage factor When the loose material is worked into a compacted state, the shrinkage factor (SF) is the ratio of the volume of compacted material to the volume of borrow material.

site The particular lot or parcel of land where grading or other development is performed.

slope An inclined ground surface. For graded slopes, the steepness is generally specified as a ratio of horizontal:vertical (e.g., 2:1 slope). Common types of slopes include natural (unaltered) slopes, cut slopes, false slopes (temporary slopes generated during fill compaction operations), and fill slopes.

slough Loose, noncompacted fill material generated during grading operations. Slough can also refer to a shallow slope failure, such as sloughing of the slope face.

slump In the placement of concrete, the slump is a measure of consistency of freshly mixed concrete as measured by the slump test. In geotechnical engineering, a slump could also refer to a slope failure.

slurry seal In the construction of asphalt pavements, a slurry seal is a fluid mixture of bituminous emulsion, fine aggregate, mineral filler, and water. A slurry seal is applied to the top surface of an asphalt pavement in order to seal its surface and prolong its wearing life.

soil stabilization The treatment of soil to improve its properties. There are many methods of soil stabilization such as adding gravel, cement, or lime to the soil. The soil could also be stabilized by using geotextiles, by drainage, or by compaction.

specification A precise statement in the form of specific requirements. The requirements could be applicable to a material, product, system, or engineering service.

stabilization fill Similar to a buttress fill, the configuration of which is typically related to slope height and is specified by the standards of practice for enhancing the stability of locally adverse conditions. A stabilization fill is normally specified by minimum key width and depth and by maximum backcut angle. A stabilization fill usually has a backdrainage system. See App. B, Standard Detail No. 3.

staking During grading, staking is the process in which a land surveyor places wood stakes that indicate the elevation of existing ground surface and the final proposed elevation according to the grading plans.

structure A structure is defined as that which is built or constructed, an edifice or building of any kind, or any piece of work artificially built up or composed of parts joined together in some definite manner.

subdrain (for canyons) A pipe-and-gravel or similar drainage system placed in the alignment of canyons or former drainage channels. After placement of the subdrain, structural fill is placed on top of the subdrain (see App. B, Standard Detail No. 1).

subgrade For roads and airfields, the subgrade is defined as the underlying soil or rock that supports the pavement section (subbase, base, and wearing surface). The subgrade is also referred to as the basement soil or foundation soil.

substructure and superstructure The substructure is the foundation and the superstructure is the portion of the structure located above the foundation (includes beams, columns, floors, and other structural and architectural members).

sulfate (SO_4) A chemical compound occurring in some soils that, at above certain levels of concentration, has a corrosive effect on ordinary portland cement concrete and some metals.

sump A small pit excavated in the ground or through the basement floor to serve as a collection basin for surface runoff or groundwater. A sump pump is used to periodically drain the pit when it fills with water.

tack coat In the construction of asphalt pavements, the tack coat is a bituminous material that is applied to an existing surface to provide a bond between different layers of the asphalt concrete.

tailings In terms of grading, tailings are nonengineered fill which accumulates on or adjacent to equipment haul roads.

terrace A relatively level step constructed in the face of a graded slope surface for drainage control and maintenance purposes.

underpinning Piles or other types of foundations built to provide new support for an existing foundation. Underpinning is often used as a remedial measure.

vibrodensification The densification or compaction of cohesionless soils by imparting vibrations into the soil mass so as to rearrange soil particles, resulting in less voids in the overall mass.

walls

bearing wall Any metal or wood stud walls that support more than 100 lb per linear foot of superimposed load. Any masonry or concrete wall that supports more than 200 lb per linear foot of superimposed load or is more than one story (*Uniform Building Code* 1997).

cutoff wall The construction of tight sheeting or a barrier of impervious material extending downward to an essentially impervious lower boundary to intercept and block the path of groundwater seepage. Cutoff walls are often used in dam construction.

retaining wall A wall designed to resist the lateral displacement of soil or other materials.

water-cement ratio For concrete mix design, the ratio of the mass of water (exclusive of that part absorbed by the aggregates) to the mass of cement.

well point During the pumping of groundwater, the well point is the perforated end section of a well pipe where the groundwater is drawn into the pipe.

windrow A string of large rock buried within engineered fill in accordance with guidelines set forth by the geotechnical engineer or governing agency requirements. See App. B, Standard Detail No. 7.

workability of concrete The ability to manipulate a freshly mixed quantity of concrete with a minimum loss of homogeneity.

GLOSSARY 5

TERMINOLOGY FOR SEWAGE DISPOSAL AND PERCOLATION TESTS

bell-and-spigot pipe A type of pipe that is expanded at one end (bell) to receive the unexpanded end (spigot) of the next section. With the assistance of a proper sealing material, a tight joint is formed at the union of the two sections.

cesspool A term which is often used interchangeably with seepage pit. A cesspool is usually defined as a pit that receives unsettled sewage.

community well A well used for domestic water supply by more than one dwelling.

digestion The natural biological process that breaks down the organic matter into its components or basic parts.

distribution box A container that distributes or divides the flow of effluent from a septic tank into several seepage pits.

domestic well A well used for drinking and cooking purposes.

effluent The liquid portion of sewage that flows out of a septic tank.

field lateral Lengths of pipe joined together to form a line.

impervious soil For sewage disposal systems, impervious soil refers to compact or clayey soil that has a very low percolation rate or does not permit the infiltration of water.

invert The floor, bottom, or lowest portion of the internal cross section of a closed conduit.

leaching lines Tile or perforated pipes that permit the percolation (leaching) of effluent into the ground.

open joint tile Pipes without tight joints.

percolation The movement of effluent into or through the soil.

percolation test A field test that is used to determine the percolation rate of water according to specified test procedures (see App. C).

porous soil For sewage disposal systems, porous soil has a high permeability that readily permits percolation of effluent into and through the soil.

seepage pit A hole in the ground that is loosely walled and receives sewage liquids from a septic tank. The seepage pit permits the sewage liquids to filter into the ground.

sewage spring Underground sewage water that reaches the surface of the ground or flows out of the ground surface.

sidewalls The sides of the seepage pit, where most of the percolation takes place.

sludge The solid part of sewage which settles out and digests in a septic tank.

soil absorption system Tile line or a seepage pit that permits the sewage liquids to filter into the ground.

starters or activators Substances or compounds that assist in carrying out digestion in a septic tank.

tight line A solid-walled pipe (i.e., no perforations) that has tight joints (such as a bell-and-spigot pipe).

tile line Plain end tile line laid with loose joints in a trench that has been partly filled with gravel or crushed stone. The tile is also surrounded and covered with loose gravel or crushed stone and permits sewage liquid to filter into the ground over a large area.

APPENDIX B
EXAMPLE OF GRADING SPECIFICATIONS

A. GENERAL

Al. The enclosed document consists of grading recommendations and standard details. This information should be considered to be a part of the project specifications.

A2. The contractor should not vary from these specifications without prior recommendation by the geotechnical engineer and the approval of the client or the authorized representative. Recommendations by the geotechnical engineer and/or client should not be considered to preclude requirements issued by the local building department.

A3. These grading specifications may be modified and/or superseded by recommendations contained in the text of the preliminary geotechnical report and/or subsequent reports.

A4. If disputes arise out of the interpretation of these grading specifications, the geotechnical engineer shall provide the governing interpretation.

B. OBLIGATIONS OF PARTIES

B1. The geotechnical engineer should provide observation and testing services and should make evaluations to advise the client on geotechnical matters. The geotechnical engineer should report the findings and recommendations to the client or the authorized representative.

B2. The client should be chiefly responsible for all aspects of the project. The client or authorized representative has the responsibility of reviewing the findings and recommendations of the geotechnical engineer. The client shall authorize or cause to have authorized the contractor and/or other consultants to perform work and/or provide services. During grading the client or the authorized representative should remain on-site or should remain reasonably accessible to all concerned parties in order to make decisions necessary to maintain the flow of the project.

B3. The contractor should be responsible for the safety of the project and satisfactory completion of all grading and other associated operations on construction projects, including, but not limited to, earthwork in accordance with the project plans, specifications and controlling agency requirements. During grading, the contractor or the authorized representative should remain on-site. Overnight and on days off, the contractor should remain accessible.

C. SITE PREPARATION

C1. The client, prior to any site preparation or grading, should arrange and attend a meeting among the grading contractor, the design structural engineer, the geotechnical engineer, representatives of the local building department, as well as any other concerned parties. All parties should be given at least 48 hours notice.

C2. Clearing and grubbing should consist of the removal of vegetation such as brush, grass, woods, stumps, trees, roots of trees, and otherwise deleterious natural materials from the areas to be graded. Clearing and grubbing should extend to the outside of all proposed excavation and fill areas.

C3. Demolition should include removal of buildings, structures, foundations, reservoirs, utilities (including underground pipelines, septic tanks, leach fields, seepage pits, cisterns, mining shafts, tunnels, etc.) and other man-made surface and subsurface improvements from the areas to be graded. Demolition of utilities should include proper capping and/or rerouting pipelines at the project perimeter and cutoff and capping of wells in accordance with the requirements of the local building department and the recommendations of the geotechnical engineer at the time of demolition.

C4. Trees, plants, or man-made improvements not planned to be removed or demolished should be protected by the contractor from damage or injury.

C5. Debris generated during clearing, grubbing, and/or demolition operations should be wasted from areas to be graded and disposed off-site.

Clearing, grubbing, and demolition operations should be performed under the observation of the geotechnical engineer.

C6. The client or contractor should obtain the required approvals from the local building department for the project prior, during, and/or after demolition, site preparation, and removals. The appropriate approvals should be obtained prior to proceeding with grading operations.

D. SITE PROTECTION

Dl. Protection of the site during the period of grading should be the responsibility of the contractor. Unless other provisions are made in writing and agreed upon among the concerned parties, completion of a portion of the project should not be considered to preclude that portion or adjacent areas from the requirements for site protection until such time as the entire project is complete as identified by the geotechnical engineer, the client, and the local building department.

D2. The contractor should be responsible for the stability of all temporary excavations. Recommendations by the geotechnical engineer pertaining to temporary excavations (e.g., backcuts) are made in consideration of stability of the completed project and, therefore, should not be considered to preclude the responsibilities of the contractor.

D3. Precautions should be taken during the performance of site clearing, excavations, and grading to protect the work site from flooding, ponding, or inundation by poor or improper surface drainage. Temporary provisions should be made during the rainy season to adequately direct surface drainage away from and off the work site. Where low areas can not be avoided, pumps should be kept on hand to continually remove water during periods of rainfall.

D4. During periods of rainfall, plastic sheeting should be kept reasonably accessible to prevent unprotected slopes from becoming saturated. Where necessary during periods of rainfall, the contractor should install checkdams, desilting basins, riprap, sand bags, or other devices or methods necessary to control erosion and provide safe conditions.

D5. During periods of rainfall, the geotechnical engineer should be kept informed by the contractor as to the nature of remedial or preventive work being performed (e.g., pumping, placement of sandbags or plastic sheeting, other labor, dozing, etc.).

D6. Following periods of rainfall, the contractor should contact the geotechnical engineer and arrange a walk-through of the site in order to visually assess rain-related damage. The geotechnical engineer may also recommend excavations and testing in order to aid in the assessments. At the request of the geotechnical engineer, the contractor shall make excavations in order to evaluate the extent of rain-related damage.

D7. Rain-related damage should be considered to include, but may not be limited to, erosion, silting, saturation, swelling, structural distress and other adverse conditions identified by the geotechnical engineer. Soil adversely affected should be classified as unsuitable materials and should be subject to overexcavation and replacement with compacted fill or other remedial grading as recommended by the geotechnical engineer.

D8. Relatively level areas, where saturated soils and/or erosion gullies exist to depths of greater than 1.0 foot, should be overexcavated to unaffected, competent material. Where less than 1.0 foot in depth, unsuitable materials may be processed in place to achieve near-optimum moisture conditions, then thoroughly recompacted in accordance with the applicable specifications. If the desired results are not achieved, the affected materials should be overexcavated, then replaced in accordance with the applicable specifications.

D9. In slope areas, where saturated soil and/or erosion gullies exist to depths of greater than 1.0 foot, they should be overexcavated and replaced as compacted fill in accordance with the applicable specifications. Where affected materials exist to depths of 1.0 foot or less below proposed finished grade, remedial grading by moisture conditioning in place, followed by thorough recompaction in accordance with these grading specifications, may be attempted. If the desired results are not achieved, all affected materials should be overexcavated and replaced as compacted fill in accordance with the slope repair recommendations herein. As field conditions dictate, other slope repair procedures may be recommended by the geotechnical engineer.

E. EXCAVATIONS

E1. Unsuitable Materials

E1.1. Materials which are unsuitable should be excavated under observation and recommendations of the geotechnical engineer. Unsuitable materials include, but may not be limited to: (1) dry, loose, soft, wet, organic, or compressible natural soils; (2) fractured, weathered, or soft bedrock; (3) nonengineered fill; and (4) other deleterious fill materials.

E1.2. Material identified by the geotechnical engineer as unsatisfactory because of its moisture conditions should be overexcavated, watered or dried as needed, and thoroughly blended to a uniform near-optimum moisture condition prior to placement as compacted fill.

E2. Cut Slopes

E2.1. Unless otherwise recommended by the geotechnical engineer and approved by the local building department, permanent cut slopes should not be steeper than 2:1 (horizontal:vertical).

E2.2. If excavations for cut slopes expose loose, cohesionless, significantly fractured or otherwise unsuitable material, overexcavation and replacement of the unsuitable materials with a compacted stabilization fill should be accomplished as recommended by the geotechnical engineer. Unless otherwise specified by the geotechnical engineer, stabilization fill construction should conform to the requirements of Standard Detail No. 3.

E2.3. The geotechnical engineer should review cut slopes during excavation. The geotechnical engineer should be notified by the contractor prior to beginning slope excavations.

E2.4. If, during the course of grading, adverse or potentially adverse geotechnical or geologic conditions are encountered which were not anticipated in the preliminary report, the geotechnical engineer or engineering geologist should explore, analyze, and make recommendations to treat these problems.

E2.5. When cut slopes are made in the direction of the prevailing drainage, a nonerodible diversion swale (brow ditch) should be provided at the top-of-cut.

E3. Pad Areas

E3.1. All lot pad areas having cut-fill transitions in the building footprint should be overexcavated to provide for a minimum of 3 feet of compacted fill over the entire pad area (see Standard Detail No. 6). Cut areas exposing significantly varying material types should also be overexcavated to provide for at least a 3-foot-thick compacted fill blanket. Geotechnical conditions may require greater depth of overexcavation. The actual depth should be determined by the geotechnical engineer during grading.

E3.2. For pad areas created above cut or natural slopes, positive drainage should be established away from the top-of-slope. This may be accomplished by utilizing a berm and/or an appropriate pad gradient. A gradient in soil areas away from the top-of-slopes of 2 percent or greater is recommended.

F. COMPACTED FILL

All fill materials should be compacted as specified below or by other methods specifically recommended by the geotechnical engineer. Unless otherwise specified, the minimum degree of compaction (relative compaction) should be 90 percent of the laboratory maximum dry density (Modified Proctor).

F1. Placement

F1.1. Prior to placement of compacted fill, the contractor should request a review by the geotechnical engineer of the exposed ground surface. Unless otherwise recommended, the exposed ground surface should then be scarified (6 inches minimum), watered or dried as needed, thoroughly blended to achieve near-optimum moisture conditions, then thoroughly compacted to a minimum of 90 percent of the maximum dry density (Modified Proctor). The review by the geotechnical engineer should not be considered to preclude requirement of inspection and approval by the local building department.

F1.2. Compacted fill should be placed in thin horizontal lifts not exceeding 8 inches in loose thickness prior to compaction. Each lift should be watered or dried as needed, thoroughly blended to achieve near-optimum moisture conditions then thoroughly compacted by mechanical methods to a minimum of 90 percent of laboratory maximum dry density (Modified Proctor). Each lift should be treated in a like manner until the desired finished grades are achieved.

F1.3. The contractor should have suitable and sufficient mechanical compaction equipment and watering apparatus on the job site to handle the amount of fill being placed in consideration of moisture retention properties of the materials. If necessary, excavation equipment should be shut down temporarily in order to permit proper compaction of fills. Earth-moving equipment should be considered only a supplement and not substituted for conventional compaction equipment.

F1.4. In placing fill in horizontal lifts adjacent to areas sloping steeper than 5:1 (horizontal:vertical), horizontal keys and vertical benches should be excavated into the adjacent slope area. Keying and benching should be sufficient to provide at least 6-foot-wide benches and a minimum of 4 feet of vertical bench height within the firm natural ground, firm bedrock, or engineered compacted fill. No compacted fill should be placed in an area after keying and benching until the area has been reviewed by the geotechnical engineer. Material generated by the benching operation should be moved far enough away from the bench area to allow for the recommended review of the horizontal bench prior to placement of fill. Typical keying and benching details have been included in Standard Details Nos. 1 and 2.

F1.5. Within a single fill area where grading procedures dictate two or more separate fills, temporary slopes (false slopes) may be created. When placing fill adjacent to a false slope, benching should be conducted in the same manner as described above. At least a 3-foot vertical bench should be established within the firm core of adjacent approved compacted fill prior to placement of additional fill. Benching should proceed in at least 3-foot vertical increments until the desired finished grades are achieved.

F1.6. Fill should be tested for compliance with the recommended relative compaction and moisture conditions. Field density testing should conform to ASTM Method of Test D 1556 (Sand Cone), D 2922 (Nuclear Method), and D 2937 (Drive-Cylinder). Tests should be provided for about every 2 vertical feet or 1000 cubic yards of fill placed. Actual test intervals may vary as field conditions dictate. Fill found not to be in conformance with the grading recommendations should be removed or otherwise handled as recommended by the geotechnical engineer.

F1.7. The contractor should assist the geotechnical engineer or field technician by digging test pits for removal determinations or for testing compacted fill.

F1.8. As recommended by the geotechnical engineer, the contractor should shut down or remove grading equipment from an area being tested.

F1.9. The geotechnical engineer should maintain a plan with estimated locations of field tests. Unless the client provides for actual surveying of test locations, the estimated locations by the geotechnical engineer should be considered only rough estimates and should not be utilized for the purpose of preparing cross sections showing test locations or in any case for the purpose of after-the-fact evaluation of the sequence of fill placement.

F2. Moisture

F2.1. For field testing purposes, "near-optimum" moisture will vary with material type and other factors including compaction procedure. Near optimum may be specifically recommended in preliminary investigation reports and/or may be evaluated during grading. As a preliminary guideline, near optimum should be considered from 1 percent below to 3 percent above optimum.

F2.2. Prior to placement of additional compacted fill after an overnight or other grading delay, the exposed surface or previously compacted fill should be processed by scarification, watered or dried as needed, thoroughly blended to near-optimum moisture conditions, then recompacted to a minimum of 90 percent of laboratory maximum dry density (Modified Proctor). Where wet, dry, or other unsuitable materials exist to depths of greater than 1 foot, the unsuitable materials should be overexcavated.

F2.3. After a period of flooding, rainfall, or overwatering by other means, no additional fill should be placed until damage assessments have been made and remedial grading has been performed.

F3. Fill Material

F3.1. Excavated on-site materials which are acceptable to the geotechnical engineer may be utilized as compacted fill, provided trash, vegetation, and other deleterious materials are removed prior to placement.

F3.2. Where import materials are required for use on site, the geotechnical engineer should be notified at least 72 hours in advance of importing, in order to sample and test materials from proposed borrow sites. No import materials should be delivered for use on-site without prior sampling and testing by the geotechnical engineer.

F3.3. Where oversized rock or similar irreducible material is generated during grading, it is recommended, where practical, to waste such material off-site, or on-site in areas designated as "nonstructural rock disposal areas." Rock placed in disposal areas should be placed with sufficient fines to fill voids. The rock should be compacted in lifts to an unyielding condition. The disposal area should be covered with at least 3 feet of compacted fill that is free of oversized material. The upper 3 feet should be placed in accordance with these specifications for compacted fill.

F3.4. Rocks 12 inches in maximum dimension and smaller may be utilized within the compacted fill, provided they are placed in such manner that nesting of the rock is avoided. Fill should be placed and thoroughly compacted over and around all rock. The amount of rock should not exceed 40 percent by dry weight passing the ¾-inch sieve size. The 12-inch and 40 percent recommendations herein may vary as field conditions dictate.

F3.5. During the course of grading operations, rocks or similar irreducible materials greater than 12 inches maximum dimension (oversized material), may be generated. These rocks should not be placed within the compacted fill unless placed as recommended by the geotechnical engineer.

F3.6. Where rocks or similar irreducible materials of greater than 12 inches but less than 4 feet of maximum dimension are generated during grading, or otherwise desired to be placed within an engineered fill, special handling in accordance with Standard Detail No. 7 is recommended. Rocks greater than 4 feet should be broken down or disposed off-site. Rocks up to 4 feet maximum dimension should be placed below the upper 10 feet of any fill and should not be closer than 20 feet to any slope face. These recommendations could vary as locations of improvements dictate.

Where practical, the rocks should not be placed below areas where structures or deep utilities are proposed. The rocks should be placed in windrows on a clean, overexcavated or unyielding compacted fill or firm natural ground surface. Select native or imported granular soil (SE = 30 or higher) should be placed, and thoroughly flooded, over and around all windrowed rock, such that voids are filled. Windrows of large rocks should be staggered so that successive strata of the large rocks are not in the same vertical plane.

The contractor should be aware that the placement of rock in windrows will significantly slow the grading operation and may require additional equipment or special equipment.

F3.7. It may be possible to dispose of individual larger rock as field conditions dictate and as recommended by the geotechnical engineer at the time of placement.

F3.8. Material that is considered unsuitable by the geotechnical engineer should not be utilized in the compacted fill.

F3.9. During grading operations, placing and mixing the materials from the cut or borrow areas may result in soil mixtures which possess unique physical properties. Testing may be required of samples obtained directly from the fill areas in order to verify conformance with the specifications. Processing of these additional samples may take two or more working days. The contractor may elect to move the operation to other areas within the project, or may continue placing compacted fill, pending laboratory and field test results. Should the contractor use the second alternative, the fill may need to be removed and recompacted depending on the outcome of the laboratory and field tests.

F3.10. Any fill placed in areas not previously reviewed and evaluated by the geotechnical engineer may require removal and recompaction at the contractor's expense. Determination of overexcavation should be made upon review of field conditions by the geotechnical engineer.

F4. Fill Slopes

F4.1. Unless otherwise recommended by the geotechnical engineer and approved by the local building department, permanent fill slopes should not be steeper than 2:1 (horizontal:vertical).

F4.2. Except as specifically recommended otherwise, or as otherwise provided for in these grading specifications, compacted fill slopes should be overbuilt and cut back to grade, exposing the firm, compacted fill inner core. The actual amount of overbuilding may vary as field conditions dictate. If the desired results are not achieved, the existing slopes should be overexcavated and reconstructed according to the recommendation of the geotechnical engineer. The degree of overbuilding shall be increased until the desired compacted slope surface condition is achieved. Care should be taken by the contractor to provide thorough mechanical compaction to the outer edge of the overbuilt slope surface.

F4.3. Although no construction procedure produces a slope free from risk of future movement, overfilling and cutting back of slope to a compacted inner core is, given no other constraints, the most desirable procedure. Other constraints, however, must often be considered. These constraints may include property line situations, access, the critical nature of the development, and cost. Where such constraints are identified, slope face compaction on slopes of 2:1 or flatter may be attempted as a second-best alternative by conventional construction procedures including backrolling techniques upon specific recommendation by the geotechnical engineer.

Fill placement should proceed in thin lifts (i.e., 6- to 8-inch loose thickness). Each lift should be moisture-conditioned and thoroughly compacted. The desired moisture condition should be maintained or reestablished, where necessary, during the period between successive lifts. Selected lifts should be tested to ascertain that desired compaction is being achieved. Care should be taken to extend compactive effort to the outer edge of the slope.

Each lift should extend horizontally to the desired finished slope surface or more, as needed to ultimately establish desired grades. Grade during construction should not be allowed to roll off at the edge of the slope. It may be helpful to elevate slightly the outer edge of the slope. Slough resulting from the placement of individual lifts should not be allowed to drift down over previous lifts. At intervals not exceeding 4 feet in vertical slope height or the capability of available equipment, whichever is less, fill slopes should be thoroughly backrolled by a conventional sheepsfoot-type roller. Care should be taken to maintain the desired moisture conditions, and/or reestablishing them, as needed prior to backrolling. Upon achieving final grade, the slopes should again be moisture-conditioned and thoroughly backrolled. The use of a side-boom roller will probably be necessary and vibratory methods are strongly recommended. Without delay, so as to avoid (if possible) further moisture conditioning, the slopes should then be grid-rolled to achieve a relatively smooth surface and uniformly compact condition.

In order to monitor slope construction procedures, moisture and density tests should be taken at regular intervals. Failure to achieve the desired results will likely result in a recommendation by the geotechnical engineer to overexcavate the slope surfaces followed by reconstruction of the slopes utilizing overfilling and cutback procedures or further attempts at the conventional backrolling approach. Other recommendations may also be provided which would be commensurate with field conditions.

F4.4. Where placement of fill above a natural slope or above a cut slope is proposed, the fill slope configuration as presented in Standard Detail No. 2 should be adopted.

F4.5. For pad areas above fill slopes, positive drainage should be established away from the top-of-slope. This may be accomplished by utilizing a berm and pad gradients of at least 2 percent in soil areas. See Standard Detail No. 9.

F5. Off-Site Fill

F5.1. Off-site fill should be treated in the same manner as recommended in these specifications for site preparation, excavation, and compaction.

F5.2. Off-site canyon fill should be placed in preparation for future additional fill, as shown in the accompanying Standard Detail No. 5.

F5.3. Off-site fill subdrains temporarily terminated (up canyon) should be surveyed for future relocation and connection.

G. DRAINAGE

G1. Canyon subdrain systems specified by the geotechnical engineer should be installed in accordance with Standard Detail No. 1.

G2. Typical subdrains for a compacted fill buttress or slope stabilization fill should be installed in accordance with the specifications of Standard Details Nos. 3 or 4.

G3. Roof, pad, and slope drainage should be directed away from slopes and areas of structures to suitable disposal areas by nonerodible devices (i.e., gutters, downspouts, concrete swales).

G4. For drainage over soil areas immediately away from structures (i.e., within 4 feet), a minimum of 4 percent gradient should be maintained. Pad drainage of at least 2 percent should be maintained over soil areas as shown in Standard Detail No. 9. Pad drainage may be reduced to at least 1 percent for flatland projects. Flatland projects are defined as those projects where no natural or man-made slopes exist that are greater than 10 feet in height or steeper than 2:1 (horizontal:vertical) in slope ratio.

G5. Drainage patterns established at the time of fine grading should be maintained throughout the life of the project. Property owners should be made aware that altering drainage patterns can be detrimental to slope stability and foundation performance.

H. STAKING

H1. In all fill areas, the fill should be compacted prior to the placement of the stakes. This particularly is important on fill slopes. Slope stakes should not be placed until the slope is thoroughly compacted (backrolled). If stakes must be placed prior to the completion of compaction procedures, it must be recognized that they will be removed and/or demolished at such time as compaction procedures resume.

H2. In order to allow for remedial grading operations, which could include overexcavations or slope stabilization, appropriate staking offsets should be provided. For finished slope and stabilization backcut areas, at least a 10-foot setback is recommended from proposed toes and tops-of-cut.

I. MAINTENANCE

I1. Landscape Plants

In order to enhance surficial slope stability, slope planting should be accomplished at the completion of grading. Slope planting should consist of deep-rooting vegetation. Plants native to the area of grading are generally desirable. A landscape architect would be the best person to consult regarding types of plants and planting configuration.

I2. Irrigation

I2.1. Irrigation pipes should be anchored to slope faces, not placed in trenches excavated into slope faces.

I2.2. Slope irrigation should be minimized. If automatic timing devices are utilized on irrigation systems, provisions should be made for interrupting normal irrigation during periods of rainfall.

I2.3. Though not a requirement, consideration should be given to the installation of near-surface moisture-monitoring control devices. Such devices can aid in the maintenance of relatively uniform and reasonably constant moisture conditions.

I2.4. Property owners should be made aware that overwatering of slopes is detrimental to slope stability.

I3. Maintenance

I3.1. Periodic inspections of landscaped slope areas should be planned and appropriate measures should be taken to control weeds and enhance growth of the landscape plants. Some areas may require occasional replanting or reseeding.

I3.2. Terrace drains and downdrains should be periodically inspected and maintained free of debris. Damage to drainage improvements should be repaired immediately.

I3.3. Property owners should be made aware that burrowing animals can be detrimental to slope stability. A preventive program should be established to control burrowing animals.

I3.4. As a precautionary measure, plastic sheeting should be readily available, or kept on hand, to protect all slope areas from saturation by periods of heavy or prolonged rainfall. This measure is strongly recommended, beginning before landscape planting.

I4. Repairs

I4.1. If slope failures occur, the geotechnical engineer should be contacted for a field review of site conditions and development of recommendations for evaluation and repair.

I4.2. If slope failures occur as a result of exposure to periods of heavy rainfall, the failure area and currently unaffected areas should be covered with plastic sheeting to protect against additional saturation.

I4.3. Standard Detail No. 8 suggests appropriate repair procedures for surficial slope failures.

J. TRENCH BACKFILL

J1. Utility trench backfill should, unless otherwise recommended, be compacted by mechanical means. Unless otherwise recommended, the degree of compaction should be a minimum of 90 percent of the laboratory maximum dry density (Modified Proctor).

J2. As an alternative, granular material (sand equivalent greater than 30) may be thoroughly jetted in place. Jetting should be considered to apply only to trenches no greater than 2 feet in width and 4 feet in depth. Following jetting operations, trench backfill should be thoroughly mechanically compacted and wheel-rolled from the surface.

J3. Backfill of exterior and interior trenches extending below a 1:1 projection from the outer edge of foundations should be mechanically compacted to a minimum of 90 percent of the laboratory maximum dry density (Modified Proctor).

J4. Within slab areas, but outside the influence of foundations, trenches up to 1 foot wide and 2 feet deep may be backfilled with sand and densified by jetting, flooding, or by mechanical means. If on-site materials are utilized, they should be wheel-rolled, tamped, or otherwise compacted to a firm condition. For minor interior trenches, density testing may be deleted or spot testing may be elected if deemed necessary, on the basis of a review of backfill operations during construction.

J5. If utility contractors indicate that it is undesirable to use compaction equipment in close proximity to a buried conduit, the contractor may elect to use lightweight mechanical compaction equipment or to shade the conduit with clean granular material, which should be thoroughly jetted in place above the conduit, before initiating mechanical compaction procedures. Other methods of utility trench compaction may also be appropriate, on review by the geotechnical engineer at the time of construction.

J6. In cases where clean granular materials are proposed for use in lieu of native materials or where flooding or jetting is proposed, the procedures should be considered subject to review by the geotechnical engineer.

J7. Clean granular backfill and/or bedding are not recommended in slope areas unless provisions are made for a drainage system to mitigate the potential buildup of seepage forces.

K. STATUS OF GRADING

Prior to proceeding with any grading operation, the geotechnical engineer should be notified at least 2 working days in advance in order to schedule the necessary observation and testing services.

K1. Prior to any significant expansion or cutback in the grading operation, the geotechnical engineer should be provided with adequate notice (i.e., 2 days) in order to make appropriate adjustments in observation and testing services.

K2. Following completion of grading operations or between phases of a grading operation, the geotechnical engineer should be provided with at least 2 working days' notice in advance of commencement of additional grading operations.

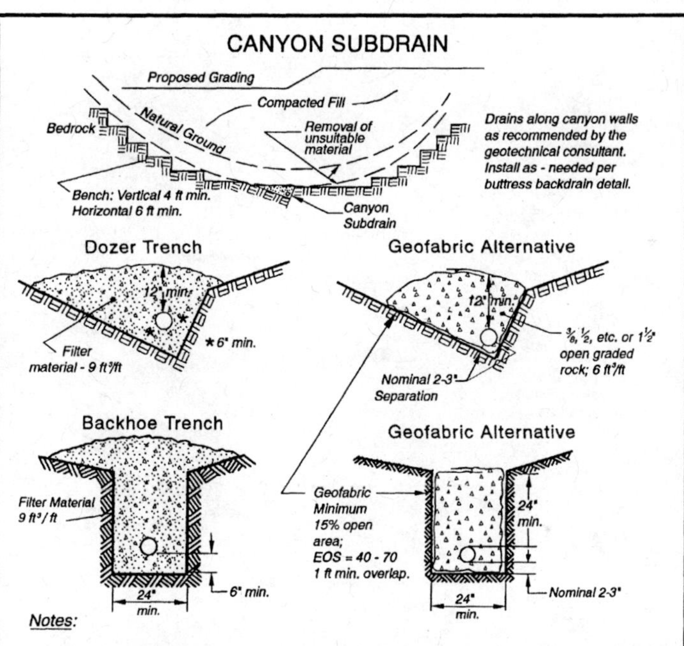

Notes:

1- Pipe be 4" min. diameter, 6" min. for runs of 500 ft to 1000 ft, 8" min. for runs of 1000ft. or greater.

2- Pipe should be schedule 40 PVC or similiar. Upstream ends should be capped.

3- Pipe should have 8 uniformly spaced 3/8" perforations per foot placed at 90° offset on underside of pipe. Final 20 feet of pipe should be nonperforated.

4- Filter material should be Calif. Class 2 Permable Material.

5- Appropriate gradiant should be provided for drainage; 2% minimum is recommended.

6- For the Geofabric Alternatives and gradients of 4% or greater, pipe may be omitted from the upper 500 ft. For runs of 500, 1000, and 1500 ft or greater 4", 6", and 8" pipe, respectively, should be provided.

STANDARD DETAIL NO. 1

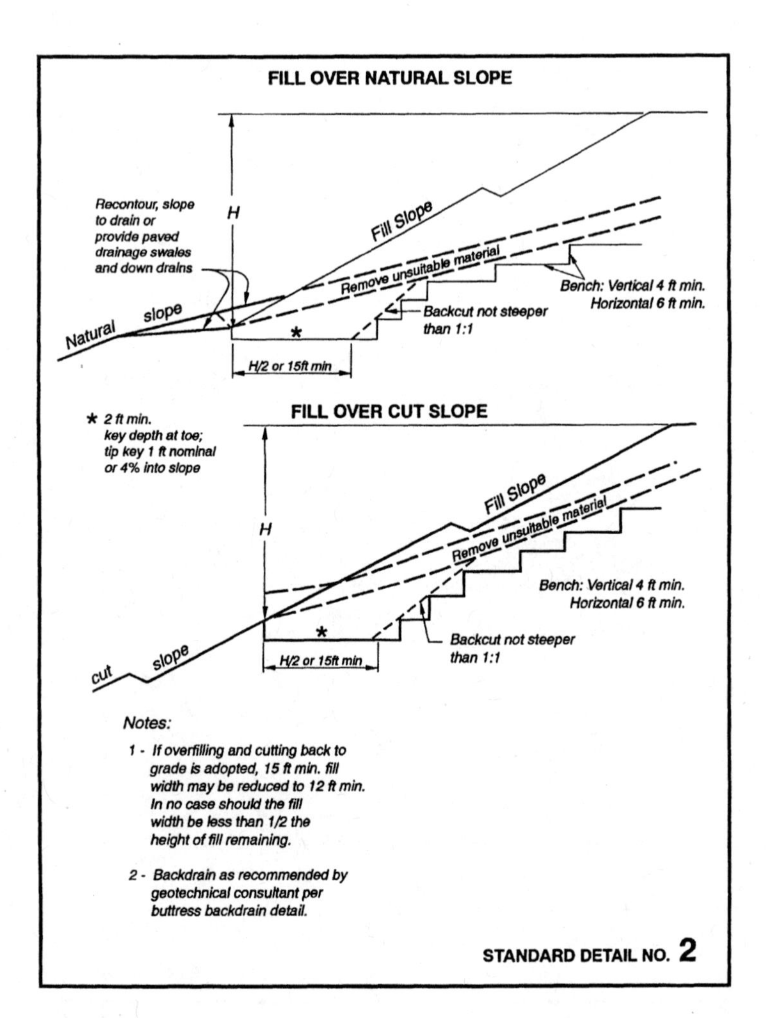
FILL OVER NATURAL SLOPE
Recontour, slope to drain or provide paved drainage swales and down drains
H
Fill Slope
Remove unsuitable material
Natural slope
*
Backcut not steeper than 1:1
Bench: Vertical 4 ft min. Horizontal 6 ft min.
H/2 or 15ft min
* 2 ft min. key depth at toe; tip key 1 ft nominal or 4% into slope
FILL OVER CUT SLOPE
H
Fill Slope
Remove unsuitable material
Bench: Vertical 4 ft min. Horizontal 6 ft min.
cut slope
*
H/2 or 15ft min
Backcut not steeper than 1:1
Notes:
1 - If overfilling and cutting back to grade is adopted, 15 ft min. fill width may be reduced to 12 ft min. In no case should the fill width be less than 1/2 the height of fill remaining.
2 - Backdrain as recommended by geotechnical consultant per buttress backdrain detail.
STANDARD DETAIL NO. 2

STABILIZATION FILL

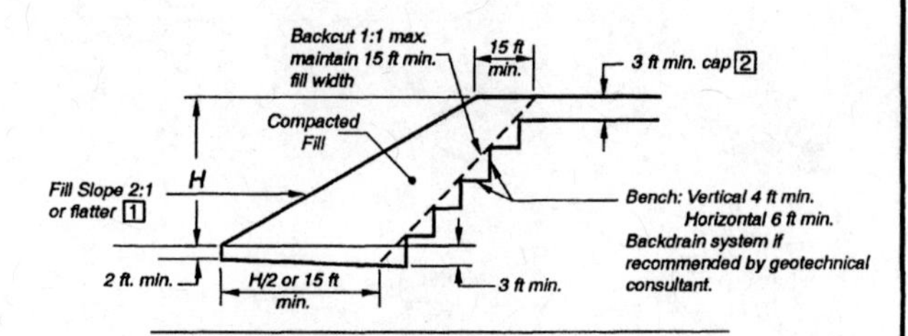

BUTTRESS FILL

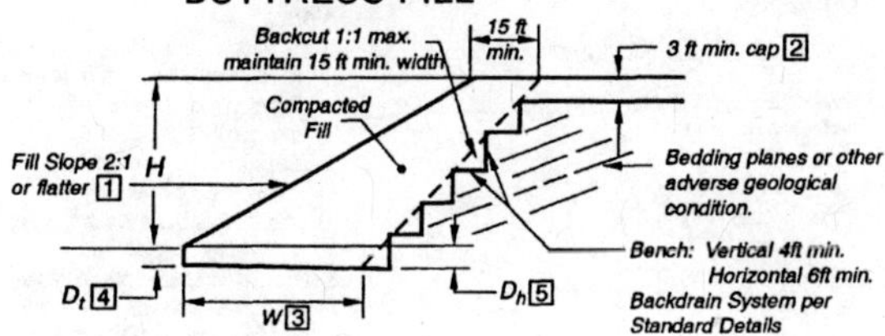

Notes:

1 - *If overfilling and cutting back to grade is adopted, 15 ft may be reduced to 12 ft. In no case should the fill width be less than 1/2 the fill height remaining.*

2 - *A 3 ft blanket fill shall be provided above stabilization and buttress fills. The thickness may be greater as recommended by the geotechnical engineer.*

3 - *W = designed width of key.*

4 - *D_t = designed depth of key at toe*

5 - *D_h = depth of key at heel; unless otherwise specified, $D_h = D_t + 1ft$*

STANDARD DETAIL NO. 3

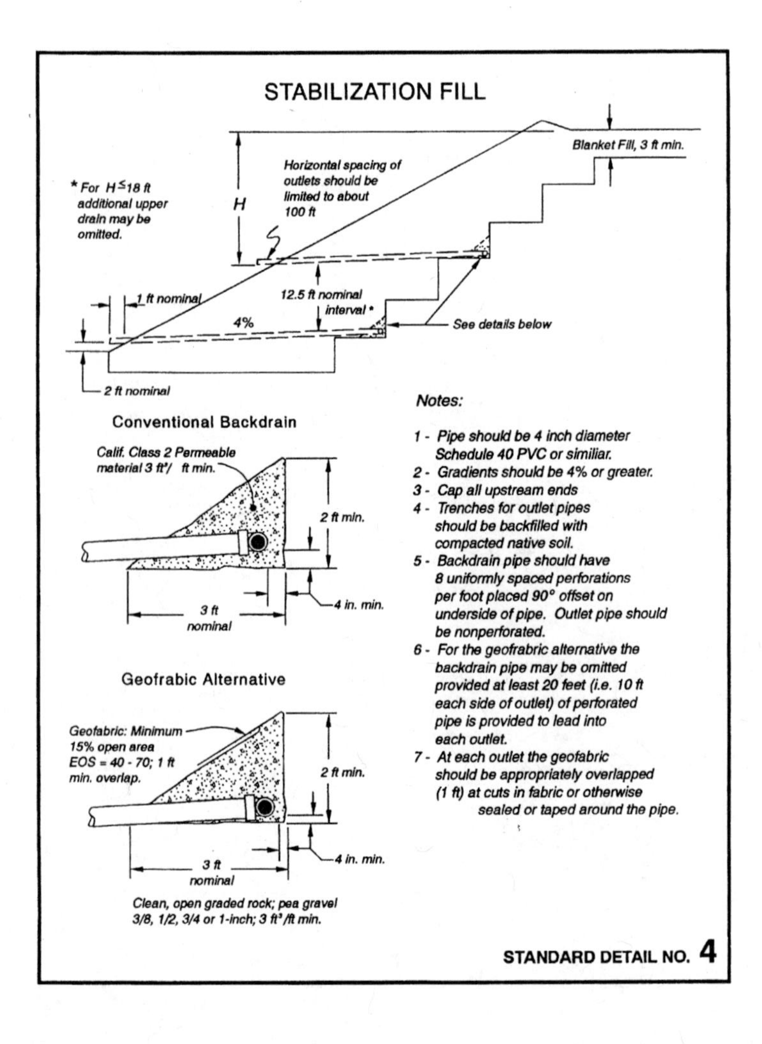
STABILIZATION FILL
Blanket Fill, 3 ft min.
Horizontal spacing of outlets should be limited to about 100 ft
* For H ≤ 18 ft additional upper drain may be omitted.
H
1 ft nominal
12.5 ft nominal interval *
4%
See details below
2 ft nominal
Conventional Backdrain
Calif. Class 2 Permeable material 3 ft³/ ft min.
2 ft min.
4 in. min.
3 ft nominal
Geofrabic Alternative
Geofabric: Minimum 15% open area EOS = 40 - 70; 1 ft min. overlap.
2 ft min.
4 in. min.
3 ft nominal
Clean, open graded rock; pea gravel 3/8, 1/2, 3/4 or 1-inch; 3 ft³/ft min.
Notes:
1 - Pipe should be 4 inch diameter Schedule 40 PVC or similiar.
2 - Gradients should be 4% or greater.
3 - Cap all upstream ends
4 - Trenches for outlet pipes should be backfilled with compacted native soil.
5 - Backdrain pipe should have 8 uniformly spaced perforations per foot placed 90° offset on underside of pipe. Outlet pipe should be nonperforated.
6 - For the geofrabric alternative the backdrain pipe may be omitted provided at least 20 feet (i.e. 10 ft each side of outlet) of perforated pipe is provided to lead into each outlet.
7 - At each outlet the geofabric should be appropriately overlapped (1 ft) at cuts in fabric or otherwise sealed or taped around the pipe.
STANDARD DETAIL NO. 4

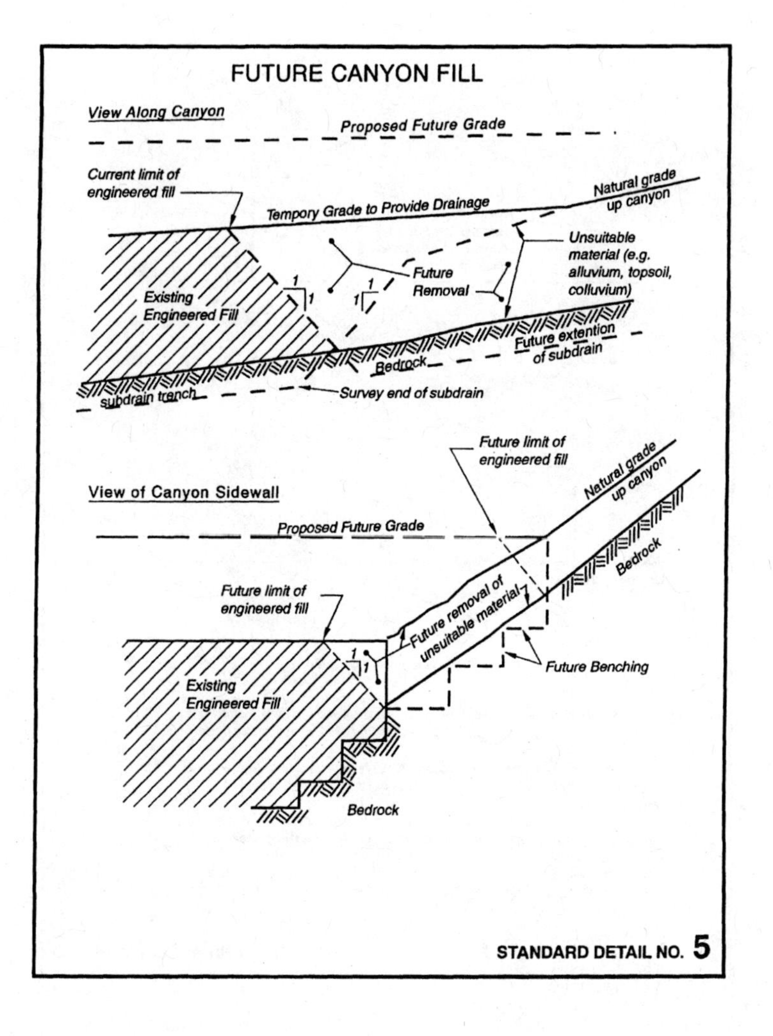
FUTURE CANYON FILL
View Along Canyon
Proposed Future Grade
Current limit of engineered fill
Tempory Grade to Provide Drainage
Natural grade up canyon
Unsuitable material (e.g. alluvium, topsoil, colluvium)
Future Removal
Existing Engineered Fill
Bedrock
Future extention of subdrain
subdrain trench
Survey end of subdrain
View of Canyon Sidewall
Future limit of engineered fill
Natural grade up canyon
Proposed Future Grade
Bedrock
Future limit of engineered fill
Future removal of unsuitable material
Future Benching
Existing Engineered Fill
Bedrock
STANDARD DETAIL NO. 5

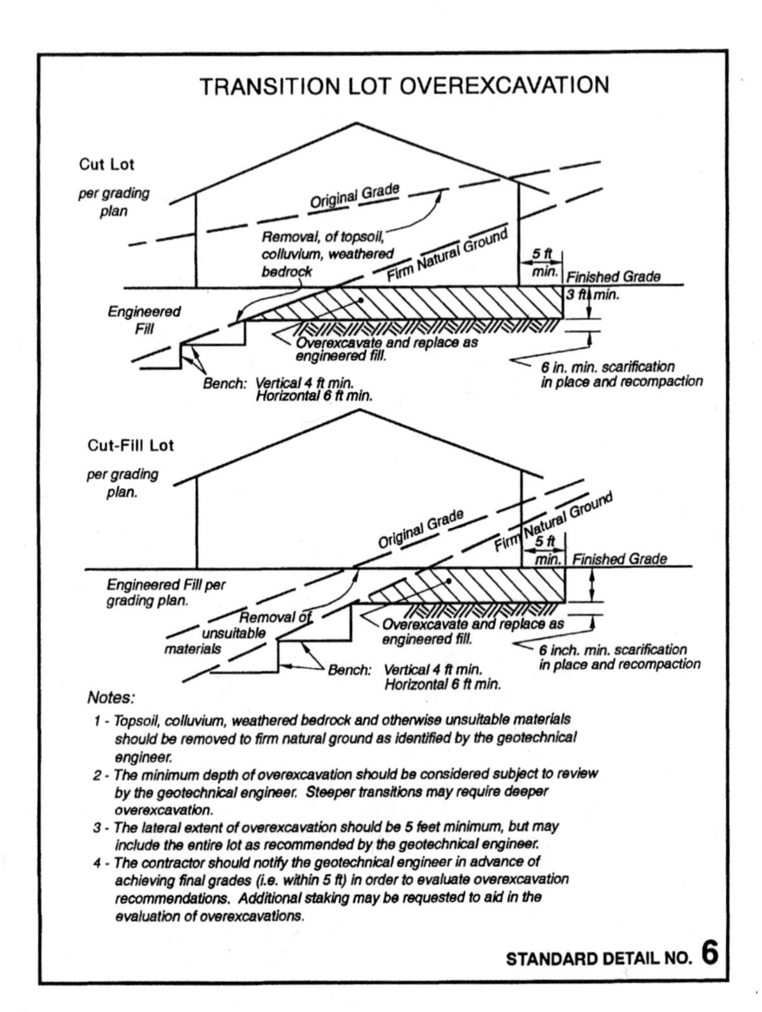

TRANSITION LOT OVEREXCAVATION
Cut Lot
per grading plan
Original Grade
Removal, of topsoil, colluvium, weathered bedrock
Firm Natural Ground
5 ft min.
Finished Grade
3 ft min.
Engineered Fill
Overexcavate and replace as engineered fill.
6 in. min. scarification in place and recompaction
Bench: Vertical 4 ft min. Horizontal 6 ft min.
Cut-Fill Lot
per grading plan.
Original Grade
Firm Natural Ground
5 ft min.
Finished Grade
Engineered Fill per grading plan.
Removal of unsuitable materials
Overexcavate and replace as engineered fill.
6 inch. min. scarification in place and recompaction
Bench: Vertical 4 ft min. Horizontal 6 ft min.
Notes:
1 - Topsoil, colluvium, weathered bedrock and otherwise unsuitable materials should be removed to firm natural ground as identified by the geotechnical engineer.
2 - The minimum depth of overexcavation should be considered subject to review by the geotechnical engineer. Steeper transitions may require deeper overexcavation.
3 - The lateral extent of overexcavation should be 5 feet minimum, but may include the entire lot as recommended by the geotechnical engineer.
4 - The contractor should notify the geotechnical engineer in advance of achieving final grades (i.e. within 5 ft) in order to evaluate overexcavation recommendations. Additional staking may be requested to aid in the evaluation of overexcavations.
STANDARD DETAIL NO. 6

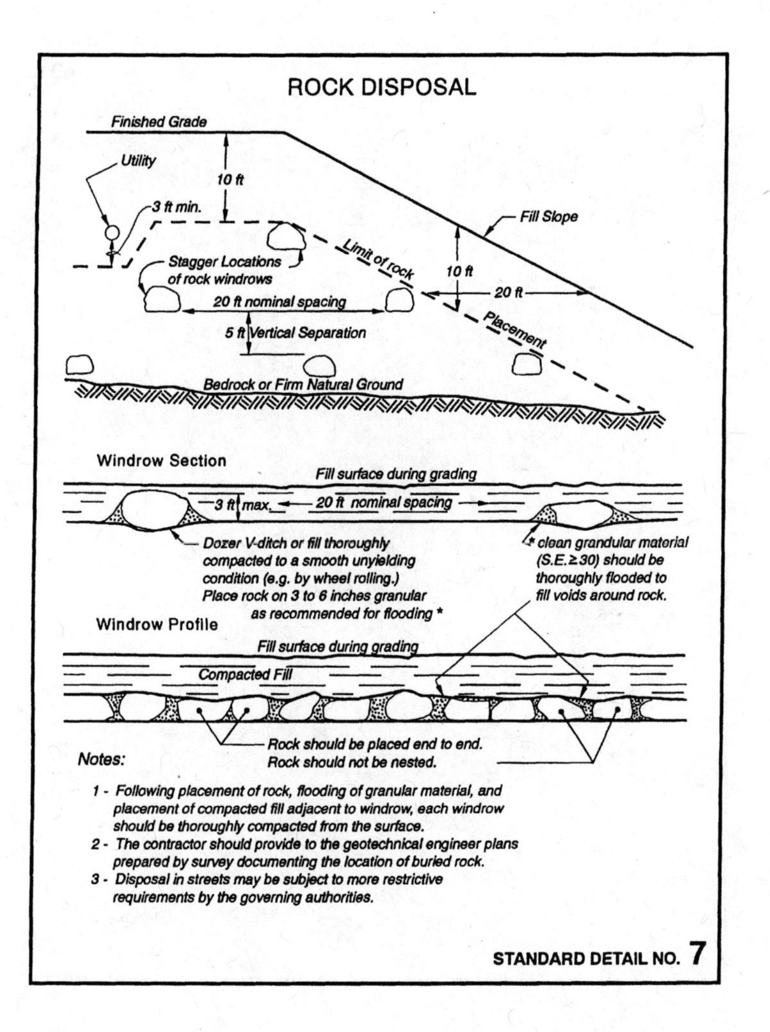
ROCK DISPOSAL
Finished Grade
Utility
10 ft
3 ft min.
Fill Slope
Limit of rock
Stagger Locations of rock windrows
10 ft
20 ft
20 ft nominal spacing
Placement
5 ft Vertical Separation
Bedrock or Firm Natural Ground
Windrow Section
Fill surface during grading
3 ft max.
20 ft nominal spacing
Dozer V-ditch or fill thoroughly compacted to a smooth unyielding condition (e.g. by wheel rolling.) Place rock on 3 to 6 inches granular as recommended for flooding *
* clean granular material (S.E.≥30) should be thoroughly flooded to fill voids around rock.
Windrow Profile
Fill surface during grading
Compacted Fill
Rock should be placed end to end.
Rock should not be nested.
Notes:
1 - Following placement of rock, flooding of granular material, and placement of compacted fill adjacent to windrow, each windrow should be thoroughly compacted from the surface.
2 - The contractor should provide to the geotechnical engineer plans prepared by survey documenting the location of buried rock.
3 - Disposal in streets may be subject to more restrictive requirements by the governing authorities.
STANDARD DETAIL NO. 7

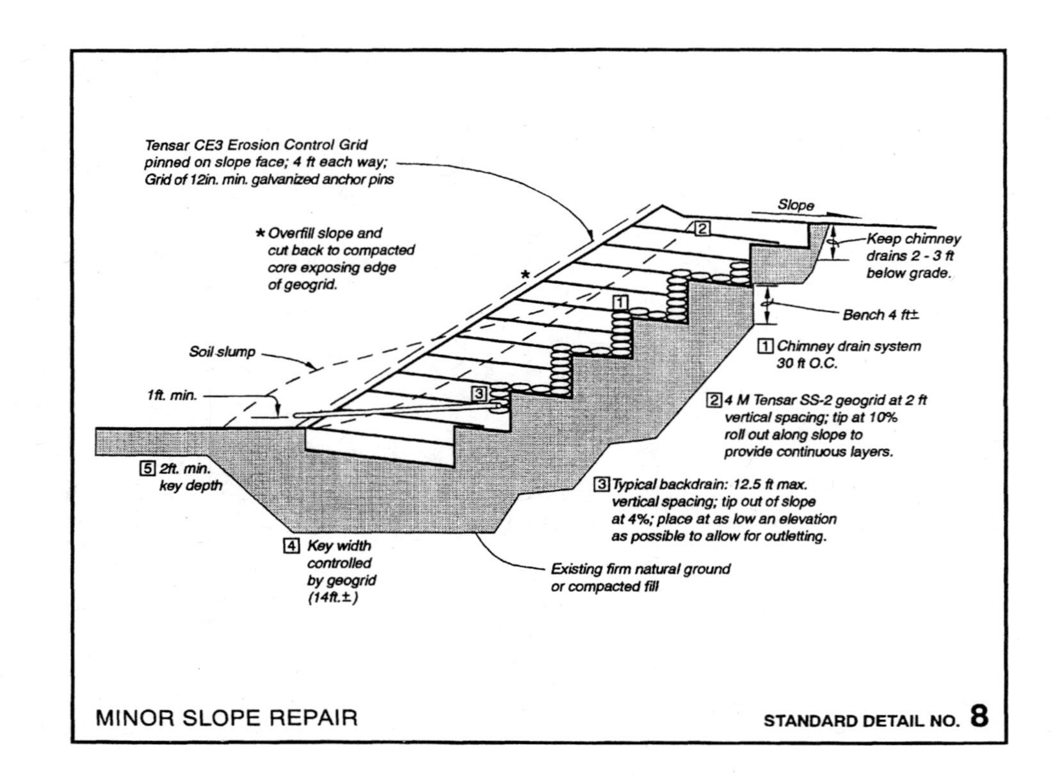
Tensar CE3 Erosion Control Grid pinned on slope face; 4 ft each way; Grid of 12in. min. galvanized anchor pins
* Overfill slope and cut back to compacted core exposing edge of geogrid.
Slope
Keep chimney drains 2 - 3 ft below grade.
Bench 4 ft±
1 Chimney drain system 30 ft O.C.
2 4 M Tensar SS-2 geogrid at 2 ft vertical spacing; tip at 10% roll out along slope to provide continuous layers.
3 Typical backdrain: 12.5 ft max. vertical spacing; tip out of slope at 4%; place at as low an elevation as possible to allow for outletting.
Soil slump
1ft. min.
5 2ft. min. key depth
4 Key width controlled by geogrid (14ft.±)
Existing firm natural ground or compacted fill
MINOR SLOPE REPAIR
STANDARD DETAIL NO. 8

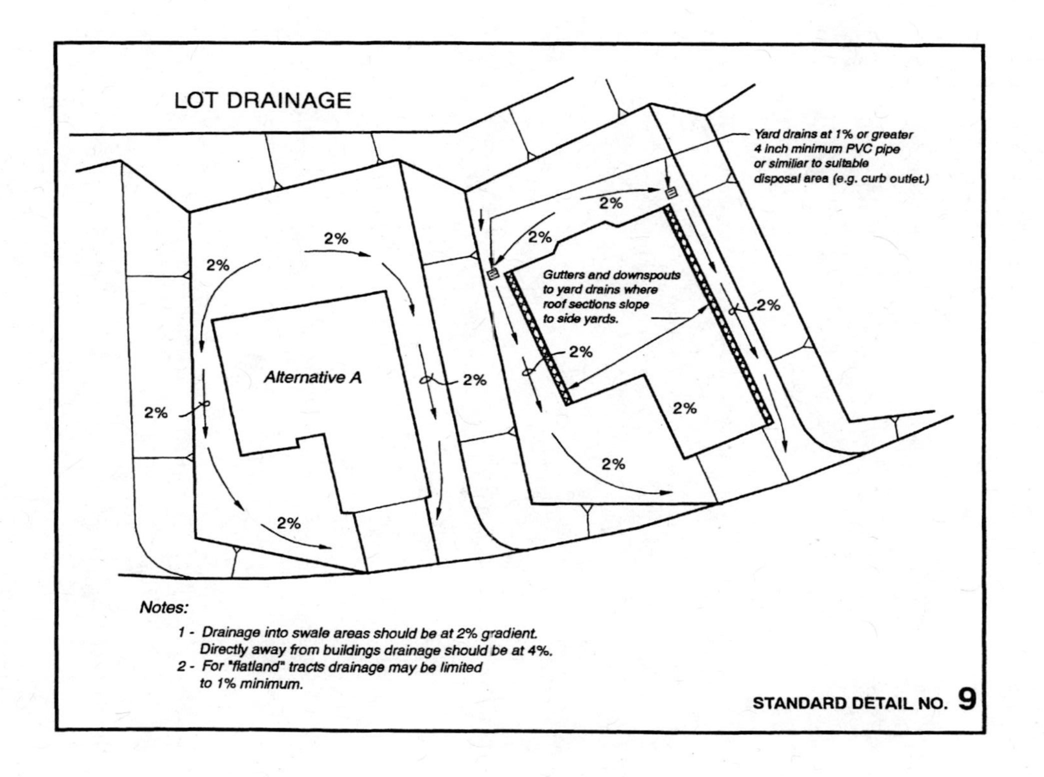
LOT DRAINAGE
Yard drains at 1% or greater
4 inch minimum PVC pipe
or similiar to suitable
disposal area (e.g. curb outlet.)
Gutters and downspouts
to yard drains where
roof sections slope
to side yards.
Alternative A
2%
Notes:
1 - Drainage into swale areas should be at 2% gradient.
Directly away from buildings drainage should be at 4%.
2 - For "flatland" tracts drainage may be limited
to 1% minimum.
STANDARD DETAIL NO. 9

APPENDIX C

PERCOLATION TEST PROCEDURES*

PURPOSE

To establish clear direction and methodology for percolation testing in San Diego County.

BACKGROUND

"Hybridization" of percolation test procedures, concerns expressed by industry regarding "perc" tests, and inconsistency in observed test methods have created a need to standardize a method that can be performed easily and will have results that can be duplicated on the basis of a single, county-wide test.

OBJECTIVE OF THE TEST

Determine the area necessary to properly treat and maintain sewage underground, to size the disposal system with adequate infiltration surface on the basis of expected hydraulic conductivity of the soil and the rate of loading, and to provide for a system with long-term expectation of satisfactory performance.

*Reproduced from County of San Diego (1991). Document has been edited for content.

POLICY

All percolation testing in the county of San Diego shall follow procedures described herein. Any deviation shall be authorized only after receiving written approval by the department of environmental health services.

SAN DIEGO COUNTY PERCOLATION TEST PROCEDURES

1. Test Holes

A. Number of Test Holes

- A minimum of four test holes are required when percolation rates are less than 60 min/in.
- A minimum of six test holes are required when the average percolation rate is more than 60 min/in. For those soils having an average percolation rate greater than 60 min/in., the requirements of Table C.2 must be met.
- Additional test holes may be necessary for adequate and sufficient information on a site-specific basis for reasons that include, but are not limited to, unacceptable or failed tests, areas of the disposal field requiring defined limits for exclusion, a disposal system located outside the concentrated area, and soil conditions that are variable or inconsistent.

B. Depth of Testing

- Test holes shall be representative of the leach line installation depth.
- Conditions which may require testing deeper than leach line depth include shallow consolidated rock, impervious layers, shallow groundwater, slope gradient that exceeds 25%, and other factors as might be determined by sound engineering practices.

C. Soil Classification

- All test holes and deep borings shall have adequate definition of soil types, consolidation, and rock.
- All borings are to be reported, including any which encountered groundwater or refusal.

D. Location of Test Holes

Test holes shall be representative of the disposal area, demonstrating site conditions throughout the entire leach field, equal consideration of primary and reserve, and distances between test holes that are adequate to describe the disposal area.

TABLE C.1 Primary Disposal Trench Length (feet) Based on the Average Percolation Rate (minutes per inch)

Average percolation rate (minutes per inch)	Primary disposal trench length (feet)				
	Columns below indicate the number of bedrooms in the house				
	2	3	4	5	6
<3	200	240	270	280	300
5	240	290	320	320	340
10	275	330	370	410	420
15	300	360	400	450	470
20	315	380	430	470	530
25	330	400	450	500	600
30	350	420	470	540	650
35	365	440	490	590	710
40	380	460	520	630	750
45	400	480	540	670	800
50	415	500	560	700	840
55	430	520	580	740	890
60	450	540	610	770	930
65	500	590	660	780	940
70	550	640	710	790	950
75	600	690	760	810	960
80	650	740	810	860	970
85	700	790	860	910	980
90	755	845	915	965	1005
95	830	920	990	1040	1080
100	905	995	1065	1115	1155
105	980	1070	1140	1190	1230
110	1055	1145	1215	1265	1305
115	1130	1220	1290	1340	1380
120	1210	1300	1370	1420	1460
125	1310	1400	1470	1520	1560
130	1410	1500	1570	1620	1660
135	1510	1600	1670	1720	1760
140	1610	1700	1770	1820	1860
145	1710	1800	1870	1920	1960
150	1810	1900	1970	2020	2060

Notes: An 18-inch trench is the maximum allowable width to be used in determining the total linear drain line footage. Septic tank sizes 2–3 bedrooms require 1000-gallon, 4 bedrooms require 1200-gallon, and 5–6 bedrooms require 1500-gallon capacity.

TABLE C.2 Standards and Requirements for Design and Installation of Subsurface Sewage Disposal Systems for Soils Having Poor Percolation (i.e., over 60 mpi)

Average percolation rate (minutes per inch)	Required minimum lot size (acres)	Available expansion area (percent)
61 to 90	3	200
91 to 120	5	300
121 to 150	7	400
Over 150	10	500

Notes: This table is for soils having an average percolation rate over 60 mpi. Percolation rates must be made at six different locations on the site of the proposed subsurface sewage disposal field. There shall be a minimum of 10 feet of soil above any impervious formation such as rock, clay, adobe, and/or groundwater table. Fractured and hard rock will not be considered as soil. Deep testing can be required to ensure uniform conditions exist below the disposal area. The land on which the subsurface sewage disposal system will have to be installed shall not have a slope gradient greater than 25%.

E. *Identification of Test Holes*

Stake and flag test holes, as needed, so holes can be located. Identify test holes with a test hole number or letter and depth of the test boring.

F. *Drilling or Boring of Test Holes*

- Diameter of each test hole shall be a minimum of 6 inches and a maximum of 10 inches.
- If a backhoe excavation is used, a test hole that is 12 to 14 inches in depth should be excavated into the bottom of the trench.

G. *Preparation of Test Holes*

- The sides and bottom of the holes shall be scarified so as to remove the areas that became smeared by the auger or other tool used to develop the hole. Remove all loose material from the hole.
- Place 2 inches of open graded gravel in the bottom of the hole, prior to running the percolation test (maximum size of open graded gravel is $^1/_2$ inch).

2. Presoaking

A. *Filling the Test Hole*

Carefully fill the test hole with 12 to 14 inches of clear water over the 2 inches of open graded gravel.

B. Time and Duration of Presoaking

- Maintain 12 to 14 inches of clear water over the 2 inches of open graded gravel for a minimum of 4 hours. After 4 hours, allow the water column to drop overnight. Testing must be done within 15 to 30 hours after the initial 4-hour presoak.
- Overnight option: If clay soils are present, it is recommended that the 12- to 14-inch water column be maintained overnight. A siphon can be used to maintain the supply at a constant level.
- If overnight presoak option is used, it will eliminate the normal 1-hour presoak on the following day.
- In highly permeable sandy soils with no clay and/or silt, the presoak procedure may be modified. If, after the hole is filled twice with 12 to 14 inches of clear water, the water seeps completely away in less than 30 minutes, proceed immediately to Case 2 in Item 3, "Determination of the Percolation Rate," and refill to 6 inches above the open graded gravel. If the test is done the following day, a presoak will be necessary.

C. Saturation and Swelling

- Saturation means that the void spaces between the soil particles are full of water. This can be accomplished in a short period of time.
- Swelling is caused by the intrusion of water into the void spaces between the soil particles. This is a slow process, especially in clay-type soil and is the reason for requiring a prolonged soaking.

D. Use of Inserts

- If side walls are not stable or sloughing results in changing depth, the test hole may be abandoned or retested after means are taken to shore up the sides. The holes shall be recleaned prior to restart of the test.
- Options for shoring or maintaining test hole stability include using a hardware cloth ($\frac{1}{8}$-inch grid) or perforated pipe or containers.

3. Determination of the Percolation Rate

Case 1. The water remains overnight following the 4-hour presoak.

Case 2. Fast soil with two columns of 12 to 14 inches of water percolating in less than 30 minutes during second presoak period.

Case 3. No water remains 15 to 30 hours after 4-hour presoak.

A. *Case 1 Testing Procedure*

- Adjust depth of water to 6 inches over the gravel.
- Take two readings at 30-minute intervals and report percolation rate as the slowest of the two readings. If a minimum amount of water remains because of a damaged hole or silting, the hole may be cleaned out and tested under Case 3, starting with the presoak.

B. *Case 2 Testing Procedure*

- After filling the hole twice with 12 to 14 inches of clear water, observe to see if the water will seep away in less than 30 minutes. If so, then proceed with this test procedure. If not, go to Case 3.
- Refill to 6 inches above the open graded gravel.
- Measure from a fixed reference point at 10-minute intervals over a period of 1 hour to the nearest 1/16 inch. Add water at each 10-minute time interval.
- Continue 10-minute readings as long as necessary to obtain a "stabilized" rate with the last two rate readings not varying more than 1/16 inch. The last water drop will be considered the percolation rate.

C. *Case 3 Testing Procedure*

- Maintain a static head of 12 to 14 inches of clean water over the open graded gravel for a period of 1 hour.
- Adjust water depth to 6 inches above the 2 inches of open graded gravel and measure from a fixed reference point at 30-minute intervals to the nearest 1/16 inch.
- Refill the hole as necessary to maintain a 6-inch column of water over the 2 inches of open graded gravel. A fall of 1 inch can be allowed before refilling; for example, if fall is less than 1 inch, allow test to continue to next 30-minute-reading interval.
- Continue 30-minute readings as long as necessary to obtain a "stabilized" rate, with the last two rate readings not varying more than 1/16 inch. The test shall run a minimum of 3 hours after the 1-hour presoak.
- The last water level drop is used to calculate the percolation rate.

4. Calculations and Measurements

A. *Calculation Example*

Percolation rate is reported in minutes per inch; for example, a 30-minute time interval with a 3/4-inch fall would be 30 minutes divided by 3/4 inch, or 40 minutes per inch (mpi).

B. Measurement Principles

- The time intervals for readings are to reflect the actual times and are to be maintained as near as possible to the intervals outlined for the test (i.e., 10 or 30 minutes).
- Measurements to the nearest $^1/_{16}$ inch should be adjusted to the slowest rate. For example, a reading observed between $^3/_8$ inch and $^5/_{16}$ inch (80 mpi and 96 mpi) would be reported as 96 mpi.
- Measurements on an engineering scale (tenths of an inch) should follow the same principle. For example, a reading observed between 0.4 inch and 0.3 inch (75 and 100 mpi) would be reported as 100 mpi.

C. Measurement and Special Considerations

- Measurement from a fixed reference point shall be from a platform that is stable and represents the center of the hole.
- Accurate measuring devices are encouraged for the water level measurements, especially when the test depth is greater than 60 inches. A description of the measurement and clean-out methodology may be required.

5. Reports

All test data and required information shall be submitted on approved Environmental Health Service forms (four copies). Reports shall be signed with an original signature by the consultant. San Diego County Code requires all percolation testing to be done by a civil engineer, geologist, or environmental health specialist, registered in the state of California. These consultants are required to be on the approved list on file with the Environmental Health Services.

The percolation test is only one critical factor in siting an on-site disposal system. Site conditions may require special evaluation by a consultant qualified to technically address issues such as high groundwater, steep slopes, nitrate impacts, cumulative impacts, mounding, and horizontal transmissibility.

Companies whose consultants employ a technician are responsible for the work performed by the technician. It is incumbent upon consultants to properly train, equip, and supervise anyone performing work under their direction and license.

APPENDIX D
CONVERSION FACTORS

From	Multiply by*	Converts to†
Area, acre	4046.9	square meters
Area, square yard	0.8361	square meters
Area, square foot	0.0929	square meters
Area, square inch	0.0006451	square meters
Bending moment, lb-force-foot	1.3558	newton-meter
Density, pounds/cubic yard	0.5932	kilograms/cubic meter
Density, pounds/cubic foot	16.0185	kilograms/cubic meter
Force, kips	4.4482	kilonewton
Force, pounds	4.4482	newtons
Length, miles	1609.344	meter
Length, yards	0.9144	meter
Length, feet	0.3048	meter
Length, inches	0.0254	meter
Force/length, pounds/foot	14.5939	newtons/meter
Force/length, pounds/inch	175.127	newtons/meter
Mass, ton	907.184	kilogram
Mass, pound	0.4536	kilogram
Mass, ounce	28.35	gram
Pressure or stress, pounds/square foot	47.8803	pascal
Pressure or stress, pounds/square inch	6.8947	kilopascal
Temperature, °F	$(t_F^{\circ} - 32)/1.8 = t_C^{\circ}$	°C
Volume, cubic yard	0.7646	cubic meters
Volume, cubic foot	0.02831	cubic meters
Volume, cubic inch	1.6387×10^{-5}	cubic meters

*The precision of a measurement converted to other units can never be greater than that of the original. To go from SI units to U.S. customary units, divide by the given constant. ASTM E 380 provides guidance on use of the SI system.

†The common SI prefixes are

mega	M	1,000,000
kilo	k	1,000
centi	c	0.01
milli	m	0.001
micro	μ	0.000001

APPENDIX E
REFERENCES

AASHTO (1996). *Standard Specifications for Highway Bridges,* 16th ed. American Association of State Highway and Transportation Officials (AASHTO), Washington, D.C.

ACI (1982). *Guide to Durable Concrete.* ACI Committee 201, American Concrete Institute, Detroit, Mich.

ACI Manual of Concrete Practice, Part 1, Materials and General Properties of Concrete (1990). American Concrete Institute, Detroit, Mich.

ASCE (1964). *Design of Foundations for Control of Settlement.* American Society of Civil Engineers, New York, 592 pp.

ASCE (1972). "Subsurface Investigation for Design and Construction of Foundations of Buildings." Task Committee for Foundation Design Manual. Part I, *Journal of the Soil Mechanics and Foundations Division,* ASCE, vol. 98, no. SM5, pp. 481–490; Part II, no. SM6, pp. 557–578; Parts III and IV, no. SM7, pp. 749–764.

ASCE (1976). *Subsurface Investigation for Design and Construction of Foundations of Buildings.* Manual no. 56. American Society of Civil Engineers, New York, 61 pp.

ASCE (1978). *Site Characterization and Exploration.* Proceedings of the Specialty Workshop at Northwestern University, C. H. Dowding, ed. New York, 395 pp.

ASCE (1982). *Gravity Sanitary Sewer Design and Construction.* Manuals and Reports on Engineering Practice, no. 60, Joint Publication: American Society of Civil Engineers and Water Pollution Control Federation, New York, 275 pp.

ASTM (1970). "Special Procedures for Testing Soil and Rock for Engineering Purposes." ASTM Special Technical Publication 479, Philadelphia, 630 pp.

ASTM (1971). "Sampling of Soil and Rock." ASTM Special Technical Publication 483, Philadelphia, 193 pp.

ASTM (1998). *Annual Book of ASTM Standards: Concrete and Aggregates,* vol. 04.02. Standard no. C 127-93, "Standard Test Method for Specific Gravity and Absorption of Coarse Aggregate," West Conshohocken, Pa., pp. 64–68.

ASTM (1997). *Annual Book of ASTM Standards: Concrete and Aggregates,* vol. 04.02. Standard no. C 881-90, "Standard Specification for Epoxy-Resin-Base Bonding Systems for Concrete," West Conshohocken, Pa., pp. 436–440.

ASTM (1997). *Annual Book of ASTM Standards: Road and Paving Materials; Vehicle-Pavement Systems,* vol. 04.03. Standard no. D 5340-93, "Standard Test Method for Airport Pavement Condition Index Surveys," West Conshohocken, Pa., pp. 546–593.

ASTM (1997). *Annual Book of ASTM Standards: Road and Paving Materials; Vehicle-Pavement Systems,* vol. 04.03. Standard no. E 1778-96a, "Standard Terminology Relating to Pavement Distress," West Conshohocken, Pa., pp. 843–845.

ASTM (1998). *Annual Book of ASTM Standards,* vol. 04.08, *Soil and Rock* (*I*). Standard no. D 420-93, "Standard Guide to Site Characterization for Engineering, Design, and Construction Purposes," West Conshohocken, Pa., pp. 1–7.

ASTM (1998). *Annual Book of ASTM Standards,* vol. 04.08, *Soil and Rock* (*I*). Standard no. D 422-90, "Standard Test Method for Particle-Size Analysis of Soils," West Conshohocken, Pa., pp. 10–20.

ASTM (1998). *Annual Book of ASTM Standards,* vol. 04.08, *Soil and Rock* (*I*). Standard no. D 427-93, "Standard Test Method for Shrinkage Factors of Soils by the Mercury Method," West Conshohocken, Pa., pp. 21–24.

ASTM (1998). *Annual Book of ASTM Standards,* vol. 04.08, *Soil and Rock* (*I*). Standard no. D 653-97, "Standard Terminology Relating to Soil, Rock, and Contained Fluids." Terms prepared jointly by the American Society of Civil Engineers and ASTM, West Conshohocken, Pa., pp. 42–76.

ASTM (1998). *Annual Book of ASTM Standards,* vol. 04.08, *Soil and Rock* (*I*). Standard no. D 698-91, "Standard Test Method for Laboratory Compaction Characteristics of Soil Using Standard Effort," West Conshohocken, Pa., pp. 77–87.

ASTM (1998). *Annual Book of ASTM Standards,* vol. 04.08, *Soil and Rock* (*I*). Standard no. D 854-92, "Standard Test Method for Specific Gravity of Soils," West Conshohocken, Pa., pp. 88–91.

ASTM (1998). *Annual Book of ASTM Standards,* vol. 04.08, *Soil and Rock* (*I*). Standard no. D 1143-94, "Standard Test Method for Piles under Static Axial Compressive Load," West Conshohocken, Pa., pp. 95–105.

ASTM (1998). *Annual Book of ASTM Standards,* vol. 04.08, *Soil and Rock* (*I*). Standard no. D 1196-97, "Standard Test Method for Nonrepetitive Static Plate Load Tests of Soils and Flexible Pavement Components, for Use in Evaluation and Design of Airport and Highway Pavements," West Conshohocken, Pa., pp. 111–112.

ASTM (1998). *Annual Book of ASTM Standards,* vol. 04.08, *Soil and Rock* (*I*). Standard no. D 1556-96, "Standard Test Method for Density and Unit Weight of Soil in Place by the Sand-Cone Method," West Conshohocken, Pa., pp. 120–125.

ASTM (1998). *Annual Book of ASTM Standards,* vol. 04.08, *Soil and Rock* (*I*). Standard no. D 1557-91, "Standard Test Method for Laboratory Compaction Characteristics of Soil Using Modified Effort," West Conshohocken, Pa., pp. 126–133.

ASTM (1998). *Annual Book of ASTM Standards,* vol. 04.08, *Soil and Rock* (*I*). Standard no. D 1586-92, "Standard Test Method for Penetration Test and Split-Barrel Sampling of Soils," West Conshohocken, Pa., pp. 137–141.

ASTM (1998). *Annual Book of ASTM Standards,* vol. 04.08, *Soil and Rock* (*I*). Standard no. D 1587-94, "Standard Practice for Thin-Walled Tube Geotechnical Sampling of Soils," West Conshohocken, Pa., pp. 142–144.

ASTM (1998). *Annual Book of ASTM Standards,* vol. 04.08, *Soil and Rock* (*I*). Standard no. D 1883-94, "Standard Test Method for CBR (California Bearing Ratio) of Laboratory-Compacted Soils," West Conshohocken, Pa., pp. 159–165.

ASTM (1998). *Annual Book of ASTM Standards,* vol. 04.08, *Soil and Rock* (*I*). Standard no. D 2166-91, "Standard Test Method for Unconfined Compressive Strength of Cohesive Soil," West Conshohocken, Pa., pp. 172–176.

ASTM (1998). *Annual Book of ASTM Standards,* vol. 04.08, *Soil and Rock* (*I*). Standard no. D 2216-92, "Standard Test Method for Laboratory Determination of Water (Moisture) Content of Soil and Rock," West Conshohocken, Pa., pp. 188–191.

ASTM (1998). *Annual Book of ASTM Standards,* vol. 04.08, *Soil and Rock* (*I*). Standard no. D 2434-94, "Standard Test Method for Permeability of Granular Soils (Constant Head)," West Conshohocken, Pa., pp. 202–206.

ASTM (1998). *Annual Book of ASTM Standards,* vol. 04.08, *Soil and Rock* (*I*). Standard no. D 2435-96, "Standard Test Method for One-Dimensional Consolidation Properties of Soils," West Conshohocken, Pa., pp. 207–216.

ASTM (1998). *Annual Book of ASTM Standards,* vol. 04.08, *Soil and Rock* (*I*). Standard no. D 2487-93, "Standard Classification of Soils for Engineering Purposes (Unified Soil Classification System)," West Conshohocken, Pa., pp. 217–227.

ASTM (1998). *Annual Book of ASTM Standards,* vol. 04.08, *Soil and Rock* (*I*). Standard no. D 2573-94, "Standard Test Method for Field Vane Shear Test in Cohesive Soil," West Conshohocken, Pa., pp. 239–241.

ASTM (1998). *Annual Book of ASTM Standards,* vol. 04.08, *Soil and Rock* (*I*). Standard no. D 2844-94, "Standard Test Method for Resistance *R*-Value and Expansion Pressure of Compacted Soils," West Conshohocken, Pa., pp. 246–253.

ASTM (1998). *Annual Book of ASTM Standards,* vol. 04.08, *Soil and Rock* (*I*). Standard no. D 2850-95, "Standard Test Method for Unconsolidated-Undrained Triaxial Compression Test on Cohesive Soils," West Conshohocken, Pa., pp. 260–264.

ASTM (1998). *Annual Book of ASTM Standards,* vol. 04.08, *Soil and Rock* (*I*). Standard no. D 2922-96, "Standard Test Methods for Density of Soil and Soil-Aggregate in Place by Nuclear Methods (Shallow Depth)," West Conshohocken, Pa., pp. 268–272.

ASTM (1998). *Annual Book of ASTM Standards,* vol. 04.08, *Soil and Rock* (*I*). Standard no. D 2937-94, "Standard Test Method for Density of Soil in Place by the Drive-Cylinder Method," West Conshohocken, Pa., pp. 275–278.

ASTM (1998). *Annual Book of ASTM Standards,* vol. 04.08, *Soil and Rock* (*I*). Standard no. D 2974-95, "Standard Test Methods for Moisture, Ash, and Organic Matter of Peat and Other Organic Soils," West Conshohocken, Pa., pp. 285–287.

ASTM (1998). *Annual Book of ASTM Standards,* vol. 04.08, *Soil and Rock* (*I*). Standard no. D 3080-90, "Standard Test Method for Direct Shear Test of Soils under Consolidated Drained Conditions," West Conshohocken, Pa., pp. 300–305.

ASTM (1998). *Annual Book of ASTM Standards,* vol. 04.08, *Soil and Rock* (*I*). Standard no. D 3441-94, "Standard Test Method for Deep, Quasi-Static, Cone and Friction-Cone Penetration Tests of Soil," West Conshohocken, Pa., pp. 348–354.

ASTM (1998). *Annual Book of ASTM Standards,* vol. 04.08, *Soil and Rock (I).* Standard no. D 3689-90, "Standard Test Method for Individual Piles under Static Axial Tensile Load," West Conshohocken, Pa., pp. 366–376.

ASTM (1998). *Annual Book of ASTM Standards,* vol. 04.08, *Soil and Rock (I).* Standard no. D 3966-90, "Standard Test Method for Piles under Lateral Loads," West Conshohocken, Pa., pp. 389–403.

ASTM (1998). *Annual Book of ASTM Standards,* vol. 04.08, *Soil and Rock (I).* Standard no. D 4253-96, "Standard Test Methods for Maximum Index Density and Unit Weight of Soils Using a Vibratory Table," West Conshohocken, Pa., pp. 498–510.

ASTM (1998). *Annual Book of ASTM Standards,* vol. 04.08, *Soil and Rock (I).* Standard no. D 4254-96, "Standard Test Method for Minimum Index Density and Unit Weight of Soils and Calculation of Relative Density," West Conshohocken, Pa., pp. 511–518.

ASTM (1998). *Annual Book of ASTM Standards,* vol. 04.08, *Soil and Rock (I).* Standard no. D 4318-95, "Standard Test Method for Liquid Limit, Plastic Limit, and Plasticity Index of Soils," West Conshohocken, Pa., pp. 519–529.

ASTM (1998). *Annual Book of ASTM Standards,* vol. 04.08, *Soil and Rock (I).* Standard no. D 4429-93, "Standard Test Method for CBR (California Bearing Ratio) of Soils in Place," West Conshohocken, Pa., pp. 605–608.

ASTM (1998). *Annual Book of ASTM Standards,* vol. 04.08, *Soil and Rock (I).* Standard no. D 4452-95, "Standard Test Methods for X-Ray Radiography of Soil Samples," West Conshohocken, Pa., pp. 618–629.

ASTM (1998). *Annual Book of ASTM Standards,* vol. 04.08, *Soil and Rock (I).* Standard no. D 4546-96, "Standard Test Methods for One-Dimensional Swell or Settlement Potential of Cohesive Soils," West Conshohocken, Pa., pp. 663–669.

ASTM (1998). *Annual Book of ASTM Standards,* vol. 04.08, *Soil and Rock (I).* Standard no. D 4647-93, "Standard Test Method for Identification and Classification of Dispersive Clay Soils by the Pinhole Test," West Conshohocken, Pa., pp. 757–766.

ASTM (1998). *Annual Book of ASTM Standards,* vol. 04.08, *Soil and Rock (I).* Standard no. D 4648-94, "Standard Test Method for Laboratory Miniature Vane Shear Test for Saturated Fine-Grained Clayey Soil," West Conshohocken, Pa., pp. 767–772.

ASTM (1998). *Annual Book of ASTM Standards,* vol. 04.08, *Soil and Rock (I).* Standard no. D 4767-95, "Standard Test Method for Consolidated Undrained Triaxial Compression Test for Cohesive Soils," West Conshohocken, Pa., pp. 850–859.

ASTM (1998). *Annual Book of ASTM Standards,* vol. 04.08, *Soil and Rock (I).* Standard no. D 4829-95, "Standard Test Method for Expansion Index of Soils," West Conshohocken, Pa., pp. 860–863.

ASTM (1998). *Annual Book of ASTM Standards,* vol. 04.09, *Soil and Rock (II), Geosynthetics.* Standard no. D 4943-95, "Standard Test Method for Shrinkage Factors of Soils by the Wax Method," West Conshohocken, Pa., pp. 1–5.

ASTM (1998). *Annual Book of ASTM Standards,* vol. 04.09, *Soil and Rock (II), Geosynthetics.* Standard no. D 4945-96, "Standard Test Method for High-Strain Dynamic Testing of Piles," West Conshohocken, Pa., pp. 10–16.

ASTM (1998). *Annual Book of ASTM Standards,* vol. 04.09, *Soil and Rock (II), Geosynthetics.* Standard no. D 5084-97, "Standard Test Method for Measurement of Hydraulic Conductivity of Saturated Porous Materials Using a Flexible Wall Permeameter," West Conshohocken, Pa., pp. 62–69.

ASTM (1998). *Annual Book of ASTM Standards,* vol. 04.09, *Soil and Rock (II), Geosynthetics.* Standard no. D 5333-96, "Standard Test Method for Measurement of Collapse Potential of Soils," West Conshohocken, Pa., pp. 224–226.

ASTM (1998). *Annual Book of ASTM Standards,* vol. 04.09, *Soil and Rock (II), Geosynthetics.* Standard no. D 5780-95, "Standard Test Method for Individual Piles in Permafrost under Static Axial Compressive Load," West Conshohocken, Pa., pp. 596–609.

ASTM (1998). *Annual Book of ASTM Standards,* vol. 04.09, *Soil and Rock (II), Geosynthetics.* Standard no. D 4439-97, "Standard Terminology for Geosynthetics," West Conshohocken, Pa., pp. 1219–1222.

Aberg, B. (1996). "Grain-Size Distribution for Smallest Possible Void Ratio." *Journal of Geotechnical Engineering,* ASCE, vol. 122, no. 1, pp. 74–77.

Abramson, L. W., Lee, T. S., Sharma, S., and Boyce, G. M. (1996). *Slope Stability and Stabilization Methods.* John Wiley & Sons, New York, 629 pp.

Al-Homoud, A. S., Basma, A. A., Husein Malkawi, A. I., and Al Bashabsheh, M. A. (1995). "Cyclic Swelling Behavior of Clays." *Journal of Geotechnical Engineering,* ASCE, vol. 121, no. 7, pp. 562–565.

Al-Homoud, A. S., Basma, A. A., Husein Malkawi, A. I., and Al Bashabsheh, M. A. (1997). Closure of "Cyclic Swelling Behavior of Clays." *Journal of Geotechnical and Geoenvironmental Engineering,* ASCE, vol. 123, no. 8, pp. 783–786.

Allen, L. R., Yen, B. C., and McNeill, R. L. (1978). "Stereoscopic X-ray Assessment of Offshore Soil Samples." *Offshore Technology Conference,* vol. 3, pp. 1391–1399.

Alpan, I. (1967). "The Empirical Evaluation of the Coefficients K_o and K_{oR}." *Soils and Foundations,* vol. 7, no. 1, pp. 31–40.

Ambraseys, N. N. (1960). "On the Seismic Behavior of Earth Dams." *Proceedings of the 2nd World Conference on Earthquake Engineering,* vol. 1, Tokyo and Kyoto, pp. 331–358.

American Geological Institute (1982). *AGI Data Sheets for Geology in the Field, Laboratory, and Office.* American Geological Institute, Falls Church, Va., 61 data sheets.

Anderson, S. A., and Sitar, N. (1995). "Analysis of Rainfall-Induced Debris Flows." *Journal of Geotechnical Engineering,* ASCE, vol. 121, no. 7, pp. 544–552.

Anderson, S. A., and Sitar, N. (1996). Closure of "Analysis of Rainfall-Induced Debris Flows." *Journal of Geotechnical Engineering,* ASCE, vol. 122, no. 12, pp. 1025–1027.

Asphalt Institute (1984). *Thickness Design—Asphalt Pavements for Highways and Streets.* The Asphalt Institute, College Park, Md., 80 pp.

Association of Engineering Geologists (1978). *Failure of St. Francis Dam.* Southern California Section.

Atkins, H. N. (1983). *Highway Materials, Soils, and Concretes,* 2nd ed. Reston Publishing, Reston, Va., 377 pp.

Atterberg, A. (1911). "The Behavior of Clays with Water, Their Limits of Plasticity and Their Degrees of Plasticity." *Kungliga Lantbruksakademiens Handlingar och Tidskrift,* vol. 50, no. 2, pp. 132–158.

Attewell, P. B., Yeates, J., and Selby, A. R. (1986). *Soil Movements Induced by Tunneling and Their Effects on Pipelines and Structures.* Chapman and Hall, New York, 325 pp.

Baldwin, J. E., Donley, H. F., and Howard, T. R. (1987). "On Debris Flow/Avalanche Mitigation and Control, San Francisco Bay Area, California." *Debris Flow/Avalanches: Process, Recognition, and Mitigation,* The Geological Society of America, Boulder, Colo., pp. 223–236.

Bates, R. L., and Jackson, J. A. (1980). *Glossary of Geology.* American Geological Institute, Falls Church, Va., 751 pp.

Bell, F. G. (1983). *Fundamentals of Engineering Geology.* Butterworths, London, 648 pp.

Bellport, B. P. (1968). "Combating Sulphate Attack on Concrete on Bureau of Reclamation Projects." *Performance of Concrete, Resistance of Concrete to Sulphate and Other Environmental Conditions,* University of Toronto Press, Toronto, pp. 77–92.

Best, M. G. (1982). *Igneous and Metamorphic Petrology.* W. H. Freeman, San Francisco.

Biddle, P. G. (1979). "Tree Root Damage to Buildings—An Arboriculturist's Experience." *Arboricultural Journal,* vol. 3, no. 6, pp. 397–412.

Biddle, P. G. (1983). "Patterns of Soil Drying and Moisture Deficit in the Vicinity of Trees on Clay Soils." *Geotechnique,* London, vol. 33, no. 2, pp. 107–126.

Bishop, A. W. (1955). "The Use of the Slip Circle in the Stability Analysis of Slopes." *Geotechnique,* London, vol. 5, no. 1, pp. 7–17.

Bishop, A. W., and Henkel, D. J. (1962). *The Measurement of Soil Properties in the Triaxial Test,* 2nd ed. Edward Arnold, London, 228 pp.

Bjerrum, L. (1963). "Allowable Settlement of Structures." *Proceedings of the 3rd European Conference on Soil Mechanics and Foundation Engineering,* vol. 2, Wiesbaden, Germany, pp. 135–137.

Bjerrum, L. (1967a). "The Third Terzaghi Lecture: Progressive Failure in Slopes of Overconsolidated Plastic Clay and Clay Shales." *Journal of the Soil Mechanics and Foundations Division,* ASCE, vol. 93, no. SM5, part 1, pp. 1–49.

Bjerrum, L. (1967b). "Engineering Geology of Norwegian Normally Consolidated Marine Clays as Related to Settlements of Buildings." Seventh Rankine Lecture, *Geotechnique,* vol. 17, no. 2, London, pp. 81–118.

Bjerrum, L. (1972). "Embankments on Soft Ground." *Proceedings of the ASCE Specialty Conference on Performance of Earth and Earth-Supported Structures.* Purdue University, West Lafayette, Ind., vol. 2, pp. 1–54.

Bjerrum, L. (1973). "Problems of Soil Mechanics and Construction on Soft Clays and Structurally Unstable Soils." Session 4, *Proceedings of the 8th International Conference on Soil Mechanics and Foundation Engineering,* Moscow, vol. 3, pp. 111–159.

Bjerrum, L., and Simons, N. E. (1960). "Comparison of Shear Strength Characteristics of Normally Consolidated Clays." *Proceedings of the ASCE Research Conference on the Shear Strength of Cohesive Soils,* Boulder, Colo., pp. 711–726.

Blight, G. E. (1965). "The Time-Rate of Heave of Structures on Expansive Clays." *Moisture Equilibria and Moisture Changes in Soils beneath Covered Areas,* G. D. Aitchison, ed. Butterworths, Sydney, Australia, pp. 78–88.

Boardman, B. T., and Daniel, D. E. (1996). "Hydraulic Conductivity of Desiccated Geosynethic Clay Liners." *Journal of Geotechnical Engineering,* ASCE, vol. 122, no. 3, pp. 204–208.

Bonilla, M. G. (1970). "Surface Faulting and Related Effects." Chapter 3 of *Earthquake Engineering,* R. L. Wiegel, coordinating ed. Prentice-Hall, Englewood Cliffs, N.J., pp. 47–74.

Boone, S. T. (1996). "Ground-Movement-Related Building Damage." *Journal of Geotechnical Engineering,* ASCE, vol. 122, no. 11, pp. 886–896.

Boone, S. T. (1998). Closure to "Ground-Movement-Related Building Damage." *Journal of Geotechnical and Geoenvironmental Engineering,* ASCE, vol. 124, no. 5, pp. 463–465.

Boscardin, M. D., and Cording, E. J. (1989). "Building Response to Excavation-Induced Settlement." *Journal of Geotechnical Engineering,* ASCE, vol. 115, no. 1, pp. 1–21.

Bourdeaux, G., and Imaizumi, H. (1977). "Dispersive Clay at Sabradinho Dam." *Dispersive Clays, Related Piping, and Erosion in Geotechnical Projects,* STP 625, American Society for Testing and Materials, Philadelphia, pp. 13–24.

Boussinesq, J. (1885). *Application des Potentiels à L'Étude de L'Équilibre et du Mouvement des Solides Élastiques,* Gauthier-Villars, Paris.

Bowles, J. E. (1982). *Foundation Analysis and Design,* 3rd ed. McGraw-Hill, New York, 816 pp.

Brewer, H. W. (1965). "Moisture Migration—Concrete Slab-on-Ground Construction." *Journal of the PCA Research and Development Laboratories,* May, pp. 2–17.

Bromhead, E. N. (1984). *Ground Movements and Their Effects on Structures,* chap. 3, "Slopes and Embankments." P. B. Attewell and R. K. Taylor, eds. Surrey University Press, London, p. 63.

Brooker, E. W., and Ireland, H. O. (1965). "Earth Pressures at Rest Related to Stress History." *Canadian Geotechnical Journal,* vol. 2, no. 1, pp. 1–15.

Brown, D. R., and Warner, J. (1973). "Compaction Grouting." *Journal of the Soil Mechanics and Foundations Division,* ASCE, vol. 99, no. SM8, pp. 589–601.

Brown, R. W. (1990). *Design and Repair of Residential and Light Commercial Foundations.* McGraw-Hill, New York, 241 pp.

Brown, R. W. (1992). *Foundation Behavior and Repair, Residential and Light Commercial.* McGraw-Hill, New York, 271 pp.

Bruce, D. A., and Jewell, R. A. (1987). "Soil Nailing: The Second Decade." *International Conference on Foundations and Tunnels,* London, pp. 68–83.

Burland, J. B., Broms, B. B., and deMello, V. F. B. (1977). "Behavior of Foundations and Structures." State-of-the-Art-Report, *Proceedings of the 9th International Conference on Soil Mechanics and Foundation Engineering,* Japanese Geotechnical Society, Tokyo, vol. 1, pp. 495–546.

Butt, T. K. (1992). "Avoiding and Repairing Moisture Problems in Slabs on Grade." *The Construction Specifier,* December, pp. 17–27.

California Department of Water Resources (1967). "Earthquake Damage to Hydraulic Structures in California." California Department of Water Resources, Bulletin 116-3, Sacramento.

California Division of Highways (1973). *Flexible Pavement Structural Section Design Guide for California Cities and Counties,* Sacramento, 44 pp.

California Plain Language Pamphlet of the Professional Engineers Act and the Board Rules (1995). Board of Registration for Professional Engineers and Land Surveyors, Department of Consumer Affairs, State of California, Sacramento, 60 pp.

Casagrande, A. (1932). Discussion of "A New Theory of Frost Heaving," by A. C. Benkelman and F. R. Ohlmstead, *Proceedings of the Highway Research Board,* vol. 11, pp. 168–172.

Casagrande, A (1932). "Research on the Atterberg Limits of Soils." *Public Roads,* vol.13, pp. 121–136.

Casagrande, A. (1936). "The Determination of the Pre-Consolidation Load and Its Practical Significance." Discussion D-34, *Proceedings of the 1st International Conference on Soil Mechanics and Foundation Engineering,* Cambridge, Mass., vol. 3, pp. 60–64.

Casagrande, A. (1940). "Seepage through Dams." *Contributions to Soil Mechanics, 1925–1940.* Boston Society of Civil Engineers, Boston, pp. 295–336, originally published in the *Journal of the New England Water Works Association,* June, 1937.

Casagrande, A. (1948). "Classification and Identification of Soils." *Transactions,* ASCE, vol. 113, pp. 901–930.

Cashman, P. M., and Harris, E. T. (1970). *Control of Groundwater by Water Lowering.* Conference on Ground Engineering, Institute of Civil Engineers, London.

Caterpillar Performance Handbook (1997). 28th ed. Caterpillar, Inc., Peoria, Ill., 1006 pp.

Cedergren, H. R. (1989). *Seepage, Drainage, and Flow Nets,* 3rd ed., John Wiley & Sons, New York, 465 pp.

Cernica, J. N. (1995a). *Geotechnical Engineering: Soil Mechanics.* John Wiley & Sons, New York, 453 pp.

Cernica, J. N. (1995b). *Geotechnical Engineering: Foundation Design.* John Wiley & Sons, New York, 486 pp.

Cheeks, J. R. (1996). "Settlement of Shallow Foundations on Uncontrolled Mine Spoil Fill." *Journal of Performance of Constructed Facilities,* ASCE, vol. 10, no. 4, pp. 143–151.

Chen, F. H. (1988). *Foundations on Expansive Soils,* 2nd ed. Elsevier Scientific, New York, 463 pp.

Cheney, J. E., and Burford, D. (1975). "Damaging Uplift to a Three-Story Office Block Constructed on a Clay Soil Following Removal of Trees." *Proceedings, Conference on Settlement of Structures,* Cambridge, England. Pentech Press, London, pp. 337–343.

City of San Diego Standard Drawings (1986). Document no. 769374, City of San Diego, San Diego, Calif., 285 pp.

Cleveland, G. B. (1960). "Geology of the Otay Clay Deposit, San Diego County, California." California Division of Mines Special Report 64, Sacramento, 16 pp.

Clough, G. W., and Davidson, R. R. (1977). "Effects of Construction on Geotechnical Performance." Specialty Session no. 3, Relationship between Design and Construction in Soil Engineering, *9th International Conference on Soil Mechanics and Foundation Engineering,* Tokyo, vol. 3, pp. 479–485.

Coduto, D. P. (1994). *Foundation Design, Principles and Practices.* Prentice Hall, Englewood Cliffs, N.J., 796 pp.

Collepardi, M. (1999). "Damage by Delayed Ettringite Formation." *Concrete International,* American Concrete Institute, vol. 21, no. 1, Farmington Hills, Mich., pp. 69-74.

Collins, A. G., and Johnson, A. I. (1988). *Ground-Water Contamination, Field Methods,* Symposium papers published by American Society for Testing and Materials, Philadelphia, 491 pp.

Committee Report for the State (1928). "Causes Leading to the Failure of the St. Francis Dam." California Printing Office.

Compton, R. R. (1962). *Manual of Field Geology.* John Wiley & Sons, New York, pp. 255–256.

Converse Consultants Southwest, Inc. (1990). *Soil and Foundation Investigation, Proposed 80-Acre Clayton-Alexander Parcel, North Las Vegas, Nevada.* CCSW Project no. 90-33264-01, Las Vegas, Nev., 46 pp.

Corns, C. F. (1974). "Inspection Guidelines—General Aspects." *Safety of Small Dams. Proceedings of Engineering Foundation Conference,* Henniker, N.H., ASCE, New York, pp. 16–21.

County of San Diego (1983). "Below Slab Moisture Barrier Specifications." DPL no. 65B, Department of Planning and Land Use, County of San Diego, Calif.

County of San Diego (1991). "Percolation Test Procedure." Department of Environmental Health Services and Land Use Division, County of San Diego, Calif.

County of San Diego (undated). *Sewage Disposal for Individual Homes, Small Industries and Institutions.* Department of Public Health, Division of Sanitation, County of San Diego, Calif., 23 pp.

Cousins, B. F. (1978). "Stability Charts for Simple Earth Slopes." *Journal of the Geotechnical Engineering Division,* ASCE, vol. 104, no. GT2, pp. 267–279.

Cox, J. B. (1968). "A Review of the Engineering Characteristics of the Recent Marine Clays in South East Asia." Asian Institute of Technology, Research Report no. 6, Bangkok.

Cutler, D. F., and Richardson, I. B. K. (1989). *Tree Roots and Buildings,* 2nd ed. Longman Scientific & Technical, Essex, England, pp. 1–67.

David, D., and Komornik, A. (1980). "Stable Embedment Depth of Piles in Swelling Clays." *4th International Conference on Expansive Soils,* ASCE, vol. 2, Denver, pp. 798–814.

Day, R. W. (1980). *Engineering Properties of the Orinoco Clay.* Thesis for the master of science in civil engineering (MCE) and civil engineer (CE) degrees. Massachusetts Institute of Technology, Cambridge, Mass., 140 pp.

Day, R. W. (1989). "Relative Compaction of Fill Having Oversize Particles." *Journal of Geotechnical Engineering,* ASCE, vol. 115, no. 10, pp. 1487–1491.

Day, R. W. (1990a). "Differential Movement of Slab-on-Grade Structures." *Journal of Performance of Constructed Facilities,* ASCE, vol. 4, no. 4, pp. 236–241.

Day, R. W. (1990b). "Index Test for Erosion Potential." *Bulletin of the Association of Engineering Geologists,* vol. 27, no. 1, pp. 116–117.

Day, R. W. (1991a). Discussion of "Collapse of Compacted Clayey Sand." *Journal of Geotechnical Engineering,* ASCE, vol. 117, no. 11, pp. 1818–1821.

Day, R. W. (1991b). "Expansion of Compacted Gravelly Clay." *Journal of Geotechnical Engineering,* ASCE, vol. 117, no. 6, pp. 968–972.

Day, R. W. (1992a). "Effective Cohesion for Compacted Clay." *Journal of Geotechnical Engineering,* ASCE, vol. 118, no. 4, pp. 611–619.

Day, R. W. (1992b). "Swell Versus Saturation for Compacted Clay." *Journal of Geotechnical Engineering,* ASCE, vol. 118, no. 8, pp. 1272–1278.

Day, R. W. (1992c). "Walking of Flatwork on Expansive Soils." *Journal of Performance of Constructed Facilities,* ASCE, vol. 6, no. 1, pp. 52–57.

Day, R. W. (1992d). "Moisture Migration through Concrete Floor Slabs." *Journal of Performance of Constructed Facilities,* ASCE, vol. 6, no. 1, pp. 46–51.

Day, R. W. (1993a). "Expansion Potential According to the Uniform Building Code." *Journal of Geotechnical Engineering,* ASCE, vol. 119, no. 6, pp. 1067–1071.

Day, R. W. (1993b). "Surficial Slope Failure: A Case Study." *Journal of Performance of Constructed Facilities,* ASCE, vol. 7, no. 4, pp. 264–269.

Day, R. W. (1994a). "Inorganic Soil Classification System Based on Plasticity." *Bulletin of the Association of Engineering Geologists,* vol. 31, no. 4, pp. 521–527.

Day, R. W. (1994b). Discussion of "Evaluation and Control of Collapsible Soils." *Journal of Geotechnical Engineering,* ASCE, vol. 120, no. 5, pp. 924–925.

Day, R. W. (1994c). "Performance of Slab-on-Grade Foundations on Expansive Soil." *Journal of Performance of Constructed Facilities,* ASCE, vol. 8, no. 2, pp. 129–138.

Day, R. W. (1994d). "Surficial Stability of Compacted Clay: Case Study." *Journal of Geotechnical Engineering,* ASCE, vol. 120, no. 11, pp. 1980–1990.

Day, R. W. (1994e). "Weathering of Expansive Sedimentary Rock Due to Cycles of Wetting and Drying." *Bulletin of the Association of Engineering Geologists,* vol. 31, no. 3, pp. 387–390.

Day, R. W. (1994f). "Moisture Migration through Basement Walls." *Journal of Performance of Constructed Facilities,* ASCE, vol. 8, no. 1, pp. 82–86.

Day, R. W. (1994g). "Swell-Shrink Behavior of Compacted Clay." *Journal of Geotechnical Engineering,* ASCE, vol. 120, no. 3, pp. 618–623.

Day, R. W. (1995a). Discussion of "Numerical Analysis of Drained Direct and Simple Shear Tests." *Journal of Geotechnical Engineering,* ASCE, vol. 121, no. 2, pp. 223–227.

Day, R. W. (1995b). "Effect of Maximum Past Pressure on Two-Dimensional Immediate Settlement." *Journal of Environmental and Engineering Geoscience,* vol. 1, no. 4, pp. 514–517.

Day, R. W. (1995c). "Pavement Deterioration: Case Study." *Journal of Performance of Constructed Facilities,* ASCE, vol. 9. no. 4, pp. 311–318.

Day, R. W. (1995d). "Reactivation of an Ancient Landslide." *Journal of Performance of Constructed Facilities,* ASCE, vol. 9, no. 1, pp. 49–56.

Day, R. W. (1995e). "Engineering Properties of Diatomaceous Fill." *Journal of Geotechnical Engineering,* ASCE, vol. 121, no. 12, pp. 908–910.

Day, R. W. (1996a). "Study of Capillary Rise and Thermal Osmosis." *Journal of Environmental and Engineering Geoscience,* AEG and GSA, vol. 2, no. 2, pp. 249–254.

Day, R. W. (1996b). "Effect of Gravel on Pumping Behavior of Compacted Soil." *Journal of Geotechnical Engineering,* ASCE, vol. 122, no. 10, pp. 863–866.

Day, R. W. (1997a). Discussion of "Grain-Size Distribution for Smallest Possible Void Ratio." *Journal of Geotechnical and Geoenvironmental Engineering,* vol. 123, no. 1, p. 78.

Day, R. W. (1997b). "Hydraulic Conductivity of a Desiccated Clay Upon Wetting." *Journal of Environmental and Engineering Geoscience,* Joint Publication, AEG and GSA, vol. 3, no. 2, pp. 308–311.

Day, R. W. (1998). Discussion of "Ground-Movement-Related Building Damage." *Journal of Geotechnical and Geoenvironmental Engineering,* ASCE, vol. 124, no. 5, pp. 462–463.

Day, R. W. (1999). *Geotechnical and Foundation Engineering: Design and Construction.* McGraw-Hill, New York, 800 pp.

Day, R. W., and Axten, G. W. (1989). "Surficial Stability of Compacted Clay Slopes." *Journal of Geotechnical Engineering,* ASCE, vol. 115, no. 4, pp. 577–580.

Day, R. W., and Axten, G. W. (1990). "Softening of Fill Slopes Due to Moisture Infiltration." *Journal of Geotechnical Engineering,* ASCE, vol. 116, no. 9, pp. 1424–1427.

Day, R. W., and Marsh, E. T. (1995). "Triaxial *A*-Value versus Swell or Collapse for Compacted Soil." *Journal of Geotechnical Engineering,* ASCE, vol. 121, no. 7, pp. 566–570.

Day, R. W., and Poland, D. M. (1996). "Damage Due to Northridge Earthquake Induced Movement of Landslide Debris." *Journal of Performance of Constructed Facilities,* ASCE, vol. 10, no. 3, pp. 96–108.

Day, R. W., and Thoeny, S. (1998). "Reactivation of a Portion of an Ancient Landslide." *Journal of the Environmental and Engineering Geoscience,* AEG and GSA, vol. 4, no. 2, pp. 261–269.

DeGaetano, A. T., Wilks, D. S., and McKay, M. (1997). "Extreme-Value Statistics for Frost Penetration Depths in Northeastern United States." *Journal of Geotechnical and Geoenvironmental Engineering,* ASCE, vol. 123, no. 9, pp. 828–835.

de Mello, V.F.B. (1971). "The Standard Penetration Test." State-of-the-Art Report, *Fourth Panamerican Conference on Soil Mechanics and Foundation Engineering,* San Juan, Puerto Rico, vol. 1, pp. 1–86.

Department of the Army (1970). *Engineering and Design, Laboratory Soils Testing* (Engineer Manual EM 1110-2-1906). U.S. Army Engineer Waterways Experiment Station, published by the Department of the Army, Washington, D.C., 282 pp.

Departments of the Army and the Air Force (1979). *Soils and Geology, Procedures for Foundation Design of Buildings and Other Structures (Except Hydraulic Structures).* TM 5/818-1/AFM 88-3, chap. 7, Washington, D.C.

Design and Control of Concrete Mixtures (1994), 4th printing of 13th ed. Portland Cement Association, Skokie, Ill., 205 pp.

Diaz, C. F., Hadipriono, F. C., and Pasternack, S. (1994). "Failures of Residential Building Basements in Ohio." *Journal of Performance of Constructed Facilities,* ASCE, vol. 8, no. 1, pp. 65–80.

Dounias, G. T., and Potts, D. M. (1993). "Numerical Analysis of Drained Direct and Simple Shear Tests." *Journal of Geotechnical Engineering,* ASCE, vol. 119, no. 12, pp. 1870–1891.

Driscoll, R. (1983). "The Influence of Vegetation on the Swelling and Shrinking of Clay Soils in Britain." *Geotechnique,* London, vol. 33, no. 2, pp. 1–67.

Dudley, J. H. (1970). "Review of Collapsing Soils," *Journal of the Soil Mechanics and Foundations Division,* ASCE, vol. 96, no. SM3, pp. 925–947.

Duke, C. M. (1960). "Foundations and Earth Structures in Earthquakes." *Proceedings of the 2nd World Conference on Earthquake Engineering,* vol. 1, Tokyo and Kyoto, pp. 435–455.

Duncan, J. M. (1993). "Limitations of Conventional Analysis of Consolidation Settlement." *Journal of Geotechnical Engineering,* ASCE, vol. 119, no. 9, 1331–1359.

Duncan, J. M. (1996). "State of the Art: Limit Equilibrium and Finite-Element Analysis of Slopes." *Journal of Geotechnical Engineering,* ASCE, vol. 122, no. 7, pp. 577-596.

Duncan, J. M., and Buchignani, A. L. (1976). "An Engineering Manual for Settlement Studies." *Geotechnical Engineering Report,* University of California at Berkeley, 94 pp.

Duncan, J. M., Williams, G. W., Sehn, A. L., and Seed, R. B. (1991). "Estimation Earth Pressures Due to Compaction." *Journal of Geotechnical Engineering,* ASCE, vol. 117, no. 12, pp. 1833–1847.

Dyni, R. C., and Burnett, M. (1993). "Speedy Backfilling for Old Mines." *Civil Engineering Magazine,* ASCE, vol. 63, no. 9, pp. 56–58.

Earth Manual (1985), 2nd ed. A Water Resources Technical Publication, U.S. Department of the Interior, Bureau of Reclamation, Denver, 810 pp.

Ehlig, P. L. (1986). "The Portuguese Bend Landslide: Its Mechanics and a Plan for Its Stabilization." *Landslides and Landslide Mitigation in Southern California.* 82nd Annual Meeting of the Cordilleran Section of the Geological Society of America, Los Angeles, pp. 181–190.

Ehlig, P. L. (1992). "Evolution, Mechanics, and Migration of the Portuguese Bend Landslide, Palos Verdes Peninsula, California." *Engineering Geology Practice in Southern California.* B. W. Pipkin and R. J. Proctor, eds. Star Publishing, Association of Engineering Geologists, Southern California section, Special Publication No. 4, pp. 531–553.

Ellen, S. D., and Fleming, R. W. (1987). "Mobilization of Debris Flows from Soil Slips, San Francisco Bay Region, California." *Debris Flows/Avalanches: Process, Recognition, and Mitigation,* The Geological Society of America, Boulder, Colo., pp. 31–40.

Engineering Geology Field Manual (1987). U.S. Department of the Interior, Bureau of Reclamation, Washington, D.C., 598 pp.

Evans, D. A. (1972). *Slope Stability Report,* Slope Stability Committee, Department of Building and Safety, Los Angeles.

Feld, J. (1965). "Tolerance of Structures to Settlement." *Journal of the Soil Mechanics and Foundations Division* , ASCE, vol. 91, no. SM3, pp. 63–77.

Feld, J., and Carper, K. L. (1997). *Construction Failure,* 2nd ed. John Wiley & Sons, New York, 512 pp.

Fellenius, W. (1936). "Calculation of the Stability of Earth Dams." *Proceedings of the 2d Congress on Large Dams,* vol. 4, Washington, D.C., pp. 445–463.

Fields of Expertise (undated). Joint Civil Engineers/Engineering Geology Committee, appointed by the Professional Engineers and Geologist and Geophysicist Boards of Registration, Calif., 5 pp.

Flaate, K., and Preber, T. (1974). "Stability of Road Embankments." *Canadian Geotechnical Journal,* vol. 11, no. 1, pp. 72–88.

Florensov, N. A., and Solonenko, V. P., eds. (1963). "Gobi-Altayskoye Zemletryasenie." *Iz. Akad. Nauk SSSR.*; also 1965, *The Gobi-Altai Earthquake,* U.S. Department of Commerce (English translation), Washington, D.C.

Fluet, J. E. (1988). "Geosynthetics for Soil Improvement: A General Report and Keynote Address." *Geosynthetics for Soil Improvement,* R. D. Holtz, ed. Geotechnical Special Publication No. 18, American Society of Civil Engineers, New York.

Foott, R., and Ladd, C. C. (1981). "Undrained Settlement of Plastic and Organic Clays." *Journal of the Geotechnical Engineering Division,* ASCE, vol. 107, no. GT8, pp. 1079–1094.

Foshee, J., and Bixler, B. (1994). "Cover-Subsidence Sinkhole Evaluation of State Road 434, Longwood, Florida." *Journal of Geotechnical Engineering,* ASCE, vol. 120, no. 11, pp. 2026–2040.

Foster, C. R., and Ahlvin, R. G. (1954). "Stresses and Deflections Induced by a Uniform Circular Load." *Proceedings of the Highway Research Board,* vol. 33, Washington, D.C., pp. 467–470.

Fourie, A. B. (1989). "Laboratory Evaluation of Lateral Swelling Pressure." *Journal of Geotechnical Engineering,* ASCE, vol. 115, no. 10, pp. 1481–1486.

Fredlund, D. G. (1983). "Prediction of Ground Movements in Swelling Clays." *31st Annual Soil Mechanics and Foundation Engineering Conference,* ASCE, invited lecture, University of Minnesota.

Fredlund, D. G., and Rahardjo, H. (1993). *Soil Mechanics for Unsaturated Soil.* John Wiley & Sons, New York, 517 pp.

Geologist and Geophysicist Act (1986). Board of Registration for Geologists and Geophysicists, Department of Consumer Affairs, State of California, Sacramento, 55 pp.

Geo-Slope (1991). *User's Guide, SLOPE/W for Slope Stability Analysis.* Version 2, Geo-Slope International, Calgary, Canada, 444 pp.

Geo-Slope (1992). *User's Guide, SEEP/W for Finite Element Seepage Analysis.* Version 2, Geo-Slope International, Calgary, Canada, 349 pp.

Geo-Slope (1993). *User's Guide, SIGMA/W for Finite Element Stress/Deformation Analysis.* Version 2, Geo-Slope International, Calgary, Canada, 407 pp.

Gill, L. D. (1967). "Landslides and Attendant Problems." *Mayor's Ad Hoc Landslide Committee Report,* Los Angeles.

g*INT* (1991). *Geotechnical Integrator Software Manual.* Computer software and manual developed by geotechnical computer applications (GCA), Santa Rosa, Calif., 800 pp.

Goh, A. T. C. (1993). "Behavior of Cantilever Retaining Walls." *Journal of Geotechnical Engineering,* ASCE, vol. 119, no. 11, pp. 1751–1770.

Goldman, S. J., Jackson, K., and Bursztynsky, T. A. (1986). *Erosion and Sediment*

Control Handbook. McGraw-Hill, New York, 454 pp.

Graf, E. D. (1969). "Compaction Grouting Techniques," *Journal of the Soil Mechanics and Foundations Division,* ASCE, vol. 95, no. SM5, pp. 1151–1158.

Grant, R., Christian, J. T., and Vanmarcke, E. H. (1974). "Differential Settlement of Buildings." *Journal of the Geotechnical Engineering Division,* ASCE, vol. 100, no. GT9, pp. 973–991.

Grantz, A., Plafker, G., and Kachadoorian, R. (1964). *Alaska's Good Friday Earthquake, March 27, 1964.* Department of the Interior, Geological Survey Circular 491, Washington, D.C.

Gray, R. E. (1988). "Coal Mine Subsidence and Structures." *Mine Induced Subsidence: Effects on Engineered Structures Division,* Geotechnical Special Publication 19, ASCE, New York, pp. 69–86.

Greenfield, S. J., and Shen, C. K. (1992). *Foundations in Problem Soils.* Prentice-Hall, Englewood Cliffs, N.J., 240 pp.

Griffin, D. C. (1974). "Kentucky's Experience with Dams and Dam Safety." *Safety of Small Dams. Proceedings of Engineering Foundation Conference,* Henniker, N.H., ASCE, New York, pp. 194–207.

Gromko, G. J. (1974). "Review of Expansive Soils." *Journal of the Geotechnical Engineering Division,* ASCE, vol. 100, no. GT6, pp. 667–687.

Hammer, M. J., and Thompson, O. B. (1966). "Foundation Clay Shrinkage Caused by Large Trees." *Journal of the Soil Mechanics and Foundations Division,* ASCE, vol. 92, no. SM6, pp. 1–17.

Hansbo, S. (1975). *Jordmateriallära,* Almqvist & Wiksell, Stockholm, Sweden, 218 pp.

Hansen, W. R. (1965). *Effects of the Earthquake of March 27, 1964 at Anchorage, Alaska.* Geological Survey Professional Paper 542-A, U.S. Department of the Interior, Washington, D.C.

Harr, M.E. (1962). *Groundwater and Seepage.* McGraw-Hill, New York, 315 pp.

Hawkins, A. B., and Pinches, G. (1987). "Expansion Due to Gypsum Crystal Growth." *Proceedings of the 6th International Conference on Expansive Soils,* vol. 1, New Delhi, India, pp. 183–188.

Hawkins, A. B., and Privett, K. D. (1985). "Measurement and Use of Residual Shear Strength of Cohesive Soils." *Ground Engineering,* vol. 18, no. 8, pp. 22–29.

Hollingsworth, R. A., and Grover, D. J. (1992). "Causes and Evaluation of Residential Damage in Southern California." *Engineering Geology Practice in Southern California,* Pipkin, B. W., and Proctor, R. J., eds. Star Publishing, Belmont, Calif., pp. 427–441.

Holtz, R. D., and Holm, G. (1979). "Test Embankment on an Organic Silty Clay." *Proceedings of the 7th European Conference on Soil Mechanics and Foundation Engineering,* Brighton, England, vol. 3, pp. 79–86.

Holtz, R. D., and Kovacs, W. D. (1981). *An Introduction to Geotechnical Engineering,* Prentice-Hall, Englewood Cliffs, N.J., 733 pp.

Holtz, W. G. (1959). "Expansive Clays—Properties and Problems." *Quarterly of the Colorado School of Mines,* vol. 54, no. 4, pp. 89–125.

Holtz, W. G. (1984). "The Influence of Vegetation on the Swelling and Shrinking of

Clays in the United States of America." *The Influence of Vegetation on Clays,* Thomas Telford, London, pp. 69–73.

Holtz, W. G., and Gibbs, H. J. (1956). "Engineering Properties of Expansive Clays." *Transactions,* ASCE, vol. 121, pp. 641–677.

Holtz, W. G., and Gibbs, H. J. (1956). "Shear Strength of Pervious Gravelly Soils." *Proceedings,* ASCE, paper no. 867.

Horn, H. M., and Deere, D. U. (1962). "Frictional Characteristics of Minerals." *Geotechnique,* vol. 12, London, pp. 319–335.

Hough, B. K. (1969) *Basic Soils Engineering.* Ronald Press, New York, 634 pp.

Housner, G. W. (1970). "Strong Ground Motion." Chapter 4 of *Earthquake Engineering,* Robert L. Wiegel, coordinating ed. Prentice-Hall, Englewood Cliffs, N.J., pp. 75–92.

Houston, S. L., and Walsh, K. D. (1993). "Comparison of Rock Correction Methods for Compaction of Clayey Soils." *Journal of Geotechnical Engineering,* ASCE, vol. 119, no. 4, pp. 763–778.

Hurst, W. D. (1968). "Experience in the Winnipeg Area with Sulphate-Resisting Cement Concrete." *Performance of Concrete, Resistance of Concrete to Sulphate and Other Environmental Conditions,* University of Toronto Press, Toronto, pp. 125–134.

Hvorslev, M. J. (1949). *Subsurface Exploration and Sampling of Soils for Civil Engineering Purposes.* Waterways Experiment Station, Vicksburg, Miss., 521 pp.

Hvorslev, M. J. (1951). "Time Lag and Soil Permeability in Ground Water Measurements." Bulletin 36, Corps of Engineers Waterways Experiment Station, Vicksburg, Miss., 50 pp.

Ishihara, K. (1985). "Stability of Natural Deposits during Earthquakes." *Proceedings of the 11th International Conference on Soil Mechanics and Foundation Engineering,* vol.1, San Francisco, pp. 321–376.

Ishihara, K. (1993). "Liquefaction and Flow Failure during Earthquakes." *Geotechnique,* vol. 43, no. 3, London, pp. 351–415.

Jaky, J. (1944). "The Coefficient of Earth Pressure at Rest" (in Hungarian). *Journal of the Society of Hungarian Architects and Engineers,* vol. 78, no. 22, pp. 355–358.

Jaky, J. (1948). "Earth Pressure in Silos." *Proceedings of the Second International Conference on Soil Mechanics and Foundation Engineering,* Rotterdam, Netherlands, vol. 1, pp. 103–107.

Jamiolkowski, M., Ladd, C. C., Germaine, J. T., and Lancellotta, R. (1985). "New Developments in Field and Laboratory Testing of Soils." *Proceedings of the 11th International Conference on Soil Mechanics and Foundation Engineering,* San Francisco, vol. 1, pp. 57–153.

Janbu, N. (1957). "Earth Pressure and Bearing Capacity Calculation by Generalized Procedure of Slices." *Proceedings of the 4th International Conference on Soil Mechanics and Foundation Engineering,* London, vol. 2, pp. 207–212.

Janbu, N. (1968). "Slope Stability Computations." *Soil Mechanics and Foundation Engineering Report,* The Technical University of Norway, Trondheim.

Jennings, J. E. (1953). "The Heaving of Buildings on Desiccated Clay." *Proceedings of the 3rd International Conference on Soil Mechanics and Foundation Engineering,* vol. 1, Zurich, pp. 390-396.

Jennings, J. E., and Knight, K. (1957). "The Additional Settlement of Foundations Due to a Collapse Structure of Sandy Subsoils on Wetting." *Proceedings of the 4th International Conference on Soil Mechanics and Foundation Engineering,* vol. 1, London, pp. 316–319.

Johnpeer, G. D. (1986). "Land Subsidence Caused by Collapsible Soils in Northern New Mexico." *Ground Failure,* National Research Council, Committee on Ground Failure Hazards, vol. 3, Washington, D.C., 24 pp.

Johnson, A. M., and Hampton, M. A. (1969). "Subaerial and Subaqueous Flow of Slurries." Final Report, U.S. Geological Survey (USGS) Contract no. 14-08-0001-10884, USGS, Boulder, Colo.

Johnson, A. M., and Rodine, J. R. (1984). "Debris Flow." Chapter 8 of *Slope Instability,* John Wiley & Sons, New York, pp. 257–361.

Johnson, L. D. (1980). "Field Test Sections on Expansive Soil." *4th International Conference on Expansive Soils,* Denver, ASCE, pp. 262–283.

Jones, D. E., and Jones, K. A. (1987). "Treating Expansive Soils." *Civil Engineering Magazine,* ASCE, vol. 57, no. 8, August.

Jones, D. E., and Holtz, W. G. (1973). "Expansive Soils—The Hidden Disaster." *Civil Engineering Magazine, ASCE,* vol. 43, no. 8, pp. 49–51.

Jubenville, D. M., and Hepworth, R. C. (1981). "Drilled Pier Foundations in Shale, Denver Colorado Area." *Proceedings of the Session on Drilled Piers and Caissons,* ASCE, New York, pp. 66–81.

Kaplar, C. W. (1970). "Phenomenon and Mechanism of Frost Heaving." *Highway Research Record 304,* pp. 1–13.

Kassiff, G., and Baker, R. (1971). "Aging Effects on Swell Potential of Compacted Clay." *Journal of the Soil Mechanics and Foundations Division,* ASCE, vol. 97, no. SM3, pp. 529–540.

Kay, B. L. (1983). "Straw as an Erosion Control Mulch." Agronomy Progress Report no. 140, University of California, Davis, Agricultural Experiment Station Cooperative Extension, Davis, Calif.

Kayan, R. E., Mitchell, J. K., Seed, R. B., Lodge, A., Nishio, S., and Coutinho, R. (1992). "Evaluation of SPT-, CPT-, and Shear Wave-Based Methods for Liquefaction Potential Assessments Using Loma Prieta Data." *Proceedings, 4th Japan-U.S. Workshop on Earthquake Resistant Design of Lifeline Facilities and Countermeasures for Soil Liquefaction*; NCEER-92-0019, National Center for Earthquake Engineering, Buffalo, N.Y., pp. 177–192.

Kennedy, M. P. (1975). "Geology of the Western San Diego Metropolitan Area, California." Bulletin 200, California Division of Mines and Geology, Sacramento, 39 pp.

Kennedy, M. P., and Tan, S. S. (1977). "Geology of National City, Imperial Beach and Otay Mesa Quadrangles, Southern San Diego Metropolitan Area, California." Map Sheet 29, California Division of Mines and Geology, Sacramento, 1 sheet.

Kenney, T. C. (1959). Discussion of "Geotechnical Properties of Glacial Lake Clays," by T. H. Wu, *Journal of the Soil Mechanics and Foundations Division,* ASCE, vol. 85, no. SM3, pp. 67–79.

Kenney, T. C. (1964). "Sea-Level Movements and the Geologic Histories of the Post-Glacial Marine Soils at Boston, Nicolet, Ottawa, and Oslo." *Geotechnique,* vol. 14, no. 3, London, pp. 203–230.

Koerner, R. M. (1990). *Designing with Geosynthetics,* 2nd ed. Prentice-Hall, Englewood Cliffs, N.J.

Kratzsch, H. (1983). *Mining Subsidence Engineering.* Springer-Verlag, Berlin, 543 pp.

Ladd, C. C. (1971). "Strength Parameters and Stress-Strain Behavior of Saturated Clays." Massachusetts Institute of Technology, MIT Research Report R71-23.

Ladd, C. C. (1973). "Settlement Analysis for Cohesive Soil." MIT Research Report R71-2, no. 272, Department of Civil Engineering, Massachusetts Institute of Technology, Cambridge, Mass., 115 pp.

Ladd, C. C. (1975). *Foundation Design of Embankments Constructed on Connecticut Valley Varved Clays.* MIT Research Report R75-7, no. 343, Cambridge, Mass., 439 pp.

Ladd, C. C., Azzouz, A. S., Martin, R. T., Day, R. W., and Malek, A. M. (1980). *Evaluation of Compositional and Engineering Properties of Offshore Venezuelan Soils,* vol. 1, MIT Research Report R80-14, no. 665, Cambridge, Mass., 286 pp.

Ladd, C. C., and Foott, R. (1974). "A New Design Procedure for Stability of Soft Clays." *Journal of the Geotechnical Engineering Division,* ASCE, vol. 100, no. GT7, pp. 763–786.

Ladd, C. C., Foott, R., Ishihara, K., Schlosser, F., and Poulos, H. G. (1977). "Stress-Deformation and Strength Characteristics." State-of-the-Art Report, *Proceedings, 9th International Conference on Soil Mechanics and Foundation Engineering,* Japanese Society of Soil Mechanics and Foundation Engineering, Tokyo, vol. 2, pp. 421–494.

Ladd, C. C., and Lambe, T. W. (1961). "The Identification and Behavior of Expansive Clays." *Proceedings, 5th International Conference on Soil Mechanics and Foundation Engineering,* Paris, vol. 1.

Ladd, C. C., and Lambe T. W. (1963). "The Strength of 'Undisturbed' Clay Determined from Undrained Tests," STP 361, ASTM, Philadelphia, pp. 342–371.

Lambe, T. W. (1951). *Soil Testing for Engineers.* John Wiley & Sons, New York, 165 pp.

Lambe, T. W. (1958a). "The Structure of Compacted Clay." *Journal of the Soil Mechanics and Foundations Division,* ASCE, vol. 84, no. SM2, pp. 1654-1 to 1654-34.

Lambe, T. W. (1958b). "The Engineering Behavior of Compacted Clay." *Journal of the Soil Mechanics and Foundations Division,* ASCE, vol. 84, no. SM2, pp. 1655-1 to 1655-35.

Lambe, T. W. (1967). "Stress Path Method." *Journal of the Soil Mechanics and Foundations Division,* ASCE, vol. 93, no. SM6, pp. 309–331.

Lambe, T. W., and Whitman, R. V. (1969). *Soil Mechanics.* John Wiley & Sons, New York, 553 pp.

Lane et al. (1947). "Modified Wentworth Scale." *Transactions of the American Geophysical Union,* vol. 28, pp. 936–938.

La Rochelle, P., Trak, B., Tavenas, F., and Roy, M. (1974). "Failure of a Test Embankment on a Sensitive Champlain Clay Deposit." *Canadian Geotechnical Journal,* vol. 11, no. 1, pp. 142–164.

Lawson, A. C., et al. (1908). *The California Earthquake of April 18, 1906—Report of the State Earthquake Investigation Commission,* vol. 1, part 1, pp. 1–254; part 2, pp. 255–451. Carnegie Institution of Washington, Publication 87.

Lawton, E. C. (1996). "Nongrouting Techniques." *Practical Foundation Engineering Handbook.* Robert W. Brown, ed. McGraw-Hill, New York, Sec. 5, pp. 5.3–5.276.

Lawton, E. C., Fragaszy, R. J., and Hardcastle, J. H. (1989). "Collapse of Compacted Clayey Sand." *Journal of Geotechnical Engineering,* ASCE, vol. 115, no. 9, pp. 1252–1267.

Lawton, E. C., Fragaszy, R. J., and Hardcastle, J. H. (1991). "Stress Ratio Effects on Collapse of Compacted Clayey Sand." *Journal of Geotechnical Engineering,* ASCE, vol. 117, no. 5, pp. 714–730.

Lawton, E. C., Fragaszy, R. J., and Hetherington, M. D. (1992). "Review of Wetting-Induced Collapse in Compacted Soil." *Journal of Geotechnical Engineering,* ASCE, vol. 118, no. 9, pp. 1376–1394.

Lea, F. M. (1971). *The Chemistry of Cement and Concrete,* 1st American ed., Chemical Publishing Company, New York.

Leonards, G. A. (1962). *Foundation Engineering.* McGraw-Hill, New York, 1136 pp.

Leonards, G. A., and Altschaeffl, A. G. (1964). "Compressibility of Clay." *Journal of the Soil Mechanics and Foundations Division,* ASCE, vol. 90, no. SM5, pp. 133–156.

Leonards, G. A., and Ramiah, B. K. (1959). "Time Effects in the Consolidation of Clay." Special Technical Publication No. 254, ASTM, Philadelphia, pp. 116–130.

Lin, G., Bennett, R. M., Drumm, E. C., and Triplett, T. L. (1995). "Response of Residential Test Foundations to Large Ground Movements." *Journal of Performance of Constructed Facilities,* ASCE, vol. 9, no. 4, pp. 319–329.

Lowe, J., and Zaccheo, P. F. (1975). "Subsurface Explorations and Sampling," Chap. 1 of *Foundation Engineering Handbook,* Hans F. Winterkorn and Hsai-Yang Fang, eds. Van Nostrand Reinhold, New York, pp. 1–66.

Lytton, R. L., and Dyke, L. D. (1980). "Creep Damage to Structures on Expansive Clay Slopes." *4th International Conference on Expansive Soils,* ASCE, vol. 1, New York, pp. 284–301.

Mabsout, M. E., Reese, L. C., and Tassoulas, J. L. (1995). "Study of Pile Driving by Finite-Element Method." *Journal of Geotechnical Engineering,* ASCE, vol. 121, no. 7, pp. 535–543.

Mahtab, M. A., and Grasso, P. (1992). *Geomechanics Principles in the Design of Tunnels and Caverns in Rock.* Developments in Geotechnical Engineering, 72, Elsevier Science, New York, 250 pp.

Maksimovic, M. (1989a). "On the Residual Shearing Strength of Clays." *Geotechnique,* London, vol. 39, no. 2, pp. 347–351.

Maksimovic, M. (1989b). "Nonlinear Failure Envelope for Soils." *Journal of Geotechnical Engineering,* ASCE, vol. 115, no. 4, pp. 581–586.

Marino, G. G., Mahar, J. W., and Murphy, E. W. (1988). "Advanced Reconstruction for Subsidence-Damaged Homes." *Mine Induced Subsidence: Effects on Engineered Structures.* H. J. Siriwardane, ed. ASCE, New York, pp. 87–106.

Marsh, E. T., and Walsh, R. K. (1996). "Common Causes of Retaining-Wall Distress: Case Study." *Journal of Performance of Constructed Facilities,* ASCE, vol. 10, no. 1, pp. 35–38.

Marston, A. (1930). "The Theory of External Loads on Closed Conduits in the Light of the Latest Experiments." Bulletin no. 96, Iowa Engineering Experiment Station.

Massarsch, K. R. (1979). "Lateral Earth Pressure in Normally Consolidated Clay." *Proceedings of the Seventh European Conference on Soil Mechanics and Foundation Engineering,* Brighton, England, vol. 2, pp. 245–250.

Massarsch, K. R., Holtz, R. D., Holm, B. G., and Fredricksson, A. (1975). "Measurement of Horizontal In Situ Stresses." *Proceedings of the ASCE Specialty Conference on In Situ Measurement of Soil Properties,* Raleigh, N.C., vol. 1, pp. 266–286.

Mather, B. (1968). "Field and Laboratory Studies of the Sulphate Resistance of Concrete." *Performance of Concrete, Resistance of Concrete to Sulphate and Other Environmental Conditions,* University of Toronto Press, Toronto, pp. 66–76.

McCarthy, D. F. (1977). *Essentials of Soil Mechanics and Foundations.* Reston Publishing, Reston, Va., 505 pp.

McElroy, C. H. (1987). "The Use of Chemical Additives to Control the Erosive Behavior of Dispersed Clays." *Engineering Aspects of Soil Erosion, Dispersive Clays and Loess,* Geotechnical Special Publication No. 10. C. W. Lovell and R. L. Wiltshire, eds. ASCE, New York, pp. 1–16.

Meehan, R. L., and Karp, L. B. (1994). "California Housing Damage Related to Expansive Soils." *Journal of Performance of Constructed Facilities,* ASCE, vol. 8, no. 2, pp. 139–157.

Mehta, P. K. (1976). Discussion of "Combating Sulfate Attack in Corps of Engineers Concrete Construction" by Thomas J. Reading, *ACI Journal Proceedings,* vol. 73, no. 4, pp. 237–238.

Merfield, P. M. (1992). "Surfical Slope Failures: the Role of Vegetation and Other Lessons from Rainstorms." *Engineering Geology Practice in Southern California,* B. W. Pipkin and R. J. Proctor, eds. Star Publishing, Association of Engineering Geologists, Southern California section, Special Publication No. 4, pp. 613–627.

Meyerhof, G. G. (1951). "The Ultimate Bearing Capacity of Foundations." *Geotechnique,* vol. 2, no. 4, pp. 301–332.

Meyerhof, G. G. (1953). "Bearing Capacity of Foundations under Eccentric and Inclined Loads." *Proceedings of the 3rd International Conference on Soil Mechanics and Foundation Engineering,* vol. 1, Zurich, pp. 440–445.

Meyerhof, G. G. (1961). Discussion of "Foundations Other Than Piled Foundations." *Proceedings of the Fifth International Conference on Soil Mechanics and Foundation Engineering,* vol. 3, Paris, p. 193.

Meyerhof, G. G. (1965). "Shallow Foundations." *Journal of the Soil Mechanics and Foundations Division,* ASCE, vol. 91, no. SM2, pp. 21–31.

Middlebrooks, T. A. (1953). "Earth Dam Practice in the United States." *Transactions, American Society of Civil Engineers,* centennial volume, pp. 697.

Milligan, V. (1972). Discussion of "Embankments on Soft Ground," *Proceedings of the ASCE Specialty Conference on Performance of Earth and Earth-Supported Structures,* Purdue University, vol. 3, pp. 41–48.

Mitchell, J. K. (1970). "In-Place Treatment of Foundation Soils," *Journal of the Soil Mechanics and Foundations Division,* ASCE, vol. 96, no. SM1, pp. 73–110.

Mitchell, J. K. (1976). *Fundamentals of Soil Behavior.* John Wiley & Sons, New York, 422 pp.

Mitchell, J. K. (1978). "*In Situ* Techniques for Site Characterization." *Site Characterization & Exploration. Proceedings of Specialty Workshop,* Northwestern University, C. H. Dowding, ed. ASCE, New York, pp. 107–130.

Monahan, E. J. (1986). *Construction of and on Compacted Fills.* John Wiley & Sons, New York, 200 pp.

Morgenstern, N. R., and Price, V. E. (1965). "The Analysis of the Stability of General Slip Surfaces." *Geotechnique,* vol. 15, no.1, London, pp. 79–93.

Mottana, A., Crespi, R., and Liborio, G. (1978). *Rocks and Minerals.* Simon & Schuster, New York, 607 pp.

Munsell Soil Color Charts (1975). Munsell Color Division of Kollmorgen Corp., Baltimore.

Myslivec, A., and Kysela, Z. (1978). *The Bearing Capacity of Building Foundations,* Developments in Geotechnical Engineering 21, Elsevier Scientific, New York, 237 pp.

Nadjer, J., and Werno, M. (1973). "Protection of Buildings on Expansive Clays." *Proceedings of the 3rd International Conference on Expansive Soils,* Haifa, Israel, vol. 1, pp. 325–334.

National Coal Board (1975). *Subsidence Engineers Handbook.* National Coal Board Mining Department, National Coal Board, London, 111 pp.

National Research Council (1985). *Reducing Losses from Landsliding in the United States.* Committee on Ground Failure Hazards, Commission on Engineering and Technical Systems, National Academy Press, Washington, D.C., 41 pp.

NSF (1992). "Quantitative Nondestructive Evaluation for Constructed Facilities." *Announcement Fiscal Year 1992.* National Science Foundation, Directorate for Engineering, Division of Mechanical and Structural Systems, Washington, D.C.

NAVFAC DM-7 (1971). *Soil Mechanics, Foundations, and Earth Structures,* Design Manual DM-7, Department of the Navy, Naval Facilities Engineering Command, Alexandria, Va., 280 pp.

NAVFAC DM-7.1 (1982). *Soil Mechanics,* Design Manual 7.1, Department of the Navy, Naval Facilities Engineering Command, Alexandria, Va., 364 pp.

NAVFAC DM-7.2 (1982). *Foundations and Earth Structures,* Design Manual 7.2, Department of the Navy, Naval Facilities Engineering Command, Alexandria, Va., 253 pp.

NAVFAC DM-7.3 (1983). *Soil Dynamics, Deep Stabilization, and Special Geotechnical Construction,* Design Manual 7.3, Department of the Navy, Naval Facilities Engineering Command, Alexandria, Va., 106 pp.

NAVFAC DM-21.3 (1978). *Flexible Pavement Design for Airfields,* Design Manual 21.3, Department of the Navy, Naval Facilities Engineering Command, Alexandria, Va., 98 pp.

Neary, D. G., and Swift, L. W. (1987). "Rainfall Thresholds for Triggering a Debris Avalanching Event in the Southern Appalachian Mountains." *Debris Flows/Avalanches: Process, Recognition, and Mitigation,* The Geological Society of America, Boulder, Colo., pp. 81–92.

Nelson, J. D., and Miller, D. J. (1992). *Expansive Soils, Problems and Practice in Foundation and Pavement Engineering.* John Wiley & Sons, New York, 259 pp.

Newmark, N. M. (1935). "Simplified Computation of Vertical Pressures in Elastic Foundations." University of Illinois Engineering Experiment Station Circular 24, Urbana, Ill., 19 pp.

Newmark, N. M. (1942). "Influence Charts for Computation of Stresses in Elastic Foundations." University of Illinois Engineering Experiment Station Bulletin, Series no. 338, vol. 61, no. 92, Urbana, Ill., reprinted 1964, 268 pp.

Nichols, H. L., and Day, D. A. (1999). *Moving the Earth: The Workbook of Excavation.* 4th ed., McGraw-Hill, New York, 1400 pp.

Oldham, R. D. (1899). "Report on the Great Earthquake of 12th June, 1897." India Geologic Survey Memorial, Publication 29, 379 pp.

Oliver, A. C. (1988). *Dampness in Buildings.* Internal and Surface Waterproofers, Nichols Publishing, New York, 221 pp.

O'Neill, M. W., and Reese, L. C. (1970). *Behavior of Axially Loaded Drilled Shafts in Beaumont Clay,* Research Report 89-8, Part 1, State of the Art Report, Center for Highway Research, University of Texas at Austin.

Orange County Grading Manual (1993). Part of the *Orange County Grading and Excavation Code,* prepared by Orange County, Calif.

Ortigao, J. A. R., Loures, T. R. R., Nogueiro, C., and Alves, L. S. (1997). "Slope Failures in Tertiary Expansive OC Clays." *Journal of Geotechnical and Geoenvironmental Engineering,* ASCE, vol. 123, no. 9, pp. 812–817.

Osterberg, J. O. (1957). "Influence Values for Vertical Stresses in a Semi-infinite Mass Due to an Embankment Loading." *Proceedings of the Fourth International Conference on Soil Mechanics and Foundation Engineering,* vol. 1, London, pp. 393–394.

Peck, R. B., Hanson, W. E., and Thornburn, T. H. (1974). *Foundation Engineering,* John Wiley & Sons, New York, 514 pp.

Peckover, F. L. (1975). "Treatment of Rock Falls on Railway Lines." *American Railway Engineering Association,* Bulletin 653, Chicago, pp. 471–503.

Peng, S. S. (1986). *Coal Mine Ground Control,* 2nd ed., John Wiley & Sons, New York, 491 pp.

Peng, S. S. (1992). *Surface Subsidence Engineering.* Society for Mining, Metallurgy and Exploration, Littleton, Colo.

Perloff, W. H., and Baron, W. (1976). *Soil Mechanics, Principles and Applications.* John Wiley & Sons, New York, 745 pp.

Perry, E. B. (1987). "Dispersive Clay Erosion at Grenada Dam, Mississippi." *Engineering Aspects of Soil Erosion, Dispersive Clays and Loess,* Geotechnical Special Publication No. 10. C. W. Lovell and R. L. Wiltshire, eds. ASCE, New York, pp. 30–45.

Petersen, E. V. (1963). "Cave-in!" *Roads and Engineering Construction,* November, pp. 25–33.

Piteau, D. R., and Peckover, F. L. (1978). "Engineering of Rock Slopes." *Landslides, Analysis and Control,* Special Report 176, Transportation Research Board, National Academy of Sciences, chap. 9, pp. 192–228.

Post-Tensioning Institute (1996). "Design and Construction of Post-Tensioned Slabs-on-Ground," report, 2nd ed. Phoenix, 101 pp.

Poulos, H. G., and Davis, E. H. (1974). *Elastic Solutions for Soil and Rock Mechanics.* John Wiley & Sons, New York, 411 pp.

Pradel, D., and Raad, G. (1993). "Effect of Permeability on Surficial Stability of Homogeneous Slopes." *Journal of Geotechnical Engineering,* ASCE, vol. 119, no. 2, pp. 315–332.

Prentis, E. A., and White, L. (1950). *Underpinning,* 2nd ed. Columbia University Press, New York.

Price, N. J. (1966). *Fault and Joint Development in Brittle and Semi-Brittle Rock.* Pergamon Press, Oxford, England.

Proctor, R. R. (1933). "Fundamental Principles of Soil Compaction." *Engineering News-Record,* vol. 111, Nos. 9, 10, 12, and 13.

Raschke, S. A., and Hryciw, R. D. (1997). "Vision Cone Penetrometer for Direct Subsurface Soil Observation." *Journal of Geotechnical and Geoenvironmental Engineering,* ASCE, vol. 123, no. 11, pp. 1074-1076.

Raumann, G. (1982). "Outdoor Exposure Tests on Geotextiles." *Proceedings of the 2nd International Conference on Geotextiles,* Industrial Fabrics Association International, St. Paul, Minn.

Ravina, I. (1984). "The Influence of Vegetation on Moisture and Volume Changes." *The Influence of Vegetation on Clays,* Thomas Telford, London, pp. 62–68.

Reading, T. J. (1975). "Combating Sulfate Attack in Corps of Engineering Concrete Construction." *Durability of Concrete,* SP47, American Concrete Institute, Detroit, pp. 343–366.

Reed, M. A., Lovell, C. W., Altschaeffl, A. G., and Wood, L. E. (1979). "Frost Heaving Rate Predicted from Pore Size Distribution," *Canadian Geotechnical Journal,* vol. 16, no. 3, pp. 463–472.

Reese, L. C. (1978). "Design and Construction of Drilled Shafts." *Journal of the Geotechnical Engineering Division,* ASCE, vol. 104, no. GT1, pp. 95–116.

Reese, L. C., Owens, M., and Hoy, H. (1981). "Effects of Construction Methods on Drilled Shafts." *Drilled Piers and Caissons.* M. W. O'Neill, ed. ASCE, New York, pp. 1–18.

Reese, L. C., and Tucker, K. L. (1985). "Bentonite Slurry in Constructing Drilled Piers." *Drilled Piers and Caisson II.* C. N. Baker, ed. ASCE, New York, pp. 1–15.

Rice, R. J. (1988). *Fundamentals of Geomorphology,* 2nd ed. John Wiley & Sons, New York.

Richardson, G. N., and Koerner, R. M. (1990). *A Design Primer: Geotextiles and Related Materials,* Industrial Fabrics Association, St. Paul, Minn.

Richardson, G. N., and Wyant, D. C. (1987). "Geotextiles Construction Criteria." *Geotextile Testing and the Design Engineer,* J. E. Fluet, ed., ASTM Special Technical Publication 952, American Society for Testing and Materials, Philadelphia, pp. 125–138.

Ritchie, A. M. (1963). "Evaluation of Rockfall and Its Control." *Highway Research Record 17,* Highway Research Board, Washington, D.C., pp. 13–28.

Robertson, P. K., and Campanella, R. G. (1983). "Interpretation of Cone Penetration Tests: Parts 1 and 2." *Canadian Geotechnical Journal,* vol. 20, pp. 718–745.

Rogers, J. D. (1992). "Recent Developments in Landslide Mitigation Techniques." Chapter 10 of *Landslides/Landslide Mitigation,* J. E. Slosson, G. G. Keene, and J. A. Johnson, eds. The Geological Society of America, Boulder, Colo., pp. 95–118.

Rollings, M. P., and Rollings, R. S. (1996). *Geotechnical Materials in Construction.* McGraw-Hill, New York, 525 pp.

Rollins, K. M., Rollins, R. L., Smith, T. D., and Beckwith, G. H. (1994). "Identification and Characterization of Collapsible Gravels." *Journal of Geotechnical Engineering,* ASCE, vol. 120, no. 3, pp. 528–542.

Ross, C. S., and Smith, R. L. (1961). *Ash-Flow Tuffs, Their Origin, Geologic Relations and Identification,* U.S. Geological Survey Professional Paper 366: U.S. Geological Survey, Denver.

Rutledge, P. C. (1944). "Relation of Undisturbed Sampling to Laboratory Testing," *Transactions,* ASCE, vol. 109, pp. 1162–1163.

Sandström, G. E. (1963a). *History of Tunneling.* Barrie and Rockcliff, London, 427 pp.

Sandström, G. E. (1963b). *Tunnels.* Holt, Rinehart, and Winston, New York, 427 pp.

Sanglerat, G. (1972). *The Penetrometer and Soil Exploration.* Elsevier Scientific, New York, 464 pp.

Savage, J. C., and Hastie, L. M. (1966). "Surface Deformation Associated with Dip-Slip Faulting." *Journal of Geophysical Research,* vol. 71, no. 20, pp. 4897–4904.

Saxena, S. K., Lourie, D. E., and Rao, J. S. (1984). "Compaction Criteria for Eastern Coal Waste Embankments." *Journal of Geotechnical Engineering,* ASCE, vol. 110, no. 2, pp. 262–284.

Schlager, N. (1994). *When Technology Fails.* "St. Francis Dam Failure." Gale Research, Detroit, pp. 426–430.

Schmertmann, J. H. (1970). "Static Cone to Compute Static Settlement over Sand." *Journal of the Soil Mechanics and Foundations Division,* ASCE, vol. 96, no. SM3, pp. 1011–1043.

Schmertmann, J. H. (1975). "Measurement of In Situ Shear Strength." State-of-the-Art Report, *Proceedings of the ASCE Specialty Conference on In Situ Measurement of Soil Properties,* Raleigh, N.C., vol. 2, pp. 57–138.

Schmertmann, J. H. (1977). *Guidelines for Cone Penetration Test, Performance and Design.* U.S. Department of Transportation, Federal Highway Administration, Washington, D.C., 145 pp.

Schmertmann, J. H., Hartman, J. P., and Brown, P. R. (1978). "Improved Strain Influence Factor Diagrams." *Journal of the Geotechnical Engineering Division,* vol. 104, no. GT8, pp. 1131–1135.

Schuster, R. L. (1986). *Landslide Dams: Processes, Risk, and Mitigation,* Geotechnical Special Publication No. 3. *Proceedings of Geotechnical Session,* Seattle. ASCE, New York, 164 pp.

Schuster, R. L., and Costa, J. E. (1986). "A Perspective on Landslide Dams." *Landslide Dams: Processes, Risk, and Mitigation,* Geotechnical Special Publication No. 3. *Proceedings of Geotechnical Session,* Seattle. ASCE, New York, pp. 1–20.

Schutz, R. J. (1984). "Properties and Specifications for Epoxies Used in Concrete Repair." *Concrete Construction Magazine,* Concrete Construction Publications, Addison, Ill., pp. 873–878.

Seed, H. B. (1970). "Soil Problems and Soil Behavior." Chapter 10 of *Earthquake Engineering,* Robert L. Wiegel, coordinating ed. Prentice-Hall, Englewood Cliffs, N.J., pp. 227–252.

Seed, H. B., and De Alba (1986). "Use of SPT and CPT Tests for Evaluating the Liquefaction Resistance of Sands." *Proceedings, In Situ 1986, ASCE Specialty Conference on Use of In Situ Testing in Geotechnical Engineering,* Special Publication No. 6, ASCE, New York.

Seed, H. B., and Idriss, I. M. (1971). "Simplified Procedure for Evaluating Soil Liquefaction Potential." *Journal of the Soil Mechanics and Foundations Division,* ASCE, vol. 97, no. SM9, 1249–1273.

Seed, H. B., Idriss, I. M., and Arango, I. (1983). "Evaluation of Liquefaction Potential Using Field Performance Data." *Journal of Geotechnical Engineering,* ASCE, vol. 109, no. 3, pp. 458–482.

Seed, H. B., Tokimatsu, K., Harder, L. F., and Chung, R. (1985). "Influence of SPT Procedures in Soil Liquefaction Resistance Evaluations." *Journal of Geotechnical Engineering,* ASCE, vol. 111, no. 12, pp. 1425–1445.

Seed, H. B., and Whitman, R. V. (1970). *Design of Earth Structures for Dynamic Loads.* Lateral Stresses in the Ground and Design of Earth Retaining Structures, ASCE, Cornell University.

Seed, H. B., Woodward, R. J., and Lundgren, R. (1962). "Prediction of Swelling Potential for Compacted Clays." *Journal of the Soil Mechanics and Foundations Division,* ASCE, vol. 88, no. SM3, pp. 53–87.

Seismic Safety Study (1995). City of San Diego, Development Services Department, San Diego.

Shannon and Wilson, Inc. (1964). *Report on Anchorage Area Soil Studies, Alaska, to U.S. Army Engineer District, Anchorage, Alaska.* Seattle.

Sherard, J. L. (1972). "Study of Piping Failures and Eroding Damage from Rain in Clay Dams in Oklahoma and Mississippi." U.S. Department of Agriculture, Soil Conservation Service, Washington, D.C.

Sherard, J. L., Decker, R. S., and Ryker, N. L. (1972). "Piping in Earth Dams of Dispersive Clay." *Proceedings of the Specialty Conference on Performance of Earth and Earth-Supported Structures.* Vol. 1, part 1, cosponsored by ASCE and Purdue University, Lafayette, Ind., pp. 589–626.

Sherard, J. L., Woodward, R. J., Gizienski, S. F., and Clevenger, W. A. (1963). *Earth and Earth-Rock Dams.* John Wiley & Sons, New York, 725 pp.

Sinha, R. S. (1989). *Underground Structures, Design and Instrumentation.* Developments in Geotechnical Engineering, 59A, Elsevier Science, New York, 480 pp.

Sinha, R. S. (1991). *Underground Structures, Design and Construction.* Developments in Geotechnical Engineering, 59B, Elsevier Science, New York, 529 pp.

Skempton, A. W. (1953). "The Colloidal Activity of Clays." *Proceedings of the Third International Conference on Soil Mechanics and Foundation Engineering,* Zurich, Switzerland, vol. 1, pp. 57–61.

Skempton, A. W. (1954). "The Pore-Pressure Coefficients *A* and *B*." *Geotechnique,* vol. 4, no. 4, pp. 143–147.

Skempton, A. W. (1961). "Effective Stress in Soils, Concrete and Rock." *Proceedings of Conference on Pore Pressure and Suction in Soils,* Butterworths, London, pp. 4–16.

Skempton, A. W. (1964). "Long-term Stability of Clay Slopes." *Geotechnique,* London, vol. 14, no. 2, pp. 75–101.

Skempton, A. W. (1985). "Residual Strength of Clays in Landslides, Folded Strata and the Laboratory." *Geotechnique,* London, vol. 35, no. 1, pp. 3–18.

Skempton, A. W. (1986). "Standard Penetration Test Procedures and the Effects in Sands of Overburden Pressure, Relative Density, Particle Size, Aging and Overconsolidation." *Geotechnique,* vol. 36, no. 3, pp. 425–447.

Skempton, A. W., and Bjerrum, L. (1957). "A Contribution to the Settlement Analysis of Foundations on Clay." *Geotechnique,* vol. 7, no. 4, pp. 168–178.

Skempton, A. W., and Henkel, D. J. (1953). "The Post-Glacial Clays of the Thames Estuary at Tilbury and Shellhaven." *Proceedings of the 3rd International Conference on Soil Mechanics and Foundation Engineering,* Switzerland, vol. 1, p. 302–308.

Skempton, A. W., and Hutchinson, J. (1969). "State of-the-art Report: Stability of Natural Slopes and Embankment Foundations." *Proceedings of the 7th International Conference on Soil Mechanics and Foundation Engineering,* Mexico, pp. 291–340.

Skempton, A. W., and MacDonald, D. H. (1956). "The Allowable Settlements of Buildings." *Proceedings of the Institution of Civil Engineers,* part III. The Institution of Civil Engineers, London, no. 5, pp. 727–768.

Slope Indicator Company (1998). *Geotechnical and Structural Instrumentation,* prepared by Slope Indicator Company, Bothell, Wash., 98 pp.

Smith, D. D., and Wischmeier, W. H. (1957). "Factors Affecting Sheet and Rill Erosion." *Transactions of the American Geophysical Union,* vol. 38, no. 6, pp. 889–896.

Smith, R. L. (1960). *Zones and Zonal Variations in Welded Ash Flows,* U.S. Geological Survey Professional Paper 354-F: U.S. Geological Survey, Denver.

Snethen, D. R. (1979). *Technical Guidelines for Expansive Soils in Highway Subgrades.* U.S. Army Engineering Waterway Experiment Station, Vicksburg, Miss., Report no. FHWA-RD-79-51.

Soils and Geology, Procedures for Foundation Design of Buildings and Other Structures (Except Hydraulic Structures) (1979). Departments of the Army and Air Force, TM 5-818-1/AFM 88-3, Chap. 7, Washington, D.C.

Sowers, G. B., and Sowers, G. F. (1970). *Introductory Soil Mechanics and Foundations,* 3rd ed. Macmillan, New York, 556 pp.

Sowers, G. F. (1962). "Shallow Foundations." Chapter 6 from *Foundation Engineering,* G. A. Leonards, ed. McGraw-Hill, New York.

Sowers. G. F. (1974). "Dam Safety Legislation: A Solution or a Problem." *Safety of Small Dams. Proceedings of Engineering Foundation Conference,* Henniker, N.H. ASCE, New York, pp. 65–100.

Sowers, G. F. (1979). *Soil Mechanics and Foundations: Geotechnical Engineering,* 4th ed. Macmillan, New York.

Sowers, G. F. (1997). *Building on Sinkholes: Design and Construction of Foundations in Karst Terrain,* ASCE Press, New York.

Sowers, G. F., and Royster, D. L. (1978). "Field Investigation," chap. 4 of *Landslides, Analysis and Control,* Special Report 176, Transportation Research Board, National Academy of Sciences, R. L. Schuster and R. J. Krizek, eds. Washington, D.C., pp. 81–111.

Spencer, E. (1967). "A Method of Analysis of the Stability of Embankments Assuming Parallel Interslice Forces." *Geotechnique,* vol. 17, no.1, London, pp. 11–26.

Spencer, E. (1968). "Effect of Tension on Stability of Embankments." *Journal of the Soil Mechanics and Foundations Division,* ASCE, vol. 94, no. SM5, pp. 1159–1173.

Spencer, E. W. (1972). *The Dynamics of the Earth, an Introduction to Physical Geology.* Thomas Y. Crowell, New York, 649 pp.

Standard Specifications for Public Works Construction (1997). 11th ed., BNi Building News, Anaheim, Calif., commonly known as the "Greenbook," 761 pp.

Standards Presented to California Occupational Safety and Health Standard Board, Secs. 1504 and 1539–1547 (1991). California Occupational Safety and Health Standard Board, San Francisco, July.

Stapledon, D. H., and Casinader, R. J. (1977). "Dispersive Soils at Sugarloaf Dam Site near Melbourne, Australia." *Dispersive Clays, Related Piping, and Erosion in Geotechnical Projects.* STP 623, American Society for Testing and Materials, Philadelphia, pp. 432–466.

Stark, T. D., and Eid, H. T. (1994). "Drained Residual Strength of Cohesive Soils." *Journal of Geotechnical Engineering,* ASCE, vol. 120, no. 5, pp. 856–871.

Stark, T. D., and Olson, S. M. (1995). "Liquefaction Resistance Using CPT and Field Case Histories." *Journal of Geotechnical Engineering,* ASCE, vol. 121, no. 12, pp. 856–869.

State of California Special Studies Zones Maps (1982). Prepared by the State of California according to the Alquist-Priolo Special Studies Zones Act.

Steinbrugge, K. V. (1970). "Earthquake Damage and Structural Performance in the United States." Chapter 9 of *Earthquake Engineering,* Robert L. Wiegel, coordinating ed. Prentice-Hall, Englewood Cliffs, N.J., pp. 167–226.

Stokes, W. L., and Varnes, D. J. (1955). *Glossary of Selected Geologic Terms.* Colorado Scientific Society Proceedings, vol. 16, Denver, 165 pp.

Szechy, K. (1973). *The Art of Tunneling.* Akkdemiaikido, Budapest, 1097 pp.

Tadepalli, R., and Fredlund, D. G. (1991). "The Collapse Behavior of a Compacted Soil During Inundation." *Canadian Geotechnical Journal,* vol. 28, no. 4, pp. 477–488.

Taylor, D. W. (1937). "Stability of Earth Slopes." *Journal of the Boston Society of Civil Engineers,* vol. 24, no. 3, pp. 197–246.

Taylor, D. W. (1948). *Fundamentals of Soil Mechanics.* John Wiley & Sons, New York, 700 pp.

Terzaghi, K. (1925). *Erdbaumechanik.* Franz Deuticke, Vienna, 399 pp.

Terzaghi, K. (1938). "Settlement of Structures in Europe and Methods of Observation," *Transactions,* ASCE, vol. 103, p. 1432.

Terzaghi, K. (1943). *Theoretical Soil Mechanics.* John Wiley & Sons, New York, 510 pp.

Terzaghi, K., and Peck, R. B. (1967). *Soil Mechanics in Engineering Practice,* 2nd ed. John Wiley & Sons, New York, 729 pp.

Thompson, L. J., and Tanenbaum, R. J. (1977). "Survey of Construction Related Trench Cave-Ins." *Journal of the Construction Division,* ASCE, vol. 103, no. CO3, September.

Thorburn, S., and Hutchison, J. F., eds. (1985). *Underpinning.* Surrey University Press, London, 296 pp.

Tokimatsu, K., and Seed, H. B. (1984). *Simplified Procedures for the Evaluation of Settlements in Sands due to Earthquake Shaking.* Report no. UCB/EERC-84/16, Report sponsored by the National Science Foundation, Earthquake Engineering Research Center, College of Engineering, University of California, Berkeley, 41 pp.

Tomlinson, M. J. (1986). *Foundation Design and Construction,* 5th ed. Longman Scientific & Technical, Essex, England, 842 pp.

Transportation Research Board (1977). *Rapid-Setting Materials for Patching Concrete.* National Cooperative Highway Research Program Synthesis of Highway Practice 45, National Academy of Sciences, Washington, D.C.

"Trench and Excavation Safety Guide." (1984). Publication S-358, California Department of Industrial Relations/California Occupational Safety and Health Administration, San Francisco.

Tucker, R. L., and Poor, A. R. (1978). "Field Study of Moisture Effects on Slab Movements." *Journal of the Geotechnical Engineering Division,* ASCE, vol. 104, no. GT4, pp. 403–414.

Turnbull, W. J., and Foster, C. R. (1956). "Stabilization of Materials by Compaction." *Journal of the Soil Mechanics and Foundations Division,* ASCE, vol. 82, no. SM2, pp. 934.1–934.23.

Tuthill, L. H. (1966). "Resistance to Chemical Attack-Hardened Concrete." *Significance of Tests and Properties of Concrete and Concrete-Making Materials,* STP-169A, ASTM, Philadelphia, pp. 275–289.

Uniform Building Code (1997). International Conference of Building Officials, 3 Volumes, Whittier, Calif.

United States Army (1960). *The Unified Soil Classification System.* Technical memorandum no. 3-357. U.S. Army Engineer Waterways Experiment Station, Vicksburg, Miss.

United States Department of Agriculture (1975). *Agriculture Handbook No. 436.* U.S. Department of Agriculture, Washington, D.C.

United States Department of Agriculture (1977). *Guides for Erosion and Sediment Control in California.* U.S. Department of Agriculture, SCS, Davis, Calif.

USGS (1975). *Encinitas Quadrangle, San Diego, California.* Topographic map, U.S. Geological Survey, Denver, Colo.

USGS (1997). *Index of Publications of the Geological Survey,* Department of the Interior, Washington, D.C.

USS Steel Sheet Piling Design Manual (1984). U.S. Department of Transportation, Washington, D.C., 132 pp.

Van der Merwe, C. P., and Ahronovitz, M. (1973). "The Behavior of Flexible Pavements on Expansive Soils." *3rd International Conference on Expansive Soil,* Haifa, Israel.

Varnes, D. J. (1978). "Slope Movement Types and Processes." *Landslides, Analysis and Control,* Transportation Research Board, National Academy of Sciences, Washington, D.C., Special Report 176, chap. 2, pp. 11–33.

Vesić, A. S. (1963). "Bearing Capacity of Deep Foundations in Sand." *Highway Research Record,* 39, National Academy of Sciences, National Research Council, Washington, D.C., pp. 112–153.

Vesić, A. S. (1967). "Ultimate Loads and Settlements of Deep Foundations in Sand." *Proceedings of the Symposium on Bearing Capacity and Settlement of Foundations,* Duke University, Durham, N.C., p. 53.

Vesić, A. S. (1975). "Bearing Capacity of Shallow Foundations." Chapter 3 of *Foundation Engineering Handbook,* Hans F. Winterkorn and Hsai-Yang Fang, eds., Van Nostrand Reinhold, New York, pp. 121–147.

Virginia Soil and Water Conservation Commission (1980). *Virginia Erosion and Sediment Control Handbook,* 2nd ed. Richmond, Va.

Waddell, J. J., and Dobrowolski, J. A. (1993). *Concrete Construction Handbook,* 3rd ed., McGraw-Hill, New York.

Wahls, H. E. (1994). "Tolerable Deformations." *Vertical and Horizontal Deformations of Foundations and Embankments,* Geotechnical Special Publication No. 40, ASCE, New York, pp. 1611–1628.

Wahlstrom, E. E. (1973). *Tunneling in Rock.* Developments in Geotechnical Engineering 3, Elsevier Scientific, New York, 250 pp.

Waldron, L. J. (1977). "The Shear Resistance of Root-Permeated Homogeneous and Stratified Soil." *Soil Science Society of America,* vol. 41, no. 5, pp. 843–849.

Warner, J. (1978). "Compaction Grouting—A Significant Case History." *Journal of the Geotechnical Engineering Division,* ASCE, vol. 104, no. GT7, pp. 837–847.

Warner, J. (1982). "Compaction Grouting—The First Thirty Years." *Proceedings of the Conference on Grouting in Geotechnical Engineering,* W. H. Baker, ed. ASCE, New York, pp. 694–707.

Watry, S. M., and Ehlig, P. L. (1995). "Effect of Test Method and Procedure on Measurements of Residual Shear Strength of Bentonite from the Portuguese Bend Landslide." *Clay and Shale Slope Instability,* W. C. Haneberg and S. A. Anderson, eds. Geological Society of America, Reviews in Engineering Geology, vol. 10, Boulder, Colo., pp. 13–38.

Wellington, A. M. (1888). "Formulae for Safe Loads of Bearing Piles." *Engineering News,* no. 20, pp. 509–512.

Westergaard, H. M. (1938). "A Problem of Elasticity Suggested by a Problem in Soil Mechanics: A Soft Material Reinforced by Numerous Strong Horizontal Sheets." *Contributions to the Mechanics of Solids, Stephen Timoshenko 60th Anniversary Volume,* Macmillan, New York, pp. 268–277.

WFCA (1984). "Moisture Guidelines for the Floor Covering Industry." WFCA Management Guidelines, Western Floor Covering Association, Los Angeles.

Whitaker, T. (1957). "Experiments with Model Piles in Groups." *Geotechnique,* London.

Whitman, R. V., and Bailey, W. A. (1967). "Use of Computers for Slope Stability Analysis." *Journal of the Soil Mechanics and Foundations Division,* ASCE, vol. 93, no. SM4, pp. 475–498.

Williams, A. A. B. (1965). "The Deformation of Roads Resulting from Moisture Changes in Expansive Soils in South Africa." *Symposium Proceedings, Moisture Equilibria and Moisture Changes in Soils Beneath Covered Areas,* G. D. Aitchison, ed., Butterworths, Australia, pp. 143–155.

Winterkorn, H. F., and Fang, H. (1975). *Foundation Engineering Handbook.* Van Nostrand Reinhold, New York, 751 pp.

Wischmeier, W. H., and Smith, D. D. (1965). *Predicting Rainfall Erosion Losses from Cropland East of the Rocky Mountains.* Agriculture Handbook No. 282, U.S. Department of Agriculture, Washington D.C.

Wischmeier, W. H., and Smith, D. D. (1978). *Predicting Rainfall Erosion Losses—A Guide to Conservation Planning.* Agriculture Handbook No. 537, U.S. Department of Agriculture, Science and Education Administration, Washington, D.C.

Wolman, M. G., and Schick, A. P. (1967). "Effects of Construction on Fluvial Sediment, Urban and Suburban Areas of Maryland." *Water Resources Research,* vol. 3, pp. 451–464.

Woodward, R. J., Gardner, W. S., and Greer, D. M. (1972). "Design Considerations," *Drilled Pier Foundations,* D. M. Greer, ed. McGraw-Hill, New York, pp. 50–52.

Wu, T. H., Randolph, B. W., and Huang, C. (1993). "Stability of Shale Embankments." *Journal of Geotechnical Engineering,* ASCE, vol. 119, no. 1, pp. 127–146.

Yong, R. N., and Warkentin, B. P. (1975). *Soil Properties and Behavior.* Elsevier Scientific, New York, 449 pp.

Zaruba, Q., and Mencl, V. (1969). *Landslides and Their Control.* Elsevier, New York, 205 pp.

Zornberg, J. G., Sitar, N., and Mitchell, J. K. (1998). "Limit Equilibrium as Basis for Design of Geosynthetic Reinforced Slopes." *Journal of Geotechnical and Geoenvironmental Engineering,* ASCE, vol. 124, no. 8, pp. 684–698.

INDEX

A-line, **4.**5–**4.**8
A-value, **7.**27–**7.**29
AASHTO soil classification system, **4.**16–**4.**17
Absorbed water layer, **4.**4
Absorption, **A.**10
Acid mine water, **2.**44, **17.**6
Acidic environment, **12.**39
Active earth pressure (*see* Retaining wall analysis)
Activity of clay, **4.**5–**4.**6, **A.**10
Adhesion, **10.**28–**10.**31, **A.**15
Adobe, **A.**4
Aeolian, **A.**4
Aerial photographs, **2.**9
Aerobic conditions, **12.**39
Aggregate, **11.**3, **19.**19, **A.**22
Aging, **14.**4
Agreement (*see* Contract; Proposal)
Agronomy, **1.**4
Allowable bearing pressure (*see* Bearing capacity)
Alluvial fans, **4.**25
Alluvium (*see* Soil deposits)
American Society for Testing and Materials (ASTM), **3.**1, **3.**3
Amphibolite, **4.**31
Andesite, **4.**28
Angle of repose, **13.**31
Angular distortion (*see* Settlement)
Anhydrite, **20.**4
Animal wastes, **17.**6
Anisotropic soil, **7.**37, **8.**10, **8.**14, **A.**15
Anthracite coal, **4.**28
Appalachians, **13.**48
Apparent opening size, **22.**7, **A.**22
Approval, **A.**22
Approved plans, **A.**22
Approved testing agency, **A.**22
Aquiclude, **A.**4
Aquifer, **A.**4
Arching, **A.**15
Area ratio, **2.**24–**2.**25
Artesian condition, **2.**42, **8.**7, **17.**7, **A.**4
As-graded (definition), **A.**22
Ash (volcanic), **4.**24–**4.**25, **A.**4
Asphalt (*see* Pavements)
Asphalt concrete (*see* Pavements)
Atterberg limits, **3.**10, **4.**4–**4.**5, **A.**10

B-value, **7.**27, **7.**29
Bacillariophyceae, **4.**26
Backdrain (definition), **A.**22
Backfill, **16.**5, **16.**18–**16.**19, **16.**37, **A.**22
Backhoe, **2.**19
Badlands, **15.**3, **A.**4
Barium chloride, **17.**3
Barium sulfate, **17.**3
Barriers for moisture migration, **22.**3, **22.**9
Basalt, **4.**28, **4.**31
Base (*see* Pavements)
Basement walls (*see* Walls)
Bearing capacity, **10.**1–**10.**34
  adjustment for groundwater table, **10.**13
  allowable bearing pressure, **10.**8–**10.**11, **A.**15
  bearing capacity factors, **10.**9–**10.**12, **10.**20–**10.**21

Bearing capacity (*Cont.*):
causes of bearing capacity failures, **10.**8
for cohesionless soil, **10.**13–**10.**14, **10.**24–**10.**26
for cohesive soil, **10.**15–**10.**17, **10.**28–**10.**30
for deep foundations, **10.**2, **10.**22–**10.**32
depth of bearing capacity failure, **10.**7–**10.**8
earthquake loading, **10.**19
eccentric loads, **10.**18
end bearing, **10.**24–**10.**27
example problems, **10.**33–**10.**34
failure (definition), **10.**1, **A.**15
field load tests, **10.**23
for footings at the top of slopes, **10.**19
frequency of bearing capacity failures, **10.**7
frictional resistance, **10.**9, **10.**25–**10.**26
general shear failure, **10.**3–**10.**4
inclined base of footing, **10.**19
lateral loads, **10.**18
load settlement curve, **10.**3–**10.**6
local shear failure, **10.**3, **10.**6
moments, **10.**18
other considerations, **10.**18–**10.**19
for pile groups, **10.**26, **10.**29–**10.**30, **10.**32
plane strain condition, **10.**11
punching shear failure, **10.**3, **10.**5
for retaining wall footings, **16.**8, **16.**12
for roads (*see* Pavements)
for shallow foundations, **10.**1–**10.**21
soil rupture, **10.**3
for square footings, **10.**10–**10.**11
for strip footings, **10.**9–**10.**11
Terzaghi bearing capacity equation, **10.**9–**10.**11
ultimate bearing capacity (definition), **10.**3, **A.**15
Bedding, **A.**4
Bedrock, **A.**4
Bell, **A.**15
Bell-and-spigot pipe, **A.**30
Bench excavation, **A.**23
Bentonite, **4.**25–**4.**26, **22.**2, **A.**4
Berm, **A.**23
Bernoulli's energy equation, **8.**7
Biological activity, **12.**39
Biotite, **3.**8
Bishop method of slices, **13.**27
Bit, **A.**4
Blasting (of rock), **21.**7
Blowout condition, **8.**15
Bog, **A.**5
Boiling condition, **8.**15
Boring, **2.**1, **2.**11–**2.**18
air track, **2.**14
auger, **2.**11
camera, **2.**13
definition of boring, **A.**5
depth of borings, **2.**17–**2.**18
dynamic soundings, **2.**13
hollow stem auger, **2.**11
layout of borings, **2.**16
log (*see* Logs)
number of borings, **2.**15
percussion drilling, **2.**14
rotary coring, **2.**12
soil samples (*see* Soil sampling)
wash type, **2.**11
Borrow (definition), **A.**23
Boulders, **4.**11, **A.**5
Braced excavation (*see* Temporary retaining walls)
Breccia, **4.**28, **4.**31
Bromhead ring shear apparatus, **3.**19
Brooming, **A.**23
Building official, **A.**23
Buildings:
allowable lateral movement, **13.**6–**13.**7
definition, **A.**28
type of projects, **1.**12
Bulking, **A.**23
Bull's liver, **4.**24
Bureau of Workers' Compensation, **16.**37
Buttress, **13.**46, **19.**7, **A.**23

Caissons, **1.**7, **10.**22, **16.**41, **A.**23
CAL/OSHA, **16.**38
Calcite, **3.**7
Calcium aluminate, **17.**3
Calcium carbonate, **4.**25, **9.**21
Calcium hydroxide, **17.**3
Calcium sulfate, **17.**3
Calcium sulfoaluminate, **17.**3
Caliche, **4.**25
California, **2.**10, **3.**5, **9.**55, **12.**4, **13.**48, **15.**12, **18.**3, **19.**12, **21.**19
California bearing ratio (*see* Pavements)
Canada, **17.**9
Canyon, **9.**9, **9.**14
Capillarity (definition), **A.**10
Capillary action, **6.**5, **12.**19
Capillary break, **17.**9, **17.**11
Capillary rise, 6.5
forming ice lenses, **17.**9–**17.**11
Capillary tension, **6.**5
Casing, **A.**5

Cation exchange capacity, **4**.4
Cement (for concrete), **17**.4–**17**.5
Cement treatment (expansive soil), **12**.28
Center lift (*see* Expansive soil)
Cesspool, **A**.30
Chert, **4**.28, **4**.31
Chicago, Illinois, **17**.10
Chlorite, **4**.7–**4**.8
Civil engineer, **A**.2
Civil engineering, **1**.4, **A**.2
Classification (*see* Soil classification; Rock classification)
Clay:
 activity, **4**.5–**4**.6, **A**.10
 attractive forces, **4**.4
 backfill for retaining walls, **16**.18–**16**.19
 bearing capacity failure, **10**.15–**10**.17, **10**.28–**10**.32
 classification of, **4**.11–**4**.18
 consistency, **4**.9–**4**.10, **A**.11
 desiccated, **12**.1–**12**.2, **12**.13–**12**.17
 dispersive, **13**.55–**13**.56, **15**.3, **15**.5
 expansion of (*see* Expansive soil)
 fraction, **4**.5
 Leda clay, **4**.26
 London clay, **7**.26, **7**.29, **13**.14
 Mexico City, **9**.55
 mineralogy, **4**.4–**4**.7
 permeability, **3**.26, **8**.3
 quick (*see* Quick clay)
 repulsive forces, **13**.55
 sensitive, **4**.26, **14**.14, **21**.9
 shear strength (*see* Shear strength)
 shrinkage (*see* Expansive soil)
 size of the particles, **4**.11, **A**.10
 stiff-fissured, **21**.9
 sulfate content of clay, **17**.6
Clayey gravel, **12**.10
Claystone, **4**.28
Clearing, brushing, and grubbing, **19**.7, **19**.9, **A**.23
Clients:
 agreement (*see* Contract; Proposal)
 contractors, **1**.12
 design professionals, **1**.12
 governmental agencies, **1**.12
 owner (definition), **A**.26
 professional developers, **1**.12
 property owners, **1**.12
 types, **1**.12
Clogging, **A**.23
Close sheeting, **16**.37, **16**.40
Coal, **4**.28
Coal mines, **9**.54–**9**.55, **9**.58
Coarse-grained soil, **4**.11–**4**.13, **A**.10
Cobbles, **4**.11, **A**.5
Codes (*see Uniform Building Code*)
Coefficient of consolidation (*see* Consolidation)
Coefficient of curvature, **4**.12
Coefficient of earth pressure at rest, **6**.6, **16**.2
Coefficient of permeability, **3**.22–**3**.27, **8**.1–**8**.7, **A**.12
Coefficient of uniformity, **4**.12
Cofferdams, **16**.31–**16**.32
Cohesion (*see* Shear strength)
Cohesionless soil (definition), **A**.5
Cohesive soil (definition), **A**.5
Collapse:
 of mines, **9**.54–**9**.55, **9**.58
 of tunnels, **9**.54–**9**.55, **9**.58
 of utility trenches, **16**.36–**16**.37
Collapse potential, **9**.17–**9**.18
Collapsible formations, **A**.15
Collapsible soil, **9**.2, **9**.15–**9**.23
 alluvium, **9**.15–**9**.16
 colluvium, **9**.15–**9**.16
 debris fill, **9**.15
 deep fill, **9**.15
 deep foundation system, **9**.20
 definition, **9**.15, **A**.15
 design and construction on collapsible soil, **9**.19–**9**.20
 due to soluble soil particles, **9**.21–**9**.22
 dumped fill, **9**.15
 foundation options, **9**.20
 laboratory testing, **9**.17–**9**.18
 pipe break, **9**.15
 removal and replacement, **9**.19
 settlement analyses, **9**.19–**9**.20
 stabilization, **9**.19
 uncontrolled fill, **9**.15
 variables that govern collapse potential, **9**.15
 wetting front, **9**.15–**9**.16, **9**.19
Colloidal soil particles, **A**.11
Colluvium (*see* Soil deposits)
Colorado, **12**.4
Community well, **A**.30
Compacted soil:
 base, **11**.3
 decomposed granite, **5**.4
 London clay, **7**.26, **13**.14
 lowest void ratio, **5**.4
 subgrade, **11**.3–**11**.5

Compaction:
backfill compaction, **16.**5, **16.**18–**16.**19, **16.**37
checking field compaction, **19.**12–**19.**18
drive cylinder test, **19.**12
guidelines, **19.**12–**19.**18
nuclear method, **19.**12
sand cone test, **19.**12
chopper wheels, **19.**11
compaction energy, **19.**10
definition, **19.**4, **A.**23
for expansive soil, **16.**18
factors that affect compaction:
compaction energy, **19.**10
material gradation, **19.**10
soil type, **19.**10
water content, **19.**10
field compaction, **19.**10–**19.**18
field density tests, **19.**11
fundamentals, **19.**4, **19.**10–**19.**22
impact or sharp blow, **19.**10–**19.**11
kneading action, **19.**10–**19.**11
for mechanically stabilized earth retaining walls, **16.**23
one-point Proctor test, **19.**17, **19.**22
oversize particles, **19.**10, **19.**17–**19.**18
pumping, **19.**13–**19.**14
relative compaction, **19.**12–**19.**14, **A.**27
for retaining wall backfill, **16.**18–**16.**19
role of technicians, **19.**13, **19.**15
specifications for compaction, **B.**1-**B.**23
static weight or pressure, **19.**10–**19.**11
utility trench compaction, **11.**10, **11.**16, **19.**8
vibration or shaking, **19.**10–**19.**11
(*See also* Grading; Fill)
Compaction equipment:
bulldozer, **19.**11
definition, **A.**23
scraper, **19.**11
water trucks, **19.**11
Compaction grouting, **19.**5
Compaction production, **A.**23
Compaction test (in the laboratory):
compaction curve, **3.**28–**3.**29, **3.**31, **A.**11
compaction energy, **3.**28
definition of, **3.**28, **A.**11
maximum dry density, **3.**28–**3.**29, **3.**31, **A.**11
modified Proctor, **3.**28–**3.**29, **3.**31
optimum moisture content, **3.**29, **3.**31, **19.**10, **A.**12
standard Proctor, **3.**28
zero air voids curve, **3.**29, **A.**14
Compensation (*see* Payment)
Compressibility:
definition, **A.**16
of subgrade, **11.**5
Compression index (*see* Consolidation)
Computer programs:
for gross slope stability analysis, **13.**25, **13.**32
for landslide slope stability analysis, **13.**41–**13.**42, **13.**44–**13.**45
for seepage analysis, **8.**10
for settlement analysis, **9.**50
Concrete:
admixtures, **17.9**
concrete pavements, **11.**3, **12.**37–**12.**38
cracking [*see* Cracks (concrete)]
definition of concrete, **A.**23
deterioration of concrete:
alkali-silica reaction (ASR), **17.**1
corrosion of reinforcing steel, **17.**1
freeze-thaw cycles, **17.**1, **17.**9
sulfate attack, **17.**1, **17.**3–**17.**6
discoloration, **17.**3
durability of concrete to resist freeze-thaw cycles, **17.9**
moisture migration through concrete floor slabs, **20.**4, **20.**13–**20.**14, **20.**17
sulfate resistance of concrete:
curing, **17.4**
concentration of sulfates, **17.4**, **17.6**
consolidation of concrete, **17.4**
quality of concrete, **17.4**
surface preparation of concrete, **17.4**
types of Portland cements, **17.4**–**17.5**
water cement ratio, **17.4**–**17.5**
Cone penetration test, **2.**28–**2.**30, **2.**36, **A.**5
cone resistance, **2.**29–**2.**30, **2.**36
electric cone, **2.**36, **A.**5
friction ratio, **2.**30
mechanical cone, **2.**36, **A.**5
mechanical-friction cone, **2.**36, **A.**5
piezocone, **2.**36, **A.**5
used to identify soil types and conditions, **2.**29–**2.**30
used to predict drained modulus, **9.**50
used to predict friction angle, **7.**10
used to predict liquefaction potential, **14.**6, **14.**10–**14.**11
Conglomerate, **4.**28, **4.**31
Consistency of clay, **4.**9–**4.**10, **A.**11
Consolidation, **9.**2, **9.**27–**9.**45
average degree of consolidation settlement, **9.**36–**9.**45

Consolidation (*Cont.*):
average degree of consolidation versus time factor, **9**.40
Casagrande construction technique, **9**.30–**9**.31, **9**.42
coefficient of compressibility, **9**.35
coefficient of consolidation, **9**.35–**9**.39, **A**.10
compression index, **9**.32–**9**.33, **A**.11
consolidation curve, **9**.30, **9**.41–**9**.42
consolidation ratio, **9**.36, **9**.44
definition of primary consolidation, **9**.2, **9**.27, **A**.16
double drainage, **9**.36, **9**.38
effects of sample disturbance (*see* Sample disturbance)
excess pore water pressure, **9**.35–**9**.36
laboratory testing, **9**.30–**9**.32
limitations of consolidation theory, **9**.38–**9**.39
load increment ratio, **9**.30, **9**.38
maximum past pressure, **9**.27, **9**.30–**9**.34, **9**.42
mechanisms causing an overconsolidated soil, **9**.29
modified compression index, **9**.34
modified recompression index, **9**.34
normally consolidated, **9**.28, **9**.33, **A**.18
overconsolidated, **9**.28, **9**.33–**9**.34, **A**.18
overconsolidation ratio, **9**.27, **A**.12
preconsolidation pressure, **9**.27, **9**.30–**9**.34, **9**.42, **A**.18
primary consolidation, **9**.27–**9**.45
rate of one-dimensional consolidation, **9**.35–**9**.39
recompression curve, **9**.32–**9**.33, **9**.42
recompression index, **9**.32
settlement analyses, **9**.33–**9**.34
single drainage, **9**.36, **9**.38
stress history, **9**.31, **9**.43
Terzaghi one-dimensional consolidation equation, **9**.35–**9**.39
test (definition), **9**.30, **A**.11
time factor, **9**.36, **A**.21
underconsolidated, **9**.27, **9**.33, **A**.21
virgin consolidation curve, **9**.32–**9**.33, **9**.42
Consolidometer (*see* Oedometer)
Construction:
excavation (*see* Excavations)
field load tests [*see* Load test (for piles)]
geosynthetics (*see* Geosynthetics)
grading (*see* Grading)
groundwater control (*see* Groundwater)
Construction (*Cont.*):
instrumentation (*see* Instrumentation)
percolation tests (*see* Percolation tests)
site improvement methods (*see* Site improvement methods)
underpinning (*see* Underpinning)
Contract:
disclaimer of warranties, **1**.17
jurisdiction, **1**.17
limitation of liability, **1**.17
modification of, **1**.17
ownership of documents, **1**.17
payment, **1**.17
retainers, **1**.17
safety, **1**.17
signature page, **1**.17
termination of, **1**.17
time limit, **1**.17
(*See also* Proposal)
Contraction (during shear), **7**.4, **A**.11
Contractors, **A**.24
Conversion factors, **D**.1
Core drilling, **2**.11–**2**.12, **A**.6
Core recovery, **A**.6
Corrosive environment, **17**.6
Cost estimating sheet, **1**.15–**1**.16
Coulomb, **3**.17
Cousins slope stability charts, **13**.30–**13**.31, **13**.35–**13**.37
Cracks (asphalt), alligator, **12**.37
Cracks (concrete):
at cut-fill transitions, **9**.9
due to sulfate attack, **17**.3
flaking, **17**.3
flatwork, **12**.37–**12**.38
repair, **21**.5–**21**.6, **21**.8–**21**.9
severity, **9**.11
spalling, **17**.3
width, **9**.11
x-type crack pattern, **12**.38
Cracks (in outlet pipes), **13**.55
Cracks (soil):
dam, **13**.55, **14**.2
desiccation, **12**.13
due to earthquakes, **14**.2
due to landslide movement, **13**.40, **13**.43
ground, **12**.13
Creep (of geogrids), **22.5**
Creep (of slopes), **13**.50–**13**.52
definition, **13**.50, **A**.16
depth of creep, **13**.51
due to seasonal moisture changes, **13**.51
effect on top of slope retaining walls, **16**.17
method of analysis, **13**.51

Creep (*Cont.*):
primary or transient, **13.**51–**13.**52
secondary or steady-state, **13.**51–**13.**52
tertiary, **13.**51–**13.**52
Creep (secondary compression), **9.**46
Critical height or critical slope, **A.**16
Crown, **A.**16
Crushing of aggregate, **19.**19
Curbs (*see* Pavements)
Cut, **9.**9, **19.**7
Cut-fill transition, **9.**9, **19.**7, **A.**24

Damage:
caused by frost action, **16.**5, **17.**9
classification, **9.**11
due to lateral movement, **13.**6–**13.**7
due to settlement, **9.**11
Dams:
causes of failure, **13.**55–**13.**56
definition, **A.**24
dispersive clays, **13.**55–**13.**56
flow net, **8.**13
homogeneous dams, **A.**24
hydrological design, **13.**53
landslide dams, **13.**54
large dams, **13.**53
maintenance, **13.**53
overtopping, **13.**55
pinhole test, **13.**56
piping, **13.**55–**13.**56
sand boils, **13.**55
slope instability, **13.**56
small dams, **13.**53
St. Francis dam failure, **13.**53
stability analysis, **13.**56
surficial failures, **13.**11
zoned dams, **A.**24
Darcy's law, **3.**22, **8.**1, **8.**7, **8.**9
Dead load, **9.**4, **10.**2, **A.**16
Death Valley, California, **3.**5
Debris, **4.**10, **19.**7, **A.**24
Debris flow, **4.**25, **13.**2, **13.**48–**13.**49
definition, **13.**2, **13.**48, **A.**16
depositional area, **13.**48
example, **13.**48
historical method, **13.**48
main tract, **13.**48
measures to protect structures, **13.**49
mobilization from a surficial failure, **13.**11
prediction of a debris flow, **13.**48
rainfall intensity, **13.**48
source area, **13.**48
Decomposed granite, **5.**4, **19.**10
Decomposition, **9.**56
Deep foundation (*see* Foundation design)
Degree of saturation, **4.**10, **5.**4
Densification of soil (*see* Compaction)
Density:
definition, **5.**2, **A.**11
density relationships, **5.**6
Deposition of soil, **A.**6
Desiccated clay (definition), **A.**16
Design (*see* Engineering analyses)
Deterioration, **17.**1–**17.**11
caused by frost:
formation of ice lenses, **17.**2, **17.**9–**17.**11
freezing of water in cracks, **17.**2, **17.**9
permafrost, **17.**2, **A.**8
of concrete:
alkali-silica reaction (ASR), **17.**1
corrosion of reinforcing steel, **17.**1
freeze-thaw cycles, **17.**1, **17.**9
sulfate attack, **17.**1–**17.**6
of pavements:
heavy traffic loads, **17.**7
measures to reduce deterioration, **17.**7–**17.**8
pavement section thickness, **17.**7–**17.**8
pavement strength parameters, **17.**8
water trapped in pavements, **17.**7
Detritus, **A.**6
Deviator stress (*see* Shear strength)
Dewatering, **20.**8–**20.**11, **21.**7, **21.**12, **A.**24
Diabase, **4.**31
Diaphragm wall, **21.**14
Diatomaceous earth, **4.**26–**4.**27, **A.**6
Diatomaceous rock, **4.**26–**4.**27
Diatomite, **4.**26–**4.**27
Diatoms, **4.**26–**4.**27, **9.**55
Differential settlement (*see* Settlement)
Digestion (sewage disposal), **A.**30
Dilation (during shear), **7.**4, **A.**11
Diorite, **4.**28, **4.**31
Direct shear apparatus, **3.**21
Direct shear test, **3.**17–**3.**18
Dispersing agent, **2.**9
Dissolution, **9.**21–**9.**22
Distribution box, **A.**30
Document review, **2.**1, **2.**4–**2.**10
aerial photographs, **2.**9
building codes, **2.**10
for construction phase of project, **2.**8
correspondence, **2.**8
data on the history of the site, **2.**9
for design phase of project, **2.**8
field change orders, **2.**8

Document review (*Cont.*):
geologic data, **2.**4–**2.**5, **2.**9
information bulletins, **2.**8
plans, **2.**8
preliminary design information, **2.**9
reference materials, **2.**8
reports, **2.**8
seismic data, **2.**8
special study data, **2.**7, **2.**10
specifications, **2.**8
standard drawings, **2.**8
topographic maps, **2.**6, **2.**10
Dolomite, **3.**7, **4.**31
Double layer, **4.**4, **A.**11
Downdrag, **10.**22, **10.**28, **21.**18, **A.**17
Dozer, **A.**24
Drag effect, **9.**9
Drainage:
canyon subdrain, **19.**7
definition, **A.**24
ditches (of slopes), **15.**14
drainage buttress, **20.**4
galleries, **20.**3
horizontal drains, **20.**4
of pavements, **17.**7, **20.**2–**20.**3
of retaining wall backfill, **16.**5
of slopes, **13.**16, **20.**3–**20.**4, **20.**12
sump pumping, **20.**8, **20.**11
system, **16.**5, **19.**7, **20.**8, **20.**11
trenches, **19.**23
using geosynthetics, **22.**3
Drainage properties of soil, **3.**26
Drained modulus, **9.**49–**9.**50
Drained shear strength (*see* Shear strength)
Drawdown, **A.**24
Drought, **12.**18, **12.**21
Dry density (*see* Density)
Dynamite, **4.**27

Earth flow, **13.**48
Earth material (definition), **A.**24
Earth pressure (*see* Retaining wall analysis)
Earthquakes:
Alaskan (1964), **14.**3
Alquist-Priolo special studies zone map, **14.**2
Assam (1897), **14.**2
Chile (1960), **14.**3
duration, **14.**4
effect on earth dams, **14.**2
effect on landslides, **14.**13
effect on retaining walls, **14.**13
effect on sensitive soil, **14.**13
Earthquakes (*Cont.*):
epicenter, **14.**4
fault displacement, **14.**2
fault zones, **14.**2
flow slides, **14.**3
foundation behavior, **14.**3
Gobi-Altai (1957), **14.**2
ground movement, **14.**2
ground rupture, **14.**2
intensity, **14.**4
liquefaction (*see* Liquefaction)
Niigata (1964), **14.**3
peak acceleration, **14.**5
pseudo-static analysis, **14.**13
San Francisco (1906), **13.**53
seismic shear stress ratio, **14.**5–**14.**11, **14.**13–**14.**14
seismic study maps, **14.**2
settlement induced by the earthquake, **14.**12–**14.**14
slope movement, **14.**3
soil strength loss, **14.**12–**14.**13
surface faulting, **14.**2
Easements, **19.**7
Economic losses:
collapse of mines and tunnels, **9.**54
due to saturation and seepage, **20.**1
expansive soil, **12.**4
frost action, **17.**9
landslides, **13.**38
pavements, **17.**9
St. Francis Dam failure, **13.**53
Edge lift (*see* Expansive soil)
Effective stress:
calculation of, **6.**4
definition, **6.**4, **A.**17
Effective stress analysis:
for bearing capacity, **10.**16
for dam design, **13.**56
discussion of, **7.**40–**7.**41
for gross stability, **13.**22–**13.**24, **13.**26–**13.**27, **13.**31
for landslides, **13.**41–**13.**42, **13.**44–**13.**45
for pier walls to stabilize slopes, **16.**43–**16.**45, **16.**47
for piles in cohesive soil, **10.**29
for retaining walls, **16.**8
for sheet pile walls, **16.**33
shear strength for effective stress analysis, **3.**13
for surficial stability, **13.**13–**13.**15
Efflorescence, **20.**13, **20.**16
Effluent (definition), **A.**30

El Niño, **15.**7
Elasticity (theory of), **6.**11, **9.**24, **9.**26, **9.**49–**9.**50
Electroosmosis, **13.**46, **19.**23, **A.**24
Elimination method, **19.**17
Encinitas, California, **2.**10
Engineer (*see* Geotechnical engineer)
Engineering analyses:
  bearing capacity (*see* Bearing capacity)
  design load, **9.**4, **10.**7, **A.**16
  deterioration (*see* Deterioration)
  earthquakes (*see* Earthquakes)
  effective stress analysis (*see* Effective stress analysis)
  erosion (*see* Erosion)
  expansive soil (*see* Expansive soil)
  groundwater (*see* Groundwater)
  moisture migration (*see* Moisture migration)
  phase relationships (*see* Phase relationships)
  retaining walls (*see* Retaining walls)
  settlement (*see* Settlement)
  short-term analysis (*see* Total stress analysis)
  slope stability (*see* Slope stability analysis)
  total stress analysis (*see* Total stress analysis)
  unusual soil (*see* Unusual soil)
Engineering geologist:
  analysis for debris flow, **13.**49
  analysis for landslides, **13.**41
  analysis for potential rockfall, **13.**8
  analysis for slope creep, **13.**51
  areas of responsibility, **1.**11
  classifying rock, **4.**3
  definition, **1.**5, **A.**2
  fields of expertise, **1.**9–**1.**10
  preparation of logs, **2.**42
  role of, **1.**9–**1.**10
  soil classification, **4.**1
Engineering geology (definition), **A.**2
Engineering News formula, **10.**23
Epoxy, **21.**19, **21.**23
Equipotential line, **8.**9–**8.**11, **A.**17
Equivalent fluid pressure (*see* Retaining wall analysis)
Equivalent wheels loads (*see* Pavements)
Erosion, **15.**1–**15.**15
  of badlands, **15.**3, **A.**4
  definition of erosion, **A.**6
  design to reduce soil erosion, **15.**12–**15.**13
  of dispersive clays, **15.**3
Erosion (*Cont.*):
  drainage ditches for slopes, **15.**14
  effect of vegetation on erosion, **15.**13, **15.**19
  erosion control devices (permanent), **A.**26
  erosion control devices (temporary), **A.**24
  erosion control fabric (*see* Geosynthetics)
  erosion control system, **A.**24
  erosion potential, **15.**2
  erosion prone landforms, **15.**3
  factors causing erosion, **15.**2
  factors governing the amount of soil loss, **15.**2–**15.**11
    ground cover, **15.**2, **15.**9
    length of slope, **15.**2, **15.**5–**15.**6
    management, **15.**2, **15.**9
    rainfall, **15.**2, **15.**4, **15.**10–**15.**11
    slope gradient, **15.**2, **15.**5–**15.**6
    soil type, **15.**2, **15.**8
  gullies, **15.**2
  jugs, **15.**3
  levels of erosion, **15.**2
  principles of erosion control, **15.**12–**15.**13
  ravines, **15.**3
  rills, **15.**2
  rillwash, **15.**2
  runoff, **15.**12
  of sand, **15.**3, **15.**8
  of sea cliffs, **15.**3
  sediment basin, **15.**13, **15.**15
  sediment control, **22.**3
  sediment trap, **15.**13, **15.**15
  sheet erosion, **15.**2
  silt fences, **15.**13
  slope wash, **15.**2, **A.**8
  of slopes, **15.**2–**15.**12
  soil loss, **15.**2, **15.**4–**15.**11
  stream flow, **15.**2
  underground erosion [*see* Piping (of soil)]
  universal soil loss equation, **15.**4–**15.**11
    combined slope length and steepness factor, **15.**5–**15.**6
    erosion control practice factor, **15.**6, **15.**9
    example problems, **15.**6–**15.**7
    limitations, **15.**7
    rainfall erosion index, **15.**4, **15.**10–**15.**11
    soil erodibility factor, **15.**4–**15.**5, **15.**8
    soil loss, **15.**4–**15.**11
    vegetative cover factor, **15.**6, **15.**9
  unusual landforms, **15.**3
Ettringite, **17.**3
Evaporites, **17.**6, **20.**4

Examples:
bearing capacity of a strip footing on cohesionless soil, **10.**33
building load determination, **9.**59
combined slope length and steepness factor, **15.**6
consolidation of a clay deposit, **9.**59
earthquake-induced settlement, **14.**13
end-bearing capacity of a pile, **10.**33
exit gradient using a flow net, **8.**17
expansion soil foundation heave, **12.**22–**12.**23
field permeability, **8.**16
foundation, selection of foundation type, **18.**15
frictional capacity of a pile, **10.**33–**10.**34
gross slope stability analysis, **13.**31
liquefaction analysis, **14.**7
mechanically stabilized earth retaining wall, **16.**24
pavement design, **11.**7
phase relationships, **5.**7
pipeline loadings, **11.**11–**11.**12
pore water pressure using a flow net, **8.**16–**8.**17
retaining wall analysis, **16.**14–**16.**16
seepage into a foundation pit, **8.**16
shear strength, **7.**42–**7.**43
sheet pile wall, **16.**30
soil classification, **4.**20–**4.**21
stress distribution, **6.**7, **6.**21
surficial slope stability analysis, **13.**15
time for end of primary consolidation, **9.**60
universal soil loss equation, **15.**7
wedge slope stability analysis, **13.**23–**13.**24
Excavation slopes:
dewatering, **20.**8–**20.**11, **21.**7
engineering evaluation, **21.**10
factors governing the stability of excavation slopes, **21.**9–**21.**10
grading, **21.**7
problem soils:
loess, **21.**9
loose sands, **21.**9
residual soil, **21.**9
sensitive clay, **21.**9
stiff-fissured clays, **21.**9
talus, **21.**9
stability of excavation slopes, **21.**7–**21.**8
Excavations:
braced excavations (*see* Temporary retaining walls)
Excavations (*Cont.*):
deep excavations:
dewatering, **20.**8–**20.**11, **21.**14
factors involved in the choice of a support system, **21.**14
definition, **21.**1, **21.**14, **A.**24
during grading operations, **19.**8, **21.**1
for footing excavations, **21.**2, **21.**4
open excavations, **21.**2, **21.**7–**21.**14
for pier excavations, **21.**2, **21.**5–**21.**6
tunnels (*see* Tunnels)
Exit gradient (*see* Groundwater)
Expansion index, **12.**6, **12.**9–**12.**10, **12.**12
Expansion index test, **12.**9–**12.**10, **12.**12
Expansion potential, **12.**5–**12.**11
Expansive rock, **12.**39
expansion due to physical factors, **12.**39
expansion due to weathering of rock, **12.**39
rebound, **12.**39
Expansive soil, **12.**1–**12.**38
active zone, **12.**13–**12.**14
bearing capacity of, **12.**37
calculating foundation heave, **12.**22–**12.**25
capillary action, **12.**19
cement treatment, **12.**28
center-lift, **12.**18–**12.**20, **12.**31, **12.**35
chemical injection, **12.**28
classification charts, **12.**7–**12.**8
clay size particles, **12.**4–**12.**5
coefficient of swell, **12.**10
compaction control, **12.**26, **12.**29
correlations with index tests, **12.**6
cost of damage, **12.**4
cyclic heave and shrinkage, **12.**18
depth of seasonal moisture changes, **12.**13–**12.**14, **A.**16
depth of the active zone, **12.**13–**12.**14
desiccated clay, **12.**1–**12.**2, **12.**13–**12.**17
desiccation cracks, **12.**13
downward displacement of foundation, **9.**4–**9.**5
drought, **12.**18, **12.**21
edge lift, **12.**18, **12.**31, **12.**35
edge moisture variation distance, **12.**31, **12.**35
effect of roots, **12.**21
effect of vegetation, **12.**21
effect on concrete pavements, **12.**37
effect on retaining walls, **12.**18
example problem, **12.**22–**12.**23
factors causing expansion, **12.**4
flatwork, **12.**37–**12.**38

Expansive soil (*Cont.*):
- fly ash treatment, **12.**28
- foundation design, **12.**30–**12.**36
- foundation repair, **21.**18–**21.**19, **21.**22–**21.**23
- heave caused by water from pipe breaks, **20.**23
- horizontal barriers, **12.**29
- hydraulic conductivity, **12.**15–**12.**17
- identification, **12.**13
- laboratory testing, **12.**9–**12.**12
- lateral movement, **12.**18
- lime treatment, **12.**27–**12.**28
- mitigation options, **12.**26–**12.**29
- moisture variation, **12.**14
- pavements, **12.**37–**12.**38
- prewetting, **12.**26–**12.**27
- progressive swelling, **12.**18–**12.**19, **12.**31
- rate of swell, **12.**10
- removal and replacement, **12.**26
- salt treatment, **12.**28
- shrinkage, **12.**15–**12.**16
- slaking, **12.**16, **A.**8
- soil properties versus expansion potential, **12.**6
- suction pressure, **12.**16
- surcharge pressure, **12.**5, **12.**26
- swell test, **12.**11, **12.**22, **12.**24
- swelling pressure, **12.**11, **12.**22
- thermal osmosis, **12.**18–**12.**19
- total heave, **12.**22–**12.**25
- treatment alternatives, **12.**26–**12.**29
- types of expansive soil movement, **12.**18–**12.**21, **12.**35
- vertical movement, **12.**18–**12.**20
- walking of flatwork, **12.**38
- wetting and drying, **12.**18
- (*See also* Swell)

Expansive soil analysis, **12.**22–**12.**25
Experience, **1.**8
Exploratory logs (*see* Logs)

Fabric (of soil), **A.**12
Factor of safety:
- for bearing capacity failure, **10.**7, **10.**10, **10.**25–**10.**26, **10.**29
- for gross stability, **13.**22
- for landslides, **13.**41–**13.**42
- for mechanically stabilized earth retaining walls, **16.**23–**16.**24
- for pier and grade beam support (expansive soil), **12.**32
- for pier walls, **16.**43–**16.**44
- for pipeline design, **11.**12
- for quicksand condition, **8.**15
- for retaining walls, **16.**8–**16.**9, **16.**12–**16.**13, **16.**16
- for surficial stability, **13.**13–**13.**15
- for temporary retaining walls, **21.**12–**21.**13
- for toe kick-out of sheet pile walls, **16.**29–**16.**30

Faults (definition), **A.**6
Fee schedule, **1.**14
Feldspar, **3.**7
Fellenius method, **13.**25
Felsite, **4.**31
Fertilizers, **17.**6
Field exploration, **2.**1–**2.**54
- document review (*see* Document review)
- exploratory logs (*see* Logs)
- field testing (*see* Field testing)
- geophysical techniques (*see* Geophysical techniques)
- problem conditions requiring special consideration, **2.**44
- soil sampling (*see* Soil sampling)
- subsoil profile, **2.**3, **2.**51–**2.**54, **A.**9
- subsurface exploration (*see* Subsurface exploration)

Field lateral, **A.**30
Field services (*see* Construction)
Field testing:
- cone penetration test (*see* Cone penetration test)
- Iowa borehole shear test, **2.**37–**2.**38, **A.**7
- pressuremeter test, **2.**37–**2.**38, **A.**8
- screw plate compressometer, **2.**37–**2.**38, **A.**8
- standard penetration test (*see* Standard penetration test)
- vane shear testing (*see* Vane shear test)

Fill:
- cut-fill transitions, **9.**9
- debris fill, **2.**43
- definition, **A.**25
- dumped fill, **2.**43, **19.**15
- engineered fill, **2.**43, **19.**15
- field density tests, **19.**11–**19.**18
- hydraulic fill, **14.**4, **A.**25
- placement of fill, **19.**7–**19.**22, **B.**1-**B.**23
- structural fill:
  - borrow with oversize particles, **19.**10, **19.**17–**19.**18
  - definition, **2.**43, **19.**15
  - mixed borrow, **19.**17
  - select import, **19.**16, **19.**19
  - uniform borrow, **19.**16–**19.**17

Fill (*Cont.*):
uncompacted fill, **2.**43, **19.**15
(*See also* Compaction; Grading)
Filter fabric (*see* Geosynthetics)
Filtration, **22.**3, **22.**7
Fine-grained soil, **4.**11, **4.**14, **A.**12
Fines, **4.**13–**4.**14, **7.**2, **A.**6
Finite element analyses (definition), **A.**17
Fire, **13.**11
Flatwork, **12.**37–**12.**38
Flocculation, **3.**9, **A.**12
Flow (laminar), **8.**7
Flow line, **8.**9–**8.**10, **A.**17
Flow net:
for anisotropic soil, **8.**14
construction of a flow net, **8.**10–**8.**11
definition, **A.**17
examples of flow nets, **8.**12–**8.**14
exit hydraulic gradient, **8.**15
flow of water from one soil strata to another, **8.**14
(*See also* Groundwater)
Flow slides, **14.**3
Flowing soil, **21.**16
Flows, **13.**48
Fold (of rock layers), **A.**6
Footing (*see* Foundation types)
Foraminifera, **4.**27
Force equilibrium equations, **13.**25
Forms, **A.**25
Foundation construction:
field load tests, **21.**3
footings excavations, **21.**2, **21.**4
footing observations, **21.**4
Foundation (definition), **1.**5, **A.**25
Foundation design:
ability to resist lateral movement, **13.**6
adequate depth, **18.**2
adequate strength, **18.**2
adjacent top of slopes, **10.**19
adverse soil changes, **18.**2
bearing capacity failure (*see* Bearing capacity)
calculating foundation heave (*see* Expansive soil)
footing dimensions, **10.**7
for expansive soil (*see* Expansive soil)
quality of concrete, **18.**2
required specifications, **18.**2
seismic forces, **18.**2
selection of foundation type, **10.**22, **18.**1–**18.**15
settlement of the foundation (*see* Settlement)
Foundation design (*Cont.*):
sulfate attack of concrete (*see* Sulfate attack of concrete)
undermining due to erosion, **9.**4
Foundation, selection of foundation type, **10.**22, **18.**1–**18.**15
example problems, **18.**15
Foundation types:
deep foundations:
basement types, **1.**7
caissons, **1.**7, **10.**22, **16.**41, **A.**23
definition, **A.**25
floating foundation, **1.**7, **18.**8
mat supported by piers or piles, **1.**7, **18.**8
pier and grade beam, **1.**7, **12.**32–**12.**33, **12.**36
piers (*see* Piers)
piles (*see* Piles)
raft foundation, **1.**7, **18.**8
selection of foundation type, **10.**22, **18.**1–**18.**15
shallow foundations:
California slab, **12.**30
cantilever footings, **18.**3, **18.**5
combined footing, **1.**6, **18.**3, **18.**5
definition, **A.**25
eccentric loaded footing, **18.**3, **18.**5
footings (definition), **A.**25
mat foundation (*see* Mat foundations)
octagonal footing, **18.**3, **18.**5
posttensioned, **1.**6, **12.**30–**12.**31, **18.**3–**18.**4
raised wood floor, **1.**6, **18.**4
ribbed foundation, **12.**30
slab-on-grade, **1.**6, **9.**7–**9.**8, **12.**30, **18.**3–**18.**4
spread footings, **1.**6, **18.**3
strap footings, **18.**3, **18.**5
strip footings, **1.**6, **18.**3
wall footings, **1.**6, **18.**3
Fracture (of rock), **17.**9, **A.**6
Fredonia, New York, **17.**9
Freeze (of piles), **10.**28, **A.**25
Freezing of soil (*see* Site improvement methods)
Friars formation, **4.**9
Friction angle (*see* Shear strength)
Frost:
depth of frost action, **17.**10–**17.**11
effect on buildings, **17.**9–**17.**10
effect on concrete, **17.**9
effect on pavements, **11.**4
effect on retaining walls, **16.**5, **21.**12
effect on rock slopes, **17.**9

Frost (*Cont.*):
expansive forces of freezing water, **17.**9
frost action, **17.**9–**17.**10
ice lenses, **17.**9–**17.**11
permafrost, **17.**2, **A.**8
spring thaw, **17.**9
Frustules, **4.**26

Gabbro, **4.**28, **4.**31
Gas, **5.**1, **5.**3
Geofabric (*see* Geosynthetics)
Geogrid (*see* Geosynthetics)
Geologic:
faults [*see* Faults (definition)]
features, **2.**4–**2.**5
hazards, **2.**7
maps, **2.**4–**2.**7
Geologist (*see* Engineering geologist)
Geology, **13.**5
Geomembranes (*see* Geosynthetics)
Geonets (*see* Geosynthetics)
Geophysical techniques, **2.**2, **2.**49–**2.**50
definition, **A.**6
drop in potential, **2.**50
E-logs, **2.**50
electrical methods, **2.**50
gravity measurements, **2.**50
high-resolution refraction, **2.**49
magnetic methods, **2.**50
resistivity, **2.**50
seismic refraction, **2.**49
seismic wave velocity, **2.**49
uphole, downhole, and cross-hole surveys, **2.**49
vibration, **2.**49
Geostatic condition, **6.**4
Geosynthetic clay liners (*see* Geosynthetics)
Geosynthetics:
definition, **22.**1, **A.**25
geocomposites, **22.**2–**22.**3, **22.**10
geogrid, **19.**25, **22.**1, **22.**3–**22.**5
limitations of geogrid, **22.**5
geomembranes, **19.**25, **22.**2–**22.**3, **22.**8–**22.**9
limitations of geomembranes, **22.**9
puncture resistance, **22.**9
sealing of seams, **22.**9
slide resistance, **22.**9
visqueen, **22.**2
geonets, **19.**25, **22.**2–**22.**3, **22.**10
geosynthetic clay liners, **4.**26, **22.**2–**22.**3
geotextiles, **19.**25, **22.**1–**22.**3, **22.**6–**22.**8
Geosynthetics, geotextiles (*Cont.*):
definition, **A.**25
equivalent opening size, **22.**7
erosion control fabric, **13.**16
filter fabric, **22.**1
flow capacity, **22.**7
geofabric, **22.**1
limitations of geotextiles, **22.**7–**22.**8
nonwoven geotextiles, **22.**6–**22.**7
soil retention capability, **22.**7
woven geotextiles, **22.**6–**22.**7
polymeric material, **22.**1
Geotechnical engineer:
areas of responsibility, **1.**11
definition, **A.**25
fields of expertise, **1.**9–**1.**10
qualifying experience, **1.**8
Geotechnical engineering:
analysis (*see* Engineering analyses)
contract for services, **1.**17
cost estimating sheet, **1.**15–**1.**16
definition, **1.**4–**1.**5, **A.**2
project requirements, **1.**13
schedule of fees, **1.**14
types of projects, **1.**12
Geotextile (*see* Geosynthetics)
Gibbsite sheet, **4.**6
Glacier, **5.**4
Glass, **4.**10
Gneiss, **4.**28, **4.**31
Grade:
existing grade, **A.**25
finished grade or final grade, **19.**9, **A.**25
lowest adjacent grade, **A.**25
natural grade, **A.**25
rough grade, **A.**25.
Grading:
adjacent property damage, **19.**8–**19.**9
back-rolling technique, **19.**7
benching, **13.**16, **19.**7, **19.**9
blasting of rock, **19.**7
buttress fill, **19.**7
canyon subdrain, **19.**7
cleanout, **19.**7
clearing, brushing, and grubbing, **19.**7, **19.**9
cut, **9.**9, **19.**17
cut-fill transition, **9.**9, **B.**5, **B.**20
definition, **19.**3, **A.**25
earthwork operations, **19.**11–**19.**18
equipment (*see* Compaction equipment)
final grade, **19.**9
fine grading, **19.**8

Grading (*Cont.*):
for fill slopes, **19.**7
hillside site, **A.**26
inspection during grading, **19.**9
key grading activities, **19.**9
overbuilding and cutting back slopes faces, **19.**7
pregrading meeting, **19.**9
process of grading, **19.**7–**19.**8
recompaction, **19.**7
revision of grading operations, **19.**7
ripping of rock, **19.**7, **19.**9
rough grading operations, **19.**9
scarifying, **19.**7
shear key, **19.**7
specifications for grading, **19.**9, **B.**1-**B.**23
stabilization fill, **13.**16, **19.**7
steps in the grading operation, **19.**7–**19.**8
terms and definitions, **A.**22-**A.**29
windrow, **19.**7, **A.**29, **B.**8-**B.**9, **B.**21
(*See also* Compaction; Fill)
Grading contractor, **A.**26
Grading permit, **A.**26
Grain size analysis (*see* Soil classification)
Granite, **4.**28, **4.**31
Gravel, **4.**11–**4.**13, **4.**15–**4.**16, **17.**6
Gravel equivalent factor (*see* Pavements)
Gravel-size particles, **4.**11, **A.**12
Gross slope stability analysis, **13.**2, **13.**22–**13.**37
charts for gross slope stability analysis, **13.**34–**13.**37
different soil layers, **13.**28
effective stress analysis, **13.**22–**13.**24, **13.**26–**13.**27, **13.**31
example problem, **13.**31
factor of safety, **13.**22
method of slices, **13.**25–**13.**27
for nonplastic soil, **13.**31
for plane strain condition, **13.**28
pore water pressure, **13.**26–**13.**27
progressive failure, **13.**28–**13.**29
slip surfaces, **13.**25, **13.**28
total stress analysis, **13.**22–**13.**23, **13.**25–**13.**26
with surcharge loads, **13.**28
with tension cracks, **13.**28
using computer programs, **13.**25, **13.**32
using nonlinear shear strength, **13.**28
wedge analysis, **13.**22–**13.**24
Groundwater, **8.**1–**8.**17, **20.**1–**20.**23
critical hydraulic gradient, **8.**15
effect on bearing capacity, **10.**13

Groundwater (*Cont.*):
effect on landslides, **13.**42
effect on liquefaction, **14.**4
effect on pavements, **11.**4, **17.**7, **20.**2–**20.**3
effect on retaining walls, **21.**12
effect on sewage disposal system, **C.**7
effect on shear strength, **8.**2
effect on slopes, **13.**5, **13.**26–**13.**27
effect on vegetation, **20.**4
evaporation at ground surface, **20.**4
exit hydraulic gradient, **8.**15, **A.**17
flow net (*see* Flow net)
flow of groundwater in utility trenches, **17.**7, **20.**3
groundwater table (definition), **A.**6
laboratory testing, **8.**1
laminar flow, **8.**7
lowering of groundwater table, **20.**2, **20.**8–**20.**11
methods to control groundwater, **20.**2, **20.**8–**20.**11
phreatic surface (definition), **20.**1, **A.**6
piezometers (*see* Instrumentation)
piping of soil [*see* Piping (of soil)]
pore water pressure (*see* Pore water pressure)
pumping of groundwater, **20.**8–**20.**11
quantity of flow, **8.**9–**8.**10
quicksand condition, **8.**15
seepage beneath sheet pile walls, **16.**32, **16.**35
seepage (definition), **A.**19
seepage forces, **8.**8, **A.**19
seepage pressures, **8.**15
seepage velocity, **8.**7, **A.**19
settlement related to groundwater extraction, **9.**55–**9.**56
for slope stability analyses, **13.**26–**13.**27
steady-state flow conditions, **13.**13
subsurface exploration, **8.**1
sulfate concentration, **17.**3–**17.**6
superficial velocity, **8.**7
velocity of flow, **8.**7
(*See also* Permeability)
Grouting (*see* Site improvement methods)
Gunite, **13.**16
Gypsiferous soil, **9.**21
Gypsum, **3.**8, **4.**28, **9.**21, **12.**39, **17.**3, **17.**6

Halite, **4.**28, **9.**21
Halloysite, **4.**7
Head (fluid flow), **3.**22, **8.**7, **A.**17
Heave (definition), **A.**17

Heavy media separator, **19**.19
Height of capillary rise, **6**.5
Hematite, **4**.28
Highway, **11**.6
Historic structures (monitoring of), **23**.3
Historical method (*see* Debris flow)
History of slope changes, **13**.5–**13**.6
Homogeneous soil, **A**.18
Horizon (of soil), **A**.6-A.7
Hornfels, **4**.28
Humidity, **12**.13
Hydraulic conductivity (*see* Coefficient of permeability)
Hydraulic gradient, **8**.7–**8**.9, **A**.18
Hydrogen bonding, **4**.4
Hydrometer analysis, **3**.9
Hydrostatic pore water pressure (*see* Pore water pressure)

Ibrid, Jordan, **12**.13
Ice lenses (*see* Frost)
Igneous rock, **4**.28
Illite, **3**.8, **4**.6–**4**.7
Immediate settlement, **9**.2, **9**.24–**9**.26
  definition, **9**.24, **A**.16
  plastic flow, **9**.24
  plate load tests, **9**.25
  Poisson's ratio, **9**.24
  stress path method, **9**.25
  theory of elasticity, **9**.24
  types of loading, **9**.24
  undrained modulus, **9**.24
Impervious soil, **A**.30
In situ (definition), **A**.7
Inclinometer (*see* Instrumentation)
Index tests:
  Atterberg limits, **3**.10, **A**.10
  expansion index test, **12**.9–**12**.10, **12**.12
  for collapse potential, **9**.17–**9**.18
  listed on boring logs, **2**.40–**2**.41
  moisture content, **3**.5
  particle size distribution, **3**.9
  specific gravity tests, **3**.7–**3**.8
  water content, **3**.5
  wet density determinations, **3**.6
  (*See also* Laboratory testing; Phase relationships)
Initial settlement (*see* Immediate settlement)
Inorganic Soil Classification System Based on Plasticity (ISBP), **4**.15
Inside clearance ratio, **2**.24–**2**.25
Instrumentation, **23**.1–**23**.12
  Avongard crack monitor, **23**.4, **23**.11
  beam sensors, **23**.4
  borehole and tape extensometers, **23**.4
  crack monitoring devices, **23**.3–**23**.4, **23**.11
  crack pins, **23**.3
  inclinometers, **23**.2, **23**.5, **A**.7
  piezometers, **23**.2, **23**.6–**23**.8, **A**.8
  pressure and load cells, **23**.2, **23**.6–**23**.8
  settlement monuments or cells, **23**.2, **23**.9
  soil strainmeters, **23**.4, **23**.12
  strain gauges, **23**.4
  tiltmeters, **23**.4
International System of Units (SI), **3**.6, **5**.2, **6**.1, **D**.1
Invert, **A**.30
Irrigation, **13**.11–**13**.12
Isobars, **6**.19
Isomorphous substitution, **4**.7
Isotropic soil (definition), **A**.18

Janbu method of slices, **13**.27
Jetting, **A**.26
Jib benefaction, **19**.19
Joints in rocks, **4**.30, **12**.39
Jute netting, **15**.9

Kaolinite, **3**.8, **4**.6, **12**.4
Karst topography, **9**.54, **A**.7
Kelly, **A**.7
Key, **16**.23, **19**.9, **A**.26
Keyway, **A**.26

Laboratory testing, **3**.1–**3**.32
  Atterberg limits, **3**.10, **A**.10
  buoyant unit weight, **3**.6
  common laboratory tests, **3**.4
  compressibility, **3**.11
  consolidated drained triaxial compression test, **3**.16
  consolidated undrained triaxial compression test, **3**.15
  consolidation test, **9**.30–**9**.32
  constant head permeameter, **3**.22, **3**.24
  direct shear test, **3**.17–**3**.18, **3**.21
  drained residual shear strength, **3**.19–**3**.20
  dry unit weight, **3**.6
  expansion index test, **12**.9–**12**.10, **12**.12
  falling head permeameter, **3**.22–**3**.23, **3**.25
  general discussion, **3**.3
  hydraulic conductivity, **3**.22–**3**.23
  hydrometer analysis, **3**.9
  index tests, **3**.4–**3**.10

Laboratory testing (*Cont.*):
laboratory tests for collapsible soil, **9.**17–**9.**18
laboratory tests for expansive rock, **12.**39
laboratory tests for pavements, **3.**32
maximum dry density, **3.**28–**3.**29
oedometer, **3.**11–**3.**12
optimum moisture content, **3.**29
particle size analysis, **3.**9
particle size distribution, **3.**9
plasticity, **3.**10
rock classification (*see* Rock classification)
sample disturbance (*see* Sample disturbance)
saturated unit weight, **3.**6
shear strength tests, **3.**13–**3.**21, **A.**13
sieve analysis, **3.**9
soil classification (*see* Soil classification)
specific gravity, **3.**7–**3.**8, **A.**14
sulfate content, **17.**3–**17.**6
swell tests, **12.**11
testing program, **3.**3
total density, **3.**6
total unit weight, **3.**6
triaxial apparatus, **3.**21
unconfined compression test, **3.**14
unconsolidated undrained triaxial compression test, **3.**15
vane shear test (*see* Vane shear test)
water content tests, **3.**5, **A.**12, **A.**14
Laminar flow, **8.**7, **A.**18
Landfill, **9.**56, **22.**2, **22.**9
Landfill liners (*see* Geosynthetics)
Landscape architect, **13.**11
Landslide, **13.**2, **13.**38–**13.**47
active landslides, **13.**38
ancient landslides, **13.**38
computer analysis, **13.**41, **13.**44–**13.**45
cross section of landslide, **13.**44–**13.**45, **13.**47
crown of the landslide, **13.**40, **13.**43
debris, **A.**7
definition, **13.**38, **A.**7
described on boring log, **2.**40–**2.**41
destabilizing effects, **13.**39
displaced landslide material, **13.**40, **13.**43
example, **13.**44–**13.**45
flank of the landslide, **13.**40, **13.**43
foot of the landslide, **13.**40, **13.**43
fossil landslides, **13.**38
groundwater, **13.**42
head of the landslide, **13.**40, **13.**43
main body, **13.**40, **13.**43
main scarp, **13.**40, **13.**43
Landslide (*Cont.*):
method of analysis, **13.**41–**13.**42
minor scarp, **13.**40, **13.**43
nomenclature, **13.**40
Portuguese Bend, **21.**19
residual shear strength, **13.**42
rotational landslide, **13.**38
rupture surface, **13.**40, **13.**43
shear key, **13.**46–**13.**47, **A.**27
stabilization, **13.**46
surface of landslide separation, **13.**40, **13.**43
tip of the landslide, **13.**40, **13.**43
toe of the landslide, **13.**40, **13.**43
top of the landslide, **13.**40, **13.**43
translational landslide, **13.**38
transverse ridges, **13.**38, **13.**43
triggering conditions, **13.**39
zone of landslide accumulation, **13.**40, **13.**43
zone of landslide depletion, **13.**40, **13.**43
Lapilli, **4.**24
Lateral fill extension, **13.**50
Lateral movement (*see* Slope movement; Slope stability analysis)
Leaching, **4.**6, **A.**7
Leaching lines, **A.**30
Leakage, **20.**13
Leyte, Philippines, **13.**48
Lift (of fill), **A.**26
Limestone, **4.**28, **4.**31, **9.**54
Liquefaction:
analysis, **14.**5–**14.**11
definition, **14.**3, **A.**18
effect of aging, **14.**4
effect of confining pressure, **14.**4
effect of drainage conditions, **14.**4
effect of groundwater table, **14.**4
effect of particle size gradation, **14.**4
effect of placement conditions, **14.**4
effect of relative density, **14.**4
example problem, **14.**7
factors causing liquefaction, **14.**4
soil type susceptible to liquefaction, **14.**4
(*See also* Earthquakes)
Liquid limit, **3.**10, **4.**4, **A.**10
Liquidity index, **4.**4–**4.**5
Liquids, **5.**1, **5.**3
Live load, **9.**4, **10.**7, **A.**18
Load cells (*see* Instrumentation)
Load test (for piles), **10.**23, **21.**3, **21.**24
Loading of soil (*see* Stress distribution)
Loess, **4.**25, **15.**3, **21.**9, **A.**7
Log washers, **19.**19

Logs:
definition, **2.**2, **A.**5
description of geologic soil deposits, **2.**43
description of man-made soil deposits, **2.**43
example of, **2.**40–**2.**41
preparation of, **2.**2, **2.**40–**2.**44
type of information recorded on, **2.**42
Long Beach naval shipyard, **9.**55
Longwall mining, **9.**54
Los Altos Hills, California, **13.**48
Louisiana, **12.**30

Marble, **4.**28, **4.**31
Mass, **5.**3, **5.**5
Mat foundations:
basement walls as part of mat foundation, **18.**6
beam and slab, **18.**6
description of mat foundation, **1.**6, **18.**3
flat plate, **18.**6
plate thickened under columns, **18.**6
plate with pedestals, **18.**6
reasons for using mat foundation, **18.**3
Maximum dry density (*see* Compaction test)
Mechanically stabilized earth retaining walls (MSE walls), **16.**2, **16.**23–**16.**25
construction of MSE walls:
compacted fill, **16.**23
drainage system, **16.**23
soil reinforcement, **16.**23
wall facing element, **16.**23
design analysis for MSE walls:
external stability, **16.**23–**16.**25
internal stability, **16.**24
example problem, **16.**24
geogrid (soil reinforcement), **16.**2, **16.**23–**16.**24, **22.**3–**22.**5
geosynthetic (soil reinforcement), **16.**2, **16.**23–**16.**24, **22.**1, **22.**3–**22.**5
movement of MSE walls, **16.**24
Metamorphic rock, **4.**28
Metaquartzite, **4.**28
Method of slices, **13.**25–**13.**27
Mexico City, **9.**55
Mica, **7.**6
Microfossils, **9.**55
Mildew, **20.**13
Mineral, **3.**7–**3.**8, **A.**7
Mines:
collapse of, **9.**54, **9.**58
spoil, **9.**55
strip mining, **9.**54
Modified Proctor (*see* Proctor)
Modified Wentworth Scale, **4.**28
Modulus of elasticity of soil, **9.**24
Modulus of subgrade reaction, **9.**50–**9.**51, **9.**53
Mohr circle, **6.**3, **6.**22, **A.**18
Moisture content (*see* Water content)
Moisture migration: **20.**4, **20.**13–**20.**17
barriers, **22.**2–**22.**3, **22.**8–**22.**9
through basement walls, **20.**15–**20.**17
capillary break, **20.**13–**20.**14, **20.**17
construction details, **20.**13–**20.**17
damage caused by, **20.**13, **20.**16
due to capillary action, **20.**13, **20.**15
due to hydrostatic pressure, **20.**13, **20.**15
due to leakage, **20.**13
due to water vapor, **20.**13, **20.**15
through floor slabs, **20.**13–**20.**14, **20.**17
waterproofing, **20.**15–**20.**17
Monitoring devices (*see* Instrumentation)
Montmorillonite, **3.**8, **4.**6, **4.**25–**4.**26, **12.**4, **16.**5
Morgenstern-Price method of slices, **13.**27
Mud flow, **13.**48
Mud slide, **13.**48
Mulch, **15.**9, **15.**12
Muscovite, **3.**7

National Research Council, **13.**38
National Science Foundation, **17.**1
Necking, **A.**26
Negative skin friction (*see* Downdrag)
New Orleans, **15.**4
Nitroglycerin, **4.**27
Nonplastic soil, **3.**30, **4.**15
Normally consolidated (*see* Consolidation)

Observational method, **19.**6, **19.**28–**19.**29
Obsidian, **4.**28
Oedometer, **3.**11–**3.**12
Ohio, **16.**37
Open joint tile, **A.**30
Optimum moisture content (*see* Compaction test)
Ordinary method of slices, **13.**25, **13.**27
Organic soil:
classification of organic soil, **4.**22–**4.**23
erosion potential, **15.**5
matter, **4.**24
peat (*see* Peat)
range of laboratory test results, **4.**22–**4.**23
settlement of organic soil, **4.**24, **9.**2–**9.**3, **9.**24–**9.**44
soil, **4.**22–**4.**23

Orinoco clay:
consolidation properties of Orinoco clay, **9.**30–**9.**31, **9.**41
factors that effect the undrained shear strength, **7.**12, **7.**37–**7.**38
radiographs of, **2.**21–**2.**23
undrained shear strength of, **7.**18, **7.**38
Orinoco River, Venezuela, **7.**12
Ottawa, Ontario, **4.**26
Overburden (definition), **A.**7
Overconsolidated soil (*see* Consolidation)
Overconsolidation ratio (*see* Consolidation)
Oversize particles, **3.**5, **19.**10, **19.**17–**19.**18
Overtopping (*see* Dams)
Owner (*see* Clients)

Palos Verdes, California, **21.**19
Particle size, **4.**11–**4.**12
Particle size distribution (*see* Soil classification)
Passive earth pressure (*see* Retaining wall analysis)
Pavements, **11.**1–**11.**9
aggregates, **11.**3
asphalt, **11.**3, **A.**22
asphalt concrete, **11.**3, **A.**22
asphalt overlays, **22.**3
base course, **11.**3, **17.**7–**17.**8, **A.**22
bearing capacity, **11.**3
black top, **11.**3
California bearing ratio, **2.**38, **3.**32, **11.**1, **11.**5, **17.**8, **19.**14, **19.**20, **A.**5
California method of design, **11.**2, **11.**6–**11.**9
cement treated base, **11.**3
concrete, **11.**3
crushed stone, **11.**3
design, **11.**6–**11.**9
design life of pavements, **11.**6
deterioration, **17.**7–**17.**9
drainage properties of the subgrade, **11.**5
effect of expansive soil, **12.**27–**12.**28
effect of groundwater, **17.**7, **20.**2–**20.**3
equivalent wheel loads, **11.**6
example problem, **11.**7
on expansive soil, **12.**37–**12.**38
flexible pavement, **11.**1, **11.**3
frost, **11.**4
gravel, **11.**3
gravel equivalent factor, **11.**6–**11.**7
hot mix, **11.**3
laboratory tests, **3.**32
macadam, **11.**3
mineral filler, **11.**3

Pavements (*Cont.*):
R-value, **3.**32, **11.**6–**11.**7, **11.**9, **17.**8
rigid pavement, **11.**1, **11.**3
rutting, **11.**3
subbase, **11.**3
subgrade (*see* Subgrade)
subgrade modulus (*see* Subgrade modulus)
surface course, **11.**3
traffic index, **11.**6–**11.**8
traffic loads, **11.**6
Payment, **1.**17
Peat:
bogs, **4.**24
classification, **4.**12, **4.**14, **4.**22
composition, **4.**24
definition, **A.**7
moors, **4.**24
water content of, **3.**5
(*See also* Organic soil)
Pebbles, **4.**28
Penetration resistance (definition), **A.**7
Percolation (definition), **A.**30
Percolation test procedures, **C.**1-**C.**7
Percolation tests, **20.**4–**20.**5, **20.**18–**20.**19, **A.**30, **C.**1-C.7
Percussion drilling, **2.**14, **A.**7
Peridotite, **4.**31
Permafrost, **17.**2, **A.**8
Permanent erosion control devices, **A.**26
Permeability, **8.**1–**8.**17
coefficient of permeability, **3.**22–**3.**27, **8.**3, **A.**12
coefficient of permeability versus drainage property, **3.**26
coefficient of permeability versus void ratio, **3.**27
of concrete, **17.**4
Darcy's law, **3.**22, **8.**1, **8.**7, **8.**9
definition, **A.**12
degree of permeability, **3.**22–**3.**23
determined from field measurements, **8.**4–**8.**6
determined from laboratory tests, **3.**22–**3.**25
effect of layering of soil on permeability, **8.**3
effect of particle size distribution on permeability, **8.**3
effect of soil imperfections or discontinuities on permeability, **8.**3
effect of soil structure on permeability, **8.**3
effect of void ratio on permeability, **8.**3

Permeability (*Cont.*):
effect on sewage disposal system, **20.**18–**20.**19
effect on surficial stability, **13.**10
factors that effect permeability, **8.**3
hydraulic conductivity, **3.**22–**3.**23
of loess, **4.**25
of soil, **3.**22–**3.**27
of swelling clay, **12.**17
of trench backfill, **20.**3
(*See also* Groundwater)
Permeameter test:
constant head permeameter, **3.**22, **3.**24
falling head permeameter, **3.**22–**3.**23, **3.**25
Permit, **A.**26
Phase relationships, **5.**1–**5.**7
buoyant unit weight, **5.**6
degree of saturation, **4.**10, **5.**4
density, **3.**6, **5.**2
direct relationships, **5.**2
dry unit weight, **5.**6
elements of soil, **5.**3
example problems, **5.**7
gas, **5.**1, **5.**3
indirect phase relationships, **5.**2, **5.**4
liquids, **5.**1, **5.**3
mass and volume relationships, **5.**5
porosity, **5.**4
relative density, **2.**35, **3.**30
saturated unit weight, **5.**6
soil element, **5.**1, **5.**3
solids, **5.**1, **5.**3
specific gravity, **3.**7–**3.**8, **5.**2
total density, **3.**6, **5.**2
total unit weight, **5.**6
unit weight, **3.**6, **5.**2
unit weight relationships, **5.**6
useful relationships, **5.**4–**5.**6
void ratio, **5.**4
water content, **3.**5, **5.**2
(*See also* Index tests; Laboratory testing)
Phreatic surface (definition), **A.**6
Phyllite, **4.**28
Pier walls, **16.**3, **16.**41–**16.**47
advantages of pier walls, **16.**41
construction of pier walls, **16.**41–**16.**45
design of pier walls:
for deep excavations, **21.**14
earth pressure theory, **16.**44, **16.**47
end bearing, **16.**45
for landslide stabilization, **13.**46
pier spacing, **16.**45
skin friction, **16.**45
Pier walls, design of pier walls (*Cont.*):
for slope stabilization, **16.**41–**16.**47
tieback anchors, **16.**45–**16.**46
wedge method, **16.**43–**16.**44, **16.**47
disadvantages of pier walls, **16.**41–**16.**42
soil arching, **16.**44
Piers:
definition, **1.**7, **10.**22, **18.**8, **A.**26
excavation of piers, **21.**2, **21.**5–**21.**6
for expansive soil, **12.**32–**12.**33, **12.**36
load test, **10.**23
for underpinning, **21.**18
(*See also* Pier walls)
Piezometer (*see* Instrumentation)
Pile groups, **10.**26, **10.**29–**10.**30, **10.**32
Piles:
batter pile, **10.**22, **18.**7, **A.**26
cast-in-place piles, **18.**7–**18.**13
combination end-bearing and friction pile, **10.**22, **18.**7, **A.**26
definition, **1.**7, **10.**22, **A.**26
driving resistance, **10.**23
end-bearing pile, **10.**22, **18.**7, **A.**27
experience, **10.**23
field load tests, **10.**23, **21.**3, **21.**24
friction pile, **10.**22, **18.**7, **A.**27
group efficiency, **10.**29, **10.**32
high displacement piles, **18.**7
load tests, **10.**23
low displacement piles, **18.**7
material types, **18.**7–**18.**14
mixed in place, **10.**22
pile cap, **10.**26
pile characteristics and uses, **18.**9–**18.**14
pile configurations, **18.**14
pile driving equations, **10.**23
prestressed concrete piles, **18.**8, **18.**11–**18.**12, **18.**14
steel pipe piles, **18.**8–**18.**13
uplift capacity, **10.**25
Pinhole test, **13.**56
Pinnacles, **4.**25
Pipe breaks, **20.**23
Pipelines, **11.**2, **11.**10–**11.**16
bedding constant, **11.**15
cast-in-place pipeline, **11.**11
cast iron pipeline, **11.**11
corrugated metal pipe (CMP), **11.**15
design factors, **11.**2
ductile iron pipeline, **11.**15
flexible pipeline, **11.**2, **11.**15
minimum design load, **11.**11
modulus of soil resistance, **11.**15

Pipelines (*Cont.*):
  placement conditions:
    embankment, **11**.2, **11**.10–**11**.11, **11**.13
    jacked, **11**.2, **11**.10–**11**.12, **11**.14
    trench, **11**.2, **11**.10–**11**.12, **11**.14
  polyvinyl chloride (PVC), **11**.15
  precast pipeline, **11**.15
  rigid pipeline, **11**.2, **11**.10–**11**.14
  specifications for storm drain backfill, **11**.16
Piping (of soil), **8**.15, **13**.55–**13**.56, **21**.12, **A**.18
Pit (*see* Test pits)
Plastic equilibrium, **A**.18
Plastic limit, **3**.10, **A**.10
Plastic soil, **3**.10, **4**.15
Plasticity, **4**.4, **A**.12
Plasticity chart, **4**.8
Plasticity index, **3**.10, **4**.4, **A**.13
Plate load test, **2**.38–**2**.39, **9**.25, **9**.48
Poisson's ratio, **9**.24
Poplars, **12**.21
Pore water pressure:
  below groundwater table, **6**.4, **8**.1
  causing liquefaction, **14**.3
  causing slope softening, **13**.50
  definition, **6**.4, **A**.18
  determined from a flow net, **8**.15
  dissipation of excess pore water pressure, **9**.27, **9**.35–**9**.36, **10**.28
  effect on landslides, **13**.42
  excess, **3**.15–**3**.18, **A**.18
  excess caused by earthquakes, **14**.3
  hydrostatic, **6**.4, **A**.18
  hydrostatic uplift, **18**.3, **20**.2
  monitoring of pore water pressure, **23**.2, **23**.6–**23**.8
  negative, **3**.15–**3**.18
  pore water pressure parameters A and B, **7**.27–**7**.29, **7**.31
  positive, **3**.15–**3**.18
  shear induced, **3**.15–**3**.18, **7**.12
  for slope stability analysis, **13**.26–**13**.27
  unbalanced pore water pressure acting on sheet pile walls, **16**.32, **16**.35
Pore water pressure ratio, **13**.27, **13**.30–**13**.31, **13**.35–**13**.37
Porosity, **5**.4, **A**.18
Porous soil, **A**.30
Portuguese Bend Landslide, **21**.19
Post-Tensioning Institute, **12**.30, **18**.4
Pozzolan, **4**.24, **12**.27, **A**.27
Precise grading permit, **A**.27
Precompression of soft soil (*see* Site improvement methods)
Preconsolidation pressure (*see* Consolidation)
Presoaking (for expansive soil), **12**.26–**12**.27
Presoaking (for percolation tests), **20**.18
Pressure, **6**.1–**6**.22, **A**.19
Primary consolidation (*see* Consolidation)
Principal planes, **6**.22, **A**.19
Principal stresses, **6**.22, **7**.14, **A**.19
Proctor, **3**.28–**3**.29, **3**.31
Progressive failure, **13**.28–**13**.29, **A**.19
Project:
  client, **1**.12
  scope of work, **1**.12
Project types:
  commercial, **1**.12
  condominiums, **1**.12
  essential facilities, **1**.12
  flatland, **1**.12
  hillside, **1**.12
  industrial, **1**.12
  private sector, **1**.12
  public works, **1**.12
  single-family dwellings, **1**.12
  special considerations, **1**.12
Proposal:
  cost estimating sheet, **1**.15–**1**.16
  schedule of fees, **1**.14
  scope of services, **1**.17
  (*See also* Contract)
Pseudostatic earthquake analysis, **14**.13
Pumice, **4**.24–**4**.25, **4**.28
Pumping:
  of groundwater, **9**.55–**9**.56
  of oil, **9**.55
  of soil during compaction, **19**.13–**19**.14
Pyritic shale, **2**.44, **12**.39
Pyroclastic rock, **4**.24

Quartz, **3**.7
Quartzite, **4**.28, **4**.31
Quick clay, **4**.26, **14**.5, **21**.3, **A**.19
Quick condition (quicksand), **8**.15, **A**.19

Radiograph, **2**.21–**2**.23
Radiolarians, **4**.27
Rainfall, **13**.5, **13**.10, **15**.2, **15**.4, **15**.10–**15**.11
Rankine earth pressure states (*see* Retaining wall analysis)
Raveling, **9**.54, **9**.57, **21**.16
Ravines (*see* Erosion)
Refusal (during drilling), **A**.8
Relative compaction (*see* Compaction)

Relative density, **2.**35, **3.**30, **14.**4, **A.**19
Repair:
for concrete cracks, **21.**18–**21.**19, **21.**22–**21.**23
foundation strengthening, **21.**18–**21.**19, **21.**21
for gross slope failures, **13.**46
for landslides, **13.**46
strip replacement, **21.**18, **21.**32
for surficial slope failures, **B.**22
Reservoir, **13.**53
Residual shear strength (*see* Shear strength)
Residual soil (*see* Soil deposits)
Restrained retaining walls, **16.**2
Retaining wall analysis, **16.**1–**16.**47
active earth pressure, **16.**6, **16.**11, **A.**17
active earth pressure coefficient, **16.**6
active wedge, **16.**7, **16.**10
adhesion, **16.**9
at-rest earth pressure, **16.**2, **A.**17
back-cut, **16.**19
backfill, **16.**5, **16.**18–**16.**19, **16.**37
bearing capacity failure, **16.**8, **16.**12
bearing pressure of footing, **16.**8, **16.**12
common causes of failure, **16.**18–**16.**19
compaction of backfill, **16.**19
construction, **16.**17–**16.**19
construction at the top of slopes, **16.**17
Coulomb's equation, **16.**11, **16.**13
drainage system, **16.**5
earth pressure (definition), **A.**17
effect of earthquakes, **16.**18
effect of expansive soil, **12.**18, **16.**18
effect of frost, **16.**5
effective stress analysis, **16.**8
equivalent fluid pressure, **16.**6, **A.**17
example of retaining wall analysis, **16.**14–**16.**16
with expansive soil backfill, **16.**18
factor of safety, **16.**8–**16.**9, **16.**12–**16.**13, **16.**16
inclined slope, **16.**13
overturning analysis, **16.**9, **16.**13, **16.**22
passive earth pressure, **16.**7–**16.**8, **16.**12, **16.**17, **A.**17
passive earth pressure coefficient, **16.**8, **16.**17
passive wedge, **16.**8, **16.**10
plane strain condition, **16.**5
Rankine earth pressure states, **16.**6, **16.**8
reduction factor for passive pressure, **16.**8
shear strength of backfill, **16.**5
sliding analysis, **16.**8–**16.**9, **16.**12–**16.**13, **16.**16
Retaining wall analysis *(Cont.)*:
slope creep, **16.**17
slope stability analysis, **13.**29, **16.**19
surcharge, **16.**7
translation of the footing, **16.**7–**16.**8, **16.**10
with wall friction, **16.**11–**16.**16
without wall friction, **16.**6–**16.**10
Retaining walls:
cofferdams (*see* Cofferdams)
definition, **A.**29
mechanically stabilized earth retaining walls (*see* Mechanically stabilized)
pier walls (*see* Pier walls)
restrained retaining walls (*see* Restrained retaining walls)
retaining wall analysis (*see* Retaining wall analysis)
sheet pile walls (*see* Sheet pile walls)
temporary retaining walls (*see* Temporary retaining walls)
types of retaining walls (*see* Walls)
Rhyolite, **4.**28
Rillwash (*see* Erosion)
Rippability, **2.**45–**2.**48, **A.**27
Riprap, **A.**27
Rock:
chemical stability, **4.**31
classification, **4.**28–**4.**31, **A.**8
classification of rock for tunnel excavations, **21.**15
crushed shape, **4.**31
definition, **A.**8
durability, **4.**31
engineering properties, **4.**31
expansion of rock due to weathering, **12.**39
expansive rock, **12.**39, **21.**15
hardness, **4.**29
impurities, **4.**31
intact rock, **21.**15
joint spacing, **4.**30
jointed rock, **21.**15
mechanical strength, **4.**31
physical description, **4.**30
quality, **4.**30
rebound of rock, **12.**39
rock quality designation (RQD), **4.**30
rock surface (during breakage), **4.**30
squeezing rock, **21.**15
stratified rock, **21.**15
topples, **13.**4
types:
igneous, **4.**28

Rock, types. (*Cont.*):
metamorphic, **4**.28
sedimentary, **4**.28
weathering characteristics, **4**.30
Rock flour, **4**.24
Rock mechanics (definition), **1**.5, **A**.2
Rock salt (halite), **4**.28
Rockfall, **13**.8–**13**.9
definition, **13**.8
design criteria for ditches, **13**.9
factors that govern a rockfall, **13**.8
remedial measures, **13**.8
Roots, **12**.21
Rose Canyon fault zone, **2**.9
Rotary drilling, **2**.11–**2**.13, **A**.8
Rotary scrubber, **19**.19
Rubble, **A**.8
Running soil or running ground, **16**.40, **21**.16, **A**.27

Sabkhas, **9**.21
Salinas, **9**.21
Salt, **9**.21
Salt marshes, **9**.21
Salt playas, **9**.21
Sample disturbance:
altered soil, **2**.24
area ratio, **2**.24–**2**.25
disturbed soil, **2**.24–**2**.25
due to gas coming out of solution, **2**.26
due to tube friction, **2**.26
effect on coefficient of consolidation, **9**.38, **9**.41
effect on consolidation test, **9**.30–**9**.31, **9**.41
effect on shear strength, **7**.12–**7**.13, **7**.37
effect on swelling of clay, **12**.11
inside clearance ratio, **2**.24–**2**.25
of Orinoco clay, **2**.21–**2**.23, **2**.26
of quick clay, **4**.26
soil cracks, **2**.26
turning of edges, **2**.26
undisturbed soil, **2**.24–**2**.25
voids, **2**.26
Samplers:
Acker, **2**.11
definition, **A**.9
Denison, **2**.11
Pitcher, **2**.11
rotary coring, **2**.11–**2**.12
Swedish foil, **2**.13
thin wall tube, **2**.12
(*See also* Soil sampling)
Sampling (*see* Soil sampling)
San Diego, California, **2**.7
San Diego County, California, **20**.14, **C**.1
Sand:
boils, **13**.55, **A**.27
classification of, **4**.11–**4**.18
erosion potential of sand, **15**.3, **15**.18
liquefaction of, **14**.3–**14**.11
loss of shear strength, **14**.13
size particles, **4**.11, **A**.13
Sand classifying unit, **19**.19
Sand cone tests, **19**.12
Sand equivalent, **A**.13
Sand-size particles, **4**.11, **A**.13
Sandstone, **4**.28, **4**.31
Saturation (degree of), **4**.10, **5**.4, **A**.19
Schedule of fees, **1**.14
Schist, **4**.28, **4**.31
Scoria, **4**.28
Screening, **19**.19
Screw classifier, **19**.19
Sea cliffs (erosion of), **15**.3
Seasonal moisture change (*see* Expansive soil)
Seawater, **17**.3, **17**.6
Secondary compression, **9**.2, **9**.46–**9**.47, **A**.16
Secondary compression ratio, **9**.46
Sedimentary rock, **4**.28
Seep, **A**.8
Seepage (*see* Groundwater)
Seepage pit, **A**.30
Seismic activity (*see* Earthquakes)
Sensitivity, **4**.26, **A**.19
Serpentine (mineral), **3**.8
Serpentine (rock), **4**.28, **4**.31
Settlement, **9**.1–**9**.60
allowable settlement, **9**.1, **9**.6–**9**.14
angular distortion, **9**.6–**9**.14
caused by groundwater extraction, **9**.55–**9**.56
caused by oil extraction, **9**.55
caused by organic decomposition, **9**.56
caused by undermining, **9**.4
of cohesionless soil, **9**.3, **9**.48–**9**.53
of cohesive soil, **9**.2, **9**.24–**9**.45
collapse of underground mines and tunnels, **9**.54–**9**.55
collapsible soil (*see* Collapsible soil)
component of lateral movement, **9**.9
compression features, **9**.9, **9**.14, **9**.58
consolidation (*see* Consolidation)
cut-fill transition, **9**.9
definition, **9**.1, **A**.20
of diatoms, **4**.27

Settlement (*Cont.*):
differential settlement, **9.**1, **9.**6–**9.**13, **A.**20
drag effect, **9.**9
due to earthquakes, **9.**4, **14.**12–**4.**13
due to secondary influences, **9.**4
due to weight of the structure, **9.**4
elastic method, **9.**24, **9.**26, **9.**49–**9.**50
empirical correlations, **9.**49–**9.**53
example problems, **9.**59–**9.**60
immediate (*see* Immediate settlement)
landfills, **9.**56
limestone cavities, **9.**54
of organic soil, **9.**2, **9.**24–**9.**44
of pile groups, **10.**26, **10.**30
plate load test, **2.**38–**2.**39, **9.**48
rate of settlement, **9.**1, **9.**7
of sands, **9.**3, **9.**48–**9.**53
Schmertmann's method, **9.**50
secondary compression, **9.**2, **9.**46–**9.**47, **A.**16
selection of foundation type, **18.**1–**18.**15
severity of cracking damage, **9.**11
sinkholes, **9.**54, **9.**57
of soluble soil particles (*see* Soluble soil particles)
subgrade modulus, **9.**51, **9.**53
subsidence, **9.**4
tensional features, **9.**9, **9.**14
tilting, **9.**8, **9.**10
total settlement, **9.**1, **9.**3, **9.**47, **A.**20
two- or three-dimensional, **9.**24
types of settlement, **9.**4–**9.**5
Settlement monuments (*see* Instrumentation)
Setup (*see* Freeze)
Sewage disposal system: **20.**2–**20.**4, **20.**18–**20.**22, **C.**1-**C.**7
design, **20.**2–**20.**4, **20.**18–**20.**22, **C.**1-**C.**7
disposal field, **20.**18–**20.**22
effluent, **20.**18, **A.**30
leach field, **20.**18–**20.**20, **20.**22
lot size, **20.**19, **C.**4
nitrate impacts, **20.**19
percolation rate, **20.**18–**20.**19, **C.**1-**C.**7
percolation test procedures, **20.**18–**20.**19, **C.**1-**C.**7
percolation tests, **20.**18–**20.**19, **C.**1-**C.**7
primary disposal trench length, **20.**19, **C.**3
septic tank, **20.**18, **20.**21, **C.**4
terms and definitions, **A.**30-**A.**31
Sewage spring, **A.**30
Sewer, **20.**23
Shafts, **10.**22, **16.**41
Shale, **4.**28, **4.**31, **15.**3
Shallow foundation (*see* Foundation)
Shear failure (definition), **A.**20
Shear key, **13.**46–**13.**47, **19.**7, **A.**27
Shear plane (also known as slip surface or rupture surface), **13.**13–**13.**14, **13.**25–**13.**26, **13.**28, **A.**20
Shear strain (due to earthquakes), **14.**4
Shear strength, **7.**1–**7.**43
$\phi = 0$ concept, **7.**15
clay consistence versus undrained shear strength, **4.**9–**4.**10
cohesion, **3.**13, **7.**1–**7.**43, **A.**11
of cohesionless soil, **7.**2, **7.**4–**7.**11
of cohesive soil, **7.**2, **7.**12–**7.**40
of compacted London clay, **7.**26
consolidated drained triaxial compression test, **3.**13
consolidated undrained triaxial compression test, **3.**13, **7.**16–**7.**17, **7.**20–**7.**22
contraction during shear, **7.**4
curved (nonlinear) failure envelope, **7.**23–**7.**24, **7.**26
definition, **A.**13
deviator (or deviatoric) stress, **7.**14, **A.**11
dilation during shear, **7.**4
direct shear, **7.**4–**7.**5, **7.**7–**7.**8
displacement rate, **7.**37
drained, **3.**13–**3.**20, **7.**1–**7.**3, **7.**21–**7.**35
for earthquake analysis, **14.**13
effect of anisotropy, **7.**37
effect of strain rate, **7.**37
effect of water on friction angle, **7.**4
effective cohesion, **7.**5, **7.**23–**7.**24
effective friction angle, **7.**5, **7.**23–**7.**25
effective stress analysis, **3.**13, **7.**1, **7.**40–**7.**41, **A.**13
estimated from soil type versus dry density or relative density, **7.**11
estimated from the cone penetration test, **7.**9
estimated from the standard penetration test, **7.**10
example problems, **7.**42–**7.**43
factors that cause a reduction in shear strength, **7.**12, **7.**37
factors that affect shear strength, **7.**3, **7.**6, **7.**23–**7.**24, **7.**37–**7.**38
failure envelope, **7.**5, **7.**16
failure surfaces, **7.**5, **7.**16, **7.**37
friction angle, **3.**13, **7.**1–**7.**43, **A.**12
laboratory testing, **3.**13–**3.**21
maximum obliquity, **7.**21, **7.**29

Shear strength (*Cont.*):
maximum shear stress, **7.**14
Mohr-Coulomb failure law, **7.**4, **7.**16
normalized undrained shear strength, **7.**12–**7.**13
peak, **7.**5–**7.**8, **A.**12
pore water pressure parameters A and B, **7.**27–**7.**29, **7.**31
remolded, **7.**37–**7.**38
residual, **3.**13, **3.**19–**3.**20, **7.**3, **7.**36
ring shear, **3.**19–**3.**20
root-permeated, **13.**11
sample disturbance (*see* Sample disturbance)
at soil-pile interface, **10.**26
strain rate, **7.**37–**7.**38
stress paths, **7.**30, **7.**32–**7.**33, **7.**35
summary of triaxial test results, **7.**34–**7.**35
tests, **3.**13–**3.**21
total stress analysis, **3.**13, **7.**1, **7.**40, **A.**13
ultimate shear strength, **7.**5, **7.**7–**7.**8
unconfined compression test, **3.**14
unconsolidated undrained triaxial compression test, **3.**13, **7.**14–**7.**15
undisturbed, **7.**37–**7.**38
undrained, **3.**13–**3.**18, **7.**1–**7.**3, **7.**12–**7.**22
vane tests (*see* Vane shear test)
(*See also* Laboratory testing)
Shear stress (definition), **A.**20
Sheet pile walls, **16.**2, **16.**26–**16.**35
design analysis:
active earth pressure on sheet pile wall, **16.**28, **16.**34
anchor pull, **16.**29
design analysis for cohesive soil, **16.**33
design charts, **16.**30, **16.**34
factors increasing the stability of the wall, **16.**33
important design considerations, **16.**32–**16.**33
loading conditions, **16.**32–**16.**33
maximum moment in the sheet pile wall, **16.**30, **16.**34
passive earth pressure on the sheet pile wall, **16.**28–**16.**29, **16.**34
penetration depth, **16.**29, **16.**32, **16.**34
plane strain condition, **16.**33
soil layers, **16.**32
surcharge loads, **16.**32
toe kick-out, **16.**29
unbalanced hydrostatic and seepage forces, **16.**32, **16.**35
water pressure, **16.**29, **16.**32, **16.**35
Sheet pile walls (*Cont.*):
example problem, **16.**30
steel sheet piling sections, **16.**26–**16.**27
types of sheet pile walls:
anchored sheet pile walls, **16.**28–**16.**29, **16.**34
cantilevered sheet pile walls, **16.**30, **16.**34
cofferdams, **16.**31
Sheffield Dam, **14.**2
Shells, **4.**10
Shoring (*see* Temporary retaining walls)
Shotcrete, **A.**27
Shrinkage factor, **A.**27
Shrinkage limit, **3.**10, **12.**13, **A.**10
Shrinkage of soil (*see* Expansive soil)
Sidewalls, **A.**31
Sieve, **3.**9, **A.**13
Sieve analysis (*see* Soil classification)
Silicates, **3.**7–**3.**8
Silt, **4.**11–**4.**18
Silt-size particles, **4.**11, **A.**13
Siltstone, **4.**28
Silty sand, **7.**5, **7.**7–**7.**8, **9.**17, **9.**22–**9.**23
Sinkhole activity, **9.**54, **9.**57
Site, **A.**27
Site improvement methods:
geosynthetics (*see* Geosynthetics)
grouting methods, **19.5**
compaction grouting, **19.5**
deep mixing, **19.**25
definition, **A.**26
injection of grout, **19.**5, **19.**25
mudjacking, **19.5**
soil replacement methods, **19.**4, **19.**23
displacement, **19.**4, **19.**23
removal and replacement, **19.**4, **19.**23
soil stabilization, **19.**4–**19.5**
soil strengthening, **19.**5, **19.**24
dynamic compaction, **19.**24
vibrocompaction, **19.**24
vibrodensification (definition), **A.**29
vibrodisplacement, **19.**24
vibroreplacement, **19.**24
thermal methods, **19.**5, **19.**25
water removal methods, **19.**4–**19.**5, **19.**23
electroosmosis, **19.**23
precompression, **19.**23
precompression with vertical drains, **19.**23
trenching, **19.**23
Site investigation (*see* Investigation)
Slab-on-grade (*see* Foundation)

Slag, **4.**10
Slaking, **12.**16, **A.**8
Slate, **4.**28, **4.**31
Slickensides, **A.**8
Slides, **13.**2
Slip surface (*see* Shear plane)
Slope (definition), **A.**27
Slope Indicator Company, **23.**2–**23.**3, **23.**5–**23.**12
Slope movement:
  allowable, **13.**6–**13.**7
  causes, **13.**4
  creep (*see* Creep of slopes)
  of dams (*see* Dams)
  debris flow (*see* Debris flow)
  due to freezing of water, **17.**9
  due to gross instability (*see* Gross slope stability analysis)
  due to liquefaction of seams, **14.**3
  due to reservoir drawdown (*see* Dams)
  due to rockfall (*see* Rockfall)
  due to slope softening (*see* Slope softening)
  due to surficial instability (*see* Surficial slope stability)
  during construction, **23.**1–**23.**2
  effect of groundwater, **13.**26–**13.**27, **20.**3–**20.**4, **20.**12
  landslides (*see* Landslides)
  rockfall (*see* Rockfall)
  stabilization measures, **13.**46
  types of slope movement, **13.**4
Slope softening, **13.**50–**13.**51
  definition, **13.**50
  indications of slope softening, **13.**50
  lateral fill extension, **13.**50
  method of analysis, **13.**51
  pore water pressure, **13.**50
Slope stability analysis:
  allowable lateral movement, **13.**6–**13.**7
  checklist for slope stability analysis, **13.**5–**13.**6
  debris flow analysis (*see* Debris flow)
  different types of analyses, **A.**20
  excavation slopes (*see* Excavation slopes)
  gross slope stability analysis (*see* Gross slope stability analysis)
  landslide stability analysis (*see* Landslides)
  pseudo-static analysis, **14.**13
  beneath retaining walls (*see* Retaining walls)
  rockfall analysis (*see* Rockfall)
  slope creep analysis (*see* Creep of slopes)
Slope stability analysis (*Cont.*):
  slope softening analysis (*see* Slope softening)
  slope stability for earth dams (*see* Dams)
  surficial stability analysis (*see* Surficial slope stability)
Slope wash (*see* Erosion)
Slough, **A.**28
Sludge, **A.**31
Slumps, **13.**17, **A.**28
Slurry seal, **A.**28
Soapstone, **4.**28
Sodium hexametaphosphate, **3.**9
Soil absorption system, **A.**31
Soil behavior (*see* Engineering analyses)
Soil cementing agents, **15.**5
Soil classification, **4.**1–**4.**27
  color, **4.**9
  consistency, **4.**9–**4.**10
  density conditions, **4.**10
  descriptive terms, **4.**10
  grain size analysis, **3.**9
  group symbols, **4.**12–**4.**15
  hydrometer tests, **3.**9
  moisture condition, **4.**10
  particle size distribution, **3.**9, **4.**19
  plasticity, **4.**4
  porosity, **4.**9
  sieve analysis, **3.**9
  structure, **4.**9
  texture, **4.**9
Soil classification systems:
  AASHTO soil classification system, **4.**16–**4.**17
  classification of soil for tunnel excavations, **21.**16
  comparison of plastic soil for USCS and AASHTO, **4.**18
  Inorganic Soil Classification System Based on Plasticity (ISBP), **4.**15
  organic soil classification system, **4.**22–**4.**23
  soil classification examples, **4.**20
  Unified Soil Classification system (USCS), **4.**11–**4.**14
  USDA textural classification, **4.**2, **4.**18
Soil containing gypsum, **9.**21
Soil containing soluble minerals, **9.**21
Soil containing sulfate (*see* Sulfate)
Soil (definition), **1.**4, **A.**8
Soil deposits:
  aeolian, **2.**43, **A.**4
  alluvium, **2.**43, **A.**4

Soil deposits (*Cont.*):
  artificial fill, **2.**43
  colluvium, **2.**43, **A.**5
  debris fill, **2.**43
  glacial, **2.**43
  gypsiferous, **9.**21
  lacustrine, **2.**43
  marine, **2.**43
  municipal dump, **2.**43
  organic, **2.**43
  pyroclastic, **2.**43
  residual, **2.**43, **21.**9, **A.**8
  structural fill, **2.**43
  uncompacted fill, **2.**43
Soil matrix, **3.**5, **19.**17–**19.**18
Soil mechanics (definition), **1.**4, **A.**3
Soil nailing, **13.**46
Soil profile (*see* Subsoil profile)
Soil reinforcement (with geosynthetics), **22.**3–**22.**5
Soil replacement (*see* Site improvement methods)
Soil samples:
  altered soil, **2.**24
  definition, **A.**9
  disturbed samples, **2.**24
  overdriven samples, **2.**24
  undisturbed soil samples, **2.**24–**2.**25
Soil sampling:
  block samples, **2.**13
  bulk samples, **2.**19
  drive cylinders, **2.**11–**2.**12
  standard penetration test (*see* Standard penetration test)
  (*See also* Samplers)
Soil scientists, **4.**1
Soil stabilization, **A.**28
Soil structure, **4.**9, **A.**13
  different types of soil structure, **4.**9, **A.**13
Soils engineer (*see* Geotechnical engineer)
Soils engineering (*see* Geotechnical engineering)
Solids, **5.**1, **5.**3
Soluble soil particles, **9.**21
  design and construction for soluble soil, **9.**21
  example of settlement analysis, **9.**21
  percent soluble soil particles, **9.**21
  settlement analysis, **9.**21
  types of soluble soil particles, **9.**21
South Carolina, **15.**7
Specific gravity, **3.**7–**3.**8, **A.**14
Specification, **A.**28
Spencer method of slices, **13.**27, **13.**42
Spicules of sponges, **4.**27
Spillway, **13.**55
St. Francis Dam, **13.**53
Stabilization fill, **A.**28
Staking, **A.**28
Standard penetration test:
  correction factors for field testing procedures, **2.**32
  definition, **2.**32, **A.**9
  description, **2.**32
  factors that influence the N-value, **2.**33–**2.**34
  N-value, **2.**32–**2.**35
  used to predict density condition, **2.**35
  used to predict drained modulus, **9.**50
  used to predict earthquake induced settlement, **14.**12–**14.**14
  used to predict friction angle, **7.**9
  used to predict liquefaction potential, **14.**5–**14.**9
  used to predict settlement, **9.**49–**9.**50, **9.**52
Standard Proctor (*see* Proctor)
Standpipe piezometer (*see* Instrumentation)
Starters or activators, **A.**31
Stereoscope, **2.**9
Stiffening beams, **12.**31
Stokes law, **3.**9
Stone columns, **10.**22
Strain (definition), **A.**20
Stress, **6.**1–**6.**22, **A.**19
Stress concentrations (in geogrid), **16.**23
Stress distribution, **6.**2, **6.**7–**6.**20
  Boussinesq equations, **6.**10–**6.**11
  for layered soil, **6.**14, **6.**19
  Newmark charts, **6.**13, **6.**18–**6.**20
  for one-dimensional loading, **6.**2, **6.**7
  plane strain condition, **6.**11
  pressure bulbs, **6.**12, **6.**19
  stress distribution examples, **6.**7, **6.**21
  theory of elasticity:
    charts, **6.**3, **6.**12–**6.**20
    tables, **6.**3, **6.**10–**6.**11
  for three-dimensional loading, **6.**2
  for two-dimensional loading, **6.**2
  2:1 approximation, **6.**2, **6.**8–**6.**9
Stress path method, **9.**25
Stress paths, **7.**30, **7.**32–**7.**33
Strike and dip, **A.**9
Structural engineer, **9.**6, **10.**7, **12.**31
Structure (*see* Buildings)
Subdrain, **A.**28

Subgrade:
  California bearing ratio for subgrade soils, **11.**5
  clay subgrade, **12.**37–**12.**38
  compaction equipment for subgrade soils, **11.**5
  compressibility of subgrade soils, **11.**5
  definition, **11.**3, **A.**28
  drainage properties of subgrade soils, **11.**5
  frost action of subgrade soils, **11.**4
  subgrade modulus, **9.**50–**9.**51, **9.**53, **11.**5, **A.**9
  subgrade properties, **11.**4–**11.**5
  typical dry densities of compacted sub grade, **11.**5
  (*See also* Pavements)
Subsidence, **9.**4, **9.**55, **A.**20
Subsoil profile, **2.**3, **2.**51–**2.**54, **A.**9
Substructure, **A.**28
Subsurface exploration:
  borings (*see* Boring)
  cone penetration tests, **2.**28–**2.**30, **2.**36
  depth of subsurface exploration, **2.**17
  dozer cuts, **2.**19
  soil sampling (*see* Soil sampling)
  standard penetration test (*see* Standard penetration test)
  test pits, **2.**19, **A.**8
  trenches, **2.**19
  (*See also* Field exploration)
Suction pressure (*see* Expansive soil)
Sulfate attack of concrete:
  chemical reactions, **17.**3
  design recommendations, **17.**6
  formation of ettringite, **17.**3
  physical growth of crystals, **17.**3
  requirements for concrete exposed to soluble sulfate, **17.**5
  sulfate resistance of concrete (*see* Concrete)
  (*See also* Deterioration)
Sulfate concentration at the site:
  corrosive environments, **17.**6
  natural variation in sulfate content, **17.**6
  obtaining soil and groundwater samples, **17.**6
  soil containing gypsum, **17.**6
Sulfates (definition), **A.**28
Sump, **20.**8, **20.**11, **A.**29
Sump pump, **20.**8, **20.**11
Superficial velocity (*see* Groundwater)
Superstructure, **A.**28
Surface drainage, **20.**23
Surficial slope stability, **13.**2, **13.**10–**13.**21
  cause of surficial failure, **13.**10
  for cut slopes, **13.**10
  damage caused by surficial failures, **13.**11
  design for surficial stability, **13.**13–**13.**15, **13.**18–**13.**20
  for earth dams, **13.**11
  effect of seepage, **13.**10
  effect of vegetation, **13.**11–**13.**12
  example of stability calculations, **13.**15
  failure mechanism, **13.**10–**13.**11, **13.**17–**13.**18
  for fill slopes, **13.**10
  methods to increase factor of safety:
    erosion control fabric, **13.**16
    flatten slope inclination, **13.**16
    gunite facing, **13.**16
    soil reinforcement, **13.**16, **13.**21
    stabilization fill, **13.**16
  for natural slopes, **13.**11
  repair of, **B.**22
  stability analysis, **13.**13–**13.**15, **13.**18–**13.**20
  surficial failures, **13.**10–**13.**11, **13.**17
Swedish circle method, **13.**25
Sweetwater formation, **4.**9
Swell:
  definition, **12.**15, **A.**20
  effect on surficial stability, **13.**10
  laboratory swell tests, **12.**11
  primary swell, **12.**15, **12.**17
  progressive swell, **12.**19, **12.**31
  secondary swell, **12.**15, **12.**17
  steady-state swell condition, **12.**15, **12.**17
  total swell, **12.**22
  (*See also* Expansive soil)
Swelling index, **12.**22
Swelling pressure, **12.**11, **12.**22
Syenite, **4.**31

Tack coat, **A.**29
Tailings (from grading operation), **A.**29
Tailings (from mining operation), **14.**4
Taylor slope stability chart, **13.**30, **13.**34
Technician, **3.**3, **19.**13, **19.**15
Temperature, **12.**13, **12.**39, **21.**12
Temporary retaining walls, **16.**3, **16.**36–**16.**40
  construction sequence, **21.**12
  design analysis for temporary retaining walls:
    braced retaining walls, **21.**2, **21.**11–**21.**14
    dewatering and recharge, **21.**12
    earth pressure distribution, **16.**36, **16.**39

Temporary retaining walls (*Cont.*):
effects of groundwater table, **16.**36
factors of safety, **21.**12–**21.**13
important design factors, **16.**36–**16.**40, **21.**12–**21.**13
movement of the wall, **21.**12
short term analysis (total stress analysis), **16.**36
surcharge loads, **16.**36, **16.**40, **21.**12
tieback anchors, **16.**36, **16.**40, **21.**12
excavation cave-ins, **16.**36
types of temporary retaining walls:
anchor or tieback wall, **21.**11
braced retaining walls, **16.**36, **16.**39
cantilever walls, **21.**11
close sheeting, **16.**37, **16.**39
cross-lot braced wall, **21.**11
earth berm support, **21.**11
raker system, **21.**11
sheet pile with struts, **16.**36
steel I-beam and wood lagging, **16.**36
utility trench shoring, **16.**36–**16.**38, **16.**40
utility trench backfill technique, **16.**38
Tensile test (geosynthetic), **A.**14
Terrace, **A.**29
Test pits, **2.**19, **A.**8
Testing:
borings (*see* Boring)
cone penetration tests (*see* Cone penetration test)
core drilling (*see* Core drilling)
field load tests (*see* Field exploration; subsurface exploration)
laboratory (*see* Laboratory testing)
in-place testing (*see* Field exploration; subsurface exploration)
subsurface exploration (*see* Field exploration; Subsurface exploration)
test pits (*see* Test pits)
trenches (*see* Trenches)
Testing requirements (*see* Laboratory tests)
Texas, **12.**4, **12.**30
Texture (of soil), **A.**14
Thermal osmosis, **12.**18–**12.**19
Thermal soil stabilization (*see* Site improvement methods)
Thixotropy, **A.**14
Tieback anchors, **16.**36, **16.**40, **21.**11–**21.**12, **23.**10
Tight line, **A.**31
Tile line, **A.**31
Till, **5.**4, **A.**9
Time factor (*see* Consolidation)
Topographic map, **2.**6, **2.**10
Topography, **2.**6, **2.**10, **13.**5
Topple (of rocks), **13.**4
Topsoil, **A.**9
Torvane, **3.**14, **7.**12, **7.**18
Total stress (definition), **6.**4, **A.**21
Total stress analysis:
for bearing capacity, **10.**15–**10.**16
for dam stability, **13.**56
discussion of, **7.**40–**7.**41
for gross slope stability, **13.**22–**13.**26, **13.**30–**13.**31, **13.**34
for piles in cohesive soil, **10.**28–**10.**29
shear strength tests based on total stress, **3.**13
for temporary retaining walls, **16.**36
Traffic index (*see* Pavements)
Transportation engineer, **11.**1
Trench shoring (*see* Temporary retaining walls)
Trenches (*see* Subsurface exploration)
Triaxial apparatus, **3.**21
Triaxial tests, **3.**15–**3.**16, **7.**34–**7.**35, **A.**14
Tricalcium aluminate, **17.**4
Tropical Storm Thelma, **13.**48
Tuff, **4.**24–**4.**25, **4.**28
Tunnels:
classification of rock for tunnel excavations, **21.**15
classification of soil for tunnel excavations, **21.**16
collapse of, **9.**54–**9.**55, **21.**2
coring and boring machines, **21.**2–**21.**3
Turbidity, **17.**3
Type of project, **1.**12

U-line, **4.**5, **4.**8
Ultimate bearing capacity (*see* Bearing capacity)
Ultraviolet light, **22.**5, **22.**7
Unconfined compressive strength:
definition, **A.**14
shear strength, **7.**12, **7.**18, **7.**37–**7.**38
soil, **3.**14
rock, **4.**29
Underconsolidated (*see* Consolidation)
Underpinning:
alternate methods to foundation underpinning, **21.**18–**21.**19
common reasons for underpinning, **21.**17
definition, **A.**29

Underpinning (*Cont.*):
repair of existing cracks, **21**.18–**21**.19, **21**.22–**21**.23
shoring for underpinning work:
needle beam method, **21**.17, **21**.20
notched wall method, **21**.17, **21**.20
types of underpinning:
underpinning with a deep foundation system, **21**.18
underpinning with a shallow foundation system, **21**.18, **21**.21
unique projects, **21**.17
Undrained modulus, **9**.24
Undrained shear strength (*see* Shear strength)
Unified Soil Classification System (USCS), **4**.11–**4**.14
group symbols, **4**.12–**4**.14
Uniform Building Code, **10**.7–**10**.8, **14**.12, **16**.8, **19**.12
Unit weight, **3**.6, **5**.6, **A**.14
United States Customary System units (USCS), **3**.6, **5**.2, **6**.1, **D**.1
United States Department of Agriculture (USDA) soil classification, **4**.2, **4**.18
United States Geological Survey (USGS), **2**.6, **2**.9–**2**.10
Universal soil loss equation (*see* Erosion)
Unusual soil, **4**.3, **4**.24–**4**.27
bentonite, **4**.25–**4**.26
bull's liver, **4**.24
caliche, **4**.25
diatomaceous earth, **4**.26–**4**.27
loess, **4**.25
nonwelded tuff, **4**.24–**4**.25
peat, **4**.24
quick clay, **4**.26
rock flour, **4**.24
sensitive clay, **4**.26
varved clay, **4**.25
volcanic ash, **4**.24–**4**.25
Uplift loads, **14**.13
Utility trenches, **11**.14, **11**.16, **16**.36–**16**.38, **17**.7, **19**.8
Utility trench shoring (*see* Temporary retaining walls)
UUC tests (unconfined compression tests), **7**.18

Vane shear test:
definition, **A**.9
illustration of field vane test, **2**.31
laboratory shear strength tests, **3**.14
miniature vane, **3**.14
Vane shear test (*Cont.*):
tapered vane, **2**.37
torvane, **2**.37, **3**.14
undrained shear strength, **2**.36–**2**.37, **7**.1, **7**.12, **7**.18
Varved clay, **4**.25, **A**.9
Varved silt, **A**.9
Vegetation, **12**.21, **15**.6, **15**.9, **19**.7
Vernal pool, **5**.7
Vertical pressure (*see* Stress distribution)
Vibrations, **13**.5
Vibro columns, **10**.22
Vibrodensification methods (*see* Site improvement methods)
Vibroflotation, **10**.22
Void ratio, **3**.30, **5**.4, **A**.21
Volcanic:
ash, **4**.24–**4**.25
dust, **4**.24
eruptions, **4**.24
rock, **4**.24
Volume, **5**.5
Volumetric strain (caused by an earthquake), **14**.12–**14**.14

Walls:
basement, **16**.2
bearing wall, **A**.29
bridge abutment walls, **16**.4
buttressed wall, **16**.4
cantilevered, **16**.4–**16**.5, **16**.21
counterfort, **16**.4–**16**.5, **16**.21
crib walls, **16**.4–**16**.5
cutoff wall, **A**.29
definition, **A**.29
gravity, **16**.4–**16**.5, **16**.20
mechanically stabilized earth retaining walls (*see* Mechanically stabilized)
pier walls (*see* Pier walls)
restrained retaining walls (*see* Restrained retaining walls)
semigravity, **16**.4, **16**.20
sheet pile walls (*see* Sheet pile walls)
temporary retaining walls (*see* Temporary retaining walls)
utility trench shoring (*see* Temporary retaining walls)
(*See also* Retaining walls)
Water cement ratio, **17**.4, **A**.29
Water content, **3**.5, **A**.12, **A**.14
Water content profile, **2**.52–**2**.54
Water table (*see* Groundwater)
Water vapor, **20**.13, **20**.15

Waterproofing:
membrane, **20.**16–**20.**17
mastic, **20.**16
primer, **20.**16
protection board, **20.**16
self-adhering, **20.**16
Weather, **13.**5
Weathering:
causing surficial failures, **13.**10–**13.**11, **13.**18
of rock, **4.**30
Wedge method, **13.**22–**13.**24, **13.**32, **16.**43–**16.**44, **16.**47
Welding, **4.**24–**4.**25
Well point, **20.**8, **20.**10, **A.**29
Wells, **20.**8, **20.**10
Wetland (definition), **A.**9
Wetting (*see* Expansive soil; Collapsible soil)
Windrow, **19.**7, **A.**29, **B.**8-B.9, **B.**21
Workability of concrete, **A.**29
Wyoming, **12.**4

X-ray diffraction tests, **4.**6
X-ray radiograph, **2.**27
X-rays, **2.**27

Yonkers, New York, **17.**9

Zero air voids curve, **3.**29, **A.**14

ABOUT THE AUTHOR

Robert W. Day is a leading geotechnical engineer and the Chief Engineer at American Geotechnical in San Diego, California. The author of over 200 published technical papers and the textbooks *Forensic Geotechnical and Foundation Engineering* and *Geotechnical and Foundation Engineering: Design and Construction*, he serves on advisory committees for several professional associations, including ASCE, ASTM, and NCEES. He holds four college degrees: two from Villanova University (bachelor's and master's degrees majoring in structural engineering), and two from the Massachusetts Institute of Technology (master's and the Civil Engineer degree[highest degree] majoring in geotechnical engineering). He is also a registered civil engineer in several states and a registered geotechnical engineer in California.